Lecture Notes in Computer Science   9833

Commenced Publication in 1973
Founding and Former Series Editors:
Gerhard Goos, Juris Hartmanis, and Jan van Leeuwen

Advanced Research in Computing and Software Science
Subline of Lecture Notes in Computer Science

More information about this series at http://www.springer.com/series/7407

Pierre-François Dutot · Denis Trystram (Eds.)

Euro-Par 2016:
Parallel Processing

22nd International Conference
on Parallel and Distributed Computing
Grenoble, France, August 24–26, 2016
Proceedings

 Springer

Editors
Pierre-François Dutot
Université Grenoble Alpes
Grenoble
France

Denis Trystram
Université Grenoble Alpes
Grenoble
France

ISSN 0302-9743 ISSN 1611-3349 (electronic)
Lecture Notes in Computer Science
ISBN 978-3-319-43658-6 ISBN 978-3-319-43659-3 (eBook)
DOI 10.1007/978-3-319-43659-3

Library of Congress Control Number: 2016945998

LNCS Sublibrary: SL1 – Theoretical Computer Science and General Issues

Printed on acid-free paper

This Springer imprint is published by Springer Nature
The registered company is Springer International Publishing AG Switzerland

Preface

It is our pleasure and privilege to introduce this new Euro-Par proceedings volume. Over the years, Euro-Par has become a major event of our well-established parallel and distributed community. This was the 22nd edition of a successful series of conferences which started in Stockholm in 1995.

Today, it is evidence that the field of parallelism is still very active. The traditional topics have been renewed year after year, following the successive generations of machines (including parallel architectures, algorithms, performance, languages, compilers, runtimes, parallel programming, foundations of parallelism, numerical applications) and some new topics have emerged (accelerator computing, hierarchical platforms, clouds, interactive computing, data management analytics). We now face the challenge of building extreme-scale high-performance computing platforms, which are the only way to address the hard societal challenges in health, security, or the climate. All the historical disciplines of parallelism are impacted by such deep changes and by strong interactions between the various topics (e.g., energy optimization, scalability, or fault-tolerance issues). Euro-Par keeps the original organization into independent topics covering all these aspects.

The conference is organized in parallel sessions that reflect the diversity of the topics. We received 176 full submissions and accepted 47 papers after an 8-week review period in a full-day Program Committee (PC) meeting involving all the topic chairs or local chairs, which led to the selective rate of acceptance of less than 27 %. This high quality is also assessed by three distinguished papers. Most papers received four reviews while a few received three or five evaluations. Two additional presentations given by industrial sponsors have been included in the program. We would like to warmly thank all the Scientific Committee members of the various topics. The team of chairpersons had the enormous task of bringing many innovative ideas, which finally led to a very attractive and timely program. The whole evaluation process ran smoothly thanks to the involvement of each of them.

In addition, we are pleased to present three keynote talks of well-recognized colleagues, namely, Susanne Albers ("Energy-Efficient Algorithms"), Walfredo Cirne ("Improving Cloud Effectiveness"), and Dror Feitelson ("Resampling with Feedback—A New Paradigm of Using Workload Data for Performance Evaluation") plus an invited talk of Arnold Rosenberg ("Scheduling DAGs Opportunistically: The "Dream and the Reality Circa 2016"). This exciting program was complemented by two days of dedicated workshops and tutorials on more specialized themes. The huge work of managing workshops and tutorials was conducted very efficiently by Frédéric Desprez, always in a positive mood. As with the previous editions, the papers selected in the various workshops are published in a separate proceedings volume after the conference.

We would like to emphasize that Euro-Par is not only a premium scientific meeting, but it is also a very social event. The organization of such a conference involves a great amount of work, completed under the guidance and support of the Steering Committee members and especially Christian Lengauer and Luc Bougé. However, it was also a great opportunity for our team in Grenoble to strengthen our internal links in order to provide the best for participants coming from all over the world. We are sincerly grateful to Sophie Azzaro and Annie Simon at Inria and to the other members of the organization team for their constant help.

We would also like to warmly thank our academic institutions and their staff, namely, Université Grenoble Alpes (and particularly the Institut Universitaire de Technologie 2), Grenoble Institute of Technology, and Inria for their logistical support at many levels, as well as our industrial partners.

June 2016

Pierre-François Dutot
Denis Trystram

Organization

Steering Committee

Full Members

Christian Lengauer (Chair)	University of Passau, Germany
Luc Bougé (Vice-Chair)	ENS Rennes, France
Emmanuel Jeannot	LaBRI-Inria, Bordeaux, France
Christos Kaklamanis	Computer Technology Institute, Patras, Greece
Paul Kelly	Imperial College, London, UK
Thomas Ludwig	University of Hamburg, Germany
Emilio Luque	Autonomous University of Barcelona, Spain
Tomàs Margalef	Autonomous University of Barcelona, Spain
Wolfgang Nagel	Dresden University of Technology, Germany
Rizos Sakellariou	University of Manchester, UK
Fernando Silva	University of Porto, Portugal
Henk Sips	Delft University of Technology, The Netherlands
Domenico Talia	University of Calabria, Italy
Jesper Larsson Träff	TU Vienna, Austria
Felix Wolf	TU Darmstadt, Germany

Honorary Members

Ron Perrott	Oxford e-Research Centre, UK
Karl Dieter Reinartz	University of Erlangen-Nürnberg, Germany

Observers

Francisco Rivera	CiTIUS, Santiago de Compostela, Spain
Denis Trystram	Grenoble Institute of Technology, France

Euro-Par 2016 Organization

Chair

Denis Trystram

Proceeding

Pierre-François Dutot

Workshop

Frédéric Desprez

Local Organization

Annie Simon
Sophie Azzaro
Grégory Mounié
Frédéric Wagner

Industrial Chair

Laurent Colombet
Pierre Neyron

Web and Publicity

Christophe Cérin
David Glesser

Program Committee

Topic 1: Support Tools and Environments

Chair

Tomàs Margalef   Universitat Autònoma de Barcelona, Spain

Local Chair

Olivier Richard   University of Grenoble Alpes, France

Members

Siegfried Benkner   University of Vienna, Austria
Joao Cardoso   University of Porto and INESC-TEC, Portugal
Michael Gerndt   Technical University of Munich, Germany
Martin Schulz   Lawrence Livermore National Laboratory, USA
Ana Lucia Varbanescu   University of Amsterdam, The Netherlands

Topic 2: Performance and Power Modeling, Prediction, and Evaluation

Chair

Laura Carrington   San Diego Supercomputer Center, USA

Local Chair

Arnaud Legrand   CNRS/University of Grenoble, France

Members

Roy Campbell	DOD High Performance Computing Modernization Program, USA
Kirk Cameron	Virginia Tech, USA
Marcos Dias de Assucao	LIP/ENS Lyon, France
Georg Hager	Friedrich-Alexander-Universität, Germany
Sascha Hunold	Vienna University of Technology, Austria
Darren Kerbyson	Pacific Northwest National Laboratory, USA
Shirley Moore	The University of Texas at El Paso, USA

Topic 3: Scheduling and Load Balancing

Chair

Rizos Sakellariou	University of Manchester, UK

Local Chair

Fanny Pascual	Université Pierre et Marie Curie, France

Members

Luiz Fernando Bittencourt	University of Campinas, Brazil
Florina Ciorba	University of Basel, Switzerland
Julita Corbalan	Barcelona Supercomputing Center, Spain
Ivan Rodero	Rutgers University, USA
Krzysztof Rzadca	University of Warsaw, Poland
Uwe Schwiegelshohn	TU Dortmund, Germany

Topic 4: High-Performance Architectures and Compilers

Chair

Henri Bal	Vrije Universiteit, The Netherlands

Local Chair

Sid Touati	University of Nice Sophia Antipolis, France

Members

Eduard Ayguadé	Technical University of Catalonia, Spain
Pedro Diniz	University of South of California, USA
Thomas Fahringer	University of Innsbruck, Austria
David Gregg	Trinity College, Ireland
Wolfgang Karl	Karlsruhe Institut für Technologie, Germany
Hand Vandierendonck	Queen's University Belfast, UK

Topic 5: Parallel and Distributed Data Management and Analytics

Chair

Tom Peterka   Argonne National Laboratory, USA

Local Chair

Bruno Raffin   Inria, France

Members

Christopher Carothers   Rensselaer Polytechnic Institute, USA
Toni Cortes   BSC, Spain
Matthieu Dorier   Argonne National Laboratory, USA
Wolfgang Frings   JSC, Germany
Patrick Martin   Queen's University, Kingston, Canada
Yang-Sae Moon   Kangwon National University, Korea

Topic 6: Cluster and Cloud Computing

Chair

Adrien Lèbre   Inria, France

Members

Ivona Brandic   Vienna University of Technology, Austria
Fabien Hermenier   University Nice Sophia Antipolis, France
Peter Pietzuch   Imperial College London, UK
Ioan Raicu   Illinois Institute of Technology, USA
Xuanhua Shi   Huazhong University of Science and Technology,
                                 China
Wenhong Tian   University of Electronic Science and Technology
                                 of China, China
Srikumar Venugopal   IBM, Ireland

Topic 7: Distributed Systems and Algorithms

Chair

Domenico Talia   Università della Calabria, Italy

Local Chair

Stéphane Devismes   University of Grenoble Alpes, France

Members

Adriana Iamnitchi   University of South Florida, USA
Alexandru Iosup   TU Delft, The Netherlands

Stefan Schmid   TU Berlin/Telekom Innovation Laboratories, Germany
Josef Widder   TU Wien, Austria

Topic 8: Parallel and Distributed Programming, Interfaces, Languages

Chair

Philippe O.A. Navaux   INF-UFRGS, Brazil

Local Chair

Christian Perez   Inria Grenoble, France

Members

Abhinav Bhatele   Lawrence Livermore National Laboratory, USA
Alba Cristina De Melo   University of Brasilia, Brazil
Raymond Namyst   University of Bordeaux - Inria, France
Celso Mendes   University of Illinois, USA
Marco Danelutto   University of Pisa, Italy
Esteban Meneses   Costa Rica Institute of Technology, Costa Rica
Kuan-Ching Li   Providence University, Taiwan

Topic 9: Multicore and Manycore Parallelism

Chair

Michela Taufer   University of Delaware, USA

Local Chair

Renaud Lachaize   University Grenoble Alpes, France

Members

Michela Becchi   University of Missouri, USA
Sunita Chandrasekaran   University of Delaware, USA
Anne Elster   NTNU and University of Texas at Austin, USA
Naoya Maruyama   RIKEN, Japan
Dimitrios Nikolopoulos   Queen's University of Belfast, UK
Sabela Ramos Garea   ETH Zurich, Switzerland
Guido Juckeland   Helmholtz Zentrum Dresden Rossendorf, Germany
Graham Lopez   Oak Ridge National Laboratory, USA
Vania Marangozova-Martin   Grenoble University, France

Topic 10: Theory and Algorithms for Parallel Computation and Networking

Chair

Klaus Jansen   University of Kiel, Germany

Local Chair

Bora Ucar   CNRS and ENS Lyon, France

Members

Petra Berenbrink   Simon Fraser University, Canada
Christos Kaklamanis   University of Patras and CTI Diophantus, Greece
Nicole Megow   Technische Universität München, Germany
Erik Saule   University of North Carolina at Charlotte, USA
Christian Scheideler   Universität Paderborn, Germany
Jiri Sgall   Charles University, Czech Republic

Topic 11: Parallel Numerical Methods and Applications

Chair

Matthias Bolten   Universität Kassel, Germany

Local Chair

Laurent Philippe   FEMTO-ST, France

Members

Peter Arbenz   ETH Zurich, Switzerland
El Mostafa Daoudi   Université Mohammed Premier-Oujda, Morroco
Maya Neytcheva   Uppsala University, Sweden
Marian Vajtersic   Universität Salzburg, Austria

Topic 12: Accelerator Computing

Chair

Enrique S. Quintana-Orti   Universidad Jaume I, Spain

Local Chair

Samuel Thibault   University of Bordeaux, France

Members

Taisuke Arai Boku   University of Tsukuba, Japan
Esteban Clua   Universidade Federal Fluminense, Brazil
Hatem Ltaief   King Abdullah University of Science and Technology,
                                     Saudi Arabia
Jeff Hammond   Intel, USA
John Stone   University of Illinois at Urbana-Champaign, USA
Robert Strzodka   Universität Heidelberg, Germany

Euro-Par 2016 Reviewers

Euro-Par is grateful to all the reviewers for their willingness and their effort in providing good feedback to authors and topic committees. All external reviewers are listed and hereby thanked.

Ahmad Abdelfattah	Carmela Comito	Matthew Hammer
Sangeetha Abdu Jyothi	Daniel Cordeiro	Paul Harvey
José Ignacio Aliaga	Helene Coullon	Khalid Hasanov
Pedro Alonso	Eduardo Cruz	Ahmad Hassan
Karine Altisen	Daniele D'Agostino	Timo Heister
Frederico Alvares De	Patrizio Dazzi	Oscar Hernandez
Oliveira	Steven Derrien	Tetsuya Hoshino
Damian Alvarez Mallon	Frederic Desprez	Kuo-Chan Huang
Andrew Anderson	Javier Diaz-Montes	Thomas Huckle
José M. Andión	Kiril Dichev	Andra-Ecaterina Hugo
Hartwig Anzt	Matthias Diener	Shadi Ibrahim
Mahwish Arif	Romain Dolbeau	Francisco D. Igual
Olivier Aumage	Ali Dorostkar	Siddhartha Jana
Ahsan Javed Awan	Matthieu Dreher	Tong Jin
Marc Baboulin	Fernando Duarte	Josep Jorba
Jose Badia	Anaïs Durand	Herbert Jordan
Denis Barthou	Juan J. Durillo	Hirotsugu Kakugawa
Cédric Bastoul	Yehia Elshater	Jeffrey Kelling
Thomas Becker	Joseph Emeras	Darren Kerbyson
Vicenç Beltran	Christian Engwer	Christoph Kessler
Anne Benoit	Roberto R. Expósito	Dounia Khaldi
Josep Lluis Berral-García	Pablo Ezzatti	Shadi Khalifa
Philipp Birken	Thomas L. Falch	Mario Kicherer
George Bosilca	Mathieu Faverge	Mauricio Kischinhevsky
Michael Bromberger	Michael Feldmann	Peter Kling
François Broquedis	Gianluigi Folino	Christina Kolb
Sebastian Buchwald	Tom Friedetzky	Ivana Kolingerova
Alexandra Carpen-Amarie	Ryan Friese	Igor Konnov
Márcio Castro	Stratis Gallopoulos	Charalampos
Eduardo Cesar	Thierry Gautier	Konstantopoulos
Eugenio Cesario	João Gazolla	Harald Köstler
Wei Chang	Giorgis Georgakoudis	Alessandro Kraemer
Vincent Chau	Domingo Gimenez	Moritz Kreutzer
Renjie Chen	Robert Gmyr	Pascal Lafourcade
Nathanaël Cheriere	Jorge	Felix Land
Kallia Chronaki	González-Domínguez	Bruno Lang
Terry Cojean	Clemens Grelck	Vincent Lanore
Guillaume Colin de	Philipp Gschwandtner	Alexey Lastovetsky
Verdiere	Georg Hager	Jonathan Lejeune

Euro-Par 2016 Invited Talks

Resampling with Feedback — A New Paradigm of Using Workload Data for Performance Evaluation

Dror Feitelson, Hebrew University of Jerusalem, Israel

Reliable performance evaluations require representative workloads. This has led to the use of accounting logs from production systems as a source for workload data in simulations. I will survey 20 years of ups and downs in the use of workload logs, culminating with the idea of resampling with feedback. It all started with the realization that using workload logs directly suffers from various deficiencies, such as providing data about only one specific situation, and lack of flexibility, namely the inability to adjust the workload as needed. Creating workload models solves some of these problems but creates others, most notably the danger of missing out on important details that were not recognized in advance, and therefore not included in the model. Resampling solves many of these deficiencies by combining the best of both worlds. It is based on partitioning the workload data into basic components (e.g. the jobs contributed by different users), and then generating new workloads by sampling from this pool of basic components. This allows analysts to create multiple varied (but related) workloads from the same original log, all the time retaining much of the structure that exists in the original workload. However, resampling should not be applied in an oblivious manner. Rather, the generated workloads need to be adjusted dynamically to the conditions of the simulated system using a feedback loop. Resampling with feedback is therefore a new way to use workload logs which benefits from the realism of logs while eliminating many of their drawbacks. In addition, it enables evaluations of throughput effects that are impossible with static workloads.

Improving Cloud Effectiveness

Walfredo Cirne, Google, USA

Cloud computing has emerged in the last decade as a very cost effective way to do computing. Consumers avoid the fixed cost and slow deployment of running their own computers, gaining the ability to massively scale their computational ability.

This talk discusses what is need to make the Cloud even more effective. Part of it is to further increase the scope of Cloud, by making it better cover the demands of specialized, large users. The other part is how to make the Cloud more efficient. How can we manage the data center as to increase its utilization? In particular, we show how to providing a different SLOs enables us to better utilize our datacenter, as well as the impact such strategies have in the user experience, from reliability and performance to scalability and prices.

Energy-Efficient Algorithms

Susanne Albers, Technical University of Munich, Germany

We survey algorithmic techniques for energy savings. So far the algorithms literature focuses mostly on the system and device level: How can we save energy in a given computational device? More specifically, (a) power-down mechanisms and (b) dynamic speed scaling have been explored.

Power-down mechanisms: Consider a single device that is equipped with two states, an active state and a sleep state. These states have individual power consumption rates. Moreover, transitions between the states consume energy. We first present simple algorithms that specify state transitions in idle periods where the device is not in use. In the offline setting, the length of an idle period is known in advance. In the online setting, this information is not available.

Furthermore, we review results for the more advanced setting that the device has several low-power states. Again we show offline and online algorithms. Moreover, we study the challenging scenario that a large set of parallel devices/processors is given. The processors are heterogeneous in that each one has an individual set of low-power states with associated power consumption rates. Over a time horizon the processing demands vary. We give algorithms for constructing state transition schedules that minimize the total energy consumed by all the processors.

Dynamic speed scaling: This technique is based on the fact that many modern microprocessors can run at variable speed. High speeds imply high performance but also high energy consumption. Low speed levels save energy but the performance degrades. The general goal is to execute a set of jobs on variable-speed processors so as to optimize energy and, possibly, a second objective.

We first review basic results for a single processor. We consider classical deadline-based scheduling where each job is specified by an arrival time, a deadline and a processing volume. Offline and online strategies are presented. We also study a second setting where jobs are not labeled with deadlines and, instead, the objective is to minimize the total cost consisting of job response times and energy. Additionally, we review results for parallel processing environments where a set of homogeneous or heterogeneous processors is given. Last not least we address an advanced problem setting in which dynamic speed scaling and power-down mechanisms are combined.

Euro-Par 2016 Topics Overview

Topic 1: Support Tools and Environments

Tomàs Margalef, Olivier Richard, Siegfried Benkner, Joao Cardoso,
Michael Gerndt, Martin Schulz, Ana Lucia Varbanescu

High performance computing systems are becoming more and more complex. They feature large node counts and each node often contains multicore microprocessors combined with hardware accelerators and a complex multilevel memory hierarchy shared among different components of the system. Faced with such complexity, correctness and the performance of the applications are crucial, yet difficult issues that are far from being solved.

The Support Tools and Environments Topic focuses on tools and techniques addressing challenges of parallel and distributed systems related to programmability, portability, correctness, reliability, scalability, efficiency and energy/power consumption. This year, a diversity of papers proposing interesting and valuable research contributions was submitted to this topic. As a result of the reviewing process, three papers were accepted for publication.

We would like to thank all the authors who submitted papers to this topic as well as the external reviewers, for their contribution to the success of the conference.

Topic 2: Performance and Power Modeling, Prediction and Evaluation

Laura Carrington, Arnaud Legrand, Roy Campbell, Kirk Cameron,
Marcos Dias de Assucao, Georg Hager, Sascha Hunold, Darren Kerbyson,
Shirley Moore

In recent years, a range of novel methods and tools have been developed for the evaluation, design, and modeling of parallel and distributed systems and applications. At the same time, the term 'performance' has broadened to also include scalability and energy efficiency, and touching reliability and robustness in addition to the classic resource-oriented notions. The aim of this topic is to gather researchers working on different aspects of performance modeling, evaluation, and prediction, be it for systems or for applications running on the whole range of parallel and distributed systems (multi-core and heterogeneous architectures, HPC systems, grid and cloud contexts etc.). Authors are invited to submit novel research in all areas of performance modeling, prediction and evaluation, and to help bring together current theory and practice.

This track received 18 submissions, all of which received at least 4 reviews, from the 10 PC members or from the 8 additional subreviewers. The papers and reviews were discussed extensively. As a result, four submissions have been accepted (22 %

acceptance rate). This track has thus been particularly selective and we have tried to provide the authors with the most valuable and constructive feedback.

Topic 3: Scheduling and Load Balancing

Rizos Sakellariou, Fanny Pascual, Luiz Fernando Bittencourt, Florina Ciorba, Julita Corbalan, Ivan Rodero, Krzysztof Rzadca, Uwe Schwiegelshohn

As parallelism now permeates all levels of modern computer systems, it opens up new opportunities for improving application performance but, at the same time, it also increases the complexity of the resource management challenge. In this environment, the importance of scheduling and load balancing as a key research topic in parallel computing continues to grow. In addition to long standing problems, hitherto unexplored scenarios emerge in which scheduling and load balancing may need to be addressed in the presence of multiple and/or conflicting optimization objectives where, for instance, energy consumption may have to be traded with performance.

This topic covered all aspects related to scheduling and load balancing on parallel and distributed machines, from theoretical foundations for modelling and designing efficient and robust strategies to experimental studies, applications and practical tools and solutions. This applies to multi-core processors, servers, heterogeneous systems, HPC clusters as well as distributed systems such as clouds and global computing platforms. Concrete areas of interest included scheduling algorithms for homogeneous or heterogeneous platforms, theoretical foundations of scheduling algorithms, robustness of scheduling algorithms, multi-objective scheduling, scheduling at extreme scale, on-line scheduling, energy awareness in scheduling and load balancing, workload characterization and modelling, workflow scheduling, performance models for scheduling and load balancing, and resource management and awareness.

Seven papers were selected for presentation following a rigorous review process with four independent reviews per paper.

Topic 4: High Performance Architectures and Compilers

Henri Bal, Sid Touati, Eduard Ayguadé, Pedro Diniz, Thomas Fahringer, David Gregg, Wolfgang Karl, Hand Vandierendonck

This topic deals with hardware architecture design, languages, and compilation for parallel and high performance systems. The areas of interest range from microprocessors to large-scale parallel machines (including multi and many-core, possibly heterogeneous, processor architectures); from general-purpose to specialised hardware platforms (e.g., graphic coprocessors, low-power embedded systems); and from architecture design to compiler technology and language design.

On the compilation side, topics of interest include programmer productivity issues, concurrent and/or sequential language aspects, vectorisation, program analysis, program transformation, automatic discovery and/or management of parallelism at all levels, autotuning and feedback directed compilation, and the interaction between the compiler and the operating system at large. On the machine architecture side, the scope spans system architectures, processor architecture and micro-architecture, memory hierarchy, multi-threading, architectural support for parallelism, and the impact of emerging hardware technologies.

This track received 12 submissions, all of which received 4 reviews, from the 8 PC members or from the 25 subreviewers. The papers and reviews were discussed extensively. As a result, four submissions have been accepted, including one that was nominated as distinguished paper.

Topic 5: Parallel and Distributed Data Management and Analytics

Tom Peterka, Bruno Raffin, Christopher Carothers, Toni Cortes, Matthieu Dorier, Wolfgang Frings, Patrick Martin, Yang-Sae Moon

Many areas of science, industry, and commerce are producing extreme scale data that must be processed—stored, managed, analyzed—in order to extract useful knowledge. This topic seeks papers in all aspects of distributed and parallel data management and data analysis. For example, HPC in situ data analytics, cloud and grid data-intensive processing, parallel storage systems, and scalable data processing workflows are all in the scope of this topic.

Focus:

- Parallel, replicated, and highly-available distributed databases
- Data-intensive clouds and grids
- HPC scientific data analytics
- Middleware for processing large-scale data
- Programming models for parallel and distributed data analytics
- Workflow management for data analytics
- Coupling HPC simulations with in situ data analysis
- Parallel data visualization
- Distributed and parallel transaction and query processing and information retrieval
- Internet-scale data-intensive applications
- Sensor network data management
- Cloud and HPC storage architectures and systems
- Parallel data streaming and data stream mining
- Parallel and distributed knowledge discovery and data mining
- New storage hierarchies in distributed data systems based on NVRAM technologies

Thirteen full-length papers were submitted, and each paper received four reviews. After discussion with the reviewers and track chairs, five papers were selected for publication. Topics ranged from Spark and MapReduce applications, graph analytics, coupling tasks in workflows, and kernels for advanced architectures.

Topic 6: Cluster and Cloud Computing

Adrien Lebre, Ivona Brandic, Fabien Hermenier, Peter Pietzuch, Ioan Raicu,
Xuanhua Shi, Wenhong Tian, Srikumar Venugopal

The success of Cloud Computing solutions such as the ones provided by Amazon or Google has driven the advent of the Utility Computing (UC) paradigm. The use of massive storage and computing resources accessible remotely in a seamless way has become essential for many applications in various areas. While significant progresses have been achieved in the past decade, the complete adoption of the UC paradigm is still facing important challenges. Beyond the scene, most of Cloud Computing solutions rely on federations of large-scale clusters where well-known but still unsolved challenges related to performance, reliability and energy efficiency of the infrastructures should be addressed by research. Moreover, Cloud Computing emphasized the importance of fundamental capabilities and services that are required to achieve the goal of user-friendly, security and service guarantees.

Topic 6 sought papers covering many aspects of Cluster and Cloud Computing dealing with infrastructure layer challenges (performance/energy optimizations, security enhancements, Edge Computing) as well as how different kinds of applications (scientific workflows, HPC, mobile) can benefit from such infrastructures.

22 papers have been submitted in Topic 6. Each submission was reviewed by at least four reviewers. Finally three papers have been selected.

We would like to sincerely thank all the authors for their submissions, the Euro-Par 2016 Organizing Committee for their valuable help and the reviewers for their excellent review work. All of them have contributed to make this topic and EuroPar an excellent forum to discuss Cluster and Cloud Computing challenges.

Topic 7: Distributed Systems and Algorithms

Domenico Talia, Stéphane Devismes, Adriana Iamnitchi, Alexandru Iosup,
Stefan Schmid, Josef Widder

Parallel computing is heavily related to the developments and challenges of distributed systems. Problems including load balancing, asynchrony, failures, malicious and selfish behavior, long latencies, network partitions, disconnected operations, distributed computing models and concurrent data structures, and heterogeneity are representative of typical distributed issues that often appear along the design of parallel applications.

This track of Euro-Par provides a forum for both theoretical and practical research, of interest to both academia and industry, on distributed computing, distributed algorithms, distributed systems, distributed data structures, and parallel processing on distributed systems, in particular in relation to efficient high performance computing.

This year, 10 complete papers have been submitted to this track. After a bidding phase, each paper has been evaluated by 4 reviewers with high expertise. Overall, 18 experts have been involved into the review process. Finally, despite the high quality of the submitted papers, only two papers have been accepted for publications.

The PC chairs, Domenico Talia (Università della Calabria, Italy) and Stéphane Devismes (Université Grenoble Alpes, France), are very grateful to all researchers that have participated to the review process and permitted to select two high-quality papers.

Topic 8: Parallel and Distributed Programming, Interfaces, Languages

Philippe O.A. Navaux, Christian Perez, Abhinav Bhatele, Alba Cristina De Melo, Raymond Namyst, Celso Mendes, Marco Danelutto, Esteban Meneses, Kuan-Ching Li

Parallel and distributed applications requires adequate programming abstractions and models, efficient design tools, parallelization techniques and practices. This topic was open for presentations of new results and practical experience in this domain. efficient and effective parallel languages, interfaces, libraries and frameworks, as well as solid practical and experimental validation. It provides a forum for research on high-performance, correct, portable, and scalable parallel programs via adequate parallel and distributed programming model, interface and language support. Contributions that assess programming abstractions, models and methods for usability, performance prediction, scalability, self-adaptation, rapid prototyping and fault-tolerance, as needed, for instance, in dynamic heterogeneous parallel and distributed infrastructures, were accepted.

All fifteen papers of this topic received three reviews that were further discussed among all nine PC members. As a result, three strong papers were accepted for the conference, covering important topics.

Topic 9: Multi- and Many-core Programming

Michela Taufer, Renaud Lachaize, Michela Becchi, Sunita Chandrasekaran, Anne Elster, Naoya Maruyama, Dimitrios Nikolopoulos, Sabela Ramos Garea, Guido Juckeland, Graham Lopez, Vania Marangozova-Martin

The intrinsic complexity of emerging many- and multi-core architectures requires the deployment of software solutions capable of dealing with hybrid and heterogeneous systems, from multi- and many-core systems to stand-alone systems with large numbers of cores such as GPUs and accelerators. Proposed software solutions pursue better programmability, performance portability, and levels of abstraction of these modern parallel architectures. Deployed approaches tackle challenges in the architectures' programming models, algorithms, languages, compilers, libraries, runtime and analysis tools.

The topic on multi- and many-core programming explores these software solutions and associated approaches, providing the attendees with a common platform for discussion of the state of the art and future directions in the field. Novel research and solutions proposed in accepted papers includes: crucial data structures such as concurrent search trees for supporting fine-grained, high-concurrent data locality in an energy-efficient manner; scheduling strategies for task-based workflows to improve runtime performance of applications on NUMA architectures; algorithmic improvements based on recursive formulation of triangular matrix-matrix multiplication (TRMM) and the triangular solve (TRSM) on GPUs; memory management techniques to map threads and data to increase locality of memory accesses; studies of relationships between performance and energy in concurrent programs; and runtime support for energy efficiency in parallel pipelines on heterogeneous multiprocessing architectures.

Six papers were selected for presentation, one as distinguished paper.

Topic 10: Theory and Algorithms for Parallel Computation and Networking

Klaus Jansen, Bora Ucar, Petra Berenbrink, Christos Kaklamanis, Nicole Megow, Erik Saule, Christian Scheideler, Jiri Sgall

Parallel computing is everywhere, on smartphones, laptops; at online shopping sites, universities, computing centers; behind the search engines. Efficiency and productivity at these scales and contexts are only possible by scalable parallel algorithms.

Theoretical tools enabling scalability, modeling and understanding parallel algorithms, and data structures for exploiting parallelism are more important than ever. Topic 10 addressed this general topic of theory and algorithms for parallel computation including communication and network algorithms. We have received 14 submissions in very relevant topics including graph algorithms, data structures, interconnection networks, distributed algorithms, and scientific computing algorithms.

The submissions were evaluated by a committee of eight members from diverse background and geographical regions. All submissions received at least four reviews. In most cases, the reviews were equivocal; in two cases the consensus were reached after a lively discussion. At the end of the PC meeting, three papers were accepted.

We thank the authors who submitted papers and congratulate those whose papers were accepted. We are grateful to the PC members and the referees who provided us with carefully written, constructive and informative reviews. We also thank the conference organizers for answering our questions and smoothly running the PC meeting that took place in Grenoble.

Topic 11: Parallel Numerical Methods and Applications

Matthias Bolten, Laurent Philippe, Peter Arbenz, El Mostafa Daoudi, Maya Neytcheva, Marian Vajtersic

A large amount of compute time used on HPC systems is used for numerical simulations run by scientists from engineering and the sciences. Most of these simulations rely on the availability of numerical algorithms that on the one hand provide the accuracy that is needed to answer the scientific questions and that on the other hand are scalable such that the available computer architecture is used as efficient as possible. The latter implies a challenge for applied mathematicians and computer scientists, as the fast growth of core numbers and the implications of modern architectures require a careful algorithmic design and implementation.

The topic covers a wide range of aspects, from algorithms for basic linear algebra problems to methods for differential equations, dealing with the avoidance of communication, implementation details, and detection of faults.

Numerous papers have been submitted to this topics from international researchers. After four reviews for each paper have been received the topic committee proposed a selection of papers to be presented to the Euro-Par program committee that based on these suggestions selected four papers for presentation at Euro-Par 2016 in Grenoble.

The selected papers highlight important aspects of parallel numerical algorithms that are currently under investigation.

The topic committee wishes to thank all authors who contributed to this important topic by submitting a paper to Euro-Par, as well as all referees for providing the reviews on time. Further, we like to thank the organizing committee for the effort spent in order to allow the participants of this year's Euro-Par to present and discuss the latest results and findings in parallel computing in Grenoble.

Topic 12: Accelerator Computing

Enrique S. Quintana-Orti, Samuel Thibault, Taisuke -Arai- Boku, Esteban Clua, Hatem Ltaief, Jeff Hammond, John Stone, Robert Strzodka

Different co-processor technology promise today a potential for accelerating large-scale applications by leveraging their much higher hardware concurrency and/or customization. Current examples range from graphics processors (GPUs) to "many-core" general-purpose processors, such as the Intel Xeon Phi, as well as custom devices, FPGA-based systems, and streaming data-flow architectures.

This topic aims to explore new avenues for actually realizing this potential, promoting significant advances and solutions in areas related to accelerators, and in particular, in architectures, algorithms, languages, compilers, libraries, runtime systems, coordination of accelerators and CPU, and debugging and profiling tools.

The topic received 16 contributions, and 3 of these were accepted for presentation.

Contents

Scheduling and Load Balancing

High Performance Architectures and Compilers

Parallel and Distributed Data Management and Analytics

Cluster and Cloud Computing

Distributed Systems and Algorithms

Parallel and Distributed Programming, Interfaces, Language

Multicore and Manycore Parallelism

Theory and Algorithms for Parallel Computation and Networking

Parallel Numerical Methods and Applications

Accelerator Computing

Invited Papers

Resampling with Feedback — A New Paradigm of Using Workload Data for Performance Evaluation

Dror G. Feitelson[✉]

School of Computer Science and Engineering,
The Hebrew University of Jerusalem, 91904 Jerusalem, Israel
feit@cs.huji.ac.il

Abstract. Reliable performance evaluations require representative workloads. This has led to the use of accounting logs from production systems as a source for workload data in simulations. But using such logs directly suffers from various deficiencies, such as providing data about only one specific situation, and lack of flexibility, namely the inability to adjust the workload as needed. Creating workload models solves some of these problems but creates others, most notably the danger of missing out on important details that were not recognized in advance, and therefore not included in the model. Resampling solves many of these deficiencies by combining the best of both worlds. It is based on partitioning real workloads into basic components (e.g. the jobs contributed by different users), and then generating new workloads by sampling from this pool of basic components. The generated workloads are adjusted dynamically to the conditions of the simulated system using a feedback loop, which may adjust the throughput. Using this methodology analysts can create multiple varied (but related) workloads from the same original log, all the time retaining much of the structure that exists in the original workload. Resampling with feedback thus provides a new way to use workload logs which benefits from the realism of logs while eliminating many of their drawbacks. In addition, it enables evaluations of throughput effects that are impossible with static workloads.

This paper was written to accompany a keynote address at EuroPar 2016. It summarizes my and my students' work and reflects a personal view. The goal is to show the big picture and the building and interplay of ideas, at the possible expense of not providing a full overview of and comparison with related work.

1 Introduction

Performance evaluation is a basic element of experimental computer science. It is used to compare design alternatives when building new systems, to tune parameter values of existing systems, and to assess capacity requirements when setting up systems for production use. Lack of adequate performance evaluations can lead to bad decisions, which imply either not being able to accomplish mission

© Springer International Publishing Switzerland 2016
P.-F. Dutot and D. Trystram (Eds.): Euro-Par 2016, LNCS 9833, pp. 3–21, 2016.
DOI: 10.1007/978-3-319-43659-3_1

objectives or inefficient use of resources. A good evaluation study, on the other hand, can be instrumental in the design and realization of an efficient and useful system.

It is widely accepted that the performance of a computer system depends on its design and implementation. This is why performance evaluations can be used to judge designs and assess implementations. But performance also depends on the workload to which the system is subjected. Evaluating a system with the wrong workload will most probably lead to erroneous results, that cannot be relied upon [9,15]. It is therefore imperative to use representative and reliable workloads to drive performance evaluation studies. However, workloads can have complicated structures and distributions, so workload characterization can be a hard task to perform [4].

The problem is exacerbated by the fact that workloads may interact with the system and even with the performance metrics in non-trivial ways [11,12]. Thus it may not be enough to use a workload model that is generally correct, and it may be important to get minute details correct too. But it is not always clear in advance which details are the important ones. This suggests that workload should be comprehensive and include *all* possible attributes [22].

In the field of parallel job scheduling, the workload is the sequence of jobs submitted to the system. Early research in this field, in the 1980s, lacked data on which to base workloads. Instead studies were based on what were thought to be reasonable assumptions, or compared possible distributions — for example, a uniform distribution of job sizes, a distribution over powers of 2, and a harmonic distribution [19,20]. But it was not known which of these is the most realistic.

Since the mid 1990s workload logs became available (starting with [18]) and were collected in the Parallel Workloads Archive [24,29]. This enabled the substitution of assumptions with hard data [14]. In particular, using logged data to drive simulations became the norm in evaluations of parallel job schedulers. But experience with this methodology exposed problems, especially in the context of matching the workload to the simulated system. This is described below in Sect. 3.

The suggested solution to these problems is to use resampling with feedback, as described in Sect. 4. The idea is to partition the workload into basic components, and sample from this pool of components to create multiple alternative workloads [42]. At the same time, feedback from the simulated system is used to pace the workload generation process as would occur in reality [30,32,44]. The resulting methodology enables evaluations that are not possible when using logs as they were recorded. And it applies to any system type, not only to the context of parallel job scheduling.

2   Background

Our discussion is couched in the domain of parallel job scheduling. Parallel jobs are composed of multiple interacting processes which run on distinct processors. They can therefore be modeled as rectangles in *processors* × *time* space, where the height of the rectangle represents the number of processors used, and its width represents the duration of use.

Scheduling parallel jobs is the decision of when each job will run. The simplest scheduling algorithm is First-Come-First-Serve (FCFS), which simply schedules the jobs in the order that they are submitted to the system (Fig. 1). An alternative is EASY, named after the Extensible Argonne Scheduling sYstem which introduced it [26, 28]. The idea here is to optimize the schedule by taking small jobs from the back of the queue, and using them to fill in holes that were left in the schedule, an operation known as *backfilling*. This reduces fragmentation and improves throughput.

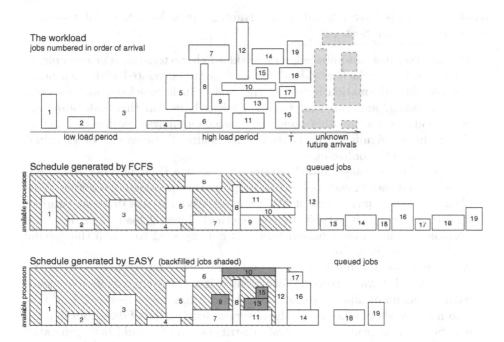

Fig. 1. Illustration of a sequence of parallel jobs (the workload) and how it would be scheduled by FCFS and EASY up to time T.

But note that the utility of backfilling depends on the workload. For example, if all the jobs require more than half the processors, two jobs can never run at the same time, and backfilling cannot be used. Thus if EASY is evaluated with such a workload, the result would be that the backfilling optimization is useless, but if real workloads actually do include many small jobs then this conclusion would be wrong. Therefore workloads used in evaluations must be representative of real workloads. Our work is about how to achieve this goal.

3   Using Workload Logs and Models to Drive Simulations

There are two common ways to use a measured workload to analyze or evaluate a system design: (1) use the logged workload directly to drive a simulation, or (2) create a model from the log and use the model for either analysis or simulation. As we'll show, both have deficiencies that may lead to problems in evaluations. The idea of resampling can be thought of as combining the two in order to enjoy the best of both worlds.

3.1   Workload Modeling

Workload models have a number of advantages over logs. Some of the most salient ones are [15, Sect. 1.3.2]:

- The modeler has full knowledge of workload characteristics. For example, it is easy to know which workload parameters are correlated with each other because this information is part of the model. Such knowledge increases our understanding, and can lead to new designs based on this understanding. Workload logs, on the other hand, may include unknown features that nevertheless have a significant influence on the results. These cannot be exploited and may lead to confusion.
- It is possible to change model parameters one at a time, in order to investigate the influence of each one, while keeping other parameters constant. This allows for direct measurement of system sensitivity to the different parameters. In particular, it is typically easy to check different load levels. It is also possible to select model parameters that are expected to match the specific workload at a given site.
- A model is not affected by policies and constraints that are particular to the site where a log was recorded. For example, if a site configures its job queues with a maximum allowed duration of 4 h, it forces users to break long jobs into multiple short jobs. Thus, the observed distribution of durations in a log will be different from the "natural" distribution users would have generated under a different policy, and the log — despite being "real" — is actually unrepresentative.
- Logs may be polluted by bogus data. For example, a log may include records of jobs that were killed because they exceeded their resource bounds. Such jobs impose a transient load on the system, and influence the arrival process. However, they may be replicated a number of times before completing successfully, and only the successful run represents "real" work. In a model, such jobs can be avoided (but they can also be modeled explicitly if so desired).
- Models have better statistical properties: they are usually stationary, so evaluation results converge faster [8], and they allow multiple statistically equivalent simulations to be run so as to support the calculation of confidence intervals. Logs, on the other hand, provide only a single data point, which may be based on an unknown mixture of conditions.

These advantages have led to the creation and use of several workload models (e.g. [2,3,25,27]), and even a quest for a general, parameterized workload model that can serve as a canonical workload in all evaluations [21].

3.2 Problems with Models

But models include only what you know about in advance, and decide to incorporate in the model. Over the years several examples of important attributes that were missed have been uncovered.

Perhaps the most interesting feature of parallel job workloads — in terms of its unexpected importance — is user runtime estimates. Many schedulers (including EASY) require users to provide estimates of job runtime when submitting a job; these estimates are then used by the scheduler to plan ahead. But simulations often assumed that perfect estimates are available. This turned out to be wrong on two counts: first, estimates are actually very inaccurate (Fig. 2) [28], and second, it actually matters [12,38]. In retrospect we can now fully understand the interactions between estimates and other features of the workload, and the conditions under which one scheduler is better than the other. We can also model realistic (inaccurate) estimates [35]. But the more important result is the demonstration that performance evaluation results may be swayed by innocent-looking workload details, and that a very detailed analysis is required in order to uncover such situations.

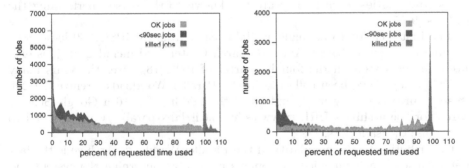

Fig. 2. Histograms of user runtime estimates as a fraction of the actual runtimes, from logs from the CTC and KTH SP2 machines [28]. The peak at 100 % is jobs killed because they exceeded their estimate; for other jobs except the very shortest ones the histogram is flat. (Color figure online)

Another example is that real workloads are obviously non-stationary: they have daily, weekly, and even yearly cycles. In many cases this is ignored in performance evaluations, with the justification that only the high load at prime time is of interest. While this is reasonable in the context of network communication, where the individual workload items (packets) are very small, it is very dubious in the context of parallel jobs, that may run for many hours. And in fact we have found that optimizing schedulers may actually depend on the existence of the daily cycle, because they try to delay non-critical jobs submitted during prime time and execute them at night [22]. If there is no daily cycle there is no non-prime time, and thus no alternative to executing these jobs at once.

Yet another effect that is prevalent in logs but usually absent from models is locality [13]. The locality properties of real workloads are especially important for the evaluation of adaptive and predictive systems (for example, it may be possible to predict job runtimes and compensate for inaccurate estimates [36]). Such features are becoming more commonplace with the advent of self-tuning and self-management. The idea is that the system should be able to react to changing conditions, without having to be reconfigured by a human operator [17]. But in order to study such systems, we need workloads with changing conditions as in real workload logs. A model based on random sampling from a distribution will not do, as it creates a stationary workload. This can be solved by employing "localized sampling" from the distribution [13], but a better solution is to use user-based modeling (or resampling, as described below).

3.3 Using Logs Directly

The perception that workload models may be over-simplified and unjustified has led many researchers to prefer real workload logs. The advantage of using a traced log directly as the input to a simulation is that it is the most "real" test of the simulated system: the workload reflects a real workload precisely, with all its complexities, even if they are not known to the person performing the analysis [9,15].

The first such log to be made available came from the iPSC/860 hypercube machine installed at NASA Ames Research Center, and included all jobs executed on the system in the fourth quarter of 1993 [18]. Over the years many additional logs have been collected in the Parallel Workloads Archive [24,29]. This resource is widely used, and as of the middle of 2016 a Google Scholar search for the archive's URL (www.cs.huji.ac.il/labs/parallel/workload) led to nearly one thousand hits.

Contributing to the popularity of the Parallel Workloads Archive is the fact that each log is accompanied by copious metadata concerning the system and the logged data. In addition, all the logs are converted to a "standard workload format" [1]. Thus if a simulator can read this format, it can immediately run simulations using all the logs in the archive.

3.4 Drawbacks of Using Logs

While using logs "as is" avoids the problems associated with models, logs too have their drawbacks. The most noticeable ones are as follows:

- Each log reflects only one specific workload, and can only provide a single data point to the evaluation. But evaluations often require multiple simulations with related workloads. For example, the calculation of confidence intervals is best done by running multiple simulations with distinct but statistically identical workloads. This is easy with a workload model but impossible with a log.

– More specifically, it is not possible to manipulate logs to adjust the workload to the simulated system and conditions, and even when it is possible, it can be problematic. In particular, it is often desirable to evaluate the performance of a system under different load conditions, e.g. to check its stability or the maximal load it can handle before saturating. Thus a single load condition (as provided by a log) is not enough, and we need a tunable parameter that allows for the generation of different load conditions.

In log-based simulations it is common practice to increase the load on the system by reducing the average interarrival time. For example, if a log represents a load of 70 % of system capacity, multiplying all interarrival times by a factor of $7/8 = 0.875$ will increase the load to 80 %. But this practice has the undesirable consequence of shrinking the daily load cycle as well. The alternative of increasing the runtime to increase the load is not much better: jobs that originally came one after the other, and maybe even depended on each other, may now overlap. And increasing the number of processors to increase load is even worse. For example, if job sizes tend to be powers of 2 (which they are) then they pack well together. Increasing them by say 10 % is not always possible (a 4-processor job can only be increased in increments of 25 %), and when possible it has detrimental effects on the packing of the jobs onto processors.

– Another drawback is the need for workload cleaning. Real workloads sometimes include unrepresentative activity, like huge short-lived surges of activity by individual users (flurries, Fig. 3). While the existence of flurries is not uncommon (many logs exhibit them up to a few times a year), they are very different from the normal workload between them, and also different from each other. They should therefore be removed from the workload logs before they are analyzed and used in simulations [23,37].

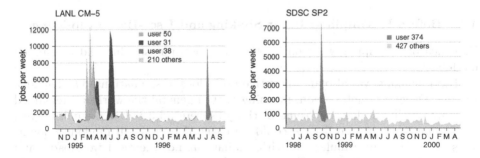

Fig. 3. Arrivals per week on two parallel supercomputers, showing flurries of activity due to single users [23,37]. (Color figure online)

At a deeper level, we find that logged workloads actually contain a "signature" of the logged system. In other words, there is no such thing as a "real" workload which is the right one for general use: *every workload observed on a real system is the result of the interaction between that particular system and its users.* If the system behaves differently, the users change their behavior as well.

This has grave implications. It means that using a workload from one system to evaluate a different system is wrong, because the workload will not fit the simulation conditions [30, 31]. We demonstrated this using a pair of cross-simulations of two different schedulers. The first is the well known FCFS scheduler, which is inefficient and leads to wasted resources as processors are left idle until the first queued job can run. The second is the optimizing EASY scheduler, which optimizes the schedule by taking small jobs from down the queue and backfilling them into holes left between earlier jobs. This allows EASY to sustain a heavier load. And indeed, simulation of FCFS using a workload generated by an EASY simulation (using the same underlying system) led to system saturation and overloading: FCFS could not handle the load that was generated when users interacted with EASY. Conversely, simulation of EASY using a workload generated by a FCFS simulation failed to show that EASY had any advantage, because the workload was not challenging enough.

Taken together, all these problems seem to imply that workload logs are actually not any better than workload models. Resampling and feedback are designed to solve the problems and facilitate reliable evaluations.

4   Resampling and Feedback

The root cause of many of the problems with using logs is that a log represents unique conditions that were in effect when it was recorded, and may not be suitable for the desired evaluation. At the same time logs contain significant structure that we want to retain. Resampling is a way to provide flexibility while preserving the structure. And feedback adds just the necessary level of adjustment to the conditions that hold during the evaluation.

4.1   Before Resampling: Input Shaking and User-Based Modeling

The idea of resampling grew out of the ideas of input shaking and user-based modeling.

Input shaking was also an innovative use of logs in simulations [39]. The idea was to "shake" the job stream, meaning that in general the workload remained the same as in the log, but some of the jobs were adjusted to a small degree. For example, their arrival time could be changed by a small random amount. This enabled many simulations with similar but not identical workloads, and facilitated the identification of situations where the original results were actually due to some artifact and therefore not representative.

User-based modeling is a generative approach to creating workloads, which was first proposed as a mechanism to generate locality of sampling [10]. The idea was to view the workload as being composed from the activities of many individual users, each of which submits jobs with different characteristics [4,7]. Since the community of active users changes over time, the number of active users in a given week — and the number of different programs they run — will be relatively small. The short-term workload will therefore tend to include repetitions of similar jobs, and will consequently tend to have more locality and be more predictable. But over a longer period this will change, because the set of active users has changed.

The essence of user-based modeling is an attempt to capture this structure using a multi-level model of the user population and the behavior of individual users. The top level is a model of the user population, including the arrival of new users and the departure of previous users. The second level models the activity of individual users as a sequence of sessions synchronized with the time of day (again based on data extracted from logs [41]). The lowest level includes repetitions of jobs within each session.

Note, however, that user-based modeling is not easy, as we typically do not have any explicit information about a user's motivation and considerations. But still some information can be gleaned by analyzing logs. For example, the question of what annoys users more and causes them to abort their interactive work has been investigated by tabulating the probability to submit another job as a function of the previous job's response time or its slowdown [31]. The result was that response time was the more meaningful metric (Fig. 4).

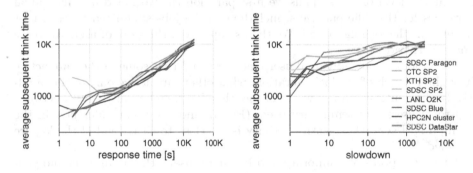

Fig. 4. A job's performance as measured by the response time is a better predictor of subsequent behavior (think time till the next job) than the job's slowdown. (Color figure online)

Remarkably, user-based modeling makes significant progress towards solving the problems outlined above:

– The workload will naturally have locality provided that the job models of different users are different from each other. During the tenure of each set of users the job stream will reflect the behavior of those users.

– The load on the system can be modified by changing the number of active users, or in other words, by changing parameters of the user population model. More users would generate higher load, but do it "in the right way".
– The generated workload can include non-stationary elements such as a daily cycle, by virtue of the model of when users engage in sessions of activity [32].
– As a special case, unique events such as workload flurries can be included or excluded at will, by including or excluding users with such unique behaviors.
– By using heavy-tailed session durations (and inter-session breaks) one can generate self similarity [40], which has been found in many types of workloads including parallel jobs [34].

But on the other hand, maybe all this modeling is too far removed from the original log data? Resampling was designed to retain the original data as much as possible, and modify only whatever is needed for a specific purpose.

4.2   Resampling from a Log

Resampling is a powerful technique for statistical reasoning in situations where not enough empirical data is available [5,6]. The idea is to use the available data sample as an approximation of the underlying population, and resample from it. Applying this to workloads, we partition a workload log into its basic components and re-group them in different ways to achieve the desired effects. In the context of parallel job scheduling, we suggest that the resampling be done at the level of users. Thus we first partition the workload into individual subtraces for the different users, including all the jobs submitted by each user throughout the logging period. We then sample from this pool of users to create a new workload [42].

When looking at individual user traces, we find that some of then are active throughout much of the log's duration, while others are active only during a relatively short interval (a few weeks or months). We therefore distinguish between long-term users and temporary users (Fig. 5), and use them differently in the resampling. Users whose entire activity is too close to either end of the log are excluded.

Given the pools of temporary and long-term users, the resampling and generation of a new workload is done as follows:

– **Initialization:** We initialize the active users set with some temporary users and some long-term users. The defaults are the number of long-term users in the original log, and the average number of temporary users present in a single week of the original log. Users are not started with their first job from the trace, because we are trying to emulate a workload that was recorded over an arbitrary timespan, and there is no reason to assume that the beginning of the logging period should coincide with the beginning of a user's activity. Therefore each user is started in some arbitrary week of his traced activity. However, care is taken that jobs start on the same day of the week and time of the day in the simulation as in the original log.

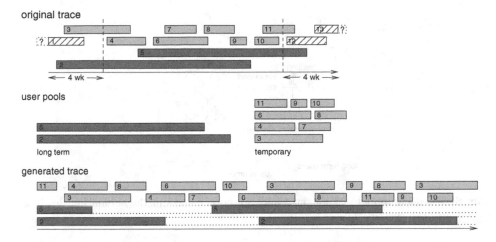

Fig. 5. Conceptual framework of dividing users into long-term and temporary, and reusing them in a generated workload [42]. Each rectangle represents the full extent of activity by a certain user.

- **Temporary users:** In each new week of the simulation, a certain number of new temporary users are added (and a similar number are expected to leave, on average). The exact number is randomized around the target number, which defaults to the average rate at which temporary users arrived in the original log. The selected users are started from their first traced jobs. A user can be selected from the pool multiple times, but care is taken not to select the same user twice in the same week.
- **Long-term users:** The population of long-term users is constant and consists of those chosen in the initialization. When the traced activity of a long-term user is finished, it is simply regenerated after a certain interval. Naturally the regenerations are also synchronized correctly with the time and day.

Each active user submits jobs to the system exactly as in the log (except that their timing may vary to reflect feedback as explained below). The flow of the simulation is shown in Fig. 6.

4.3   Adding Feedback

Computer systems are not closed systems. Rather, they interact with their environment, and in particular with their users. We therefore suggest that it is not enough to simulate the computer system in isolation — we should also simulate the system's environment, namely the users who interact with the system, create its input, and wait for its response [30]. With resampling we introduce this explicitly by including a changing user community in the simulation. It is these (simulated) users who create the (simulated) jobs submitted to the (simulated) system.

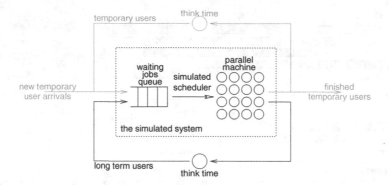

Fig. 6. Queueing model of long term and temporary users in the simulation, leading to a semi-open system [44].

The fact that jobs are submitted by simulated users might seem innocuous at first, but in fact it has grave implications. When users wait for the system before deciding that to do next they introduce a feedback loop (Fig. 7). And such feedback implies a pacing of the workload — it is a stabilizing *negative* feedback, where extra load causes the generation of additional load to be throttled [16]. This reduces the risk of system saturation.

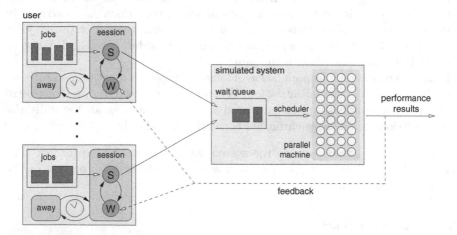

Fig. 7. Illustration of a user-based simulation with feedback. When users are in session, they alternate between submitting jobs (S) and waiting for feedback regarding previous jobs (W).

The problem with modeling the effect of feedback is that accounting logs used as data sources do not include explicit information regarding the dependencies between jobs. We therefore need to identify user sessions and extract dependencies between the jobs in each session [30,43]. These dependencies are then used during the simulation to pace the job submittal rate. Additional jobs will be submitted (by the simulated users) only after the jobs that they depend on have terminated (on the simulated system).

In other words, when we want to evaluate a new scheduling policy using a representative workload, the workload should reflect the user-level logic and not just parrot a previous workload. This logic is embodied in the dependencies between jobs. We argue that *it is more important to preserve the logic of the users' behavior than to repeat the exact timestamps that appear in the original log.*

The way to integrate such considerations into log-driven simulations is by manipulating the timing of job arrivals. In other words, the *sequence* of jobs submitted by each user stays the same, but the *submittal times* are changed [43]. Specifically, each job's submit time is adjusted to reflect feedback from the system performance to the user's behavior.

However, a job cannot arrive immediately when all its constraints are removed. Rather, its arrival should reflect reasonable user behavior (for example, users often go to sleep at night). One possible model of user behavior is the "fluid" user model. The idea of this model is to retain the original session times of the users, but allow jobs to flow from one session to another according to the feedback. To do that, we keep each session's start and end timestamps from the original log. The think times between successive jobs are also retained from the original log. But if a job's execution is delayed in the simulation, leading to the next arrival falling beyond the end of the session, the next job will be delayed even more and arrive only at the beginning of the next session [43]. Contrariwise, if jobs terminate sooner in the simulation, jobs that were submitted originally in the next session may flow forward to occur in the current one.

4.4 Applications and Benefits

So what can we do with this new tool of workload resampling with feedback? Here are some results that would be hard or impossible to achieve with conventional simulations that just replay an existing log.

The first and foremost is to validate simulation results. Simulating with a given log provides a single data point. But with resampling we can get a distribution based on statistically similar workloads. In most cases this distribution is centered on the value which is obtained using a conventional simulation, and the result is verified (Fig. 8). But in some cases (e.g. the Blue log on the right) the distribution is shifted, indicating a mismatch between the behavior of the users in the original log and the expected behavior in the simulated system.

Using results from resampling and feedback for verification hinges on the claim that such simulations are more valid to begin with. As noted above, using a log to drive a simulation suffers from the possible mismatch between the behavior of the users in the logged system and the behavior that would be appropriate for the simulated system. In particular, if the simulated system is more powerful, the users would be expected to submit more jobs, and vice versa. In simulations with feedback this indeed happens automatically, as demonstrated in Fig. 9.

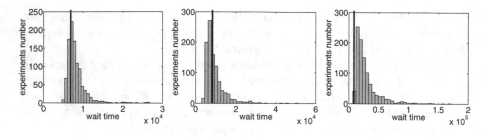

Fig. 8. Histograms of the average waiting time in a thousand simulations of EASY on resampled workloads, compared to a simulation using the original logs (vertical line). Left to right: CTC SP2, SDSC Datastar, and Blue Horizon logs. (Color figure online)

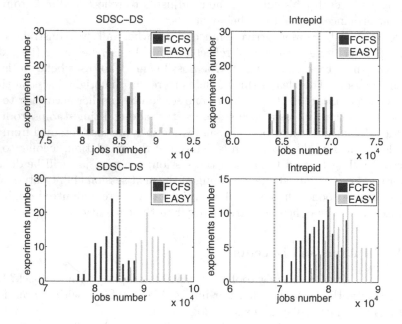

Fig. 9. Histograms of the throughput achieved in one hundred simulations of the EASY and FCFS schedulers using the SDSC DataStar and Intrepid logs [44]. Top: in simulation without feedback the throughput is determined by the input, and a better scheduler has no effect. Bottom: in simulation with feedback EASY is seen to support a higher throughput than FCFS. Vertical line represents the throughput in the original log. (Color figure online)

Other benefits are more technical in nature. One of them is the ability to extend a log and create a longer workload, with more jobs in total, so as to facilitate better convergence of the results and reduce the detrimental effects of the initial "warmup" period. This is easy to achieve: we just continue resampling on and on for as long as we wish.

We can also change the load on the system (as has been noted above) by increasing or decreasing the number of active users. This is done by changing the number of long-term users in the initialization, and the number of new temporary users which arrive in each simulated week. Such a manipulation facilitates the study of how the system responds to load, and enables the generation of response curves similar to those obtained from queueing analyses.

However, note that increasing the load on the system is at odds with the throttling effect that comes with feedback. As the load increases, the system will naturally take longer to process each job. Simulated users who are waiting for a job to terminate before submitting their next job will therefore be delayed. So having more users will eventually cause each of these users to produce additional work at a slower rate, and the load will cease to increase! This is good because it is assumed that such an effect exists in real systems, where users abandon the system if it is too slow. But it frustrates our attempt to control the load and increase it at will.

Nevertheless, adding users to increase the load has two additional benefits. One is the ability to make low-load logs usable. Some logs were recorded on very low-load systems, with a utilization of only 25 % or capacity or so. These workloads are not interesting as they do not tax the system to any appreciable degree. But by using resampling to increase their load they become interesting.

The other and more important benefit is the ability to measure the maximal load that can be sustained before the system saturates. Resampling and user-based modeling support the use of throughput as a performance metric, because the users become part of the simulation. But every system has a maximal capacity, and if the load exceeds this capacity the system saturates. Identifying this maximal capacity is an important part of a performance evaluation. Likewise, if we add system abandonment to the user behavior model, we can add the metric of number of frustrated users.

In a related vein, simulations based on resampling and feedback can be used to evaluate adaptive systems that are designed to enhance throughput (and thus productivity) rather than response time [32,44]. For example, we can design a scheduler that prioritizes jobs based on the elapsed time since the same user's previous job has terminated. The idea is that if this interval is short, there is a good chance that the user is waiting for this job. Therefore prioritizing the awaited job will enhance the productivity of this user. Trying to evaluate this idea with conventional workloads and simulations is useless — such simulations cannot evaluate productivity, and might even show that average response time is actually increased. But with a dynamic user-based simulation we can do away with such averages, and focus on the users and the service they receive.

Finally, we note that once we partition the workload into individual users we can also look and different user classes in isolation. One example noted before is the workload flurries occasionally produced by some users. We can then evaluate the effect of such flurries by oversampling these users, and thus causing more and more flurries to occur.

5 Conclusions

Resampling with feedback provides a new way to use workload logs in simulations, enabling the generation of varied and dynamically adjusted workloads that are specifically suited to evaluate the simulated system. This combines the realism of real log data with the flexibility of models. Basing the simulations as closely as possible on real logs reflects the importance of using hard data rather than assumptions. Adjusting the workload to the specific conditions during the simulation reflects the importance of the interaction between the users and the system. Without this, evaluation results are of unknown relevance, and might pertain to only irrelevant situations which do not occur in practice.

Particularly, by applying resampling and feedback to real workloads we achieve the following:

– Retain the all important (but possibly unknown) details of the workload as they exist in logs recorded from production systems, with as little modifications as possible.
– Enable evaluations of throughput and user satisfaction in addition to (or instead of) being limited to the response time and slowdown metrics. This also leads to natural support for assessing the saturation limit of the system.
– Provide a new interpretation of the goal of "comparing alternatives under equivalent conditions": this is not to process exactly the same job stream, but rather to face the same workload generation process (users). This acknowledges the realization that there is no such thing as a generally correct workload — rather, the workload depends on system.

Workload manipulations such as those embodied in resampling with feedback are important tools in the performance analyst's toolbox, that have not received due attention in terms of methodological research. As a result, inappropriate manipulations are sometimes used, which in turn has led to some controversy regarding whether *any* manipulations of real workloads are legitimate. By increasing our understanding of resampling-based manipulations we hope to bolster the use of this important tool, allowing new types of manipulations to be applied to workload logs, and enabling researchers to achieve better control over their properties, as needed for different evaluation scenarios.

Naturally, there are many opportunities for additional research regarding resampling and feedback. One element that is still largely missing is the user population model, and especially the issue of leaving the system when performance is inadequate. Another is the distribution of user types and behaviors. Resolving these issues requires not only deep analysis of workload logs, but also a collaboration with researchers in psychology and cognition [33]. After all, computer systems are used by humans.

Acknowledgments. The work described here was by and large performed by several outstanding students, especially Edi Shmueli, Netanel Zakay, and Dan Tsafrir. Our work was supported by the Israel Science Foundation (grants no. 219/99 and 167/03) and the Ministry of Science and Technology, Israel.

References

1. Chapin, S.J., Cirne, W., Feitelson, D.G., Jones, J.P., Leutenegger, S.T., Schwiegelshohn, U., Smith, W., Talby, D.: Benchmarks and standards for the evaluation of parallel job schedulers. In: Feitelson, D.G., Rudolph, L. (eds.) JSSPP 1999. LNCS, vol. 1659, pp. 67–90. Springer, Heidelberg (1999). doi:10.1007/3-540-47954-6_4
2. Cirne, W., Berman, F.: A comprehensive model of the supercomputer workload. In: 4th Workshop on Workload Characterization, pp. 140–148, December 2001. doi:10.1109/WWC.2001.990753
3. Downey, A.B.: A parallel workload model and its implications for processor allocation. Cluster Comput. 1(1), 133–145 (1998). doi:10.1023/A:1019077214124
4. Downey, A.B., Feitelson, D.G.: The elusive goal of workload characterization. Perform. Eval. Rev. 26(4), 14–29 (1999). doi:10.1145/309746.309750
5. Efron, B.: Bootstrap methods: another look at the jackknife. Ann. Statist. 7(1), 1–26 (1979). doi:10.1214/aos/1176344552
6. Efron, B., Gong, G.: A leisurely look at the bootstrap, the jackknife, and cross-validation. Am. Stat. 37(1), 36–48 (1983). doi:10.2307/2685844
7. Feitelson, D.G.: Memory usage in the LANL CM-5 workload. In: Feitelson, D.G., Rudolph, L. (eds.) JSSPP 1997. LNCS, vol. 1291, pp. 78–94. Springer, Heidelberg (1997). doi:10.1007/3-540-63574-2_17
8. Feitelson, D.G.: Metrics for parallel job scheduling and their convergence. In: Feitelson, D.G., Rudolph, L. (eds.) JSSPP 2001. LNCS, vol. 2221, pp. 188–205. Springer, Heidelberg (2001). doi:10.1007/3-540-45540-X_11
9. Feitelson, D.G.: The forgotten factor: facts; on performance evaluation and its dependence on workloads. In: Monien, B., Feldmann, R.L. (eds.) Euro-Par 2002. LNCS, vol. 2400, pp. 49–60. Springer, Heidelberg (2002). doi:10.1007/3-540-45706-2_4
10. Feitelson, D.G.: Workload modeling for performance evaluation. In: Calzarossa, M.C., Tucci, S. (eds.) Performance 2002. LNCS, vol. 2459, pp. 114–141. Springer, Heidelberg (2002). doi:10.1007/3-540-45798-4_6
11. Feitelson, D.G.: Metric and workload effects on computer systems evaluation. Computer 36(9), 18–25 (2003). doi:10.1109/MC.2003.1231190
12. Feitelson, D.G.: Experimental analysis of the root causes of performance evaluation results: a backfilling case study. IEEE Trans. Parallel Distrib. Syst. 16(2), 175–182 (2005). doi:10.1109/TPDS.2005.18
13. Feitelson, D.G.: Locality of sampling and diversity in parallel system workloads. In: 21st International Conference on Supercomputing, pp. 53–63, June 2007. doi:10.1145/1274971.1274982
14. Feitelson, D.G.: Looking at data. In: 22nd International Parallel & Distributed Processing Symposium, April 2008. doi:10.1109/IPDPS.2008.4536092
15. Feitelson, D.G.: Workload Modeling for Computer Systems Performance Evaluation. Cambridge University Press, Cambridge (2015)
16. Feitelson, D.G., Mu'alem, A.W.: On the definition of "on-line" in job scheduling problems. SIGACT News 36(1), 122–131 (2005). doi:10.1145/1052796.1052797
17. Feitelson, D.G., Naaman, M.: Self-tuning systems. IEEE Softw. 16(2), 52–60 (1999). doi:10.1109/52.754053
18. Feitelson, D.G., Nitzberg, B.: Job characteristics of a production parallel scientific workload on the NASA Ames iPSC/860. In: Feitelson, D.G., Rudolph, L. (eds.) JSSPP 1995. LNCS, vol. 949, pp. 337–360. Springer, Heidelberg (1995). doi:10.1007/3-540-60153-8_38

19. Feitelson, D.G., Rudolph, L.: Distributed hierarchical control for parallel processing. Computer **23**(5), 65–77 (1990). doi:10.1109/2.53356
20. Feitelson, D.G., Rudolph, L.: Evaluation of design choices for gang scheduling using distributed hierarchical control. J. Parallel Distrib. Comput. **35**(1), 18–34 (1996). doi:10.1006/jpdc.1996.0064
21. Feitelson, D.G., Rudolph, L.: Metrics and benchmarking for parallel job scheduling. In: Feitelson, D.G., Rudolph, L. (eds.) JSSPP 1998. LNCS, vol. 1459, pp. 1–24. Springer, Heidelberg (1998). doi:10.1007/BFb0053978
22. Feitelson, D.G., Shmueli, E.: A case for conservative workload modeling: parallel job scheduling with daily cycles of activity. In: 17th Modeling, Analysis & Simulation of Computer and Telecommunication Systems, September 2009. doi:10.1109/MAS-COT.2009.5366139
23. Feitelson, D.G., Tsafrir, D.: Workload sanitation for performance evaluation. In: IEEE International Symposium on Performance Analysis of Systems & Software, pp. 221–230, March 2006. doi:10.1109/ISPASS.2006.1620806
24. Feitelson, D.G., Tsafrir, D., Krakov, D.: Experience with using the Parallel Workloads Archive. J. Parallel Distrib. Comput. **74**(10), 2967–2982 (2014). doi:10.1016/j.jpdc.2014.06.013
25. Jann, J., Pattnaik, P., Franke, H., Wang, F., Skovira, J., Riodan, J.: Modeling of workload in MPPs. In: Feitelson, D.G., Rudolph, L. (eds.) JSSPP 1997. LNCS, vol. 1291, pp. 95–116. Springer, Heidelberg (1997). doi:10.1007/3-540-63574-2_18
26. Lifka, D.: The ANL/IBM SP scheduling system. In: Feitelson, D.G., Rudolph, L. (eds.) JSSPP 1995. LNCS, vol. 949, pp. 295–303. Springer, Heidelberg (1995). doi:10.1007/3-540-60153-8_35
27. Lublin, U., Feitelson, D.G.: The workload on parallel supercomputers: modeling the characteristics of rigid jobs. J. Parallel Distrib. Comput. **63**(11), 1105–1122 (2003). doi:10.1016/S0743-7315(03)00108-4
28. Mu'alem, A.W., Feitelson, D.G.: Utilization, predictability, workloads, and user runtime estimates in scheduling the IBM SP2 with backfilling. IEEE Trans. Parallel Distrib. Syst. **12**(6), 529–543 (2001). doi:10.1109/71.932708
29. Parallel Workloads Archive. http://www.cs.huji.ac.il/labs/parallel/workload/
30. Shmueli, E., Feitelson, D.G.: Using site-level modeling to evaluate the performance of parallel system schedulers. In 14th Modeling, Analysis & Simulation of Computer and Telecommunication Systems, pp. 167–176, September 2006. doi:10.1109/MAS-COTS.2006.50
31. Shmueli, E., Feitelson, D.G.: Uncovering the effect of system performance on user behavior from traces of parallel systems. In 15th Modeling, Analysis & Simulation of Computer and Telecommunication Systems, pp. 274–280, October 2007. doi:10.1109/MAS-COTS.2007.67
32. Shmueli, E., Feitelson, D.G.: On simulation and design of parallel-systems schedulers: are we doing the right thing? IEEE Trans. Parallel Distrib. Syst. **20**(7), 983–996 (2009). doi:10.1109/TPDS.2008.152
33. Snir, M.: Computer and information science and engineering: one discipline, many specialties. Comm. ACM **54**(3), 38–43 (2011). doi:10.1145/1897852.1897867
34. Talby, D., Feitelson, D.G., Raveh, A.: Comparing logs and models of parallel workloads using the co-plot method. In: Feitelson, D.G., Rudolph, L. (eds.) JSSPP 1999. LNCS, vol. 1659, p. 43. Springer, Heidelberg (1999). doi:10.1007/3-540-47954-6_3
35. Tsafrir, D., Etsion, Y., Feitelson, D.G.: Modeling user runtime estimates. In: Feitelson, D.G., Frachtenberg, E., Rudolph, L., Schwiegelshohn, U. (eds.) JSSPP 2005. LNCS, vol. 3834, pp. 1–35. Springer, Heidelberg (2005). doi:10.1007/11605300_1

36. Tsafrir, D., Etsion, Y., Feitelson, D.G.: Backfilling using system-generated predictions rather than user runtime estimates. IEEE Trans. Parallel Distrib. Syst. **18**(6), 789–803 (2007). doi:10.1109/TPDS.2007.70606

37. Tsafrir, D., Feitelson, D.G.: Instability in parallel job scheduling simulation: the role of workload flurries. In: 20th International Parallel & Distributed Processing Symposium, April 2006. doi:10.1109/IPDPS.2006.1639311

38. Tsafrir, D., Feitelson, D.G.: The dynamics of backfilling: Solving the mystery of why increased inaccuracy may help. In: IEEE International Symposium on Workload Characterization, pp. 131–141, October 2006. doi:10.1109/IISWC.2006.302737

39. Tsafrir, D., Ouaknine, K., Feitelson, D.G.: Reducing performance evaluation sensitivity and variability by input shaking. In: 15th Modelling, Analysis & Simulation of Computer and Telecommunication Systems, pp. 231–237, October 2007. doi:10.1109/MAS-COTS.2007.58

40. Willinger, W., Taqqu, M.S., Sherman, R., Wilson, D.V.: Self-similarity through high-variability: statistical analysis of Ethernet LAN traffic at the source level. In: ACM SIGCOMM Conference, pp. 100–113 (1995)

41. Zakay, N., Feitelson, D.G.: On identifying user session boundaries in parallel workload logs. In: Cirne, W., Desai, N., Frachtenberg, E., Schwiegelshohn, U. (eds.) JSSPP 2012. LNCS, vol. 7698, pp. 216–234. Springer, Heidelberg (2013). doi:10.1007/978-3-642-35867-8_12

42. Zakay, N., Feitelson, D.G.: Workload resampling for performance evaluation of parallel job schedulers. Concurrency Comput. - Pract. Exp. **26**(12), 2079–2105 (2014). doi:10.1002/cpe.3240

43. Zakay, N., Feitelson, D.G.: Preserving user behavior characteristics in trace-based simulation of parallel job scheduling. In: 22nd Modelling, Analysis & Simulation of Computer and Telecommunication Systems, pp. 51–60, September 2014. doi:10.1109/MAS-COTS.2014.15

44. Zakay, N., Feitelson, D.G.: Semi-open trace based simulation for reliable evaluation of job throughput and user productivity. In: 7th IEEE International Conference on Cloud Computing Technology and Science, pp. 413–421, November 2015. 10.1109/CloudCom.2015.35

Scheduling DAGs Opportunistically: The Dream and the Reality Circa 2016

Arnold L. Rosenberg[✉]

Computer Science, Northeastern University, Boston, MA, USA
rsnbrg@ccs.neu.edu

Abstract. A broad-brush tour of a platform-oblivious approach to scheduling DAG-structured computations on platforms whose resources can change dynamically, both in availability and efficiency. The main focus is on the IC-scheduling and Area-oriented scheduling paradigms—the motivation, the dream, the implementation, and initial work on evaluation.

Keywords: Area-oriented DAG-scheduling · Dynamically changing platforms · IC-DAG-scheduling · Opportunistic DAG-scheduling

1 Prehistory

Early this century, Fran Berman, then-director of the San Diego Supercomputing Center (SDSC), gave a distinguished lecture at my then-home institution, UMass-Amherst. During a subsequent one-on-one, Fran educated me about a Grid-consortium that SDSC participated in, jointly with several kindred centers. The consortium "contract" allowed any member institution to submit computing jobs to any other. There was a guarantee that submitted jobs would be completed—but not when. When I asked what kind of computations SDSC performed using this paradigm, I was shocked to learn that the computations had dependencies among subcomputations that constrained the order in which work could be done. (As I recall, these were wavefont-structured dependencies.) I asked Fran how her team coped with the possibility that work could grind to a halt pending the completion of jobs that had been deployed within the consortium but not yet completed. Fran responded that they used heuristics that seemed to work well—but that she did not know of any mathematical setting that would allow one to think about this situation rigorously. The challenge was irresistible!

2 The Dream of Opportunistic Scheduling

2.1 An Informal Overview

Many modern computing platforms—notably including clouds [26,27], desktop grids [2], and volunteer-computing projects [11,15]—exhibit extreme levels of

© Springer International Publishing Switzerland 2016
P.-F. Dutot and D. Trystram (Eds.): Euro-Par 2016, LNCS 9833, pp. 22–33, 2016.
DOI: 10.1007/978-3-319-43659-3_2

dynamic heterogeneity. The availability and relative efficiencies of such platforms' computing resources can change at unexpected times and in unexpected ways. Scheduling a computation for efficient execution on such a platform can be quite challenging, particularly when there are dependencies among the computation's constituent *chores*[1] (jobs, tasks, etc.). We wanted to take up this challenge for the traditional scheduling setting of computations whose dependencies had the structure of DAGs (*directed acyclic graphs*).

The nodes of a computation-DAG $\mathcal{G}$ represent chores to be executed; $\mathcal{G}$'s arcs (directed edges) represent inter-chore dependencies that constrain the order in which chores can be executed. Specifically, a node v cannot be executed until all of its *parents* have been: these are the nodes that have arcs *into* v. Once all of v's parents have been executed, v becomes *eligible* (for execution) and remains so until it is executed. $\mathcal{G}$ has one or more *sources*—nodes that have no parents, hence are immediately eligible—and one or more *sinks*—nodes that have no "children." Clearly, executing a non-sink renders new nodes eligible. The execution of $\mathcal{G}$ terminates once all nodes have been executed.

2.2   Opportunistic DAG-Execution via Platform-Oblivious Scheduling

Recent studies have proposed seeking high performance and low cost within platforms that are dynamically heterogeneous and/or elastic by scheduling computations in a *platform-oblivious* manner. One compensates for ignoring platform details by carefully exploiting the detailed characteristics of one's computation. The central thesis motivating this approach is that, particularly with the targeted platforms, one always benefits computationally with DAG-structured workflows *by enhancing the likelihood of having as many eligible chores as possible.* Such scheduling enhances the likelihood of having work available as (advantageous) resources become available, hence being able to exploit resources *opportunistically.* Platform-oblivious scheduling can be advantageous for the targeted platforms because it exploits unchanging, perfectly-known characteristics of one's computation rather than attempting to adapt to characteristics of the platform, which are at best imperfectly known and, indeed, may change dynamically.

As we have pursued the dream of high-performing platform-oblivious schedules, we have found it technically advantageous to follow the lead of work-centric systems such as CHARM++ [14], by refining input DAGs before scheduling. We thereby can focus on scheduling *fine-grained* DAGs whose chores are all of (roughly) equal complexity. This focus extrapolates easily to DAGs that represent heterogeneous workloads: one simply models large chores as chains of "unit-size" ones with sequential dependencies, in the manner discussed in [9].

3   The Reality

3.1   Formalizing the Dream

A schedule Σ for a DAG $\mathcal{G}$ is a *topological sort* [10] of $\mathcal{G}$, i.e., a linear ordering of $\mathcal{G}$'s nodes in which all parents of each node v lie to the left of v. The schedule prescribes

---

[1] We use the granularity-neutral "chore" for the units that form the computation.

the order in which $\mathcal{G}$'s nodes are selected for execution. For any schedule Σ for $\mathcal{G}$ and any integer $T \in [0..N_\mathcal{G}]$,[2] $E_\Sigma(T)$ denotes the number of nodes of $\mathcal{G}$ that are eligible for execution at step T when Σ executes $\mathcal{G}$.

A. ICO Quality and Optimality [21]. Our first quality measure for DAG-schedules embodies the strictest possible interpretation of "eligible-node enhancement." We measure the *IC quality* of an execution of $\mathcal{G}$ by the number of nodes that are eligible after each node-execution—the more, the better. (Note that *we measure time in an event-driven manner*, as the number of nodes that have been executed to that point.) Our goal is to execute $\mathcal{G}$'s nodes in an order that maximizes the production rate of eligible nodes *at every step of the execution*, i.e., to craft a schedule Σ^* such that

$$(\forall t) \; E_{\Sigma^*}(t) = \max_{\Sigma \text{ a schedule for } \mathcal{G}} \{E_\Sigma(t)\}. \tag{1}$$

A schedule for $\mathcal{G}$ that achieves this demanding goal is *IC optimal* (*ICO*, for short).

In Sect. 3.2.A, we discuss ICO schedules for many classes of significant "real" computations—surprisingly many, given the strictness of the condition in Eq. 1.

B. AREA Quality and Optimality [3]. As we detail in Sect. 3.2.A, the demands of Eq. 1 are so stringent that many DAGs do not admit ICO schedules. This led us to weaken the IC-scheduling paradigm in [3], by introducing the *Area-oriented* DAG-scheduling paradigm.

Let Σ be a schedule for DAG $\mathcal{G}$. The *Area*, $Area(\Sigma)$, of Σ, is the sum

$$Area(\Sigma) = E_\Sigma(0) + E_\Sigma(1) + \cdots + E_\Sigma(N_\mathcal{G}).$$

Note that schedule Σ's normalized Area—obtained by dividing $AREA(\Sigma)$ by the number of nodes in $\mathcal{G}$—is the average number of nodes that are eligible as Σ executes $\mathcal{G}$. (The term *Area* is by analogy with Riemann sums approximating integrals.) Our goal is to find, for each DAG $\mathcal{G}$, an Area-maximal schedule, i.e., a schedule Σ^* for $\mathcal{G}$ such that

$$Area(\Sigma^*) = \max_{\Sigma \text{ a schedule for } \mathcal{G}} Area(\Sigma). \tag{2}$$

A schedule for $\mathcal{G}$ that achieves this goal is *Area-optimal* (*A-O*, for short).

Easily, every DAG admits an A-O schedule. Importantly for our dream, the A-O scheduling paradigm is a strict extension of the ICO paradigm, in the following sense.

Theorem 1 ([3]). *If DAG $\mathcal{G}$ admits an ICO schedule Σ, then every ICO schedule for $\mathcal{G}$ is A-O, and vice versa.*

---

[2] $[a..b]$ denotes the set of integers $\{a, a+1, \ldots, b\}$.

C. Optimal Schedules via DAG-duality. An important "meta-scheduling" contribution appears in [6] for ICO scheduling and in [3] for A-O scheduling. In both cases, one finds an algorithm that converts an optimal ICO (resp., A-O) schedule for a DAG $\mathcal{G}$ to an optimal ICO (resp., A-O) schedule for $\mathcal{G}$'s *dual* DAG $\widehat{\mathcal{G}}$. $\widehat{\mathcal{G}}$ is obtained from $\mathcal{G}$ by reversing all of $\mathcal{G}$'s arcs (e.g., the evolving mesh and reduction-mesh in Fig. 1(a) are dual to each other, as are the expansion-tree and reduction-tree in Fig. 1(b)).

3.2 Finding High-Quality Schedules

A. Schedules with High ICO Quality. The stringent demands of IC-*optimality*—the *maximum* number of eligible nodes at *every* step of a DAG-execution; cf. Eq. 1—raises the specter that ICO schedules exist only for a very constrained class of DAGs. Our first goal was to refute this possibility. We derived the following results.

(1) *ICO schedules for specific families of DAGs and computations.* In [6, 21, 22], we developed ICO scheduling strategies for many familiar classes of DAGs, including

- *evolving meshes* and *reduction-meshes*; see Fig. 1(a)
- *expansion-trees* and *reduction-trees*; see Fig. 1(b).
- *butterfly-structured, convolutional* DAGs; see Fig. 1(c, right).

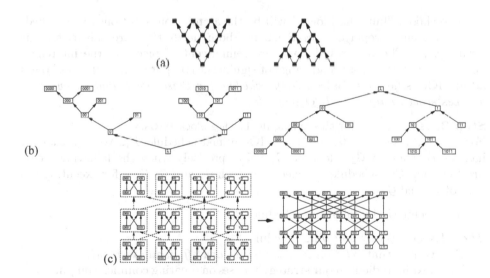

Fig. 1. Familiar DAGs that admit ICO schedules

In [5], we expanded the abstract, DAG-oriented, perspective of the preceding sources, to develop ICO scheduling strategies for many familiar classes of *computations*, including

- *convolutions*—e.g., the Fast Fourier Transform, polynomial multiplication
- *expansion-reductions*—e.g., numerical integration, comparator-based sorting
- many "named" computations—e.g., *Discrete Laplace Transform, matrix multiply.*

(2) *ICO schedules via* DAG *decomposition.* Careful analysis of our *ad hoc* schedules enabled us, in [19], to develop efficient—i.e., quadratic-time—algorithms that produce ICO schedules for a broad range of DAGs, based on structural decomposition. When the strategy succeeds in decomposing a DAG $\mathcal{G}$ in the prescribed manner, one can "read off" an ICO schedule for $\mathcal{G}$ from the decomposition. The strategy has two major steps.

Step 1. Select a set of bipartite[3] "building-block" DAGs that admit ICO schedules.

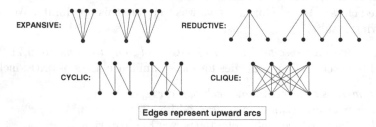

Fig. 2. A sampler of small instances of useful bipartite "building-block" DAGs.

The chosen "building blocks" will be the atomic computations in the schedule. The sample repertoire in Fig. 2 fits both needs that are salient for our strategy. (*a*) The illustrated DAGs are reminiscent of pieces of the interchore dependency-DAGs for a broad range of significant computations. (*b*) These DAGs admit ICO schedules. Indeed, *any schedule for these DAGs that executes all sources sequentially is an ICO schedule.*

Step 2. Establish ▷-priorities among the building-block DAGs.
For $i = 1, 2$, let DAG $\mathcal{G}_i$ admit an IC-optimal schedule Σ_i. We say that $\mathcal{G}_1$ has ▷-*priority* over $\mathcal{G}_2$—denoted $\mathcal{G}_1 \triangleright \mathcal{G}_2$—precisely when the following recipe produces an ICO schedule for executing both $\mathcal{G}_1$ and $\mathcal{G}_2$ (i.e., for executing the *sum* of $\mathcal{G}_1$ and $\mathcal{G}_2$):

First: Execute $\mathcal{G}_1$ by following schedule Σ_1.

Then: Execute $\mathcal{G}_2$ by following schedule Σ_2.
    One verifies that *relation* ▷ *is transitive and efficiently tested* [7].
    The next ingredient in our strategy focuses on creating complex computation-DAGs by *composing* simple computation-DAGs. One *composes* DAG $\mathcal{G}_1$ with DAG $\mathcal{G}_2$ by *merging/identifying* some k sources of $\mathcal{G}_2$ with some k sinks of $\mathcal{G}_1$: the resulting DAG is *composite of type* $\mathcal{G}_1 \Uparrow \mathcal{G}_2$. (Easily, DAG-composition composes

---

[3] A bipartite DAG's nodes are partitioned into sets X and Y, with every arc going from X to Y.

the function specified by $\mathcal{G}_1$ with the one specified by $\mathcal{G}_2$.) The following sample composition illustrates DAG-composition and its associativity.

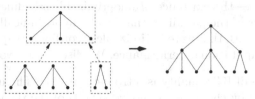

We can now announce the major contribution of our decomposition-based strategy.

Theorem 2 ([19]). *Focus on a* DAG $\mathcal{G}$ *that is composite of type* $\mathcal{G}_1 \Uparrow \mathcal{G}_2 \Uparrow \cdots \Uparrow \mathcal{G}_n$. *Say that*
— *each* DAG $\mathcal{G}_i$ *admits the IC-optimal schedule* Σ_i;
— $\mathcal{G}_1 \rhd \mathcal{G}_2 \rhd \cdots \rhd \mathcal{G}_n$.
Then, the following schedule for $\mathcal{G}$ *is IC optimal:*
Use the schedules $\{\Sigma_i\}$ *to execute the* DAGs $\{\mathcal{G}_i\}$ *seriatim, in order of* $\rhd$-*priority.*

Efficient algorithms implement Theorem 2 on a large variety of "well-structured" DAGs. In particular, the two core processes in the theorem are computationally efficient:
— "parsing" DAG $\mathcal{G}$ into $\mathcal{G}_1, \ldots, \mathcal{G}_n$ (when such a parsing exists)
— testing $\rhd$-priorities among the $\mathcal{G}_i$.
Two clarifications will help illuminate Theorem 2.

1. *A* DAG *can have very nonlinear structure, even though it is composed from small* DAGs *that obey a linear chain of* $\rhd$-*priorities.*
   Butterfly DAGs provide an example. Every butterfly DAG $\mathcal{B}$ is composed from many copies of the bipartite butterfly DAG $\mathcal{B}_2$: symbolically, $\mathcal{B}$ is composite of type $\mathcal{B}_2 \Uparrow \mathcal{B}_2 \Uparrow \cdots \Uparrow \mathcal{B}_2$ (see Fig. 1(c)). One verifies easily that $\mathcal{B}_2$ has "self $\rhd$-priority"—i.e., $\mathcal{B}_2 \rhd \mathcal{B}_2$—so that $\mathcal{B}$ admits a linear chain of $\rhd$-priorities: $\mathcal{B}_2 \rhd \cdots \rhd \mathcal{B}_2$.
2. *Many* DAGs *that admit ICO schedules are quite nonuniform in a graph-structure sense:*

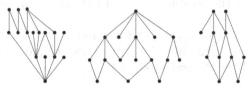

The "well-structuredness" exploited in Theorem 2 is algebraic in nature, in terms of composition and $\rhd$-priority.

(3) *A weakness in the IC-scheduling paradigm.* Using Theorem 2 and *ad hoc* techniques, we developed ICO—i.e., *optimal* eligible-node-enhancing—schedules for many popular families of DAGs, including "butterflies," "meshes," "trees."

But, with little difficulty, we also discovered "cousins" of these "well-structured" DAGs that *do not admit any ICO schedule* [19]. This deficiency in the IC-scheduling paradigm—the existence of unoptimizable schedules—led us to seek "weakened" versions of the paradigm that would algorithmically produce schedules for *every* input DAG, that were optimizable according to a quality metric that correlated with computational performance. We discovered two such paradigms.

1. A *batched* notion of ICO quality is introduced in [17]. The underlying idea is to execute a DAG by choosing successive *subsets* of the then-eligible nodes.
2. An *averaged* notion of ICO quality underlies the *Area* quality metric of Sect. 3.1.B and [3]. The underlying quest is for schedules that maximize the *average* number of nodes that are eligible at each step of a DAG-execution.

Both the batched-ICO and Area quality measures admit optimal schedules for every DAG—but the general versions of both optimization problems are NP-Complete [17,20]. In the case of the Area measure, we were able to craft two readily computable associated heuristics (AO and SIDNEY) that are (empirically) computationally beneficial—as discussed at length in Sect. 3.2.B. Regrettably, we have not yet succeeded in finding such an associated heuristic for the batched version of the IC-scheduling paradigm. We leave the attractive challenges related to the batched paradigm to the interested reader.

B. Schedules with High AREA Quality. In contrast to the IC-scheduling paradigm, our major accomplishments with Area-oriented scheduling involved *heuristics* inspired by the paradigm. We begin our discussion with theoretical developments.

(1) *A-O schedulers for specific DAG-families.* In [3], we developed A-O schedulers for several classes of DAGs, including

- *monotonic tree-DAGs*: each DAG is either an *expansion-tree*—a DAG having one source, in which each nonsource has one parent—or the dual of an expansion-tree.
- *expansion-reduction DAGs*: each DAG is obtained by composing a k-sink expansion-tree with a k-source reduction-tree. (Imagine, e.g., that we match up the sources of the righthand tree in Fig. 1(b) with the sinks/leaves of the lefthand tree.)
- *compositions of bipartite cycle- and clique-DAGs*. (The "building-block" DAGs of Fig. 2(bottom) exemplify the cycles and cliques; the butterfly-DAG of Fig. 1(c) exemplifies the end product.)

Among the family-specific A-O schedulers that we developed, one stands out for its DAG-scheduling consequences. This is the efficient algorithm developed in [8], that produces A-O schedules for *series-parallel DAGs* (*SP-DAGs*, for short). (SP-DAGs have a rich history in the design of logic circuits. More recently, they have been used to model multi-threaded parallel computations; cf. [1].) This algorithm decomposes an input SP-DAG $\mathcal{G}$ according to the following recursive recipe for generating SP-DAGs, and then it "reads off" an A-O schedule from the resulting "parse" of $\mathcal{G}$.

A *(2-terminal) series-parallel* DAG $\mathcal{G}(SP\text{-}DAG)$ is produced by a sequence of the following operations.

1. **Create.** Form a DAG $\mathcal{G}$ that has:
   (a.) two nodes, a *source* s and a *target* t, which are jointly $\mathcal{G}$'s *terminals*,
   (b.) one arc, $(s \to t)$, directed from s to t.
2. **Compose.** SP-DAGs, $\mathcal{G}'$ with terminals s', t', and $\mathcal{G}''$, with terminals s'', t''.
   (a.) *Parallel composition.* Form the SP-DAG $\mathcal{G} = \mathcal{G}' \Uparrow \mathcal{G}''$ from $\mathcal{G}'$ and $\mathcal{G}''$ by merging s' with s'' to form a new source s and t' with t'' to form a new target t.
   (b.) *Series composition.* Form the SP-DAG $\mathcal{G} = (\mathcal{G}' \to \mathcal{G}'')$ from $\mathcal{G}'$ and $\mathcal{G}''$ by merging t' with s''. $\mathcal{G}$ has the single source s' and the single target t''.

(2) *The NP-completeness of AREA-maximization.* After we developed the ICO schedules for specific DAG-families discussed in Sect. 3.2.A(1), we were able to detect commonalities in reasoning that ultimately culminated in a proof for Theorem 2. In contrast, we found that our A-O schedules for specific DAG-families discussed in Sect. 3.2.B(1) relied in a fundamental way on the specific structures of the specific DAG-families. We were, therefore, not surprised to learn, in [20], that the general problem of computing A-O schedules is NP-complete. The proof in [20] reduces the 0–1 *Minimum Weighted-Completion-Time* Problem for bipartite DAGs, which is known to be NP-complete [25], to the A-O scheduling problem. This result shifted our focus entirely to the development of scheduling heuristics that (empirically) produced schedules with large Areas. We now describe the main heuristics that we have developed.

(3) *Area-oriented scheduling heuristics.* We have developed three scheduling heuristics that are "Area-centric," in the sense that they exploit Area-related structural properties of the DAG being scheduled.

(a) Heuristic D-G [3]. The *dynamic-greedy scheduling heuristic* D-G crafts a schedule for a DAG $\mathcal{G}$ by organizing $\mathcal{G}$'s eligible chores in a list structure that is (partially) ordered by chores' *yields,* with ties broken randomly. The *yield* $v(t)$ of eligible chore v at step t is the number of non-eligible chores that would be rendered eligible if v were executed now. The yield of a chore u can change at each step, and the execution of u can change the yields of many other chores, specifically, those that share children with u. Thus, in contrast with our other schedulers, the schedules produced by D-G change at each step—which gives D-G time-complexity commensurate with our other heuristics.

Note. D-G's successive choices of the next node to execute are *locally optimal*— except for its (nonexistent) tie-breaking mechanism.

(b) Heuristic AO [8]. The *Area-oriented scheduling heuristic* AO builds on two facts: (i) We have access to an efficient A-O scheduler for SP-DAGs; cf. Section 3.2.B(1). (ii) Every DAG $\mathcal{G}$ can be transformed efficiently to an SP-DAG $\sigma(\mathcal{G})$ that retains both $\mathcal{G}$'s inter-chore dependencies and (roughly) its degree

of inherent parallelism. Several sources describe "SP-izing" transformations; a perspicuous version from [12] is invoked in [8]. Heuristic AO produces a high-Area schedule for a DAG $\mathcal{G}$ in three steps.

Step 1. Transform $\mathcal{G}$ to an SP-DAG $\sigma(\mathcal{G})$, using an algorithm from [12].

Step 2. Produce an A-O schedule $\widetilde{\Sigma}$ for $\sigma(\mathcal{G})$, via the algorithm in [8].

Step 3. "Filter" schedule $\widetilde{\Sigma}$ to remove the "auxiliary" nodes added when SP-izing $\mathcal{G}$.

(c) Heuristic SIDNEY. The SIDNEY scheduling heuristic of [20] inherits both its name and its algorithmic underpinnings from a sophisticated DAG-decomposition scheme from [23]. It schedules an input DAG $\mathcal{G}$ in four steps.

Step 1. Transform $\mathcal{G}$ to its associated 0–1 *version* $\mathcal{G}_{0,1}$.

> The nodes of $\mathcal{G}_{0,1}$ are obtained by splitting each node v of $\mathcal{G}$ into two nodes, v_0 and v_1. Give each node of $\mathcal{G}_{0,1}$ that has a 0 subscript (the 0 nodes) a *processing time* of 0 and a *weight* of 1; give each node of $\mathcal{G}_{0,1}$ that has a 1 subscript (the 1-nodes) a processing time of 1 and a weight of 0. Finally, give $\mathcal{G}_{0,1}$ an arc $(u_1 \rightarrow v_0)$ for each arc $(u \rightarrow v)$ of $\mathcal{G}$ and an arc $(u_0 \rightarrow u_1)$ for each node u of $\mathcal{G}$.

Step 2. Use a max-flow computation to perform a Sidney decomposition of $\mathcal{G}_{0,1}$, via the algorithm in [23].

Step 3. Say that the decomposition of $\mathcal{G}_{0,1}$ produces DAGS $\mathcal{G}_1, \ldots, \mathcal{G}_k$.

a. Remove all 0-nodes from every $\mathcal{G}_i$.

b. Use heuristic D-G to produce a schedule Σ_i for each $\mathcal{G}_i$.

Step 4. Output schedule $\Sigma \overset{\text{def}}{=} \Sigma_1 \cdots \Sigma_k$, the concatenation of the k subschedules.

At the cost of somewhat more computation than needed for heuristics D-G and AO, SIDNEY empirically produces schedules whose Areas are within 85 % of maximal [20].

3.3   The Benefits of Opportunistic Scheduling

A. Benefits Exposed via Simulation Experiments. Simulation-based studies of the opportunistic scheduling paradigms we have discussed appear in [3, 4, 13, 16, 24]. Rather than reproduce material that appears in great detail in those sources, I have decided to summarize here the major *messages* of those studies.

B1. One observes in all of the cited sources that there are two circumstances under which all (oblivious) scheduling paradigms are essentially equivalent in performance.

(*a*) When computing resources are plentiful, then the inherently sequential critical path of a DAG is the only constraint on the speed of executing the DAG.

(b) When computing resources are really meager, then there are no opportunities for efficiency-enhancing concurrency.

B2. One observes in [13,16] situations where ICO schedules outperform a variety of platform-oblivious competitor-schedules by as much a 10–20%. The tested workloads in [16] were real scientific computations; the ones in [13] were synthetic, but with structures that approximated those of real scientific computations.

B3. Since the Area quality-metric is a weakening of IC-quality, it is not surprising that the benefits of A-O schedules observed in [3] are more modest than those observed in the IC-frameworks of [13,16]. That said, one still often observes A-O schedules outperforming a variety of platform-oblivious competitor-schedules by double-digit percentages. Indeed, the same type of performance is observed even in [4] with heuristic AO. One observes that A-O schedules outperform those produced by heuristic AO, but only by percentages that may not justify the computational cost of producing the A-O schedule.

B4. The experiments in [4] suggest that the schedules produced by heuristic AO perform best when computing resources become available according to distributions that have low variances.

B5. The experimental settings in [3,4] (involving, respectively, A-O schedules and schedules produced by heuristic AO) posit that computing resources become available according to low-variance distributions. It is observed experimentally in [4] that, within such settings, the Areas of generated schedules inversely track the makespans of the schedules' DAG-executions—i.e., larger Areas correlate with smaller makespans.

B6. In contrast to heuristic AO, the very high-Area schedules produced by heuristic SIDNEY seem to favor situations wherein the distributions governing computing-resource availability do not have low variances [20]. This suggests that SIDNEY's schedules may be desirable in settings such as enterprise clouds, where the user can tailor the purchase of computing resources based on the varying numbers of eligible nodes produced over time by one's DAG-schedule.

B7. The experiments reported in [24] seem to validate Observations 4 and 6: schedules produced by heuristic AO are observed to perform very well in "single-instance" enterprise clouds, wherein there is a single block of computing resources that are available at any moment. In fact, the static heuristic AO i sobserved to compete well with dynamic competitor schedules.

B. Two Major Open Issues. We close with a two open issues regarding opportunistic DAG-scheduling. The benefits we have already uncovered—and enumerated in this section—explain our belief in the potential significance of success in addressing these issues.

Q1. The discovery in [19] that many DAGs do not admit ICO schedules led to three weakened version of IC-scheduling: batched IC-scheduling [18], a version based on weakening the ▷-priority relation of Theorem 2 [16], and Area-oriented scheduling [3]. Of these alternatives, only Area-oriented scheduling has been studied in any detail. The other alternatives certainly deserve more attention than they have received.

Q2. Our study of opportunistic DAG-scheduling began with a focus on *dynamically heterogeneous* computing platforms—and it has largely retained that focus. The benefits of eligible-node-enhancing DAG-schedules should be significant also in other domains:

(*a*) Opportunistic DAG-schedulers may be valuable when pursuing cost-effective computing within an enterprise cloud. Having access to large numbers of eligible nodes should alow a user to maximally exploit available cost-effective resources. This benefit is hinted at in [24], but it deserves careful study.

(*b*) In a similar vein, opportunistic DAG-schedulers may be beneficial in power-aware computing environments. Their schedules may enable one to maximally exploit low-power resources as they become available. This possibility, too, deserves careful study.

Acknowledgments. It is a pleasure to acknowledge the invaluable contributions of my collaborators on the work discussed here: Gennaro Cordasco, Rosario De Chiara, Ian Foster, Robert Hall, Greg Malewicz, Rajmohan Rajaraman, Scott Roche, Mark Sims, Michela Taufer, Arun Venkataramani, Mike Wilde, Matt Yurkewych. Our work on opportunistic DAG-scheduling has been supported in part by several grants from the US National Science Foundation, most recently Grant CSR-1217981.

References

1. Blumofe, R.D., Joerg, C.F., Kuszmaul, B.C., Leiserson, C.E., Randall, K.H., Zhou, Y.: Cilk: an efficient multithreaded runtime system. In: 5th ACM SIGPLAN Symposium on Principles and Practices of Parallel Programming (PPoPP 1995) (1995)
2. Casanova, H., Dufossé, F., Robert, Y., Vivien, F.: Scheduling parallel iterative applications on volatile resources. In: 25th IEEE International Parallel and Distributed Processing Symposium (2011)
3. Cordasco, G., De Chiara, R., Rosenberg, A.L.: On scheduling DAGs for volatile computing platforms: area-maximizing schedules. J. Parallel Distrib. Comput. **72**(10), 1347–1360 (2012)
4. Cordasco, G., De Chiara, R., Rosenberg, A.L.: An AREA-oriented heuristic for scheduling DAGs on volatile computing platforms. IEEE Trans. Parallel Distrib. Syst. **26**(8), 2164–2177 (2015)
5. Cordasco, G., Malewicz, G., Rosenberg, A.L.: Applying IC-scheduling theory to some familiar classes of computations. In: Workshop on Large-Scale, Volatile Desktop Grids (PCGrid 2007) (2007)
6. Cordasco, G., Malewicz, G., Rosenberg, A.L.: Advances in IC-scheduling theory: scheduling expansive and reductive DAGs and scheduling DAGs via duality. IEEE Trans. Parallel Distrib. Syst. **18**, 1607–1617 (2007)
7. Cordasco, G., Malewicz, G., Rosenberg, A.L.: Extending IC-scheduling via the Sweep algorithm. J. Parallel Distrib. Comput. **70**, 201–211 (2010)

8. Cordasco, G., Rosenberg, A.L.: On scheduling series-parallel DAGs to maximize AREA. Int. J. Found. Comput. Sci. **25**(5), 597–621 (2014)
9. Cordasco, G., Rosenberg, A.L., Sims, M.: On clustering DAGs for task-hungry computing platforms. Cent. Eur. J. Comput. Sci. **1**, 19–35 (2011)
10. Cormen, T.H., Leiserson, C.E., Rivest, R.L., Stein, C.: Introduction to Algorithms, 2nd edn. MIT Press, Cambridge (1999)
11. Estrada, T., Taufer, M., Reed, K.: Modeling job lifespan delays in volunteer computing projects. In: 9th IEEE International Symposium on Cluster, Cloud, and Grid Computing (CCGrid) (2009)
12. González-Escribano, A., van Gemund, A.J.C., Cardeñoso-Payo, V.: Mapping unstructured applications into nested parallelism. In: Palma, J.M.L.M., Sousa, A.A., Dongarra, J., Hernández, V. (eds.) VECPAR 2002. LNCS, vol. 2565, pp. 407–420. Springer, Heidelberg (2003)
13. Hall, R., Rosenberg, A.L., Venkataramani, A.: A comparison of DAG-scheduling strategies for internet-based computing. In: 21st IEEE International Parallel and Distributed Processing Symposium (IPDPS) (2007)
14. Kale, L.V., Bhatele, A. (eds.): Parallel Science and Engineering Applications: The Charm++ Approach. New York, Taylor & Francis Group, CRC Press (2013)
15. Korpela, E., Werthimer, D., Anderson, D., Cobb, J., Lebofsky, M.: SETI@home: massively distributed computing for SETI. In: Dubois, P.F (ed.) Computing in Science and Engineering. IEEE Computer Society Press (2000)
16. Malewicz, G., Foster, I., Rosenberg, A.L., Wilde, M.: A tool for prioritizing DAG-Man jobs and its evaluation. J. Grid Comput. **5**, 197–212 (2007)
17. Malewicz, G., Rosenberg, A.L.: Batch-scheduling dags for internet-based computing. In: Cunha, J.C., Medeiros, P.D. (eds.) Euro-Par 2005. LNCS, vol. 3648, pp. 262–271. Springer, Heidelberg (2005)
18. Malewicz, G., Rosenberg, A.L.: A pebble game for internet-based computing. In: Goldreich, O., Rosenberg, A.L., Selman, A.L. (eds.) Theoretical Computer Science. LNCS, vol. 3895, pp. 291–312. Springer, Heidelberg (2006)
19. Malewicz, G., Rosenberg, A.L., Yurkewych, M.: Toward a theory for scheduling DAGs in internet-based computing. IEEE Trans. Comput. **55**, 757–768 (2006)
20. Roche, S.T., Rosenberg, A.L., Rajaraman, R.: On constructing DAG-schedules with large AREAs. Concurrency Comput. Pract. Experience **27**(16), 4107–4121 (2015)
21. Rosenberg, A.L.: On scheduling mesh-structured computations for internet-based computing. IEEE Trans. Comput. **53**, 1176–1186 (2004)
22. Rosenberg, A.L., Yurkewych, M.: Guidelines for scheduling some common computation-DAGs for internet-based computing. IEEE Trans. Comput. **54**, 428–438 (2005)
23. Sidney, J.B.: Decomposition algorithms for single-machine sequencing with precedence relations and deferral costs. Oper. Res. **23**(2), 283–298 (1975)
24. Taufer, M., Rosenberg, A.L.: Scheduling DAG-based workflows on single cloud instances: high performance and cost effectiveness with a static scheduler. Int. J. High Perform. Comput. Appl. (2015). doi:10.1177/1094342015594518
25. Woeginger, G.J.: On the approximability of average completion time scheduling under precedence constraints. Discrete Appl. Math. **131**(1), 237–252 (2003)
26. Yao, S., Lee, H.-H.S.: Using mathematical modeling in provisioning a heterogeneous cloud computing environment. IEEE Comput. **44**, 55–62 (2011)
27. Zaharia, M., Konwinski, A., Joseph, A.D., Katz, R., Stoica, I.: Improving MapReduce performance in heterogeneous environments. In: 7th USENIX Symposium on Operating System Design and Implementation (2008)

Support Tools and Environments

Synchronization Debugging
of Hybrid Parallel Programs

Olaf Krzikalla$^{(\boxtimes)}$, Ralph Müller-Pfefferkorn, and Wolfgang E. Nagel

Technische Universität, Dresden, Germany
{olaf.krzikalla,ralph.mueller-pfefferkorn,wolfgang.nagel}@tu-dresden.de

Abstract. In this paper we address the problem of locating race conditions among synchronization primitives in execution traces of hybrid parallel programs. In hybrid parallel programs collective and point-to-point synchronization can't be analyzed separately. We introduce a model for synchronization primitives and formally define synchronization races with respect to the model. Based on these concepts we present an algorithm which accurately detects synchronization races and yields a task graph of the execution trace. The task graph represents the guaranteed ordering of events across thread and process boundaries. It is an essential core element for the further analysis (e.g. a data race detection) of a program.

Depending on the synchronization model task graph construction can be an NP-hard problem. Our model allows to construct an algorithm with sub-quadratic time complexity. Thus programs adhering to the principles of our model are provable against race conditions. Therefore we argue, that our model should be used as a foundation for the design and implementation of synchronization functions.

1 Introduction

Exascale systems are expected to exhibit a hybrid architecture. Even contemporary systems are clusters of shared memory nodes. On such systems several levels of parallelism exist, e.g., the node level, the core level, and the SIMD level. In this paper we consider a thread the smallest execution element of a program parallelization. A process consists of a number of threads, with each thread able to call distributed synchronization and communication functions. A hybrid program in turn consists of a set of such processes.

Hybrid programs raise new challenges to debugging and correctness tools. Consider two processes each executing a barrier call twice (Fig. 1a). A tool analyzing the execution traces of the two processes can enumerate the barrier calls of each process and by this means compute the matching barrier calls. Identifying the relation between barrier calls becomes difficult in the presence of a hybrid parallel execution (Fig. 1b). Let's assume process 1 consists of two threads each executing the barrier once. Thread 2 sends a message to thread 1 in-between the two barrier executions. Thread 1 waits for that message before it executes its barrier. Thus the execution order of the two barriers is determined.

© Springer International Publishing Switzerland 2016
P.-F. Dutot and D. Trystram (Eds.): Euro-Par 2016, LNCS 9833, pp. 37–50, 2016.
DOI: 10.1007/978-3-319-43659-3_3

However, in order to compute the order it is necessary to take the point-to-point synchronization into account, which happens between thread 2 and thread 1. Without that point-to-point synchronization a synchronization race would arise. It would be undetermined, whether the first barrier call of process 2 matches the barrier call of thread 1 or of thread 2. In practice a concurrent call to the same barrier is often forbidden (e.g. by MPI or GASPI [1,7]). Due to its non-deterministic nature such an error could cause an untimely program abortion. The other side of the problem is illustrated in Fig. 1c. In this case two point-to-point and one collective synchronization occur. Due to the barrier the first wait at process 2 will wait for the post of thread 1 leading to a determined execution order again. But it is also necessary to take the collective synchronization into account in order to compute the order of point-to-point synchronizations. The conclusion is that an algorithm computing the order of events on the basis of a hybrid parallel program trace cannot handle point-to-point and collective synchronization in two independent steps. Only a consolidated computation of both types of synchronization can yield a task graph, which represents the guaranteed ordering of events.

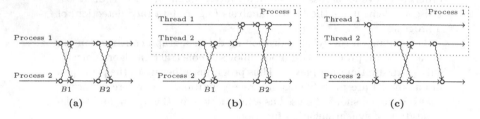

Fig. 1. Interaction of collective and point-to-point synchronization in hybrid parallel program executions

In the following we will introduce a model, which can be used to describe both point-to-point and collective synchronization. Based on that model we formally explore how races can be detected and how a task graph of a given program trace can be efficiently constructed. Our work is novel as it unifies the handling of point-to-point and collective synchronizations. The major result of our work is an algorithm to analyze the synchronization operations of the trace of an application's execution to compute its guaranteed orderings. The algorithm requires $\mathcal{O}(|T|^2)$ time, where $|T|$ is the number of traced synchronization operations. The algorithm reports synchronization races, which are sequences of synchronization operations leading to a non-deterministic program behavior. In addition we present an important optimization, which decreases the time complexity of the algorithm down to sub-quadratic and makes it highly scalable. We have implemented and evaluated our concept as a tool capable of analyzing hybrid GASPI/OpenMP/Pthreads programs. The task graphs generated by the tool visualize the synchronization relations among the threads and processes in terms of necessity.

2   Model

We derive our model from the classic point-to-point or event-style synchronization model [6] and extend it, that it can handle collective synchronization too. The basic concept is the *event*. An event has two states: posted and cleared. In the classic model three operations can be performed on an event: *POST* sets the state of the event to *posted*; *WAIT* suspends the calling thread until the state of the event is *posted*; and *CLEAR* sets the state of the event to *cleared*.

Typically, point-to-point synchronizations use simple flags as events. These flags are shared among the threads of a process. Events of collective primitives are handled similar. Every executing element (i.e. a thread or a process) participating in a collective has its own event. A thread being part of an OpenMP barrier has a thread-local event for that barrier. A process participating in an MPI barrier shares the corresponding event among its threads. When a blocking collective is entered by a thread, a *POST* operation is performed on the corresponding event first. Afterward, a *WAIT* operation waits until all participating executing elements have entered the collective and set their respective events to *posted*. Finally, a *CLEAR* operation is performed before the execution returns from the collective.

In a blocking collective the three primitives *POST*, *WAIT* and *CLEAR* are tied together and executed in that order. In a non-blocking collective (e.g. a *split-phase barrier* [4]) the *POST* operation is swapped out to a dedicated enter routine (e.g. upc_notify). *WAIT* and *CLEAR* remain tied together in one routine (e.g. upc_wait).

The coupling of *WAIT* and *CLEAR* is important. In our model it is not only used for collective but also for point-to-point synchronization. Thus we reduce the classic model to two principal operations:

- post(e) or P: sets the state of the event e to *posted*.
- wait(e) or W: suspends the executing thread until the state of the event e is *posted*. If e belongs to a collective, then W waits until all participating elements have set their respective events to *posted*. Upon exiting the state of e is set to *cleared*.

Performing *WAIT* and *CLEAR* in one operation is a common practice. It is used on a regular basis in collective synchronization. Another example is the GASPI standard, which resembles the *WAIT,CLEAR* sequence in the gaspi_notify_reset function. This function resets an event and returns its former state. A caller can choose the further execution path by means of the function result.

A *program execution* $\mathcal{P} = \langle E, \prec \rangle$ represents a particular execution of a parallel program. E is a finite set of tasks and $\prec$ is the *happens-before relation* defined over E [9]. $\mathcal{P}$ constitutes a directed acyclic graph with E being the nodes and $\prec$ being the edges. We assume a trace of a program as input representing a partial task graph $\mathcal{P}^T = \langle E, \prec^T \rangle$. A task in E can be either a post(e) or a wait(e) operation. The event e is part of the input and contains information about the synchronization type. The $\prec^T$ relation denotes the execution order of the tasks in a thread. It is implicitly given by the input trace. The challenge is to compute the $\prec^S$ relations, which are induced among threads by the synchronization tasks. If this computation leads to a uniquely determined program execution $\mathcal{P} = \langle E, \prec^T \cup \prec^S \rangle$, then the input trace $\mathcal{P}^T$ is free of synchronization races.

3   Synchronization Races

Parallel programs can exhibit various forms of non-deterministic behavior, which is caused by race conditions at different levels. Value non-determinacies are the most fundamental race conditions – data races. Data races are generally considered a programming error. However, there are also benign and even intended data races, for example to implement synchronization operations.

Static non-determinacy is a property of the program control flow, which is typically intended and built in the source code. An example are programs where the threads adjust their execution according to the content of received messages (where content may refer to the sender, the message type or the actual data). Stencil codes are representative: halos are processed in the order, in which they are received from neighboring threads. Another form of non-determinacy is mutual exclusion, where two or more synchronization operations intentionally race toward the acquisition of the same resource. Unlike point-to-point synchronization mutual exclusion does not establish directed synchronization relations.

Our notion of synchronization races lead to a form of non-determinacy, which conceptually differs from other forms of non-determinacy. A synchronization race can only occur among point-to-point synchronization operations accessing the same event. In Fig. 2a process 2 issues a P operation, but it is unclear, whether thread 1, thread 2, or both will perceive the posted event and reset it. This depends on the point in time, at which the execution of thread 1 and thread 2 reaches the respective W operations. Figure 2b depicts a race of two posts toward the same wait. If process 1 has entered the wait operation before process 2 executes P_1, then process 1 can proceed after P_1 and eventually the state of the event is *posted* after the execution of P_2. However, if process 1 doesn't enter W before process 2 has executed P_2, then the state of the event is eventually *cleared*. Figure 2c is an extension of Fig. 2b. At first glance the execution order seems well defined, since $P_1 \prec W_1$ and $P_2 \prec W_2$. But if process 2 has executed P_1 and P_2 before process 1 enters W_1, then the second post gets lost and process 1 will be stuck in the second wait. This may lead to an unpredictable dead-lock.

We formally define a synchronization race as a specific global program state. A global program state can be seen as a *frontier* drawn across all threads in between tasks of a task graph [3]. All tasks before the frontier were already executed. Tasks immediately after the frontier are just about to be executed. We call such tasks *active*. A *consistent* global state is an execution point, at which all threads *could* have simultaneously arrived.

Definition 1. *A synchronization race* exists in a program execution $\mathcal{P}$, iff a consistent global state exists such that a wait task on an event e is active and

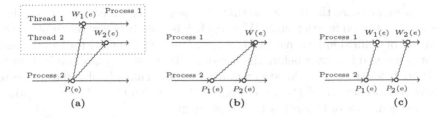

Fig. 2. Different types of synchronization races

1. *another wait task on* **e** *is active or*
2. *at least two post tasks on* **e** *exist before the frontier and none of them is connected to a wait task before the frontier.*

A frontier of a consistent global state can only be crossed by arrows toward the direction of the program execution. Thus a task after a consistent frontier can never happen before a task before the frontier. Figure 3a resembles Fig. 2c and illustrates the concept. The frontier belongs to a consistent global state – all arrows cross the frontier onward. This case constitutes a synchronization race by Definition 1: W_1 is active, P_1 and P_2 are before the frontier and none of them has triggered a wait before the frontier. On the contrary the frontier in Fig. 3b is not consistent any more, since it is crossed by an arrow backwards from $P(x)$ to $W(x)$. In this case it is indeed not possible to construct a consistent frontier such that a synchronization race could be constituted according to Definition 1. Figure 3c applies the frontier concept to a collective synchronization operation in a hybrid environment. The shown frontier separates the enter and leave events (post and wait operations resp.) of the barrier calls $B1$ and $B2$. Thus W_{B1} at thread 2 and W_{B2} at thread 1 are both active. But this frontier is not consistent, since it is crossed by an arrow backwards due to a point-to-point synchronization from thread 2 to thread 1. Again, a construction of a consistent frontier fulfilling all requirements of Definition 1 is not possible.

The examples give us a hint, how synchronization races can be detected. If $P(e)$ happens after $W(e)$, then these two tasks can never form a synchronization race.

Theorem 1. *Let P be a post task triggering a wait task W; P_r another post task on the same event; and $P_r \nprec P$. A synchronization race exists between W and P_r, iff $W \nprec P_r$.*

Proof. According to Definition 1, pt.2 we try to construct a consistent frontier such that W is active and both P and P_r are located before the frontier.

$\Rightarrow$: Since W is active, it lies after the frontier. If $W \prec P_r$, then P_r lies after the frontier too. Thus it is not possible to construct a consistent frontier with P_r being located before the frontier. The conditions of Definition 1 can't be met.

$\Leftarrow$: Let $Next(P_r)$ be the event immediately following P_r. We place the frontier between P_r and $Next(P_r)$, so that any wait triggered by P_r is after the frontier.

Furthermore we place the frontier so that W is active. This step requires no shift of the already placed frontier, since $W \not\prec P_r$. If P is already before the frontier, the conditions of Definition 1 are met: W is active, P and P_r lie before the frontier and are not connected to a wait before the frontier. Otherwise we place the frontier so that P lies before it. Again, this step requires no shift of already placed frontiers to preserve consistency: $W \not\prec P$ since P triggers W, but also $P_r \not\prec P$ by assumption. Thus the conditions of Definition 1 are met again. $\qquad\square$

Definition 1 requires that the sequence of wait operations on a particular event is totally ordered in a race-free task graph. Theorem 1 reveals how we can check this property: whenever a post task P is encountered it is checked against the last wait task W on the same event that has been triggered. If $W \not\prec P$ then a synchronization race has been found.

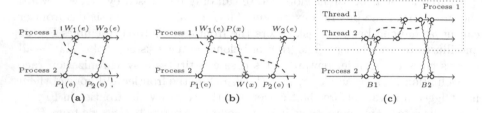

Fig. 3. Frontiers of consistent (a) and non-consistent (b,c) global program states

We can also prove, that Definition 1 is feasible to identify nondeterminism in a program execution.

Theorem 2. *If a program execution $\mathcal{P}$ has no synchronization races, then $\mathcal{P}$ is deterministic.*

Proof. We assume a program execution $\mathcal{P} = \langle E, \prec \rangle$ free of synchronization races. If $\mathcal{P}$ is non-deterministic, then another execution $\dot{\mathcal{P}} = \langle \dot{E}, \dot{\prec} \rangle$ with the same input could exhibit the same synchronization events and relations up to some point, after which they differ. Let W be the first wait event at which $\mathcal{P}$ and $\dot{\mathcal{P}}$ differ. We distinguish two cases:

1. Let P_1 and P_2 be different post events, which trigger W in $\mathcal{P}$ and $\dot{\mathcal{P}}$ respectively. Then $W \not\prec P_1$, since P_1 triggers W in $\mathcal{P}$. In addition $W \not\prec P_2$ in $\mathcal{P}$, since P_2 triggers W in $\dot{\mathcal{P}}$ and all events and relations before W are the same in $\mathcal{P}$ and $\dot{\mathcal{P}}$. Hence we can construct a consistent frontier in $\mathcal{P}$, such that W is active and P_1 and P_2 are both before the frontier. W.l.o.g. we assume $P_1 \not\prec P_2$ in $\mathcal{P}$, since $P_1 \prec P_2 \land P_2 \prec P_1$ cannot hold. Then the conditions of Theorem 1 are met with $P = P_2$ and $P_r = P_1$. But this contradicts the initial assumption, that $\mathcal{P}$ is free of synchronization races.

2. W.l.o.g. we assume that W is not triggered in $\mathcal{P}$, but triggered in $\dot{\mathcal{P}}$ by P. Then there is a task W_x, which has cleared the event posted by P before W in $\mathcal{P}$. Thus $W \not\prec W_x$ in $\mathcal{P}$, since W_x is executed, but W is not triggered. If $W_x \prec W$ in $\mathcal{P}$, then W_x would be included in the set of events, which are the same in $\mathcal{P}$ and $\dot{\mathcal{P}}$. Then $W_x \prec W$ in $\dot{\mathcal{P}}$ and W_x would be triggered in $\dot{\mathcal{P}}$ by P. But P has triggered W in $\dot{\mathcal{P}}$ too, which is not possible if $W_x \prec W$. Thus $W_x \not\prec W$ in $\mathcal{P}$. The conditions of Definition 1, pt.1 are met. Again this contradicts the initial assumption, that $\mathcal{P}$ is free of synchronization races. □

Theorem 2 is literally taken from [12]. We have adapted the proof to our model and extended it in order to deal with the possibility of concurrent wait tasks in hybrid parallel programs. Theorem 2 implies, that exactly one resulting task graph $\mathcal{P}$ exists for a race-free input trace $\mathcal{P}^T$. Moreover, no race-free task graph $\mathcal{P}$ can exist for an input trace containing synchronization races.

Unlike other non-determinacies we consider synchronization non-determinacy always a programming error. In the case covered by Theorem 1 both P and P_r might be executed before W. As a result one of these post events is lost, a subsequent wait might never trigger, and at least one thread never finishes. But even in the case, that superfluous post events prevent such a kind of deadlock, no reliable happens-before relation is established. We only have $P \prec W \vee P_r \prec W$, but this also means, that anyone of P or P_r may happen after W. This behavior contradicts the notion of point-to-point synchronization, whose purpose is to create happens-before relations.

4   The Replay Algorithm

The following algorithm to analyze synchronization operations is based on a replay approach. It performs a mock-up execution of the traced input tasks. Due to Theorem 2 our algorithm can replay the tasks in any order, which preserves the semantics of the synchronization primitives. During the replay the algorithm checks for the occurrence of synchronization races according to Theorem 1. If no races are found, the result is a race-free task graph $\mathcal{P}$. This graph contains all happens-before relations induced by the traced synchronization primitives.

Listing 1 is a condensed version of our actual implementation, which demonstrates the unified handling of blocking collective and point-to-point operations. The function `replay_tasks` replays the traced tasks of one thread consecutively until there are no more traced events or an untriggered wait is encountered. Depending on the type of the processed task T the variable e (line 3) denotes the flag number (point-to-point operation), the process group (GASPI collective) or the thread team (OpenMP barrier). Also depending on the type of T the index r (line 4) denotes the particular position of the thread of T inside e. This index is always 0 for point-to-point operations, it refers to a process index for a GASPI collective, and to the thread index for an OpenMP barrier. Every event is assigned a data structure PWP. PWP.Wait stores the active wait task, PWP.PreviousWait stores the last wait task that has been triggered. PWP.Post stores an already replayed post task, which hasn't been connected to one or more wait tasks yet. Race conditions are checked at line 8 (Definition 1, pt.2), at line 10 (Theorem 1) and at line 14 (Definition 1, pt.1). The lines 18–26 handle triggered wait tasks. If all members of a synchronization operation (a point-to-point operation has only one member) have set their respective events to *posted*, then a happens-before relation is added from the respective post tasks to all active wait tasks (line 20). At line 25 the execution of formerly suspended threads is resumed. If the current task is a wait task, then the thread is suspended at line 27. Note however, that by this time the thread might be already further processed at line 25. If the current task is a post task, then the replay of the thread just proceeds (line 28). The handling of non-blocking collectives is omitted for brevity. They require a special handling, since it is not possible to wait until all wait tasks of such a collective are encountered (line 18).

```
1   function replay_tasks (Task T) {
2     while T != nil {
3       let e = event of T
4       let r = index of T in e
5       let PWP = map[e]
6       switch type (T) {
7         case Post:
8           if PWP[r].Post != nil
9             abort and report post/post race
10          if PWP[r].PreviousWait != nil and not PWP[r].PreviousWait ≺ T
11            abort and report wait/post race
12          PWP[r].Post = T
13        case Wait:
14          if PWP[r].Wait != nil
15            abort and report wait/wait race
16          PWP[r].Wait = T
17      }
18      if (∀ x: PWP[x].Post != nil and PWP[x].Wait != nil) {
19        ∀ x: ∀ y:
20          add PWP[x].Post ≺ PWP[y].Wait
21        ∀ x:
22          PWP[x].PreviousWait = PWP[x].Wait
23          PWP[x].Post = PWP[x].Wait = nil
24        ∀ x:
25          replay_tasks(next_Task(PWP[x].PreviousWait))
26      }
27      if (type (T) == Wait) return
28      T = next_Task(T)
29    }
30  }
```

Listing 1. The replay algorithm

The performance-critical part of our algorithm is the reachability test at line 10, which we have implemented using depth-first-search (DFS). Therefore the complexity of the algorithm is $\mathcal{O}(|T|^2)$ with $|T|$ being the total number of tasks. However we have optimized the reachability test by leveraging the fact, that the replay order of the tasks is topological sorted. Albeit the worst case complexity would remain $\mathcal{O}(|T|^2)$, in practice large portions of the search space are cut off reducing the complexity of our replay algorithm to sub-quadratic time. In addition, the topological sorting helps in further analysis tasks (e.g. data race detection), which perform reachability tests too.

Since the replay order of tasks doesn't matter due to Theorem 2, the algorithm can be easily parallelized. The function `replay_tasks` can be executed in parallel for tasks of multiple threads. The access to the PWP map must be synchronized. Instead of the recursive call at line 27 a queue should be used, from which analysis threads fetch tasks, which are ready to be replayed.

5   Practical Evaluation

We have implemented the replay algorithm in a tool capable of analyzing post-mortem execution traces of hybrid programs using GASPI on the process level and OpenMP/Pthreads at the thread level. The tool combines this work with the model introduced in [8] in order to obtain task graphs of GASPI programs. The execution traces are generated by recording function enter and function leave events, their respective arguments, and return values using the dynamic binary instrumentation framework Pin [11]. Thus, the analysis doesn't require a recompilation of the source code.

With our replay algorithm we are able to generate a task graph of a GASPI program run out of an execution trace. Since such a task graph contains the happens-before relations in terms of necessity, it reveals the logic connections among the threads. As such, our algorithm opens up a complete new perspective to a parallel program. A programmer can visualize, understand and also easily teach the interactions of the asynchronous weak synchronization operations exhibited by a GASPI program.

In the following figures the time line is top-down and ranks are ordered from left to right (starting with rank 0). Collective synchronization is not visualized for clarity. Figure 4 depicts a detail of a task graph visualizing an one-sided broadcast implemented as a binary tree. Rank 0 sends the data to Rank 1,2,4, and 8 via the asynchronous one-sided `gaspi_write_notify` function. After rank 2,4, and 8 have received the data, they redistribute it.

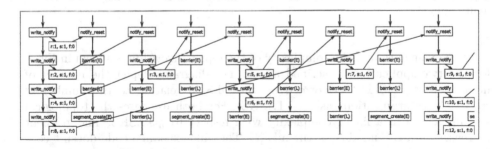

Fig. 4. Asynchronous one-sided broadcasting in a binary tree

Figure 5 shows two iterations of an one-dimensional halo-exchange code in a ring of 4 processes. The code uses double-buffering and switches back and forth between two data segments. A particular event e is defined by its rank r, its segment s and its flag number f. The notify_reset nodes enclosed in the two dotted rectangles are a case of static non-determinism. In the first iteration rank 0 receives its data first from rank 3 and then from rank 1. In the second iteration the receiving order changes, now rank 0 receives its data first from rank 1 and from rank 3 afterward. The dashed red line marks the happens-before relation between a wait operation (notify_reset) and a subsequent asynchronous post operation (issued by write_notify) on the same event (rank 1, segment 0, flag number 0). Thus the requirement imposed by Theorem 1 holds. During the construction of the task graph the replay algorithm has checked this requirement for all post/wait chains on all events. The program run doesn't contain any synchronization races. Thus, while the program itself is statically non-deterministic, the analyzed program run doesn't contain any problematic non-determinacies with respect to Theorem 2.

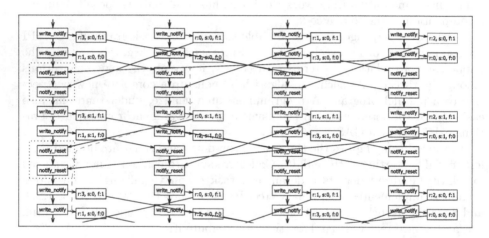

Fig. 5. Synchronization relations of an one-dimensional halo-exchange code (Color figure online)

The examination of the complexity of the replay algorithm is shown in Fig. 6 for the two applications described above. The diagrams depict the number of node visits (#VISITS) performed by the DFS in relation to the number of replayed synchronization tasks (#TASKS). For both use-cases the topological sorting results in a linear complexity with respect to #TASKS. The gradient increases with the number of threads $|t|$. The halo-exchange code doesn't contain a collective operation in its main computational loop. The complexity is $\mathcal{O}(|T| * |t|)$. The binary broadcast code performs number of collective operations. The additional edges thus introduced raise the complexity to $\mathcal{O}(|T| * |t|^2)$. However, the influence of $|t|$ can be mitigated by the already outlined parallelization, since more threads allow more tasks to be replayed in parallel.

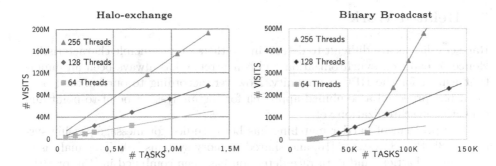

Fig. 6. Actual complexity of the task graph generation

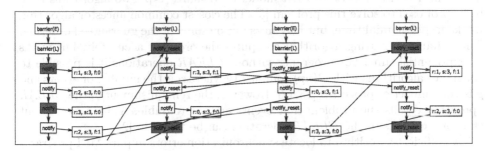

Fig. 7. Synchronization race

Figure 7 depicts the task graph of a program, which sometimes got stuck. The analyzed execution trace was recorded of a successful program run. Nevertheless, our replay algorithm revealed a post/post collision and marked the corresponding nodes in the output graph. The problem was introduced by a program optimization, where a collective reduction in the initialization phase was replaced by a more efficient binary broadcast routine similar to the one depicted in Fig. 4. That routine was taken from another program, where it had worked. The problem was, that the flag range used by the initialization routine overlapped with the flag range of the worker phase. The first marked notify_reset node at rank 3 was meant to receive the notification from rank 0. However, it could also receive a notification from rank 2, which already belongs to the worker phase. In such a case our tool marks the colliding notifications and connects them with the notify_reset node. If the replay algorithm finishes, it also marks still untriggered wait operations, e.g. the second notify_reset node at rank 3. With this information we could fix the bug by assigning different numbers to the flags of the initialization routine. This removed the overlap and freed the program of synchronization races.

6   Related Work

Race conditions are difficult to detect due to their irreproducible characteristics. Hence research on synchronization and concurrency has always been an important aspect for the HPC community. However, according to our knowledge no work has proposed a combined approach for the analysis of point-to-point and collective synchronization yet.

The problem of barrier matching has been studied for message-passing systems [18], PGAS systems [16] and shared-memory systems [10]. The analysis of split-phase barriers and data race detection has been combined in [15] for UPC programs. The problem of computing all guaranteed orderings in a program trace using the $POST$, $WAIT$, $CLEAR$ model is NP-hard [14]. Two algorithms have been proposed to solve this problem [5]. The closest common ancestor algorithm works in polynomial time, but may miss out on some of the guaranteed ordering. The exhaustive-pairing algorithm computes the ordering accurately, but works in exponential time. For programs without $CLEAR$ operations it is possible to construct algorithms with $\mathcal{O}(np)$ complexity where n is the number of events and p is the number of processes [13,17]. However, the relinquishment of the $CLEAR$ operation entails the problem that events are not reusable. A discussion about the consequences of the $CLEAR$ operation can be found in [2].

An efficient algorithm to locate synchronization errors in pure MPI programs is described in [12]. While this approach does not handle collective synchronization, the theoretical background presented there is similar to our approach. Theorem 2 appears in our work in a more generalized context.

7   Conclusion

This paper makes two important contributions. First, we have extended the event-style synchronization model to collectives. By doing so we are able to handle point-to-point and collective synchronization in a unified manner. This enables us to reason about the execution order of events in hybrid parallel programs. Second, we have condensed the event-style synchronization paradigm to two operations – `post` and `wait`. Our `wait` operation is a concatenation of the classic $WAIT$ and $CLEAR$ operations. This simplification has the important effect, that task graph construction is not NP-hard any more. Thus programs using our synchronization paradigm are testable for race conditions of various kinds. Our paradigm is used by the collective and the point-to-point synchronization routines defined by the GASPI standard. Even `MPI_Send` and `MPI_Recv` can be regarded as `post` and `wait` respectively.

Our model does not require an atomic coupling of $WAIT$ and $CLEAR$. A programmer could also manually perform a $CLEAR$ after a $WAIT$. For instance, the OpenShmem function `shmem_int_wait` forces a thread to wait until an integer is no longer equal to a certain value. One could reset the respective integer to that value as soon as `shmem_int_wait` has returned and thus achieve the functionality required by our model. A function `shmem_int_wait_and_clear` would lead to programs, which are implicitly testable against race conditions. That's why we think, that our work should be considered, whenever decisions have to be made during the design of parallel programming APIs. Point-to-point synchronization using post and wait makes the reasoning about the correctness of programs easier than the *POST*, *WAIT*, *CLEAR* paradigm.

While we have introduced the algorithm in the context of post-mortem analysis, an adaption to an on-the-fly approach is possible. As discussed in Sect. 3 tasks can be replayed in the order of their delivery. On-the-fly techniques can cope with much longer program runs than post-mortem techniques, since tasks can be discarded once they are evaluated. An interesting research topic is the question, which tasks can be discarded so that the test of Theorem 1 (PWP[r].PreviousWait $\prec$ T) on line 10 of Listing 1 is not affected.

References

1. GASPI - Global Adress Space Programming Interface (2011). http://www.gaspi.de. Accessed 29 Jan 2015
2. Callahan, D., Kennedy, K., Subhlok, J.: Analysis of event synchronization in a parallel programming tool. In: ACM SIGPLAN Notices, vol. 25, pp. 21 30. ACM (1990)
3. Chandy, K.M., Lamport, L.: Distributed snapshots: determining global states of distributed systems. ACM Trans. Comput. Syst. (TOCS) **3**(1), 63–75 (1985)
4. El-Ghazawi, T., Carlson, W., Sterling, T., Yelick, K.: UPC: Distributed Shared Memory Programming. Wiley Series on Parallel and Distributed Computing. Wiley, New York (2005)
5. Emrath, P., Ghosh, S., Padua, D.: Detecting nondeterminacy in parallel programs. IEEE Softw. **9**(1), 69–77 (1992)
6. Emrath, P.A., Ghosh, S., Padua, D.A.: Event synchronization analysis for debugging parallel programs. In: Proceedings of Supercomputing 1989, pp. 580–588 (1989)
7. Grünewald, D., Simmendinger, C.: The GASPI API specification and its implementation GPI 2.0. In: 7th International Conference on PGAS Programming Models, vol. 243 (2013)
8. Krzikalla, O., Knüpfer, A., Müller-Pfefferkorn, R., Nagel, W.E.: On the modelling of one-sided communication systems. In: Proceedings of the 7th International Conference on PGAS Programming Models, Edinburgh, UK, October 2013, pp. 41–53 (2013)
9. Lamport, L.: Time, clocks, and the ordering of events in a distributed system. Commun. ACM **21**(7), 558–565 (1978)
10. Lin, Y.: Static nonconcurrency analysis of OpenMP programs. In: Mueller, M.S., Chapman, B.M., de Supinski, B.R., Malony, A.D., Voss, M. (eds.) IWOMP 2005 and IWOMP 2006. LNCS, vol. 4315, pp. 36–50. Springer, Heidelberg (2008)
11. Luk, C.K., Cohn, R., Muth, R., Patil, H., Klauser, A., Lowney, G., Wallace, S., Reddi, V.J., Hazelwood, K.: Pin: building customized program analysis tools with dynamic instrumentation. SIGPLAN Not. **40**(6), 190–200 (2005)
12. Netzer, R.H., Brennan, T.W., Damodaran-Kamal, S.K.: Debugging race conditions in message-passing programs. In: Proceedings of the SIGMETRICS Symposium on Parallel and Distributed Tools, pp. 31–40. ACM (1996)

13. Netzer, R.H., Ghosh, S., et al.: Efficient race condition detection for shared-memory programs with Post/Wait synchronization. University of Wisconsin-Madison, Computer Sciences Department (1992)
14. Netzer, R.H., Miller, B.P.: On the complexity of event ordering for shared-memory parallel program executions. In: Proceedings of the 1990 International Conference on Parallel Processing, pp. 93–97 (1990)
15. Park, C.S., Sen, K., Hargrove, P., Iancu, C.: Efficient data race detection for distributed memory parallel programs. In: Proceedings of 2011 International Conference for High Performance Computing, Networking, Storage and Analysis, SC 2011, pp. 51:1–51:12. ACM, New York (2011)
16. Pophale, S., Hernandez, O., Poole, S., Chapman, B.M.: Extending the OpenSHMEM analyzer to perform synchronization and multi-valued analysis. In: Poole, S., Hernandez, O., Shamis, P. (eds.) OpenSHMEM 2014. LNCS, vol. 8356, pp. 134–148. Springer, Heidelberg (2014)
17. Ramanujam, J., Mathew, A.: Analysis of event synchronization in parallel programs. In: Pingali, K.K., Gelernter, D., Padua, D.A., Banerjee, U., Nicolau, A. (eds.) LCPC 1994. LNCS, vol. 892, pp. 300–315. Springer, Heidelberg (1995)
18. Zhang, Y., Duesterwald, E.: Barrier matching for programs with textually unaligned barriers. In: Proceedings of the 12th ACM SIGPLAN Symposium on Principles and Practice of Parallel Programming, pp. 194–204. ACM (2007)

Nasty-MPI: Debugging Synchronization Errors in MPI-3 One-Sided Applications

Roger Kowalewski[✉] and Karl Fürlinger

MNM-Team, Ludwig-Maximilians-Universität München,
Oettingenstr. 67, Munich, Germany
{kowalewski,fuerling}@mnm-team.org

Abstract. The Message Passing Interface (MPI) specifies a one-sided interface for Remote Memory Access (RMA), which allows one process to specify all communication parameters for both the sending and receiving side by providing support for asynchronous reads and updates of distributed shared data. While MPI RMA communication can be highly efficient, proper synchronization of possibly conflicting accesses to shared data is a challenging task.

This paper presents a novel debugging tool that supports developers in finding latent synchronization errors. It dynamically intercepts RMA calls and reschedules them into pessimistic executions which are valid in terms of the MPI-3 standard. Given an application with a latent synchronization error, we force a manifestation of this error which can easily be detected with the help of program invariants. An experimental evaluation shows that the tool can uncover synchronization errors which would otherwise likely go unnoticed for a wide range of scenarios.

Keywords. Bug detection · MPI · One-sided communication

1 Introduction

Modern remote direct memory access (RDMA) network interconnects leverage efficient MPI one-sided communication, also known as MPI RMA (remote memory access), as an important communication paradigm. In contrast to traditional message passing, RMA conceptually decouples data transfer and synchronization, enabling superior performance potential [2]. Furthermore, while message passing is natural for some problems, it can be cumbersome to use for applications with irregular communication patterns. However, the non-blocking nature of MPI RMA poses several challenges. Programmers must understand the complex synchronization model to maintain memory consistency between possibly conflicting asynchronous data accesses. Latent synchronization bugs may lead to an erroneous state manifested during one execution that may not be triggered during another execution due to the underlying MPI library, network interconnect facilities, thread interleaving, etc. Often, errors remain unnoticed for a long period of time and only occur in large-scale scenarios [1] or after porting the application to a different HPC platform [3].

© Springer International Publishing Switzerland 2016
P.-F. Dutot and D. Trystram (Eds.): Euro-Par 2016, LNCS 9833, pp. 51–62, 2016.
DOI: 10.1007/978-3-319-43659-3_4

As an example, MPI RMA provides only weak ordering guarantees. In Fig. 1b, a `MPI_Put` modifies a remote memory buffer x which is subsequently accessed by a `MPI_Get` call. Due to the non-blocking semantics, both RMA calls may complete in any order. Furthermore, they are non-atomic allowing the `MPI_Get` to fetch a partially written value by the preceding `MPI_Put`. While this semantic flexibility is a major strength of MPI and necessary to achieve high performance, it complicates the task to write portable and correct programs.

For this purpose, we present Nasty-MPI, a runtime debugging tool to support programmers in finding latent synchronization bugs within their applications. Like many other tools, it is based on the MPI profiling interface (PMPI) and can therefore be used with any MPI application. The approach takes the RMA semantics into account to schedule *pessimistic executions* of the issued RMA communications which force a manifestation of latent synchronization bugs. Because each application may have numerous of such pessimistic executions, we provide external configuration parameters to control the scheduling process of Nasty-MPI, enabling easy integration into any environment. Since Nasty-MPI has no correctness model of the target applications, the only requirement on programmers is to supply program invariants (e.g., `assert` statements) which uncover possibly forced synchronization errors.

Such a tool supports programmers already during early development stages. We have integrated the concept into the DASH library distribution [6]. DASH is a C++ template library which adopts hierarchical PGAS[1] concepts to important standard containers (arrays, matrices, etc.) and is published along with extensive unit test suites to validate the implemented containers and algorithms. Applying Nasty-MPI improves these unit tests as it significantly increases the chance to uncover latent synchronization bugs in the low-level MPI RMA communications.

In the reminder of this paper, we first summarize the MPI RMA synchronization semantics and present a formalism to model memory consistency in Sect. 2. Section 3 elaborates the concept and strategies of Nasty-MPI to uncover synchronization bugs. An experimental evaluation in Sect. 4 with small test cases compares the behavior of applications with latent synchronization bugs on different HPC platforms. It reveals that the presented approach can manifest these synchronization bugs which would otherwise likely go unnoticed. Finally, Sect. 5 summarizes related work and Sect. 6 concludes the results of this paper.

2   MPI-3 One-Sided Communication

MPI RMA can be applied only on a point-to-point basis, i.e., an origin process remotely accesses memory on a target process. All communication actions (puts, gets, accumulates) operate in the context of a *window*, abstracting the distributed memory between MPI processes, and are grouped into synchronization phases, called *access epochs*. No RMA operation may be issued before opening an access epoch and no completion guarantees, neither local nor remote, are available before closing an access epoch.

---

[1] Partitioned global address space.

MPI RMA offers two synchronization modes which are called the *active target* and *passive target* mode. In the scope of this paper, we focus on passive target mode as only the origin is involved in synchronization, which closely matches the requirements of one-sided communication. Passive target synchronization relies on a single *lock/unlock* model to open and close an access epoch, respectively. Nevertheless, we can adopt the proposed techniques to the active target mode, as well as to other one-sided programming models.

2.1 Challenges in MPI RMA

In MPI RMA, all communication primitives and the *lock* routine to open an access epoch are, in fact, non-blocking. Thus, memory operations within an access epoch, whether RMA or native loads/stores, may conflict with each other. In particular, we have to consider three critical properties:

Completion: RMA communication operations are not guaranteed to complete before the surrounding access epoch is explicitly synchronized. For example in Fig. 1a, the receive buffer (buf) for the MPI_Get is subsequently accessed by a native load. Both memory accesses conflict, resulting in undefined behavior.

Ordering: In general, MPI provides no ordering guarantees for RMA operations, i.e., the order in which they are applied within an access epoch is

```
MPI_Win_lock(target);
MPI_Get(&buf, ..., target);

if (buf % 2 == 0)
    buf++;
MPI_Win_unlock(target);
```

```
MPI_Win_lock(target);
int s = 10, r = 0;
MPI_Put(&s, ..., x, target);
MPI_Get(&r, ..., x, target);
assert(r == 10);
MPI_Win_unlock(target);
```

```
int buf[100];
/* init buf */

MPI_Win_lock(target);
MPI_Put(&buf, 100 ..., target);
MPI_Win_unlock(target);
```

(a) Conflicting native load and MPI_Get.   (b) Unsynchronized Put-Get sequence.   (c) Non-atomic Put.

Fig. 1. Application samples with synchronization bugs.

```
MPI_Win_lock(target);
...
if (buf % 2 == 0)
    buf++;

/* Defer Get*/
MPI_Get(&buf, ..., target);

MPI_Win_unlock(target);
```

```
MPI_Win_lock(target);
int s = 10, r = 0;

/* Reverse order */
MPI_Get(&r, ..., x, target);
MPI_Win_flush(target);
MPI_Put(&s, ..., x, target);

assert(r == 10); /* Fails */
MPI_Win_unlock(target);
```

```
/* init buf[100] */
MPI_Win_lock(target);

/* Splitting */
MPI_Put,(&buf, ..., target);
MPI_Put,(&(buf + 1), ..., target);
...
MPI_Put,(&(buf + 99), ..., target);

MPI_Win_unlock(target);
```

(a) Deferred MPI_Get.   (b) Reordered Put-Get sequence.   (c) Split non-atomic Put.

Fig. 2. Exemplified modifications by Nasty-MPI.

unspecified. An exception is made for *accumulates* directed to the same target, and with the same operation and basic data type. In Fig. 1b, two RMA calls read (MPI_Put) and write (MPI_Get) on a single memory buffer, respectively. Since the operations may complete in any order, they conflict with each other.

Atomicity: In general, RMA operations are non-atomic, except *accumulates*, which guarantee element-wise atomic reads and writes to a single target if they use the same basic data type. Figure 1c shows an example where an origin copies an array, consisting of 100 integers, to a target memory. This MPI_Put is non-atomic and may conflict with any memory accesses, operating concurrently on the target memory location.

These guarantees are crucial in even simple concurrent protocols. Thus, writing portable and well-defined programs requires properly synchronized memory accesses on overlapping locations. MPI RMA specifies dedicated synchronization primitives for this purpose. Beside the common approach of synchronizing by distinct access epochs, MPI additionally provides *local_flush* and *flush* primitives to locally or remotely complete pending RMA operations within an access epoch. While local completion guarantees consistent memory buffers only on the origin process, remote completion guarantees memory consistency of the target memory as well [16].

2.2 Modeling Memory Consistency in MPI RMA

To model and analyze the RMA operations issued by an application, we use a formalism based on a paper written by the MPI RMA Working Group [10].

Two memory accesses a and b conflict if they target overlapping memory and are not synchronized by both a happens-before ($\xrightarrow{hb}$) [11] and a consistency edge ($\xrightarrow{co}$) [10]. The happens-before order may either be the program order, if both operations occur in a single process, or the synchronization order between two MPI processes, such as blocking send-receive pairs. A consistency edge between two operations (i.e., $a \xrightarrow{co} b$) implies that the memory effects of a may be observed by b. Consistency edges are established by the RMA synchronization primitives, as described earlier.

Utilizing this notation, we derive an execution model of all issued RMA communications in an MPI program P. All executions E over the set of RMA calls in P may be modeled as a partially ordered *happens-before graph*, formed by the transitive closure of $\xrightarrow{hb}$ and $\xrightarrow{co}$ edges. Two executions e_1 and e_2 in E are semantically equivalent if they result in the same happens-before graph. If a and b are not synchronized, they are contained in a parallel region. For example, Fig. 3 represents a happens-before graph, derived from the program in Fig. 1b. Since both RMA operations operate on overlapping memory and are within a parallel region, the program includes a synchronization bug. If we want to guarantee that both operations remotely complete in program order, one valid solution is to synchronize by an additional *flush*, which establishes the required $\xrightarrow{cohb}$ edge, as depicted in Fig. 4.

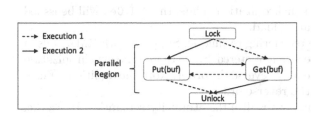

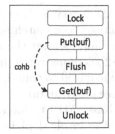

Fig. 3. Unsynchronized (two executions).

Fig. 4. Synchronized execution.

The next section explains how we exploit this formalism to uncover latent synchronization bugs.

3 Forcing Synchronization Errors with Nasty-MPI

This section describes an effective approach to support programmers in debugging MPI programs with improperly synchronized RMA communications. Suppose an MPI program P contains a latent synchronization bug. Given further that P has a predefined correctness model in the form of included program invariants, as illustrated by the **assert** statements in Fig. 1b. Based on the presented memory consistency model we are able to explore different execution paths in the happens-before graph of P with the objective of finding at least one execution which forces a manifestation of this bug.

3.1 Conceptual Overview

By exploiting the PMPI interface we intercept all RMA communication actions at runtime and initially buffer them, instead of handing them over to the MPI library. This enables to dynamically construct a happens-before graph and, in particular, tracking all its parallel regions. The approach relies on the RMA completion semantics, allowing to defer the execution of communication actions to a matching synchronization call. When the application issues a synchronization action, it triggers a three-stage rescheduling process:

(1) **Completion Stage:** We consider only those communication actions which are necessarily required to complete, as specified by the synchronization action.

(2) **Atomicity Stage:** We break non-atomic communication actions into a set of smaller requests in such a way that the memory semantics are identical.

(3) **Reordering Stage:** We reorder communication actions which do not conceptually give any ordering guarantees within the synchronized access epoch.

Figure 2 illustrates the rescheduling techniques when applying Nasty-MPI to the programs in Fig. 1 in the form of source code modifications that are equivalent to the effects achieved by the dynamic interception and rescheduling process.

In Fig. 2a, Nasty-MPI exploits the completion semantics and defers communication actions to a matching synchronization. Thus, the MPI_Get will be issued to the MPI library after the native load.

Figure 2b demonstrates the reordering technique. Suppose both RMA calls in Fig. 1b are required to complete as encountered. Since there is no synchronization to guarantee program order, we may reverse the order. Note the additional flush, issued by Nasty-MPI to force the reverse order.

The last example depicts how we utilize the atomicity semantics. In Fig. 2c, we split one single MPI_Put into 100 separate MPI_Put calls. While both variants have identical semantics, splitting RMA operations can effectively force errors which result from non-atomic memory access on overlapping locations.

In the next section, we explain the rescheduling process in more detail and discuss how the tool uses the full semantic flexibility, given by the MPI standard, to schedule pessimistic executions.

3.2   Nasty-MPI Rescheduling Process

Suppose Nasty-MPI receives a synchronization action, triggering the rescheduling process on buffered communication actions. The three stages of this rescheduling process are described in the following.

Completion Stage. Nasty-MPI first distinguishes between local and remote completion. If the issued synchronization action has remote completion semantics (i.e., *unlock* or *flush*), we filter all buffered RMA calls which are necessarily required to complete. A synchronization action can complete either all pending RMA calls within a window or to a specific target rank [16].

In the case of local completion (i.e., *flush_local*), however, all MPI_Put calls remain in the buffer and are not issued to the MPI library. This approach is allowed, because local completion only guarantees memory consistency of local buffers. However, because local completion creates a consistency edge between two consecutive memory access (i.e., $a \xrightarrow{co} b$), we have to copy the source buffer of a to keep it internally until remote completion is forced. This approach is applicable to RMA *accumulates* as well. However, because accumulates are conceptually ordered under certain conditions [16], we have to make sure that there are no subsequent correlated accumulates which atomically fetch data from remote memory. In this case, we are not allowed to further postpone the first accumulate operation. This strategy is useful because several experiments revealed that some MPI libraries do not distinguish between local and remote completion, i.e., they always apply remote completion.Table 1 lists two parameters for the completion stage to control, whether Nasty-MPI should apply local completion semantics (NASTY_LOCAL_COMPLETION_ENABLED) or even bypass the completion stage (NASTY_SKIP_COMPLETION_STAGE).

Table 1. Nasty-MPI configuration parameters.

Parameter	Options
NASTY_SKIP_COMPLETION_STAGE	0*, 1
NASTY_LOCAL_COMPLETION_ENABLED	0, 1*
NASTY_SKIP_ATOMICITY_STAGE	0, 1*
NASTY_SUBMIT_ORDER	see Table 2
NASTY_ADD_FLUSH_ENABLED	0, 1*
NASTY_ADD_LATENCY	unit32 range**

*default value **default value: 0

Table 2. Options for NASTY_SUBMIT_ORDER.

Option	Description
random	Random (default)
reverse_po	Reverse program order
put_before_get	Schedule *put* before *get* calls
get_before_put	Schedule *get* before *put* calls

Atomicity Stage. While fast RMA data transfers (i.e., *put, get*) are non-atomic, accumulates guarantee it only on a per element granularity. Thus, we apply a splitting technique to break a single RMA call into a set of many smaller RMA calls which have identical memory semantics. We first analyze the count and datatype parameters which are contained in the signature of each RMA call. If the count parameter is specified with at least 2 elements (i.e., count >= 2), we further determine the *extent* of a single datatype element. Based on these two parameters, one RMA call can be split into many single-element calls. For example, in Fig. 1c, count is 100 and the extent of MPI_INT is 4 bytes. This results in 100 MPI_Put calls, each having a source buffer which starts at increasing 4 bytes offsets relative to the original buffer address (see Fig. 2c).

RMA *put* and *get* calls can be even split into 1-byte RMA operations. However, we are restricted by the *displacement unit* in MPI *windows* which defines the minimum size of a single element. Thus, this approach applies only if the displacement unit is specified with a size of MPI_BYTE at window creation.

The atomicity stage may skipped by setting the NASTY_SKIP_ATOMICITY_STAGE parameter to 1, as listed in Table 1.

Reordering Stage. Passing the first two stages gives a set of RMA calls which are (*a*) required to remotely complete; and (*b*) split into many small RMA calls in order to explore the minimal completion and atomicity semantics. Before we hand over these RMA calls to the native MPI library, they are finally reordered. The only restriction applies to accumulates. We can interleave them with any other communication action, however, their syntactic order has to be preserved. The default reordering approach is to randomly shuffle buffered communication actions. More fine-grained control is provided by the configuration parameter NASTY_SUBMIT_ORDER which can be set to any of the options in Table 2. However, simply reorder RMA operations does not guarantee that the native MPI library obeys the scheduled order. Similar to Nasty-MPI, MPI libraries are free to reorder or even apply additional optimizations, such as merging of RMA calls [7]. Thus, we must explicitly force the scheduled ordering. One option is to simulate communication latency between consecutive communication actions, giving the MPI library a chance to asynchronously process an RMA operation before the next call is issued. However, if the MPI library does not facilitate asynchronous progress mechanisms or applies lazy execution, this approach has no effect. An effective solution is to issue additional *flush* operations which is semantically valid, as we modify only parallel regions in the original happens-before graph.

The reordering stage can be further controlled by two parameters in order to configure the simulation of communication latency (NASTY_ADD_LATENCY) and to configure whether Nasty-MPI is allowed to inject additional *flush* synchronizations (NASTY_ADD_FLUSH_ENABLED).

4   Experimental Evaluation

The experiments were conducted on two HPC platforms: The NERSC Edison Cray XC 30 supercomputer [17] and SuperMUC [12] at the Leibniz Supercomputing Centre. The Cray machine is interconnected by an Aries network and provides its own MPI library and compiler, included in Cray's Message Passing Toolkit. SuperMUC facilitates a fully non-blocking Infiniband network and supports three MPI libraries: IBM (v9.1.4), Intel(v5.0) and open MPI(v1.8). The corresponding compiler is Intel icc (v15.0.4). A prototypical implementation of Nasty-MPI is publicly accessible on Github[2].

4.1   Methodology

All experiments include at least two MPI processes which communicate by improperly synchronized RMA operations. The correctness model of these applications is defined by included assert statements in the source code to uncover the synchronization errors.

Each experiment is evaluated with all MPI libraries in 4 scenarios, which are based on two parameters. The first parameter determines process locality, i.e., the origin and target process reside either on a single node or on two distant nodes. Process locality is an important property, because MPI libraries may hide communication latency in MPI RMA calls by utilizing shared memory semantics. The second parameter depends on whether Nasty-MPI is linked to the target application. If Nasty-MPI is linked, all applications are repeatedly executed with distinct combinations of the Nasty-MPI configuration parameters, listed in Table 1. The assumption is that, if Nasty-MPI is not linked, the MPI libraries can successfully execute the applications, i.e., the assert statements manifest no errors. In this case, there has to be at least one configuration for Nasty-MPI which forces a pessimistic execution to uncover the synchronization bug.

4.2   Effectiveness of Nasty-MPI

The first test case is a binary tree broadcast algorithm which was described by Luecke et al. [13]. The code relies on MPI_Get being a blocking MPI call because there is no synchronization action which actually completes it. The relevant

---

[2] https://github.com/rkowalewski/nasty-MPI.

snippet is shown in Fig. 5. Executing this program setup leads to different results, depending on the test setup. If the communicating processes, involved in the MPI_Get, reside on distant nodes no MPI library can successfully terminate this program due to an infinite loop. But the situation changes, if both processes reside on the same node. While IBM MPI and open MPI again cannot exit from the polling loop, the implementations of Intel (SuperMUC) and Cray (NERSC Edison) can complete the RMA call. This demonstrates that process locality may impact the behavior of RMA communications, depending on the underlying MPI library. If Nasty-MPI is linked and the completion stage is not skipped (i.e., NASTY_SKIP_COMPLETION_STAGE = 0), the MPI library does never receive the MPI_Get request, because no synchronization action completes the buffered RMA call.

```
MPI_Win_lock(target);
double check = 0;
...
while (check == 0)
{
  MPI_Get(&check, ..., target);
  /* Missing Synchronization */
}
...

MPI_Win_unlock(target);
```

```
MPI_Win_lock_all(win);

MPI_Accumulate(...,
                    predecessor, ..., win);
do {
  MPI_Fetch_and_op(..., self, ...,win);

  MPI_Win_flush(self);
} while (flag);

MPI_Win_unlock_all(win);
```

Fig. 5. Non-completed MPI_Get   **Fig. 6.** Improperly synchronized Acc.

The second test case is an implementation of the MCS lock [15] which can be implemented using MPI RMA primitives [10]. In the code for acquiring the lock (Fig. 6), a requesting process issues two RMA calls which are directed to different targets. For test purposes, we have injected a synchronization error in such a way that only one target is synchronized. As listed in Table 3, all MPI libraries, except Intel, can successfully execute this program. This observation confirms that some MPI libraries always complete all pending RMA calls, regardless of the synchronization target. In Nasty-MPI, however, only the second RMA call reaches the native MPI library, while the first MPI_Accumulate is rejected in the completion stage, causing a manifestation of the synchronization bug.

The third test case is a slight modification from the example in Fig. 1b. The MPI_Put modifies a remote memory location x and is only locally completed by a *local_flush*. All MPI libraries pass the assert statement, i.e., the MPI_Get fetches the modified value by the MPI_Put. If Nasty-MPI is linked and the parameter NASTY_LOCAL_COMPLETION_ENABLED is set to 1, it defers the MPI_Get to the *unlock* call, leading to a manifestation of the synchronization error.

Program 4 tests the ordering guarantees of the MPI libraries. It requires that two consecutive remote writes, one MPI_Put followed by an MPI_Accumulate, are completed in target memory as encountered by the program order. Still, there is no synchronization action to ensure this order. If the origin and target processes reside on a single node, all MPI libraries, except Intel, complete both RMA calls in program order. Nasty-MPI can easily force the synchronization bug by setting NASTY_SUBMIT_ORDER to reverse_po.

Finally, Nasty-MPI helped to detect a synchronization bug in the DASH library, while it was applied to a large test suite. The root cause was to pass a memory buffer, located on the stack frame, to a MPI_Put. However, the matching synchronization call was outside of the method scope, causing the memory buffer to be invalid if the RMA call is deferred to this synchronization call.

Table 3. Results of the experiments without linking Nasty-MPI.

		Edison	SuperMUC		
No	Test Program	Cray	IBM	Intel	oMPI
1	Binary Broadcast [13]	×	✓	×	✓
2	MCS lock [15]	×	×	✓	×
3	Local completion	×	×	×	×
4	Unordered Put calls	×	×	✓	×

✓ Synchronization error manifested
× Synchronization error not manifested

5   Related Work

There is a large number of approaches for automatic bug detection in *two-sided* MPI [4,5,20,21], however, they cannot be applied to one-sided MPI due to the contrary synchronization model.

A tool, called MC-Checker [3], is closely related to this paper. It can detect memory consistency errors by profiling both MPI RMA and native memory accesses, i.e., loads and stores. Based on the MPI semantics, it effectively finds potential data races even across different origins which concurrently access overlapping target memory. However, MC-Checker only covers the MPI-2 standard which follows different synchronization semantics compared to MPI-3. Moreover, Nasty-MPI follows a different approach, since it forces synchronization errors, rather than detecting them. MUST [9] focuses on deadlocks and semantic parameter checking, which is not the scope of Nasty-MPI. However, both tools can complement each other to debug memory consistency and semantic parameter errors. Scalasca [8] detects inefficient wait states in MPI RMA applications. Another approach applies model checking [19] for deadlock detection in MPI RMA programs. Furthermore, there are tools from other PGAS languages. UPC-Thrill [18] uncovers data races in UPC programs. Significant semantic differences between UPC and MPI RMA distinguish this tool from Nasty-MPI.

6   Conclusion and Future Work

This paper discusses the semantic challenges of MPI-3 RMA and presents Nasty-MPI, a novel approach to support the detection of latent synchronization bugs in MPI applications. Based on the complex RMA semantics, we apply a systematic strategy to force latent errors, which may be easily manifested with the help of program invariants. An experimental evaluation has demonstrated that we can uncover synchronization errors which would be otherwise go unnoticed for a wide range of synchronization scenarios. Furthermore, the tool detected a synchronization bug in the DASH library [6].

We currently evaluate to track native memory accesses by using tools, such as Pin [14]. This enables to more effectively force synchronization errors between MPI RMA and native memory accesses. Another challenge are RMA communications which use complex MPI data types (e.g., structs). Currently, we cannot apply the full potential of Nasty-MPI to such RMA operations, as it requires to understand the memory layout of complex MPI data types.

Finally, Nasty-MPI may be used by any programmer who wants to verify the semantic correctness of a given MPI RMA program.

Acknowledgments. We gratefully acknowledge funding by the German Research Foundation (DFG) through the German Priority Programme 1648 Software for Exascale Computing (SPPEXA).

References

1. Arnold, D., Ahn, D., de Supinski, B., Lee, G., Miller, B., Schulz, M.: Stack trace analysis for large scale debugging. In: IEEE International Parallel and Distributed Processing Symposium, IPDPS 2007, pp. 1–10, March 2007
2. Bell, C., Bonachea, D., Nishtala, R., Yelick, K.: Optimizing bandwidth limited problems using one-sided communication and overlap. In: Proceedings of the 20th International Conference on Parallel and Distributed Processing, IPDPS 2006, p. 84. IEEE Computer Society, Washington, DC (2006)
3. Chen, Z., Dinan, J., Tang, Z., Balaji, P., Zhong, H., Wei, J., Huang, T., Qin, F.: MC-Checker: detecting memory consistency errors in MPI one-sided applications. In: Proceedings of the International Conference for High Performance Computing, Networking, Storage and Analysis, pp. 499–510. IEEE Press (2014)
4. Chen, Z., Li, X., Chen, J.Y., Zhong, H., Qin, F.: SyncChecker: detecting synchronization errors between MPI applications and libraries. In: Proceedings of the 2012 IEEE 26th International Parallel and Distributed Processing Symposium, IPDPS 2012, pp. 342–353. IEEE Computer Society, Washington, DC (2012)
5. DeSouza, J., Kuhn, B., de Supinski, B.R., Samofalov, V., Zheltov, S., Bratanov, S.: Automated, scalable debugging of MPI Programs with intel message checker. In: Proceedings of the Second International Workshop on Software Engineering for High Performance Computing System Applications, SE-HPCS 2005, pp. 78–82, New York (2005)
6. Fürlinger, K., et al.: DASH: data structures and algorithms with support for hierarchical locality. In: Lopes, L., et al. (eds.) Euro-Par 2014, Part II. LNCS, vol. 8806, pp. 542–552. Springer, Heidelberg (2014)

7. Gropp, W.D., Thakur, R.: An evaluation of implementation options for MPI one-sided communication. In: Di Martino, B., Kranzlmüller, D., Dongarra, J. (eds.) EuroPVM/MPI 2005. LNCS, vol. 3666, pp. 415–424. Springer, Heidelberg (2005)
8. Hermanns, M.A., Miklosch, M., Böhme, D., Wolf, F.: Understanding the formation of wait states in applications with one-sided communication. In: Proceedings of the 20th European MPI Users' Group Meeting, pp. 73–78. ACM (2013)
9. Hilbrich, T., Protze, J., Schulz, M., de Supinski, B.R., Müller, M.S.: MPI runtime error detection with MUST: advances in deadlock detection. In: Proceedings of the International Conference on High Performance Computing, Networking, Storage and Analysis, SC 2012, pp. 30:1–30:11. IEEE Computer Society Press, Los Alamitos (2012)
10. Hoefler, T., Dinan, J., Thakur, R., Barrett, B., Balaji, P., Gropp, W., Underwood, K.: Remote memory access programming in MPI-3. ACM Trans. Parallel Comput. **2**(2), 9:1–9:26 (2015)
11. Lamport, L.: Time, clocks, and the ordering of events in a distributed system. Commun. ACM **21**(7), 558–565 (1978)
12. Leibniz Supercomputing Centre, Munich, Germany: SuperMUC Petascale System. https://www.lrz.de/services/compute/supermuc/systemdescription/
13. Luecke, G.R., Spanoyannis, S., Kraeva, M.: The performance and scalability of SHMEM and MPI-2 one-sided routines on a SGI Origin 2000 and a Cray T3E–600: performances. Concurr. Comput. Pract. Exper. **16**(10), 1037–1060 (2004)
14. Luk, C.K., Cohn, R., Muth, R., Patil, H., Klauser, A., Lowney, G., Wallace, S., Janapa, V., Hazelwood, R.K.: Pin: building customized program analysis tools with dynamic instrumentation. In: Proceedings of the 2005 ACM SIGPLAN conference on Programming language design and implementation, PLDI 2005, pp. 190–200. ACM Press (2005)
15. Mellor-Crummey, J.M., Scott, M.L.: Algorithms for scalable synchronization on shared-memory multiprocessors. ACM Trans. Comput. Syst. **9**(1), 21–65 (1991)
16. MPI Forum: MPI: A Message-Passing Interface Standard. Version 3.0, September 2012. http://www.mpi-forum.org
17. National Energy Research Center, United States: Edison System Configuration. https://www.nersc.gov/users/computational-systems/edison/configuration/
18. Park, C.S., Sen, K., Hargrove, P., Iancu, C.: Efficient data race detection for distributed memory parallel programs. In: Proceedings of 2011 International Conference for High Performance Computing, Networking, Storage and Analysis, SC 2011, pp. 51:1–51:12. ACM, New York (2011)
19. Pervez, S., Gopalakrishnan, G.C., Kirby, R.M., Thakur, R., Gropp, W.D.: Formal verification of programs that use MPI one-sided communication. In: Mohr, B., Träff, J.L., Worringen, J., Dongarra, J. (eds.) PVM/MPI 2006. LNCS, vol. 4192, pp. 30–39. Springer, Heidelberg (2006)
20. Vakkalanka, S.S., Sharma, S., Gopalakrishnan, G., Kirby, R.M.: ISP: a tool for model checking MPI programs. In: Proceedings of the 13th ACM SIGPLAN Symposium on Principles and Practice of Parallel Programming, PPopp 2008, pp. 285–286. ACM, New York (2008)
21. Vetter, J.S., de Supinski, B.R.: Dynamic software testing of MPI applications with umpire. In: Proceedings of the 2000 ACM/IEEE Conference on Supercomputing, SC 2000. IEEE Computer Society, Washington, DC (2000)

Automatic Benchmark Profiling Through Advanced Trace Analysis

Alexis Martin[1,2,3](✉) and Vania Marangozova-Martin[1,2,3](✉)

[1] CNRS, LIG, 38000 Grenoble, France
[2] University Grenoble Alpes, LIG, 38000 Grenoble, France
[3] Inria, Grenoble, France
alexis.martin@inria.fr, vania.marangozova-martin@imag.fr

Abstract. Benchmarking has proven to be crucial for the investigation of system behavior and performances. However, the choice of relevant benchmarks still remains a challenge. To help the process of comparing and choosing among benchmarks, we propose a solution for automatic benchmark profiling. It computes unified profiles reflecting benchmarks' duration, function repartition, stability, CPU efficiency, parallelization and memory usage. It identifies the needed system information for profile computation, collects it from execution traces and produces profiles through automatic and reproducible trace analysis. The paper presents the design, the implementation and the evaluation of the approach.

1 Introduction

System performance is a major preoccupation during system design and implementation. Even if some performance aspects may be guaranteed by design using formal methods [1], all systems undergo a testing phase during which their execution is evaluated. The evaluation typically consists in quantifying performance metrics or in checking behavior correction in a set of use cases. In many cases, system performance evaluation does not only consider absolute measures for performance metrics but is completed by *benchmarks*. The point is to use well-known and accepted test programs to compare the target system against competitor solutions.

Constructing a benchmark is a difficult task [2] as it needs to capture relevant system behaviors, under realistic workloads and provide interesting performance metrics. This is why benchmarks evolve with the maturation of a given application domain and new benchmarks appear as new system features need to be put forward. Developers frequently find themselves confronted with the challenge of choosing *the right* benchmark among the numerous available. To do so, they need to understand under which conditions the benchmark is applicable, what system characteristics it tests, how its different parameters should be configured and how to interpret the results. In most cases, the choice naturally goes to the

This work is funded by the SoC-Trace FUI project http://soc-trace.minalogic.net.

P.-F. Dutot and D. Trystram (Eds.): Euro-Par 2016, LNCS 9833, pp. 63–74, 2016.
DOI: 10.1007/978-3-319-43659-3_5

most popular benchmarks. Unfortunately, these are not suitable for all use cases and an incorrect usage may lead to irrelevant results.

In this paper, we present our solution for automatic profiling of benchmarks. The profiles characterize the runtime behavior of benchmarks using well defined metrics and thus help benchmark comparison and developers' choices. The profile computation uses information contained in execution traces and is structured as a deterministic trace analysis workflow. The contributions of the paper can be summarized as follows:

- *definition of unified profiles for benchmarks.* We define profiles in terms of execution duration, function repartition, stability, CPU efficiency, parallelization and memory usage. These are standard metrics and can be easily understood and interpreted by developers.
- *definition of the tools needed to compute the profiles.* We structure the computation as a reproducible workflow with parallel and streaming features. The final workflow is automatic and may be easily applied to different benchmarks or to different configurations of the same benchmark.
- *definition of the data needed for profile computation.* We use system tracing and extract useful data in application-agnostic manner. The application source code is not needed.
- *profiling of the Phoronix benchmarks.* We use our solution to profile the Phoronix Test Suite [3]. The results are obtained on different embedded and desktop platforms.

The paper is organized as follows. Section 2 presents the design ideas behind our solution. Section 3 discusses our implementation by considering the Phoronix use case. Section 4 distinguishes our proposal from related work. Finally, Sect. 5 presents the conclusions and the perspectives of this work.

2   Automatic Profiling of Benchmarks

This section presents the benchmark profiles (Sect. 2.1), the needed data for their computation (Sect. 2.2) and the computation process itself (Sect. 2.3).

2.1   Benchmark Profiles Definition

The profile considered for a benchmark is independent of its semantics and is composed of the following features:

- *Duration.* This metric gives the time needed to run the benchmark. It allows developers to estimate the time-cost of a benchmarking process and to choose between short and long-running benchmarks.
- *CPU Occupation.* This metric characterizes the way a benchmark runs on the target system's available processors. It gives information about the CPU usage, as well as about the benchmark's parallelization.

– *Kernel vs User Time*. This metric gives the time distribution between the benchmark-specific (user) and kernel operations. It gives initial information on the parts of the system that are stressed by the benchmark.
– *Benchmark Type*. The type of a benchmark is defined by the part of the system which is stressed during the benchmarking process. Namely, we distinguish between benchmarks that stress the processor (CPU-intensive), the memory (memory-intensive), the system, the disk, the network or the graphical devices. The motivation behind this classification is that it is application-agnostic and may be applied to all kinds of benchmarks.
– *Memory Usage*. This part of the profile provides information about the memory footprint of the benchmark, as well as the memory allocation variations.
– *Stability*. This metric reflects the execution determinism of a benchmark, namely the possible variations of the above metrics across multiple runs.

2.2   Initial Profile Data

The computation of the above metrics needs detailed data about the execution of a benchmark. It needs timing information both about the benchmark's global execution and about its fine-grained operations. It also needs information about the number, the type and the scheduling of the different execution events.

To collect this data, we decide to use system tracing and work with a historical log containing timestamped information about the different execution events. To ensure minimal system intrusion, we propose to use LTTng [4,5][1] which is a *de facto* standard for tracing Linux systems. Indeed, LTTng is capable of tracing hardware counters, the kernel and user-level operations. Hardware counters are available through the profiling *perf* kernel API, while kernel and user operations are traced with *tracepoints*. For the kernel, predefined tracepoints refer to context switches, interruptions, system calls, memory management and I/O.

2.3   Profile Computation

The profile computation is a two-phase process which respectively analyzes the kernel and the user-level traces.

The analysis of the kernel trace is implemented as a VisTrails [6] workflow. It thus benefits from its reusability, efficiency and reproducibility features. Indeed, the workflow takes as input a benchmark trace and automatically computes the corresponding profile. The same workflow may be reused for the analysis of a different benchmark or a different configuration of the same benchmark.

The kernel trace analysis workflow follows the standard logic where traces are first captured and stored, and then analyzed offline (Fig. 1a). The first (top) step of the workflow imports the kernel trace into a relational database. The database's generic trace format may represent not only LTTng traces but also other formats [7]. This step also reconstructs processes' states (not active or active and executing a given function).

---

[1] http://lttng.org.

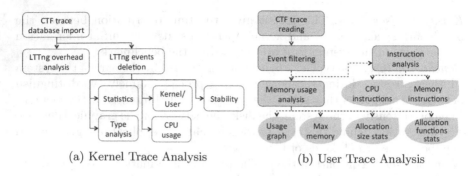

(a) Kernel Trace Analysis   (b) User Trace Analysis

Fig. 1. Profile computation process

The intermediary steps focus on LTTng tracing. The second step character-izes LTTng's tracing overhead in terms of number of traced events and execution slowdown. The third step filters out LTTng-related events to focus only on the benchmark's performances. All computations are done via SQL requests.

The last steps provide execution statistics in terms of number of execution events and execution duration, categorize the execution events to characterize the type of the benchmark, compute the repartition between kernel and user time, analyze the CPU occupation and explore benchmark stability.

For the user-level trace analysis, the trace data is read and streamed to the analysis treatments that operate in parallel and on-the-fly (Fig. 1b). This approach is motivated by the fact that the database-oriented store-and-later-analyze approach does not scale in the case of big execution traces. Indeed, execution traces may easily size up to several GB and their database import and subsequent analysis is costly in terms of storage and computation time.

The user-level trace analysis comprises several modules that process the trace in a pipeline. The first module reads directly the initial trace and transfers it to the events filtering module. The filtering module forwards to the subsequent modules only the information related to the benchmark processes. The memory usage analysis module provides information about the variations in the bench-mark's allocated memory, about the size of the allocated chunks of memory and about the frequency of usage of the memory allocation functions. The trace also provides information about the hardware counters which are used to quantify the user-level computations and the memory accesses.

3   Profiling the Phoronix Test Suite

This section details our benchmark profiling in the Phoronix Test Suite case.

3.1   The Phoronix Test Suite

The *Phoronix Test Suite (PTS)* [3] provides a set of benchmarks targeting differ-ent aspects of a system. *PTS* is available on multiple platforms including Linux, MacOS, Windows, Solaris and BSD.

PTS comes with some 200 open-source test programs. It includes hardware benchmarks typically testing battery consumption, disk performance, processor efficiency or memory consumption. It also targets diverse environments including OpenGL, Apache, compilers, games and many others.

PTS provides little information about benchmarks' logic and internals. Even if each benchmark is tagged as one of Disk, Graphics, Memory, Network, Processor and System, supposedly to indicate which system part is tested, there is no further information on how this tag has been decided or how exactly the benchmark tests this system part.

The repartition of the benchmarks is highly irregular. If we consider that PTS benchmarks having the same tag form a benchmark family, the Network family contains only one test, while the Processor family contains around 80 tests. If, in the first case, a developer has no choice, in the second case, he/she will need to know more about the benchmarks to choose the most relevant.

Table 1. Tagging LTTng Kernel Tracepoints

Family	Events
Processor	timer_*; hrtimer_*; itimer_*; power_*; irq_*; softirq_*;
Memory	kemem_*; mm_*
System	workqueue_*; signal_*; sched_*; module_*;rpm_*; lttng_*; rcu_*;
	regulator_*; regmap_*;regcache_*; random_*; console_*;gpio_*;
Graphics	v4l2_*; snd_*;
Disk	scsi_*; jbd2_*; block_*;
Network	udp_*; rpc_*; sock_*; skb_*; net_*; netif_*;napi_*;

3.2 Tracing Phoronix with LTTng

By enabling all LTTng kernel tracepoints, we collect information about scheduling decisions, process management (exec, clone, switch and exit function calls) and kernel usage (syscalls). Associated with each traced event is a hardware (CPU) and a software (PID) provenance context.

To analyze which parts of the target system are tested and thus deduce the type of a benchmark, we have analyzed the types of kernel events and mapped them to the PTS family tags. For example, the hmm_page_aloc and mm_page_free are clearly events related to Memory-related activity, while power_cpu_idle and htimer_expire are related to the Processor. Table 1 gives the mapping between events and system activity.

User-level tracing is highly dependent of the application to trace and PTS benchmarks' are highly heterogenous. To provide a generic tracing solution, we focus on the interface that is commonly used by all benchmarks, namely the standard C library (libc). Redefining the LD_PRELOAD environment variable and overloading the libc functions, it is easy to obtain the information about the memory management functions (malloc, calloc, realloc and free) needed for the computation of benchmarks' memory profiles.

Another aspect we are interested in is to characterize the user level behavior of a benchmark in terms of CPU or memory-related activity. To do so, we use the information provided by hardware performance counters. In particular we use the `Instruction` counter, which gives the total number of instructions executed. We also use the `L1-dcache-loads` and `L1-dcache-stores` counters that provide the total number of L1 cache reads and L1 cache writes. As all data access go through the L1 cache, the sum of those two values gives the total number of data related instructions. To get the number of computation related instructions, we use the difference `Instruction` − (`L1-dcache-stores` + `L1-dcache-stores`).

3.3   Experimental Setup

We have worked with 10 *PTS* bechmarks, namely `compressgzip`, `ffmpeg`, `scimark2`, `stream`, `ramspeed`, `idle`, `phpbench`, `pybench`, `network-loopback` and `dbench`. The set is part of the *PTS recommended* benchmarks which are the most popular ones as determined by the number of downloads and available results [8]. We have used three benchmarks from the *Processor* family, three from the *System* family, two from *Memory*, one *Disk* and one *Network* benchmark.

Each benchmark is run with its default options as defined by the *PTS* system except for the number of runs. Instead of 3 times we run benchmarks 32 times to ensure statistically reliable results [9]. The score for each benchmark, which is benchmark-specific, is computed as the mean value of the 32 obtained scores.

The experiments have been run on three different platforms which helped validate the fact that benchmarks have similar executions, hence profiles, whatever the platform. We have used one UDOO board[2], one Juno board[3] and a desktop machine. In the following we show results from the UDOO and the Juno boards. The UDOO has an i.MX 6 4core ARM CPU at 1 GHz, a Cortex-M3 coprocessor and 1 GB of RAM. It runs the multi-platform Debian kernel for ARM `armmp3.16`.The Juno board has one 2core CortexA57 and one 4core CortexA53 processors with 2 GB of RAM. It runs a Debian kernel (`4.3.0-1-arm64`).

3.4   LTTng Overhead and Benchmark Stability

Our analysis starts with an evaluation of LTTng's perturbation of the target system. In terms of execution duration, both for kernel and user traces, LTTng's overhead is negligible as it is less that 1 %. The only notable exception is the case of the `phpbench` benchmark slowed down by 78 % by user-level tracing because of its heavy use of memory operations.

In terms of collected events, LTTng-related events account for 10 % to 26 % in kernel traces and between $100K$ and $200K$ in user traces. To prevent bias in statistics metrics computations, these events are filtered out and ignored during trace analysis. Finally, considering benchmarks' results, scores from executions with and without tracing do not differ more than 2.5 %.

---

[2] http://www.udoo.org.
[3] http://www.arm.com/products/tools/development-boards/.

Table 2. Information on Phoronix Benchmarks (UDOO board)

Benchmark	Exec.(m)	Size(GB)	Idle	SD	User / Kernel ratio	
compress-gzip	94	5.18	76	0.03	78%	22%
ffmpeg	221.90	62	62	0.01	99%	01%
scimark2	22.41	0.28	76	0.00	99%	01%
stream	7.00	0.35	33	0.01	99%	01%
ramspeed	2019.25	30.20	24	0.00	83%	17%
idle	1.09	0.01	99	0.60	48%	52%
phpbench	649.72	15.35	75	0.00	97%	03%
pybench	267.71	1.60	75	0.00	99%	01%
dbench	915.72	58.20	0	0.03	01%	99%
network-loopback	90.92	25.50	54	0.00	11%	89%

Table 2 summarizes information about the 32 runs of the considered benchmarks. Namely we have the global execution time, the corresponding trace size, the relative time spent in idle mode (`idle` stands out), the standard deviation for benchmarks' duration and the ratio between user and kernel time. There are important differences, even between benchmarks belonging to the same family. A simple recommendation to developers would be to use shorter benchmarks.

We have investigated benchmark stability over the 32 runs. Having reverse-engineered the Phoronix launching process and identified the trace parts about the 32 distinct runs, we evaluated the stability of the considered profile metrics (number of events, execution time, kernel and user time, number of cores). Considering the benchmark execution time, for example, we have computed the mean value, the maximum, the minimum and the standard deviation. The latter is close to zero meaning that the benchmarks are stable. The only exception is `idle` whose variation may be explained by its short execution time ($6\,ms$). The analysis of the other parameters shows similar results.

3.5 Benchmark Types

A first simple classification of Phoronix benchmark is to consider the ratio of kernel versus user operation. Table 2 gives this ratio and shows that there are only 4 benchmarks spending significant time in kernel mode. It is worth noting that the ratio here is computed over benchmarks' useful execution time and ignoring the idle time. For `idle`, for example, this represents only 1 % of its total execution time. We can conclude that the others are either CPU- or memory intensive.

If we use the classification of kernel events introduced in Sect. 3.2 and use the number of traced events, we obtain the kernel profiles shown in Fig. 2. We can clearly see that there is no benchmark testing the graphics subsystem (no graphics events) and that `network-loopback` and `dbench` respectively test the network and the disk. Indeed, they are the only ones with a significant amount of respectively *Network* and *Disk* events.

If we consider the benchmarks tagged as *Memory* within Phoronix, `stream` has an important kernel activity and its kernel profile confirms the frequent usage

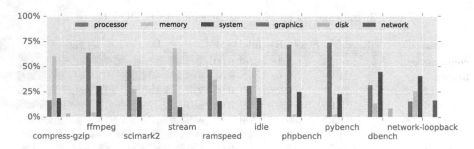

Fig. 2. Kernel Operation (UDOO board)

of memory-related functions. However, the profile of `scimark2` is different and to verify whether the benchmark is indeed memory-intensive one should consider its execution in user mode.

In the *Processor* family, `ffmpeg` and `scimark2` have the expected kernel profiles with predominant *Processor* events. `compress-gzip`, however, shows an important memory management activity so the profile computation should consider the user-level information.

Our analysis of the *System* Phoronix family made us understand that it includes various benchmarks testing different software systems (or layers, or middleware) and it does not necessarily focus on the operating system level. `idle` does test the operating system and quantifies the execution time of a program doing nothing. However, `phpbench` and `pybench`, which, by the way, have similar kernel profiles, respectively test the performances of PHP and Python code.

3.6   CPU Usage and Parallelization

An interesting aspect we have investigated is the way benchmarks use the available processors. In the case of the UDOO platform, we can see that benchmarks have quite different parallelization schemes (Fig. 3). The `idle` benchmark does not use the CPU, as expected. The `pybench` benchmark uses only 3 CPU out of 4. The other benchmarks do use the 4 processors but only `dbench`, `stream`

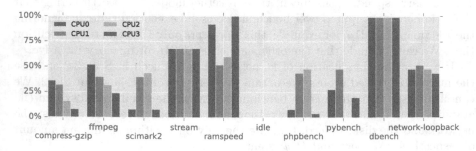

Fig. 3. CPU Usage and Parallelization (UDOO board)

and `network-loppback` are totally balanced. `dbench` uses the CPUs at 100 %, `stream` at 68 % and `network-loppback` 45 %.

Another interesting observation is that there are two couples with quite similar CPU usage profiles. These are `scimark2` and `phpbench`, on one hand, and `compress-gzip` and `ffmpeg`, on the other. However `scimark2` and `phpbench` belong to the *Processor* and *System* family respectively. As for `compress-gzip` and `ffmpeg`, the two being of the *Processor* family, it may be better to consider the `ffmpeg` benchmark which runs longer but makes a more efficient usage of the platform processors.

Table 3. Maximum Memory and Instructions repartition (Juno board)

Benchmark	Memory (KB)	Instructions repartition (Compute / Memory)	
compress-gzip	138	72%	28%
ffmpeg	5 210	71%	29%
scimark2	16 779	64%	36%
stream	9	58%	42%
ramspeed	3 456 108	49%	51%
idle	2	56%	44%
phpbench	3 225	54%	46%
pybench	1 827	56%	44%
dbench	225 608	77%	23%
network-loopback	1 123	55%	45%

3.7 Memory Usage Profile

User-level tracing yields interesting information about the differences between benchmarks. The maximum memory allocated by benchmarks, for example, is quite varying (Table 3). Indeed, only `ramspeed` with 3.45 GB uses almost all virtual memory available on the UDOO board (4 s). `stream` and `idle` are memory light-weight as they consume respectively 9 KB and 2 KB.

The results from profiling of user-level operations and classifying them into computational and memory-related are given in the third row of Table 3. These confirm that the *Memory* benchmarks, `ramspeed` and `stream` do use intensively the memory as expected. The other benchmarks which reveal to be memory-intensive are the *System* ones we have selected, namely `phpbench` and `pybench`. As for `idle` and `network-loopback`, these are not representative as the time spend in user mode is very short (0.6 % of the total execution time for `idle` and 5 % for `network-loopback`).

Figure 4 gives the evolution of the memory usage of benchmarks over time and shows that there are various behaviors.

4 Related Work

Current benchmark-oriented efforts [3,10,11] focus on the problems of providing a set of benchmarks which is to be *complete*, *portable* and *easy* to use. Indeed, the

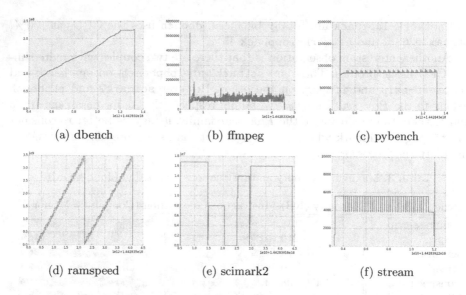

(a) dbench   (b) ffmpeg   (c) pybench

(d) ramspeed   (e) scimark2   (f) stream

Fig. 4. Memory usage (bytes) over time (ns). (UDOO board)

goal is to provide benchmarks that cover different performance aspects, support different platforms and can be automatically downloaded, installed and executed. However, the classification of benchmarks is *ad hoc* and there is no detailed information about their functionality. Our proposal is a step forward as it allows benchmark comparison through automatic profiling.

Our work is motivated by the need of identifying relevant and representative benchmarks and is thus related to [12]. The authors use simulation to obtain source-code-related statistics which are further treated with cluster-based methods to identify a minimal representative benchmark set. Our solution is complementary as it considers real-world settings, applies to binary programs and, most of all, automates the analysis. It can benefit from the learning techniques to identify the representative set of Phoronix benchmarks.

Our proposal can be seen as a profiling tool for benchmarks. However, existing profiling tools [13–16] typically provide detailed low-level information on a particular system aspect. Moreover, they are system dependent. Our proposal is applicable to all types of benchmarks, on different platforms and provides a macroscopic vision of their behavior.

The major aspects of our profile computation are trace analysis and workflow management. Concerning trace analysis, most existing tools are system and format specific [17,18] and limit themselves to time-chart visualizations and basic statistics. In our work, trace analysis is brought to a higher level of abstraction. It is based on generic data representation of traces and thus may be applied to execution traces (hence benchmarks) from different systems. It is not organized as a set of predefined and thus limited treatments but may be configured and enriched to better respond to the user needs. Finally, its structuration in terms of a deterministic workflow allows for automation and reproducibility of the analysis process.

As for workflow-oriented tools [19,20], they focus on formal specification, automation, optimisation and reproducibility of computations aspects. Generic trace analysis and especially the problem of huge traces have not been considered.

5   Conclusion and Ongoing Work

We have presented in this paper a workflow-based tool for automatic profiling of benchmarks. The result is a unified profile which characterizes a benchmark, allows the comparison among benchmarks and thus facilitates the choices of a system performance analyst. We have illustrated our approach with the Phoronix Test Suite and have experimented with embedded and desktop Linux-based platforms. We have successfully produced the profiles for several benchmarks exhibiting their different characteristics. Our experimentation puts forward the fact that the initially provided description is far from sufficient for understanding the way benchmarks test the target system.

In our work we have taken advantage of workflows' useful features such as automation, result caching and reproducibility. However, most workflow systems do not properly address the data management issues when it comes to manipulating big data sets. In this regard, we have shown that workflow tools should provide for pipelining, streaming and parallel computations. An ongoing collaboration with the VisTrails team brings these features to the VisTrails tool.

The benchmark profiles we provide are easy to compute, to understand, and to compare. All metrics observed, tools used, and profiles drawn, do not depend on any specifics of the benchmarks.

A long term research objective would be to provide generic means for reflecting the benchmark specifics into the profile and thus help even more the performance evaluation work of an analysis.

References

1. Almeida, J.B., Frade, M.J., Pinto, J.S., de Sousa, S.M.: Rigorous Software Development: An Introduction to Program Verification. Springer, London (2011)
2. A Practitioner's Guide. Cambridge University Press, New York (2000)
3. Phoronix Test Suite. http://www.phoronix-test-suite.com/
4. Desnoyers, M.: Low-Impact Operating System Tracing. PhD thesis (2009)
5. Chen, K.Y., Chang, Y.H., Liao, P.S., Yew, P.C., Cheng, S.W., Chen, T.F.: Selective Profiling for OS scalability study on multicore systems. In: IEEE 6th International Conference on Service-Oriented Computing and Applications, pp. 174–181 (2013)
6. Callahan, S.P., Freire, J., Santos, E., Scheidegger, C., Silva, C.T., Vo, H.T.: VisTrails: visualization meets data management. In: Proceedings of the ACM SIGMOD International Conference on Management of Data, pp. 745–747 (2006)
7. Pagano, G., et al.: Trace management and analysis for embedded systems. In: IEEE 7th International Symposium on Embedded Multicore Socs, September 2013

8. Phoronix Test Suite, User Manual. http://www.phoronix-test-suite.com/documentation/
9. Jain, R.: The Art of Computer Systems Performance Analysis (1991)
10. Industry-Standard Benchmarks for Embedded Systems. http://www.eembc.org/benchmark/products.php/
11. Future Benchmarks and Performance Tests. http://www.futuremark.com/
12. Joshi, A., Phansalkar, A., Eeckhout, L., John, L.K.: Measuring benchmark similarity using inherent program characteristics. IEEE Trans. Comput. **55**(6), 769–782 (2006)
13. Yamamoto, M., Ono, M., Nakashima, K., Hirai, A.: Unified performance profiling of an entire virtualized environment. In: Second International Symposium on Computing and Networking (CANDAR), pp. 106–115, December 2014
14. Gouigoux, J.-P.: Practical Performance Profiling: Improving the Efficiency of .NET Code. Red gate books, United Kingdom (2012)
15. Seward, J., Nethercote, N., Weidendorfer, J.: Valgrind 3.3 - Advanced Debugging and Profiling for GNU/Linux Applications. Network Theory Ltd., Bristol (2008)
16. Janjusic, T., Kartsaklis, C.: Glprof: a Gprof inspired, callgraph-oriented per-object disseminating memory access multi-cache profiler. In: Procedia Computer Science, International Conference on Computational Science, ICCS Computational Science at the Gates of Nature, vol. 51, pp. 1363–1372 (2015)
17. Knüpfer, A., et al.: Score-p: a joint performance measurement run-time infrastructure for periscope, scalasca, tau, and vampir. In: Tools for High Performance Computing 2011. ZIH, Dresden, pp. 79–91, September 2011
18. Prada-Rojas, C., Santana, M., De-Paoli, S., Raynaud, X.: Summarizing embedded execution traces through a compact view. In: Conference on System Software, SoC and Silicon Debug S4D (2010)
19. Cohen-Boulakia, S., Leser, U.: Search, adapt and reuse: the future of scientific workflows. SIGMOD Records **40**(2), 1187–1189 (2011)
20. Qin, J., Fahringer, T.: Scientific Workflows, Programming, Optimization, and Synthesis with ASKALON and AWDL. Springer, Heidelberg (2012). ISBN 978-3-642-30714-0

Performance and Power Modeling, Prediction and Evaluation

Addressing Materials Science Challenges Using GPU-accelerated POWER8 Nodes

Paul F. Baumeister[1]([✉]), Marcel Bornemann[2], Markus Bühler[3], Thorsten Hater[1], Benjamin Krill[3], Dirk Pleiter[1], and Rudolf Zeller[2]

[1] Jülich Supercomputing Centre, Forschungszentrum Jülich, Jülich 52425, Germany
p.baumeister@fz-juelich.de
[2] Institute for Advanced Simulation, Forschungszentrum Jülich,
Jülich 52425, Germany
[3] IBM Germany Research and Development, Böblingen 71032, Germany

Abstract. Materials research is an area that is expected to strongly benefit from the growing performance capabilities of future supercomputers towards exascale. Density functional theory (DFT) has become one of the most important methods for numerical materials science. In this paper we present results of a performance model based analysis of a particular, scalable DFT-based application on GPU-accelerated compute nodes with POWER8 processors. These technologies are part of a future roadmap for pre-exascale architectures. With power consumption becoming a major design constraint, we also determine the energy required for executing the most performance critical kernel.

1 Introduction

Density Functional Theory (DFT) is a key method for addressing challenges in materials science that require an accurate description of the electronic properties of a material. The complexity of calculating the full wave function of the many-electron system is avoided by considering a single-particle picture with an effective potential [15], giving rise to the Kohn-Sham equation $\hat{H}\Psi = E\Psi$. The solutions can take the form of either a set of eigenstates of the Hamiltonian $\hat{H}$ as realised in wave function based implementations [3,13,21] or the Green function $\hat{G}(E) = (E - \hat{H})^{-1}$ as proposed by Korringa, Kohn and Rostoker (KKR) [5,14,16]. Here, the energy E is continued into the complex plane with a non-vanishing imaginary part in order to prevent the inversion of a singular operator. A suitable representation allows for casting the problem into a matrix inversion maintaining high accuracy via a full-potential description. Despite, the matrix dimension only grows as $16\,N_{\text{atom}}$ assuming a truncation of angular momenta beyond $\ell = 3$. The screened KKR method allows for finding a short ranged formulation and, hence, the equivalent operator $\hat{G}$ becomes block-sparse [26]. In large systems, where the number of atoms $N_{\text{atom}} \gg 1000$, the Green function formulation can be approximated by systematically truncating long-ranged interactions between well-separated atoms. This reduces the overall complexity of the method from cubic to linear and large systems with $100,000$ atoms

© Springer International Publishing Switzerland 2016
P.-F. Dutot and D. Trystram (Eds.): Euro-Par 2016, LNCS 9833, pp. 77–89, 2016.
DOI: 10.1007/978-3-319-43659-3_6

and more become thus feasible. KKRnano is a DFT application implementing the original cubic method as well as the linear-scaling approach [23, 27]. It has been proven to scale to massively-parallel architectures leveraging MPI and OpenMP programming models. Central to its performance is an iterative solver for the linear system and the application of the block-sparse operator.

Massively-parallel computing resources are required to facilitate high throughput for medium-sized problems as well as to address large-scale challenges. The former will, e.g., be required to scan parameter spaces and evaluate high-dimensional phase diagrams. The latter involves problems where large N_{atom} are required, e.g. when effects that occur at the length-scale of several nanometers need to be understood and investigated. Ideal atomic geometries can be analysed using a workstation to run a DFT code that exploits symmetries. In contrast, realistic samples of a material are hardly ever perfect crystals with full translational symmetry or isolated molecules in vacuum. Addressing these challenges requires dealing with broken symmetries, i.e. crystals with impurities, random alloys or amorphous materials and thus result in calculations with $N_{atom} \gg 10,000$.

Due to the end of Dennard scaling the level of parallelism in HPC systems will become even more extreme to offer an increase in the number of floating-point operations per time unit. In order to minimise power consumption, low-clocked, but highly parallel compute devices like GPUs have become increasingly popular. Operating at clock speeds below 1 GHz means that more than 10^8 floating-point operations per clock cycle are required to reach a pre-exascale performance level of about 100 PFlop/s. In case of KKRnano the exploitable parallelism scales with N_{atom}, enabling exploitation of such massively-parallel architectures.

This article makes the following contributions:

1. We present a performance analysis for highly optimised implementations of the main kernel of the application KKRnano on both, IBM POWER8 processors and NVIDIA K40 GPUs.
2. To enable analysis of performance as well as scalability properties a simple performance model is developed. We use this model to explore scalability of the application for (not yet existing) large-scale systems based on this processor and accelerator technologies.
3. Finally, we evaluate energy-to-solution of our implementation with and without GPUs based on power consumption measurements of the system.

In this section and Sect. 2 we provide background on the application domain and relevant technology. After presenting an analysis of the application's performance characteristics in Sect. 3, we outline the main features of our implementation and provide a performance analysis for the kernel on POWER8 and the GPU in Sect. 4 and 5, respectively. In Sect. 6 we present our performance model and use it to explore the scalability of the application. We continue with a power consumption analysis in Sect. 7. Before concluding in Sect. 9, we provide an overview on related work in Sect. 8.

2 GPU-accelerated POWER Architectures

We evaluate application performance on commercially available POWER8 824
47 L servers [8], comprising two POWER8 sockets, 256 GiByte of memory and
one NVIDIA K40m GPU per socket.

The POWER8 processors in the considered system are dual-chip modules,
where each module comprises 5 cores, i.e. there are 20 cores per node. Each
core offers two sets of the following instruction pipelines: fixed point, floating-
point, pure load (LU) and a load-store unit (LSU). Instructions are processed
out-of-order to increase instruction level parallelism. The cores support 8-way
Simultaneous Multithreading. For the HPC workloads, as considered here, a
few details are of special interest. The floating point unit, called the Vector
Scalar Unit (VSU), supports two- or four-way SIMD for single-precision and two-
way SIMD for double-precision floating-point instructions. Fused multiply-add
instructions are provided. In case of floating-point instructions, the operands
have to be present in VSU registers; load to these are processed in the LU
exclusively. Further, stores from VSU are issued both to the LSU and the VSU
internally.

Per cycle up to eight double-precision floating-point operations can be performed in the form of two fused multiply-add instructions on 128 bit vector
registers, providing 29 GFlop/s per core or 590 GFlop/s per node at the peak
clock of 3.69 GHz. Each core has a private L1 data cache of 64 kByte, a private
L2 cache of 512 kiByte and a segment of 8 MiByte associated to it in the shared
L3 cache (total 80 MiByte). In concert with a set of external memory buffers
– called the Centaur chip – the POWER8 CPU provides a maximum read and
write bandwidth of 256 GByte/s and 128 GByte/s per socket, respectively. The
memory system can provide up to two 16 Byte loads and one 16 Byte store per
cycle at L1.

Each of the POWER8 sockets is connected to an NVIDIA K40m GPU via an
x16 PCIe GEN3 link. The K40m is based on GK110 GPUs of the Kepler generation running at 745 MHz. With a total of 15 streaming multi-processors it has a
peak performance of 1430 GFlop/s. Each GPU can either write or read to or from
its 12 GiByte of GDDR5 memory with a nominal bandwidth of 288 GByte/s.

Both compute devices, the POWER8 processor as well as the K40m
GPU, thus offer significantly different hardware performance capabilities. The
POWER8 processor features very high memory bandwidth at moderate floating-
point operations throughput and operates at relatively high clock speed. In con-
strast, the K40m has a much higher concurrency to provide very high floating-
point operation throughput at moderate clock speed and a memory bandwidth
that is relatively small compared to its compute capabilities.

3 Application Performance Characteristics

We focus on a single iteration of the KKR algorithm, comprising the solution
of a linear system locally and, afterwards, setup of a new system for the next

iteration. Solving the local problem is approached using a variant of the Quasi Minimal Residual (QMR) method, an iterative solver [10]. In the case at hand, simultaneous solutions of a set of right-hand sides are sought. Given Λ and ω we have the following problem to solve

$$\Lambda\gamma = \omega \qquad (1)$$

where the elements Λ_{ij} are operators describing the interaction between an atom i with its direct neighbours j. We fix the number of columns to $N_{cl} = 13$ entries from here on, which corresponds to a close packed lattice structure ($N_{cl}=13$ for hcp or fcc, while for bcc, $N_{cl}=15$ is a good choice). The number of rows corresponds to the number of atoms in a truncation cluster, we primarily use $N_{tr} = 1000$. The elements of Λ are small dense square matrices over $\mathbb{C}$. The size of these entries b corresponds to the order to which the expansion in the angular momentum is truncated, we pick the current default, namely $b = 16$. Since Λ is sparse, the operator is compressed in memory by dropping the zero elements in each row and carrying the appropriate index list. The runtime of the solver is dominated by the application of the operator Λ, which consumes around 90 % of the solver's runtime. KKRnano operates on double-precision complex numbers so 16 Byte are assumed per number and 8 Flop are required to perform a complex fused multiply-accumulate operation.

The parallelisation strategy of KKRnano foresees one MPI rank per atom, i.e. the number of tasks per node is given by N_{atom}/N_{node}, where N_{atom} and N_{node} are the number of atoms and nodes, respectively. Each task has to solve Eq. (1) using the iterative solver which does not require any inter-node communication. After solving the linear system, the operator Λ needs to be updated, which involves communication with N_{tr} other tasks. We analyse the relevant kernel using an information exchange approach [7,19], which models the hardware as a graph of data stores connected by edges representing communication links or processing pipelines. We choose a simple model for the processor consisting of two data stores, the external main memory and the on-chip memory, representing register file and caches.

The performance of the overall kernel is driven by accumulating dense matrix products when applying the operator Λ. In the following we assume that the solver always performs a fixed number of iterations N_{iter} with two applications of Λ per iteration. At node level we therefore can characterise the kernel by the following information exchange functions:

$$I_{fp} = 2N_{iter} \cdot \frac{N_{atom}}{N_{node}} \cdot N_{tr}N_{cl} \cdot b^3 \cdot 8\,\text{Flop}, \qquad (2)$$

$$I_{ld} = 2N_{iter} \cdot \frac{N_{atom}}{N_{node}} \cdot N_{tr}N_{cl} \cdot b^2 \cdot 16\,\text{Byte}, \qquad (3)$$

$$I_{st} = 2N_{iter} \cdot \frac{N_{atom}}{N_{node}} \cdot N_{cl} \cdot b^2 \cdot 16\,\text{Byte}, \qquad (4)$$

where I_{fp} is the number of floating-point operations required to solve Eq. (1) for all atoms on one node. I_{ld} and I_{st} account for the input and output operands that need to be loaded and stored, respectively. We furthermore assume that all other numerical subtasks - which scale as $N_{\text{tr}} \cdot b^2$ - within the solver can be ignored. No assumptions are made about exploitation of data reuse outside the complex multiplications. The information exchange functions can be used to compute the arithmetic intensity

$$AI = \frac{I_{\text{fp}}}{I_{\text{st}} + I_{\text{ld}}} \underset{\approx}{^{N_{\text{cl}} \gg 1}} \frac{b}{4} \frac{\text{Flop}}{\text{Byte}} \overset{b=16}{=} 4 \frac{\text{Flop}}{\text{Byte}}. \tag{5}$$

Following the roofline performance model approach [25] we thus expect the maximum attainable performance of the application to be limited by the throughput of double-precision floating-point operations on the POWER8 processors, while on the K40 GPU the nominal memory bandwidth limits the attainable performance to 80 % of the nominal floating-point performance. Our previous investigations showed memory bandwidths on the CPU of more than 280 GByte/s [2], while on the K40 210 GByte/s (ECC active) were achievable, resulting in the same expected performance for the host and a reduced expectation of 840 GFlop/s on the K40m.

When the solver is executed on the GPU, additional data transfers are needed. Before launching the solver, Λ and ω need to be transfered from host to device. After completion the result vector γ has to be transfered from device to host. Both vectors are stored as dense arrays of N_{tr} blocks. Thus, we write for this sub-task:

$$I_{\text{acc}} = \frac{N_{\text{atom}}}{N_{\text{node}}} (2N_{\text{tr}} + N_{\text{cl}} N_{\text{tr}}) \, b^2 \, 16 \, \text{Byte}. \tag{6}$$

The full vector is required to be present on the device, if even the operator application might only utilise a subset, consequently, the full transfer is accounted for.

After solving Eq. (1), Λ is updated. In the worst case all N_{tr} pairing atoms are located on other nodes, i.e. all information needs to be communicated over the network. This information exchange is captured by by the information exchange function

$$I_{\text{net}} = \frac{N_{\text{atom}}}{N_{\text{node}}} N_{\text{cl}} N_{\text{tr}} \, b^2 \, 16 \, \text{Byte}. \tag{7}$$

4   Application Performance Analysis on Processor

To simplify adaption of the code, we extracted the performance critical part of the code in a benchmark, i.e. the $2N_{\text{iter}}$ applications of operator Λ. While the original code is implemented in Fortran, in case of the benchmark we choose C++. The benchmark retains only the block sparse operator application from the original solver, however, this part is reproduced in full. The omission is limited to parts scaling as $b^2 N_{\text{tr}}$ in arithmetic operations, compared to $b^3 N_{\text{tr}} N_{\text{cl}}$ for the operator. The reduction to this core can increase the effectiveness of data caches, due to the smaller working set size and higher temporal locality.

Λ is stored in compressed block sparse row format. The kernel traverses the per-row index list π to accumulate the required blocks of the result. Multiple rows are processed in parallel using OpenMP threads. We compute the result vector in terms of its individual blocks, each corresponding to a row of the operator Λ. Each row i is processed by one thread which utilises the indices $\pi(i,j)$ to compute $\omega_i \leftarrow \Lambda_{ij}\gamma_{\pi(i,j)}$. The core of the algorithm is the dense matrix product in $\mathbb{C}^{b\times b}$.

Based on the analysis presented in Sect. 3 we expect the performance of the benchmark to be limited by the floating-point throughput. To maximize this throughput it is necessary to exploit 2-way SIMD. These expectations are confirmed by our observations. To enable the compiler to use SIMD instructions we changed the data layout. While the original code follows an array-of-structure design with arrays of complex numbers, the benchmark employs a structure-of-array separating real and imaginary parts numbers into different arrays.

In Table 1 we show a performance counter analysis for the full solver taken from the original code as well as our optimised benchmark. The parameters of both runs have been chosen such that the same number of inner matrix-matrix multiplications is performed. More specifically, the run was for a single atom on a single node, i.e. $N_{\mathrm{atom}} = N_{\mathrm{node}} = 1$. To obtain stable numbers, a single pinned core per atom was utilised and measured. Furthermore, we have set $N_{\mathrm{tr}} = 1000$, $N_{\mathrm{cl}} = 13$, $b = 16$ and $N_{\mathrm{iter}} = 1000$. As not all performance counters can be measured during a single run, Table 1 combines the results obtained from multiple runs. For each performance counter we have repeated the same run 10 times and use only the minimum value for our analysis.

Table 1. Selected performance counters for the full solver mini-application and the performance optimized benchmark, which mimics the behaviour of this solver. The parameters of these runs are discussed in the text. Cycles in which the core is waiting for completion of a group of finished instructions are marked as *completing*, those in which another thread blocked the completion port are marked *thread*. Stores are counted twice by the hardware counters, as they are issued to *both* the LSU and VSU.

	Solver	Benchmark		Solver	Benchmark
Cycles (10^9)	850	188	Instructions (10^9)	1718	487
—Running	332	116	—Branch	11	7
—Completing	118	28	—Integer	30	88
—Thread	16	4	—Arithmetic	943	232
—Stalled	378	38	—Scalar	656	0
—VSU	366	24	—Vector	7	217
—LSU	10	14	—(Stores)	280	15
Transfer (GByte)	211	213	—Memory	1015	180

Using Eq. (2) we find $I_{\mathrm{fp}} = 852\,\mathrm{GFlop}$. The number of floating-point instructions would be minimized if the application could be mapped to 2-way SIMD fused multiply-add instructions, i.e. $N_{\mathrm{vfp}} = I_{\mathrm{fp}}/4$. In practice, we find an overhead of less than 1 % in the number of arithmetic vector instructions. For the original code we observe no vector instructions and the number of scalar arithmetic instructions $N_{fp} \gg I_{fp}/2$ due to a lack of fused multiply-add operations, which is confirmed by an inspection of the assembly. Over the runtime of the benchmark a total volume of 211 GByte is loaded and stored, while the full solver transfers 213 GByte. Note that both numbers are slightly lower than the estimated value from Eqs. 3 and 4, which we attribute to the large L3 cache, which could in theory hold one full problem set. Thus, the ratio of required floating point operations to actually transferred bytes is larger than four. The two programs utilise 1.3 GByte/s and 4.6 GByte/s of memory bandwidth.

Assuming that memory instructions and arithmetic instructions can be perfectly overlapped and distributed over at least 2 pipelines, we would expect that the minimum time-to-solution in units of clock cycles is equal to $N_{vfp}/2 \simeq I_{fp}/8 = 106 \cdot 10^9$. In practice, we observe that due to a significant number of stall cycles the number of clock cycles spent in the solver Δt_{solver} to be almost 80 % larger. In summary, using a benchmark version of the application kernel, we are able to reach on a single core a floating-point efficiency $\epsilon_{fp} = I_{fp}/(8 \cdot \Delta t_{solver}) = 56 \%$.

5 Kernel Acceleration on GPU

We investigate the viability of GPU acceleration for KKRnano by porting the complete benchmark version of the solver. For the GPU implementation CUDA is used. The porting efforts are significantly reduced as the block sparse matrix-vector multiplication can be implemented using the cuSPARSE library.

With GPUs featuring extreme levels of parallelism, the obtained performance can in practice strongly depend on the level of parallelism of the problem solved on the GPU. Additionally, kernel launch times can have a non negligible effect. In Fig. 1 we therefore explore both kernel execution time as well as performance as a function of N_{tr} (the other parameters are the same as in the previous section). We observe that the performance saturates for $N_{tr} \gtrsim 1000$. Maximum performance is obtained for $N_{tr} = 3000$. From Eq. (2) we obtain $I_{fp} = 2.55$ TFlop, 8 s to execute on a single K40. This corresponds to a performance of about 320 GFlop/s, which is far below the maximum attainable performance as expected from the roofline model. We analysed the resulting performance using GPU hardware counters and the NVIDIA profiling tools and observed the bandwidth to the shared memory being almost fully used. This could indicate that the bandwidth to the shared memory in the cuSPARSE implementation is the limiter and not the external memory bandwidth, as it was expected from the analysis in Sect. 3.

In order to improve the resource utilisation on the GPU, we investigated how performance changes when multiple tasks running on the CPU use the GPU simultaneous for solving Eq. (1). This is possible using the multi-process service mps. The performance gain can be quantified by a weak-scaling efficiency $\epsilon_{par}(n) = n\Delta t_s/\Delta t_p(n)$, where Δt_s is the serial solver execution time for a single solver instance without mps and $\Delta t_p(n)$ is the time required for n concurrent calls of the solver. The results for $1 \leq n \leq 10$ are shown in Fig. 1. The upper

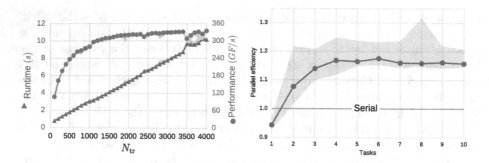

Fig. 1. Benchmark performance results obtained using $N_{\text{iter}} = 1000$ using a single (left) and multiple tasks for $N_{\text{tr}} = 1000$ (right) on a single K40m.

limit corresponds to one task per core of the processor, to which the GPU is attached. A gain of 17% in efficiency is observed.

6   Performance Model Analysis

To enable an assessment of the performance of KKRnano on not yet existing larger systems based on GPU-accelerated nodes with POWER8 processors, we employ a performance modeling approach used in [4], which combines the information exchange analysis with semi-empirical performance analysis [12]. For this we assume that time-to-solution depends linearly on the information exchange. Furthermore, we assume that arithmetic operations and memory transfers can be perfectly overlapped. In case the solver is executed on the POWER8 processor, the performance can be expected to be limited by the floating-point operation throughput and we thus make the following ansatz:

$$\Delta t_{\text{solver}}^{\text{CPU}} = a_0^{\text{CPU}} + a_{1,\text{fp}}^{\text{CPU}} I_{\text{fp}}, \tag{8}$$

where I_{fp} is defined in Eq. (2). The coefficients a_0^{CPU}, $a_{1,\text{fp}}^{\text{CPU}}$ are determined by fitting Eq. (8) to timing measurements for different application parameters.

If the solver is executed on the GPU, we assume performance to be limited by memory bandwidth. Additionally we have to take the time into account that is required to data transfer from host to device and vice versa. This results in a slightly more complex ansatz using I_{ld}, I_{st} and I_{acc} from Eq. (3), (4) and (6), respectively:

$$\Delta t_{\text{solver}}^{\text{GPU}} = a_0^{\text{GPU}} + a_{1,\text{mem}}^{\text{GPU}}(I_{\text{ld}} + I_{\text{st}}) + a_{1,\text{acc}}^{\text{GPU}} I_{\text{acc}}. \tag{9}$$

To determine the model parameters we have performed multiple runs with fixed $N_{\text{atom}} = 20$, $N_{\text{node}} = 1$, $N_{\text{cl}} = 13$, $b = 16$, and different N_{iter} as well as N_{tr}. The runs are repeated multiple times for the same parameter setting and the minimal value is used. Error bounds are established by *k-fold* cross-validation with $k = 100$. Due to the size of the problem, the constant terms turned out to be insignificant and have been ignored.

The final contribution to our model is the update of the operator Λ, which requires a local computation of one row (neglected) and assembling the remote rows into the full operator. Applying the same approach as before we have

$$\Delta t_{\text{upd}} = c_{0,\text{net}} + c_{1,\text{net}} I_{\text{net}}, \tag{10}$$

where I_{net} is defined in Eq. (7). To determine the coefficients $c_{0,\text{net}}$ and $c_{1,\text{net}}$ we used the OSU micro-benchmarks [1] to measure the bandwidth between two POWER8 systems interconnected via a Mellanox EDR Infiniband network. Since for realistic parameter settings the effect of the constants a_0^{CPU}, a_0^{GPU} and c_0 is negligible, we focus on the linear term only. In Fig. 2 we show the inverse values for the coefficients of the linear terms to facilitate comparison with the bandwidth and throughput parameters of the hardware.

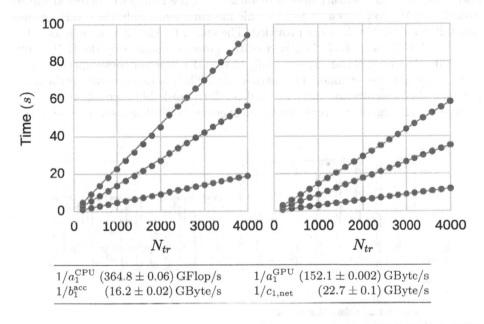

| $1/a_1^{\text{CPU}}$ | (364.8 ± 0.06) GFlop/s | $1/a_1^{\text{GPU}}$ | (152.1 ± 0.002) GByte/s |
| $1/b_1^{\text{acc}}$ | (16.2 ± 0.02) GByte/s | $1/c_{1,\text{net}}$ | (22.7 ± 0.1) GByte/s |

Fig. 2. Data points used to determine the model parameters for CPU (left) and GPU (right) and predictions for $N_{\text{iter}} = 200, 600, 1000$. The parameters are tabulated below with their respective errors.

The model allows us to assess whether KKRnano, which scales with good efficiency on a 28-rack Blue Gene/Q system, could scale on a hypothetical system comprising nodes that have a similar architecture as the one considered in this paper. We would need at least 2100 nodes to reach a similar peak performance. For an efficient utilization of the resources of a single node, we assume $N_{\text{atom}}/N_{\text{node}} \geq 20$, i.e. $N_{\text{atom}} \geq 42000$ for $N_{\text{node}} = 2100$. This matches the target problem size of this application area. From the performance model we find that $\Delta t_{\text{upd}} \ll \Delta t_{\text{solver}}$, even were we to assume much smaller values of $1/c_{1,\text{net}}$ due to network congestion.

7   Energy Efficiency Analysis

Let us finally consider the energy-to-solution for a single execution of the solver
on the considered architecture. The POWER8 processor provides an on-chip controller (OCC) to measure a set of sensors in the hardware. The data is available out-
of-band via a service processor and can be read out by the AMESTER tool [9,17].
The measurement granularity depends on the number of sensors, each requires an
additional latency of typically 200 ms. The data is, therefore, gathered in irregular
intervals. We resample it to a set of regular 1 s measurement points. The incoming
data represents the current power consumption of the component corresponding
to the sensor. To calculate the overall energy consumption, we use thresholding
of the data to detect active phases, sum the power consumption measurements P_i
and scale with the measurement interval Δt and the number of detected solver
executions. We do not report all available measurements, only the total of memory, CPU and GPU values are provided. The sensor for the 12 V domain includes
different I/O devices, including part of the power consumed by the GPUs. We
attribute the values of these sensors fully to the GPU's power consumption, which
leads to a slight overestimate of the actual value. The power consumed by the cooling fans shows significant variation and no distinguishable correlation with the
workload. The signal was replaced by its average. We utilise a setup close to the

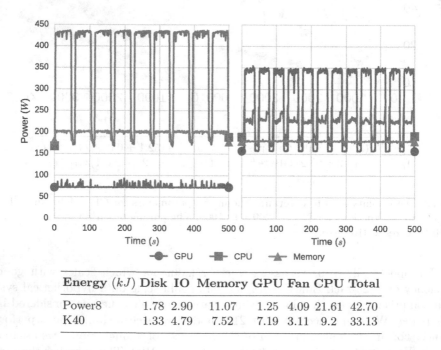

Energy (kJ)	Disk	IO	Memory	GPU	Fan	CPU	Total
Power8	1.78	2.90	11.07	1.25	4.09	21.61	42.70
K40	1.33	4.79	7.52	7.19	3.11	9.2	33.13

Fig. 3. Power consumption of the linear solver, CPU (left) and GPU (right). Below, we
report the averaged total energy to solution for the corresponding benchmarks. (Color
figure online)

configuration used in production runs of KKRnano, that is $N_{cl} = 13$, $N_{tr} = 1000$, $b = 16$ and $N_{iter} = 1000$ iterations inside the solver. The number of iteration is chosen as the maximum number allowed in KKRnano, on average numbers are typically $\mathcal{O}(100)$. Per core one instance of the problem is solved, for a total of 20; from prior analysis we have the requirement of 17 TFlop. The power consumption over multiple invocations of the CPU and GPU implementations of solver is shown in Fig. 3. Only about 20 W of additional power is utilised by the memory system, as the solver is not very memory intensive. We report energy metrics of the full node in Fig. 3 for the solution of one instance of the problem per socket or GPU respectively. Power consumption is averaged over multiple invocations of the solver. Since power required for an idle system is quite high, much of the total energy required for the solution is explained by the base cost. Thus, this metric likely favors fast implementations of the solver.

8 Related Work

Recent efforts to accelerate DFT methods leveraging GPU-based systems can be found in literature. GPU acceleration has been achieved for wave function based DFT methods, e.g. plane wave methods, wavelets, grid and local orbitals [11,20,22,24]. Closer to this work are projects in the class of linear-scaling methods like SIESTA or CP2K [6,21]. As node architectures based on POWER8 processors are relatively new and, in particular, the GPU-accelerated versions not yet widely available, only few performance investigations related to scientific applications have been published. Applications based on the Lattice Boltzmann method, a brain simulator as well as an application based on the Finite Difference Time Domain method are considered [2,4]. The authors of [18] focus on server workloads as well as big data, analytics, and cloud workloads.

9 Conclusions and Future Work

In this paper we presented results for a highly scalable materials science application based on the Density Functional Theory (DFT) method. Typically, most of the computational resources are spent in an iterative solver. We could demonstrate that for this kernel a high or at least good floating-point efficiency could be obtained on the POWER8 processors and K40m GPUs, respectively.

To explore the scalability properties of this application on future systems based on GPU-accelerated compute nodes with POWER processors, which could provide a performance of $\mathcal{O}(10)$ PFlop/s, we designed a simple performance model. From this we could conclude that assuming a network technology that is state-of-the-art as of today a good scalability is achievable. An analysis of the energy-to-solution for the relevant kernel revealed that, although much higher floating-point operation efficiency can be obtained on the POWER8 processors, the energy-to-solution is significantly smaller when using GPUs.

This work leaves multiple opportunities for future work. First, the model analysis suggests that a specialised implementation of the sparse operator application GPU kernel could outperform the cuSPARSE library for this concrete problem. Second, with the upcoming availability of large scale POWER8 based systems employing a high performance interconnect, we will investigate the validity of the developed models. Finally, the application might benefit from a flexible distribution of work among processor and accelerator, as the application kernel runs efficiently on both.

Acknowledgements. This work was done in the framework of the POWER Acceleration and Design Center. We acknowledge the support from Charles Lefurgy (IBM) and Willi Homberg (JSC) on performing power consumption measurements using AMESTER. Furthermore, we thank Jiri Kraus (NVIDIA) for many helpful discussions.

References

1. OSU Micro-Benchmarks. http://mvapich.cse.ohio-state.edu/benchmarks/
2. Adinetz, A.V., Baumeister, P.F., Böttiger, H., Hater, T., Maurer, T., Pleiter, D., Schenck, W., Schifano, S.F.: Performance evaluation of scientific applications on POWER8. In: Jarvis, S.A., Wright, S.A., Hammond, S.D. (eds.) PMBS 2014. LNCS, vol. 8966, pp. 24–45. Springer, Heidelberg (2015)
3. Baumeister, P.F.: Real-Space Finite-Difference PAW Method for Large-Scale Applications on Massively Parallel Computers. Ph.D. thesis, RWTH Aachen (2012)
4. Baumeister, P.F., Hater, T., Kraus, J., Pleiter, D., Wahl, P.: A performance model for GPU-accelerated FDTD applications. In: 2015 IEEE 22nd International Conference on High Performance Computing (HiPC), pp. 185–193 (2015)
5. Beeby, J.: The density of electrons in a perfect or imperfect lattice. In: Proceedings of the Royal Society of London A: Mathematical, Physical and Engineering Sciences, vol. 302. The Royal Society (1967)
6. Ben, M.D., Hutter, J., VandeVondele, J.: Second-order Møller-Plesset perturbation theory in the condensed phase. J. Chem. Theory. Comput. **8**(11), 4177–4188 (2012)
7. Bilardi, G., Pietracaprina, A., Pucci, G., Schifano, F., Tripiccione, R.: The potential of on-chip multiprocessing for QCD machines. In: Bader, D.A., Parashar, M., Sridhar, V., Prasanna, V.K. (eds.) HiPC 2005. LNCS, vol. 3769, pp. 386–397. Springer, Heidelberg (2005)
8. Caldeira, A.B., et al.: IBM Power System S824L technical overview and introduction (2014). redbooks.ibm.com/Redbooks.nsf/RedbookAbstracts/redp5139.html
9. Floyd, M., et al.: Introducing the adaptive energy management features of the POWER7 chip. IEEE Micro **31**(2), 60–75 (2011)
10. Freund, R.W., Nachtigal, N.: QMR: a quasi-minimal residual method for non-Hermitian linear systems. Numer. Math. **60**(1), 315–339 (1991)
11. Hakala, S., Havu, V., Enkovaara, J., Nieminen, R.: Parallel electronic structure calculations using multiple graphics processing units (GPUs). In: Manninen, P., Öster, P. (eds.) PARA. LNCS, vol. 7782, pp. 63–76. Springer, Heidelberg (2013)
12. Hoefler, T., Gropp, W., Kramer, W., Snir, M.: Performance modeling for systematic performance tuning. In: State of the Practice Reports. SC 2011. ACM (2011)
13. Hutter, J., Iannuzzi, M., Schiffmann, F., VandeVondele, J.: CP2K: atomistic simulations of condensed matter systems. Comp. Mol. Sci. **4**(1), 15–25 (2014)

14. Kohn, W., Rostoker, N.: Solution of the Schrödinger equation in periodic lattices with an application to metallic Lithium. Phys. Rev. **94**, 1111–1120 (1954)
15. Kohn, W., Sham, L.J.: Self-consistent equations including exchange and correlation effects. Phys. Rev. **140**, A1133–A1138 (1965)
16. Korringa, J.: On the calculation of the energy of a Bloch wave in a metal. Physica **13**(6), 392–400 (1947)
17. Lefurgy, C., Wang, X., Ware, M.: Server-level power control. In: Fourth International Conference on Autonomic Computing, 2007. ICAC 2007, pp. 4–4, June 2007
18. Mericas, A.E.A.: IBM POWER8 performance features and evaluation. IBM J. Res. Dev. **59**(1), 6:1–6:10 (2015)
19. Pleiter, D.: Parallel computer architectures. In: 45th IFF Spring School 2014 "Computing Solids Models, ab-initio Methods and Supercomputing", Schriften des Forschungszentrums Jülich, Reihe Schlüsseltechnologien, vol. 74 (2014)
20. Solcà, R., Kozhevnikov, A., et al.: Efficient implementation of quantum materials simulations on distributed CPU-GPU systems. In: SC 2015 Conference on Proceed, pp. 10:1 (2015)
21. Soler, J.M., et al.: The SIESTA method for ab initio order-N materials simulation. J. Phys.: Condens. Matter **14**(11), 2745 (2002)
22. Spiga, F., Girotto, I.: phiGEMM: a CPU-GPU library for porting Quantum ESPRESSO on hybrid systems. In: 2012 20th Euromicro International Conference on Parallel, Distributed and Network-Based Processing, PDP 2012, pp. 368–375, February 2012
23. Thiess, A., et al.: Massively parallel density functional calculations for thousands of atoms: KKRnano. Phys. Rev. B **85**, 235103 (2012)
24. Videau, B., Marangozova-Martin, V., Genovese, L., Deutsch, T.: Optimizing 3D convolutions for wavelet transforms on CPUs with SSE units and GPUs. In: Wolf, F., Mohr, B., an Mey, D. (eds.) Euro-Par 2013. LNCS, vol. 8097, pp. 826–837. Springer, Heidelberg (2013)
25. Williams, S., Waterman, A., Patterson, D.: Roofline: an insightful visual performance model for multicore architectures. Commun. ACM **52**(4), 65–76 (2009)
26. Zeller, R., et al.: Theory and convergence properties of the screened Korringa-Kohn-Rostoker method. Phys. Rev. B **52**, 8807–8812 (1995)
27. Zeller, R.: Towards a linear-scaling algorithm for electronic structure calculations with the tight-binding Korringa-Kohn-Rostoker Green function method. J. Phys.: Condens. Matter **20**(29), 294215 (2008)

Performance Prediction and Ranking of SpMV Kernels on GPU Architectures

Christoph Lehnert[1], Rudolf Berrendorf[1(✉)],
Jan P. Ecker[1], and Florian Mannuss[2]

[1] Computer Science Department, Bonn-Rhein-Sieg University of Applied Sciences,
Sankt Augustin, Germany
{christoph.lehnert,rudolf.berrendorf,jan.ecker}@h-brs.de
[2] EXPEC Advanced Research Center,
Saudi Arabian Oil Company, Dhahran, Saudi Arabia
florian.mannuss@aramco.com

Abstract. Predicting the runtime of a sparse matrix-vector multiplication (SpMV) for different sparse matrix formats and thread mappings allows the dynamic selection of the most appropriate matrix format and thread mapping for a given matrix. This paper introduces two new generally applicable performance models for SpMV – for linear and non-linear relationships – based on machine learning techniques. This approach supersedes the common manual development of an explicit performance model for a new architecture or for a new format based on empirical data. The two new models are compared to an existing explicit performance model on different GPUs. Results show that the quality of performance prediction results, the ranking of the alternatives, and the adaptability to other formats/architectures of the two machine learning techniques is better than that of the explicit performance model.

Keywords: SpMV · Performance prediction · Linear regression · Gradient-boosting · KNN · Instance-based learning

1 Introduction

The sparse matrix-vector multiplication (SpMV) is the most time-consuming operation in iterative solvers [14]. Improving the efficiency of these operations is therefore important in many application fields [2], and many papers have been published on different sparse matrix formats and related SpMV implementations. Besides handling the sparsity as such, some formats try to utilize additional structural properties of a matrix. For example, work has been done on formats that do not rely on certain properties of a matrix and are therefore generally applicable, e.g., CSR [14], SELL-C-σ [8]. Additional research has been conducted on other formats that take advantage of the matrix structure (e.g., BCSR [7]) and/or a target architecture (e.g., supported formats in the Intel MKL and Nvidia cuSPARSE). Even if one format and target architecture is fixed, some formats/architectures have additional parameters, e.g., a slice size in the SELL-C-σ format or grid/block size on

© Springer International Publishing Switzerland 2016
P.-F. Dutot and D. Trystram (Eds.): Euro-Par 2016, LNCS 9833, pp. 90–102, 2016.
DOI: 10.1007/978-3-319-43659-3_7

Graphics Processor Units (GPUs) that need to be optimized for a given matrix or an architecture.

However no sparse matrix format performs best for all matrices and all processor architectures. Instead the formats differ significantly in their performance between formats for certain matrix structures or when switching between architectures. Choosing the right data format, parameter values, and possibly even the target architecture for a given matrix is therefore extremely important.

Simple and fast heuristics have been developed (see related work in Sect. 2) that try to predict which format or parameter values should be used for a given matrix based on few parameters than can be efficiently determined with low overhead. These heuristics have to be adapted to new formats, possibly to new sparsity patterns, and if possible to new architectures. Instead of manually developing new heuristics for new configurations, the question is whether it is possible to develop more general techniques to rank SpMV alternatives through runtime prediction. These techniques could then be used as a basis for deciding on a configuration for a given matrix.

This paper investigates three modeling techniques to predict the runtime of an SpMV operation on a GPU. Based on such predictions, a ranking of alternatives is possible that lets a program choose the best ranked alternative or, dependent on external parameters (prefer architecture X), choose the best ranked alternative after applying a filter at runtime.

One runtime prediction technique used in this work is based on the work of Guo et al. [6] and uses special benchmark matrices that determine two parameters in a simple linear model specific to some basic matrix formats. These performance models themselves are specific to a matrix format and are therefore not transferable to other formats. The other two newly developed models proposed in this paper use machine learning techniques and are more general. These two models differentiate between whether a mostly linear relationship exists between features and the SpMV runtime (using regression techniques) or whether non-linear relationships have a non-neglecting influence on the runtime (then using KNN). The three different approaches are evaluated using a set of sparse matrices and 5 different matrix formats: 3 basic ones that were used already in other papers (COO, CSR, ELL [14]) and 2 complex ones not used in such investigations before (SELL-C-σ [8], ELL-BRO [17]).

The paper is organized as follows. Section 2 gives an overview on related work. Section 3 introduces the three performance modeling methods. Then Sect. 4 discusses in detail the evaluation and reliability of the runtime prediction/ranking. The paper closes with a summary.

2   Related Work

Some papers [1,3,8,9,13] deal with simple explicit heuristics or provide empirical modeling for a Central Processing Unit (CPU) or GPU. This allows a program to choose at runtime which matrix format and/or which parameter values to use. This type of system is mainly used for autotuning. Those papers work with empirically derived fixed heuristics that work reasonable well for the supported simple matrix formats/architectures but need to be adapted for new and more complex matrix formats and/or architectures.

Xu et al. [18] claim that the SpMV is a memory-bounded problem. They suggest a prediction concept that estimates the memory access times needed to load and store the matrix and vector data for the SpMV operation. Although SpMV is memory bound, there are others factors that influence the runtime of this operation as well. Li et al. [10] published a probabilistic approach that takes into account the distribution of non-zero elements in each row of the sparse matrix. They use the model for the runtime prediction of the (simple) CSR, COO, ELL and HYB formats. Sedaghati et al. [15] use decision tree algorithms to select the most suitable format for a specific sparse matrix. Similar to our work, they use machine learning techniques, but with important differences. Their matrix formats are those supported by the vendor library and are fairly regular/simple. We used in addition more complex formats (SELL-C-σ, ELL-BRO), that have a more complex performance behavior. Additionally we allow a choice not only between the matrix formats but also between format/architecture parameter values for a specific format.

Guo et al. [6] uses profiled data for benchmark matrices and a simple 2-parameter linear model that is discussed and evaluated in detail in Sects. 3.1 and 4. Offline benchmark matrices and a heuristic performance model was also used in Lee et al. [9] for the CSR matrix format only.

3   Performance Modeling

This work aim to develop performance models that are not explicitly specific to a certain matrix format or a processor architecture but are more generally applicable and do not need to be reworked for new configurations. We describe in more detail below an existing benchmarking-based approach and then introduce a linear regression technique and the k-nearest neighbors approach.

Table 1 summarizes all features of the platform and matrices we found to be relevant for predicting performance in any of the described models on GPUs. Further features were analyzed, but no relevance for the runtime prediction could be identified. For some formats, there are several format specific parameters that may also influence the runtime of the SpMV operation. Although we have not included such format parameters, our work could be extended in that direction.

3.1   Benchmarking-Based Approach

The benchmarking approach we use in this paper is based on the work of Guo et al. [6]. Their prediction method consists of two major phases. In the instrumentation phase, platform information is gathered including the number of available streaming multiprocessors (SMs) and the maximum number of threads that can be processed by each SM at once. These are used to compute the so-called strip size. A strip is a maximum submatrix that can be computed by the

Table 1. Platform and matrix features that are relevant for the presented approaches.

Feature	Description
blocksize	The CUDA blocksize
nRows	The number of rows (the dimension of the square matrix)
nnz	The overall number of non zeros of the matrix
minNnz	The minimum number of non-zero elements per row among all rows
maxNnz	The maximum number of non-zero elements per row among all rows
modeNnz	The statistic mode of the number of non-zero elements per row among all rows
medianNnz	The median of the number of non-zero elements per row among all rows
minDist	The minimum distance between non-zero elements per row among all rows
maxDist	The maximum distance between non-zero elements per row among all rows
bandwidth	The maximum distance of non-zero elements from the diagonal
dispersion	The standard deviation of the numbers of non-zero elements among all rows
density	The fraction of the non-zero elements in the total number of array elements

GPU in one iteration if the full parallelism is used. The number of matrix rows that fit into a strip is therefore different for each of the used formats. According to Guo et al., one strip can be calculated in one step (simplified); therefore the number of strips a matrix consists of has a major influence on the runtime. Another important attribute in this model is the average number of non-zero elements in the matrix rows. The execution of synthetical special benchmark matrices is used to determine the relation between the number of strips and the average number of non-zero elements per row to the execution time. The SpMV operation is executed on multiple benchmark matrices:

– with a fixed size and an increasing number of non-zeros per row
– with a fixed number of non-zeros per row and an increasing number of strips
– multiple runs of matrices with (different) fixed numbers of non-zeroes per row and an increasing number of strips (ELL format only)

The information gathered from these SpMV executions is used to create a 2-parameter linear model [6]. In a program run, this model is parametrized with the information from the target matrix, and the actual runtime is predicted. This model is different for every matrix format and for a new matrix format an appropriate linear relationship has to be explicitly specified.

3.2   Linear Regression Techniques

Algorithms based on machine learning techniques are used in various scientific fields. One such machine learning approach is linear regression. It is a technique that relates a response vector of training instances, for example the measured SpMV runtimes of several matrices, to the features of the matrix by assuming that a linear relationship exists. The goal is to determine the relevance of these predictors and apply linear coefficients to each of them so that the best approximation of the responses for all (training) instances can be obtained. This goal can be achieved by iteratively refining the linear regression model to minimize the residual error between the modeled values and the actual responses [12].

Basic Model. In this step, the linear relation for the first training data record (feature values and SpMV runtime of the first matrix) is established. The second record is added, and the coefficients that have been identified in the first step are adjusted until the squared error between the estimations of both instances (by using the common coefficients) and the actual response is minimal. This process is repeated for all training data records. The result of this training phase is a linear model that leads to the best approximated responses for all training instances. Its particular coefficients ω_i can be utilized to predict the result T_{test} for all new test instances with feature values α_i. This prediction is obtained by using Eq. (1) for an instance with m features. A linear intercept ω_0 also results from the procedure described in [12].

$$T_{\text{test}} = \omega_0 + \omega_1 \times \alpha_1 + \omega_2 \times \alpha_2 + ... + \omega_m \times \alpha_m \qquad (1)$$

Model Enhancements. This regression technique is applicable if linear relations exist between the features. For a SpMV operation, many but not all features have such properties. An open question at the beginning of the research was whether the linear features dominate the runtime or whether a linear model is not suitable because the influence of non-linear features is too high.

The relationships between the matrix formats are certainly not linear, and the thread mapping also behaves non-linearly. Therefore in the training phase, distinct models are generated for different matrix formats and thread mappings. First investigations have also shown that selecting proper features per format (and only those) is an essential step. Table 2 presents the selected subset of features for each format. Other features did not positively influence the quality of the predictions. In a next step in the training phase, the feature values were logarithmized. These steps had proved to be sufficient to accurately predict SpMV runtimes for the simple formats COO, CSR and ELL.

Gradient-Boosting. For the more complex matrix formats SELL-C-σ and ELL-BRO, the prediction quality was less good. To be able to predict with a high accuracy the SpMV runtimes even in cases where (1) some non-linear relationship exists between the matrix features and SpMV execution times for a specific

Table 2. Features selection for the approaches linear regression/gradient-boosting (LR) and k-Nearest Neighbors (KNN) and the selected formats.

Feature	COO		CSR		ELL		SELL-C-σ		ELL-BRO	
	LR	KNN	LR	KNN	LR	KNN	LR	KNN	LR	KNN
blocksize	X	X	X	X	X	X	X	X	X	X
nRows	X		X	X	X					
nnz	X	X		X	X	X	X	X	X	X
minNnz	X		X			X	X	X	X	X
maxNnz	X				X	X	X		X	
modeNnz	X		X				X		X	
medianNnz	X		X	X	X		X		X	
minDist	X					X	X		X	
maxDist			X		X	X				
bandwidth	X		X		X		X			
dispersion	X				X	X	X	X	X	
density			X	X			X		X	

format or (2) important features for this format have not been identified and can thus not be represented in the modeling process, a gradient-boosting technique was chosen as an alternative to the linear regression model. This technique was then used for the formats SELL-C-σ and ELL-BRO.

Gradient-boosting [5] is also a regression technique. To determine the responses of completely new instances, a general function is approximated using all available training samples. Here, the search is for the concrete function that leads to a minimal estimated error among all training instances. As explained in detail in [5], gradient-boosting approaches start with a simple and often weak approximating model that only fits a small number of the training instances and iteratively refine it by applying the same *base learner* to the previous intermediate results. In our approach, the base learner is a regression tree, parametrized using several factors including its depth and the splitting rules at each non-terminal node. The model is built by roughly fitting the training instances by initially using a simple tree. The result is iteratively refined by applying further regression trees on the particular residual errors of the prior iteration and combining the intermediate results to a final complex model.

3.3 k-Nearest Neighbors

The k-Nearest Neighbors (KNN) [11] approach belongs to the instance-based learning methods that are suitable for both regression and classification tasks. The general idea behind such techniques is not to determine and store a concrete calculated function but to compare a new instance to all training records or a subset of them. The desired response value for the new instance is then retrieved by identifying similar training instances and accounting for their individual responses.

The KNN algorithm compares the feature values of the new instance with those of the training instances. A distance measure is chosen to select the k training records that are closest to the new instance, presuming that they are the most similar ones among all training instances. For a regression problem, the response value is (for example) retrieved by computing the mean of the k neighbors responses [11].

To successfully use the KNN algorithm, an appropriate distance function as well as a proper k-value have to be accommodated to different problem areas. Further enhancements of the KNN technique are the distance-weighted and the feature-weighted KNN approach. When using the first one, the calculated distances to the neighbors of a new instance are weighted. While neighbors with a low distance have a high impact on the calculated value, the impact decreases the greater the distances are [11]. The second technique applies weights to each feature.

Model Specifics. We achieved the most accurate runtime predictions by using k-values of 4 for the COO, CSR, ELL and BRO-ELL formats and 2 for the SELL-C-σ format. These differences are caused by the provided training sets per kernel. Using a k-value of 4 for the SELL-C-σ format leads to higher inaccuracies for a few test matrices, because only few neighbors for this kernel exists for those matrices. Taking more neighbors into account diminishes the quality of the runtime estimations. This problem can be dealt with by providing a bigger set of training instances with a higher variety of features. As a distance function, the euclidean distance was used. Manually derived weights were used for the preselected features shown in Table 2. The KNN approach has been shown to be much more sensitive to the correct selection of features compared to the linear regression approach. Furthermore, the feature values have been 0-1-normalized since the data ranges of the features are too different. The distance weighting was realized by using the inverse of the euclidean distance for computing the neighbors contribution to the calculation of the response variables, i.e., the estimated runtimes.

4   Evaluation

In this section, all three approaches are evaluated. First the evaluation methodology will be explained, followed by the evaluation of the runtime prediction quality and the ranking quality.

4.1   Evaluation Methodology

The training and evaluation of the prediction approaches requires the measurement of the runtimes of the different SpMV kernels on a GPU. The runtimes do not include the data transfer times to the GPU. The measurements are performed on GPUs with two different architectures (Nvidia M2050 and K80). Only one of the two GPUs of the K80 is used. To minimize measurement inaccuracies and startup overhead, all SpMV operations are executed 200 times and the median is always used as the actual runtime.

Table 3. Set of used test matrices with some additional structure information.

#	Matrix	Rows/columns	nnz	Bandwidth	nnz per row			
					min	max	mode	med
1	Ga41As41H72	268096	18488476	22519	18	702	37	37
2	PR02R	161070	8185136	84250	1	92	66	66
3	Si34H36	97569	5156379	18908	17	494	37	37
4	crankseg_2	63838	14148858	61047	48	3423	195	195
5	nd6k	18000	6897316	16766	130	514	468	416
6	TSOPF_RS_b2383	38120	16171169	33353	2	983	4	6
7	tmt_sym	726713	5080961	1921	3	9	7	7
8	af_1_k101	503625	17550675	859	15	35	35	35
9	af_shell1	504855	17588875	4909	20	40	35	35
10	gsm_106857	589446	21758924	588744	12	106	32	32
11	matrix_spe1Ref_a	900000	18612000	17999	12	21	21	21
12	boneS10	914898	55468422	8969	12	81	81	66
13	atmosmodd	1270432	8814880	21904	4	7	7	7
14	kkt_power	2063494	14612663	2046911	1	96	3	3
15	memchip	2707524	14810202	1647939	2	27	4	5
16	Flan_1565	1564794	117406044	20702	24	81	81	81
17	circuit5M_dc	3523317	19194193	2832158	1	27	4	5
18	matrix_spe10_dpdp_a	3506080	50928264	3378961	2	16	16	16

A set of 74 square matrices is used for the training and evaluation process. Table 3 shows the 18 test matrices that are used as target- or test matrices where the runtime must be predicted. The other matrices are the training matrices. The matrices show a wide variety of feature values. All matrices originate from the University of Florida Sparse Matrix Collection [4] or the SPE Comparative Solution Project [16]. For the SELL-C-σ format, a fixed C-value of 512 and σ-value of 2048 are used for all measurements. Likewise for the ELL-BRO format, a slice size of 256 and symbol size of 64 bit are used.

An exhaustive search on the available formats and valid thread mappings was performed for each matrix to determine the best achievable runtime for that matrix and the corresponding format and thread mapping. Only the three basic formats CSR, ELL and COO are used for the benchmarking-based approach, while all five formats are used for the two machine learning approaches.

4.2 Prediction Quality

The prediction quality is determined by comparing the predicted runtime with the actual measured runtime for the same format and same thread mapping.

This was done with all matrices of Table 3 using all supported formats and a large set of thread mappings, in total about 8600 test instances for the machine learning approaches and about 6100 ones for the benchmarking approach. Figures 1a and b present the divergence between the predicted and measured runtimes over all test instances of all three approaches on the M2050 and K80. They show that the median divergence of the machine learning approaches is very low at around 10 %. The average divergence of both these approaches is higher due to some inaccurate predictions.

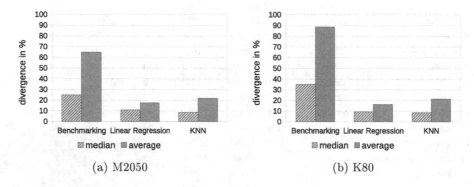

(a) M2050   (b) K80

Fig. 1. Comparison of the predicted runtimes and the actual measured runtimes.

The median divergence of the linear regression approach is higher compared to the KNN approach, but its average is significantly better. This difference is caused by an overall smaller number of predictions with greater inaccuracy. The few high over- and underestimations of the KNN approach can be found in a small number of matrices. Examples are matrices nd6k, circuit5M_dc and kkt_power. They have in common that some of their features have the smallest values among all available training instances. The number of rows of matrix nd6k, for example, is much smaller than those of all training matrices. Since this value is crucial for estimating the CSR-runtime, training instances with a significantly higher number of rows may be chosen as the nearest neighbors, resulting in an imprecise estimation. This problem can be dealt with by extending the training set and including smaller matrices.

The figures also clearly show that the performance of the benchmarking-based approach is much worse. The median divergence is about 25 % and the average accuracy only reaches 65 %. On the newer architecture of the K80, the benchmarking-based approach performs even worse, which could indicate that the relatively simple model is less suitable for the newer and more complex Kepler architecture. In summary, both machine learning approaches deliver a very high prediction quality with median divergences of around 10 %.

4.3   Ranking Quality

The runtime prediction is used as a tool, for ranking SpMV alternatives for a given matrix. Even with a inaccurate prediction, correct ranking could still be possible, e.g., with a continuous over estimation as long as the ranking order is still correct. The evaluation of the quality of the ranking is done by comparing the measured runtime of the predicted first ranked configuration with the overall best measured runtime for that matrix over all configurations (formats and thread mappings).

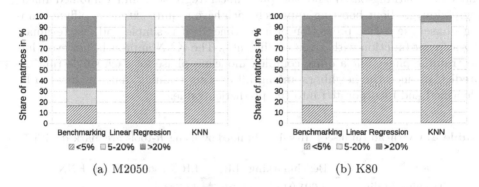

(a) M2050   (b) K80

Fig. 2. Share of matrices where the divergence between the reached runtime and the overall best runtime is between certain values.

Figure 2a presents the share of predictions where the divergence is below 5 %, between 5 % and 20 %, and over 20 % for all approaches on the M2050. The figure shows that the machine learning approaches again deliver in total very good results, and much better results than the benchmarking-based approach. For the majority of matrices, the predictions of the machine learning approaches result in runtimes that are only less than 5 % slower than the best possible runtimes. There was also no prediction of the linear regression/gradient-boosting approach, which was more than 20 % slower than the optimal runtime. The real runtimes of the benchmarking approach are more than 20 % slower than the optimum in most of the cases.

Figure 2b presents the same comparison for the K80 GPU. The linear regression on the K80 performs slightly worse than that on the M2050 and the KNN, and the benchmarking approach perform slightly better than that on the M2050. A more detailed comparison of the data revealed more details not easily presentable in plots:

- The benchmarking approach delivers very inconsistent results, and the runtimes for the same matrix on different architectures vary greatly.
- The quality of the predictions for a specific matrix is very consistent for the linear regression and the KNN approaches and mostly independent of the used architecture.

- The KNN approach delivers very good predictions for the majority of matrices, but also some highly inaccurate predictions with up to around 160 % slower runtimes. The linear regression delivers better average predictions and no such extreme outliers.

4.4   Other Aspects

Table 4 shows the durations of the training-, modeling- and ranking-phases for all approaches. The ranking was done for the matrix af_1_k101. While the benchmarking-based and the pure linear regression and combined linear regression/gradient-boosting approach each have a quite extensive offline training phase, the online prediction of a runtime itself is simple and fast (evaluating a linear function with given coefficients). The KNN approach, however, has no training phase but a quite complex modeling phase for each single runtime prediction since the neighbors among all existing training matrices have to be identified and incorporated into the prediction value.

Table 4. Overhead (in msec). Results obtained by using R-tools are marked with *.

Phase	Benchmarking	LR	LR/Grad. boosting	KNN
Training (offline)	2,601,042	130.7*	4,078*	-
Modeling & ranking	0.001	0.001	0.013	153.4*

Table 5. SpMV-runtimes (in msec) per format of matrix af_1_k101

Format	BRO-ELL	SELL-C-σ	ELL	CSR	COO
Runtime (msec)	1.029	1.187	1.362	2.861	7.571

Table 5 shows the measured SpMV execution times for the same matrix using different formats. A comparison of the two tables shows that the modeling/ranking-phases for all approaches other than KNN are even faster than one SpMV execution with the most appropriate kernel for the given matrix. Besides predicting an adequate format, the approaches can also be used for selecting a proper thread mapping, simply by calculating the predictions for a set of reasonable thread mappings. Regarding the prediction overhead, this process is suitable for the benchmarking-based and linear regression techniques, whereas the overhead for the KNN approach is quite high.

5 Summary and Outlook

We developed two new general models for performance prediction and ranking of SpMV kernels on GPUs and compared these to a known explicit linear model. The two new models both show better prediction and ranking results than the simple explicit model. The linear regression model is most appropriate if linearity of parameters dominates. This was the case for simple/regular sparse matrix formats. The gradient-boosting regression technique could also handle the two more complex formats. If parameters show non-linearity, a KNN model is more suitable. This model delivers better results for formats with more complex performance behavior, but it has a higher runtime overhead. Based on the prediction, we were also able to use ranking to determine the most suitable format and architecture parameters for a given matrix.

Our high quality results were only available after we fitted the general model more specifically to the concrete problems; this procedure is common when using these techniques. The procedure includes the proper selection of relevant features and weights and the separation of models with respect to formats and thread mapping. The approaches themselves are transferable to other formats and (GPU) architectures.

There are several opportunities to further improve our models. For the linear regression, we used a least-square/gradient-boosting approach. Here different regression techniques might deliver an even better quality. For the KNN approach, the weights could be determined by the system itself. Applying the models to a CPU-based systems would also be of interest.

Acknowledgements. We would like to thank the CMT team at Saudi Aramco EXPEC ARC for their support and input. Especially we want to thank Ali H. Dogru for making this research project possible. Additionally we appreciate the discussions on modeling with Marlis von der Hude and Peter Becker.

References

1. Ashari, A., Sedaghati, N., Eisenlohr, J., Sadayappan, P.: An efficient two-dimensional blocking strategy for sparse matrix-vector multiplication on GPUs. In: Proceedings of the 28th ACM International Conference on Supercomputing (ICS 2014), pp. 273–282. ACM (2014)
2. Berrendorf, R., Weierstall, M., Mannuss, F.: Program optimization strategies to improve the performance of SpMV-operations. In: Proceedings of the 8th International Conference on Future Computational Technologies and Applications, pp. 34–40 (2016)
3. Choi, J.W., Singh, A., Vuduc, R.W.: Model-driven autotuning of sparse matrix-vector multiply on GPUs. In: Proceedings of Principles and Practices of Parallel Programming (PPoPP 2010), pp. 115–125. ACM, January 2010
4. Davis, T.A., Hu, Y.: The University of Florida Sparse Matrix Collection. ACM Trans. Math. Softw. **38**(1), 1:1–1:25 (2010)
5. Friedman, J.H.: Greedy function approximation: a gradient boosting machine. Ann. Stat. **29**, 1189–1232 (2000)

6. Guo, P., Wang, L., Chen, P.: A performance modeling and optimization analysis tool for sparse matrix-vector multiplication on a GPUs. IEEE Trans. Parallel Distrib. Syst. **25**(5), 1112–1123 (2014)

7. Im, E.J., Yelick, K., Vuduc, R.: Sparsity: optimization framework for sparse matrix kernels. Int. J. High Perform. Comput. Appl. **18**(1), 135–158 (2004)

8. Kreutzer, M., Hager, G., Wellein, G., Fehske, H., Bishop, A.R.: A unified sparse matrix data format for efficient general sparse matrix-vector multiply on modern processors with wide SIMD units. SIAM J. Sci. Comput. **26**(5), C401–C423 (2014)

9. Lee, B.C., Vuduc, R.W., Demmel, J.W., Yelick, K.A.: Performance models for evaluation and automatic tuning of symmetric spare matrix-vector multiply. In: Proceedings of the International Conference on Parallel Processing, vol. 1, pp. 169–176. IEEE (2004)

10. Li, K., Yang, W., Li, K.: Performance analysis and optimization for SpMV on GPU using probalistic modeling. IEEE Trans. Parallel Distrib. Syst. **26**(1), 196–205 (2015)

11. Mitchell, T.M.: Machine Learning, vol. 1. McGraw-Hill, Singapore (1997)

12. Murphy, K.P.: Machine Learning: A Probabilistic Perspective, vol. 1. The MIT Press, Cambridge (2012)

13. Neelima, B., Reddy, G., Raghavendra, P.: Predicting an optimal sparse matrix format for SpMV computation on GPU. In: Proceedings of International Parallel & Distributed Processing Symposium Workshops (IPDPSW 2014), pp. 1427–1436. IEEE (2014)

14. Saad, Y.: Iterative Methods for Sparse Linear Systems, 2nd edn. SIAM, Philadelphia (2003)

15. Sedaghati, N., Mu, T., Pouchet, L.N., Parthasarathy, S., Sadayappan, P.: Automatic selection of sparse matrix representation on GPUs. In: Proceedings of the 25th International Conference on Supercomputing (ICS 2015). ACM (2015)

16. Society of Petroleum Engineers. http://www.spe.org/web/csp/: SPE Comparative Solution Project

17. Tang, W., Tan, W., Ray, R., Wong, Y., Chen, W., Kuo, S., Goh, R., Turner, S., Wong, W.: Accelerating sparse matrix-vector multiplication on GPUs using bit-representation-optimized schemes. In: Proceedings of Intrnational Conference on High Performance Computing, Networking, Storage and Analysis (SC 2013). ACM (2013) (article no. 26)

18. Xu, S., Xue, W., Lin, H.X.: Performance modeling and optimization of sparse matrix-vector multiplication on NVIDIA CUDA platform. J. Supercomput. **63**(3), 710–721 (2011)

The Impact of Voltage-Frequency Scaling for the Matrix-Vector Product on the IBM POWER8

Sandra Catalán[1(✉)], A. Cristiano I. Malossi[2],
Costas Bekas[2], and Enrique S. Quintana-Ortí[1]

[1] Dpto. de Ingeniería y Ciencia de Computadores,
Universidad Jaume I, 12071 Castellón, Spain
{catalans,quintana}@uji.es
[2] IBM Research–Zurich, Foundations of Cognitive Solutions,
8803 Rüschlikon, Switzerland
{acm,bek}@zurich.ibm.com

Abstract. The physical limitations of CMOS miniaturization have promoted understanding the interplay between performance and energy into a primary challenge. In this paper we contribute towards this goal by assessing the effect of voltage and frequency scaling (VFS) on the energy consumption of the dense and sparse matrix-vector products. The optimization of the sparse kernel, from the perspective of both performance and energy efficiency, is especially difficult due to its irregular memory access pattern, but the potential benefits are remarkable because of its varied applications.

Our experiments with a small synthetic training set show that it is possible to build a general classification of sparse matrices that governs the optimal VFS level from the point of view of energy efficiency. More importantly, this characterization can be leveraged to tune VFS for a major portion of the University of Florida Matrix Collection, when executed on the IBM Power8, yielding significant gains with respect to a (power-hungry) configuration that simply favours performance.

Keywords: Energy efficiency · Voltage-frequency scaling · Performance prediction · Performance metrics · Matrix-vector product · IBM POWER8

1 Introduction

The matrix-vector product is an important numerical kernel as well as one of the 7+ dwarfs [3] proposed for the evaluation of parallel programming models and architectures. In particular, the sparse instance of the matrix-vector multiplication (SpMV) underlies the HPCG benchmark [8], and is also a crucial kernel for

---

IBM POWER8: IBM, the IBM logo, and ibm.com are trademarks or registered trademarks of International Business Machines Corp. in the United States, other countries, or both. Other product and service names might be trademarks of IBM or other companies.

© Springer International Publishing Switzerland 2016
P.-F. Dutot and D. Trystram (Eds.): Euro-Par 2016, LNCS 9833, pp. 103–116, 2016.
DOI: 10.1007/978-3-319-43659-3_8

the solution of linear systems and eigenvalue problems arising in many scientific and engineering applications [21]. Furthermore, the connection between sparse linear algebra and graph algorithms has been recently exploited by a new class of algorithms, based among others on SPMV, to tackle the vast volume of information that is common in social networks and other data analytic processes [5,14].

In this paper, we perform a complete experimental analysis of a generic implementation of the matrix-vector product on a current multi-threaded architecture with support for dynamic voltage-frequency scaling (VFS). Our study focuses on the energy efficiency using the energy-per-flop metric as reference. This analysis is timely because even though the matrix-vector product kernel has been extensively analyzed and optimized from the point of view of performance, the number of studies that investigate its energy consumption is limited. This is especially relevant since power/energy are factors which constrain the performance of current processor designs for the high performance computing systems running the aforementioned numerical and graph-related applications [9,17].

In particular, this work makes the following specific contributions with respect to the energy optimization of SPMV:

- We analyze several critical parameters with respect to matrix size, sparsity degree, and non-zero clustering of the sparse matrix that drive the energy efficiency of this kernel on a modern multithreaded architecture.
- We derive a simple recipe to optimize VFS for a SPMV operation involving any given sparse matrix that exploits the aforementioned relevant parameters from the graph representing the sparse matrix.
- We demonstrate the robustness of our VFS optimization recipe by applying it to optimize the energy performance of the entire University of Florida Sparse Matrix Collection (UFMC) [1] on the IBM Power8.
- We evaluate the energy gains that an appropriate adaption of VFS can yield with respect to an energy-oblivious approach that considers performance as the only optimization goal.

Our driving motivation is the design of an easy-to-use and widely-applicable strategy to significantly reduce energy consumption of SPMV without the need of a per-case analysis. Indeed, finding the optimal VFS for each individual sparse problem is unrealistic as well as unaffordable in practical production scenarios, where the sparsity pattern may change from one application to the next, and time cannot be spent on preliminary fine tuning energy-benchmarks.

The rest of the paper is structured as follows. In Sect. 2 we briefly review some related works and next, in Sect. 3, we describe the experimental setup. In Sects. 4 and 5 we analyze the energy consumption of the sparse matrix-vector kernel with dense and sparse matrices, respectively. We close the paper with a few concluding remarks in Sect. 6.

2   Related Work

There exists a large volume of work addressing the performance optimization of SPMV; see, e.g., [5,16,22,23] and the references therein. Among others, pOSKI [6] is a multithreaded library for SpMV that leverages automatic search over multiple implementations and sparse data layouts to optimize performance on multicore processors. Model-driven optimization of SpMV for data-parallel accelerators has been studied in [7,10].

The number of efforts dedicated to the energy modeling and optimization of SpMV is significantly more reduced. In [20] we introduced a systematic methodology to derive reliable time and power models for algebraic kernels employing a bottom-up approach. However, the recipe resulted a bit cumbersome to leverage, requiring a large number of calibration tests. In [19] we devised a systematic machine learning algorithm to classify and predict the performance and energy costs of the SpMV kernel, but did not consider the effect of VFS. The work in [2] presents an extensive experimental study of the interactions occurring in the triangle performance-power-energy for the execution of a pivotal numerical algorithm, the iterative Conjugate Gradient (CG) method, on an ample collection of parallel multithreaded architectures. However, that work does not produce a recipe to optimize VFS for any given sparse problem. Other work related to modeling sparse linear algebra operations can be found in [11,13,18].

3   Experimental Setup

3.1   Hardware

The target platform for our analysis is the IBM Power System S812L. The POWER8 processor in this system, fabricated on 22 nm silicon, features 2 sockets, each with 6 cores, offering hardware support for up to 8-way simultaneous multi-threading (SMT) as well as dynamic VFS. Each core in the IBM POWER8 is furnished with a 512-KB L2 cache. Furthermore, the chip contains a shared L3 cache of 8 MB per core, and 16 MB of L4 cache per buffer, with up to 8 buffers per socket [4]. The server was also equipped with 64 GB of DDR RAM.

Table 1 displays the frequency-voltage configuration pairs and the idle power dissipated by the system, measured during the execution of a sleep test,[1] for two scenarios: socket+DDR ("SD") only vs the full server ("Node"). Power measures were obtained using AMESTER [15]. This tool runs on a separate server and connects to the service processor of the node in order to obtain voltage/frequency/power-per-core samples from several sensors, while avoiding interference with the workload. Using this information, we calculate the time-per-flop and *net* power-per-core (without the idle power), yielding the *net* energy per-flop-and-core from their product. All energy consumption values reported next refer to the net energy-per-flop and core.

The codes were compiled using IBM's mpcc 13.01.0003.0000, with the flags: -O3 -qprefetch=dscr=0 -qhot -qstrict -qsmp=noauto:omp -qthreaded -qsimd= auto -qaltivec -q64 -qarch=pwr8. Each test was repeated for at least 60 s,

---

[1] Although the idle power could be determined with higher accuracy by via an extrapolation to the power usage with 0 cores, we believe that the sleep-based test provides enough precision for our purposes.

Table 1. Voltage-frequency pairs and idle power in the IBM POWER8 processor.

Config.	Frequency F (GHz)	Voltage V (mV)	Idle power, P^{IDLE} (W)	
			SD	Node
C_1 ●	2.13	875.0	147.9	358.3
C_2 ●	2.53	931.3	151.9	367.7
C_3 ●	2.96	987.5	158.1	380.1
C_4 ●	3.36	1,037.5	165.3	392.6
C_5 ●	3.79	1,093.8	173.2	411.2
C_6 ●	4.22	1,187.5	192.6	451.9

and the results average the values from these runs. The experiments targeted a single "socket" of the IBM POWER8 chip (i.e., 6 cores), with either 1 thread or 8 threads per core (1-SMT or 8-SMT, respectively), and all threads/cores collaborating to compute one instance of SpMV.

3.2 Kernel and Implementation

We analyze an implementation of the SpMV $y := A \cdot x$, with sparse matrix $A \in \mathbb{R}^{n \times n}$ and dense vectors $x, y \in \mathbb{R}^n$, based on the CSR (compressed sparse row) storage format [21]. For sparse matrices, CSR offers a fair balance between compression efficiency (as it is one of the most efficient formats for generic sparse matrices on cache-based microprocessors) and architecture-independent performance (since it does not directly exploit graph characteristics that may emerge from the specific physical problem) [22]. The CSR data layout employs a real array for the values of A (A_val), and two auxiliary integer arrays (col_ind and row_ptr) to maintain (respectively) the column indices of the nonzero entries in A and the initial/final index of each row of A within the other two arrays [21]. All our experiments employ double precision floating-point arithmetic so that the values of A, x, y occupy $s_d = 8$ bytes each. Each component of the indexing integer arrays occupies $s_i = 4$ bytes. Therefore, storing an $n \times n$ matrix with n_z nonzero entries in this format requires $M_S = n_z(s_d + s_i) + (n + 1)s_i$ bytes, and x, y occupy $V_S = ns_d$ bytes each.

The implementation of SpMV in CSR format is illustrated in Fig. 1. The optimization is left to the compiler, except for some minor details omitted for simplicity. The parallelization strategy distributes the computation of the entries of y among the threads/cores (via the OpenMP #pragma omp directive before the outer loop).

In the operation $y := A \cdot x$, there is no reuse of the entries of A and the only opportunity to exploit data locality is in the accesses to x, y. In the CSR version of SpMV, the entries of A_val and col_ind are streamed from the memory layer where they reside into the processor register file with unit stride; each entry of y is loaded into a register once and re-used until it has been completely updated; and the re-use factor of x depends on the sparsity pattern of A.

```
1  void SpMV_CSR( int n, int * row_ptr,   int * col_ind,
2                         double * A_val, double * x,     double * y ) {
3     int i, j;      double tmp;
4
5     #pragma omp parallel for private ( tmp, j ) schedule static
6     for ( i = 0; i < n; i++ ) {
7        tmp = 0.0;
8        for ( j = row_ptr [i]; j < row_ptr [i+1]; j++ )
9           tmp += A_val [j] * x [col_ind[j]];
10       y[i] += tmp;
11    }
12 }
```

Fig. 1. SpMV based on the CSR format.

4   Tuning VFS for the Dense Matrix-Vector Product

We commence our analysis by considering an $n \times n$ *dense* matrix-vector product kernel, GEMV, computed via the code in Fig. 1. While a dense matrix can be more efficiently stored as a conventional 1-D array, in column- or row-major order, this initial study will offer us some preliminary insights on the energy behavior of this memory-bound operation. For the following experiments, we use two square dense matrices, of dimension $n = 312$ and 30, 512 (with $n_z = n^2$ nonzeros). Taking into account the CSR memory layout, and the fact that all cores collaborate in the execution of the same matrix-vector product, the data for these two problems respectively requires about 1.15 MB and 10.6 GB. Thus, the small case easily fits into the on-chip L3 cache (8 MB/core), while the larger problem can only be stored in the off-chip DDR RAM.

Scaling the voltage and frequency (VFS) can be expected to produce an effect on performance and power dissipation which, in turn, produces an impact on the energy consumption. In principle, one could expect that a change of frequency results in a proportional variation of the GFLOPS (billions of flops per second). The left-hand side plot in Fig. 2 investigates the behaviour as the frequency is increased, and the socket is populated with 1 or 8 threads per core (1-SMT and 8-SMT). To capture the theoretical linear relation between the GFLOPS and the frequency, both metrics are normalized in the figure with respect to those observed for C_1. On one hand, when operating with the off-chip problem, the GFLOPS rate attained with 1-SMT stagnates for the two higher frequency rates while, for 8-SMT, the performance does not vary with the frequency. These results show scenarios where the DDR bandwidth is saturared for our GEMV code. On the other hand, the GFLOPS rate grows linearly with the frequency for the L3 on-chip case, independently of the number of threads.

The analysis from the point of view of power dissipation is more complex. Concretely, for a given voltage-frequency configuration pair $C_A = (V_A, F_A)$, the power dissipation can be decomposed into its static and dynamic components which depend, respectively, on V_A^2 and $V_A^2 \cdot F_A$ [12]. We can assume that the idle power (see Table 1) is mostly due to leakage (static power), while a substantial fraction of the net power is due to the application's activity (dynamic power). The right-hand side plot in Fig. 2 compares the experimental net power ratio

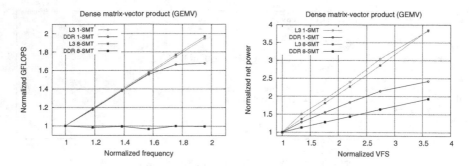

Fig. 2. Performance (left) and SD power consumption (right) for GEMV. (Color figure online)

Table 2. Normalized energy consumption with respect to C_1 for 8-SMT and different VFS levels. Best choices are highlighted in green.

Config.	SD		Node	
	L3	DDR	L3	DDR
C_1 ●	1.00	1.00	1.00	1.00
C_2 ●	0.90	1.06	0.87	1.04
C_3 ●	0.83	1.11	0.79	1.08
C_4 ●	0.79	1.22	0.73	1.16
C_5 ●	0.78	1.26	0.70	1.19
C_6 ●	0.82	1.43	0.71	1.32

P_A^{NET}, normalized with respect to that observed when running in C_1, against the theoretical VFS ratio $(V_A^2 \cdot F_A)/(V_1^2 \cdot F_1)$ during the execution of GEMV. Considering only the 8-SMT cases, there is a clear difference between the L3 case, which shows a perfect match between the theory and the experimental behaviour, and the DDR problem, for which the net power grows at a much lower pace due to the memory bottleneck.

Table 2 illustrates the combined effect of the variations in performance and power dissipation on the energy consumption, identifying the best VFS configuration depending on the problem dimension. The values there are again normalized with respect to those observed for the configuration C_1. Hereafter, we only report the results obtained with 8-SMT, as they are consistently superior to those obtained with 1-SMT in both performance and energy efficiency.

The previous experiments reveal that the execution time as well as the power consumption of GEMV can be accurately modeled when the problem data fits on-chip. This leads to a straight-forward derivation of energy-related metrics (such as our target net energy-per-flop) and, as illustrated in Table 2, paves the road to a direct optimization of VFS from the perspective of energy efficiency. The behavior of performance and power consumption is more difficult to predict for the DDR problems; however, if the goal is to optimize energy efficiency, the best strategy for GEMV simply runs this kernel at the lowest voltage-frequency pair. Compared with this energy-aware VFS configuration, an execution of GeMV that simply aims to enhance performance consumes significantly more energy: 1.43× for SD and 1.32× for the node.

5 Tuning VFS for the Sparse Matrix-Vector Product

Unfortunately, the sparsity exhibited by most real applications generally results in irregular access patterns (in CSR, to the entries of x), which may render an imbalanced workload distribution yielding the guidelines derived to adjust VFS for GeMV in Sect. 4 suboptimal for the sparse case.

In [19], we identified a reduced set of critical structural parameters which impact the performance, power, and energy consumption of the CSR implementation of SpMV. These properties were used to generate a small synthetic sparse benchmark (or training set) which was then employed to build a model that accurately predicts performance and energy consumption of any sparse problem. In the following, we investigate whether the same approach, based on a synthetic sparse training set, can produce a strategy to select a (close-to-)optimal VFS configuration for any sparse problem.

5.1 Training Set

In order to determine an appropriate configuration for SpMV, we leverage the benchmark introduced in [19], which is characterized by five parameters:

- Number of rows/columns n and nonzeros n_z.
- Block size: b_s. Many applications lead to sparse matrices where the non-zeros are clustered into a few compact dense blocks in each row. This parameter specifies the number of entries in these blocks, and determines the number of elements of x accessed with unit stride.
- Block density: $b_d = b_s/n_{zr} = b_s n/n_z \in [0, 1]$ is the inverse of the number of blocks per row. With b_s fixed, b_d defines the re-use factor for y.
- Row density: $r_d = n_{zr}/n \in [0, 1]$ is the number of non-zeros per row relative to the row size. With n_{zr} fixed, r_d is an indicator of the probability of finding an entry of x already fetched into a higher level of the cache hierarchy during the computation of a previous entry of y.

For the analysis, we distribute the matrix instances of the training set evenly in the $\log_2$ space comprised by $b_s \in \{2^0, 2^2, 2^4, \ldots, 2^{14}\}$, $b_d \in \{2^0, 2^{-2}, \ldots, 2^{-14}\}$, and $r_d \in \{2^0, 2^{-2}, \ldots, 2^{-28}\}$. Thus, for a matrix in the training set, the triplet-coordinates (b_s, b_d, r_d) identify a matrix of dimension $n = n_{zr}/r_d = b_s/(b_d r_d)$ with $n_z = n\, n_{zr}$ nonzeros. As $b_s \leq n_{zr} \leq n$, this distribution yields a total of 162 samples only, yet offers enough variability to characterize sparse matrices from real applications while avoiding a costly calibration.

Figure 3 illustrates a compact representation of the training set, with each matrix identified by a single point (b_s, b_d, r_d) in the 3-D space. The colors of the points identify the optimal VFS configuration, from the perspective of energy consumption, taking into account two scenarios: SD power only or the full node consumption. The matrix instances in the training set are divided into three categories, according to the size of the SpMV data ($P_S = M_S + 2V_S$): "L3", "DDR", and "transition". The first category includes those instances where the matrix data occupies less than 18 MB and involve small vectors x, y that easily fit into the L3 cache; the second category contains those instances requiring more than 200 MB, which therefore can only reside in the DDR; all the remaining cases are assigned to the transition category.

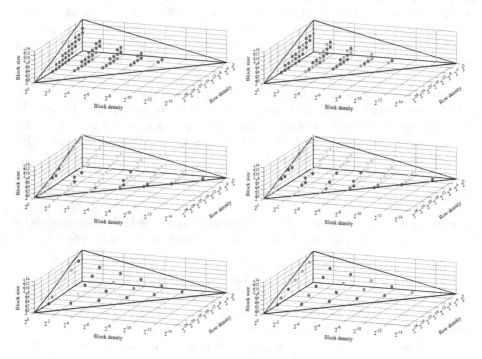

Fig. 3. Optimization of SD (left) and node (right) energy consumption via VFS for the synthetic training set, with matrices in the L3, DDR, and transition categories (top, middle, and bottom, resp.). Color codes: : C_1; ●: C_2; ●: C_3; ●: C_4; ●: C_5; ●: C_6. (Color figure online)

These results show that, for any scenario/category, there is no single configuration that optimizes VFS for all cases in it. For brevity, let us examine the energy consumed by the SD (scenario) in some detail (left column of plots):

– L3: in most cases, the C_5 configuration (●) is optimal.
– DDR: the C_1 configuration ( ) is possibly the best compromise solution. However, we can observe that, as the row density r_c decreases, the C_2 configuration (●) seems to offer a fair alternative.
– Transition: the figures expose no clear winner in this case.

The plots in Fig. 3 offer a quick glance of the impact of VFS depending on the scenario (SD or Node) and category (L3, DDR or transition). We next refine this information to quantify the overhead incurred by an approach that chooses a single configuration for each scenario/category combination. Consider in particular a category consisting of s problem instances $P = \{p_1, p_2, \ldots, p_s\}$, which are executed under a given VFS configuration C_A and scenario S, resulting in a range of values for the energy consumption: $E^S \in \{E^{SD}, E^{Node}\}$. Let us denote by $C_{OPT}(p, S)$ the configuration that minimizes energy consumption for a given problem instance/scenario p/S. With these premises, Table 3 reports the *relative deviation from the energy-optimal VFS configuration* for each problem category and scenario, given (in %) by

$$\frac{100}{s} \cdot \sum_{i=1}^{s} \left(\frac{E^S(p_i, C_A) - E^S(p_i, C_{OPT}(p_i, S))}{E^S(p_i, C_{OPT}(p_i, S))} \right).$$

The "L3", "DDR", and "Trans(ition)" columns in the table thus show the additional energy (overhead in %) that is spent if one chooses a single configuration for all problems in the category, instead of the specific optimal configuration for each problem instance/scenario. Let us analyze one of the scenarios in more detail, namely the SD energy consumption. The results in the table reveal that this excess is acceptable for the instances in the L3 and DDR categories when choosing C_5 (0.9 %) and C_1 (2.4 %), respectively, but they also expose a significant loss if a single configuration is selected for all cases in the transition category (15.3 % in the best case, corresponding to C_2).

Table 3. Deviation (in %) of energy consumption with respect to the optimal configuration for the synthetic training set. Best choices are highlighted in green.

Config.	SD					Node				
	L3	DDR	Trans.	HRD	LRD	L3	DDR	Trans.	HRD	LRD
C_1 ●	26.7	2.4	18.9	23.6	7.1	39.9	3.2	27.4	34.1	10.3
C_2 ●	14.7	5.4	15.3	19.8	4.1	24.4	5.0	21.9	28.2	6.1
C_3 ●	6.9	11.2	18.9	24.6	4.9	12.3	8.8	22.4	29.8	3.6
C_4 ●	2.2	17.2	17.1	18.8	12.8	5.2	13.6	18.7	21.9	10.4
C_5 ●	0.9	23.1	17.0	14.9	22.3	0.7	17.1	15.1	14.2	17.1
C_6 ●	6.2	38.4	21.9	20.0	26.7	2.2	28.7	15.9	15.0	17.8

Motivated by the blurry behaviour of the instances in the transition category, we next analyze these cases in more detail. Table 4 reports the SD energy consumption for the 21 instances in this problem category, normalized with respect to the configuration C_1. The values in the table show no clear relation between problem size, defined by n and P_S, and the optimal configuration for these instances. On the other hand, among the remaining three problem parameters, only the row density seems to play a role, as high row densities (HRD: $r_d > 2^{-14}$) favour high frequencies, while low row densities (LRD: $r_d \leq 2^{-14}$) benefit more from a lower frequency. This criteria pushes us to split the transition category into two subcategories, LRD and HRD, yielding the deviations in the columns labelled as "HRD" and "LRD" of Table 3, which reveal a significant

Table 4. Normalized energy consumption of SD with respect to C_1 for 8-SMT and different VFS levels. Best choices are highlighted in green.

	Problem parameters					Configuration					
	n	P_S (MB)	b_s	b_d (exponent)	r_d	C_1 ●	C_2 ●	C_3 ●	C_4 ●	C_5 ●	C_6 ●
HRD	4,096	48.1	1	-10	-2	1	0.98	1.10	0.71	0.78	0.76
	4,096	48.1	4	-8	-2	1	0.79	1.01	1.12	0.70	1.12
	4,096	48.1	16	-6	-2	1	1.01	0.76	0.96	0.75	0.77
	4,096	48.1	64	-4	-2	1	1.06	1.20	1.04	0.74	0.83
	4,096	48.1	256	-2	-2	1	0.91	0.77	0.74	1.08	1.18
	4,096	48.1	1,024	0	-2	1	0.82	1.02	0.79	0.83	0.82
	16,384	48.3	1	-8	-6	1	0.91	0.92	0.87	0.75	0.83
	16,384	48.3	4	-6	-6	1	0.93	0.84	1.05	1.09	1.19
	16,384	48.3	16	-4	-6	1	1.09	1.07	1.07	1.16	0.88
	16,384	48.3	64	-2	-6	1	0.90	1.15	1.12	0.80	1.10
	16,384	48.3	256	0	-6	1	1.04	0.96	1.15	0.94	0.78
	65,536	49.2	1	-6	-10	1	0.97	0.91	0.91	1.06	1.00
	65,536	49.2	4	-4	-10	1	0.99	1.20	0.84	1.14	1.10
	65,536	49.2	16	-2	-10	1	1.11	0.99	0.95	1.16	1.11
	65,536	49.2	64	0	-10	1	1.04	1.20	1.03	1.10	1.09
LRD	262,144	53.0	1	-4	-14	1	0.95	0.94	0.93	1.03	1.23
	262,144	53.0	4	-2	-14	1	1.08	1.00	1.18	1.23	1.25
	262,144	53.0	16	0	-14	1	1.04	1.07	1.03	1.14	1.20
	10,485,766	32.0	1	0	-20	1	0.94	0.99	1.04	1.18	1.32
	10,485,766	68.0	1	-2	-18	1	0.80	0.92	1.05	1.14	0.93
	10,485,766	68.0	4	0	-18	1	1.03	0.94	1.05	1.10	1.17

improvement for the LRD subcategory but no relevant gain for the HRD. To conclude this analysis of the transition cases, we note that LRD consist of just 6 instances which may be too small to derive a strong conclusion.

5.2 Validation with UFMC

The synthetic benchmark is only a small collection of sparse problems, that aims to provide a rough approximation of the sparsity patterns present in real applications, in order to offer some guidance on the energy-optimal VFS configuration. The motivation for this is that choosing the optimal VFS for each individual sparse problem can be unrealistic, as that may require to execute each case at each VFS level to make an appropriate choice. As an alternative, we trade off accuracy (and, therefore, energy efficiency) for flexibility, by using the classification and VFS configuration obtained with synthetic benchmark to dictate the selection for the real applications.

We validate the energy efficiency that can be attained if we base our VFS selection for real sparse matrices on the previous problem classification. For this purpose, we employ 1,202 problem instances from very different real applications in the set UFMC. Among these, 1,044 cases fit into the L3 cache, 75 can be classified as DDR cases, and the rest belong to the transition category, with 69 in HRD and 14 only in LRD.

Table 5 displays the *relative deviation from the energy-optimal VFS configuration* and the performance-optimal configuration for the problem instances in the UFMC. Regarding the energy, when comparing the optimal global VFS

Table 5. Deviation (in %) of energy consumption and performance with respect to the optimal configuration for the UFMC. Best choices are highlighted in green.

Config.	SD energy				Node energy				Performance			
	L3	DDR	HRD	LRD	L3	DDR	HRD	LRD	L3	DDR	HRD	LRD
C_1 ●	28.2	1.9	17.2	17.2	41.3	3.6	26.0	34.7	-47.91	-11.63	-37.43	-44.96
C_2 ●	15.8	4.9	10.9	10.9	24.8	4.9	16.6	23.4	-38.62	-10.09	-30.59	-37.44
C_3 ●	7.1	7.7	7.7	7.4	12.6	5.9	10.8	13.1	-28.39	-6.87	-23.50	-28.17
C_4 ●	2.4	11.6	6.5	4.3	5.1	7.9	7.0	7.3	-19.21	-4.15	-17.12	-20.50
C_5 ●	0.5	19.1	5.9	2.8	0.5	13.1	3.8	2.9	-9.95	-3.18	-9.14	-11.89
C_6 ●	5.1	32.2	11.6	5.2	1.5	22.4	5.8	1.5	-0.01	-0.65	-0.28	0

configuration in the table with those in Table 3, we observe that the training set did actually offer an appropriate guidance to select the energy-optimal VFS for the L3, DDR and HRD cases, for both the SD and node scenarios. On the other hand, the global energy-optimal VFS options for the UFMC problem instances in the LRD category are C_5 (SD) or C_6 (node) instead of those pointed out by the synthetic benchmark. With respect to performance, that when applying the energy-optimal VFS this metric decreases up to 11.63 %, except for the LRD cases where it could reach 37.44 %. Again, this is due to the fact that the training set prediction does not match the energy-optimal VFS for the UFMC cases that fall into the LRD category. At this point, it is worth reminding that the synthetic collection included only 6 problem instances in this category, with the data for five of them occupying in the range of $(P_S=)53$–68 MB. Compared with this, the UFMC set has 14 instances in the LRD category, but only 4 of this are in the same P_S-dimension range. The fact that the samples in the (synthetic and real) LRD category are few, and that the problem sizes for the synthetic set and the UFMC do not overlap, explains the different energy-optimal VFS configuration determined for each case. However, we emphasize that the LRD category contains only 14 cases out of 1,202 real problems, which is less than 1.2 %! For the remaining 98.8 % cases, the training set did actually identify a fair classification into categories as well as offer a good VFS selection.

A final question to investigate is the balance between the energy gains vs the performance loss that an energy-aware VFS configuration, based on the categories/VFS levels determined from the previous experimental studies, can yield compared with a conventional performance-oriented VFS selection that simply runs all (real) cases at the highest VFS level. Table 6 shows that, when considering the socket+DDR (SD), an energy-aware VFS configuration can yield savings between 3.0 % and 21.0 % with respect to the performance-oriented option, and more reduced if we consider the full node consumption. These savings come at a certain cost from the perspective of execution time, reporting a loss of the energy-aware VFS configuration with respect to the performance-oriented case that is between 9.6 % and 22.7 %.

Table 6. Average energy savings vs performance loss between the energy-optimal and the performance-optimal VFS configurations for the UFMC (denoted as C_E and $C_P = C_6$, respectively). From previous experiments with UFMC, $C_E = C_5$ for the problem instances in the L3 and transition categories, while $C_E = C_1$ for the DDR cases. Performance is measured in time-per-flop.

Metric	SD				Node			
	L3	DDR	HRD	LRD	L3	DDR	HRD	LRD
Energy ratio: C_E/C_P	1.046	1.210	1.054	1.030	1.011	1.097	1.019	0.995
Performance ratio: C_P/C_E	1.104	1.227	1.096	1.125	Same as SD scenario			

6   Concluding Remarks

Voltage-frequency scaling (VFS) is an energy-oriented technology present in current hardware that the operating system/programmer can leverage to adapt the execution pace of an application without modifying the code. Unfortunately, selecting the energy-optimal VFS configuration is both architecture- and application-dependent. For the (sparse) matrix-vector product kernel, our work shows that it is possible to rely on a portable benchmark, consisting of a reduced number of synthetic sparse matrices, to establish a general classification of the problems data, (according to criteria related to problem dimension and sparsity pattern,) and to determine a global energy-optimal VFS configuration for the matrices in each group. Our experiments on a multicore server equipped with an IBM POWER8 show a strong dependence between energy consumption and problem dimension, exposing an interesting trade-off between energy efficiency and performance for this particular kernel.

Our work also analyzed the energy-delay product, with similar conclusions to those presented in the paper for the energy efficiency. As part of future work, we plan to investigate the energy savings that can be attained with a limited loss in performance as well as prediction of the optimal level of concurrency throttling from the point of view of energy efficiency.

Acknowledgements. This work was supported by project Exa2Green (under grant agreement n°318793) of the Future and Emerging Technologies (FET) programme within the ICT theme of the Seventh Framework Programme for Research (FP7/2007–2013) of the European Commission. The researchers from Universidad Jaume I were supported by project TIN2014-53495-R of the MINECO and FEDER, and the FPU program of MECD.

References

1. The University of Florida Sparse Matrix Collection, January 2016. http://www.cise.ufl.edu/research/sparse/matrices/

2. Aliaga, J.I., Anzt, H., Castillo, M., Fernández, J., León, G., Pérez, J., Quintana-Ortí, E.S.: Unveiling the performance-energy trade-off in iterative linear system solvers for multithreaded processors. Concurr. Comput. Pract. Exper. **27**(4), 895–904 (2015)
3. Asanovic, K., et al.: The landscape of parallel computing research: a view from berkeley. Technical report UCB/EECS-2006-183, EECS Department, University of California, Berkeley, December 2006
4. Bergner, P., et al.: Performance Optimization and Tuning Techniques for IBM Power Systems Processors Including IBM POWER8. IBM (2015). IBM Reed Books
5. Buono, D., et al.: Optimizing sparse linear algebra for large-scale graph analytics. Computer **48**(8), 26–34 (2015)
6. Byun, J.-H., Lin, R., Yelick, K.A., Demmel, J.: Autotuning sparse matrix-vector multiplication for multicore. Technical report UCB/EECS-2012-215, EECS Dept., Univ. California, Berkeley (2012)
7. Choi, J.W., Singh, A., Vuduc, R.W.: Model-driven autotuning of sparse matrix-vector multiply on GPUs. In Proceedings of the 15th ACM SIGPLAN Symposium on Principles and Practice of Parallel Programming, PPopp 2010, pp. 115–126 (2010)
8. Dongarra, J., Heroux, M.A.: Toward a new metric for ranking high performance computing systems. Sandia report SAND2013-4744, Sandia National Laboratories, June 2013
9. Duranton, M., De Bosschere, K., Cohen, A., Maebe, J., Munk, H.: HiPEAC vision 2015 (2015). https://www.hipeac.org/publications/vision/
10. Guo, P., Wang, L., Chen, P.: A performance modeling and optimization analysis tool for sparse matrix-vector multiplication on GPUs. IEEE Trans. Parallel Distrib. Syst. **25**(5), 1112–1123 (2013)
11. Hager, G., Treibig, J., Habich, J., Wellein, G.: Exploring performance and power properties of modern multi-core chips via simple machine models. Concurr. Comput. Pract. Exper. **28**(2), 189–210 (2016)
12. Hennessy, J.L., Patterson, D.A.: Computer Architecture: A Quantitative Approach, 5th edn. Morgan Kaufmann, Waltham (2012)
13. Karakasis, V., Goumas, G., Koziris, N.: Exploring the performance-energy tradeoffs in sparse matrix-vector multiplication. In: Workshop on Emerging Supercomputing Technologies (WEST) - ICS 2011 (2011)
14. Kepner, J., Gilbert, J. (eds.): Graph Algorithms in the Language of Linear Algebra. SIAM, Philadelphia (2011)
15. Lefurgy, C., Wang, X., Ware, M.: Server-level power control. In: Proceedings of the 4th IEEE Conference on Autonomic Computing (ICAC 2007), Jacksonville, Florida, USA, 11–15 June, 2007
16. Liu, X., Smelyanskiy, M., Chow, E., Dubey, P.: Efficient sparse matrix-vector multiplication on x86-based many-core processors. In: Proceedings of the 27th International Conference on Supercomputing, Eugene, Oregon, USA, pp. 273–282, June 2013
17. Lucas, R.: Top ten Exascale research challenges (2014). http://science.energy.gov/~/media/ascr/ascac/pdf/meetings/20140210/Top10reportFEB14.pdf
18. Malkowski, K.: Co-adapting Scientific Applications and Architectures Toward Energy-efficient High Performance Computing. Ph.D. thesis, University Park, PA, USA (2008) AI3346339

19. Malossi, A.C.I., Ineichen, Y., Bekas, C., Curioni, A., Quintana-Ortí, E.S.: Performance and energy-aware characterization of the sparse matrix-vector multiplication on multithreaded architectures. In Proceedings of 43rd International Conference on Parallel Processing (ICCP), Minneapolis (MN), USA, pp. 139–148 (2014)
20. Malossi, A.C.I., Ineichen, Y., Bekas, C., Curioni, A., Quintana-Ortí, E.S.: Systematic derivation of time and power models for linear algebra kernels on multicore architectures. Sustainable Comput. Inf. Syst. **7**, 24–40 (2016)
21. Saad, Y.: Iterative Methods for Sparse Linear Systems. SIAM, Philadelphia (2003)
22. Vuduc, R.: Automatic performance tuning of sparse matrix kernels. Ph.D. dissertation, Univ. California, Berkeley, January 2004
23. Williams, S., et al.: Optimization of sparse matrix-vector multiplication on emerging multicore platforms. Parallel Comput. **35**(3), 178–194 (2009)

Power Consumption Modeling and Prediction in a Hybrid CPU-GPU-MIC Supercomputer

Alina Sîrbu[1,2(✉)] and Ozalp Babaoglu[2]

[1] Department of Computer Science, University of Pisa, Pisa, Italy
alina.sirbu@unipi.it
[2] Department of Computer Science and Engineering,
University of Bologna, Bologna, Italy

Abstract. Power consumption is a major obstacle for High Performance Computing (HPC) systems in their quest towards the holy grail of ExaFLOP performance. Significant advances in power efficiency have to be made before this goal can be attained and accurate modeling is an essential step towards power efficiency by optimizing system operating parameters to match dynamic energy needs. In this paper we present a study of power consumption by jobs in Eurora, a hybrid CPU-GPU-MIC system installed at the largest Italian data center. Using data from a dedicated monitoring framework, we build a data-driven model of power consumption for each user in the system and use it to predict the power requirements of future jobs. We are able to achieve good prediction results for over 80 % of the users in the system. For the remaining users, we identify possible reasons why prediction performance is not as good. Possible applications for our predictive modeling results include scheduling optimization, power-aware billing and system-scale power modeling. All the scripts used for the study have been made available on GitHub.

Keywords: Job power modeling · Job power prediction · High performance computing · Hybrid system · Support vector regression

1 Introduction

A major impediment for supercomputers from reaching the ExaFLOP target is power consumption. Energy efficiency of computing systems has to increase by at least one order of magnitude to achieve this goal [1]. This requires power optimization at all levels of hardware and software, including computation, networking and cooling. Numerous power modeling studies have been conducted in recent years towards these goals. Models can enable prediction of power usage under different scenarios, and indicate operating modes that optimize energy needs. Optimization can be obtained not only at low levels, e.g. through frequency and voltage scaling present in most modern CPUs, but also at higher levels, e.g. through power-aware scheduling, which has not been extensively studied.

© Springer International Publishing Switzerland 2016
P.-F. Dutot and D. Trystram (Eds.): Euro-Par 2016, LNCS 9833, pp. 117–130, 2016.
DOI: 10.1007/978-3-319-43659-3_9

In this paper, we model power needs of jobs in a hybrid CPU-GPU-MIC system (Eurora) with the aim of predicting power consumption of future jobs before they are started. Among other benefits, accurate prediction could enable development of advanced power-aware schedulers that can optimize power for the same workload. Eurora is a prototype supercomputer that topped the Green500 list in July 2013 for energy efficiency. It includes an advanced monitoring framework that collects status data in an open-access database [1], which we use to extract important features that enable prediction of job power consumption. The prediction problem is formulated as a regression task: given feature values (independent variables), compute the power consumption for a job (dependent variable). We divide this task into subproblems corresponding to each component type (CPU, GPU, MIC), and use Support Vector Regression (SVR) [16] for each individual problem. The total job power is then obtained as the sum of the individual component power consumptions.

This paper makes several contributions to power research for HPC systems. First, it identifies several features relevant to job power consumption in hybrid systems, supported by data examples. These include features not previously considered when modeling power such as application names, same-node resources used by other jobs and job running times. We model power exclusively through job-related features and do not require knowledge of CPU frequencies, load or application code structure to extract the sequence of executed operations. Second, we build power consumption models *for each user* starting from historical data and employing SVR. Models are shown to have high predictive power for most users. Third, we perform an analysis of power consumption variability for the system to provide context for the model error levels, and explain model limitations. Finally, we outline a methodology to implement the prediction framework in real time and discuss application scenarios. We used the Google BigQuery data analytics service [17] for the initial data analysis phase, while model training was done using the *scikit-learn* python package [9]. We have made all of the scripts used for our study available on GitHub [6].

2   The Eurora System and its Data

Eurora [5] is a prototype HPC system hosted at CINECA (www.cineca.it) that combines the use of CPUs, GPUs and MICs to achieve higher power efficiency. It remained in production for over 2 years from 2013 to 2015. The system consists of 64 nodes, each hosting two 8-core Intel Xeon E5 CPUs and two expansion cards that can contain either GPU or MIC accelerator modules. There are 3 different classes of CPUs based on their maximum frequencies: 2.1 GHz (the *slow* class, denoted as S and present at 24 nodes), 2.2 GHz (the *medium* class, denoted as M and present at 8 nodes) and 3.1 GHz (the *fast* class, denoted as F and present at 32 nodes). Half of the nodes mount GPUs (Nvidia Tesla Kepler) while the other half mount Intel "Knights Corner" MIC (Xeon Phi). All nodes run CentOS Linux. The workload is handled through the Portable Batch System (PBS).

Eurora contains an extensive monitoring subsystem which collects high reso-lution (5-second intervals) status data from system components, including power and cooling infrastructures [1]. Log data for the period 31 March 2014 to 11 August 2015 is available (250 GB of data in 328 tables), with several gaps due to system/monitoring errors or database migration operations. The work reported in this paper is based on these data for computing power consumption per job and building a prediction framework for estimating future job power. We lim-ited our study to the data from 2014 since the system underwent several changes in 2015 and became more unstable. *Workload information* provided the num-ber of resources used by each job on each node at 5-min resolution. Although this number is known, it is impossible to extract from the data exactly which CPU/GPU/MIC is being used out of the two available on each node. *Power logs* allowed us to compute power consumption for each component (CPUs, GPUs and MICs) on each node, again every 5 min. Power data is known only at the level of CPU, GPU and MICs but is not available for the cores.

Combining workload and power data, we computed at 5-min intervals the overall power usage of a job j as the sum of the power of each component type:

$$P^j = P_S^j + P_M^j + P_F^j + P_{GPU}^j + P_{MIC}^j \tag{1}$$

Power for each component type is computed by summing over all used nodes. For example, for the F CPU type, $P_F^j = \sum_{i \in \text{nodes}} P_F^j(i)$ where $P_F^j(i)$ is the power used by job j at a fast CPU on node i. If the job does not use node i or if CPU type F is not present at node i, the corresponding power is assumed to be 0. Denoting the number of cores used by job j on node i as $n_j(i)$ and that used by other jobs as $n_{\text{other}}(i)$, the number of free cores at node i is given by $n_{\text{idle}}(i) = 16 - n_j(i) - n_{\text{other}}(i)$ (recall that each node contains two 8-core CPUs). Let $P_F(i)$ denote the total power recorded for the fast CPU type at node i. Then, the power used by job j at a fast CPU on node i is computed based on the number of cores used by job j in relation to the total number of cores at the node, the number of cores used by other jobs and the total power recorded at the node for that CPU type:

$$P_F^j(i) = n_j(i) \frac{P_F(i) - \hat{P}_F \times n_{\text{idle}}(i)}{n_j(i) + n_{\text{other}}(i)} \tag{2}$$

where $\hat{P}_F$ denotes the average power consumed by a single F type CPU core when it is idle. We estimate this quantity from the log data by dividing the total power consumed by an idle CPU of type F by the number of cores (which is 8). The same procedure is repeated for the remaining types M, S, GPUs and MICs. This procedure may introduce some noise in calculating job power, since it assumes that when two jobs share the same node, the power usage is evenly distributed across used components (e.g., cores). It is highly unlikely that this assumption holds since jobs have different power needs, yet it is necessary in order to be able to use the entire job set in our study, since many jobs indeed share nodes with other jobs.

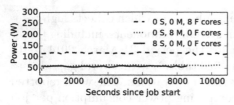

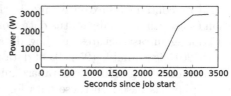

(a) Three jobs with same name and same number of cores, but allocated on different classes of CPUs.

(b) Power consumption of a single job throughout its execution.

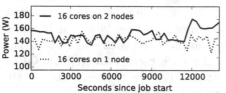

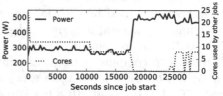

(c) Two jobs with same name and same number of cores, but allocated on different number of nodes.

(d) One job with variable number of cores in use by other users on the same nodes. The job itself uses 16 cores on 4 nodes.

Fig. 1. Power consumption for various jobs.

3 Power Model

3.1 Features

Power consumption of a job can depend on several factors. One is the number of components of each type used by the job. A job using 16 cores will most likely use more power than a job using only 8 cores. A related factor is the type of core being used, with faster class cores using more power than slower class cores (e.g. Fig. 1a). The structure of the application is also important. A job can have periods of high power usage and other periods of lower power usage, although the number of components in use remains the same (e.g. Fig. 1b). These patterns can be captured by including the runtime as a feature (i.e., the time since the job started). We also use the job name, a user-defined string, to identify the application. We performed a textual analysis, using the *CountVectorizer* class in the *scikit-learn* python package [9], checking which n-grams of 2 and 3 letters are present in the job names. This resulted in a set of numerical features that count how many times each n-gram appears.

The *distribution* of resources is also important. Figure 1c shows two jobs running the same application using 16 cores, however one job is allocated one node while the other two nodes, causing differences in power. A related factor is the load of a node that is partially used by a job. Figure 1d shows an example job using 16 cores on 4 nodes, together with the number of cores used by other jobs on the same nodes at the same time. We see a negative correlation between power and cores used by other jobs, so we include the same-node used cores as features.

To summarize, for each job we extracted the number of components of each type used (S, M, F, GPU, MIC), runtime, name (occurrence of n-grams of size 2 and 3), number of nodes and same-node components used by other jobs as *regression features*, and the power as *regression target*. These were computed at 5-min intervals, resulting in numerous data points per job. Google BigQuery [17] was used, enabling analysis of large amounts of data in a reasonable amount of time.

3.2   Regression Problem and Training Procedure

Since power is measured for each component type, we divided the problem of predicting power per job into 5 subproblems, corresponding to the terms on the right side of Eq. 1. Hence, for each user we perform 5 regression analyses, one for each component type, and then sum the predicted component powers to obtain an estimate for the *global job power*. In practice, most users use only 2 or 3 component types, so regression is only performed for those. We use SVR with Radial Basis Function (RBF) kernels [16]. To simulate the realistic scenario where power prediction is based on user histories, we train new models on a monthly basis using all past data, and then apply the models to new data for the current month. Here we show results for October 2014: models are trained on data prior to October 1st and then applied to all data recorded in October.

Training consists of two steps. First, SVR meta-parameters have to be optimized for each user and each component type. We use cross-validation to find optimal values. That is, all past data is divided into a "train" subset and a "test" subset. We use the first 80 % of the jobs of each user as training data, and the last 20 % as test data. Then multiple models are trained with a range of parameter values and the combination that produces the best results on the test data is selected. Second, a new *final model* is trained with the optimal parameter combination, using all past data in training (merging the train and test dataset). This ensures that all available past information is included in the model.

Table 1. Number of users analyzed for each component type and globally.

Component type	S	M	F	GPU	MIC	Global
Users	21	20	27	9	2	34

Once the final model is available for each user and component type, it can be applied for one month to new unseen data. Power is predicted for the individual components and then summed to obtain *global job power*. To avoid poor prediction due to limited training, we only analyzed those users for which historical data included at least 1,000 total data points, at least 10 jobs totaling at least 100 time points for each component type, and at least 10 data points to apply the model to. Table 1 shows the number of users analyzed for each component type and globally. For all users, a total of 435,079 data points from 22,130 unique jobs were used to build the model (data before October), which was then applied to 53,717 new points from 5,039 unique jobs (October data).

3.3   Evaluation

We will compare results of our multiple-SVR model with a simple *Enhanced Average Model* (EAM). For each user, the EAM computes the average power used per component unit (core, GPU, MIC), based on historic data. Then job power for each component is predicted by multiplying the number of component units by the average power per unit. For instance, if a job j of user u uses n_F^j F cores and the historical usage for one F core for the user is $\bar{P}_F^u$, then $P_F^j = n_F^j \times \bar{P}_F^u$. All other terms in Eq. 1 are computed in a similar fashion and then summed to obtain total job power. This model is an enhanced version of the so called "average model", which would compute job power just by averaging historical data, without taking into account the number of components used. Even so, it is much simpler than our multiple-SVR approach. Training is straightforward as it only requires computing averages per user, while application of the model requires knowing only the number of components used.

The models were evaluated using two standard criteria for regression: the (mean-)normalized-root-mean-squared-error (NRMSE) and R-squared (R^2):

$$\text{NRMSE} = \frac{\sqrt{(\sum_{i=1}^{N}(P_i - P_i^*)^2)/N}}{\bar{P}} \tag{3}$$

$$R^2 = 1 - \frac{\sum_{i=1}^{N}(P_i - P_i^*)^2}{\sum_{i=1}^{N}(P_i - \bar{P})^2} \tag{4}$$

where N is the number of data points considered across the jobs of the user, P_i^* and P_i are the predicted and real powers for data point i, respectively, while $\bar{P}$ is the average of the real power over all N data points.

To provide context for the errors reported, it is important to understand the natural fluctuations of power consumption at constant load — the noise levels. Power usage can vary for the same workload on the same node, due to *hardware-related noise*, such as variations in the production process which may generate different electrical behaviors across same-type cores, or adaptive mechanisms for performance optimization [12]. Additionally, there is *software-related noise*, introduced by operating system interference, external interrupts or shared resource contention [8]. Noise has a negative impact on power model performance since random fluctuations are not captured by model features, hence cannot be reproduced through regression. Thus, we cannot expect model errors to be less than the noise levels. This has been shown to *affect performance of models* by reducing the maximum accuracy they can obtain [12].

In the case of Eurora, undesired software and hardware variability *between nodes* was previously shown to reach up to 20 % (5 % software and up to 15 % hardware) [8]. Here, we look at *within-node variability* at constant load for CPUs and GPUs. Specifically, we computed the coefficient of variation (CV) of power

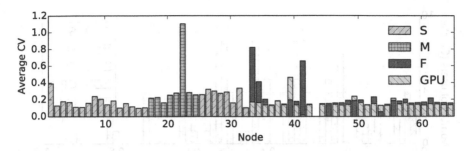

Fig. 2. Variability of power consumption. The bars represent the coefficient of variation (CV) of power at fixed load, averaged over all loads, for each node and component type in the system. (Color figure online)

at various load levels, and averaged over all loads for each component. For MICs, load information is not available so we could only analyze power at 0 load. The variability observed may come from different sources, however we evaluate them together since we are interested only in an overall value to be used as a baseline for quantifying our errors. Figure 2 shows average CV values for all 64 nodes, per component type. The M CPUs show on average largest fluctuations, with most nodes reaching over 20 %. Most S, F and GPU components have average fluctuations under 20 %. However, some nodes in all categories display much larger fluctuations, even over 100 %. For MICs (not shown in the figure since load data is unavailable), idle power fluctuates on average by 10.24 % and we expect this value to be larger at larger loads. Hence, based on our data analysis and based on previous studies, in this work we consider NRMSE values < 0.2 (20 %) to be good performance, since they are within the natural fluctuations of the individual components.

We included here both the R^2 and NRMSE evaluation criteria because they are complementary: the NRMSE looks at overall fit and gives a measure relative to the mean value, while R^2 looks at the general shape of the time series and gives a measure relative to the variations in the data. Additionally, they are affected differently by noise. For instance, if the power levels for a user are relatively flat, and vary only due to noise, the R^2 measure becomes irrelevant. This because R^2 looks at the 'shape' of the data, which in this case is entirely determined by local fluctuations, which cannot be reproduced by any model. However, a model can still capture average behavior which is the best performance possible, but which will correspond to low R^2. In this case the NRMSE provides additional information, with a NRMSE value similar to the noise level considered a good performance. Conversely, when a user has highly variable power consumption for jobs, NRMSE can be large due to a few data points, but the model can still contain useful information, reflected in the R^2 measure. In the following we will consider NRMSE > 0.2 or $R^2 > 0.5$ to be a very good result.

In our data, the distribution of job power for each user is very heterogeneous, with users ranging from those having jobs with stable power requirements to those showing very large differences across jobs and time, justifying the use of

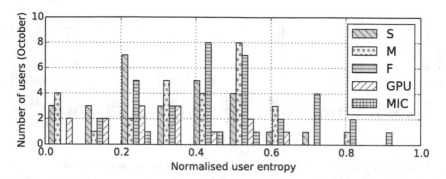

Fig. 3. Distribution of variability of power across jobs and time for each user. Variability for each user is computed as the normalized entropy of the distribution of job power levels recorded in time for that user. The plot shows a histogram of all entropies for the users active in the month of October 2014. (Color figure online)

both NRMSE and R^2. Figure 3 shows the distribution of variability in job power for all users active during the month of October 2014, for each component type. To quantify the variability for each user, we obtained the distribution of the power levels for that user, by collecting the data in bins of size 20 W in the range [0 W, 500 W], and computed the entropy of this distribution, normalized by the maximum entropy possible (logarithm of the number of bins). The normalized entropy is a measure of spread of the distribution, with 0 meaning all data fell into one bin and 1 meaning power levels are uniformly distributed across all bins. So, a user with 0 entropy has very flat power levels, while a user with high entropy has very large differences between power levels. As Fig. 3 shows, our data contains users with a wide range of entropies, hence we are dealing with a very heterogeneous user population.

4 Model Performance

Once meta-parameters are explored using cross-validation with data prior to October 1st, 2014, the best meta-parameter combination is selected and a new final model is trained on all data prior to October. One SVR model is obtained for each user and each component type, which are then combined into a global model for each user. Table 2 shows prediction performance for all users (all jobs concatenated), for each component type, throughout the month of October. For components S and GPU, both R^2 and NRMSE values are very good. For F, NRMSE is quite high, however R^2 is also very large, so the model contains useful information. The high NRMSE is due to the fact that one user has jobs that consume much more power than others (over 3 KW versus under 500 W for others), so a relatively small error in that user will produce a large overall NRMSE (due to the fact that the normalizing factor depends on all jobs of all users). If we remove the user with jobs consuming over 3KW, then we obtain $R^2 = 0.89$ and $NRMSE = 0.26$ which are very good considering the noise levels for the F CPUs shown in Fig. 2. For M CPUs, which showed highest noise in Fig. 2, performance is somewhat lower. R^2 does not reach the 0.5 threshold, albeit very close, while NRMSE is around 33 %. This shows how power fluctuations can affect model performance. Even so, the model is better than the average model (R^2 much larger than 0). For the MIC component, the amount of data is more reduced, which can be one reason for the lower performance. Only two MIC users exist, one with very good and another with lower prediction performance.

Table 2. Performance of the SVR and EAM for individual components.

	S	M	F	GPU	MIC
SVR NRMSE	0.13	0.33	0.52	0.15	0.28
SVR R^2	0.87	0.47	0.92	0.84	0.34
EAM NRMSE	0.13	0.37	1.34	0.24	0.28
EAM R^2	0.87	0.34	0.52	0.59	0.31

If we compare the SVR models with the EAM, for which results are also shown in Table 2, we note that the SVR has better performance on all component types except for S. For the S class, the SVR and EAM are comparable, meaning that jobs using this component are quite predictable and power depends mostly on the number of components used. Significantly better performance of the SVR can be seen for the F and GPU components, which are also the most used across the cluster. This increase in performance means jobs are much more complex and additional SVR features are important in predicting the power outcome.

While Table 2 shows how the model behaves on the individual components, it is total job power (global model) that interests us the most. Figure 4 shows the power time series (predicted and real) for the total job power (i.e., after applying Eq. 1), using the SVR model. Given the presence of that one user with very high job power, NRMSE is again large, however R^2 is very good. Again, by removing this user, NRMSE reduces to 21 %, meaning our model has an overall accuracy of 79 % for all other jobs of October, while R^2 stays high at 0.87. Compared to the EAM (global NRMSE = 0.91 and R^2 = 0.53), NRMSE of the SVR is 40 % that of the EAM, while R^2 is improved by 70 %.

Model performance varies also from user to user. Figure 5 plots global model NRMSE versus R^2 for each user, for both the SVR and EAM. In general, the SVR outperforms the EAM (in the plot, stars are located south-east of the corresponding circle), however there are a few users for which the EAM is better. For these, one is better off using the EAM for predictions. Out of a total of 34 users analyzed, 27 have SVR NRMSE ≤ 0.2 or $R^2 \geq 0.5$, and 7 (20 %) have lower performance. For the latter, a weak SVR model corresponds also to a weak EAM model. Poor performance could be due to noise, indicated by the fact that jobs of these users use partial node resources (i.e., 1 out of 2 MICs or 1 out of 16 cores) or run on nodes with high variability, being thus more prone to noise.

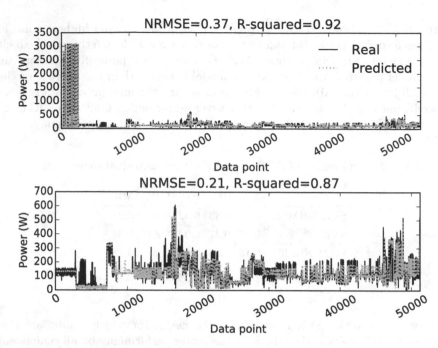

Fig. 4. Global real and predicted total power consumption (components summed together). For each job, the power was computed at 5-min intervals, with the plot showing all power values for all users and jobs. The top panel shows all users, while in the lower panel the first user with high power values was eliminated.

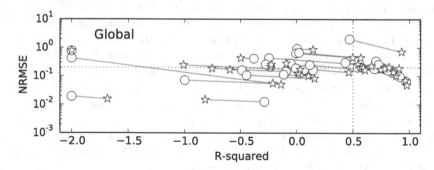

Fig. 5. Global model performance per user, NRMSE vs R^2. Circles show performance for the EAM, while stars show performance for the multiple-SVR model. Each circle/star corresponds to one separate user. For each user, the two model types (EAM and SVR) are connected by an edge. For figure readability, data points with very low negative R^2 values have been mapped to -2.0.

5   Related Work

Power monitoring, modeling and optimization have been major research concerns in recent years. Modern computing units embed advanced control mechanisms such as the Dynamic Frequency and Voltage Scaling that aim to optimize performance and can affect power levels, making modeling problematic even for a single computational unit [12]. Several models trying to explain the relation between frequency, load, hardware counters and power for single units have been introduced for multicore CPUs [7,14] and GPUs [18]. Model performance ranges widely depending on the applications running, with errors between 3.65 and 14.4 % for the CPU case, and between 1.7 % and 27.7 % for GPUs. These errors are only expected to grow when multiple units have to be combined, as is the case for HPC systems. Our approach is very different in that we are modeling power consumption per job, not per component. Additionally, our model does not require direct measures of load and frequency, which are typically not known in advance, but only workload measures which are known when the job starts.

Some work in modeling job or application power consumption has appeared recently. Performance counters are used to model application power on three small scale HPC platforms by [13]. GPU CUDA kernels are analyzed in [11], again based on job performance counters. These methods are very different from ours, since they require instrumenting the applications to extract signatures and performance counters, while we only use the number of resources required, making it much more straightforward to apply. An approach more similar to ours was recently introduced in [15] which uses the number of nodes used by applications as model input, with very good precision (errors between 0 and 5.2 % per application). A more detailed model is introduced in [4] where a Hidden Markov Model is used to represent job states and transitions. All these methods build one model per job, while we are trying to explore user patterns as well. Building one model per user has several advantages including larger training datasets and greater robustness to inaccurate use of job names by HPC users (e.g., when the user gives the same job name to different applications, or various names to the same application). Additionally, unlike other methods, our model applies to hybrid jobs using CPUs, GPUs and MICs.

On the road towards ExaFLOP performance, special attention has been given to system-level power consumption by clusters. Recent work at Google [10] describes the use of Artificial Neural Networks to model Power Usage Effectiveness using a mixture of workload and cooling features. System-level prediction of power consumption is also one application of our predictive model. In terms of power-aware scheduling, another possible application of our models, the authors in [2,3] introduce a method based on Constraint Programming, to achieve power capping on Eurora, the same HPC system analyzed here. This could benefit greatly from power prediction offered by our framework.

6   Discussion and Conclusions

We presented an analysis of *historical trace data* from Eurora, and evaluated prediction models for power consumption of jobs. The method is fully data driven — no assumptions about the model structure nor additional instrumentation of application code are required. The only application-aware feature is the job name, making our method easily applicable to any system even when application code is not available. The power of our prediction derives from user history rather than from application counters, and our results show that when enough data is available, high performance can be achieved. We employ a multiple-SVR model to estimate job power in time. One model per user is trained. An alternative would have been to build one model per application (job name) but this would have meant much less training data per model. Additionally, learning from user profiles can allow for user trends to be captured, maintaining high quality predictions even if job names are not properly employed by users (e.g., using the same name for different applications or many different names for the same application).

The multiple-SVR approach is compared to an enhanced average model (EAM) where power depends only on the number of components used. The SVR outperforms the EAM approach for most users, obtaining good prediction (error under 20 % or $R^2 \leq 0.5$) for 80 % of the users analyzed. For the rest of the users, indications are that performance is affected by noise.

The approach is intended to be used in *real time*, where predictions are made as new jobs arrive at the scheduler. Online application consists of training the model for each user, then applying it to real time data, by employing the procedure outlined in this work. Periodically, the model is updated by incorporating recent data into the training dataset. We expect monthly model updates to be sufficient in order to capture changes in job structure. available, prediction can be improved by training the multiple-SVR model.

In terms of resources, our analysis was performed on a 516-node CentOS 7.0 cluster, with 2 octa-core 2.40 GHz Intel Xeon CPUs per node. Since our problem is intrinsically parallel, we obtained each model separately on one core. Running times depended on the user (different amounts of data available) and on the meta-parameters. *Meta-parameter optimization* required a total of 185.36 core-hours for all users, with a maximum running time for one optimization run of 4.66 h. *Global model training* for all users required at total of 2.92 core-hours (maximum for one user was 1.7 h), while *model application* to all the data of October took only 6.81 min for all users. Consequently, if parallelized on a multi-core platform, the entire process incurs little overhead, especially given that the training procedure has to be repeated only once a month. We expect the method to scale to systems that are much larger than Eurora, since the analysis is performed separately for each user and can be easily parallelized.

The predictions presented here can be improved through more detailed data on job characteristics (e.g., exact application names, input datasets and parameters) and more detailed power monitoring (e.g., power per core rather than per CPU), work which we will undertake in the future after obtaining improved datasets. Furthermore, we plan to use our predictions in several applications to optimize system functionality. The first is modeling and prediction of system level power consumption, including networking equipment, IO systems and even cooling infrastructure, starting from prediction of job power. Secondly, our approach is applicable to power-aware scheduling, where the scheduler can estimate power usage for various job allocation schemes and select the best among them. Thirdly, our method can be employed by users to estimate power for their jobs before submission, which can facilitate better management of resources by the users, especially in the context of power-aware billing.

Acknowledgments. BigQuery analysis was carried out through a generous Cloud Credits grant from Google. We are grateful to Prof. L. Benini and Dr. A. Bartolini for useful discussions regarding the data and to the HPC group at CINECA, in particular Dr. E. Rossi and Dr. C. Cavazzoni for providing access to the CINECA systems. We acknowledge the CINECA ISCRA PACNA and PM-HPC awards allowing access to HPC resources and support. This work was partially funded by the European project SoBigData Research Infrastructure — Big Data and Social Mining Ecosystem under the INFRAIA-H2020 program (grant agreement 654024).

References

1. Bartolini, A., et al.: Unveiling eurora-thermal and power characterization of the most energy-efficient supercomputer in the world. In: DATE 2014 (2014)
2. Borghesi, A., et al.: MS3: a Mediterranean-stile job scheduler for supercomputers-do less when it's too hot! In: HPCS 2015, pp. 88–95 (2015)
3. Borghesi, A., Collina, F., Lombardi, M., Milano, M., Benini, L.: Power capping in high performance computing systems. In: Pesant, G. (ed.) CP 2015. LNCS, vol. 9255, pp. 524–540. Springer, Heidelberg (2015)
4. C. Storlie, C., et al.: Modeling and predicting power consumption of high performance computing jobs. arXiv preprint arXiv:14125247 (2014)
5. Cavazzoni, C.: Eurora: a european architecture toward exascale. In: Future HPC Systems: the Challenges of Power-Constrained Performance. ACM (2012)
6. Sîrbu, A., Babaoglu, O.: BigQuery and Python scripts. Github (2016). http://github.com/alinasirbu/eurora_job_power_prediction
7. Dargie, W.: A stochastic model for estimating the power consumption of a processor. IEEE Trans. Comput. **64**(5), 1311–1322 (2015)
8. Fraternali, F., et al.: Quantifying the impact of variability on the energy efficiency for a next-generation ultra-green supercomputer. In: ISLPED 2014, pp. 295–298 (2014)
9. Pedregosa, F., et al.: Scikit-learn: Machine learning in Python. J. Mach. Learn. Res. **12**, 2825–2830 (2011)
10. Gao, J.: Machine learning applications for data center optimisation. Google White Paper (2014)
11. Nagasaka, H., et al.: Statistical power modeling of GPU kernels using performance counters. In: IGCC 2010, pp. 115–122 (2010)
12. McCullough, J.C., et al.: Evaluating the effectiveness of model-based power characterization. In: USENIX ATC 2011, vol. 20 (2011)
13. Witkowski, M., et al.: Practical power consumption estimation for real life HPC applications. Future Gener. Comput. Syst. **29**(1), 208–217 (2013)
14. Gschwandtner, P., et al.: Modeling CPU energy consumption of HPC applications on the IBM Power7. In: PDP 2014, pp. 536–543 (2014)
15. Shoukourian, H., Wilde, T.: Predicting the energy and power consumption of strong and weak scaling HPC applications. Supercomp Front Innov. **1**(2), 20–41 (2014)

16. Smola, A., Vapnik, V.: Support vector regression machines. Adv. Neural Inf. Process. Syst. **9**, 155–161 (1997)
17. Tigani, J., Naidu, S.: Google BigQuery Analytics. Wiley, Hoboken (2014)
18. Ma, X., et al.: Statistical power consumption analysis and modeling for GPU-based computing. In: ACM SOSP HotPower 2009 (2009)

Scheduling and Load Balancing

Controlling and Assessing Correlations of Cost Matrices in Heterogeneous Scheduling

Louis-Claude Canon[✉], Pierre-Cyrille Héam, and Laurent Philippe

FEMTO-ST Institute/CNRS – Université de Franche-Comté/UBFC,
25000 Besançon, France
{louis-claude.canon,pierre-cyrille.heam,laurent.philippe}@univ-fcomte.fr

Abstract. This paper considers the problem of allocating independent tasks to unrelated machines such as to minimize the maximum completion time. Testing heuristics for this problem requires the generation of cost matrices that specify the execution time of each task on each machine. Numerous studies showed that the task and machine heterogeneities belong to the properties impacting heuristics performance the most. This study focuses on orthogonal properties, the average correlations between each pair of rows and each pair of columns, which is a proximity measure with uniform instances (Uniform instances are particular unrelated instances in which each execution time is proportional to the weight of the task and the cycle time of the machine.). Cost matrices generated with a novel generation method show the effect of these correlations on the performance of several heuristics from the literature. In particular, EFT performance depends on whether the tasks are more correlated than the machines and HLPT performs the best when both correlations are close to one.

1 Introduction

The problem of scheduling tasks on processors is central in parallel computing science because it supports parts of the grid, computing centers and cloud systems. Considering static scheduling, the problem is deterministic, although complex, because all the data are known a priori. In the case of independent tasks running on a heterogeneous platform and with the objective of minimizing the total execution time [13,14], the performance[1] of any scheduling algorithm depends on the properties of the input cost matrix and generating input instances is thus a crucial problem in algorithm assessment [5,7]. In a previous study [8], we have proposed heterogeneity measures and procedures to control this property when generating cost matrices. In particular, we showed that the heterogeneity was previously not properly controlled despite having a significant impact on the relative performance of scheduling heuristics. However, the proposed measures prevent tuning how the machines are related to one another in

---

[1] The performance of any algorithm for this NP-Hard problem is given by the difference between the obtained total execution time and the minimum one.

© Springer International Publishing Switzerland 2016
P.-F. Dutot and D. Trystram (Eds.): Euro-Par 2016, LNCS 9833, pp. 133–145, 2016.
DOI: 10.1007/978-3-319-43659-3_10

terms of processing time, i.e., if the execution times are proportional and depend on a task weight and a machine cycle time.

In this paper, we propose to investigate a continuum of instances between the uniform case and the unrelated case. The contribution[2] is a measure, the correlation, to explore this continuum, its analysis in existing generation methods and existing studies (Sect. 3), a new generation method with better correlation properties (Sect. 4) and its analysis on several heuristics (Sect. 5) and, last, the confrontation of the correlation to a related measure (Sect. 6).

2   Related Work

The validation of scheduling heuristics in the literature relies mainly on two generation methods: the range-based and CVB methods. The range-based method [4,5] generates n vectors of m values that follow a uniform distribution in the range $[1, R_{\mathrm{mach}}]$ where n is the number of tasks and m the number of machines. Each row is then multiplied by a random value that follows a uniform distribution in the range $[1, R_{\mathrm{task}}]$. The CVB method is based on the same principle except it uses more generic parameters and a distinct underlying distribution. In particular, the parameters consist of two CV[3] (V_{task} for the task heterogeneity and V_{mach} for the machine heterogeneity) and one expected value (μ_{task} for the tasks). The parameters of the gamma distribution used to generate random values are derived from the provided parameters. An extension has been proposed to control the consistency of any generated matrix:[4] the rows in a submatrix containing a fraction a of the initial rows and a fraction b of the initial columns are sorted.

The shuffling and noise-based methods were later proposed in [7,8]. They both start with an initial cost matrix that is equivalent to a uniform instance (any cost is the product of a task weight and a machine cycle time). The former method randomly alters the costs without changing the sum of the costs on each row and column. This step introduces some randomness in the instance, which distinguishes it from a uniform one. The latter relies on a similar principle: it inserts noise in each cost by multiplying it by a random variable with mean one. Both methods require the parameters V_{task} and V_{mach} to set the task and machine heterogeneity. In addition, the amount of noise introduced in the noise-based method can be adjusted through the parameter V_{noise}.

This study focuses on the average correlation between each pair of tasks or machines in a cost matrix. No existing work explicitly considers this property. The closest work is the consistency extension in the range-based and CVB methods mentioned above. The consistency extension could be used to generate cost matrices that are close to uniform instances because cost matrices corresponding to uniform instances are consistent. However, this mechanism modifies

---

[2] These results are also available in the companion research report [6].

[3] The Coefficient of Variation is the ratio of the standard deviation to the mean.

[4] In a consistent cost matrix, any task faster than another task on a given machine will be consistently faster than this other task on any machine.

the matrix row by row, which makes it asymmetric relatively to the rows and columns. This prevents its direct usage to control the correlation.

The TMA (Task-Machine Affinity) quantifies the specialization of a platform [1,2], i.e., whether some machines are particularly efficient for some specific tasks. This measure proceeds in three steps: first, it normalizes the cost matrix to make the measure independent from the matrix heterogeneity; second, it performs the singular value decomposition of the matrix; last, it computes the inverse of the ratio between the first singular value and the mean of all the other singular values. The normalization happens on the columns in [2] and on both the rows and columns in [1]. If there is no affinity between the tasks and the machines (as with uniform machines), the TMA is close to zero. Oppositely, if the machines are significantly specialized, the TMA is close to one. Additionally, Khemka et al. [12] claims that high (resp., low) TMA is associated with low (resp., high) column correlation. This association is however not general because the TMA and the correlation can both be close to zero. See Sect. 6 for a more thorough discussion on the TMA.

The range-based and CVB methods do not cover the entire range of possible values for the TMA [2]. Khemka et al. [12] propose a method that iteratively increases the TMA of an existing matrix while keeping the same MPH and TDH. A method that generates matrices with varying affinities (similar to the TMA) and which resembles the noise-based method is also proposed in [3]. However, no formal method has been proposed for generating matrices with a given TMA.

3   Correlation Between Tasks and Processors

As stated previously, the unrelated model is more general than the uniform model and all uniform instances are therefore unrelated instances. Let $U = (\{w_i\}_{1 \le i \le n}, \{b_j\}_{1 \le j \le m})$ be a uniform instance with n tasks and m machines where w_i is the weight of task i and b_j the cycle time of machine j. The corresponding unrelated instance is $E = \{e_{i,j}\}_{1 \le i \le n, 1 \le j \le m}$ such that $e_{i,j} = w_i b_j$ is the execution time of task i on machine j. Our objective is to generate unrelated instances that are as close as desired to uniform ones. On the one hand, all rows are perfectly correlated in a uniform instance and this is also true for the columns. On the other hand, there is no correlation in an instance generated with nm independent random values. Thus, we propose to use the correlation to measure the proximity of an unrelated instance to a uniform one.

Correlations Properties. Let $e_{i,j}$ be the execution time for task i on machine j. Then, we define the *task correlation* as follows:

$$\rho_{\text{task}} \triangleq \frac{1}{n(n-1)} \sum_{i=1}^{n} \sum_{i'=1, i' \ne i}^{n} \rho_{i,i'}^{r} \tag{1}$$

where $\rho_{i,i'}^{r}$ represents the correlation between row i and row i' as follows:

$$\rho^r_{i,i'} \triangleq \frac{\frac{1}{m}\sum_{j=1}^{m} e_{i,j}e_{i',j} - \frac{1}{m}\sum_{j=1}^{m} e_{i,j}\frac{1}{m}\sum_{j=1}^{m} e_{i',j}}{\sqrt{\frac{1}{m}\sum_{j=1}^{m} e_{i,j}^2 - \left(\frac{1}{m}\sum_{j=1}^{m} e_{i,j}\right)^2}\sqrt{\frac{1}{m}\sum_{j=1}^{m} e_{i',j}^2 - \left(\frac{1}{m}\sum_{j=1}^{m} e_{i',j}\right)^2}} \tag{2}$$

Note that any correlation between row i and itself is 1 and is hence ignored. Also, since the correlation is symmetric ($\rho^r_{i,i'} = \rho^r_{i',i}$), it is actually sufficient to only compute half of them. We define the *machine correlation*, ρ_{mach}, analogously on the columns. These correlations are the average correlations between each pair of distinct rows or columns. They are inspired by the classic Pearson definition, but adapted to the case when we deal with two vectors of costs.

There are three special cases when either one or both of these correlations are one or zero. When $\rho_{task} = \rho_{mach} = 1$, then instances may be uniform ones and the problem can be equivalent to $Q||C_{max}$ [6, Proposition 1]. When $\rho_{task} = 1$ and $\rho_{mach} = 0$, then a related problem is $Q|p_i = p|C_{max}$ where each machine may be represented by a cycle time and all tasks are identical [6, Proposition 2]. Finally, when $\rho_{mach} = 1$ and $\rho_{task} = 0$, then a related problem is $P||C_{max}$ where each task may be represented by a weight and all machines are identical [6, Proposition 3]. For any other cases, we do not have any relation to another existing problem that is more specific than scheduling unrelated instances.

Correlations of Existing Methods. Table 1 synthesises the analysis of the asymptotic correlation properties of the range-based, CVB and noise-based methods [6, Propositions 4 to 9].

Table 1. Summary of the asymptotic correlation properties of existing methods.

Method	ρ_{task}	ρ_{mach}
Range-based [4,5]	$a^2 b$	$\begin{cases} \frac{3}{7} & \text{if } a = 0 \\ b^2 + 2\sqrt{\frac{3}{7}}b(1-b) + \frac{3}{7}(1-b)^2 & \text{if } a = 1 \end{cases}$
CVB [4,5]	$a^2 b$	$\begin{cases} \frac{1}{V_{mach}^2(1+1/V_{task}^2)+1} & \text{if } a = 0 \\ b^2 + \frac{2b(1-b)}{\sqrt{V_{mach}^2(1+1/V_{task}^2)+1}} & \text{if } a = 1 \\ \quad + \frac{(1-b)^2}{V_{mach}^2(1+1/V_{task}^2)+1} \end{cases}$
Noise-based [8]	$\frac{1}{V_{noise}^2(1+1/V_{mach}^2)+1}$	$\frac{1}{V_{noise}^2(1+1/V_{task}^2)+1}$

Correlations in Previous Studies. More than 200 unique settings used for generating instances were collected from the literature and synthesized in [8]. For each of them, we computed the correlations using the formulas from Table 1. For the case when $0 < a < 1$, the correlations were measured on a single 1000×1000 cost matrix that was generated with the range-based or the CVB method as done in [8] (missing consistency values were replaced by 0 and the expected value was set to one for the CVB method).

Figure 1 depicts the values for the proposed correlation measures. The task correlation is larger than the machine correlation (i.e., $\rho_{task} > \rho_{mach}$) for only a few instances. The space of possible values for both correlations has thus been largely unexplored. Additionally, few instances have high task correlation and are thus underrepresented.

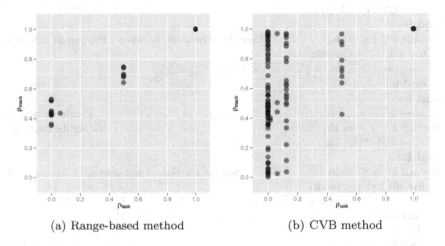

(a) Range-based method   (b) CVB method

Fig. 1. Correlation properties (ρ_{task} and ρ_{mach}) of cost matrices used in the literature.

Two matrices extracted from the SPEC benchmarks on five different machines are provided in [1]. There are 12 tasks in CINT2006Rate and 17 tasks in CFP2006Rate. The values for the correlation measures and other measures from the literature are given in Table 2. The correlations for these two benchmarks correspond to an area that is not well covered in Fig. 1. This illustrate the need for a better exploration of the correlation space when assessing scheduling algorithms.

Table 2. Summary of the properties for two benchmarks (CINT2006Rate and CFP2006Rate).

Benchmark	ρ_{task}	ρ_{mach}	$V\mu_{task}$	$V\mu_{mach}$	μV_{task}	μV_{mach}	TDH	MPH	TMA
CINT2006Rate	0.85	0.73	0.32	0.36	0.37	0.39	0.90	0.82	0.07
CFP2006Rate	0.60	0.67	0.42	0.32	0.48	0.39	0.91	0.83	0.13

4   Controlling the Correlation

Table 1 shows that the correlation properties of existing methods are determined by a combination of unrelated parameters, which is unsatisfactory. We propose

---

Algorithm 1. Combination-based cost matrix generation with gamma distribution

---

Input: n, m, r_{task}, r_{mach}, μ, V
Output: a $n \times m$ cost matrix

1: $V_{\text{col}} \leftarrow \dfrac{\sqrt{r_{\text{task}}}+\sqrt{1-r_{\text{task}}}\left(\sqrt{r_{\text{mach}}}+\sqrt{1-r_{\text{mach}}}\right)}{\sqrt{r_{\text{task}}}\sqrt{1-r_{\text{mach}}}+\sqrt{1-r_{\text{task}}}\left(\sqrt{r_{\text{mach}}}+\sqrt{1-r_{\text{mach}}}\right)}V$ {Scale variability}

2: **for all** $1 \leq i \leq n$ **do**   {Generate base column}

3:      $c_i \leftarrow G(1/V_{\text{col}}^2, V_{\text{col}}^2)$

4: **end for**

5: **for all** $1 \leq i \leq n$ **do**   {Set the correlation between each pair of columns}

6:      **for all** $1 \leq j \leq m$ **do**

7:          $e_{i,j} \leftarrow \sqrt{r_{\text{mach}}}c_i + \sqrt{1 - r_{\text{mach}}} \times G(1/V_{\text{col}}^2, V_{\text{col}}^2)$

8:      **end for**

9: **end for**

10: $V_{\text{row}} \leftarrow \sqrt{1 - r_{\text{mach}}}V_{\text{col}}$ {Scale variability}

11: **for all** $1 \leq j \leq m$ **do**   {Generate base row}

12:      $r_j \leftarrow G(1/V_{\text{row}}^2, V_{\text{row}}^2)$

13: **end for**

14: **for all** $1 \leq i \leq n$ **do**   {Set the correlation between each pair of rows}

15:      **for all** $1 \leq j \leq m$ **do**

16:          $e_{i,j} \leftarrow \sqrt{r_{\text{task}}}r_j + \sqrt{1 - r_{\text{task}}}e_{i,j}$

17:      **end for**

18: **end for**

19: **for all** $1 \leq i \leq n$ **do**   {Rescaling}

20:      **for all** $1 \leq j \leq m$ **do**

21:          $e_{i,j} \leftarrow \dfrac{\mu e_{i,j}}{\sqrt{r_{\text{task}}}+\sqrt{1-r_{\text{task}}}\left(\sqrt{r_{\text{mach}}}+\sqrt{1-r_{\text{mach}}}\right)}$

22:      **end for**

23: **end for**

24: **return** $\{e_{i,j}\}_{1\leq i\leq n, 1\leq j\leq m}$

---

a cost matrix generation method that takes the task and machine correlations as parameters. This method assumes that both these parameters are distinct from one.

Algorithm 1 presents the combination-based method. It sets the correlation between two distinct columns (or rows) by computing a linear combination between a base vector common to all columns (or rows) and a new vector specific to each column (or row). The algorithm first generates the matrix with the target machine correlation using a base column (generated on Line 3) and the linear combination on Line 7. Then, rows are modified such that the task correlation is as desired using a base row (generated on Line 12) and the linear combination on Line 16. The base row follows a distribution with a lower standard deviation, which depends on the machine correlation (Line 10). Using this specific standard deviation is essential to set the task correlation (see the proof of Proposition 1). Propositions 1 and 2 show these two steps generate a matrix with the target correlations for any value of V_{col}.

Proposition 1. *The task correlation ρ_{task} of a cost matrix generated using the combination-based method with the parameter r_{task} converges to r_{task} as $m \to \infty$.*

Proof. Given Lines 7, 16 and 21, any cost, multiplied by $\frac{1}{\mu}\left(\sqrt{r_{task}}\right.$ $+\sqrt{1-r_{task}}\left(\sqrt{r_{mach}}+\sqrt{1-r_{mach}}\right)\right)$ as it does not change $\rho_{i,i'}^r$, is: $e_{i,j} =$ $\sqrt{r_{task}}r_j + \sqrt{1-r_{task}}\left(\sqrt{r_{mach}}c_i + \sqrt{1-r_{mach}}G(1/V_{col}^2, V_{col}^2)\right)$

Let's focus on the first part of the numerator of $\rho_{i,i'}^r$ (from Eq. 2):
$\frac{1}{m}\sum_{j=1}^m e_{i,j}e_{i',j} = r_{task}^2\frac{1}{m}\sum_{j=1}^m r_j^2 + \frac{1}{m}\sum_{j=1}^m \sqrt{r_{task}}r_j\sqrt{1-r_{task}}\left(\sqrt{r_{mach}}c_i + \right.$ $\sqrt{1-r_{mach}}G(1/V_{col}^2, V_{col}^2))$ $+$ $\frac{1}{m}\sum_{j=1}^m \sqrt{r_{task}}r_j\sqrt{1-r_{task}}\left(\sqrt{r_{mach}}c_{i'}+\right.$ $\sqrt{1-r_{mach}}$ $G(1/V_{col}^2, V_{col}^2))$ $+$ $(1$ $-$ $r_{task})\frac{1}{m}\sum_{j=1}^m\left(\sqrt{r_{mach}}\ c_i\right.$ $+$ $\sqrt{1-r_{mach}}G(1/V_{col}^2, V_{col}^2))\times\left(\sqrt{r_{mach}}c_{i'} + \sqrt{1-r_{mach}}G(1/V_{col}^2, V_{col}^2)\right)$.

The first sum converges to $r_{task}(1 + (1 - r_{max})V_{col}^2)$ as $m \to \infty$ because r_j follows a gamma distribution with expected value one and standard deviation $\sqrt{1-r_{max}}V_{col}$. The second sum converges to $\sqrt{r_{task}}\sqrt{1-r_{task}}$ $\left(\sqrt{r_{mach}}c_i + \sqrt{1-r_{mach}}\right)$ as $m \to \infty$ and the third sum converges to $\sqrt{r_{task}}\sqrt{1-r_{task}}\left(\sqrt{r_{mach}}c_{i'} + \sqrt{1-r_{mach}}\right)$ as $m \to \infty$. Finally, the last sum converges to $(1 - r_{task})\left(\sqrt{r_{mach}}c_i + \sqrt{1-r_{mach}}\right)\left(\sqrt{r_{mach}}c_{i'} + \sqrt{1-r_{mach}}\right)$ as $m \to \infty$. The second part of the numerator of $\rho_{i,i'}^r$ is simpler and converges to $\left(\sqrt{r_{task}}+\sqrt{1-r_{task}}\left(\sqrt{r_{mach}}c_i + \sqrt{1-r_{mach}}\right)\right)$ $\left(\sqrt{1-r_{task}}\left(\sqrt{r_{mach}}c_{i'} + \sqrt{1-r_{mach}}\right) + \sqrt{r_{task}}\right)$ as $m \to \infty$. Therefore, the numerator of $\rho_{i,i'}^r$ converges to $r_{task}(1 - r_{max})V_{col}^2$ as $m \to \infty$.

The denominator of $\rho_{i,i'}^r$ converges to the product of the standard deviations of e_{ij} and $e_{i'j}$ as $m \to \infty$. The standard deviation of r_j (resp., $G(1/V_{col}^2, V_{col}^2)$) is $\sqrt{1-r_{mach}}V_{col}$ (resp., V_{col}). Therefore, the standard deviation of e_{ij} is $\sqrt{r_{task}(1 - r_{mach})V_{col}^2 + (1 - r_{task})(1 - r_{mach})V_{col}^2}$.

The correlation between any pair of distinct rows $\rho_{i,i'}^r$ converges thus to r_{task} as $m \to \infty$, which concludes the proof. □

Proposition 2. *The machine correlation ρ_{mach} of a cost matrix generated using the combination-based method with the parameter r_{mach} converges to r_{mach} as $n \to \infty$.*

The proof of Proposition 2 is similar to the proof of Proposition 1 [6, Proposition 14].

Finally, the resulting matrix is scaled on Line 21 to adjust its mean. The initial scaling of the standard deviation on Line 1 is necessary to ensure that the final CV (Coefficient of Variation) of the costs is V. The proof of Proposition 3 is more direct than the previous ones [6, Proposition 15].

Proposition 3. *When used with the parameters μ and V, the combination-based method generates costs with expected value μ and CV V.*

Note that the correlation parameters may be zero. However, each of them must be distinct from one. If they are both equal to one, a direct method exists by building the unrelated instance corresponding to a uniform instance. Additionally, the final cost distribution is a sum of three gamma distributions (two if either of the correlation parameters is zero and only one if both of them are zero).

Note that the previous propositions give only convergence results. For a given generated matrix with finite dimension, the effective correlation properties are distinct from the asymptotic ones.

5   Impact on Scheduling Heuristics

Controlling the task and machine correlations provides a continuum of unrelated instances that are arbitrarily close to uniform instances. This section shows how some heuristics for scheduling unrelated instances are affected by this proximity.

A subset of the heuristics from [7] were used with instances generated using the combination-based method. The three selected heuristics are based on distinct principles to emphasize how the correlation properties may have different effects on the performance. First, we selected EFT [11, E-schedule] [9, Min-Min], which relies on a greedy principle that schedules first the tasks that have the smallest duration. The second heuristic is an adaptation of LPT [10] for unrelated platforms. Since LPT is a heuristic for the $Q||C_{max}$ problem, HLPT performs as the original LPT when machines are uniform (i.e., when the correlations are both equal to 1). HLPT differs from EFT by considering first the largest tasks instead of the smallest ones based on their minimum cost on any machine. The last heuristic is BalSuff [8], which iteratively balances an initial schedule by changing the allocation of the tasks that are on the most loaded machine. The new machine that will execute it is chosen such as to minimize the increase in the task duration.

These heuristics perform identically when the task and machine correlations are arbitrarily close to one and zero, respectively. In particular, sorting the tasks for HLPT is meaningless because all tasks have similar execution times. With such instances, the problem is related to the $Q|p_i = p|C_{max}$ problem (see Sect. 3), which is polynomial. Therefore, we expect these heuristics to perform well with these instances.

In the following experiments, we rely on the combination-based method (Algorithm 1) to generate cost matrices. Instances are generated with $n = 100$ tasks and $m = 30$ machines. Without loss of generality, the mean cost μ is one (scaling a matrix by multiplying each cost by the same constant will have no impact on the scheduling heuristics). The cost CV is $V = 0.3$.

For each scenario, we compute the makespan[5] of each heuristic. We then consider the relative difference from the reference makespan: $C/C_{min} - 1$ where C is the makespan of a given heuristic and C_{min} the best makespan we obtained (we use a genetic algorithm that is initialized with all the solutions obtained by other heuristics as in [7] because the problem is NP-Complete and finding

---

[5] The makespan is the total execution time and it must be minimized.

the optimal solution would take too much time). The closer to zero, the better the performance. We assume in this study that the reference makespan closely approximates the optimal one.

The heat maps on Fig. 2 share the same generation procedure. First, 30 equidistant correlation values are considered between 0.001 and 0.999 using a probit scale (0.001, 0.002, 0.0039, 0.0071, ..., 0.37, 0.46, ..., 0.999). The probit function is the quantile function of the standard normal distribution. It highlights what happens for values that are arbitrarily close to 0 and 1 at the same time. Then, each pair of values for the task and machine correlations leads to the generation of 200 cost matrices (for a total of 180 000 instances). The actual correlations are then measured for each generated cost matrices. Any tile on the figures corresponds to the average performance obtained with the instances for which the actual correlation values lie in the range of the tile. Hence, an instance generated with 0.001 for both correlations may be assigned to another tile than the bottommost and leftmost one depending on its actual correlations. Any value outside any tile was discarded when it occurred.

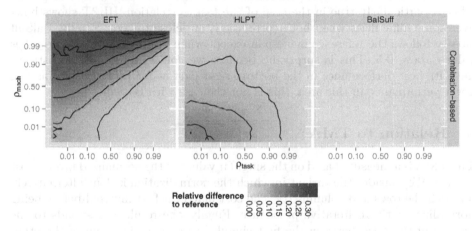

Fig. 2. Heuristic performance with 180 000 instances for the combination-based method. The cost CV V is set to 0.3. The x- and y-axes are in probit scale between 0.001 and 0.999. Each tile represents on average 200 instances. The contour lines correspond to the levels in the legend (0, 0.05, 0.1, ...).

Figure 2 compares the average performance of EFT, HLPT and Balsuff. First, EFT performance remains mainly unaffected by the task and machine correlations when they are similar. However, its performance is significantly impacted by them when one correlation is the complement of the other to one (i.e., when $\rho_{task} = 1 - \rho_{mach}$, which is the other diagonal). In this case, the performance of EFT is at its poorest on the top-left. It then continuously improves until reaching its best performance on the bottom-right (less than 5 % from the reference makespan, which is comparable to the other two heuristics for this area). This is consistent with the previous observation that this last area corresponds to

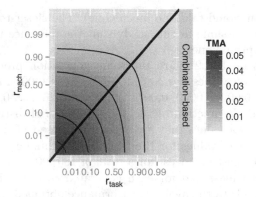

Fig. 3. TMA of the instances used in Fig. 2.

instances that may be close to $Q|p_i = p|C_{\max}$ instances, for which EFT is optimal. HLPT achieves the best performance when either correlation is close to one. This is particularly true in the case of the task correlation. HLPT shows however some difficulties when both correlations are close to zero. Finally, BalSuff closely follows the reference makespan except when the task correlation reaches values above 0.5. This is surprising because we could expect any heuristic to have its best performance in the bottom-right part as for EFT. Despite having good performance in this area, this is not the case with BalSuff.

6   Relation to TMA

The TMA is a measure based on the singular values of the normalized inverse cost matrix. We consider the variant in which the normalization is done alternatively on both the rows and columns [1]. The cost matrix is first inverted before being normalized with an iterative procedure. Finally, the result corresponds to the inverse of the ratio between the first singular value and the mean of the other singular values.

Similarly to the correlation, the TMA measures the affinities between the tasks and the machines. TMA values equal to zero means machines are uniform (no affinity) because only the first singular value is non-zero and the rank of the cost matrix is one. Oppositely, TMA values equal to one means tasks and machines have unrelated characteristics (high affinities between tasks and machines) because the cost matrix is orthogonal.

However, the correspondence with the correlation is not systematic. Let $\{e_{i,j}\}_{1\le i\le n, 1\le j\le n}$ be a cost matrix where $e_{i,j} = \epsilon$ if $i = j$ and $e_{i,j} = w_i b_j$ otherwise (with w_i the weight of task i and b_j the cycle time of machine j). The TMA of this cost matrix converges to one as $\epsilon \to 0$, which suggests a discrepancy from any uniform instance. By contrast, both its task and machine correlations converge to one as $n \to \infty$ and $m \to \infty$ (suggesting a similarity with a uniform instance). Assuming the number of tasks is greater than the number of machines (i.e., $n > m$), each task i must be scheduled on machine i for $1 \leq i \leq m$. The problem is thus equivalent to scheduling the last $n - m$ tasks, each of which has a well-defined weight. This cost matrix corresponds therefore to a uniform instance as indicated by the correlation properties. This contrived example shows that changing a few single values may impact the TMA more profoundly than the correlations. We conclude that the correlations focus on the general consistency across multiple tasks and machines, whereas the TMA stresses the specialization of a few machines for some specific tasks.

Figure 3 depicts the TMA of each of the $2 \times 30^2 \times 200$ instances generated in Sect. 5. The TMA is strongly associated with the correlations in our settings. Note that it does not reach large values given that its maximum is one, even when the correlations are close to zero.

The TMA is also symmetric relatively to the diagonal slices: it is the same when the task/machine correlations are high/low as when they are low/high. Therefore, some behaviors may not be seen with the TMA. For instance, EFT performance varies mainly relatively to the other diagonal (from the top-left to the bottom-right).

The TMA offers several advantages: its normalization procedure makes it independent from the heterogeneity and like the correlation, it is associated to the performance of the selected heuristics. However, it suffers from several drawbacks. Its value depends on the cost matrix dimension and on the cost CV. Moreover, its normalization procedure makes derivations of analytical results difficult. By contrast, the correlation has no such default but it is not independent from the heterogeneity. Also, the correlation is finer because it consists of two different values, which allow the characterization of behaviors that cannot be seen with the TMA (e.g., for EFT). Nevertheless, the TMA may be more relevant than the correlation in some specific cases. For instance, with small cost matrices, the TMA is more sensitive to individual values that may impact significantly the performance. Devising a SVD-based measure that outperforms the TMA (analytically simpler and independent from the cost matrix dimension and the cost CV) is left for future work.

7   Conclusion

This paper studies the correlations of cost matrices used to assess heterogeneous scheduling algorithms. The task and machine correlations are proposed to measure the similarity between an unrelated instance in which any cost is arbitrary ($R||C_{\max}$) and the closest uniform instance ($Q||C_{\max}$) in which any cost is proportional to the task weight and machine cycle time. We analyzed several generation methods from the literature and designed a new one to see the impact of these properties.

Even though the correlation approximates the distance between uniform and unrelated instances (a unitary correlation does not imply it corresponds to a uniform instance), our proposed generation method shows how some heuristics from the literature are affected. For instance, the closer instances are from the uniform case, the better HLPT, an adaptation of LPT to the unrelated case, performs. Additionally, the need for two correlations (for the tasks and for the machines) arise for EFT for which the performance goes from worst to best as the task and machine correlations go from zero to one and one to zero, respectively.

Although the current study highlights the importance of controlling the correlations in cost matrices, it presents some limitations. Overcoming each of them is left for future work. First, results were obtained using the gamma distribution only. However, the proposed method could use other distributions as long as the mean and standard deviation are preserved. Second, all formal derivations are in the asymptotic case only. Hence, the proposed results may be less relevant for small instances. Also, the proposed correlation measures and generation method assume that the correlations stay the same for each pair of rows and for each pair of columns: our measures average the correlations and our method is inapplicable when the correlations between each pair of rows or each pair of columns are distinct. Considering two correlation matrices that define the specific correlations between each pair of rows and each pair of columns would require the design of a finer generation method. Finally, investigating the relation with the heterogeneous properties would require the design of a method that controls both the correlation and heterogeneity properties.

Acknowledgments. Computations have been performed on the supercomputer facilities of the Mésocentre de calcul de Franche-Comté.

References

1. Al-Qawasmeh, A.M., Maciejewski, A.A., Roberts, R.G., Siegel, H.J.: Characterizing task-machine affinity in heterogeneous computing environments. In: IPDPSW (2011)
2. Al-Qawasmeh, A.M., Maciejewski, A.A., Siegel, H.J.: Characterizing heterogeneous computing environments using singular value decomposition. In: IPDPSW (2010)
3. Al-Qawasmeh, A.M., Pasricha, S., Maciejewski, A.A., Siegel, H.J.: Power and thermal-aware workload allocation in heterogeneous data centers. Trans. Comput. **64**(2), 477–491 (2013)
4. Ali, S., Siegel, H.J., Maheswaran, M., Hensgen, D.: Task execution time modeling for heterogeneous computing systems. In: HCW, pp. 185–199. IEEE (2000)
5. Ali, S., Siegel, H.J., Maheswaran, M., Hensgen, D., Ali, S.: Representing task and machine heterogeneities for heterogeneous computing systems. Tamkang J. Sci. Eng. **3**(3), 195–208 (2000)
6. Canon, L.C., Héam, P.C., Philippe, L.: Controlling and Assessing Correlations of Cost Matrices in Heterogeneous Scheduling. Technical report RR-FEMTO-ST-1191, FEMTO-ST, February 2016
7. Canon, L.-C., Philippe, L.: On the heterogeneity bias of cost matrices when assessing scheduling algorithms. In: Träff, J.L., Hunold, S., Versaci, F. (eds.) Euro-Par 2015. LNCS, vol. 9233, pp. 109–121. Springer, Heidelberg (2015)
8. Canon, L.C., Philippe, L.: On the Heterogeneity Bias of Cost Matrices when Assessing Scheduling Algorithms. Technical report RR-FEMTO-ST-8663, FEMTO-ST, March 2015

9. Freund, R.F., Gherrity, M., Ambrosius, S., Campbell, M., Halderman, M., Hensgen, D., Keith, E., Kidd, T., Kussow, M., Lima, J.D., Mirabile, F., Moore, L., Rust, B., Siegel, H.J.: Scheduling resources in multi-user, heterogeneous, computing environments with SmartNet. In: HCW, pp. 184–199. IEEE (1998)

10. Graham, R.L.: Bounds on multiprocessing timing anomalies. J. appl. math. **17**(2), 416–429 (1969)

11. Ibarra, O.H., Kim, C.E.: Heuristic algorithms for scheduling independent tasks on nonidentical processors. J. ACM **24**(2), 280–289 (1977)

12. Khemka, B., Friese, R., Pasricha, S., Maciejewski, A.A., Siegel, H.J., Koenig, G.A., Powers, S., Hilton, M., Rambharos, R., Poole, S.: Utility maximizing dynamic resource management in an oversubscribed energy-constrained heterogeneous computing system. Sustain. Comput. Inf. Syst. **5**, 14–30 (2014)

13. Luo, P., Lü, K., Shi, Z.: A revisit of fast greedy heuristics for mapping a class of independent tasks onto heterogeneous computing systems. J. Parallel Distrib. Comput. **67**(6), 695–714 (2007)

14. Maheswaran, M., Ali, S., Siegel, H.J., Hensgen, D., Freund, R.F.: Dynamic mapping of a class of independent tasks onto heterogeneous computing systems. J. Parallel Distrib. Comput. **59**(2), 107–131 (1999)

Penalized Graph Partitioning for Static and Dynamic Load Balancing

Tim Kiefer, Dirk Habich$^{(\boxtimes)}$, and Wolfgang Lehner

Dresden Database Systems Group, Technische Universität Dresden,
Dresden, Germany
{tim.kiefer,dirk.habich,wolfgang.lehner}@tu-dresden.de

Abstract. With ubiquitous parallel architectures, the importance of optimally distributed and thereby balanced work is unprecedented. To tackle this challenge, graph partitioning algorithms have been successfully applied in various application areas. However, there is a mismatch between solutions found by classic graph partitioning and the behavior of many real hardware systems. Graph partitioning assumes that individual vertex weights add up to partition weights (here, referred to as *linear graph partitioning*). This implies that performance scales linearly with the number of tasks. In reality, performance does usually not scale linearly with the amount of work due to contention on various resources. We address this mismatch with our novel *penalized graph partitioning* approach in this paper. Furthermore, we experimentally evaluate the applicability and scalability of our method.

1 Introduction

Modeling problems as graphs and balancing the load of corresponding distributed algorithms by means of graph partitioning has numerous applications in scientific computing [10,22,25]. Balanced min-cut (hyper)graph partitioning is appealing because it balances the load while at the same time minimizing communication costs. In recent years, graph partitioning was successfully used in other areas like data management as well [6,9,23]. Taking data management systems as an example, the possible applications for graph partitioning range from high-level database-as-a-service architectures [1,19] to low-level parallelism found in modern multi-socket-multi-core systems [17]. With more parallel architectures being used, the problem of optimally balancing work gains importance.

However, there is a mismatch between solutions found by classic graph partitioning and the behavior of many real hardware systems. Graph partitioning assumes that individual vertex weights add up to partition weights (here, referred to as *linear graph partitioning*). In the context of distributed systems, the assumption implies that performance scales linearly with the number of tasks. In reality however, performance does usually not scale linearly with the amount of work due to contention on hardware [3], operating system [18], or application resources [20]. We address this mismatch with *penalized graph partitioning*, a special case of non-linear graph partitioning, in this paper. The result is a load

© Springer International Publishing Switzerland 2016
P.-F. Dutot and D. Trystram (Eds.): Euro-Par 2016, LNCS 9833, pp. 146–158, 2016.
DOI: 10.1007/978-3-319-43659-3_11

balancing algorithm that shares the advantages of classic graph partitioning and that is at the same time considering the non-linear performance of real systems.

1.1 Penalized Performance Model

In this paper, we consider distributed systems where multiple (heterogeneous) tasks are executed concurrently on the various nodes. In the simplest case, loads induced by tasks are combined by summing them up to derive a node's global load. This method is referred to as the *linear model* as it models an ideal system where performance scales linearly with the amount of work that needs to be done. However, in practice, performance often depends on all kinds of workload parameters like request rates, request types, and the concurrent execution of requests. Contention on resources caused by concurrent execution may lead to performance that does not scale linearly with the amount of work. Therefore, we propose to use a *non-linear* model to combine the individual loads. To grasp the general behavior of complex systems, we assume a simplified *penalized* resource consumption model, which is a combination of the linear model and a (possibly non-linear) penalty function. Up to a certain load or degree of parallelism, the linear usage assumption often holds because the system is then underutilized and sufficient resources are available. However, when a certain load level is reached, contention occurs and the performance does not scale linearly beyond this load level. The penalty function is used to account for the contention.

While we acknowledge that modeling real systems is a challenging problem in itself, we assume here that the model, i.e., the penalty function, is given. Depending on the actual system, low-level and application-level experiments may be necessary to find a sufficiently accurate system model.

1.2 Motivating Example

To demonstrate the potential of penalized graph partitioning in presence of non-linear resources, we perform a synthetic partitioning experiment. To run the experiment, we generate a workload graph that contains 1000 heterogeneous tasks with weights following a Zipf distribution.[1] Each task in the workload graph is communicating with 0 to 10 other tasks (again Zipf distributed). To model a system, we use an exponential penalty function and assume that the underlying resource can execute 16 parallel tasks before the penalty grows with the square of the cardinality due to contention (Fig. 1a).

The workload in this experiment is partitioned into 32 balanced partitions using a standard graph partitioning library. Afterward, to estimate the actual load for each node, the penalty function is applied to each partition based on the partition cardinality (Fig. 1b). The resulting partition weights are compared to a second partitioning of the graph that was generated by our novel penalized graph partitioning algorithm (Fig. 1c).

---

[1] Comparable workloads can be found in actual systems, e.g., database-as-a-service systems [24].

The unmodified partitioning algorithm, which is unaware of the contention, tries to balance the load. The resulting relative weights show that the node with the highest partition weight receives 3.1 times the load of the node with the lowest partition weight. In contrast, the penalized partitioning algorithm leads to partition weights, and hence node utilizations, that are balanced within a tolerance of 3 %.

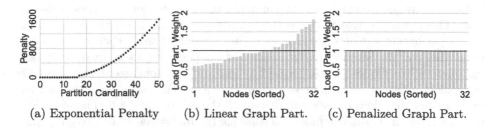

(a) Exponential Penalty   (b) Linear Graph Part.   (c) Penalized Graph Part.

Fig. 1. Partitioning experiment (loads normalized to average)

1.3   Related Work

Graph partitioning has been a topic of interest in the scientific computing community at least since the late 1990s. Early works on the multilevel graph partitioning paradigm [13] led to many papers about variations and extensions of the balanced min-cut partitioning problem, e.g., about multi-constraint partitioning [14], incremental update strategies [11], or heterogeneous infrastructures [21]. A rather recent book and a survey provide excellent overviews of the results in the field [2,4]. To the best of our knowledge, we are the first to consider penalized, i.e., non-linear, graph partitioning.

In recent years, graph partitioning was successfully used in data management applications as well [6,9,23]. These applications will most likely benefit from penalized graph partitioning due to the complex and often heterogeneous tasks and the ever-present contention on bottleneck resources.

1.4   Contributions

Our main contribution in this paper is a load balancing algorithm based on penalized graph partitioning. In detail, we recap the basics of graph partitioning (Sect. 2) before we introduce our novel method to partition graphs with penalized partition weights, i.e., vertex weights that do not sum up linearly to partition weights (Sect. 3). Thereby, we also propose an extension to the penalized graph partitioning algorithm to deal with dynamic workloads. Our experimental evaluation shows the applicability and scalability of penalized graph partitioning in Sect. 4 before we conclude the paper in Sect. 5.

2 Graph Partitioning

Given an undirected, weighted graph, the balanced k-way min-cut graph partitioning problem (GPP) refers to finding a k-way partitioning of the graph such that the total edge cut is minimized and the partitions are balanced within a given tolerance. The following definitions are used to formalize the problem and to describe its solution heuristics in detail. In this paper, we limit ourselves to graphs with a single weight per vertex. Without restriction, penalized graph partitioning works with multiple vertex weights as well (e.g., based on [14]).

Let $G = (V, E, w_V, w_E)$ be an *undirected, weighted graph* with a set of vertices V, a set of edges E, and weight functions w_V and w_E. Vertex and edge weights are positive real numbers: $w_V \colon V \to \mathbb{R}_{>0}$ and $w_E \colon E \to \mathbb{R}_{>0}$. The weight functions are naturally extended to sets of vertices and edges:

$$w_V(V') := \sum_{v \in V'} w_V(v) \text{ for } V' \subseteq V \quad \text{and} \quad w_E(E') := \sum_{e \in E'} w_E(e) \text{ for } E' \subseteq E.$$

Let $\Pi = (V_1, \dots, V_k)$ be a *partitioning* of V into k partitions $V_1, \dots, V_k$ such that: $V_1 \cup \dots \cup V_k = V$ and $V_i \cap V_j = \emptyset$ for all $i \neq j$. Given a partitioning, an edge that connects partitions is called a *cut edge* and E_c is the set of all cut edges in a graph. The objective of the GPP is to minimize the *total cut* $w_E(E_c)$, i.e., the aggregated weight of all cut edges.

A *balance constraint* demands that all partitions have about equal weights. Let μ be the average partition weight: $\mu := w_V(V)/k$. For a balanced graph partitioning it must hold that $\forall i \in \{1, \dots, k\} \colon w_V(V_i) \leq (1 + \epsilon) \cdot \mu$, where $\epsilon \in \mathbb{R}_{\geq 0}$ is a given imbalance parameter to specify a tolerable degree of imbalance (depending on the application).

2.1 Partitioning Algorithm

Partitioning a graph into k partitions of roughly equal size such that the total cut is minimized is NP-complete [12]. Heuristics, especially the multilevel partitioning framework [4, 13], are used in practice to solve the problem.

The multilevel graph partitioning framework consists of three phases: (1) *coarsening* the graph, (2) finding an *initial partitioning* of the coarse graph, and (3) *uncoarsening* the graph and projecting the coarse solution to the finer graphs. In the **coarsening phase**, a series of smaller graphs is derived from the input graph. Coarsening is commonly implemented by contracting a subset of vertices and replacing it with a single vertex. Parallel edges are replaced by a single edge with the accumulated weight of the parallel edges. Contracting vertices like this implies that a balanced partitioning on the coarse level represents a balanced partitioning on the fine level with the same total cut. Different strategies exist to select vertices to be contracted. Finding a matching is a tradeoff between using heavy edges (and hence reducing the final cut) and keeping uniform vertex weights (and hence improving partition balance). The coarsening ends when the coarsest graph is sufficiently small to be initially partitioned.

Different algorithms exist to find an **initial partitioning** [4]. Methods for the initial partitioning are either based on direct k-way partitioning or on recursive bisection. A simple but effective method to find an initial partitioning is greedy graph growing. A random start vertex is grown using breadth-first search, adding the vertex that increases the total cut the least in each step. The search is stopped as soon as half of the total vertex weight is assigned to the growing partition. Because the quality of the bisection strongly depends on the randomly selected start vertex, multiple iterations with different starts are used and the best solution is kept. The k-way extension of graph-growing starts with k random vertices and grows them in turns.

The initial partitioning is **uncoarsened** by repeatedly assigning previously contracted vertices to the same partition. Each extraction of vertices is followed by a **refinement step** to improve the total cut or the balance of the partitions. For instance, local vertex swapping is a refinement metaheuristic that can be parametrized with different strategies to select vertices to move [8,15,16].

3   Penalized Graph Partitioning

The idea of our penalized graph partitioning is to introduce a *penalized partition weight* and to modify the graph partitioning problem accordingly. We define the resulting problem as the *Penalized Graph Partitioning Problem* (P-GPP). Figure 2 shows an example graph with vertex and edge weights denoted in Fig. 2a. Solving the GPP leads to the partitioning with the total cut of 3 shown in Fig. 2b. When the cardinality of a partition is penalized linearly, the solution of the P-GPP having a total cut of 4 is shown in Fig. 2c. However, when the penalty of a partition grows with the square of the partition cardinality, the partitioning with the total cut of 4 shown in Fig. 2d is the solution to the P-GPP. The partitioning obviously depends on the performance model, i.e., the given penalty function.

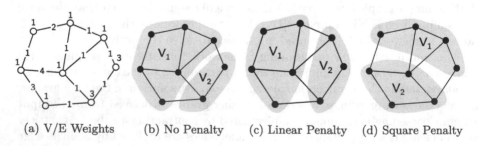

(a) V/E Weights   (b) No Penalty   (c) Linear Penalty   (d) Square Penalty

Fig. 2. Example of graph partitionings with different penalty functions

3.1   Prerequisites

Let $G = (V, E, w_V, w_E)$ be an undirected, weighted graph as in Sect. 2. Furthermore, let p be a positive, monotonically increasing penalty function that penalizes a partition weight based on the partition cardinality:

$$p \colon \mathbb{N} \to \mathbb{R}_{\geq 0} \text{ with } p(n_1) \leq p(n_2) \text{ for } n_1 \leq n_2.$$

The vertex weight function is extended to sets $V' \subseteq V$ such that it incorporates the penalty:

$$w_V(V') := \sum_{v \in V'} w_V(v) + p(|V'|).$$

The example partitioning in Fig. 2c uses a linear penalty function, i.e., $p(|V|) := |V|$. Accordingly, using the definition, the partition weights are

$$w_V(V_1) = \sum_{v \in V_1} w_V(v) + p(|V_1|) = 5 + 5 = 10 \text{ and}$$

$$w_V(V_2) = \sum_{v \in V_2} w_V(v) + p(|V_2|) = 7 + 3 = 10.$$

The example partitioning in Fig. 2d uses a square penalty function, i.e., $p(|V|) := |V|^2$. Accordingly, the partition weights are $w_V(V_1) = w_V(V_2) = 22$.

Adding penalties to partition weights invalidates some of the assumptions made in the GPP and its solution algorithms. Most fundamentally, the combined weight of two or more partitions is not equal to the weight of a partition containing all the vertices. Using the definition and two partitions V_1 and V_2:

$$w_V(V_1 \cup V_2) = w_V(V_1) + w_V(V_2) + p(|V_1 \cup V_2|) - p(|V_1|) - p(|V_2|).$$

For arbitrary penalty functions we must assume that $p(|V_1 \cup V_2|) \neq p(|V_1|) + p(|V_2|)$. It follows that in general $w_V(V_1 \cup V_2) \neq w_V(V_1) + w_V(V_2)$. Hence, the total weight of all vertices is in general not equal to the total weight of all partitions. We therefore introduce the following definitions of the two weights. Given a graph and a partitioning, the *total vertex weight* w_V is the penalized weight of all vertices, i.e.,

$$w_V := \sum_{v \in V} w_V(v) + p(|V|).$$

The *total partition weight* w_Π on the other hand is the sum of the weights of all partitions, i.e.,

$$w_\Pi := \sum_{i=1}^{k} w_V(V_i).$$

Consider the example partitioning in Fig. 2d; using the definition, $w_V = 12 + 64 = 76$ and $w_\Pi = 22 + 22 = 44$.

It follows that the total partition weight w_Π of the graph is not constant but depends on the partitioning, specifically the cardinalities of the partitions. This observation has implications in all steps of the graph partitioning algorithm, e.g., the balance constraint has to use the average total partition weight $\mu := w_\Pi / k$ instead of the average total vertex weight.

3.2 Penalized Graph Partitioning Algorithm (Static Case)

We propose modifications of the multilevel graph partitioning algorithm to solve the P-GPP. First, we describe two basic operations that need to reflect partition penalties. Then, we will detail the necessary modifications to the three building blocks of the multilevel graph partitioning framework.

During graph partitioning and refinement, it is often necessary to move a vertex between partitions or to merge partitions. For the sake of computational efficiency, the weights of the resulting partitions should be computed incrementally instead of from scratch.

Operation 1. *When a vertex v is moved from partition V_1 to partition V_2, the partition weights of the resulting partitions $V_1' := V_1 \setminus v$ and $V_2' := V_2 \cup v$ are as follows:*

$$w_V(V_1') = w_V(V_1 \setminus v) = w_V(V_1) - w_V(v) - p(|V_1|) + p(|V_1| - 1) \text{ and}$$
$$w_V(V_2') = w_V(V_2 \cup v) = w_V(V_2) + w_V(v) - p(|V_2|) + p(|V_2| + 1).$$

Operation 2. *When two partitions V_1 and V_2 are combined, the partition weight of the resulting partition $V' := V_1 \cup V_2$ can be calculated as follows:*

$$w_V(V') = w_V(V_1) + w_V(V_2) + p(|V_1| + |V_2|) - p(|V_1|) - p(|V_2|).$$

To **coarsen** the graph, a matching of vertices has to be determined and vertices have to be contracted accordingly. The heuristics introduced in Sect. 2.1 can be used to coarsen a graph with penalized partition weights. However, the vertex weight of the contracted vertex has to correctly incorporate the penalty to ensure that a balanced partitioning of the coarse graph will lead to a balanced partitioning during the uncoarsening steps. Therefore, contracted vertices are treated like partitions themselves and the weight of a contracted vertex is calculated as in Operation 2.

We use a modified version of recursive bisection and greedy region growing to find an **initial k-way partitioning** of graphs with penalized partition weights. In the region growing algorithm, moving a vertex between partitions has to use Operation 1 to calculate the resulting partition weights. Moreover, the stop condition of the region growing algorithm has to be modified to account for the new balance constraint. In the original formulation, the algorithm stopped when the growing partition reached at least half of the total vertex weight. To achieve balanced partitions and because the total vertex weight is in general not equal to the total partition weight, the latter has to be used in the stop condition. Furthermore, since the total partition weight depends on the partitioning it repeatedly has to be recalculated after vertices have been moved, again using Operation 1.

The penalties have to be considered during the **uncoarsening and refinement** of the graph. Similar to the modifications of the region growing algorithm, the local vertex swapping method has to use Operation 1 whenever a vertex is moved between partitions. Furthermore, when vertex swapping is used to balance a partitioning, the modified balance constraint has to be used. This implies that stop conditions and checks use the total partition weight instead of the total vertex weight. Since the total partition weight depends on the partitioning, it has to be recalculated after a vertex has been moved (Operation 1).

3.3 Incrementally Updating the Partitioning (Dynamic Case)

With dynamic workloads, the partitioning needs to be periodically re-evaluated to ensure balanced partitions and an optimal total cut. Updating the partitioning after changes is a tradeoff between the quality of the new partitioning and the migration costs induced by implementing the new partitioning.

The problem of incrementally updating a partitioning is known as dynamic load balancing or repartitioning and is a well studied problem for the original graph partitioning problem [5,7]. In this paper, we adapt an existing hybrid update strategy for penalized graph partitioning and show in our experimental evaluation that it performs well in the presence of penalized partition weights. Whenever the graph changes such that the balance constraint is violated, balancing and refinement steps based on local vertex swapping try to move vertices such that the partitioning is balanced again. If no balanced partitioning can be found using the local search strategy, the graph is partitioned from scratch and the new partitioning is mapped to the previous partitioning such that the migration cost is minimized. To prevent the total cut in the graph from slowly deteriorating, a new partitioning is computed in the background after a certain number of local refinement operations (even when the partitioning is still balanced). The new partitioning replaces the current one only if the new total cut justifies the migration overhead.

4   Experimental Evaluation

METIS is a set of programs for graph partitioning and related tasks based on multilevel recursive bisection, multilevel k-way partitioning, and multi-constraint partitioning.[2] We modified METIS (v5.1) to support the penalized graph partitioning methods proposed in this paper (we denote the resulting tool PENMETIS). Our modifications are based on the serial version of METIS but can also be incorporated in the parallel version of METIS in the future.

4.1   Scalability Experiments

In this section, we evaluate the overhead that penalized partition weights introduce in the partitioning process. Furthermore, we investigate how penalized graph partitioning scales with the size of the graph and the number of partitions. We use a linear penalty function and example graphs from the Walshaw Benchmark [26] to analyze penalized graph partitioning. The corresponding Graph Partitioning Archive[3] contains 34 graphs from applications such as finite element computation, matrix computation, and VLSI design. The largest graph (`auto`) contains 448695 vertices and 3314611 edges and can be considered large in the context of workload graphs.

Penalized Partitioning Overhead. In this experiment, we investigate the overhead of penalized partition weights. Figure 3 shows the absolute partitioning times for all benchmark graphs using METIS and PENMETIS.[4] The figure shows that penalized partitioning introduces only a small overhead. More specifically, PENMETIS takes on average 28 % (42 ms) more time than METIS.

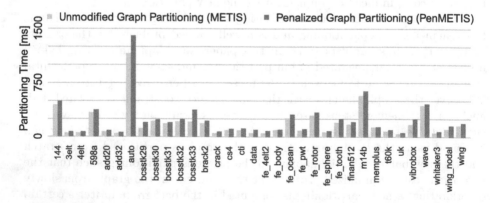

Fig. 3. Partitioning time comparison (64 partitions, 3 % imbalance) (Color figure online)

---

[2] http://glaros.dtc.umn.edu/gkhome/metis/metis/overview.

[3] http://staffweb.cms.gre.ac.uk/~c.walshaw/partition/.

[4] We use a fairly moderate AMD Opteron (Istanbul) CPU running at 2.6GHz for this experiment. As mentioned before, METIS and PENMETIS run single-threaded.

Scalability with Graph Size. Figures 4a and b show the execution times of PENMETIS charted by the number of vertices and by the number of edges. The charts indicate that the graph partitioning algorithms scale linearly with both parameters.

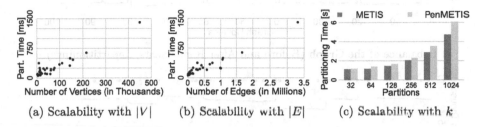

(a) Scalability with $|V|$ (b) Scalability with $|E|$ (c) Scalability with k

Fig. 4. Execution times of PENMETIS depending on the number of vertices $|V|$, edges $|E|$ (64 partitions, 3 % imbalance), and partitions k (graph `auto`) (Color figure online)

Scalability with Partition Count. In a second scalability experiment, we investigate how penalized graph partitioning scales with the number of partitions. In Fig. 4c, we show partitioning times for METIS and PENMETIS for the largest benchmark graph (`auto`) and various partition counts. Beyond 64 partitions, the partitioning time scales linearly with the number of partitions.

4.2 Incremental Update Experiment

In this experiment, we evaluate the ability of PENMETIS to react to changes in the workload. We start our experiment with the previously introduced synthetic workload graph containing 1000 vertices and the same exponential penalty function (see Sect. 1.2). We additionally generate random edge weights (between 1 and 100) to get a more realistic evaluation of the total cut. The workload graph is initially partitioned into 32 partitions with an imbalance parameter of 3 %.

To simulate a changing workload, we define two workload graph modifications. A *minor change* is implemented by updating the vertex and edge weights of 1 % of all vertices and all edges (randomly selected). A *major change* is implemented by updating the vertex and edge weights of 10 % of all vertices and all edges. The complete experiment consists of 100 workload changes where one major change follows after every 19 minor changes. Figure 5 shows the results.

After each workload change, the current partitioning is evaluated against the new workload graph. The update mechanism is triggered when the balance constraint is violated. The update strategy first tries to regain a balanced partitioning by using local refinement strategies. A complete repartitioning is only triggered when the local refinement fails. In addition, the update strategy repartitions the workload graph in the background after every ten changes. However, the new partitioning is only implemented when it leads to a total cut that is

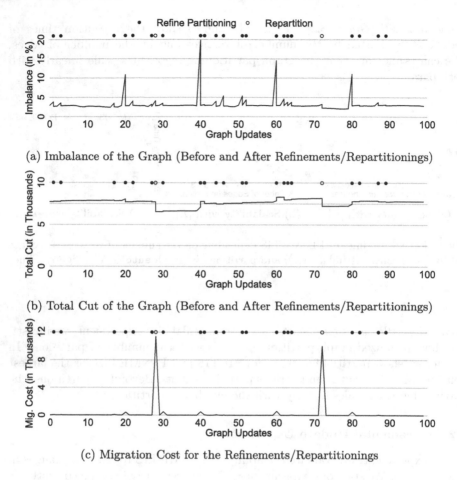

(a) Imbalance of the Graph (Before and After Refinements/Repartitionings)

(b) Total Cut of the Graph (Before and After Refinements/Repartitionings)

(c) Migration Cost for the Refinements/Repartitionings

Fig. 5. Incremental Update Experiment (32 Partitions, 3 % Imbalance)

more than 10 % better than the old cut. The evolution of the graph imbalance and the total cut are summarized in Figs. 5a and b. The results show that minor changes eventually and major changes always lead to violations of the balance constraint. However, in many cases (21 out of 23 in the experiment) the local refinement algorithm is able to regain a balanced partitioning. A complete repartitioning is triggered only in two cases, which in both cases leads to considerably better total cuts.

We report the sum of all vertex weights of vertices that are moved between partitions as the total migration cost for an update (Fig. 5c). The figure shows that partitioning the workload graph from scratch causes considerably higher migration costs than refining an existing partitioning.

5 Conclusion

In this paper, we presented penalized graph partitioning, a special case of non-linear graph partitioning. An experimental evaluation showed the applicability and scalability of penalized graph partitioning as a load balancing mechanism in the presence of non-linear performance due to contention on resources.

We believe that penalized graph partitioning is a versatile method that can be applied to many distributed systems. We showed that existing extensions for basic graph partitioning, specifically for dynamic repartitioning, can be applied to penalized graph partitioning as well. In the future, we will present results to show that the same holds for other extensions that deal with, e.g., multiple resources, heterogeneous infrastructures, or partial allocations. We will also show that the idea of penalized graph partitioning can be generalized to arbitrary non-linear performance models.

Acknowledgments. This work is partly funded by the German Research Foundation (DFG) within the Cluster of Excellence *Center for Advancing Electronics Dresden* (Orchestration Path) and under the DFG project LE 1416/22-1.

References

1. Amazon. Amazon Relational Database Service (2015)
2. Bichot, C.-E., Siarry, P. (eds.): Graph Partitioning. Wiley, Hoboken (2011)
3. Blagodurov, S., Zhuravlev, S., Fedorova, A.: Contention-aware scheduling on multicore systems. ACM Trans. Comput. Syst. **28**(4), 8 (2010)
4. Buluç, A., Meyerhenke, H., Safro, I., Sanders, P., Schulz, C.: Recent Advances in Graph Partitioning. preprint: Computing Research Repository (2013)
5. Catalyurek, U.V., et al.: Hypergraph-based dynamic load balancing for adaptive scientific computations. In: IPDPS (2007)
6. Curino, C., Jones, E.P.C., Zhang, Y., Madden, S.: Schism: a workload-driven approach to database replication and partitioning. In: VLDB (2010)
7. Devine, K.D., Boman, E.G., Heaphy, R.T., Hendrickson, B.A.: New challenges in dynamic load balancing. Appl. Numer. Math. **52**, 133–152 (2005)
8. Fiduccia, C.M., Mattheyses, R.M.: A linear-time heuristic for improving network partitions. In: DAC (1982)
9. Golab, L., Hadjieleftheriou, M., Karloff, H., Saha, B.: Distributed data placement to minimize communication costs via graph partitioning. In: SSDBM (2014)
10. Hendrickson, B., Kolda, T.G.: Graph partitioning models for parallel computing. Parallel Comput. **26**(12), 1519–1534 (2000)
11. Hendrickson, B., Leland, R., Van Driessche, R.: Enhancing data locality by using terminal propagation. In: HICSS (1996)
12. Hyafil, L., Rivest, R.L.: Graph Partitioning and Constructing Optimal Decision Trees are Polynomial Complete Problems. Technical report, IRIA (1973)
13. Karypis, G., Kumar, V.: Analysis of multilevel graph partitioning. In: SC (1995)
14. Karypis, G., Kumar, V.: Multilevel Algorithms for Multi-Constraint Graph Partitioning. Technical report, University of Minnesota (1998)
15. Karypis, G., Kumar, V.: Multilevel k-way partitioning scheme for irregular graphs. J. Parallel Distrib. Comput. **48**(1), 71–95 (1998)

16. Kernighan, B.W., Lin, S.: An efficient heuristic procedure for partitioning graphs. Bell Syst. Tech. J. **49**(2), 291–307 (1970)
17. Kissinger, T., et al.: ERIS: a NUMA-aware in-memory storage engine for analytical workloads. In: ADMS (2014)
18. Li, C., Ding, C., Shen, K.: Quantifying the cost of context switch. In: ExpCS (2007)
19. Microsoft. Microsoft Windows Azure (2015)
20. Pandis, I., Johnson, R., Hardavellas, N., Ailamaki, A.: Data-oriented transaction execution. In: VLDB (2010)
21. Pellegrini, F.: Static mapping by dual recursive bipartitioning of process and architecture graphs. In: SHPCC (1994)
22. Pothen, A.: Graph Partitioning Algorithms with Applications to Scientific Computing. Technical report, Old Dominion University (1997)
23. Quamar, A., Kumar, K.A., Deshpande, A.: SWORD: scalable workload-aware data placement for transactional workloads. In: EDBT (2013)
24. Schaffner, J., et al.: RTP: Robust tenant placement for elastic in-memory database clusters. In: SIGMOD (2013)
25. Schloegel, K., Karypis, G., Kumar, V.: Graph partitioning for dynamic. adaptive and multi-phase scientific simulations. In: CLUSTER (2001)
26. Soper, A.J., Walshaw, C., Cross, M.: A combined evolutionary search and multi-level optimisation approach to graph-partitioning. J. Global Optim. **29**(2), 225–241 (2004)

Non-preemptive Scheduling with Setup Times: A PTAS

Klaus Jansen and Felix Land[(⊠)]

Institute of Computer Science, University of Kiel, 24118 Kiel, Germany
{kj,fku}@informatik.uni-kiel.de

Abstract. Consider the following scheduling problem: a set of jobs is to be processed without preemption on m identical machines. The jobs are partitioned into classes. Before jobs from a class can be processed on a machine, a setup is required, whose duration depends on the class. The objective is to schedule all jobs while minimizing the completion time of the last job, also known as the makespan.

We present and analyze three polynomial algorithms for this problem. The first algorithm follows a next-fit strategy and has approximation ratio 3. The second is a very efficient algorithm with approximation ratio arbitrarily close to 2. The last algorithm is a polynomial time approximation scheme.

Keywords: Scheduling · Setup times · Makespan minimization · Approximation algorithms · Polynomial time approximation schemes

1 Introduction

Consider the following scheduling problem: we are given a set J of n jobs. Each job $j \in J$ has processing time $t(j) \in \mathbb{N}$. In addition, J is partitioned into c classes $C_1, \ldots, C_c$. Whenever a machine is scheduled to process a job from class k at the beginning of the schedule or after it processed a job from another class $k' \neq k$, a setup of s_k time units is required. The goal is to find a non-preemptive schedule on m identical machines, such that the makespan (the completion time of the latest job) is minimized. Note that we can group the jobs on each machine by class without increasing the makespan, and usually assume that schedules are of this form.

This model is applicable in several contexts, e.g. (1) when machines have to be configured for each class of produced items, or (2) in cloud computing, when a significant amount of data that is common for all jobs in a class needs to be transferred before starting the computation [7].

This work was supported by DFG project JA 612/14-2: Design of Efficient Polynomial Time Approximation Schemes for Scheduling and Related Optimization Problems.

© Springer International Publishing Switzerland 2016
P.-F. Dutot and D. Trystram (Eds.): Euro-Par 2016, LNCS 9833, pp. 159–170, 2016.
DOI: 10.1007/978-3-319-43659-3_12

Related Work. The considered problem is a generalization of the scheduling problem on identical machines, where no setup times are present. It is NP-hard, so we cannot hope for an exact polynomial time algorithm [4]. A *c-approximate algorithm* produces, for each input instance I, a solution with makespan at most $c\,\mathrm{OPT}(I)$, where $\mathrm{OPT}(I)$ denotes the optimum (minimum) makespan. Scheduling on identical machines is well understood in terms approximability: it admits a *polynomial time approximation scheme* (PTAS) [5], i.e. a familiy $\{A_\varepsilon \mid \varepsilon > 0\}$ of algorithms such that A_ε is a polynomial time $(1+\varepsilon)$-approximate algorithm for each $\varepsilon > 0$.

Scheduling with setup times also has been extensively studied for different machine models and objective functions [1,2]. However, as far we are aware, the exact problem we consider here is not covered by previous research. We discuss some results about similar models. Perhaps the closest related work is due to Mäcker et al. [7]. In their model, all classes have the same setup time, so our algorithms apply to their model as well. They gave a simple 2-approximate algorithm, and a $(\frac{3}{2} + \varepsilon)$-approximation for arbitrarily small $\varepsilon > 0$. The latter algorithm further requires that the total processing time of each class is bounded by $\gamma\mathrm{OPT}$ for some constant γ. Monma and Potts [8] consider a model with class-dependent setup times and preemptions. They designed two algorithms. The first one has approximation factor $\frac{3}{2} - \frac{1}{4m-4}$ if $m \leq 4$ and $\frac{5}{3} - \frac{1}{n}$ when $m > 3$ is a multiple of 3, and requires that the total processing time of each class including setup is bounded by OPT. Later, Chen [3] presented an algorithm with improved approximation guarantee $\frac{3}{2}$ for this model. Without restrictions on the value of m, Monma and Potts gave a $\left(2 - \frac{1}{\lfloor m/2 \rfloor + 1}\right)$-approximation. Schuurman and Woeginger [9] studied the setting where each class contains exactly one preemptable job, and developed both a $(\frac{4}{3} + \varepsilon)$-approximation, and a PTAS for the case that all setup times are equal.

Our Contribution. We present three algorithms for our model. The first one is a simple heuristic that employs a next-fit strategy followed by a repair step. It has approximation ratio 3 and running time $\mathrm{O}(n)$. The second algorithm follows the dual approximation paradigm and is more careful with the placement of jobs or setups that are larger than $\frac{1}{2}\mathrm{OPT}$. It achieves an approximation guarantee of $2 + \varepsilon$ for arbitrarily small $\varepsilon > 0$ in time $\mathrm{O}(n \log \frac{1}{\varepsilon})$. The final algorithm is a PTAS, improving and generalizing the result by Mäcker et al. [7]. It distinguishes between large jobs and small jobs. The large jobs are scheduled near-optimally using a dynamic program. For the small jobs, we distinguish whether the setup time is large or small compared to the optimum for each class. Small jobs in classes with small setup time are removed from the instance and are later re-added using a next-fit algorithm. In classes with large setup time, the small jobs are glued together such that they become large, and the glued jobs are placed by the dynamic program as well.

Notation. We consistently use j to denote jobs, i to denote machines, and k to denote class indices. For a set $J' \subseteq J$, we abbreviate $\mathrm{t}(J') = \sum_{j \in J'} \mathrm{t}(j)$. For $\ell \in \mathbb{N}$ we denote $[\ell] = \{1, \ldots, \ell\}$.

2   A Linear 3-Approximation

In this section we present a simple heuristic that produces a 3-approximate solution, i.e. a solution with makespan $A(I)$ such that $A(I) \leq 3\,\mathrm{OPT}(I)$ for each instance I.

Theorem 1. *There is an algorithm that finds a 3-approximate solution for non-preemptive scheduling with setup times in time* $O(n)$.

First, we need lower bounds on the optimum makespan $\mathrm{OPT} = \mathrm{OPT}(I)$. Define the values $t_{\max} = \max\{t(j) \mid j \in J\}$, $s_{\max} = \max\{s_k \mid k \in [g]\}$, and $T = \left(\sum_{j \in J} t(j) + \sum_{k=1}^{c} s_k\right)/m$. Obviously we have $t_{\max}, s_{\max}, T \leq \mathrm{OPT}$.

Now, we group the jobs by classes. Let $C_k = \{j_1^k, \ldots, j_{n_k}^k\}$. Then, beginning at machine 1, add for each group one setup and then all jobs to the schedule. Whenever the load of the current machine exceeds T, keep the last placed setup/job and proceed to the next machine. In other words, we add the items

$$s_1, j_1^1, \ldots, j_{n_1}^1, s_2, j_1^2, \ldots, j_{n_2}^2, \ldots, s_c, j_1^c, \ldots, j_{n_c}^c \tag{1}$$

using a next-fit strategy with threshold T, see also Fig. 1.

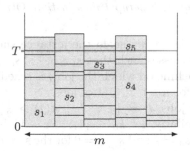

Fig. 1. Example for a next-fit schedule with five classes. The schedule is infeasible until we insert some required setups in the next step of the algorithm.

Assume that we could not add all jobs and setups in this manner. Then there is at least the job $j_{n_c}^c$ left unplaced when the last machine's load exceeds T, and the total load on all machines it greater than mT. On the other hand, the load we have to place is exactly $\sum_{j \in J} t(j) + \sum_{k=1}^{g} s_k = mT$, and we have not even placed all of that, a contradiction.

In the next step, we discard setups that were placed last on a machine. The makespan of the resulting schedule then is at most $T + t_{\max}$, but not all required setups are present. To repair the schedule, we have to add a setup at the beginning of each machine (unless it already started with a setup). This increases the makespan to at most $T + t_{\max} + s_{\max} \leq 3\mathrm{OPT}$.

This algorithm can be implemented to run in linear time $O(n)$.

3   A 2-Dual Approximation

The algorithm described in this section is a 2-dual approximation algorithm. The principle of dual approximation algorithms was introduced by Hochbaum and Shmoys [5]: a c-dual approximation algorithm accepts a number d in addition to the instance as input. It will output a solution with makespan cd, provided that a solution with makespan d exists. Otherwise, it may reject the instance.

This can be turned into a $(c + \varepsilon)$-approximate algorithm for an arbitrary small $\varepsilon > 0$ as follows: First, we obtain a 3-approximation of the optimal makespan OPT using our algorithm from Sect. 2, i.e. we find a schedule of makespan $d' \geq$ OPT such that $d' \leq 3$OPT, or equivalently, $d'/3 \leq$ OPT $\leq d'$. Then use binary search on the interval $[d'/3, d']$ to find a value d such that the algorithm is successful for d, but rejects for $d/(1+\frac{\varepsilon}{c})$ in O($\log \frac{c}{\varepsilon}$) iterations. Note that we must have $d/(1+\frac{\varepsilon}{c}) \leq$ OPT, because the algorithm cannot reject for values larger than OPT, so $d \leq (1+\frac{\varepsilon}{c})$OPT. The solution the algorithm produced for target makespan d then has makespan at most $cd \leq c(1+\frac{\varepsilon}{c})$OPT $= (c+\varepsilon)$OPT, meeting the claimed approximation guarantee.

In this section, we have $c = 2$, so we need at most O($\log \frac{1}{\varepsilon}$) iterations of our algorithm to obtain a $2 + \varepsilon$-approximate schedule:

Theorem 2. *There is an algorithm that finds a $2 + \varepsilon$-approximate solution for non-preemptive scheduling with setup times in time* O($n \log \frac{1}{\varepsilon}$)*, where $\varepsilon > 0$ is arbitrarily small.*

For notational convenience, we scale the instance such that $d = 1$ in each iteration and reorder the classes such that $s_1 \leq \cdots \leq s_c$. Also, for a schedule σ, we denote by $\sigma(j)$ the machine on which the job j is scheduled, and by extension $\sigma(J') = \{\sigma(j) \mid j \in J'\}$ for a set $J' \subseteq J$ of jobs.

We first consider the jobs $J^1 = \{j \in J \mid t(j) > \frac{1}{2}\}$ that have processing time greater than $\frac{1}{2}$. Similarly, let $k_0 \in \{1, \ldots, c+1\}$ be the index such that $s_{k_0} > \frac{1}{2} \geq s_{k_0-1}$, and let $s_0 = 1$ and $s_{c+1} = 0$ for the soundness of this definition, i.e. the classes $C_{k_0}, \ldots, C_c$ have setup times larger than $\frac{1}{2}$. Assume that there is a feasible schedule with makespan 1 and fix any such schedule σ^*. For $k \geq k_0$ define $m_k = |\sigma^*(C_k)|$. The schedule σ^* has some basic properties:

Lemma 3. *(i) The feasible schedule σ^* schedules C_k on at least $|J^1 \cap C_k|$ machines, i.e. $|\sigma^*(C_k)| \geq |C_k \cap J^1|$, for $k \in [c]$.*
(ii) We have $\sum_{k=k_0}^{c} m_k + |J^1| \leq m$.

Proof. (i) Assume that $|\sigma^*(C_k)| < |J^1 \cap C_k|$, then there must be two different jobs $j, j' \in J^1 \cap C_k$ such that $\sigma^*(j) = \sigma^*(j')$. This machine then has load greater than 1, a contradiction.
(ii) Again, σ^* cannot assign two items (jobs or setups) with processing time larger than $\frac{1}{2}$ to one machine, so the sets $\sigma^*(C_k)$, $k \geq k_0$, and $\sigma^*(\{j\})$, $j \in J^1$, are all pairwise disjoint. This implies

$$\sum_{k=k_0}^{c} m_k + |J^1| = \sum_{k=k_0}^{c} |\sigma^*(C_k)| + \sum_{j \in J^1} |\sigma^*(\{j\})| \leq m. \tag{2}$$

$\square$

The Algorithm. Consider the following procedure to stepwise construct a schedule σ:

(1) If $t(j) + s_k > 1$ for some $k \in [c]$ and $j \in C_k$, report that no schedule with makespan 1 exists.
(2) For $k \geq k_0$, schedule all jobs in C_k by using a next-fit strategy with threshold 1 and adding setups as required. If we cannot schedule all jobs on m machines, report that there is no schedule with makespan 1.
(3) Schedule all remaining jobs in J^l on different, unused machines with corresponding setups. Again, report that no schedule with makespan 1 exists if we fail to place all jobs.
(4) For each class C_k, fill the machines $\sigma(C_k \cap J^l)$ with jobs from C_k using next-fit with threshold 1.
(5) Group the remaining jobs by class and, akin to the 3-approximation, schedule one setup per class and the jobs with next-fit and threshold 1. If this fails to schedule all jobs, report that no schedule with makespan 1 exists.
(6) Add a setup at the beginning of each machine if required.

Theorem 4. *Above algorithm either yields a feasible schedule σ of makespan 2, or correctly reports that no schedule with makespan 1 exists in time $O(n)$.*

In order to prove Theorem 4, assume that a schedule σ^* with makespan 1 exists, and define m_k for $k \geq k_0$ as before. For simplicity, we also assume that the algorithm ignores the available number m of machines. We will later show that, if it requires more than m machines, there is no schedule of makespan 1. Note that in steps (2) to (5), we report that no schedule of makespan 1 exists only when the number of machines is insufficient, making this a correct decision.

We first argue that step (1) is also correct: take any job $j \in C_k$. Then σ^* must schedule both j and a setup of length s_k on one machine, and σ^* has makespan at most 1. This also implies that $t(j) \leq \frac{1}{2}$ for $j \in \bigcup_{k=k_0}^{c} C_k$.

Now consider the schedule after step (2).

Claim. Fix one class C_k, $k \geq k_0$. Step (2) uses at most m_k machines to schedule C_k, and the load of these machines is at most $\frac{3}{2}$.

Note that we do not compute the value of m_k in step (2) of the algorithm.

Proof. We have $t(C_k) \leq m_k(1 - s_k)$, because the schedule σ^* uses m_k machines with load at most 1 to schedule C_k. Let ℓ be the number of machines the next-fit procedure in step (2) used to place the jobs C_k. Assume for the sake of contradiction that $\ell > m_k$. Then we have $\ell - 1 \geq m_k$ machines with load at least 1, of which $1 - s_k$ is due to jobs and s_k due to setup. Since we also used the ℓ-th machine, we have $t(C_k) > m_k(1 - s_k)$, a contradiction.

Furthermore, step (1) ensured that $t(j) \leq \frac{1}{2}$ for $j \in C_k$, so the load of these machines is bounded by $\frac{3}{2}$. $\square$

After step (3), our schedule σ therefore uses at most $\sum_{k=k_0}^{c} m_k + |J^l|$ machines. If this is larger than m, there is no feasible schedule with makespan 1 according to

Lemma 3 (ii). In addition, the machines we used to schedule J^l have load at most 1, so the makespan of σ is still bounded by $\frac{3}{2}$.

We now bound the total load of our schedule σ after performing step (5). Note that the load of two schedules may differ due to a different amount of setups the schedule requires.

Claim. The total load of σ is at most as large as the total load of σ^*.

Proof. Fix one class C_k and let ℓ and ℓ^* be the number of setups σ and σ^* scheduled for C_k, respectively. It is sufficient to show $\ell \leq \ell^*$.
Case 1: $s_k > \frac{1}{2}$, i.e. $k \geq k_0$ We already proved that we used at most $m_k = \ell^*$ machines to schedule C_k.
Case 2: $s_k \leq \frac{1}{2}$ and $C_k \cap J^l = \emptyset$ We place jobs from C_k only in step (5), along with one setup. Remember that additional required setups are added in the next step, and should not considered here. Therefore $\ell = 1 \leq \ell^*$.
Case 3: $s_k \leq \frac{1}{2}$ and $C_k \cap J^l \neq \emptyset$ We either have $\ell = |C_k \cap J^l|$, i.e. all jobs from C_k were placed in step (4). In this case, according to Lemma 3 (i), we have $\ell = |C_k \cap J^l| \leq |\sigma^*(C_k)| = \ell^*$. Otherwise we have $\ell > |C_k \cap J^l|$, and we filled $\ell - 1$ machines with jobs from C_k, with a total load of at least $(\ell - 1)(1 - s_k)$. Therefore $(\ell - 1)(1 - s_k) < t(C_k)$. But σ^* can only schedule jobs from C_k with load $1 - s_k$ on each machine, so $t(C_k) \leq \ell^*(1 - s_k)$. It follows that $(\ell - 1)(1 - s_k) < t(C_k) \leq \ell^*(1 - s_k)$, which implies $\ell \leq \ell^*$. $\square$

If we required more than m machines in step (4) or (5), σ would use at least m machines of load 1 and at least one additional machine. Thus the total load of σ would be strictly larger than m, but the total load of σ^* is at most m, a contradiction.

Finally, we add at most one setup of length at most $\frac{1}{2}$ to each machine in step (6), increasing the makespan to at most 2.

The running time of each step is again linear in the number of jobs. The full algorithm, including the computation of the 3-approximate solution and the binary search requires $O(n \log \frac{1}{\varepsilon})$ operations.

4   A Polynomial Time Approximation Scheme

Our proposed algorithm is a 5ε-dual approximation algorithm. As described in Sect. 3, we need at most $O(\log \frac{1}{\varepsilon})$ iterations of our algorithm to obtain a $1 + 6\varepsilon$-approximate schedule.

Theorem 5. *There is a PTAS for non-preemptive scheduling with setup times.*

We assume that $\varepsilon \leq \frac{1}{2}$, and further require that $\frac{1}{\varepsilon}$ is integral. Otherwise, one can e.g. choose the unique value of the form $\frac{1}{2^i}$ in the interval $(\varepsilon/2, \varepsilon)$ as new ε. For notational convenience, we scale the instance such that $d = 1$ and reorder the classes such that $s_1 \leq \cdots \leq s_c$. Let $k_0 \in \{0, \ldots, c\}$ be the index such that $s_{k_0} \leq \varepsilon^3 < s_{k_0+1}$, and let $s_0 = 0$ and $s_{c+1} = 2$ for the soundness of this definition.

Our algorithm distinguishes between small and large jobs, which are defined differently for classes whose setup times are small compared to the target makespan 1 and classes with larger setup times. We will first describe how to handle the small jobs for both types of classes in Sects. 4.1 and 4.2. In Sect. 4.3 we then show how the large jobs can be scheduled near-optimally.

4.1 Removing Small Jobs of Classes with $s_k \leq \varepsilon^3$

Let $J_1 = \bigcup_{k=1}^{k_0} C_k$ be the set of jobs in all classes with $s_k \leq \varepsilon^3$. We consider a job j in J_1 small if $t(j) \leq \varepsilon$ and large if $t(j) > \varepsilon$. Let $J_1^s = \{j \in J_1 \mid t(j) \leq \varepsilon\}$ be the set of all small jobs in J_1 and define

$$L_s = \sum_{j \in J_1^s} t(j) + \sum_{\substack{k \in [k_0] \\ C_k \subseteq J_1^s}} s_k. \tag{3}$$

L_s is the load of all small jobs plus one setup for each class that contains only small jobs.

Lemma 6. *If there is a schedule for J with makespan at most 1, then there is a schedule for $J \setminus J_1^s$ with makespan at most 1 and total load at most $m - L_s$.*

Proof. The schedule for J with makespan 1 has total load at most m. Now simply remove the jobs in J_1^s from the schedule, and afterwards all setups that are no longer required. If a class C_k satisfies $C_k \subseteq J_1^s$, all jobs of this class were removed, and subsequently all setups for this class become superfluous. Because we assume that classes are nonempty, there has to be at least one setup for each class. Thus we removed setups of total length at least $\sum_{\substack{k \in [k_0] \\ C_k \subseteq J_1^s}} s_k$, and the resulting schedule has the desired properties. □

We will not be able to find a schedule with makespan 1 and total load at most $m - L_s$, but in Sects. 4.2 and 4.3 we will show how to find a schedule with only slightly higher makespan and load. We then can re-add the removed jobs.

Lemma 7. *If we have a schedule for $J \setminus J_1^s$ with makespan at most $M \geq 1$ and total load at most $(M - \varepsilon - 2\varepsilon^3)m - L_s$, we can compute a schedule for J with makespan $M(1 + \varepsilon)$ in linear time.*

Proof. We group the jobs in J_1^s by class, i.e. define $C_k^s = C_k \cap J_1^s$ for each $k \in \{1, \ldots, k_0\}$. For groups with $t(C_k^s) \leq \varepsilon^2$, we add them all to one arbitrary machine that already is scheduled to process a large job from the class C_k, if such a large job exists. Since on each machine there are at most $\frac{M}{\varepsilon}$ many large jobs scheduled, this increases the makespan by at most $\frac{M}{\varepsilon}\varepsilon^2 = M\varepsilon$.

After this step, we consider the other jobs. For each group C_k^s with $t(C_k^s) \leq \varepsilon^2$ and $C_k^s = C_k$, i.e. for which no large jobs exists, we instead create a container job j_k^c containing C_k and one setup. The container job has processing time $t(j_k^c) = \sum_{j \in C_k} t(j) + s_k$. Because $\varepsilon \leq \frac{1}{2}$, we have $t(j_k^c) \leq \varepsilon^2 + \varepsilon^3 \leq \varepsilon$, so the container jobs are small.

Now we add all container jobs and the remaining jobs from the groups C_k^s with $t(C_k^s) > \varepsilon^2$ in a next-fit manner such that each machine has at least load $M - \varepsilon - 2\varepsilon^3$ (and at most $M - 2\varepsilon^3$, since all added jobs are small) until we run out of jobs. The total load of the jobs we added in the first step and this step is exactly L_s.

Note that we will be able to place all jobs in this way. Assume that there is a job j we cannot place, then all machines have load at least $M - \varepsilon - 2\varepsilon^3$, i.e. the total load is at least $(M - \varepsilon - 2\varepsilon^3)m$. The original schedule had load at most $(M - \varepsilon - 2\varepsilon^3)m - L_s$, and the jobs we added caused an additional load of at most $L_s - t(j)$, making the total load in the schedule less than $(M - \varepsilon - 2\varepsilon^3)m$.

To get a feasible schedule, we may have to add setups for the jobs from the classes C_k with $t(C_k^s) > \varepsilon^2$ to some machines. Note that these machines have load at most $M - 2\varepsilon^3$. Since we grouped the jobs by class and used a next-fit strategy, there are jobs from at most $\frac{1}{\varepsilon^2} + 2$ of these classes on each machine ($\frac{1}{\varepsilon^2}$ classes with load greater ε^2 and two classes with partial load at most ε^2). Therefore the setup times generate an additional load of at most $\varepsilon + 2\varepsilon^3$ on each machine. Since the load was bounded by $M - 2\varepsilon^3$ before this step, the makespan of the final schedule will not exceed $M + \varepsilon \leq M(1 + \varepsilon)$.   □

4.2   Gluing Small Jobs of Classes with $s_k > \varepsilon^3$

In this section, we preprocess the jobs from $J_2 = \bigcup_{k=k_0+1}^{c} C_k = J \setminus J_1$. For jobs in J_2, the threshold between large and small will be $\delta = \varepsilon^4$. Define the set of all small jobs in J_2 as $J_2^s = \{j \in J_2 \mid t(j) \leq \delta\}$. Again let $C_k^s = C_k \cap J_2^s$ be the small jobs of each class C_k, $k \in \{k_0 + 1, \ldots, c\}$.

We now greedily glue small jobs in C_k^s together until their total length is at least δ (and less than 2δ). If some jobs with total length $r_k < \delta$ remain, we glue them together anyway and increase the length of the resulting job to δ. Call the resulting sets of jobs C_k^g. Now define $C_k' = (C_k \setminus C_k^s) \cup C_k^g$, $J_2' = \bigcup_{k=k_0+1}^{c} C_k'$, and $J' = (J_1 \setminus J_1^s) \cup J_2'$. We now prove that scheduling the glued jobs instead of the original jobs will not increase the optimum makespan or total work too much.

Lemma 8. *If there is a schedule for the original jobs J with makespan 1 and total load L, then there is a schedule for the jobs J' that has makespan less than $1 + 2\varepsilon$ and total load $L + \sum_{k=k_0+1}^{c}(\delta - r_k)$.*

Proof. Consider the schedule for the original jobs J and fix one class $k > k_0$. For each machine i, denote by $h_{i,k}$ the total load the small jobs C_k^s induce on machine i without setup times and remove these jobs from the schedule.

Now add jobs from C_k^g greedily to each machine i until they exceed the load $h_{i,k}$ (by at most 2δ). Note that we can always schedule all jobs from C_k^g by this procedure: if a job cannot be added, the total load of the previously added jobs must exceed $\sum_{i=1}^{m} h_{i,k} = \sum_{j \in C_k^s} t(j)$. On the other hand, the total load of the previously placed jobs must be less than $\sum_{j \in C_k^s} t(j)$, provided that we add the single job whose length was increased last.

Now consider the load of a fixed machine i after jobs from all classes have been glued and inserted. Since we considered only jobs from classes with setup times larger than ε^3, jobs from at most $\frac{1}{\varepsilon^3}$ classes are scheduled on i. The greedy procedure exceeded the load $h_{i,k}$ for each of these classes by less than 2δ, which increases the makespan by less than $\frac{1}{\varepsilon^3}2\delta = 2\varepsilon$ by the definition of δ.

Finally note that total load increases by exactly $\sum_{k=k_0+1}^{c}(\delta - r_k)$, because we increased the size of the last glued job of each class k by $\delta - r_k$.   □

4.3   Finding a Schedule for Large Jobs

The set J' contains only jobs larger than $\delta = \varepsilon^4$. We now round the length of these jobs and the setup times down to the next multiple of $U = \delta\frac{\varepsilon}{1+2\varepsilon}$, i.e. $\tilde{t}(j) = \lfloor\frac{t(j)}{U}\rfloor U$ and $\tilde{s}_k = \lfloor\frac{s_k}{U}\rfloor U$. Using rounded times does not hurt us too much:

Lemma 9. *(i) If a schedule for J' has makespan less than $1 + 2\varepsilon$ and total load L with the original processing times and setup times, then it has makespan less than $1 + 2\varepsilon$ and total load at most $L - \sum_{j\in J'}(t(j) - \tilde{t}(j))$ with the rounded times.*

(ii) If a schedule for J'_1 has makespan less than $1 + 2\varepsilon$ and total load L with the rounded times, then this schedule has makespan less than $1 + 3\varepsilon + 2\varepsilon^2$ and total load at most $L + \sum_{j\in J'}(t(j) - \tilde{t}(j)) + m(\varepsilon^2 + \varepsilon^4)$ with the original times.

Proof. (i) This is obvious, because we rounded all times down.

(ii) Since $\frac{1}{\varepsilon}$ is integral, $\delta = \frac{1+2\varepsilon}{\varepsilon}U = (\frac{1}{\varepsilon} + 2)U$ is a multiple of U. Also each job j has $t(j) > \delta$, and because we round down do the next multiple of δ, we have $\tilde{t}(j) \geq \delta$. Fix any job j, then the realtive increase of its processing time when undoing the rounding is

$$\frac{t(j) - \tilde{t}(j)}{\tilde{t}(j)} < \frac{U}{\delta} = \frac{\varepsilon}{1+2\varepsilon} < \varepsilon. \tag{4}$$

For any class k we can bound

$$\tilde{s}_k > s_k - U > \varepsilon^3 - U = \delta\left(\frac{1}{\varepsilon} - \frac{\varepsilon}{1+2\varepsilon}\right). \tag{5}$$

Since $\varepsilon \leq \frac{1}{2}$, this is further bounded by

$$\delta\left(\frac{1}{\varepsilon} - \frac{\varepsilon}{1+2\varepsilon}\right) > \delta(\frac{1}{\varepsilon} - \varepsilon) \geq \frac{3}{4}\varepsilon^3 > \varepsilon^4 = \delta, \tag{6}$$

so the setup times increase by a factor of at most $1 + \varepsilon$ as well. The load of each machine therefore is less than $(1 + 2\varepsilon)(1 + \varepsilon) = 1 + 3\varepsilon + 2\varepsilon^2$.

Now consider the load due to setups. Using (5), we can bound the number of setups on each machine. Thus, the load on each machine increases by at most

$$\frac{1 + 2\varepsilon}{\delta\left(\frac{1}{\varepsilon} - \frac{\varepsilon}{1+2\varepsilon}\right)} \times U = \varepsilon^2 + \frac{\varepsilon^4}{1 + 2\varepsilon - \varepsilon^2} \leq \varepsilon^2 + \varepsilon^4. \tag{7}$$

Finally, the total load due to jobs increases by exactly $\sum_{j \in J'}(t(j) - \tilde{t}(j))$.   □

Finally, we solve the instance with jobs J' and rounded times exactly. For this, we apply dynamic programming over the jobs. Conceptually, we have a set of feasible schedules for the previously considered jobs, and extend these schedules by adding the current job j (and its groups setup if necessary) on all machines where this does not exceed the target makespan. The key observation is, that, in order to determine where j can be scheduled, we do not need complete knowledge of the schedule. We only need to know the load of each machine, and whether it already has the setup for j's class scheduled. In fact, since machines of the same load then are indistinguishable, we only need to know how many machines of each load the schedule has, i.e. we consider two schedules equivalent if the number of machines for each load is equal. Our dynamic program only considers the equivalence classes of feasible schedules we are able to construct. Also, the load of each machine is a multiple of U and less than $1 + 2\varepsilon$, and thus takes one of the values $0, U, 2U, \ldots, \lfloor \frac{1+2\varepsilon}{U} \rfloor U$, i.e. $B = \lfloor \frac{1+2\varepsilon}{U} \rfloor + 1 = O(\frac{1}{\varepsilon^5})$ possible values.

Lemma 10. *We can find all feasible schedules (up to equivalence) of J' with rounded times and makespan less than $1 + 2\varepsilon$ in time $O(nm^{2/\varepsilon^5 + 2})$.*

Proof. Formally, the state of our dynamic program is a set S of vectors. Each vector $v = (m_0^-, m_0^+, \ldots, m_{B-1}^-, m_{B-1}^+) \in \{0, \ldots, m\}^{2B}$ represents one class of feasible schedules we were able to construct. The components m_ℓ^+ (m_ℓ^-) give the number of machines with load ℓU that already have (do not have) the setup for the current job's class scheduled. Note that we can have at most $m^{2B} + 2B = O(m^{2B})$ vectors in S: the vectors from the set $\{0, \ldots, m-1\}^{2B}$ and the $2B$ vectors of the form $(0, \ldots, 0, m, 0, \ldots, 0)$.

The state is initialized with only the class of the empty schedule, i.e. $S = \{(m, 0, \ldots, 0)\}$. To correctly keep track of which machines have which setups, we order the jobs by class. For the first job of each class, we reset the components of each vector before placing the job, i.e. we set $m_\ell^- \leftarrow m_\ell^- + m_\ell^+$ and $m_\ell^+ = 0$. This requires time $O(cm^{2B})$.

For each job j from class C_k, we create a new, initially empty set S', which will later hold the vectors corresponding to feasible schedules for all jobs up to and including j. For this, consider all vectors from S and all of the $2B$ machine types to add j to: let $v = (m_0^-, m_0^+, \ldots, m_{B-1}^-, m_{B-1}^+) \in S$, $\ell \in \{0, \ldots, B-1\}$, and $\circ \in \{+, -\}$. Let

$$\ell' = \begin{cases} \ell + \frac{\tilde{t}(j)}{U} & \text{if } \circ = + \\ \ell + \frac{\tilde{t}(j) + \tilde{s}_k}{U} & \text{if } \circ = -. \end{cases} \tag{8}$$

When $\ell' < B$, adding j to a machine with load ℓU and with/without setup for C_k (depending on $\circ$) leads to a feasible schedule. In this case, we add the vector $\hat{v} = (\hat{m}_0^-, \hat{m}_0^+, \ldots, \hat{m}_{B-1}^-, \hat{m}_{B-1}^+)$ to S' that has $\hat{m}_\ell^\circ = m_\ell^\circ - 1$, $\hat{m}_{\ell'}^+ = m_{\ell'}^+ + 1$, and is identical to v in every other component. After computing all vectors vor the job j, we replace S by S' before proceeding to the next job. This step requires $O(2Bm^{2B})$ steps for each job.

For all groups and jobs, our dynamic program requires $O(cm^{2B}+n2Bm^{2B}) = O(nm^{2B})$ operations, using that ε is a constant.

4.4   Putting it Together

We can start putting everything together for our algorithm.

Lemma 11. *If there is a schedule for J with makespan at most 1, then there is a schedule for J' that has makespan at most $1 + 2\varepsilon$, total load at most*

$$L^* = m - L_s + \sum_{k=k_0+1}^{c} (\delta - r_k) - \sum_{j \in J'} (t(j) - \tilde{t}(j)) \tag{9}$$

with the rounded times.

Proof. According to Lemma 6, there is a schedule for $J \setminus J_1^s$ with makespan at most 1 and total load at most $m - L_s$. Lemma 8 states that there is a schedule for J' with makespan at most $1+2\varepsilon$ and total load at most $m - L_s + \sum_{k=k_0+1}^{g}(\delta - r_k)$. The latter schedule has makespan at most $1 + 2\varepsilon$ and total load at most L^* by Lemma 9 (i) with the rounded times.

Note that the load of different feasible schedules do not need to be equal, because different distributions of classes to machines may require more or less setup time.

The Algorithm.

(1) Remove the jobs J_1^s from the instance.
(2) Glue the items in J_2^s together and obtain J'.
(3) Round the processing and setup times for jobs in J'.
(4) Using our dynamic program, find the schedule for J' with makespan at most $1 + 2\varepsilon$ that has the lowest total load. If no such schedule exists or its load is greater than L^*, report that no schedule for J with makespan 1 exists.
(5) Undo the rounding and replace the glued jobs with the original jobs.
(6) Add the jobs J_1^s using Lemma 7.

Theorem 12. *Above procedure either finds a schedule of makespan $1 + 4\varepsilon + 5\varepsilon^2 + 2\varepsilon^3$ or correctly decides that no schedule of makespan 1 exists.*

Proof. The decision to report that no schedule with makespan 1 exists is correct by Lemma 11. If we found a schedule that has makespan at most $1 + 2\varepsilon$ and total load at most L^* with the rounded times, then, according to Lemma 9 (ii), this schedule has makespan at most $M = 1 + 3\varepsilon + 2\varepsilon^2$ and total load at most

$$L^* + \sum_{j \in J'}(t(j) - \tilde{t}(j)) + m(\varepsilon^2 + \varepsilon^4) = m - L_s + \sum_{k=k_0+1}^{c} (\delta - r_k) + m(\varepsilon^2 + \varepsilon^4) \tag{10}$$

with the original times. Replacing the glued jobs by the original small jobs reduces the load by $\sum_{k=k_0+1}^{c}(\delta-r_k)$ to $(1+\varepsilon^2+\varepsilon^4)m-L_s \leq (M-\varepsilon-2\varepsilon^3)m-L_s$. Therefore, Lemma 7 allows us to add all the removed jobs J_1^s, and the makespan increases by most εM to $1 + 4\varepsilon + 5\varepsilon^2 + 2\varepsilon^3$. $\square$

The running time of our algorithm is dominated by the dynamic program with $O(nm^{2/\varepsilon^5+2})$ steps. Every other step can be performed in linear time $O(n)$.

5   Conclusion

We presented three algorithms for non-preemptive scheduling with setup costs on identical machines. The first two algorithms are simple to implement and extremely fast, and have approximation ratio 3 and $2 + \varepsilon$, respectively. We also presented a polynomial time approximation scheme, thus we can approximate the problem arbitrarily good.

This raises some questions: is there an efficient algorithm with approximation ratio less than 2? Can we design a faster PTAS, preferably an EPTAS? An EPTAS has a running time $f(\frac{1}{\varepsilon}) \times \text{poly}(n)$, thus avoiding large exponents.

Further research could also consider the model of uniformly related machines, where machines can have different speeds that affect the processing times. Note that, without setup times, scheduling on identical and uniformly related machines admits an EPTAS [6].

References

1. Allahverdi, A., Gupta, J.N., Aldowaisan, T.: A review of scheduling research involving setup considerations. Omega **27**(2), 219–239 (1999)
2. Allahverdi, A., Ng, C., Cheng, T., Kovalyov, M.Y.: A survey of scheduling problems with setup times or costs. Eur. J. Oper. Res. **187**(3), 985–1032 (2008)
3. Chen, B.: A better heuristic for preemptive parallel machine scheduling with batch setup times. SIAM J. Comput. **22**(6), 1303–1318 (1993)
4. Garey, M., Johnson, D.: Computers and Intractability: A Guide to the Theory of NP-Completeness. Series of Books in the Mathematical Sciences, W. H. Freeman (1979)
5. Hochbaum, D.S., Shmoys, D.B.: Using dual approximation algorithms for scheduling problems theoretical and practical results. J. ACM **34**, 144–162 (1987)
6. Jansen, K.: An EPTAS for scheduling jobs on uniform processors: Using an MILP relaxation with a constant number of integral variables. SIAM J. Discrete Math. **24**(2), 457–485 (2010)
7. Mäcker, A., Malatyali, M., Meyer auf der Heide, F., Riechers, S.: Non-preemptive scheduling on machines with setup times. In: Dehne, F., Sack, J.-R., Stege, U. (eds.) WADS 2015. LNCS, vol. 9214, pp. 542–553. Springer, Heidelberg (2015)
8. Monma, C.L., Potts, C.N.: Analysis of heuristics for preemptive parallel machine scheduling with batch setup times. Oper. Res. **41**(5), 981–993 (1993)
9. Schuurman, P., Woeginger, G.J.: Preemptive scheduling with job-dependent setup times. In: Proceedings of the Tenth Annual ACM-SIAM Symposium on Discrete Algorithms (SODA 1999), pp. 759–767. SIAM (1999)

Cuboid Partitioning for Parallel Matrix Multiplication on Heterogeneous Platforms

Olivier Beaumont[1,2], Lionel Eyraud-Dubois[1,2(✉)], and Thomas Lambert[1,2]

[1] Inria, University of Bordeaux, Bordeaux, France
Lionel.Eyraud-Dubois@inria.fr
[2] LaBRI, University of Bordeaux, Bordeaux, France

Abstract. The problem of partitioning a square into zones of prescribed areas arises when partitioning matrices for dense linear algebra kernels onto a set of heterogeneous processors, and several approximation algorithms have been proposed for that problem. In this paper, we address the natural generalization of this problem in dimension 3: partition a cuboid in a set of zones of prescribed volumes (which represent the amount of computations to perform), while minimizing the surface of the boundaries between zones (which represent the data transfers involved). This problem naturally arises in the context of matrix multiplication, and can be seen as a heterogeneous generalization of 2.5D approaches that have been proposed in this context. The contributions of this paper are twofold. We prove the NP-completeness of the general problem, and we propose a $\frac{5}{6^{2/3}} \simeq 1.51$-approximation algorithm for cube-partitioning. This is the first known approximation result for this 3D partitioning problem.

1 Introduction

In the case of homogeneous resources, the problem of partitioning data for Linear Algebra kernels in order to both balance the load throughout the computation and to minimize communications is well understood. 2D block-cyclic distributions, for instance, have been introduced in Scalapack [12] in order to achieve this goal. More recently, the problem has received a lot of attention in Communication Avoiding algorithms design (see [1,16] and [3,22] for Matrix Multiplication specifically). In this context, the goal is to partition the set of elementary computations to be performed into a minimal number of zones, each zone being able to be processed in local memory (*i.e.* both input, intermediate and output data). This corresponds to maximizing the volume of computations that can be processed with a given amount of memory.

In this paper, we concentrate on Matrix Multiplication algorithms and more specifically on Matrix Multiplication algorithms that involve N^3 elementary operations of type $C_{i,j} \leftarrow C_{i,j} + A_{i,k} B_{k,j}$, *i.e.* we ignore variants such as Strassen or Coppersmith-Winograd. Note that throughout the paper, we will assume that matrices are partitioned into blocks, whose size is chosen so as to be well adapted

© Springer International Publishing Switzerland 2016
P.-F. Dutot and D. Trystram (Eds.): Euro-Par 2016, LNCS 9833, pp. 171–182, 2016.
DOI: 10.1007/978-3-319-43659-3_13

to all types of resources (typically CPUs and GPUs). On the other hand, we consider a fully heterogeneous platform, where all nodes may have different processing capacities and we address the most general problem, where several partially aggregated copies of C can reside simultaneously in memory, such as in 2.5D algorithms [22]. In this context, the problem consists in partitioning the computational domain (the cube of N^3 points) into sub-domains allocated to the different resources. In order to balance the load between the processing units, each unit should receive a volume of computations proportional to its processing speed and the overall amount of communications, that corresponds to the overall boundary area between the zones should be minimized.

Many algorithms [6,10,14,15,17,19] have been proposed in the context of dense matrix multiplication based on Canon's-like algorithm, that corresponds to the 2D version of the problem, *i.e.* how to partition a matrix into zones of fixed area while minimizing the overall length of the boundaries. On the other hand, to the best of our knowledge, this paper is the first to consider the complexity of the 3D version of the algorithm, to prove the NP-Completeness of the underlying decision problem and to propose an approximation algorithm for it.

Related Works

The 2D version of this optimization problem has been first introduced by Lastovetsky and Kalinov in [17]. In [6], it has been proven that the problem is NP-Complete, and a first algorithm with bounded approximation ratio (1.75) has been proposed. This algorithm has been improved along two directions. On the one hand, Lastovetsky *et al.* have proposed to relax the assumption stating that the zones allocated to the processors should be single rectangles and have proposed optimal algorithms, but limited to 2 heterogeneous processors [10] and more recently to 3 heterogeneous processors [14]. On the other hand, recursive partitioning algorithms have recently been proposed where at each step, the set of processors is split into two parts. Sophisticated proof techniques enabled Nagamochi and Abe [19] to improve the approximation ratio down to 1.25. Recently, Fügenschuh *et al.* [15] improved this result to 1.15, but under the assumption that if we consider processors in decreasing order of their processing speeds, there is no abrupt change in the performance between 2 successive processors. Unfortunately, such an abrupt decrease typically happens when considering nodes consisting of CPUs and GPUs, such that Fügenschuh's algorithm is limited to the case of relatively homogeneous platforms. In [8], an algorithm based on the idea of non rectangular partitioning proposed by Lastovetsky and extended to any number of processors by adapting the recursive partitioning algorithm proposed by Nagamochi has been proposed. It achieves an approximation ratio of $\frac{2}{\sqrt{3}} \simeq 1.15$ and does not require any specific assumption on the relative speed of resources, so that it can be used in the case of nodes consisting of both regular cores and accelerators.

Besides a single heterogeneous node, this partitioning problem has been adapted to distributed hierarchical and highly heterogeneous platforms in [13], where the partitioning is applied at two levels (intra-node and inter-node), based on sophisticated performance models. The same partitioning has also been extended to finite-difference time-domain (FDTD) methods to obtain numerical solutions of Maxwell's equations in [21]. More dynamic settings have also been considered in [18]. Recently, in order to cope with resource heterogeneity and the difficulty to build optimal schedules, the use of dynamic runtime schedulers have been proposed, such as StarPU [2], StarSs [20], or PaRSEC [11]. In these systems, at runtime, the scheduler takes the scheduling and allocation decisions based on the set of ready tasks (tasks whose all data and control dependencies have been resolved), on the availability of the resources (estimated using expecting processing and communication times), and on the actual location of input data. The comparison between static scheduling strategies and runtime scheduling strategies has been considered in [7], where the analysis of the behavior of static, dynamic, and hybrid strategies highlights the benefits of introducing more static knowledge and allocation decisions in runtime libraries.

Paper Outline

The paper is organized as follows. In Sect. 2, we formally present the partitioning problem and the notations that will be used throughout the paper. In Sect. 3, the complexity of the associated decision problem in the 3D case is established and a $\frac{5}{6^{2/3}} \simeq 1.51$-approximation algorithm for cube-partitioning is proposed in Sect. 4. Conclusions and perspectives are given in Sect. 5.

2   General Context

Definition 1. *Let P be a connected polyhedron included in $[0, x] \times [0, y] \times [0, z]$. We define its covering cuboid as the smallest cuboid $Cu(P) = [x_1, x_2] \times [y_1, y_2] \times [z_1, z_2]$ that includes P. We define also its width $w(P)$ as $x_2 - x_1$, its height $h(P)$ as $y_2 - y_1$ and its length $l(P)$ as $z_2 - z_1$. Let us define $Hs(P) = h(P)l(P) + w(P)l(P) + h(P)w(P)$, $\rho(P) = \frac{max(h(P),w(P),l(P))}{min(h(P),w(P),l(P))}$ and $\rho'(P) = \frac{max(h(P),w(P),l(P))}{med(h(P),w(P),l(P))}$. Finally we denote by $V(P)$ the volume of P.*

Problem 1 (Minimizing-Surface-Cuboid-Partition (MSCuboidP)). Given a set of n numbers $\{v_1, \ldots, v_n\}$ such that $\sum v_k = xyz$, and the cuboid $Cu = [0, x] \times [0, y] \times [0, z]$, find for each v_k a polyhedron P_k of Cu such that $V(P_k) = v_k$ and $\bigcup P_k = Cu$ minimizing $\sum Hs(P_k)$.

A general lower bound for this problem has been established by Ballard et al. [4] and comes from Loomis-Whitney inequality. It simply states that a polyhedron P of volume $V(P)$ minimizes the surface of its covering cuboid if and only if it is shaped as a cube. This implies the following lower bound:

$$Hs(P) \geq 3V(P)^{\frac{2}{3}}. \tag{1}$$

3   NP-Completeness

We prove in this section the NP-completeness of the decision problem associated to MSCuboidP, MSCuboidP-DEC.

Problem 2 (MSCuboidP-DEC). Given a set of n given numbers $\{v_1, \ldots, v_n\}$ such that $\sum v_k = xyz$, a cuboid $Cu = [0, x] \times [0, y] \times [0, z]$ and a number K, is there a set of n polyhedra P_k of Cu such that $V(P_k) = v_k$, $\bigcup P_k = Cu$ and $\sum Hs(P_k) \leq K$?

We start by proving the NP-completeness of a more constrained variant in which the goal is to partition the cuboid in cubes of specified side lengths. The direct reduction to MSCuboidP-DEC is then given at the end of the section.

Problem 3 (All-Cube-Cuboid-Partition (ACCuboidP)). Given a set of p given length $\{l_1, \ldots, l_p\}$ such that $\sum l_k^3 = xyz$, and a cuboid $Cu = [0, x] \times [0, y] \times [0, z]$, is there a set of p cubes $C_k \in Cu$ such that $V(C_k) = l_k^3$ and $\bigcup C_k = Cu$?

Lemma 1. *ACCuboidP is NP-Complete.*

It is easy to check that ACCuboidP belongs to NP. We prove NP-hardness of ACCuboidP with a method inspired from the hardness proof of the equivalent 2D problem [5], by using a reduction from 2-PART-EQUAL, a variant of the well-known 2-PART problem. The NP-completeness of 2-PART-EQUAL can be proven by a reduction from 2-PART. Indeed, adding a constant C to every element of the instance of 2-PART and then adding n (where n is the size of the instance) elements of size C to the instance itself creates an instance of 2-PART-EQUAL that has a solution if and only if the original instance of 2-PART has one. Our proof consists in two steps: from an instance of 2-PART-EQUAL, we first derive another set of numbers b_i and prove that they can be partitioned into two equal size sets if and only if the 2-PART-EQUAL instance has a solution. Then, we use the b_i numbers to build an instance of ACCuboidP for which the existence of a packing is equivalent to partitioning the b_i's into two equal size sets.

Problem 4 (2-PART-EQUAL). Given a set of $2n$ integers $\{a_1, \ldots, a_{2n}\}$, is there $I \subseteq [1, n]$ such that $|I| = n$ and

$$\sum_{i \in I} a_i = \sum_{i \notin I} a_i \ ?$$

First Reduction. Let us now consider an instance of 2-PART-EQUAL, $\{a_1, \ldots, a_{2n}\}$ and let us denote $2A = \sum_{i=1}^{2n} a_i$ and $M = 6n \times \max_i a_i$. Let us suppose, without loss of generality, that n is a multiple of 120 greater than 240 and let us define a new set $\{b_1, \ldots, b_{2n}\}$ as

$$\forall i, \ b_i = a_i + 3n \times \max_i a_i + D \qquad \text{where } D = \frac{60M - (A \bmod 60M)}{n}$$

$$= a_i + \frac{M}{2} + D.$$

In addition, let us set $k = \frac{n}{120} + \frac{A+nD}{60M}$ and $S = \frac{1}{2}\sum_{i=1}^{2n} b_i$. One can prove that k is an integer (since n is a multiple of 120) and that $S = 60k \times M$. In addition, let us notice that for all i, $\frac{M}{2} < b_i$ (since $D \geq 0$ and $a_i > 0$) and $b_i \leq M$. Indeed, $D \leq \frac{60M}{n} \leq \frac{M}{4}$ and $a_i + \frac{M}{2} \leq M(\frac{1}{2} + \frac{1}{6n}) \leq \frac{4M}{6}$. Therefore $\forall i, b_i \leq \frac{11M}{12} < M$.

Let us prove now that there exists a solution to our instance of 2-PART-EQUAL if and only if there exists a set $I \subset [1,n]$ such that $\sum_{i \in I} b_i = \sum_{i \notin I} b_i$. If there exists I such that $|I| = n$ and $\sum_{i \in I} a_i = \sum_{i \notin I} a_i$, then

$$\sum_{i \in I} b_i = \sum_{i \in I} a_i + \frac{n}{2}(\frac{M}{2} + D) = \sum_{i \notin I} a_i + \frac{n}{2}(\frac{M}{2} + D) = \sum_{i \notin I} b_i.$$

Let us assume that there exists a set I such that $\sum_{i \in I} b_i = \sum_{i \notin I} b_i$. Then,

$$\sum_{i \in I} b_i - \sum_{i \notin I} b_i = \sum_{i \in I} a_i - \sum_{i \notin I} a_i + (|I| - |\bar{I}|)(\frac{M}{2} + D)$$

$$\sum_{i \notin I} a_i - \sum_{i \in I} a_i = (|I| - |\bar{I}|)(\frac{M}{2} + D).$$

Yet, $\sum_{i \notin I} a_i - \sum_{i \in I} a_i \leq 2n \times max\ a_i = \frac{M}{3}$ and $\frac{M}{2} \leq \frac{M}{2} + D$. Therefore,

$$(|I| - |\bar{I}|)\frac{M}{2} \leq \frac{M}{3} \text{ and } (|I| - |\bar{I}|) \leq \frac{2}{3} < 1.$$

By symmetry, we obtain $|I| = |\bar{I}|$ and I is a solution to 2-PART-EQUAL.

Second Reduction. In order to build the ACCuboidP instance that will be used in the reduction, we rely on a result from Walters [23] stating that it is possible to tile any cuboid with a number of cubes which is poly-logarithmic in the side lengths of the cuboid. We call the cubes in such a tiling Walters' cubes, and we denote by $WS(X, Y, Z)$ a (poly-logarithmic size) set of cubes tiling the cuboid $X \times Y \times Z$.

Let us consider the following instance of ACCuboidP:

- A cuboid of size $11M \times 15M \times S$ (with $S = 60k \times M$).
- $20k$ cubes of length $6M$.
- $24k$ cubes of length $5M$.
- $30k$ cubes of length $4M$.
- $20k$ cubes of length $3M$.
- $\forall i$, a cube of length b_i.
- $\forall i$, $WS(M - b_i, b_i, b_i)$ and $WS(M, M - b_i, b_i)$.

with M, k and the b_i's defined from the a_i's as in the first reduction described above. One can see that the reduction is polynomial, since the sizes of the Walters' cubes sets are poly-logarithmic functions of the b_i's.

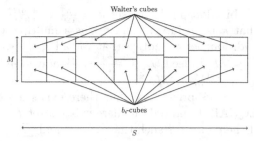

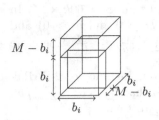

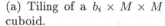

(a) Tiling of a $b_i \times M \times M$ cuboid.

(b) Side view of the $M \times M \times S$ packing.

Fig. 1. Tiling of a $M \times M \times S$ packing.

In the first part of the proof, we prove that if we can split the b_i items in two equal sets, then the above set of cubes can be packed into the cuboid.

Let us first consider, for each i, the cube of length b_i and the two associated Walters' cubes sets. Figure 1(a) shows how they can be packed into a cuboid of size $M \times M \times b_i$, where the cuboid of size $(M - b_i) \times b_i \times b_i$ and the cuboid of size $(M - b_i) \times b_i \times M$ are tiled with the cubes from $WS(M - b_i, b_i, b_i)$ and $WS(M, M - b_i, b_i)$ respectively. Stacking up such cuboids on top of one another, we can build two $M \times M \times S$ cuboids from the two sets I and $\bar{I}$, see Fig. 1(b).

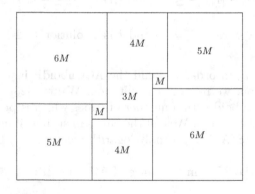

Fig. 2. Tiling of a $11M \times 15M$ rectangle.

Figure 2 shows how to tile a $11M \times 15M$ rectangle with the corresponding squares, where both $M \times M$ squares represent a slice of the $M \times M \times S$ cuboids presented above. This arrangement can be repeated for a total length of S, since $10k \times 6M = 12k \times 5M = 15k \times 4M = 20k \times 3M = 60k \times M = S$.

Hence, this provides a tiling of the whole $11M \times 15M \times S$ cuboid.

For the second part of the proof, we need to prove that if the cuboid can be tiled with the set of cubes, then a partition of the b_i values in two equal size sets exists. We start by proving that in any valid tiling of the cuboid, the $11M \times 15M$

rectangle can only be tiled as shown on Fig. 2 (or under the same pattern but with an horizontal symmetry). This can be proven by considering all possible values for the area occupied by the smallest cubes (the b_i-cubes, the Walters' cubes, and the cubes of length $3M$). Case analysis shows that this surface needs to be at least $11M^2$, which implies that any slice of the 3D tiling must intersect the same number of cubes of each type (two of each length $6M$, $5M$, $4M$ and one of length $3M$). Then one can prove that the only way to tile the $11M \times 15M$ rectangle with these squares is as shown on Fig. 2. Due to lack of space, we were not able to provide the full case analysis and we refer the interested reader to the companion research report [9] for the sake of completeness.

We have built a pattern that must appear on each slice of the cuboid, and in which the b_i cubes have to be included into two separate parts of the tiling. Let us denote by I the indexes of the b_i cubes which appear in the leftmost $M \times M \times S$ cuboid. Since by construction, $b_i > \frac{M}{2}$ for all i, these cubes have to be arranged as depicted on Fig. 1(b). This shows that $\sum_{i \in I} b_i \leq S$ and $\sum_{i \notin I} b_i \leq S$. Since the total sum is $2S$, this implies that both sums are in fact equal to S, and thus that there exists a solution to the original 2-PART-EQUAL instance.

Note that the pattern in Fig. 2 can be horizontally reversed. Furthermore, both possibles patterns can be present on the final tiling. However, even in this case, considering the b_i-cubes on the left side still yields a set I such that $\sum_{i \in I} b_i = \sum_{i \notin I} b_i = S$. $\qquad\qquad\qquad\qquad\qquad\qquad\qquad\qquad\qquad\qquad\square$

Theorem 1. *MSCuboidP-DEC is NP-complete.*

Proof. There is a reduction from ACCuboidP to MSCuboidP-DEC. Indeed, $ACCuboidP(\{l_1, \ldots, l_p\}, [0, x] \times [0, y] \times [0, z])$ is true if and only if $MSCuboidP - DEC(\{l_1^3, \ldots, l_p^3\}, [0, x] \times [0, y] \times [0, z], 3 \sum v_k^{2/3})$ is true (the bound in (1) is tight if and only if there exists a partitioning where only cubes are used). Yet, thanks to Lemma 1, ACCuboidP is NP-complete. Therefore MSCuboidP-DEC is NP-complete.

4   Approximation Algorithm

In this section, we present 3D-NRRP, an approximation algorithm for the case where the cuboid to partition is cubic, what corresponds to the multiplication of square matrices. It is inspired by the NRRP algorithm proposed in [8] and by Nagamochi *et al.* in [19].

4.1   Presentation and Correctness of 3D-NRRP

3D-NRRP (see Algorithm 1) is based on the divide and conquer principle, and its analysis relies on the following invariant: at each step, the aspect ratio of the cuboid to be partitioned is smaller than 3. In what follows, we define the aspect ratio ρ of a cuboid as the ratio of the largest length by the smallest length. We also define the second aspect ratio ρ' as the ratio of the largest length by the median length.

Algorithm 1. 3D-NRRP

Input: A set of values $\{v_1, \ldots, v_n\}$ sorted in non-decreasing order, a cuboid
$$Cu = [a, b] \times [c, d] \times [e, f] \text{ with } (b - a) \times (d - c) \times (f - e) = \sum_{i=1}^{n} v_i$$

Output: For each $i \leq n$ a polyhedron P_i such that $\bigcup P_i = Cu$ and $V(P_i) = v_i$

1   **if** $n = 1$ **then**
2   **return** Cu
3   **else**
4     $v = \sum_{i=1}^{n} v_i$;
5     $w = b - a$; $h = d - c$; $l = f - e$;
6     $\rho_1 = \frac{max(w,h,l)}{min(w,h,l)}$; $\rho_2 = \frac{max(w,h,l)}{med(w,h,l)}$;
7   **if** *there exists k such that* $\sum_{i=1}^{k-1} v_i \geq \frac{v}{3\rho_2}$ **then**
8       $k =$ the smallest such index ;
9       $v' = \sum_{i=1}^{k-1} v_i$;
10   Cut Cu along its longest edge to obtain Cu_1 and Cu_2, with respective volumes v' and $v - v'$;
11   **return** 3D-NRRP($\{v_1, \ldots, v_{k-1}\}, Cu_1$) $\cup$ 3D-NRRP($\{v_k, \ldots, v_n\}, Cu_2$) ;
12   **else**
13       $v' = \sum_{i=1}^{n-1} v_i$;
14       $Cu_1 = [a, a + \sqrt[3]{v'}] \times [c, c + \sqrt[3]{v'}] \times [e, e + \sqrt[3]{v'}]$;
15   **return** 3D-NRRP($\{v_1, \ldots, v_{n-1}\}, Cu_1$) $\cup$ ($Cu \setminus Cu_1$)

At each step of the algorithm, the current cuboid (whose aspect ratio is smaller than 3) is split into two parts, and the same routine is recursively applied to each part. To ensure that the resulting parts have an aspect ratio smaller than 3, the splitting is performed according to two modes. The first mode is the general case, in which the cuboid is partitioned in two disjoint cuboids by cutting along the largest length (Lines 7 to 11 in Algorithm 1, and Fig. 3(a)). This is possible if there exists an index k such that $\sum_{i=1}^{k-1} v_i \geq v/(3\rho_2)$. Indeed, Lemmas 2 and 3 show that this condition is sufficient to prove the invariant for both parts. More specifically, Lemma 2 states that in that case, both resulting cuboids have a total volume greater than one third of the overall volume, and Lemma 3 states that the aspect ratio of both cuboids is smaller than 3, under the assumption that the previous one had also a ratio less than 3, what ensures the correctness of the algorithm.

In the second mode, v_n is significantly larger than the other v_i's. Splitting in two cuboids would result in the smallest cuboid having an aspect ratio larger than 3. Therefore 3D-NRRP shapes the smallest part as a cube, included in the covering cuboid of the other part, which is made of one element only, namely v_n (Lines 20 to 22 of Algorithm 1, and Fig. 3(b)).

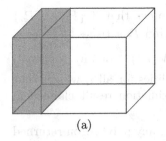
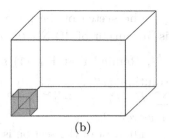

(a)   (b)

Fig. 3. The two splitting modes of 3D-NRRP. In both cases, the gray polyhedra is attributed to $\{v_1, \ldots, v_{k-1}\}$, the white one to $\{v_k, \ldots, v_n\}$.

Lemma 2. *Let $\{v_1, \ldots, v_n\}$ be a set of positive values sorted in non-decreasing order, $\rho \geq 1$, and $v = \sum_i v_i$. Let us assume that there exists an index k such that $\sum_{i=1}^{k} \frac{1}{v_i} \geq v/3\rho$, and let us consider the smallest such integer. Then $\sum_{i=k}^{n} v_i \geq v/3\rho$.*

Proof. By definition of k, $\sum_{i=1}^{k-2} v_i < v/3\rho$. Therefore, if we assume that $\sum_{i=k}^{n} v_i < v/3\rho$, we obtain $v_{k-1} = v - \sum_{i=1}^{k-2} v_i - \sum_{i=k}^{n} v_i \geq v/3\rho$ (since $\rho \geq 1$). Since $v_k \leq \sum_{i=k}^{n} v_i < v/3\rho$, we have $v_{k-1} > v_k$ which is a contradiction with the fact that the v_i's are sorted in non-decreasing order.

Lemma 3. *Let Cu be a cuboid of dimension $w \times h \times l$, with volume $V = hwl$, aspect ratio ρ, and second aspect ratio ρ'. Let us assume that Cu_1 and Cu_2 are obtained by cutting Cu along the largest length, with $V(Cu_1)$ and $V(Cu_2)$ not smaller than $\frac{V}{3\rho'}$. Then, $\rho(Cu_1) \leq max(3, \rho)$ and $\rho(Cu_2) \leq max(3, \rho)$.*

Proof. Let us suppose that $w \leq h \leq l$ without loss of generality. In this case, $w = w(Cu_1) = w(Cu_2)$, $h = h(Cu_1) = h(Cu_2)$, $\rho = l/w$ and $\rho' = l/h$. Let us denote $l_1 = l(Cu_1)$ and $l_2 = l(Cu_2)$ and let us consider cuboid Cu_1. Then, there are 3 cases to consider,

- If $h \leq l_1$, then $\rho(Cu_1) = l_1/w \leq l/w = \rho$,
- If $w \leq l_1 \leq h$, then $\rho(Cu_1) = h/w \leq l/w \leq \rho$,
- If $l_1 \leq w \leq h$, by assumption $w \times h \times l_1 = V(Cu_1) \geq \frac{V}{3\rho'} = \frac{w \times h \times l}{3\rho'}$. Therefore $l_1 \geq l/3\rho'$. Then, $\rho(Cu_1) = h/l_1 \leq \frac{3h\rho'}{l} = 3$.

Thus, in all cases, $\rho(Cu_1) \leq max(3, \rho)$. By symmetry, the same proof applies to Cu_2.

4.2 Approximation Ratio

This section is devoted to the proof of Theorem 2, which states that the 3D-NRRP achieves a $\frac{5}{6^{2/3}}$ approximation ratio ($\frac{5}{6^{2/3}} \simeq 1.51$).

Theorem 2. *3D-NRRP is a $\frac{5}{6^{2/3}}$-approximation when the given cuboid is cubic.*

The sketch of the proof is as follows. First, we prove that if $\{P_1, \ldots, P_n\}$ is an output of 3D-NRRP, then any output polyhedron P_i satisfies $\frac{Hs(P_i)}{3V(P_i)^{\frac{2}{3}}} \leq \frac{5}{6^{\frac{2}{3}}}$. Remind that Eq. (1) states that $Hs(P_i^*) \geq 3V(P_i^*)^{\frac{2}{3}}$ for any (optimal) solution $\{P_1^*, \ldots, P_n^*\}$. By summing up these inequalities for all i, we get that $\sum_i Hs(P_i) \leq \frac{5}{6^{\frac{2}{3}}} \sum_i Hs(P_i^*)$ and obtain the approximation result claimed in Theorem 2.

The rest of the section is devoted to the proof that any polyhedron returned by 3D-NRRP satisfies the property stated above. One can see that there are only two situations in which 3D-NRRP returns a polyhedron: Line 2 and Line 15. In the first case, the returned zone is a cuboid with aspect ratio less than 3. In this case Lemma 5 provides the claimed result, since $\frac{5}{3\sqrt[3]{3}} \leq \frac{5}{6^{\frac{2}{3}}}$. We first state a technical result in Lemma 4, whose proof is not included for lack of space, and can be found on the companion technical report [9].

Lemma 4. Let $f(x,y) = \frac{y+x(1+y)}{3(xy)^{2/3}}$. For $y \in [1,3]$ and $x \in [1,y]$, $f(x,y) \leq \frac{5}{3\sqrt[3]{3}}$.

Lemma 5. If P is a cuboid with $\rho(P) \leq 3$, then $\frac{Hs(P)}{3V(P)^{2/3}} \leq \frac{5}{3\sqrt[3]{3}}$.

Proof. We denote $\rho(P) = \rho$ and $V(P) = V$. We suppose that $w = w(P) \leq h = h(P) \leq l(P) = l$ without loss of generality. We denote $x = h/w$. In this case $l = \rho w$ and $V = whl = \rho x w^3$. Therefore

$$\frac{Hs(P)}{3V^{2/3}} = \frac{(x + \rho + x\rho)w^2}{3(\rho x w^3)^{2/3}} = \frac{\rho + x(1+\rho)}{3(\rho x)^{2/3}} = f(x,\rho),$$

where f is as defined in Lemma 4, what ends the proof.   □

In the other case, 3D-NRRP returns a cuboid minus a cube (Line 15 of Algorithm 1, as described on Fig. 3(b)). The bound on the volume of the cube is such that the conditions of Lemma 7 are fulfilled. As before, this relies on the technical Lemma 6, whose proof can be found on the technical report [9].

Lemma 6. Let $f(x,y) = \frac{y+x(1+y)}{3(xy-\frac{x^2}{3})^{\frac{2}{3}}}$. For $y \in [1,3]$ and $x \in [1,y]$, $f(x,y) \leq \frac{5}{6^{\frac{2}{3}}}$.

Lemma 7. If $V(P) \geq (1-\frac{1}{3\rho'(P)})V(Cu(P))$ and $\rho(P) \leq 3$, then $\frac{Hs(P)}{3V(P)^{2/3}} \leq \frac{5}{6^{\frac{2}{3}}}$.

Proof. Let us denote $\rho = \rho(P)$, $\rho' = \rho'(P)$, $V = V(P)$, and let us denote by w, h, l the dimensions of $Cu(P)$. Let us suppose that $w \leq h \leq l$ without loss of generality. Then, $l = \rho w$ and $l = \rho' h$, and let us also denote $x = h/w = \rho/\rho'$. With such notations, we get $V(Cu(P)) = \rho x w^3$ and $Hs(P) = (\rho + x(1+\rho))w^2$. Thus, the condition on $V(P)$ can be written as

$$V(P) \geq (1 - \frac{1}{3\rho'(P)})V(Cu(P)) = (1 - \frac{x}{3\rho})\rho x w^3 = (\rho x - \frac{x^2}{3})w^3,$$

what leads to $\frac{Hs(P)}{3V(P)^{2/3}} \leq \frac{(\rho + x(1+\rho))w^2}{3(\rho x - \frac{x^2}{3})^{2/3}w^2} = \frac{\rho + x(1+\rho)}{3(\rho x - \frac{x^2}{3})^{2/3}} = f(x,\rho),$

where f is as defined in Lemma 6, what provides the conclusion.   □

5   Conclusion

We introduce a model of the partitioning problem associated to the 2.5D matrix multiplication algorithm on heterogeneous resources, a problem of crucial importance in high performance computing. This corresponds to partitioning a cuboid into several polyhedra, each representing the volume of computations attributed to a resource. We provide two theoretical results: a proof of the NP-completeness for this problem, and an approximation result for 3D-NRRP, which generalizes the results obtained in the 2D case. This is the first known approximation result for this problem, and it provides a strong guarantee ($\frac{5}{6^{2/3}} \simeq 1.51$). In addition, the computational time of the algorithm is extremely low, $O(n \log n)$, where n is the number of processors, what makes it perfectly suitable for practical use. This work opens several interesting perspectives for dimensions higher than 3, what corresponds to tensor products. It would also be interesting to combine this algorithm, whose general goal is to minimize the overall volume of communications while enforcing perfect load balancing between heterogeneous resources, with algorithms that explicitly take into account memory constraints at each node.

References

1. Anderson, M., Ballard, G., Demmel, J., Keutzer, K.: Communication-avoiding QR decomposition for GPUs. In: IEEE International Parallel & Distributed Processing Symposium (IPDPS), pp. 48–58. IEEE (2011)
2. Augonnet, C., Thibault, S., Namyst, R., Wacrenier, P.A.: StarPU: a unified platform for task scheduling on heterogeneous multicore architectures. Concurrency and Comput. Practice Exp. **23**, 187–198 (2009). Special Issue: Euro-Par 2009
3. Ballard, G., Demmel, J., Holtz, O., Lipshitz, B., Schwartz, O.: Communication-optimal parallel algorithm for strassen's matrix multiplication. In: Proceedings of the Twenty-Fourth Annual ACM Symposium on Parallelism in Algorithms and Architectures, pp. 193–204. ACM (2012)
4. Ballard, G., Demmel, J., Holtz, O., Schwartz, O.: Minimizing communication in linear algebra. SIAM J. Matrix Anal. Appl. **32**(3), 866–901 (2011). arXiv: 0905.2485
5. Beaumont, O., Boudet, V., Rastello, F., Robert, Y.: Matrix multiplication on heterogeneous platforms. IEEE Trans. Parallel Distrib. Syst. **12**(10), 1033–1051 (2001)
6. Beaumont, O., Boudet, V., Rastello, F., Robert, Y., et al.: Partitioning a square into rectangles: NP-completeness and approximation algorithms. Algorithmica **34**(3), 217–239 (2002)
7. Beaumont, O., Eyraud-Dubois, L., Guermouche, A., Lambert, T.: Comparison of static and dynamic resource allocation strategies for matrix multiplication. In: Proceedings of the 26th IEEE International Symposium on Computer Architecture and High Performance Computing (SBAC-PAD), pp. 1–10. IEEE (2015)
8. Beaumont, O., Eyraud-Dubois, L., Lambert, T.: A new approximation algorithm for matrix partitioning in presence of strongly heterogeneous processors. In: 30th IEEE International Parallel and Distributed Processing Symposium (2016)
9. Beaumont, O., Eyraud-Dubois, L., Lambert, T.: Cuboid partitioning for parallel matrix multiplication on heterogeneous platforms (2016). http://hal.inria.fr/hal-01269881

10. Becker, B., Lastovetsky, A.: Towards data partitioning for parallel computing on three interconnected clusters. In: Sixth International Symposium on Parallel and Distributed Computing, ISPDC 2007, pp. 39–39. IEEE (2007)
11. Bosilca, G., Bouteiller, A., Danalis, A., Faverge, M., Hérault, T., Dongarra, J.: PaRSEC: a programming paradigm exploiting heterogeneity for enhancing scalability. Comput. Sci. Eng. **15**(6), 36–45 (2013)
12. Choi, J., et al.: ScaLAPACK: a portable linear algebra library for distributed memory computers - design issues and performance. In: Waśniewski, J., Madsen, K., Dongarra, J. (eds.) PARA 1995. LNCS, vol. 1041, pp. 95–106. Springer, Heidelberg (1996)
13. Clarke, D., Ilic, A., Lastovetsky, A., Sousa, L.: Hierarchical partitioning algorithm for scientific computing on highly heterogeneous CPU + GPU clusters. In: Kaklamanis, C., Papatheodorou, T., Spirakis, P.G. (eds.) Euro-Par 2012. LNCS, vol. 7484, pp. 489–501. Springer, Heidelberg (2012)
14. DeFlumere, A., Lastovetsky, A.: Optimal data partitioning shape for matrix multiplication on three fully connected heterogeneous processors. In: Lopes, L., et al. (eds.) Euro-Par 2014, Part I. LNCS, vol. 8805, pp. 201–214. Springer, Heidelberg (2014)
15. Fügenschuh, A., Junosza-Szaniawski, K., Lonc, Z.: Exact and approximation algorithms for a soft rectangle packing problem. Optimization **63**(11), 1637–1663 (2014)
16. Hoemmen, M.: Communication-avoiding Krylov Subspace Methods. Ph.D. thesis, University of California, Berkeley (2010)
17. Kalinov, A., Lastovetsky, A.: Heterogeneous distribution of computations solving linear algebra problems on networks of heterogeneous computers. J. Parallel Distrib. Comput. **61**(4), 520–535 (2001)
18. Mohamed, N., Al-Jaroodi, J., Jiang, H.: DDOps: dual-direction operations for load balancing on non-dedicated heterogeneous distributed systems. Cluster Comput. **17**(2), 503–528 (2014)
19. Nagamochi, H., Abe, Y.: An approximation algorithm for dissecting a rectangle into rectangles with specified areas. Discrete Appl. Math. **155**(4), 523–537 (2007)
20. Planas, J., Badia, R.M., Ayguadé, E., Labarta, J.: Hierarchical task-based programming with StarSs. Int. J. High Perform. Comput. Appl. **23**(3), 284–299 (2009)
21. Shams, R., Sadeghi, P.: On optimization of finite-difference time-domain (FDTD) computation on heterogeneous CPU and GPU clusters. J. Parallel Distrib. Comput. **71**(4), 584–593 (2011)
22. Solomonik, E., Demmel, J.: Communication-optimal parallel 2.5D matrix multiplication and LU factorization algorithms. In: Jeannot, E., Namyst, R., Roman, J. (eds.) Euro-Par 2011, Part II. LNCS, vol. 6853, pp. 90 109. Springer, Heidelberg (2011)
23. Walters, M.: Rectangles as sums of squares. Discrete Math. **309**(9), 2913–2921 (2009)

HeSP: A Simulation Framework for Solving the Task Scheduling-Partitioning Problem on Heterogeneous Architectures

Antón Rey, Francisco D. Igual$^{(\boxtimes)}$, and Manuel Prieto-Matías

Dept. Arquitectura de Computadores Y Automática,
Universidad Complutense de Madrid, Madrid, Spain
{anrey,figual,mpmatias}@ucm.es

Abstract. In this paper we describe HeSP, a complete simulation framework to study a general task scheduling-partitioning problem on heterogeneous architectures, which treats recursive task partitioning and scheduling decisions on equal footing. Considering recursive partitioning as an additional degree of freedom, tasks can be dynamically partitioned or merged at runtime for each available processor type, exposing additional or reduced degrees of parallelism as needed. Our simulations reveal that, for a specific class of dense linear algebra algorithms taken as a driving example, simultaneous decisions on task scheduling and partitioning yield significant performance gains on two different heterogeneous platforms: a highly heterogeneous CPU-GPU system and a low-power asymmetric big.LITTLE ARM platform. The insights extracted from the framework can be further applied to actual runtime task schedulers in order to improve performance on current or future architectures and for different task-parallel codes.

1 Introduction and Motivation

Task-parallel programming models have emerged as an appealing solution in order to tackle the programmability problem on both homogeneous and heterogeneous platforms. These efforts aim at reducing user intervention to manage data dependences, task allocation and data transfer management by delegating those tasks to underlying runtime task schedulers. However, the ever-increasing heterogeneity in current (and future) architectures has dramatically aggravated the challenge for runtime developers; as more types of computing resources are available, it becomes more difficult to concurrently exploit them in order to optimize co-operative parallel implementations. One of the main conceptual problems lies on how to optimally (and possibly dynamically) partition a task into sub-tasks (that is, solving a *task partitioning problem*), and how to efficiently schedule them to the most convenient resource among those available in order to maximize performance (that is, solving a *task scheduling problem*).

In this paper, we present HeSP (*Heterogeneous Scheduler-Partitioner*), a simulation framework that addresses both problems in a simultaneous fashion.

© Springer International Publishing Switzerland 2016
P.-F. Dutot and D. Trystram (Eds.): Euro-Par 2016, LNCS 9833, pp. 183–195, 2016.
DOI: 10.1007/978-3-319-43659-3_14

Based on per-task and data transfers performance models, HeSP adds an additional degree of freedom to typical task scheduling policies by considering a joint task partitioning/scheduling approach. The framework proceeds by finding a set of task partitions that divides the initial workload into a number of sub-tasks with different granularity, that better fit to the underlying hardware resources at a given execution point. The approach drives to considerable performance improvements and more efficient resource utilization. We show that the new task scheduler-partitioner paradigm is of wide appeal to increase the scheduling quality on highly heterogeneous architectures, and to gain insights that can be further applied to specific task-parallel implementations, actual runtime task schedulers, and present and future heterogeneous architectures.

Runtime task schedulers are capable of managing efficient load balancing, asynchronous out-of-order task execution and handling data across separated memory spaces, abstracting these mechanisms to the programmer. Concretely, StarPU [1], OmpSs [2] or XKaapi [4], among others, offer implicit parallel programming models with transparent data dependence analysis among tasks, and support scheduling on heterogeneous processing platforms. Efficient scheduling under this task-based perspective strongly depends on the quality of the scheduling policies implemented in the runtime, and more specifically, how they address the special features of the algorithm and the underlying architecture.

These efforts usually consider the static creation and management of equally-sized tasks operating on uniform data tiles, which naturally drives to an improper load balancing among computing resources on heterogeneous architectures, given the different processing capabilities of each type of resource. As a side effect, establishing the optimal block size, even in the homogeneous target system case, is a time-consuming effort for the developer, and strongly depends on the algorithmic properties of the target implementation and the features of the underlying architecture. Although each processor type typically reaches its performance peak for substantially different task sizes, and the chosen initial granularity exposes a fixed amount of parallelism, few strategies have been developed in order to dynamically adapt task granularity to the underlying heterogeneous hardware. Focusing on dense linear algebra implementations, [8] propose a hierarchical directed acyclic graph (DAG onwards) strategy, creating a two-level DAG hierarchy on systems featuring two types of computing platforms (CPU/GPU). Similarly, [5] proposes an offline adaptation of the task grain size to the processor type and to statically assign tasks to distributed compute nodes. On the other hand, [3] proposes an alternative approach in which computing resources are aggregated as needed in order to adapt the computing capabilities to coarse grain kernels. The *Versioning* task scheduler for the OmpSs runtime [6] defines multiple implementations per task, each one targeting a different processor type, and decides at runtime where to map them based on historical runtime information.

HeSP extends the aforementioned efforts by exploring the global impact of *arbitrary* degrees of task granularity on an arbitrary heterogeneous platform, adapting task sizes not only to the individual processor capabilities, but also to the current degree of available parallelism dictated by a specific algorithm.

1.1   A Motivating Example: Tiled Cholesky Factorization

Let us expose a motivating and illustrative example of the actual problems related with equally-sized task partitioning on heterogeneous platforms. The blocked Cholesky factorization decomposes an $n \times n$ symmetric positive definite matrix A stored by $s \times s$ blocks of dimension $b \times b$ each, into $A = LL^T$ where L is a lower triangular matrix. At runtime, the outer loop in the code depicted in Fig. 1 that calculates the Cholesky factorization divides the operation into a number of sub-tasks that, when executed under a task-parallel paradigm, generate a task DAG as that shown in Fig. 2(a). In the task DAG, nodes correspond to different tasks, and edges denote data dependencies between them.

```
void cholesky (double *A[s][s], int b, int s) {
  for (int k = 0; k < s; k++) {
    chol (A[k][k], b, b);              // Cholesky factor. (diag. block)

    for (int j = k + 1; j < s; j++)
      trsm (A[k][k], A[k][j], b, b);   // Triangular solve

    for (int i = k + 1; i < s; i++) {
      for (int j = i + 1; j < s; j++)
        gemm (A[k][i], A[k][j],        // Matrix multiplication
              A[i][j], b, b);

      syrk (A[k][i], A[i][i], b, b);   // Symmetric rank-b update
} } }
```

Fig. 1. C implementation of the blocked Cholesky factorization.

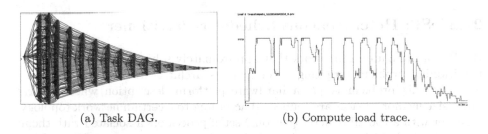

(a) Task DAG.   (b) Compute load trace.

Fig. 2. (a) Task DAG in which the computation evolves from left to right, and (b) compute load trace generated by the Cholesky factorization in Fig. 1, for a problem size $n = 16,384$, and block size $b = 1,024$.

The Cholesky factorization is an appealing example for our purposes: it exhibits different sub-task types (CHOL, SYRK, GEMM and TRSM) and complex data dependences among them, and it features different degrees of parallelism as the factorization evolves. Consider, for example, how the DAG depicted in Fig. 2(a) reduces the potential parallelism (that is, the number of tasks that can be potentially executed in parallel, typically related with the width of the DAG) at the first stages of the factorization, and (in a much larger extent) at the last stages. This is usually translated into processor load patterns like that shown in Fig. 2(b), that represents a timeline of an execution of the Cholesky factorization on a highly heterogeneous platform, composed by 28 Intel Xeon cores and 3 different GPUs. The plot represents the number of active processors as the execution proceeds. Areas with reduced load are usually due to load imbalance. Note that this phenomenon can be motivated by two different factors: different processing capabilities of each processor type, and lack of potential parallelism on specific stages of the execution. The first can be alleviated by *scheduling* heuristics (e.g. mapping tasks in the critical path to fast processors), but the second is inherent to the algorithm, and can be alleviated by dynamic task *partitioning* in order to expose additional parallelism at runtime.

Data block (tile) size is a crucial parameter in task-parallel executions, as it ultimately determines the amount of available parallelism, and the efficiency of each individual task execution. In Fig. 1, block size is determined by b; note that, typically, larger block sizes usually imply higher performance per individual task, and smaller block sizes tend to expose higher degrees of parallelism, which naturally drives to better processor occupation. In addition, different block (task) sizes are desired for different architectures, and even for different problem dimensions in the same architecture.

Altogether, these observations motivate the exploration of new techniques that explore the impact of *heterogeneous* or *non-uniform* task partitioning on the performance and resource occupation of heterogeneous architectures. In the following, we introduce HeSP, a complete framework that supports the definition of complex heterogeneous architectures and simulates simultaneous task-scheduling and task-partitioning schemes that alleviate the aforementioned problems.

2   HeSP: Heterogeneous Scheduler-Partitioner

HeSP is a simulation framework that approximately solves the task scheduling-partitioning problem targeting heterogeneous architectures. At a glance, the input to this problem is *(1)* a hardware platform description where several finite-size memory spaces are connected according to a certain network topology, together with a (possibly heterogeneous) set of processors associated with them; and *(2)* a task to be computed in that platform. A solution to this problem consists of *(1)* a set of tasks –presumably with different granularity–, related by arbitrary data dependences and equivalent to the input task, and *(2)* a task-to-processor mapping. The objective function is typically performance maximization, although energy consumption minimization is also supported by HeSP.

2.1   Features of the Scheduling-Partitioning Simulation Framework

Besides supporting recursive task partitioning, HeSP is designed to be a realistic framework that simulates not only current heterogeneous architectures, but also state-of-the-art scheduling and data management policies on task-parallel executions. In the following, we introduce its features in detail.

Task and Data Scheduling Heuristics. HeSP implements different heuristics for task-to-processor assignments. *Random* (R-P) and *Fastest* (F-P) processor selection policies consider such processor choices among idle processors at the task release time. The *Earliest Idle Time* (EIT-P) and *Earliest Finish Time* (EFT-P) policies select the processor becoming idle first, and the processor finishing first if that task is assigned to it, respectively. EFT-P estimates the finishing time accounting for eventual data transfers if needed. Task scheduling order is specified by choosing between *First-come, first-served* (FCFS) or *Priority-List* (PL) choices. In PL, a priority list is built by sorting tasks by their critical times in decreasing order. Critical times are computed by averaging task processing time for all processors, and propagating them throughout the task DAG by a backflow algorithm. The combination of *Priority-List* and EFT-P heuristics is practically identical to the well-known *HEFT* scheduling algorithm [7].

When several independent memory spaces are present, HeSP considers data movement for scheduling decisions, considering individual memory spaces of each accelerator as software caches of a main memory space, typically tied to CPUs. Common caching policies like *write-through* (WT), *write-back* (WB) or *write-around* (WA) are used. When a task is about to be scheduled to a processor, the required data transfers are issued from the source memory space to the memory space the processor is tied to using prefetching schemes.

Performance and Data Transfer Models. HeSP estimates computing or transfer times relying on analytical models extracted a priori for each task/data type and size mapped to any existing processor/interconnect in the system. These estimations are required when making both scheduling and partitioning decisions, jointly or in an isolated fashion. The quality of these models will ultimately determine the accuracy of the simulated scheduling results.

Recursive Task Partitioners. Task *partitioners*, specified for each task type willing to be partitioned, are just blocked algorithms (see, for example, Fig. 1 for the specific case of the Cholesky factorization) with an input parameter that specifies the data granularity/degree of parallelism of the following partition. On a partitioner invocation, the corresponding emergent sub-tasks are managed by HeSP by introducing them in the respective task DAG which the partitioned task belongs to. In Fig. 3, starting from initial CHOL task –Cholesky factorization–, it is illustrated how three successive task partitions –corresponding to respective CHOL, TRSM and SYRK blocked algorithms– affect the prior task DAG, and the corresponding data partitions they induce.

Note that any task can be partitioned again as long as its dependent data blocks can be divided consistently, so an extremely hierarchical task DAG can

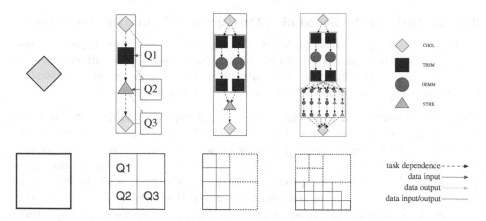

Fig. 3. Three successive task partitions and corresponding partitioned data blocks. Tasks and their related data dependences –Q1, Q2 and Q3 quadrants– are shown only for the first partition for the sake of clarity. (Color figure online)

be constructed by recursively partition its tasks. A *task cluster* is a set of tasks generated from a single task partitioning, being the source task their *parent*. We refer to the *DAG/graph depth* to indicate the maximum number of nested task clusters, and *DAG/graph width* as the maximum number of tasks that can be run in parallel. For instance, in the four task DAGs in Fig. 3, the corresponding depths are 0, 1, 2 and 2, and their widths are 1, 1, 2 and 6. Dependences between tasks, shown as dashed arrows, represent RaW, WaR and WaW constraints.

Recursive Data Partitioning and Data Coherence Management. New tasks generated after a partition reference to finer-grained input and output data dependences, which are partitions of the initial data block(s) of the parent task (see Fig. 3). HeSP implements validate/invalidate mechanisms to ensure data coherency among different memory spaces while handling asynchronous memory transfers. Since recursive task partitions induce corresponding recursive data block partitions, the existing partitioned data blocks are organized in a directed acyclic graph structure (data DAG onwards) in which nodes represent data blocks and directed links represent nesting relations between them; for example, $A \longrightarrow B$ means B is fully contained in A and A is bigger than B.

Armed with the data DAG, validations and invalidations are propagated by top-bottom and bottom-top mechanisms throughout this graph to maintain coherence. For instance, to ensure that a task can start its computation and store the result in an output block OB, not only the block OB must be invalidated on the remaining memory spaces in which the block might be present, but the hypothetical data block partitions contained in OB and the bigger blocks in which OB might be contained must be invalidated as well. Similarly, after a certain task has finished its computation updating OB, both OB and all the blocks within OB must be validated in the memory space corresponding to the processor assigned to that task.

In general, these data block partitions induced by task partitions form tree structures. However, it is possible to have a pair of blocks which intersect partially, nested within a common bigger parent block. This case shows up when two partitions of non-divisible grain sizes are applied to the same data block (for example, quadrant $Q2$ in Fig. 3). In this case, a new data block descriptor which refers to its intersection is introduced in the data DAG as a common child node of two intersecting blocks (see Fig. 4). With this mechanism, together with the validate/invalidate propagation mechanisms, data coherency is ensured for all possible data partitions and hierarchical data graphs.

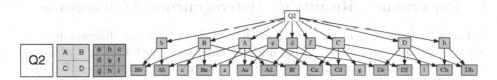

Fig. 4. A data block (Q2 quadrant) can be simultaneously divided according to different tilings (yellow and blue tilings, corresponding to TRSM and SYRK task partitions in Fig. 3). Additional data block descriptors (green) are constructed to represent partial overlaps between not nested data blocks. (Color figure online)

Iterative Solver. HeSP solves the scheduling-partitioning problem by iteratively searching for those hierarchical task DAGs which best fit to an heterogeneous processing platform—according to performance optimization—given a specific combination of the mentioned scheduling heuristics—processor selection heuristic and task ordering. A schedule stage is followed by a partition stage for each iteration, being the number of iterations a user-defined parameter.

At the partition stage, HeSP chooses a candidate task to be partitioned or a candidate task cluster to be merged back/repartitioned with a different granularity. A global analysis of the schedule-partition done in the previous iteration can provide useful information—i.e., bottleneck identification, number of idle resources, or too fine grained tasks— to help the iterative process to converge towards a better overall schedule-partition. This is the reason why we chose an iterative approach as the first implementation of HeSP instead of a *local-scoped* constructive approach, in which scheduling or partitioning decisions are made at every task arrival to the scheduling queue.

The partition procedure is based on two stages: *(1)*, task selection to build the candidate list, and *(2)*, sampling type to choose the final candidate. For *(1)*, HeSP implements three different policies: *All*, *CP* and *Shallow*. *All* selects all tasks of the previous step, *CP* selects only tasks belonging to the critical path, and *Shallow* selects those tasks whose depth (that is, number of task clusters that contain it) is minimal. All existing task clusters are candidates to be merged back or repartitioned. For each added candidate, a positive score is computed by subtracting the current cost delay by an estimated cost after its eventual partition or merge, being this estimation based on the available parallelism at its scheduling time of the previous step. For each candidate whose data dependences have a characteristic size d, a partition parameter $p \in (0, 1]$ is chosen such that new tasks created after the eventual partition will depend on data blocks of size $b = p \times d^1$. The more available parallelism is exposed, the smaller p is set in order to generate a higher amount of parallel finer-grained tasks.

In the second stage, a final selection among all candidates is done according to *Hard* or *Soft* procedures. In *Hard*, the candidate with the maximum score is chosen; in *Soft*, the candidate is randomly selected such that the selection probability equals the score divided by the sum of all scores.

3   Performance Results on Heterogeneous Architectures

In the following, we will feed HeSP with data describing two different heterogeneous architectures: BUJARUELO, a highly heterogeneous CPU-GPU architecture, composed by 28 Intel Xeon-E5 2695v3 cores running at 2.3 GHz, 2 GeForce GTX980 GPU and 1 GTX950; and ODROID, a low-power asymmetric ARM architecture with two types of processors: 4 slow Cortex-A7 and 4 fast Cortex-A15, running at 800 and 1300 MHz respectively. NVIDIA CUBLAS/CUSOLVER v7.5 and Intel MKL v11.3 were used to extract task performance models on BUJARUELO, and BLIS v0.9.1 was used on ODROID.

3.1   Framework Validation and Evaluation of Scheduling Heuristics

The goal of the first set of experiments is twofold: first, to validate the results extracted from HeSP by comparing them with an equivalent execution using a real task scheduler; second, to illustrate the impact on performance of several scheduling policies in HeSP when using *homogeneous* or *uniform* task partitions.

Each point in the OmpSs line in Fig. 5 (left) corresponds to the best scheduling performance out of 20 OmpSs executions for each grain size. These 20 trials were set to let OmpSs *Versioning* scheduler [6] improve itself by gathering enough task execution delay samples for each task type/size and processor. The other two curves—HESP-REPLICA-PM and HESP-REPLICA-RD—denote the performance attained by HeSP when applying the same task-to-processor mapping extracted from the best OmpSs trial, using our performance models and the real OmpSs task delays, respectively, for each uniform tiling.

Differences in performance between HESP-REPLICA-RD and OMPSS points are a measure of the OmpSs runtime overhead while the differences between HESP-REPLICA-PM and HESP-REPLICA-RD are mainly due to the accuracy of our performance models and possible differences between own OmpSs task delay instrumentation module and the instrumentation we used to extract our performance models. Summarizing, the differences between the replicated schedules are small enough and easily explainable to assert the validity of the following results. In general, our observations reveal a qualitative matching between real and simulated workloads for all problem sizes, with deviations that can be easily explained and do not usually affect the quality of the observations.

---

[1] A task cluster is a candidate to be merged if $p = 1$ or repartitioned if $p < 1$.

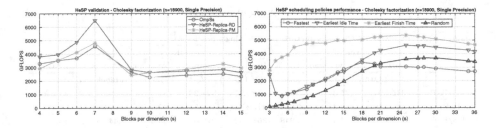

Fig. 5. Left: Comparison between OmpSs and their replicated schedules. Right: Comparison between different scheduling policies and block sizes in HeSP.

To introduce the context where our *heterogeneous* or *non-uniform* partitioning approach takes place and its potential benefits, Fig. 5 (right) reports the performance obtained by running HeSP simulations using different scheduling policies for different uniform task partitions. Some facts are remarkable: first, the optimal tile size does not only depend on the underlying architecture and problem size, but also on the selected scheduling policy; second, for every policy, performance curves follow a similar pattern, exhibiting a peak performance in a trade-off tile size that best balances potential parallelism and optimal individual task performance; third, differences in performance are relevant depending on the selected scheduling policy, being even more dramatic for large tile sizes; this gives a clue on the potential benefits that will be obtained by using a non-uniform partitioning scheme, as exposed next.

3.2 Impact of Non-uniform Partitioning on Performance

In the following, we illustrate the main performance improvements obtained with HeSP using *All/Soft* configuration for task partitioning selection. Table 1 reports performance values on BUJARUELO and ODROID using the best uniform and non-uniform partitions found by HeSP for different scheduling policies[2], together with additional metrics that clarify many of the concepts exposed hereafter, including average processor load, optimal/average block size and task DAG depth. The first point to notice is the overall improvement attained for all non-uniform task partitions found by HeSP and the overall reduction in the optimal average block size on non-uniform partitions.

Note the direct relation between the average processor occupancy and the improvements of the non-uniform partitions. For example, EIT-P with uniform partitioning yields high processor occupancies (between 91 % and 98.5 %), so the potential benefit expected from additional extracted parallelism is poor, ranging between 0.76 % and 2.02 %. Contrary, uniform partitions on EFT-P schedules yield better performance than EIT-P ones while still leaving more room for potential parallelism. Although the quality of EFT-P schedules could actually leave little room for additional improvements, the greater processor availability

---

[2] In all cases, we use WB as the caching mechanism.

Table 1. Performance comparison for BUJARUELO and ODROID.

BUJARUELO (32, 768 × 32, 768 Cholesky factorization in single precision)								
	Best Uniform			Best Found Non-uniform				
	Perf	Avg. load	Block	Perf	Improve	Avg. load	Avg	DAG
Config	(GFLOPS)	(%)	size	(GFLOPS)	(%)	(%)	block size	depth
FCFS/R-P	3453.91	75.3	1024	4189.17	21.29	82.3	991.23	2
PL/R-P	4460.30	88.4	1024	4752.43	6.55	89.4	978.33	2
FCFS/F-P	2846.78	53.4	2048	3687.93	29.55	63.6	446.52	3
PL/F-P	3381.76	68.4	2048	3614.28	6.88	66.2	1165.70	3
FCFS/EIT-P	5650.10	91.3	1024	5747.87	1.73	92.3	1002.26	2
PL/EIT-P	6096.91	93.9	1024	6206.55	1.80	95.4	1009.91	2
FCFS/EFT-P	6581.96	23.3	2048	7569.34	15.00	63.9	412.15	5
PL/EFT-P (*)	7046.87	55.9	2048	8030.50	13.96	86.9	407.41	4
ODROID (8, 192 × 8, 192 Cholesky factorization in double precision)								
	Best Uniform			Best Found Non-uniform				
	Perf	Avg. load	Block	Perf	Improve	Avg. load	Avg	DAG
Config	(GFLOPS)	(%)	size	(GFLOPS)	(%)	(%)	block size	depth
FCFS/R-P	3.75	63.9	512	4.87	29.9	70.8	458.89	2
PL/R-P	4.89	70.9	512	5.84	19.3	77.4	461.11	2
FCFS/F-P	7.59	69.7	512	8.10	6.74	73.7	335.80	3
PL/F-P	8.55	88.4	512	8.80	2.91	92.0	466.00	2
FCFS/EIT-P	8.46	98.5	256	8.52	0.76	99.1	255.19	2
PL/EIT-P	8.74	96.2	512	8.91	2.03	97.7	463.76	2
FCFS/EFT-P	8.77	89.6	512	8.96	2.20	96.2	301.23	3
PL/EFT-P (*)	8.84	91.4	512	9.08	2.75	99.0	352.07	3

they offer permits the iterative scheduler-partitioner to find finer-grained partitions (see Fig. 6(d)), attaining remarkable net improvements for BUJARUELO (between 13.96 % and 15 %). Note also that bigger performance improvements do not only correspond with lower processor occupancies, but also with higher task DAG depths (up to 5 in BUJARUELO). This observation reinforces the importance of managing arbitrary task granularity, introduced by HeSP, extending the idea of using only two degrees of granularity for two types of processors introduced in other works [8].

This reasoning also applies when comparing the highly heterogeneous BUJARUELO with the less heterogeneous ODROID since the optimal uniform tile size seems to fit better to homogeneous platforms, yielding higher occupancies for all scheduling policies tested, hence leaving less room for non-uniform partitioning improvements. Even with those limitations, HeSP does always provide improvements in all cases.

Note the even better improvements, with simpler –i.e. less deep– partitions, attained by our scheme when jointly applied with simpler schedulers –R-P/F-P– and naive FCFS task ordering. Since bad scheduling decisions exhibit a smaller worsening global impact when applied to a bigger set of smaller tasks, task partitions cooperating with a simple scheduler might alleviate its poor performance: under highly heterogeneous scenarios and available resources, it could be safer to partition a task rather than taking the risk of assigning it to the wrong processor.

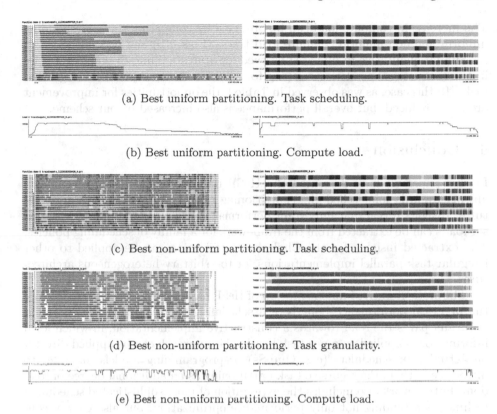

(a) Best uniform partitioning. Task scheduling.

(b) Best uniform partitioning. Compute load.

(c) Best non-uniform partitioning. Task scheduling.

(d) Best non-uniform partitioning. Task granularity.

(e) Best non-uniform partitioning. Compute load.

Fig. 6. Execution traces for the blocked Cholesky factorization on BUJARUELO (left column, $n = 32,768$) and ODROID (right column, $n = 8,192$), using PL/EFT-P. For each case, traces are adjusted to fit the longest execution. In the task scheduling traces, colors correspond to the legend in Fig. 3. (Color figure online)

Figure 6 reports execution traces for the best-performing configurations observed for both architectures[3] (marked in Table 1 with an asterisk), using uniform and non-uniform task partitioning setups. In the traces, each line corresponds to a different processor. In BUJARUELO, (25 CPUs on top, 3 GPUs on bottom), observe the amount of idle times (marked in light blue) in the early and last stages of computation; see how the corresponding best non-uniform schedule fills those gaps by exposing extra parallelism through task partitioning. Concretely, observing the task granularity trace, in which granularity is reported as a gradient (from light green for small tasks to dark blue for large tasks), it is possible to conclude that HeSP is able to refine task granularity only on those stages in which processor occupancy is scarce, improving global performance.

---

[3] Detailed trace generation is supported by HeSP using Paraver (http://www.bsc.es/computer-sciences/performance-tools/paraver).

The increase in compute load can be observed by comparing the corresponding uniform and non-uniform compute load traces (Figs. 6(b) and (e)).

Similar qualitative results are observed for ODROID, filling the gaps that arise in the same stages on slow cores (top four lines in the trace) with finer-grained tasks. In this case, as was observed in Table 1, the opportunities for improvement are more reduced, but overall performance is also increased by our scheme.

4   Conclusion

In this paper we have presented the HeSP framework and its internal mechanisms towards joint scheduling/partitioning tasks on heterogeneous architectures. Insights reveal that important performance benefits and improved processor loads can be extracted from the framework for a family of scheduling policies. The extracted insights for the Cholesky factorization can be applied to other irregular task-parallel implementations, or to arbitrary heterogeneous architectures.

The static iterative implementation of HeSP has shown to be useful to explore the practical performance bounds of a scheduling-partitioning problem, and it naturally paves the road towards a constructive implementation, in which local information is applied on a per-task basis. This approach can be applied directly on actual task schedulers (e.g. OmpSs) or programming models, in order to introduce in them the recursive task partitioning as an additional degree of freedom. Future work also includes the exploration of more sophisticated scheduling techniques attending not only performance optimization, but also energy consumption on different architectures.

Acknowledgements. This work is funded by project TIN 2015-65277-R (MINECO/ FEDER).

References

1. Augonnet, C., Thibault, S., Namyst, R., Wacrenier, P.: StarPU: a unified platform for task scheduling on heterogeneous multicore architectures. CC: PE **23**(2), 187–198 (2011)
2. Bueno, J., Planas, J., Duran, A., Badia, R.M., Martorell, X., Ayguade, E., Labarta, J.: Productive programming of GPU clusters with OmpSs. In: IPDPS 2012, pp. 557–568, May 2012
3. Cojean, T., Guermouche, A., Hugo, A., Namyst, R., Wacrenier, P.: Resource aggregation in task-based applications over accelerator-based multicore machines. Technical report, INRIA (2015)
4. Gautier, T., Ferreira Lima, J.V., Maillard, N., Raffin, B.: XKaapi: a runtime system for data-flow task programming on heterogeneous architectures. In: 27th IPDPS, Boston, May 2013
5. Haidar, A., YarKhan, A., Chongxiao, C., Luszczek, P., Tomov, S., Dongarra, J.: Flexible linear algebra development and scheduling with Cholesky factorization. HPCC **2015**, 861–864 (2015)

6. Planas, J., Badia, R.M., Ayguade, E., Labarta, J.: Self-adaptive OmpSs tasks in heterogeneous environments. In: IEEE 27th International Symposium on Parallel Distributed Processing (IPDPS), pp. 138–149, May 2013

7. Topcuoglu, H., Hariri, S., Min-You, Wu: Performance-effective and low-complexity task scheduling for heterogeneous computing. IEEE Trans. Parallel and Distrib. Syst. **13**(3), 260–274 (2002)

8. Wei, W., Bouteiller, A., Bosilca, G., Faverge, M., Dongarra, J.: Hierarchical dag scheduling for hybrid distributed systems. In: IEEE International Parallel and Distributed Processing Symposium (IPDPS), pp. 156–165, May 2015

FPT Approximation Algorithm
for Scheduling with Memory Constraints

Eric Angel[1], Cédric Chevalier[2], Franck Ledoux[2],
Sébastien Morais[1,2(✉)], and Damien Regnault[1]

[1] IBISC, Université d'Evry, Evry, France
{Eric.Angel,Sebastien.Morais,Damien.Regnault}@ibisc.univ-evry.fr
[2] CEA, DAM, DIF, 91297 Arpajon, France
{Cedric.Chevalier,Franck.Ledoux,Sebastien.Morais}@cea.fr

Abstract. In this paper we study a scheduling problem motivated by performing intensive numerical simulations on large meshes. In order to run the simulation as fast as possible, we must allocate computations on different processors such that the makespan is minimized, but also take care of the limited memory on each processor. We present a dynamic programming based algorithm that ensures that both of these objectives are satisfied, within a ratio of $1 + \varepsilon$. Our algorithm is fixed-parameter tractable (FPT) with respect to the path-width of the graph. For sake of readability, the algorithm is presented for two identical machines, but it can be generalized for a fixed number of unrelated processors.

Keywords: Scheduling · Approximation algorithm · Dynamic programming · Fixed-parameter tractable

1 Introduction

In this paper, we study a specific scheduling problem involving two types of memory constraints: each processing unit has a bounded memory capacity; the tasks to be scheduled depend on each others in a complex way, which we model using a graph structure. A motivation for this problem comes from distributed numerical simulations where most numerical schemes are based on finite elements or volume methods (FEM or VEM) [3,10]. Such approaches require the geometric domain of study Ω to be discretized into basic elements, called cells, which form a mesh. Then, each cell j is assigned a computation valued by a computation cost p_j, and data (like density, pressure, ...) valued by a memory weight m_j. Moreover, performing the computation of a cell j requires, in addition to its data, data located in its neighborhood[1], denoted $\mathcal{N}(j)$. For a distributed simulation, the problem is so to assign all the computations to processing units

---

[1] The neighborhood is most of the time topologically defined (cells sharing an edge or a face) and its depth depends on the numerical scheme used for performing the numerical simulation.

© Springer International Publishing Switzerland 2016
P.-F. Dutot and D. Trystram (Eds.): Euro-Par 2016, LNCS 9833, pp. 196–208, 2016.
DOI: 10.1007/978-3-319-43659-3_15

with bounded memory capacities, while minimizing the makespan[2] and ensuring that the following constraints are satisfied:

- the computation of each cell j is scheduled to a processing unit;
- if a processing unit performs the computation of a cell j, then it needs to locally access data from cell j and cells in $\mathcal{N}(j)$;
- the amount of data stored by a processing unit cannot exceed its capacity.

To illustrate such an assignment, let us consider Fig. 1(a), where a mesh M_Ω and its associated computations are assigned onto 3 processing units. This assignment puts each computation onto a processing unit according to the cell color. Due to the neighborhood constraint, the total amount of memory needed for each processing unit is not limited to those colored cells but extends to some adjacent cells. For an edge-based adjacency relationship, we get the configuration presented in Fig. 1(b), where the memory needed for each processing unit is equal to the memory of both white and colored cells. The white cells contain additional data needed by a processing unit to process all its assigned computation.

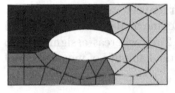 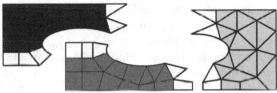

(a) Compact view of the computation assignment of M_Ω

(b) Exploded-view of M_Ω, where the white cells are cells whose data are locally known by the processing unit and the computation is not performed by the latter.

Fig. 1. Computation assignment of a mesh M_Ω onto 3 processing units in (a) and its exploded-view with neighbor memory needed (b).

In the present work, we model this problem as a scheduling problem using a graph $G(J, E)$, which we refer to as the *neighborhood graph*[3]. Computations assigned to cells correspond to *jobs* (one per cell in the mesh), modeled by the set J. Throughout this paper we will denote by n the number of jobs, i.e. $n := |J|$. Job $j \in J$ requires $p_j \in \mathbb{N}$ *units of time* to be executed (computation time) and an amount $m_j \in \mathbb{N}$ of memory. Jobs have to be assigned among k identical *machines* (i.e. processing unit), each machine l having a memory capacity M_l, for $l = 1, \ldots, k$. Moreover, each job j requires data from some *adjacent jobs*, denoted by $\mathcal{N}(j) \subseteq J$. We say that jobs $j \in J$ and $j' \in J$ are adjacent if there

---

[2] Recall that the makespan is the maximum computation time among the processing units.

[3] We draw the reader's attention on the fact that this graph is not a precedence graph, as our problem has no precedence relation between the jobs.

is an edge $(j, j') \in E$. We assume the graph is not directed, i.e. $j' \in \mathcal{N}(j)$ if and only if $j \in \mathcal{N}(j')$. For a subset of jobs $J' \subseteq J$, $\mathcal{N}(J') := \cup_{j \in J'} \mathcal{N}(j)$. When a subset of jobs $J' \subseteq J$ is scheduled on a machine, this machine needs to allocate an amount of memory equal to $\sum_{j \in J' \cup \mathcal{N}(J')} m_j$, while its processing time is $\sum_{j \in J'} p_j$. The objective is then to assign all jobs of J onto a machine, while minimizing the makespan and ensuring strong memory constraints: the amount of memory stored by each machine is smaller than or equal to its memory capacity. In the following we assume that there exists at least one feasible solution, i.e. an assignment of all the jobs such that the memory constraint on each machine is satisfied. Using the notation introduced by Graham et al. [5] we refer to our problem as $Pk|G, mem|C_{max}$. Notice that the second field doesn't contain the term *prec* as there are no precedence constraints among the jobs.

The neighborhood graph $G(J, E)$ is the main feature of $Pk|G, mem|C_{max}$ and dealing with it is the most challenging part as dynamically assigning the jobs may lead to different amount of memory needed to be allocated. As an illustration, let us consider an instance with 2 machines with the neighborhood graph depicted on Fig. 2. Suppose that the subset $J' := \{j_4, j_5, j_6\}$ is assigned to machine 1 while the subset $J'' := \{j_7\}$ is assigned to machine 2. Then the assignment of j_8 to machine 1 or 2 has a different impact in terms of memory allocation: assigning j_8 to machine 1 makes this machine to allocate an additional amount of memory equal to $m_{j_8} + m_{j_{10}}$ (see Fig. 2(a)) whereas assigning it to machine 2 makes this machine to allocate an additional amount of memory equal to $m_{j_{10}}$ only (see Fig. 2(b)).

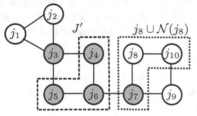

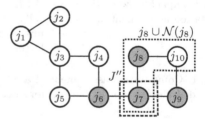

(a) The assignment of $J' = \{j_4, j_5, j_6\}$ constraints the machine to allocate an amount of memory for each (colored) job $j \in J' \cup \mathcal{N}(J') = \{j_3, j_4, j_5, j_6, j_7\}$. Assigning j_8 to this machine induces an additional amount of $m_8 + m_{10}$.

(b) The assignment of $J'' = \{j_7\}$ constraints the machine to allocate an amount of memory for each (colored) job $j \in J'' \cup \mathcal{N}(J'') = \{j_6, j_7, j_8, j_9\}$. Assigning j_8 to this machine induces an additional amount of m_{10}.

Fig. 2. A neighborhood graph $G(J, E)$ and the memory allocation induced by $J' = \{j_4, j_5, j_6\} \in J$ in (a) and $J'' = \{j_7\} \in J$ in (b).

1.1 Related Problems

When $m_j = 0$ for each job j, the problem $Rk|G, mem|C_{max}$ becomes the well-known **NP**- scheduling problem denoted by $Rk||C_{max}$. Lenstra et al. [9] gave a

2-approximation algorithm when the number of machines k is not part of the input and proved that no polynomial algorithm can achieve an approximation ratio less than $3/2$ unless $P = NP$. Their algorithm computes an optimal fractional solution to a natural LP-relaxation and then uses rounding to obtain a schedule for the discrete problem. In [4], Gairing et al. gave a faster algorithm that matches the 2-approximation quality and which is based on unsplittable flow techniques. If the number of machines is fixed, there exist a fully polynomial-time approximation scheme [14].

When the neighborhood graph has no edges, and the memory is bounded on each machine, and $m_j = 1$ for each job j, we get the so-called Scheduling Machines with Capacity Constraints problem (SMCC). In this problem, each machine k can process at most a fixed number of jobs. Saha and Srinivasan [12] gave a 2-approximation in a more general scheduling setting, i.e. Scheduling Unrelated Machines with Capacity Constraints. For the special case of two machines, Woeginger designed a FPTAS for this problem [15].

1.2 Main Contribution

As $Pk|G, mem|C_{max}$ is a generalization of those well-known scheduling problems, a reasonable question is to know whether we can get approximation algorithms, which could possibly depend on some parameters of the neighborhood graph[4]. We answer this question by providing a dynamic programming based algorithm that, assuming that there exists at least one feasible solution to our problem, returns a solution within a ratio of $(1 + \varepsilon)$ for both the optimum makespan and the memory capacity constraints. This algorithm is Fixed-Parameter Tractable (FPT) with respect to the path-width of the neighborhood graph. Notice that there cannot exist an exact FPT algorithm with respect to the path-width parameter, since when the graph has no edge (and therefore a path-width equal to 0), the problem is NP-hard (see Sect. 1.1).

1.3 Outline of the Paper

We start by briefly recalling in Sect. 2 the definitions of different notions useful for our proof. We then provide in Sect. 3 a dynamic programming based algorithm that computes all the solutions to this problem. This task is not trivial as dynamically assigning the jobs may lead to differents amounts of memory needed to be allocated. Since the time complexity of this algorithm is not polynomial in the input size, we apply the Trimming-of-the-State-Space technique [6] in Sect. 4 obtaining an approximation algorithm that is FPT with respect to the path-width of the graph. Finally, we give some concluding remarks in Sect. 5.

---

[4] Recall that an algorithm is *fixed-parameter tractable* (FPT) with respect to h if its running time is bounded by $f(h).|I|^{O(1)}$ where $|I|$ is the size of the instance and f is an arbitrary function depending only on the parameter h.

2   Definitions

Throughout this paper we consider simple, finite, undirected graphs. Let us start by defining the notions of path decomposition and path-width. They were initially introduced in the framework of graph minor theory [11]. A *path decomposition* of a graph $G(J, E)$ is a pair (P, X), where $P := (J(P), E(P))$ is a path, and $X := (X_i)_{i \in J(P)}$ is a family of subsets of J satisfying:

1. $\cup_{i \in J(P)} X_i = J$;
2. $\forall (j, j') \in E$, there exists an $i \in J(P)$ such that $\{j, j'\} \subseteq X_i$;
3. $\forall j, j', j'' \in J(P)$, if j' lies on the path from j to j'' then $X_j \cap X_{j''} \subseteq X_{j'}$.

The *width* of a path decomposition is $\max(|X_i| - 1 : i \in J(P))$ and the *path-width* of G is the minimum width of a path decomposition of G. The construction of such a path decomposition is illustrated on Fig. 3(a), and the result is presented on Fig. 3(b). When P is required to be a tree instead of being a path, previous definitions straightforwardly extend to the definitions *tree decomposition* and *tree-width* of a graph.

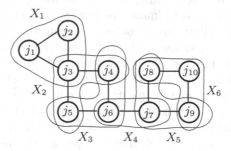

(a) Construction of $X := (X_i)_{i \in J(P)}$, a family of subsets of J satisfying the three properties of a path decomposition.

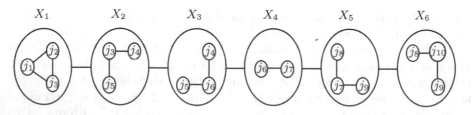

(b) Path decomposition (P, X) of the neighborhood graph $G(J, E)$.

Fig. 3. Example of a path decomposition (P, X) of the neighborhood graph $G(J, E)$, where X is composed by the subsets $X_1 = \{j_1, j_2, j_3\}$, $X_2 = \{j_3, j_4, j_5\}$, $X_3 = \{j_4, j_5, j_6\}$, $X_4 = \{j_6, j_7\}$, $X_5 = \{j_7, j_8, j_9\}$, $X_6 = \{j_8, j_9, j_{10}\}$ in (a) and (P, X) is presented in (b). This path decomposition is optimal with respect to the path-width.

In order to define the notion of vertex separation number, which is equivalent to path-width [7], let us introduce the notion of *(linear) layout* of a graph $G(J, E)$, which is simply a one-to-one mapping $L : J \rightarrow \{1, 2, \ldots, |J|\}$. For any layout L, we define

$$V_L(i) := \{j \in J \mid L(j) \leq i \text{ and } \exists j' \in J \text{ such that } (j, j') \in E \text{ and } L(j') > i\}.$$

Thus $V_L(i)$ is the set of vertices of G mapped to integers less than or equal to i and that are adjacent to vertices mapped to integers greater than i. Then the *vertex separation* of G with respect to L, $vs_L(G)$, is the maximum number of vertices in any $V_L(i)$. And eventually, the *vertex separation number* of G is the minimum, over all possible layouts L of G, of $vs_L(G)$. Formally,

$$vs_L(G) := \max_{1 \leq i \leq |J|} \{|V_L(i)|\}$$

and

$$vs(G) := min\{vs_L(G) \mid L \text{ is a linear layout of } G\}.$$

We note $L^* := \arg\min_L vs_L(G)$, i.e. L^* is a linear layout associated with the vertex separation number. Such an optimal numbering is used on Figs. 2 and 3. When G has a path-width bounded by a constant integer value h, such a layout can be obtained in polynomial time by constructing a linear decomposition (P, X) of G with width h [1], and using an algorithm presented in [7], which (by using a suitable data structure) transforms a given optimal path-decomposition into an optimal layout L^*. It has been proved that $pw(G) = O(\log(n)tw(G))$ for any graph G on n vertices [8], and therefore $vs(G) = O(\log(n)tw(G))$, where pw and tw mean path-width and tree-width respectively.

3   An Exact Algorithm Using Dynamic Programming

For sake of readability, the presentation of the algorithm is done for two machines. But it can be generalized to a constant number k of machines, with $k > 2$, as we will see in Sect. 5. In the following, we assume that the jobs have been numbered such that $L^*(j_i) = i$, for $1 \leq i \leq n$, i.e. the layout is optimal with respect to the vertex separation number.

The dynamic programming goes through n phases. Each phase i, with $i = 1, \ldots, n$, processes the job j_i and produces a set $\mathcal{S}_i$ of states. Each state in the state space $\mathcal{S}_i$ is a vector $S = [s_1, s_2, s_3, s_4, C_i] \in \mathcal{S}_i$, which encodes a partial solution for the first i jobs, i.e. an assignment of the first i jobs to the machines, and where:

1. s_1 (resp. s_2) is the total processing time on the first (resp. second) machine in the partial schedule,
2. s_3 (resp. s_4) is the total amount of memory required by the first (resp. second) machine in the partial schedule,

3. C_i is an additional structure, called *combinatorial frontier*. For a given partial solution from j_1 to j_i, it is defined as $C_i := (V_L(i), \sigma_i, \sigma_i')$ with $\sigma_i : V_L(i) \rightarrow \{1, 2\}$ and $\sigma_i' : V_L(i) \rightarrow \{0, 1\}$, such that $\sigma_i(j)$ is the machine on which j has been assigned, and $\sigma_i'(j) := 1$ if the machine on which j has not been assigned, i.e. the machine $3 - \sigma_i(j)$, has already memorized the data of j.

Notice that in the definition of C_i we use only the set of vertices $V_L(i)$, instead of the whole set $\{1, \ldots, i\}$, since these vertices are the only ones needed to compute additional amounts of memory induced by the future tasks's assignment. Moreover, the number of distinct combinatorial frontiers C_i is equal to' $4^{|V_L(i)|} \leq 4^{vs(G)}$.

The algorithm main structure is summarized in Algorithm 1.

Algorithm 1. Summary of the exact dynamic programming algorithm.

input : A graph $G(J, E)$ where $L^*(j_i) = i$, for $1 \leq i \leq n$
output: A solution vector s
1 $\mathcal{S}_1 := \{[p_1, 0, m_1 + \sum_{j \in \mathcal{N}(j_1)} m_j, 0, C_1^1], [0, p_1, 0, m_1 + \sum_{j \in \mathcal{N}(j_1)} m_j, C_1^2]\};$
2 **foreach** $i \leftarrow 2, n$ **do**
3   **foreach** $[s_1, s_2, s_3, s_4, C_{i-1}] \in \mathcal{S}_{i-1}$ **do**
4   Compute α_i^1 and C_i^1;
5          $\mathcal{S}_i \leftarrow \mathcal{S}_i \cup [s_1 + p_i, s_2, s_3 + \alpha_i^1, s_4, C_i^1];$
6   Compute α_i^2 and C_i^2;
7          $\mathcal{S}_i \leftarrow \mathcal{S}_i \cup [s_1, s_2 + p_i, s_3, s_4 + \alpha_i^2, C_i^2];$
8   **end**
9 **end**
10 $s \leftarrow [s_1, s_2, s_3, s_4, C_n] \in \mathcal{S}_n$ with $s_3 \leq M_1$ and $s_4 \leq M_2$ and such that $\max\{s_1, s_2\}$ is minimum;

Line 1 is the initialization phase. The set $\mathcal{S}_1$ contains two states, the first (resp. second) is when job 1 is assigned on machine 1 (resp. 2). $C_1^1 := (V_L(1), \sigma_1, \sigma_1')$ is the combinatorial frontier when job j_1 is assigned on machine 1. Either $V_L(1) = \{j_1\}$ or $V_L(1) = \emptyset$. In case $V_L(1) = \{j_1\}$ we have $\sigma_1(j_1) := 1$ and $\sigma_1'(j_1) := 0$. Similarly, C_1^2 is the combinatorial frontier when job j_1 is assigned on machine 2. Then at each iteration in the lines 4–7, for each state in $\mathcal{S}_{i-1}$ we add two states in $\mathcal{S}_i$: The state in line 5 (resp. 7) corresponds to the case when job j_i is assigned on machine 1 (resp. 2) and α_i^1 (resp. α_i^2) is the memory induced by this assignment. In order to show how to compute α_i^1, used in Line 5, we define two sets of jobs, namely J_1 and J_2, such that

$$J_1 := \{j \in V_L(i-1) \cap \mathcal{N}(j_i) : \sigma_{i-1}(j) \neq 1 \wedge \sigma_{i-1}'(j) = 0\},$$
$$J_2 := \{j_k \in \mathcal{N}(j_i) : k > i \wedge \forall j_l \in V_L(i-1) \cap \mathcal{N}(j_k) \ \sigma_{i-1}(j_l) \neq 1\}.$$

J_1 is the set of jobs that are in the neighborhood of j_i, have not been assigned to machine 1, and have not been memorized by machine 1 either. J_2 is the set of jobs that are in the neighborhood of j_i, not already assigned, and such that they do not have a job in their neighborhood already assigned to machine 1. Then, we have

$$\alpha_i^1 := \sum_{j \in J_1 \cup J_2} m_j + \begin{cases} m_i & \text{if } \forall j_l \in V_L(i-1) \cap \mathcal{N}(j_i), \sigma_{i-1}(j_l) \neq 1 \\ 0 & \text{otherwise} \end{cases}$$

To illustrate the sets J_1 and J_2, let us consider Fig. 4 where we want to assign j_6 to machine 1, and where $J' = \{j_1, j_2\}$ is assigned to machine 1 and $J'' = \{j_3, j_4, j_5\}$ is assigned to machine 2. We have $V_L(5) = \{j_4, j_5\}$, $\mathcal{N}(j_6) = \{j_4, j_5, j_7\}$ so $V_L(5) \cap \mathcal{N}(j_6) = \{j_4, j_5\}$. As $J'' = \{j_3, j_4, j_5\}$ is assigned to machine 2, we have $\sigma_5(j_4) \neq 1$, $\sigma_5(j_5) \neq 1$ and $\sigma_5'(j_4) = \sigma_5'(j_5) = 0$. Therefore $J_1 = \{j_4, j_5\}$, i.e. assigning j_6 to machine 1 forces this machine to allocate an amount of memory for j_4 and j_5. Moreover, we have $j_7 \in \mathcal{N}(j_6)$ such that $\forall j_l \in V_L(5) \cap \mathcal{N}(j_7), \sigma_5(j_l) \neq 1$. Therefore $J_2 = \{j_7\}$, i.e. assigning j_6 to machine 1 forces this machine to allocate an amount of memory for j_7.

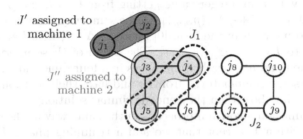

Fig. 4. An example illustrating the sets of jobs J_1 and J_2, when $i = 6$, $J' = \{j_1, j_2\}$ is assigned to machine 1 and $J'' = \{j_3, j_4, j_5\}$ is assigned to machine 2.

Value α_i^2, used in Line 7, is similarly computed. Eventually, let us show how to obtain the new combinatorial frontier $C_i := (V_L(i), \sigma_i, \sigma_i')$ in Line 5 and 7, denoted by C_i^1 and C_i^2 respectively, from $C_{i-1} := (V_L(i-1), \sigma_{i-1}, \sigma_{i-1}')$. Let us consider the first case, i.e. the vertex j_i is assigned on the first machine, and let us show how to obtain C_i^1. If $j_i \in V_L(i)$, then $\sigma_i(j_i) := 1$, and $\sigma_i'(j_i) := 1$ if $\exists j \in V_L(i-1) \cap \mathcal{N}(j_i)$ such that $\sigma_{i-1}(j) = 2$, and $\sigma_i'(j_i) := 0$ otherwise. For $j \in (V_L(i) \setminus \{j_i\}) \cap \mathcal{N}(j_i)$ we have $\sigma_i(j) := \sigma_{i-1}(j)$ and $\sigma_i'(j) := 1$ if $\sigma_{i-1}'(j) = 1$ or $\sigma_i(j) = 2$, and 0 otherwise. For $j \in V_L(i) \setminus (\{j_i\} \cup \mathcal{N}(j_i))$ we have $\sigma_i(j) := \sigma_{i-1}(j)$ and $\sigma_i'(j) := \sigma_{i-1}'(j)$. The combinatorial frontier C_i^2 can be similarly computed. Notice that in the dynamic programming algorithm, if two states S and S' have the same components, including the same combinatorial frontier, then only one of them is kept in the state space. The time complexity to test whether two states S and S' are the same, is thus $O(vs(G))$.

Let $p_{sum} := \sum_{i=1}^{n} p_i$ and $m_{sum} := \sum_{i=1}^{n} m_i$, then for each vector $S = [s_1, s_2, s_3, s_4, C_i] \in \mathcal{S}_i$, s_1 and s_2 are integers between 0 and p_{sum}, s_3 and s_4 are integers between 0 and m_{sum}, and we have $|\mathcal{S}_i| = O(p_{sum}^2 \times m_{sum}^2 \times 4^{vs(G)})$.

The time complexity of this algorithm is proportional to $\sum_{i=1}^{n} |\mathcal{S}_i|$. Thus, the overall complexity is $O(n \times vs(G) \times p_{sum}^2 \times m_{sum}^2 \times 4^{vs(G)})$.

The time complexity of this algorithm being pseudo-polynomial, we are going to transform it into an approximation FPT algorithm with respect to the path-width of the neighborhood graph.

4   Getting an Approximated Algorithm via Trimming Techniques

In this section, we propose an approximated algorithm, derived from Algorithm 1 to get an FPT approximation algorithm. The main idea is to apply the trimming-the-state technique [14] and to withdraw, during the execution of the algorithm, states that are close to each other.

We define $\Delta := 1 + \varepsilon/2n$, with $\varepsilon > 0$ a fixed constant. Let us first consider the first two coordinates of a state $S = [s_1, s_2, s_3, s_4, C_i]$. We have $0 \le s_1 \le p_{sum}$ and $0 \le s_2 \le p_{sum}$. We divide each of those intervals into intervals of the form $[0]$ and $[\Delta^l, \Delta^{l+1}]$, with l an integer value getting from 0 to $L_1 := \lceil \log_\Delta(p_{sum}) \rceil = \lceil \ln(p_{sum})/\ln(\Delta) \rceil \le \lceil (1 + \frac{2n}{\varepsilon})\ln(p_{sum}) \rceil$. In the same way, we divide the next two coordinates into intervals of the form $[0]$ and $[\Delta^l, \Delta^{l+1}]$, with l an integer value getting from 0 to $L_2 := \lceil \log_\Delta(m_{sum}) \rceil$. The union of those intervals defines a set of axis-aligned and non-overlapping boxes in a four dimensional space. If two states have the same combinatorial frontier and have their first four coordinates falling into the same box, then they encode similar solutions.

The approximation algorithm proceeds in the same way as the dynamic programming Algorithm 1, except that we add a trimming phase. The trimming phase works as follows. If in a box, there are more than one state with the same combinatorial frontier, then we keep only one of them (chosen arbitrarily). We will denote by $\mathcal{U}_i$ the (untrimmed) state space obtained before performing that trimming phase at the i-th phase of the algorithm, and $\mathcal{T}_i$ the (trimmed) state space obtained after thinning out and trimming $\mathcal{U}_i$. Algorithm 2 fully describes the approximated dynamic programming algorithm.

The worst time complexity of this algorithm is $O(n \times vs(G) \times (L_1)^2 \times (L_2)^2 \times 4^{vs(G)})$. Since the size of an instance I is $\Theta(n + |E| + \ln(p_{sum} + m_{sum}))$, this algorithm is therefore FPT with respect to the path-width. Let us also notice that if the tree-width is a constant h, then the time complexity remains polynomial since, as mentioned in Sect. 2, $vs(G) = O(\log(n)tw(G))$. Moreover, the tree-width of G being a constant h, we can construct a layout L such that $vs_L(G) = O(\log(n)h)$ in polynomial time by constructing a tree decomposition (T, X) of G with width h [1] and using works in [7,13].

Theorem 1. *There exists an FPT algorithm with respect to the path-width, which returns a solution for the problem $Pk|G, mem|C_{max}$ within a ratio of $(1 + \varepsilon)$ for the optimum makespan, where the memory capacity M_i, $1 \le i \le k$, of each machine may be exceeded by at most a factor $(1 + \varepsilon)$.*

Algorithm 2. The approximated dynamic programming algorithm.

input : A graph $G(J, E)$ where $L^*(j_i) = i$, for $1 \leq i \leq n$
output: A solution vector s

1   $\mathcal{S}_1 := \{[p_1, 0, m_1 + \sum_{j \in \mathcal{N}(j_1)} m_j, 0, C_1^1], [0, p_1, 0, m_1 + \sum_{j \in \mathcal{N}(j_1)} m_j, C_1^2]\};$

2   $\mathcal{T}_1 := \mathcal{S}_1;$

3   **foreach** $i \leftarrow 2, n$ **do**

4      $\mathcal{U}_i := \emptyset;$

5   **foreach** $[s_1, s_2, s_3, s_4, C_{i-1}] \in \mathcal{T}_{i-1}$ **do**

6   Compute α_i^1 and $C_i^1;$

7          $\mathcal{U}_i \leftarrow \mathcal{U}_i \cup [s_1 + p_i, s_2, s_3 + \alpha_i^1, s_4, C_i^1];$

8   Compute α_i^2 and $C_i^2;$

9          $\mathcal{U}_i \leftarrow \mathcal{U}_i \cup [s_1, s_2 + p_i, s_3, s_4 + \alpha_i^2, C_i^2];$

10   **end**

11   Compute a trimmed copy $\mathcal{T}_i$ of $\mathcal{U}_i;$

12   **end**

13   $s \leftarrow [s_1, s_2, s_3, s_4, C_n] \in \mathcal{T}_n$ with $s_3 \leq (1 + \varepsilon)M_1$ and $s_4 \leq (1 + \varepsilon)M_2$ and such that $\max\{s_1, s_2\}$ is minimum;

As stated before we present here the proof when $k = 2$. In the conclusion we mention the general case when k is any fixed constant. The proof of this theorem relies on the following lemma.

Lemma 1. *For each state $S = [s_1, s_2, s_3, s_4, C_i] \in \mathcal{S}_i$, there exists a state $T = [s_1^\#, s_2^\#, s_3^\#, s_4^\#, C_i] \in \mathcal{T}_i$ such that*

$$s_1^\# \leq \Delta^i s_1 \text{ and } s_2^\# \leq \Delta^i s_2 \text{ and } s_3^\# \leq \Delta^i s_3 \text{ and } s_4^\# \leq \Delta^i s_4. \tag{1}$$

Proof. The proof of this statement is by recurrence on i. By construction $\mathcal{T}_1 = \mathcal{S}_1$ so the statement is true for $i = 1$. Now, let us assume that inequalities (1) hold for some index $i - 1$, and consider an arbitrary state $S = [s_1, s_2, s_3, s_4, C_i] \in \mathcal{S}_i$. Then, S is computed from a state $[w, x, y, z, C_{i-1}] \in \mathcal{S}_{i-1}$ and either $[s_1, s_2, s_3, s_4, C_i] = [w + p_i, x, y + \alpha_i^1, z, C_i^1]$ or $[s_1, s_2, s_3, s_4, C_i] = [w, x + p_i, y, z + \alpha_i^2, C_i^2]$ must hold. We assume that $[s_1, s_2, s_3, s_4, C_i] = [w + p_i, x, y + \alpha_i^1, z, C_i^1]$ as, with similar arguments, the rest of the proof is also valid when $[s_1, s_2, s_3, s_4, C_i] = [w, x + p_i, y, z + \alpha_i^2, C_i^2]$. By the inductive assumption, there exists a vector $[w^\#, x^\#, y^\#, z^\#, C_{i-1}] \in \mathcal{T}_{i-1}$ such that

$$w^\# \leq \Delta^{i-1} w \text{ and } x^\# \leq \Delta^{i-1} x \text{ and } y^\# \leq \Delta^{i-1} y \text{ and } z^\# \leq \Delta^{i-1} z. \tag{2}$$

The trimmed algorithm generates the vector $[w^\# + p_i, x^\#, y^\# + \alpha_i^1, z^\#, C_i^1] \in \mathcal{U}_i$ and may remove it during the trimming phase, but it must leave some vector $[s_1^\#, s_2^\#, s_3^\#, s_4^\#, C_i^1] \in \mathcal{T}_i$ that is in the same box as $[w^\# + p_i, x^\#, y^\# + \alpha_i^1, z^\#, C_i^1]$. This vector $[s_1^\#, s_2^\#, s_3^\#, s_4^\#, C_i^1] \in \mathcal{T}_i$ is an approximation of $S = [s_1, s_2, s_3, s_4, C_i] \in \mathcal{S}_i$ in the sense of (2). Indeed, its first coordinate $s_1^\#$ satisfies

$$s_1^\# \leq \Delta(w^\# + p_i) \leq \Delta(\Delta^{i-1} w + p_i) \leq \Delta^i w + \Delta p_i \leq \Delta^i(w + p_i) = \Delta^i s_1, \tag{3}$$

its third coordinate $s_3^{\#}$ satisfies

$$s_3^{\#} \leq \Delta(y^{\#} + \alpha_i^1) \leq \Delta(\Delta^{i-1}y + \alpha_i^1) \leq \Delta^i y + \Delta\alpha_i^1 \leq \Delta^i(y + \alpha_i^1) = \Delta^i s_3, \quad (4)$$

and its last coordinate C_i^1 is equal to C_i. By analogous arguments, we can show that $s_2^{\#} \leq \Delta^i s_2$ and $s_4^{\#} \leq \Delta^i s_4$. Our assumption is valid during the transition from phase $i - 1$ to i, which completes the inductive proof.

□

Let us now go back to the proof of Theorem 1. At the end of phase n, the untrimmed algorithm (Algorithm 1) outputs the vector $s = [s_1, s_2, s_3, s_4, C_n]$ that minimizes the value $\max\{s_1, s_2\}$ such that $s_3 \leq M_1$ and $s_4 \leq M_2$. By Lemma 1, there exists a vector $[s_1^{\#}, s_2^{\#}, s_3^{\#}, s_4^{\#}, C_n] \in \mathcal{T}_n$ whose coordinates are at most a factor of Δ^n above the corresponding coordinates of s. We conclude that our algorithm (Algorithm 2) returns a solution such that the makespan is at most Δ^n times the optimal solution and the amount of memory for each machine is at most Δ^n its capacity. Moreover $\Delta^n \leq 1 + \varepsilon$. Indeed, if we consider functions $f(x) = (1 + x/n)^n$ and $g(x) = 1 + 2x$, with $0 \leq x \leq 1$ and $n \geq 1$, we have

$$(1 + x/n)^n \leq 1 + 2x \quad (5)$$

since f and g are respectively a convex and a linear function in x and the inequality holds true at $x = 0$ and $x = 1$.

So we have constructed an algorithm that returns a solution such that the makespan is at most $(1+\varepsilon)$ times the optimal solution and the amount of memory for each machine is at most $(1+\varepsilon)$ its capacity. It ends the proof of Theorem 1.

We provide a non-intuitive optimal solution to $P2|G, mem|C_{max}$ on Fig. 5. On that Figure the connected set of colored jobs is assigned to machine 1 while the non-connected set of white jobs is assigned to machine 2.

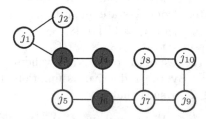

Fig. 5. Example of an optimal assignment to $P2|G, mem|C_{max}$ for the following instance : $M_l = 21, 1 \leq l \leq 2$; $p_{j_i} = m_{j_i}, 1 \leq i \leq 10$; $p_{j_1} = p_{j_2} = p_{j_3} = p_{j_5} = p_{j_7} = p_{j_9} = p_{j_{10}} = 1$; $p_{j_4} = p_{j_8} = 5$; $p_{j_6} = 9$. This optimal assignment induces a computation time of 11 to machine 1 (resp. 15 to machine 2) and a memory allocation of 21 to machine 1 (resp. 19 to machine 2). Therefore, the induced makespan is 15 and the memory capacity of each machine is satisfied.

5   Conclusion

Given 2 machines and a neighborhood graph of jobs with h-bounded linear-width or tree-width, we have constructed an algorithm that returns a solution if at least one solution exists for our scheduling problem with memory constraints. The output of this algorithm is generated in polynomial time and is such that the makespan is at most $(1 + \varepsilon)$ times the optimal solution and the amount of memory for each machine is at most $(1 + \varepsilon)$ its capacity. Moreover, we have constructed an algorithm that returns an optimal solution to our problem in pseudo-polynomial time.

This result can be extended to any constant number of machines as adding machines means increasing the number of dimensions of a state. It only requires to redefine the combinatorial frontier where $\sigma'_i(j)$ would express the machines on which j has not been assigned and which have memorized the data of j. This leads to a time complexity $O(n \times k \times vs(G) \times (L_1)^k \times (L_2)^k \times (k \times 2^k)^{vs(G)})$ where n is the number of phases; $k \times vs(G)$ is the time complexity to test whether two states S and S' are the same; $(L_1)^k \times (L_2)^k$ is the number of boxes induced by the algorithm; and $(k \times 2^k)^{vs(G)}$ is the number of distinct combinatorial frontiers. Notice that if the maximum degree of G is bounded by a constant d, we can lower the previous complexity as at most d machines can memorize the data of a task. Extending the result to unrelated machines can be easily carried over.

As the algorithm is FPT with respect to the path-width, it is particularly interesting for graphs with bounded path-width. Given the fact that such a graph does not occur naturally in large simulations on large meshes, we are wondering if there are FPT approximation algorithms with respect to more generic graph parameters such as the tree-width, and the local tree-width [2].

References

1. Bodlaender, H.L., Kloks, T.: Better algorithms for the pathwidth and treewidth of graphs. In: Leach Albert, J., Monien, B., Rodríguez-Artalejo, M. (eds.) ICALP 1991. LNCS, vol. 510, pp. 544–555. Springer, Heidelberg (1991)
2. Eppstein, D.: Diameter and treewidth in minor-closed graph families. Algorithmica **27**, 275–291 (2000)
3. Ern, A., Guermond, J.L.: Theory and Practice of Finite Elements. Applied Mathematical Sciences, vol. 159. Springer, New York (2004)
4. Gairing, M., Monien, B., Woclaw, A.: A faster combinatorial approximation algorithm for scheduling unrelated parallel machines. Theor. Comput. Sci. **380**(1–2), 87–99 (2007)
5. Graham, R., Lawler, E., Lenstra, J., Rinnooy Kan, A.: Optimization and approximation in deterministic sequencing and scheduling: a survey. Ann. Discrete Math. **5**, 287–326 (1979). Elsevier
6. Ibarra, O.H., Kim, C.E.: Fast approximation algorithms for the knapsack and sum of subset problems. J. ACM **22**(4), 463–468 (1975)
7. Kinnersley, N.G.: The vertex separation number of a graph equals its path-width. Inf. Process. Lett. **42**(6), 345–350 (1992)

8. Korach, E., Solel, N.: Tree-width, path-width, and cutwidth. Discrete Appl. Math. **43**(1), 97–101 (1993)
9. Lenstra, J.K., Shmoys, D.B., Tardos, E.: Approximation algorithms for scheduling unrelated parallel machines. Math. Program. **46**(3), 259–271 (1990)
10. LeVeque, R.J.: Finite Volume Methods for Hyperbolic Problems. Cambridge Texts in Applied Mathematics. Cambridge University Press, Cambridge (2002)
11. Robertson, N., Seymour, P.: Graph minors. I. Excluding a forst. J. Comb. Theory Ser. B **35**(1), 39–61 (1983)
12. Saha, B., Srinivasan, A.: A new approximation technique for resource-allocation problems. In: Proceeding of Innovations in Computer Science (ICS), 342–357 (2010)
13. Skodinis, K.: Construction of linear tree-layouts which are optimal with respect to vertex separation in linear time. J. Algorithms **47**(1), 40–59 (2003)
14. Woeginger, G.J.: When does a dynamic programming formulation guarantee the existence of a fully polynomial time approximation scheme (FPTAS)? INFORMS J. Comput. **12**(1), 57–74 (2000)
15. Woeginger, G.J.: A comment on scheduling two parallel machines with capacity constraints. Discret. Optim. **2**(3), 269–272 (2005)

Scheduling MapReduce Jobs Under Multi-round Precedences

D. Fotakis[1], I. Milis[2], O. Papadigenopoulos[1], V. Vassalos[2], and G. Zois[2]([✉])

[1] School of Electrical and Computer Engineering,
National Technical University of Athens, Athens, Greece
fotakis@cs.ntua.gr, opapadig@corelab.ntua.gr
[2] Department of Informatics,
Athens University of Economics and Business, Athens, Greece
{milis,vassalos,georzois}@aueb.gr

Abstract. We consider non-preemptive scheduling of MapReduce jobs consisitng of multiple map-reduce rounds so as to minimize the average weighted completion time on identical and unrelated processors. For identical processors, we present LP-based $O(1)$-approximation algorithms, while for unrelated processors the approximation ratio naturally depends on the maximum number of rounds of any job (a small constant in practice). For the single-round case, we substantially improve on previously best known approximation ratios for both identical and unrelated processors. Moreover, we conduct an experimental analysis and compare the performance of our algorithms against a fast heuristic and a lower bound on the optimal solution, thus demonstrating their promising practical performance.

1 Introduction

The sharp rise in Internet's use has boosted the amount of data stored on the web and processed daily. MapReduce [6], and its open-source implementation Hadoop, is a fundamental platform for processing data sets on large clusters. A MapReduce job starts by allocating (randomly or arbitrarily) data to a set of processors. The computation over the dataset is broken into successive rounds, where, during each round, a two-phase (map-reduce) process is executed, in which the execution of any reduce task cannot begin until all of its corresponding map tasks have finished. A key observation is that, while the map and reduce phases in each round must be executed sequentially, the tasks in each phase can be executed in parallel. In addition to the computation cost of map and reduce phases, a significant cost is the *communication cost* of transmitting the intermediate data of a job from each map task to every reduce task. Although

---

I. Milis was partially supported by the Research Center of Athens University of Economics and Business (RC-AUEB). V. Vassalos and G. Zois were supported by the European Union Seventh Framework Programme (FP7/2007-2013) under grant agreement no. 604102 (Human Brain Project).

© Springer International Publishing Switzerland 2016
P.-F. Dutot and D. Trystram (Eds.): Euro-Par 2016, LNCS 9833, pp. 209–222, 2016.
DOI: 10.1007/978-3-319-43659-3_16

MapReduce is a distributed computation model, the scheduler of such a system is operating in a centralized manner and its performance is crucial for the efficiency of large MapReduce clusters shared by many users. These clusters typically deal with many jobs that consist of many tasks and of several map-reduce rounds. In such processing environments, the quality of a schedule is typically measured by the jobs' average completion time, which for a MapReduce job takes into account the time when the last reduce task finishes its work.

In this work, we present a general model and an algorithmic framework for scheduling a set of MapReduce jobs on parallel (identical or unrelated) processors with the goal to minimize their average weighted completion time. We consider an offline setting for our model, where each job is represented by multiple successive rounds, and each round consists of multiple map and reduce tasks corresponding to the map and reduce phases respectively. Each reduce task cannot begin its execution before all map tasks of the same round are finished, while the same also holds between reduce and map tasks of two successive rounds. Moreover, the tasks are associated with positive processing times, depending on the processor environment, and each job has a positive weight to represent its priority value. Concerning the communication cost that is incurred in each round, we assume that it is incorporated in the processing times of its reduce task.

Related Work. In the distributed setting of MapReduce's architecture, two main models have been proposed for analyzing the efficiency of MapReduce algorithms with respect to the number of rounds required. Karloff et al. [12] presented a model inspired by PRAM and proved that a large class of PRAM algorithms can be efficiently (i.e., the number of processors and their memory should be sublinear and the running time in each round should be polynomial in the input size) implemented in MapReduce. Recent results in this direction [13] have proposed substantial improvements on the number of rounds for various MapReduce algorithms. Afrati et al. [1] proposed a different model that is inspired by BSP and focuses on the trade-off between communication and computation cost. The main idea is that restricting the computation cost leads to a greater amount of parallelism and to a larger communication cost between the mappers and the reducers. In this context, [2] presents multi-round MapReduce algorithms, trying to optimize the tradeoff between the communication cost and the number of rounds.

In the context of MapReduce scheduling a significant volume of work focuses on the experimental evaluation of scheduling heuristics, trying to achieve good trade-offs between various criteria (see e.g., [19]). On the other hand theoretical work (e.g., [4,8,15]) focuses on scheduling a set of MapReduce jobs on parallel processors to minimize the average (weighted) completion time, capturing the main practical insights in a MapReduce computation (e.g., task dependencies, data locality), in the restricted case where each job is executed in a single round. [4] presents approximation algorithms using simple models, equivalent to known variants of the open-shop problem, taking into account task precedences and assuming that the tasks are preassigned to processors. Moseley et al. [15] present a 12-approximation algorithm for the case of identical processors, modeling in this way MapReduce scheduling as a generalization of the so-called two-stage Flexible Flow-Shop problem. They also present a $O(1/\epsilon^2)$-competitive online algorithm, for any $\epsilon \in (0, 1)$, under $(1+\epsilon)$-speed augmentation. [8] studies the *single-round MapReduce scheduling* problem in the most general case of unrelated processors and present an LP-based 54-approximation algorithm. They also show how to incorporate the communication cost into their algorithm, with the same approximation ratio.

Contribution. Our model incorporates all the main features of the models in [1,12], aiming at an efficient scheduling and assignment of tasks in MapReduce environments. Note that, by assuming positive values for the tasks' execution times, which are polynomially bounded by the input size, we are consistent with both computation models [1,12]. We refer to our problem as the *multi-round MapReduce scheduling* problem or the *single-round MapReduce scheduling* problem (depending on the number of rounds). Our contribution is threefold. First, in terms of modeling the MapReduce scheduling process: (i) We consider the practical scenario of multi-round multi-task MapReduce jobs and capture their task dependencies, and (ii) we study both identical and unrelated processors, thus dealing with data locality. Second, in terms of algorithm design and analysis: (i) We propose an algorithmic framework for the *multi-round MapReduce scheduling* problem with proven performance guarantees, distinguishing between the case of indistinguishable and disjoint (map and reduce) sets of identical or unrelated processors, and (ii) our algorithms are based on natural LP relaxations of the problem and improve on the approximation ratios achieved in previous work [8,15]. Third, in terms of experimental analysis, we focus on the most general case of unrelated processors and show that our algorithms have an excellent performance in practice.

The rest of the paper is organized as follows. In Sect. 2, we formally define our model and provide notation. In Sect. 3, we consider the *multi-round MapReduce scheduling* problem on identical indistinguishable and disjoint processors and we design a 4-approximation and an 11-approximation algorithm, respectively. Moreover, for the *single-round MapReduce scheduling* problem on identical disjoint processors we substantially improve on the results proposed by Moseley et al. [15], presenting an LP-based 8-approximation algorithm, instead of 12-approximation. In Sect. 4, we consider the *multi-round MapReduce scheduling* problem on the most general environment of unrelated processors and we propose an LP-based $O(r_{\max})$-approximation algorithm, where $r_{\max}$ is the maximum number of rounds over all jobs. As a corollary, for the *single-round MapReduce scheduling* problem, we show a 37.87-approximation, which significantly improves on the previously proposed 54-approximation algorithm in [8]. Furthermore, we comment on the hardness of the *multi-round MapReduce scheduling* problem. In Sect. 5, we compare our algorithms via simulations of random instances with a fast heuristic, proposed in [8], as well as with a lower bound on the optimal value of the *multi-round MapReduce scheduling* problem.

2 Problem Formulation

We consider a set $\mathcal{J} = \{1, 2, \ldots, n\}$ of n *MapReduce jobs* to be scheduled on a set $\mathcal{P} = \{1, 2, \ldots, m\}$ of m parallel *processors*. Each job $j \in \mathcal{J}$ is available at time zero and comprises of $r_j \in \mathbb{N}$, $r_j \geq 1$ rounds of computation, with each round consisting of a set of map tasks and a set of reduce tasks. Moreover, each job is associated with a positive weight, let w_j, indicating its significance and, therefore, its relative *priority* to the system. Let $\mathcal{M}$, $\mathcal{R}$ be the sets of all *map* and *reduce* tasks respectively. Each task $T_{k,j} \in \mathcal{M} \cup \mathcal{R}$ of a job $j \in \mathcal{J}$, where $k \in \mathbb{N}$, is associated with a positive processing time. Note that, by assuming task processing times that are polynomially bounded by the input size we are consistent with the two above computation models [1,12]. In every round, each reduce task of a job can start its execution only after the completion of all map tasks of the same job, while similar precedence constraints hold also between the reduce and the map tasks of two successive rounds. In other words, except for the precedence constraints emerged by the existence of map and reduce phases, there are also precedence constraints between consecutive rounds, so a map task of a round $r \in \{2, \ldots r_j\}$, of a job j, cannot start its execution unless all the reduce tasks of the previous round, $r - 1$, have completed their execution. The precedence constraints of a *multi-round MapReduce job* j can be represented by an r_j-*partite-like* directed acyclic graph, as the one depicted in Fig. 1, where r_j is the number of rounds and $l_j = 2r_j - 1$ is the length of a maximal path of the tasks' precedences. Throughout the analysis, in order to upper bound the approximation ratio of our algorithms, the latter parameter is used instead of the number of rounds. Note that, in order to refer to a precedence constraint between two tasks, we use the standard notation, $T_{k,j} \prec T_{k',j}$.

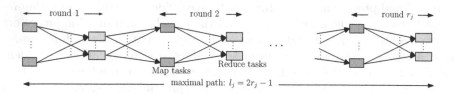

Fig. 1. A MapReduce job j of r_j rounds, and length $l_j = 2r_j - 1$.

To better capture data locality issues in task assignment, we distinguish between the standard *identical processors* environment, where the processing time of each task $T_{k,j}$, let $p_{k,j}$, is the same for every processor, and the most general *unrelated processors* environment, where there is a vector of processing times $\{p_{i,k,j}\}$, one for each processor $i \in \mathcal{P}$. Concerning the dedication of processors to either map or reduce tasks, we examine two cases: (a) The sets $\mathcal{P}_\mathcal{M}$ and $\mathcal{P}_\mathcal{R}$ are *indistinguishable* and the processors in $\mathcal{P}$ are processing both map and reduce tasks, and (b) the set $\mathcal{P}$ is divided into two *disjoint* sets $\mathcal{P}_\mathcal{M}$ and $\mathcal{P}_\mathcal{R}$, where $\mathcal{P} = \mathcal{P}_{\mathcal{M}} \cup \mathcal{P}_{\mathcal{R}}$, where the processors of $\mathcal{P}_{\mathcal{M}}$ process only map tasks, while the processors of $\mathcal{P}_{\mathcal{R}}$ process only reduce tasks.

For a given schedule we denote by C_j and $C_{k,j}$ the completion times of a job $j \in \mathcal{J}$ and a task $T_{k,j} \in \mathcal{M} \cup \mathcal{R}$ respectively. Note that, due to the task precedences along the r_j rounds of each job j, $C_j = \max_{T_{k,j} \in \mathcal{R}}\{C_{k,j}\}$. By $C_{max} = \max_{j \in \mathcal{J}}\{C_j\}$ we denote the *makespan* of the schedule, i.e. the completion time of the last finishing job. Our goal is to schedule *non-preemptively* all tasks on processors of $\mathcal{P}$, with respect to their precedences, so as to minimize the average weighted completion time, $\sum_{j \in \mathcal{J}} w_j C_j$.

3 Scheduling Tasks on Identical Processors

We first study the case of *multi-round MapReduce scheduling* on identical indistinguishable or disjoint processors. For indistinguishable processors, reducing the problem to standard job scheduling under precedence constraints, we immediately obtain a 4-approximation algorithm, a result that holds also for *single-round MapReduce scheduling*. Then, we present an 11-approximation algorithm for identical disjoint processors. For the same case, we also propose an improved 8-approximation algorithm for the *single-round MapReduce scheduling* problem, which substantially improves on the 12-approximation algorithm proposed in [15] for the same problem.

Indistinguishable Processors. We consider the *multi-round MapReduce scheduling* problem on identical indistinguishable processors. Finding an algorithm for this problem can be easily reduced to finding an algorithm for the classic problem of scheduling a set of jobs on identical processors, under precedence constraints of any kind, to minimize their average weighted completion time. More specifically, for any instance of our problem we can create an equivalent instance of the latter problem through the following transformation: For every task $T_{k,j} \in \mathcal{M} \cup \mathcal{R}$, we create a corresponding job j_k of equal processing time, $p_{k,j}$, and zero weight, $w_{j_k} = 0$. We maintain the same precedence constraints, emerged from the input of *multi-round MapReduce scheduling* problem, to the new problem, i.e. for every $T_{k,j} \succ T_{k',j}$ we set $j_k \succ j_{k'}$. For each MapReduce job j, we create a *dummy* job j_D of zero processing time and weight equal to the weight of j, i.e. $w_{j_D} = w_j$, and for every job j_k we demand that $j_k \succ j_D$. In other words, since the corresponding dummy task of a MapReduce job j has zero processing time, there exists an optimal schedule where it is executed exactly after the completion time of all corresponding jobs j_k and, therefore, indicate the completion time of the job itself in the MapReduce context. Moreover, every dummy job j_D carries the weight of the corresponding MapReduce job j. [16] shows a 4-approximation algorithm for scheduling a set of jobs on identical processors, under general precedence constraints, to minimize their average weighted completion time. Combining our transformation with this algorithm, we obtain that:

Theorem 1. *There is a 4-approximation algorithm for the* multi-round *MapReduce scheduling* problem on identical indistinguishable processors.

Disjoint Processors. Inspired by the algorithm of [10, Theorem 3.8], we present an $O(1)$-approximation algorithm which transforms a solution to an interval-indexed LP relaxation of our problem into an integral schedule by carefully applying, on each interval of execution, a variation of the well-known Graham's 2-approximation algorithm [9] for job scheduling on identical processors under precedence constraints to minimize makespan. Note that, in the following, we use the term $b \in \{\mathcal{M}, \mathcal{R}\}$ to refer to both map and reduce attributes.

For any set of tasks $S \subseteq b$, we define $p(S) = \sum_{T_{k,j} \in S} p_{k,j}$ and $p^2(S) = \sum_{T_{k,j} \in S} p_{k,j}^2$. The following (LP1) is an interval-indexed linear programming relaxation of our problem. Constraints (1) ensure that the completion time of a MapReduce job is at least the completion time of any of its tasks and that the completion time of any task is at least its processing time. Constraints (2) capture the relation of completion times of two tasks $T_{k,j} \succ T_{k',j}$. Constraints (3) have been proved [10] to hold for any feasible schedule on identical processors minimizing the average weighted completion time and give useful lower bounds to the completion times of tasks.

Let $(0, t_{\max} = \sum_{T_{k,j} \in \mathcal{M} \cup \mathcal{R}} p_{k,j}]$ be the time horizon of the schedule, where $t_{\max}$ is an upper bound on the makespan of any feasible schedule. We discretize the time horizon into intervals $[1,1], (1,2], (2,2^2], \ldots, (2^{L-1}, 2^L]$, where L is the smallest integer such that $2^{L-1} \geq t_{\max}$. Let $\mathcal{L} = \{1, 2, \ldots, L\}$. Note that, interval $[1,1]$ implies that no job finishes its execution before time 1; in fact, we can assume, w.l.o.g., that all processing times are positive integers. Let $\tau_0 = 1$ and $\tau_\ell = 2^{\ell-1}$. Our algorithm begins from a fractional solution to the LP, $(\bar{C}_{k,j}, \bar{C}_j)$, and separates tasks into intervals with respect to their completion times $\bar{C}_{k,j}$ as follows.

(LP1):   minimize $\sum_{j \in \mathcal{J}} w_j C_j$

    s.t. $C_j \geq C_{k,j} \geq p_{k,j}$ $\forall T_{k,j} \in \mathcal{M} \cup \mathcal{R}$ (1)

               $C_{k,j} \geq C_{k',j} + p_{k,j}$ $\forall T_{k',j} \prec T_{k,j}$ (2)

$$\sum_{T_{k,j} \in b} p_{k,j} C_{k,j} \geq \frac{p(S)^2 + p^2(S)}{2|\mathcal{P}_b|} \qquad b \in \{\mathcal{M}, \mathcal{R}\}, \forall S \subseteq b \quad (3)$$

Let $S(\ell) = \{T_{k,j} | \tau_{\ell-1} < \bar{C}_{k,j} \leq \tau_\ell\}$. Let also $S^M(\ell) \subseteq S(\ell)$ and $S^R(\ell) \subseteq S(\ell)$ be a partition of each set $S(\ell)$ into only map and only reduce tasks, respectively. We define $t_\ell^M = \frac{p(S^M(\ell))}{|\mathcal{P}_M|}$ and $t_\ell^R = \frac{p(S^R(\ell))}{|\mathcal{P}_R|}$ to be the average load of a map and reduce processor, respectively, for executing the map and reduce tasks of each set $S(\ell)$. Now, we can define an adjusted set of intervals as $\bar{\tau}_\ell = 1 + \sum_{k=1}^{\ell}(\tau_k + t_k^M + t_k^R)$ $\forall \ell \in \mathcal{L}$. We can schedule greedily the tasks of each set $S(\ell)$ in interval $(\bar{\tau}_{\ell-1}, \bar{\tau}_\ell]$, using the following variation of Graham's List Scheduling algorithm.

RESTRICTED-RESOURCE LIST SCHEDULING. Consider two different types of available resources, i.e. the map and the reduce processors, while each task can be scheduled only on a specific resource type. Whenever a processor becomes available, execute on it any available unscheduled task that corresponds to its type.

Lemma 1. *The tasks of $S(\ell)$ can be scheduled non-preemptively at interval $(\bar{\tau}_{\ell-1}, \bar{\tau}_{\ell}]$ by applying* RESTRICTED-RESOURCE LIST SCHEDULING.

Proof. Using the analysis of [10] we can prove that the makespan of each set $S(\ell)$ is upper bounded by the total processing time of the longest chain of precedences and the average processing time of a map (resp. reduce) processor. By definition of $S(\ell)$ and constraints (1), we know that the former value can be at most τ_{ℓ}. Therefore, if the algorithm starts by assigning tasks at time $\bar{\tau}_{\ell-1}$, it should have finished by time $C \leq \bar{\tau}_{\ell-1} + \tau_{\ell} + t_{\ell}^{M} + t_{\ell}^{R}$. Then, by definition of $\bar{\tau}_{\ell-1}$ we have that $C \leq 1 + \sum_{k=1}^{\ell-1}(\tau_k + t_k^M + t_k^R) + \tau_{\ell} + t_{\ell}^M + t_{\ell}^R \leq 1 + \sum_{k=1}^{\ell}(\tau_k + t_k^M + t_k^R) = \bar{\tau}_{\ell}.$ $\quad\square$

Note that the resulting schedule respects the tasks precedences since by (1), for any pair of tasks such that $T_{k,j} \succ T_{k',j}$, it must be the case that $T_{k,j} \in S(\ell)$ and $T_{k',j} \in S(\ell')$ with $\ell \leq \ell'$. Now we are able to prove the following theorem.

Theorem 2. *There is an 11-approximation algorithm for the* multi-round MapReduce scheduling *problem on identical disjoint processors.*

Proof. Consider the completion time $C_{k,j}$ of a task $T_{k,j} \in S(\ell)$. By constraints (1) and (2) we know that the length of any chain that ends with $T_{k,j}$ is upper bounded by $\bar{C}_{k,j}$. Therefore, using the previous lemma and since $T_{k,j} \in S(\ell)$, we can see that for its completion time it holds: $C_{k,j} \leq \bar{\tau}_{\ell-1} + t_{\ell}^M + t_{\ell}^R + \bar{C}_{k,j} = 1 + \sum_{k=1}^{\ell-1}(\tau_k + t_k^M + t_k^R) + t_{\ell}^M + t_{\ell}^R + \bar{C}_{k,j} = \tau_{\ell} + \sum_{k=1}^{\ell}(t_k^M + t_k^R) + \bar{C}_{k,j}$. Constraints (3) imply that, for the last finishing -say map- task, $T_{k',j'}$ of the set $S(\ell')$, it holds $\bar{C}_{k',j'} \geq \frac{1}{2|\mathcal{P}_M|}\sum_{k=1}^{\ell'} p(S(k))$, while the same holds for the reduce tasks. Therefore: $\sum_{k=1}^{\ell}(t_k^M + t_k^R) = \sum_{k=1}^{\ell} t_k^M + \sum_{k=1}^{\ell} t_k^R \leq \frac{1}{|\mathcal{P}_M|}\sum_{k=1}^{\ell} p(S^M(k)) + \frac{1}{|\mathcal{P}_R|}\sum_{k=1}^{\ell} p(S^R(k)) \leq 2\tau_{\ell} + 2\tau_{\ell} = 4\tau_{\ell}$. Since by definition of $S(\ell)$, $\tau_{\ell} \leq 2\bar{C}_{k,j}$ it is the case that: $C_{k,j} \leq \tau_{\ell} + 4\tau_{\ell} + \bar{C}_{k,j} \leq 11\bar{C}_{k,j}$. The theorem follows by applying the previous inequality to the objective function. $\quad\square$

Remark. A simple transformation of the previous algorithm yields a 7-approximation algorithm for indistinguishable processors. However, Theorem 1 also applies and gives a 4-approximation algorithm for the *single-round MapReduce scheduling* problem.

The Single-Round Case. For the special case of *single-round MapReduce scheduling*, we obtain an 8-approximation algorithm, improving on the 12-approximation algorithm of [15]. Our algorithm refines the idea of merging independent schedules of only map and only reduce tasks, σ_M and σ_R respectively, on their corresponding sets of processors into a single schedule, by applying a 2-approximation algorithm similar to that in [5, Lemma 6.1]. Note that [5] considers a more general case of scheduling a set of job orders, instead of jobs consisting of tasks, while the completion time of each order is specified by the completion of the job that finishes last.

$$\textbf{(LP2):} \qquad \text{minimize} \sum_{j \in \mathcal{J}} w_j C_j$$

$$\text{s.t.} \qquad C_j \geq M_{k,j} + \frac{p_{k,j}}{2} \qquad\qquad \forall T_{k,j} \in b \qquad\qquad (4)$$

$$\sum_{T_{k,j} \in S} p_{k,j} M_{k,j} \geq \frac{p(S)^2}{2|\mathcal{P}_b|} \qquad\qquad \forall S \subseteq b \qquad\qquad (5)$$

For the partial schedules σ_b of only map and only reduce tasks, since we have no precedence constraints between tasks, let $M_{k,j}$ be the midpoint of a task $T_{k,j} \in b$ in any non-preemptive schedule, i.e., $M_{k,j} = C_{k,j} - \frac{p_{k,j}}{2}$. [7] shows that in any feasible schedule on m identical processors, for every $S \subseteq b$: $\sum_{T_{k,j} \in S} p_{k,j} M_{k,j} \geq \frac{p(S)^2}{2m}$.

Now, consider the linear programming formulation (LP2). Note that, although the number of inequalities of this linear program is exponential, it is known [17] that it can be solved in polynomial time using the ellipsoid algorithm. Thus, consider an optimal solution $(\bar{M}_{k,j}, \bar{C}_j)$ to this formulation with objective value $\sum_{j \in \mathcal{J}} w_j \bar{C}_j$. If we greedily assign tasks on the processors of $\mathcal{P}_b$ in a non-decreasing order of $\bar{M}_{k,j}$ using Graham's list scheduling, then, for the resulting schedule σ_b, it holds that:

Lemma 2. *There is a 2-approximate schedule of map (resp. reduce) tasks on identical map (resp. reduce) processors to minimize their average weighted completion time.*

The second step of our algorithm is to merge the two partial schedules σ_M and σ_R into a single one. To succeed it, we can use the merging technique proposed in [15]. If we denote by $C_j^{\sigma_M}$ and $C_j^{\sigma_R}$ the completion times of a job j in σ_M and σ_R respectively, we can define the *width* of each job j to be $\omega_j = \max\{C_j^{\sigma_M}, C_j^{\sigma_R}\}$. The algorithm schedules the tasks of each job on the same processors that they have been assigned in σ_M and σ_R, in non-decreasing order of ω_j, with respect to the precedences. This merging routine is known [8, Theorem 2] to result in a schedule where the completion time of each job is at most $2 \max\{C_j^{\sigma_M}, C_j^{\sigma_R}\}$, leading to the following theorem:

Theorem 3. *There is an 8-approximation algorithm for the single-round MapReduce scheduling problem on identical disjoint processors.*

Remark. The same analysis yields an 8-approximation algorithm for *single-round MapReduce scheduling* on identical indistinguishable processors. We only have to define the width of each job to be $\omega_j = C_j^{\sigma_M} + C_j^{\sigma_R}$.

4 Scheduling Tasks on Unrelated Processors

In this section, we consider the *multi-round MapReduce scheduling* problem on unrelated processors. We present a $\mathcal{O}(l_{\max})$-approximation algorithm, where $l_{\max} = \max_{j \in \mathcal{J}} l_j$ is the maximum length over all jobs' maximal paths in the underlying precedence graph. Since $l_{max} = 2r_{\max} - 1$, our algorithm is also a $O(r_{\max})$-approximation, where $r_{\max}$ is the maximum number of rounds over all jobs. Our technique builds on ideas proposed in [8]. We formulate an interval-indexed LP relaxation for *multi-round MapReduce scheduling* so as to handle the multi-round precedences. Unlike [8,15], we avoid the idea of creating partial schedules of only map and only reduce tasks and then combining them into one. Moreover, applying the following algorithm for the *single-round MapReduce scheduling* problem, we derive a 37.87-approximation algorithm, thus improving on the 54-approximation algorithm of [8]. Even though in the following analysis, we consider the case of indistinguishable processors, we can simulate the case of disjoint processors by simply setting $p_{i,k,j} = +\infty$ for every map (resp. reduce) task $T_{k,j}$ when i is a reduce (resp. map) processor. In the sequel, we denote by $\mathcal{T} = \mathcal{M} \cup \mathcal{R}$ the set of all tasks.

We use an interval-indexed LP relaxation. Let $(0, t_{\max} = \sum_{T_{k,j} \in \mathcal{T}} \max_{i \in \mathcal{P}} p_{i,k,j}]$ be the time horizon of potential completion times, where $t_{\max}$ is an upper bound on the makespan of any feasible schedule. Similarly with (LP1), we discretize the time horizon into intervals $[1, 1], (1, (1+\delta)], ((1+\delta), (1+\delta)^2], \ldots, ((1+\delta)^{L-1}, (1+\delta)^L]$, where $\delta \in (0, 1)$ is a small constant, and L is the smallest integer such that $(1 + \delta)^{L-1} \geq t_{\max}$. Let $I_\ell = ((1 + \delta)^{\ell-1}, (1 + \delta)^\ell]$, for $1 \leq \ell \leq L$, and $\mathcal{L} = \{1, 2, \ldots, L\}$. Clearly, the number of intervals is polynomial in the size of the instance and in $\frac{1}{\delta}$.

We introduce an assignment variable $y_{i,k,j,\ell}$ indicating whether task $T_{k,j} \in \mathcal{T}$ is completed on processor $i \in \mathcal{P}$ within the interval I_ℓ. Furthermore, let $C_{k,j}$ be the completion time variable for a task $T_{k,j} \in \mathcal{T}$ and C_j be the completion time variable for a job $j \in \mathcal{J}$. (LP3) is an LP relaxation of the *multi-round MapReduce scheduling* problem, whose corresponding integer program is itself a $(1 + \delta)$-relaxation.

Algorithm 1. MULTI-ROUND MRS: An algorithm for *multi-round MapReduce scheduling* on unrelated processors

---

1: Compute a fractional solution to the LP $(\bar{y}_{i,k,j,\ell}, \bar{C}_{k,j}, \bar{C}_j)$.
2: Partition the tasks into sets $S(\ell) = \{T_{k,j} \in b \mid (1 + \delta)^{\ell-1} \leq \alpha \bar{C}_{k,j} < (1 + \delta)^\ell\}$,
3: where $\alpha > 1$ is a fixed constant.
4: **for** each $\ell = 1 \ldots L$ **do**
5:    **if** $S(\ell) \neq \emptyset$ **then**
6:       Let G_ℓ be the precedence graph of the tasks of $S(\ell)$.
7:       $V_{1,\ell}, \ldots, V_{t,\ell}, \ldots, V_{l_{\max}+1,\ell} \leftarrow$ DECOMPOSE(G_ℓ)
8:       **for** each $V_{t,\ell}$, in increasing order of t **do**
9:          Integrally assign the tasks of $V_{t,\ell}$ on $\mathcal{P}$ using [18, Theorem 2.1].
10:         Schedule tasks of $V_{t,\ell}$ on $\mathcal{P}$, as early as possible, w.r.t. their precedences.

(LP3):   minimize $\sum_{j \in \mathcal{J}} w_j C_j$

s.t. $\sum_{i \in \mathcal{P}, \ell \in \mathcal{L}} y_{i,k,j,\ell} \geq 1,$ $\forall T_{k,j} \in \mathcal{T}$ (6)

$C_j \geq C_{k,j},$ $\forall T_{k,j} \in \mathcal{T}$ (7)

$C_{k,j} \geq C_{k',j} + \sum_{i \in \mathcal{P}} p_{i,k,j} \sum_{\ell \in \mathcal{L}} y_{i,k,j,\ell},$ $\forall T_{k',j} \prec T_{k,j}$ (8)

$\sum_{i \in \mathcal{P}} \sum_{\ell \in \mathcal{L}} (1+\delta)^{\ell-1} y_{i,k,j,\ell} \leq C_{k,j},$ $\forall T_{k,j} \in \mathcal{T}$ (9)

$\sum_{T_{k,j} \in \mathcal{T}} p_{i,k,j} \sum_{t \leq \ell} y_{i,k,j,t} \leq (1+\delta)^{\ell},$ $\forall i \in \mathcal{P}, \ell \in \mathcal{L}$ (10)

$p_{i,k,j} > (1+\delta)^{\ell} \Rightarrow y_{i,k,j,\ell} = 0,$ $\forall i \in \mathcal{P}, T_{k,j} \in b, \ell \in \mathcal{L}$ (11)

$y_{i,k,j,\ell} \geq 0,$ $\forall i \in \mathcal{P}, T_{k,j} \in b, \ell \in \mathcal{L}$

Constraints (6) ensure that every task is completed on a processor of the set $\mathcal{P}$ in some time interval. Constraints (7) denote that the completion time of a job is determined by the completion time of its last finishing task. Constraints (8) describe the relation between the completion times of two jobs $T_{k,j} \succ T_{k',j}$, where the term $\sum_{i \in \mathcal{P}} p_{i,k,j} \sum_{\ell \in \mathcal{L}} y_{i,k,j,\ell}$ refers to the fractional processing time of $T_{k,j}$. Constraints (9) impose a lower bound on the completion time of each task. For each $\ell \in \mathcal{L}$, constraints (10), (11) are validity constraints which state that the total processing time of jobs executed up to an interval I_ℓ on a processor $i \in \mathcal{P}$ is at most $(1+\delta)^{\ell}$, and that if processing a task $T_{k,j}$ on a processor $i \in \mathcal{P}$ is greater than $(1+\delta)^{\ell}$, $T_{k,j}$ should not be scheduled on i, respectively.

Algorithm 1 considers a fractional solution $(\bar{y}_{i,k,j,\ell}, \bar{C}_{k,j}, \bar{C}_j)$ to (LP3) and rounds it to an integral schedule. It begins by separating the tasks into disjoint sets $S(\ell), \ell \in \mathcal{L}$ according to their fractional completion times $\bar{C}_{k,j}$. Since some of the tasks of each $S(\ell)$ may be related with precedence constraints, we proceed into a further partitioning of each set $S(\ell), \ell \in \mathcal{L}$ into pairwise disjoint sets $V_{t,\ell}, 1 \leq t \leq l_{\max} + 1$, with the following property: all the predecessors of any task in $V_{t,\ell}$ must belong either in a set $V_{t',\ell}$ with $t' < t$, or in a set $S(\ell')$ with $\ell' < \ell$. Let G be the precedence graph, given as input of the *multi-round MapReduce scheduling* problem. The above partitioning process on G can be done in polynomial time by the following simple algorithm.

DECOMPOSE(G). Identify the nodes of zero in-degree, i.e., $\delta^{-}(v) = 0$, in G. Add them in a set $V_{t,\ell}$, starting with $t = 1$, remove them from the graph, and set $t \leftarrow t + 1$. Repeat until there are no more nodes. Output the sets of tasks.

As the maximum path length in the precedence graph is $l_{\max}$, for each $\ell \in \mathcal{L}$, we could have at most $l_{\max} + 1$ sets $V_{t,\ell}$, with some of them possibly empty. Now, since there are no precedence constraints among the tasks of each set $V_{t,\ell}$, we integrally assign these tasks using the algorithm of [18, Theorem 2.1] in an increasing order of ℓ and t. The next lemmas prove an upper bound on the integral makespan of the tasks of every set $S(\ell)$ and $V_{t,\ell}$.

Lemma 3. *Suppose that we ignore any possible precedences among the tasks in $S(\ell)$, for each $\ell \in \mathcal{L}$. Then we can (fractionally) schedule them on the processors $\mathcal{P}$ with makespan at most $\frac{\alpha}{\alpha-1}(1+\delta)^{\ell}$.*

Now, since every set of tasks $V_{t,\ell}$ is a subset of $S(\ell)$, the aforementioned result on the fractional makespan of $S(\ell)$ also holds for every $V_{t,\ell} \subseteq S(\ell)$.

Lemma 4. *The tasks of every set $V_{t,\ell} \subseteq S(\ell)$ can be integrally scheduled on the processors $\mathcal{P}$ with makespan at most $(\frac{\alpha}{\alpha-1}+1)(1+\delta)^{\ell}$.*

Consider now a set of tasks $S(\ell)$ whose decomposition results in a sequence of pairwise disjoint subsets $V_{1,\ell}, \ldots, V_{t,\ell}, \ldots, V_{l_{\max}+1,\ell}$. Using the Lemma 4, we see that if we integrally schedule each subset $V_{t,\ell}$ in a time window of $(\frac{\alpha}{\alpha-1}+1)(1+\delta)^{\ell}$ and then place the schedules in an increasing order of t, the resulting schedule would respect all constraints and would have makespan at most $(l_{\max}+1)(\frac{\alpha}{\alpha-1}+1)(1+\delta)^{\ell}$. Now, we can prove the following.

Theorem 4. *Algorithm 1 is an $\alpha[(l_{\max}+1)\frac{\alpha}{\alpha-1} + l_{\max}\frac{\alpha}{\delta(\alpha-1)} + l_{\max} + 1 + \frac{l_{\max}+1}{\delta}](1+\delta)$-approximation for the multi-round MapReduce scheduling problem on unrelated processors, where $l_{\max}$ is the maximum length over all maximal paths in the precedence graph, and $\alpha > 1$, $\delta > 0$ are fixed constants.*

Proof. First, we need to note that the tasks of each set $S(\ell)$ can be scheduled integrally in the processors of $\mathcal{P}$ with makespan equal to the sum of makespans of the subsets $V_{t,\ell}, 1 \leq t \leq l_{\max}+1$. The rounding theorem of [18, Theorem 2.1] suggests that the makespan of an integral schedule of tasks in $V_{t,\ell}$ is at most the fractional assignment, $\Pi_{t,\ell} \leq \frac{\alpha}{\alpha-1}(1+\delta)^{\ell}$, of tasks to processors plus the maximum processing time on every processor, $p_{t,\ell}^{max} \leq (1+\delta)^{\ell}$. Therefore, the sets $V_{1,\ell}$ to $V_{l_{\max},\ell}$ can be scheduled with makespan at most $l_{\max}(\frac{\alpha}{\alpha-1}+1)(1+\delta)^{\ell}$, in order to respect the precedences among them. Now, consider the sets $V_{l_{\max}+1,\ell}, \forall \ell \in \mathcal{L}$. Clearly, these must include the last finishing tasks of any chain in the precedence graph. Therefore, by constraints (10), it is the case that $\sum_{t \leq \ell} \Pi_{l_{\max}+1,t} \leq \frac{\alpha}{\alpha-1}(1+\delta)^{\ell}$.

Now, let $T_{k,j} \in \mathcal{T}$ be the last finishing task of a job $j \in \mathcal{J}$ which is scheduled on a processor $i \in \mathcal{P}$. Suppose, w.l.o.g., that $T_{k,j}$ belongs to the set $S(\ell)$. By Lemma 4 and Lemma 3, taking the union of the schedules of tasks in $S(\ell')$, with $\ell' \leq \ell$, it must hold that the completion time of $T_{k,j}$ in the resulting schedule is:

$$C_{k,j} \leq \sum_{\ell' \leq \ell} [l_{\max}(\frac{a}{a-1}+1)(1+\delta)^{\ell'} + \Pi_{l_{\max}+1,\ell'} + p_{l_{\max}+1,\ell'}^{max}]$$

$$\leq \alpha\left((l_{\max}+1)\frac{\alpha}{\alpha-1} + l_{\max}\frac{\alpha}{\delta(\alpha-1)} + l_{\max} + 1 + \frac{l_{\max}+1}{\delta}\right)(1+\delta)\bar{C}_{k,j}.$$

$\square$

As for *single-round MapReduce scheduling*, for all the maximal paths of each job j in the underlying graph, $l_j = 1$. By Theorem 4 with $(\alpha, \delta) \approx (1.65, 0.80)$, we get that:

Corollary 1. *There is a 37.87-approximation algorithm for the* single-round MapReduce scheduling *problem on unrelated processors.*

A Note on the Computational Complexity. Concerning the hardness of *multi-round MapReduce scheduling* on unrelated processors, we note it is a generalization of the standard job-shop scheduling, where the precedence constraints are restricted to be a disjoint union of chains and the task assignment is given in advance, under the average weighted completion time objective. However, for the latter one, we know that it is NP-hard to obtain an $O(1)$-approximation and it does not admit an $O(\log^{1-\epsilon} lb)$-approximation algorithm for any $\epsilon > 0$, unless NP $\subseteq$ ZTIME$(2^{\log^{1/\epsilon} n})$, where lb is a standard lower bound on the makespan of any schedule [14]. Thus, the best we can expect is no more than a logarithmic improvement on our approximation ratio.

5   Simulation Results

We conclude with simulation results for *multi-round MapReduce scheduling* on unrelated processors. We compare our algorithm against the simple heuristic Fast-MR of [8] and against a lower bound derived from (LP3). We provide evidence that the empirical approximation ratio of Algorithm 1 is significantly better than the theoretical one.

Fast-MR operates in two steps. First, it computes an online assignment of tasks to processors, using the online algorithm of [3], and then, it schedules them using a variant of Weighted Shortest Processing Time first wrt. the multi-round task precedences.

Computational Experience and Results. We generate instances consisting of 30 indistinguishable processors and from 5 to 50 jobs. Each job consists of 5 rounds, where the number of map and reduce tasks in each round ranges from 20 to 35 and from 5 to 15, respectively. The weight of each job is uniformly distributed in $[1, n]$, where n is the number of jobs. Moreover, the parameters of Algorithm 1 are fixed to $\delta = 0.96$ and $\alpha = 1.69$. To better capture the unrelated nature of the processors as well as data locality issues, we generate the task processing times in each processor in a processor-task correlated way, extending on the model of [11]. Specifically, the processing times $\{p_{i,k,j}\}_{i \in \mathcal{P}}$ of each map task are equal to $b_j a_{j,i}$ plus some noise selected u.a.r. from $[0, 10]$, where b_j and $a_{j,i}$ are selected u.a.r. from $[1, 10]$, for each job $j \in \mathcal{J}$ and each processor $i \in \mathcal{P}$. The processing time of each reduce task, taking into account that is practically larger, is set to $3b_j a_{j,i}$ plus some noise selected u.a.r. from $[0, 10]$. In this context, we simulate both Algorithm 1 and Fast-MR by running 10 different trials for each possible number of jobs. Since in various applications a MapReduce computation is performed within a single round, we also simulate

Algorithm 1 in the single-round case, called SINGLE-ROUND MRS and compare it against Fast-MR. Note that in the latter case, we fix $\alpha = 1.65, \delta = 0.80$ according to Corollary 1. The instances and the results are available at http://www.corelab.ntua.gr/~opapadig/mrrounds/.

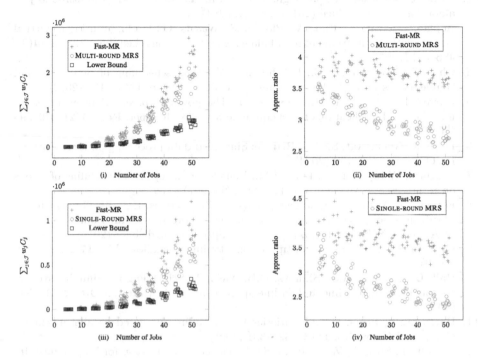

Fig. 2. Simulation results for the single-round and multi-round cases, in terms of absolute values and (empirical) approximation ratios. (Color figure online)

In Figs. 2 (i)–(ii), we note that Algorithm 1 outperforms the Fast-MR heuristic, for any simulated number of jobs. More specifically, the empirical approximation ratio of Fast-MR, ranges from 3.32 to 4.30, while the ratio of Algorithm 1 ranges from 2.57 to 3.68. More interestingly, the gap between the performance guarantee of the two algorithms is growing as the number of jobs is increasing: For $n = 5$ jobs the average ratios of the algorithms Algorithm 1 and Fast-MR are 3.43 and 3.72, while for $n = 50$, the average ratio converges to 2.71 and 3.62, respectively. Over all trials, we can see that Algorithm 1 produces up to 28.4 % better solutions. In Figs. 2 (iii)–(iv), we note that SINGLE-ROUND MRS also outperforms Fast-MR, producing up to 36.7 % better solutions. Similarly to Algorithm 1, its empirical approximation ratio ranges from 2.25 to 3.78 (vs. the ratio of Fast-MR which ranges from 2.94 to 4.44), while the gap against the approximation ratio of Fast-MR increases as the number of jobs increasing (e.g., for $n = 50$, SINGLE-ROUND MRS achieves ratio 2.37, while Fast-MR 3.40). Note that, the empirical approximation ratios in both multi-round and single-round cases of our algorithm are far from our theoretical worst-case approximation guarantees.

References

1. Afrati, F.N., Das Sarma, A., Salihoglu, S., Ullman, J.D.: Upper and lower bounds on the cost of a MapReduce computation. VLDB **6**(4), 277–288 (2013)
2. Afrati, F., Joglekar, M., Salihoglu, C.R.S., Ullman, J.D.: GYM: A multiround join algorithm in MapReduce (2014). arXiv:1410.4156
3. Aspnes, J., Azar, Y., Fiat, A., Plotkin, S., Waarts, O.: On-line routing of virtual circuits with applications to load balancing and machine scheduling. JACM **44**(3), 486–504 (1997)
4. Chen, F., Kodialam, M.S., Lakshman, T.V.: Joint scheduling of processing and shuffle phases in mapreduce systems. In: INFOCOM, pp. 1143–1151 (2012)
5. Correa, J.R., Skutella, M., Verschae, J.: The power of preemption on unrelated machines and applications to scheduling orders. Math. Oper. Res. **37**(2), 379–398 (2012)
6. Dean, J., Ghemawat, S.: MapReduce: Simplified data processing on large clusters. In: OSDI, pp. 137–150 (2004)
7. Eastman, W.L., Even, S., Iaacs, I.M.: Bounds for the optimal scheduling of n jobs on m processors. Manage. Sci. **11**, 268–279 (1964)
8. Fotakis, D., Milis, I., Papadigenopoulos, O., Zampetakis, E., Zois, G.: Scheduling MapReduce jobs and data shuffle on unrelated processors. In: Bampis, E. (ed.) SEA 2015. LNCS, vol. 9125, pp. 137–150. Springer, Heidelberg (2015)
9. Graham, R.L.: Bounds on multiprocessing timing anomalies. SIAP **17**(2), 416–429 (1969)
10. Hall, L.A., Schulz, A.S., Shmoys, D.B., Wein, J.: Scheduling to minimize average completion time: Off-line and on-line approximation algorithms. MOR **22**, 513–544 (1997)
11. Hariri, A.M., Potts, C.N.: Heuristics for scheduling unrelated parallel machines. Comp. and Oper. Res. **18**(3), 323–331 (1991)
12. Karloff, H., Suri, S., Vassilvitskii, S.: A model of computation for MapReduce. In: SODA, pp. 263-285 (2010)
13. Kumar, R., Moseley, B., Vassilvitskii, S., Vattani, A.: Fast greedy algorithms in mapreduce and streaming. In: SPAA, pp. 1–10 (2013)
14. Mastrolilli, M., Svensson, O.: Hardness of approximating flow and job shop scheduling problems. JACM **58**(5), 20 (2011)
15. Moseley, B., Dasgupta, A., Kumar, R., Sarlós, T.: On scheduling in Map-Reduce and flow-shops. In: SPAA, pp. 289–298 (2011)
16. Queyranne, M., Schulz, A.S.: Approximation bounds for a general class of precedence constrained parallel machine scheduling problems. SICOMP **35**(5), 1241–1253 (2006)
17. Queyranne, M.: Structure of a simple scheduling polyhedron. Math. Program. **58**(1), 263–285 (1993)
18. Shmoys, D.B., Tardos, É.: An approximation algorithm for the generalized assignment problem. Math. Program. **62**, 461–474 (1993)
19. Yoo, D.-J., Sim, K.M.: A comparative review of job scheduling for MapReduce. In: CCIS, pp. 353–358 (2011)

High Performance Architectures and Compilers

Code Bones: Fast and Flexible Code Generation for Dynamic and Speculative Polyhedral Optimization

Juan Manuel Martinez Caamaño$^{(\boxtimes)}$, Willy Wolff, and Philippe Clauss

INRIA CAMUS, ICube Laboratory, University of Strasbourg, Strasbourg, France
jmartinezcaamao@gmail.com

Abstract. In this paper, we present a new runtime code generation technique for speculative loop optimization and parallelization, that allows to generate on-the-fly codes resulting from any polyhedral optimizing transformation of loop nests, such as tiling, skewing, fission, fusion or interchange, without introducing a penalizing time overhead. The proposed strategy is based on the generation of *code bones* at compile-time, which are parametrized code snippets either dedicated to speculation management or to computations of the original target program. These code bones are then instantiated and assembled at runtime to constitute the speculatively-optimized code, as soon as an optimizing polyhedral transformation has been determined. Their granularity threshold is sufficient to apply any polyhedral transformation, while still enabling fast runtime code generation. This strategy has been implemented in the speculative loop parallelizing framework Apollo.

1 Introduction

The polytope model (or polyhedral model) [7] is a powerful mathematical framework for reasoning about loop nests, and for performing aggressive transformations which improve parallelism and data-locality. Although very powerful, compilers relying on this model [3,8] are restricted to a small class of compute-intensive codes that can only be handled at compile-time. However, most codes are not amenable to this model, due to dynamic data structures accessed through indirect references or pointers, which prevent a precise static dependence analysis. On the other hand, Thread-Level Speculation (TLS) [14] is a promising approach to overcome this limitation: regions of the code are executed in parallel before all the dependences are known. Hardware or software mechanisms track register and memory accesses to determine if any dependence violation occur. While traditional TLS systems implement only a straightforward loop parallelization strategy consisting of slicing the target loop into consecutive parallel threads, TLS frameworks implementing a speculative and dynamic adaptation of the polytope model have been recently proposed: VMAD [9] and Apollo [17], where parallelizing and optimizing transformations are performed for loops exhibiting a polyhedral-compliant behavior at runtime. A main limitation of these frameworks relies on the dynamic code generation mechanism:

© Springer International Publishing Switzerland 2016
P.-F. Dutot and D. Trystram (Eds.): Euro-Par 2016, LNCS 9833, pp. 225–237, 2016.
DOI: 10.1007/978-3-319-43659-3_17

for each target loop nest, some code skeletons, which are incomplete optimized code versions that will be completed at runtime, are generated at compile-time and embedded in the final executable file. This approach has several limitations: (1) Each skeleton only supports a limited set of loop optimizing transformations; for example, while a given skeleton enables a combination of skewing and interchange, it cannot support any additional transformation as tiling or fission. (2) The impact of some code transformations regarding the structure of the resulting loop nest cannot be predicted; for example, loop fission may result in an arbitrary number of loops; another example is loop unrolling, where the best unroll factor may only be known at runtime. (3) With code skeletons, the same schedule must be applied to all the statements of a target loop, while the polytope model considers scheduling per statements. (4) The complicated structure of generic code skeletons hampers the application of some compiler optimizations, as for example automatic vectorization.

In this paper, we present a dynamic code generation mechanism for speculative polyhedral optimization, that allows to apply on-the-fly any combinations of transformations to a target loop nest, similarly to what is achieved at compile-time by static polyhedral compilers as Pluto [3]. It is based on the compile-time generation of *code bones*, which are code snippets either made of instructions of the target loop nest, or of speculation verification instructions. These code bones are then instantiated and assembled at runtime, according to an optimizing transformation that has just been determined from runtime profiling. The resulting assembled code is then further optimized and compiled using the LLVM just-in-time compiler to generate the final executable code. Our contribution has been implemented in the speculative parallelization framework Apollo [17]. We show on a set of benchmark codes that this code generation technique enables: (1) significant parallel speed-ups, thanks to (2) various automatic runtime loop optimizations that are traditionally only possible at compile-time, (3) on loops that cannot be handled at compile-time.

The paper is organized as follows. An overview of Apollo is presented in Sect. 2. Section 3 details the proposed code generation mechanism (the main contribution of this paper). Section 4 present the empirical results regarding performance and time overhead of the proposed approach. Section 5 compares our mechanism against other approaches. Finally, conclusions are given in Sect. 6.

2   Speculative Parallelization

Apollo[1] [17] is a framework capable of applying polyhedral loop optimizations on any kind of loop-nest[2], even if it contains unpredictable control and memory accesses through pointers or indirections, as soon as it exhibits a polyhedral-compliant behavior at runtime. The framework is made of two components: a static compiler, whose role is to prepare the target code for speculative parallelization, and implemented as passes of the Clang-LLVM compiler [10]; and a

---

[1] Automatic POLyhedral speculative Loop Optimizer.
[2] for-loops, while-loops, do-while-loops.

runtime system, that orchestrates the execution of the code. New *virtual itera-tors*, starting at zero with step one, are systematically inserted at compile-time in the original loop nest. They are used for handling any kind of loop in the same manner, and serve as a basis for building the prediction model and for reasoning about code transformations.

Apollo's static compiler analyzes each target loop nest regarding its memory accesses, its loop bounds and the evolution of its scalar variables. It classifies these objects as being static or dynamic. For example, if the target address of a memory instruction can be defined as a linear function of the iterators, then it is considered as static. Otherwise, it is dynamic and thus requires instrumentation to be analyzed at runtime to take part of the prediction model. The same is achieved for the loop bounds and for scalars. This classification is used to build an instrumented version of the code, where instructions collecting values of the dynamic objects are inserted, as well as instructions collecting the initial values of the static objects (e.g. base addresses of regular data structures).

At runtime, Apollo executes the target loop nest in successive phases, where each phase corresponds to a slice of the outermost loop (see Fig. 1):

1. First, an on-line profiling ① phase is launched, executing only a small number of iterations, and recording memory addresses, loop-trip counts and scalar values.
2. ② From the recorded values, linear functions are interpolated to build a linear prediction model. Using this model, a loop optimizing and parallelizing transformation is determined by invoking, on-line, the polyhedral compiler Pluto. From the transformation, the corresponding parallel code is generated, with additional instructions devoted to the verification of the speculation.
3. A backup ③ of the memory regions, that are predicted to be updated during the execution of the next slice, is performed. An early detection of a mis-prediction is possible, by checking that all the memory locations that are predicted to be accessed are actually allocated to the process.
4. A large slice of iterations is executed ④ using the parallel optimized version of the code. While executing, the prediction model is also verified by comparing the actual values against their linear predictions. If a misprediction is detected, memory is restored ⑤ to cancel the execution of the current slice. Then, the execution of the slice is re-initiated using the original ⑥ serial version of the code, in order to overcome the faulty execution point. Finally, a profiling slice is launched again to capture the changing behavior and build a new prediction model. If no misprediction was detected during the run of the parallel code, a next slice of the loop nest using the same parallel code is launched.

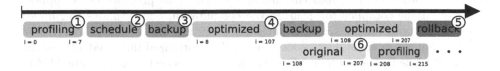

Fig. 1. Execution in slices of iterations

3 Code Generation Strategy

Until now, in order to achieve fast code generation, the Apollo framework has been using *code skeletons* [9]. Code skeletons are incomplete transformed versions of the target loop nests that are generated at compile-time, and completed at runtime as soon as the necessary information has been discovered and computed. Each of such skeletons supports a fixed combination of loop transformations, related to a fixed loop structure. This approach becomes impractical when supporting combinations of polyhedral transformations that may alter the loop structure, such as loop fission, loop unrolling or even simple statement reorderings. To cover every possible combination of loop transformations, we propose a new fast code generation strategy based on *code bones*, which are parametrized code snippets generated at compile-time, and assembled at runtime to result in the transformed code.

Any speculatively optimized code is generally composed of two types of computations: (1) computations of the original target code, whose schedule and parameters have been modified for optimization purposes; and (2) computations related to the verification of the speculation, whose role is to ensure semantic correctness and to launch a recovery process in case of wrong speculation. These computations are generated in two phases: (i) a compile-time phase where code bones of each type are built, and (ii) a runtime phase where complex transformations are determined and instantiated using the code bones.

Generation of Code Bones: At compile time, code bones are extracted from the control-flow graph (CFG) of the target loop nest. Each memory write instruction yields an associated code bone, that includes all instructions belonging to the *backward static slice* of the memory write instruction. In other words, these are all the instructions required to execute an instance of the memory write. Notice that memory read instructions are also included in code bones, since the role of any read instruction is related to the accomplishment of at least one write instruction. Starting from this first set of code bones (called *computation bones*), a second set of bones devoted to the verification of the predictions (called *verification bones*) is generated. For each memory instruction of the computation bones, that may be a write or a read, an associated verification bone is created. These verification bones contain a verification instruction comparing the actual accessed address to the generic predicting linear function that will be instantiated at runtime. Hence, the backward slice computing the target address of the verified memory instruction is also inserted in the verification bone. In the corresponding computation bone, all these instructions are removed and replaced by the computation of the predicting linear function. This provides better opportunities for the compiler to optimize the computation bones thanks to simpler address computations. Similar verification bones are also created for dynamic scalars and loop bounds. Each code bone is then optimized independently of the rest of the code by the compiler. Finally, the so-built code bones are embedded in a fat binary code in their LLVM intermediate representation form (LLVM-IR).

They will be used later by the runtime system for code generation.

Example: As an illustration, consider the loop nest in Listing 1.1. Since array A is accessed through an indirection using array B, whose values are unknown at compile-time, it is impossible to determine what elements of A will be updated. The corresponding CFG is shown in Fig. 2. To make the examples clearer for the reader, instructions are shown in a simplified SSA intermediate representation. Instructions defining original loop iterators are identified by number ①, memory accesses by ② and loop exit conditions by ③. Loop iterators are identified as phi-nodes in the header of each loop, recalling that a phi-node is an instruction used to select an incoming value depending on the predecessor of the current basic block. In order to handle this loop nest at runtime for speculative optimization, code bones are generated. Since there is only one memory write, one computation bone is built. Array B is accessed through a linear memory reference that is identified at compile-time. Hence, only one verification bone is built, which is related to the access of array A. The computation bone is shown in Fig. 3. The computations of the predictions for the original iterators and addresses are identified by number ①, while the memory access instructions using the predicted address by ②. The associated verification bone is shown in Fig. 4, where the computations of the predicted addresses are identified by number ①, number ② points out the load of B[j] using the predicted memory address, original_ptr calculates the actual address of A[B[j]], while the verification instruction is identified by ③, which compares original_ptr against the prediction stored in A.pred. Notice that this bone includes the original address computations. Variables vi.0 and vi.1 stand for the virtual iterators that are used as a basis for building the prediction model. They are passed as parameters to the code bone. The linear functions of the prediction model are interpolated in terms of these iterators. Variables coef_i.0-1, coef_j.0-2, coef_a.0-2 and coef_b.0-2 are the coefficients of the linear functions. These coefficients will be instantiated and replaced by constant values at runtime.

Runtime Composition of Code Bones: The runtime code generation process is depicted in Fig. 5. When linear interpolating functions have been successfully built from the on-line profiling phase, they are used to build the encoding of a loop nest which is compliant with the polyhedral model, using the OpenScop format [2]. This polyhedral representation is then given as input to Pluto to perform dependence analysis, and to compute an optimizing and parallelizing transformation. Pluto's result, also in OpenScop, is then passed to the code generator Cloog [1] to obtain the polyhedral scan, *i.e.*, the new loops and iterators for the statements. Then, a dedicated translation process generates LLVM-IR from Cloog's output. This translation process is straightforward: Cloog's output recalls some constructs in C code like for-loops, simple if-conditions and statement invocations. This code invokes the code bones and instantiates the embedded linear functions according to the schedule provided by Cloog. Notice that a given bone may be invoked several times, but with different parameters to instantiate the linear functions. This may happen in case of loop fission

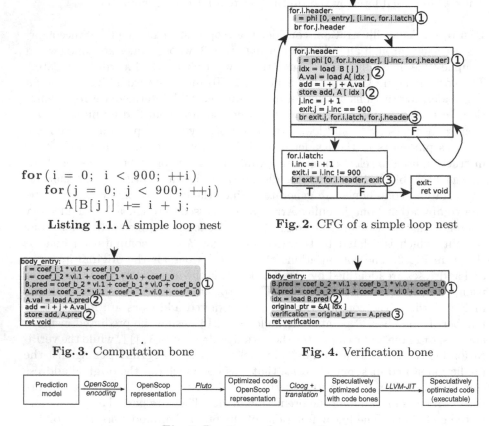

```
for( i = 0;  i < 900;  ++i )
   for( j = 0;  j < 900;  ++j )
      A[B[ j ]] += i + j;
```

Listing 1.1. A simple loop nest

Fig. 2. CFG of a simple loop nest

Fig. 3. Computation bone

Fig. 4. Verification bone

Fig. 5. Runtime code generation

for example. Finally, the resulting code is optimized further and converted into executable form using the LLVM just-in-time compiler.

Transformation Selection Overhead: The selection of a loop transformation is performed using the polyhedral compiler Pluto. However in [18], it has been shown that the execution time of Pluto increases in a roughly n^5 complexity in the number of statements in the system. In consequence, for complex kernels involving many dependences, Pluto can introduce a high time overhead which is inadequate for a runtime usage. Meanwhile the availability of a just-in-time polyhedral compiler, we bypass this issue by handling codes that may yield a high overhead of Pluto in a specific way. Kernels which are associated with more than 5 computation bones are classified as *complex*, while the others are classified as *simple*. For simple kernels, Pluto's time is masked by the execution of a slice of the original serial code. Apollo then starts executing the parallel code as soon as it is ready. For complex kernels, contiguous code bones in the same iteration domain are fused into a single bone. The execution is then equivalent to their serial executions. Notice that this fused bone can perform verification and multiple memory-writes. Simplification of complex kernels is achieved by fusing bones until the total number reaches a given threshold. From our experiments, this threshold has been fixed to 15 bones. Then Pluto is invoked to select a transformation. If the threshold cannot be reached, Pluto is used solely for dependence analysis, to determine which loops in the original code may be parallelized, without any additional transformation.

```
parallel for (t2=0;t2<=899;t2++)
    VerifBone(slice_lower ,t2);
parallel for (t2=0;t2<=899;t2++)
    for (t1=slice_lower ;t1<=slice_upper ;t1++)
        CompBone(t1 ,t2);
```

Listing 1.2. Generated code

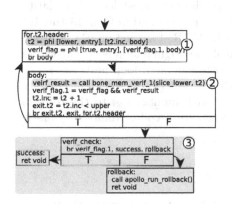

Fig. 6. CFG of the verification code

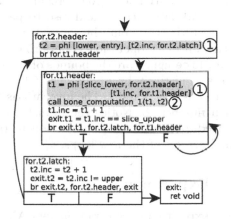

Fig. 7. CFG of the computation code

Optimization of the Verification: Verification bones exhibit three specific optimization opportunities: (1) While computation bones often participate in dependencies, verification bones may not. Such verification bones are extracted into a separate loop nest, which is run before the rest of the bones in a inspector-executor fashion. (2) For these latter verification bones, it is possible to identify dimensions of the iteration domain (*i.e.*, loop indices), for which the the predicting linear functions is invariant. This is achieved by checking the linear functions: if the coefficients multiplying the iterator of a dimension are zero, the computation remains invariant for this dimension. Thus, only the first iteration is run. (3) When some operands of the computing functions are detected as being necessarily linear, while some others require some memory accesses to be performed, outer loops embedding the memory accesses are fully executed, while the inner loops computing linear operands only require the first iteration to be executed.

Example (Continued): Let us now assume that at runtime, each array element B[j] is assigned with j. Thus, Apollo discovers, through profiling and interpolation, that addresses touched by the store instruction can be represented by a linear function of the form: $j + base\_address$. The dependence spawned is an output dependence carried by the outermost loop. The large reuse distance between elements of A is penalizing regarding temporal data locality. Since the store instruction is going to use the predicting linear function as its target address, accesses to array B and A are not dependant anymore. Notice also that all the linear functions for verifying the memory access have 0 as coefficient for the outermost loop iterator. Then, it is sufficient to execute a single iteration of this loop. Both loop nests are sent to Pluto: a first single-loop nest accessing array B and verifying that its elements yield addresses equal to the predicting linear function, and a second two-loop nest computing and storing elements of array A using the linear function as the target address. Pluto suggests to parallelize the single verification loop, to interchange loops i and j in the second nest to improve temporal data locality, and also to parallelize the outermost loop. For clarity, the resulting code built by the code generator is represented in C, in Listing 1.2, although every operation is performed on LLVM-IR. Parameters slice_lower and slice_upper are the bounds of the slice of the original loop nest that will be run speculatively by Apollo. The CFGs of both generated loop nests are represented in Figs. 6 and 7. The invocations to the respective code bones are marked with number ②. Notice the branch (marked with number ③) to the rollback procedure in case of misprediction in the verification code. Number ① marks the iterators of the new loops.

4   Experiments

Our experiments were ran on two AMD Opteron 6172 processors of 12 cores each. Reported results are obtained by averaging the outcome of three runs. The tile sizes were always set to Pluto's default (32 for each dimention). The set of benchmarks has been built from a collection of benchmark suites, such that the selected codes include a main loop nest and highlights Apollo's capabilities: SOR from the Scimark suite[3], backprop and needle from the Rodinia suite [5]; dmatmat, ispmatmat, spmatmat, djacit and pcg from the SPARK00 suite of irregular codes [13], mri-q and stencil from the Parboil suite [16], and finally seidel-2d, which is a special version of the code belonging to the Polybench suite[4], in which the arrays are allocated dynamically, thus yielding pointer aliasing issues. The input problem sizes are as follows: dmatmat, ispmatmat and SOR: 3000×3000 matrices; spmatmat (square): 2500×2500 matrices; spmatmat (diagonal): 8000×8000 matrices; spmatmat (random): 2000×2000 matrices with 3000000 non-zero elements; spmatmat (worst case scenario): 4000×4000 matrices with 1600000 non-zero elements; pcg: 1100×1100 matrices; seidel-2d: $20,000 \times 20,000$ matrices; needle: $24,000 \times 24,000$ matrices;

---

[3] http://math.nist.gov/scimark2/.
[4] http://sourceforge.net/projects/polybench.

Table 1. Number of code bones and applied transformations.

Benchmark	#comp-bones	#verif-bones	Applied transformation
needle	1	1	Tile + Skew + Vectorize + Unroll
SOR	1	6	Tile + Skew + Vectorize + Unroll
seidel-2d	1	10	Tile + Skew + Unroll
dmatmat	1	5	Tile + Unroll
ispmatmat	1	8	Tile + Unroll
spmatmat	1	11	Tile + Vectorize + Unroll
stencil	1	2	Tile + Interchange + Vectorize + Unroll
djacit	7	5	Skew + Unroll
mri-q	2	1	Interchange + Unroll
backprop	2	4	Interchange + Vectorize + Unroll
pcg	21	33	Identity + Unroll

stencil: 4000×4000 matrices; mri-q: two vectors of sizes 2048 and 262,144; backprop: a neural-network with 80,000 input units, 512 hidden units and 16 output units.

Table 1 shows the number of generated bones and the optimizing transformations that were applied on-the-fly, in addition to parallelization. For every benchmark, except spmatmat, mispredictions are detected in advance, during backup, and before launching the speculatively-optimized code, as explained in Sect. 2. Codes djacit and pcg are both classified as *complex*: code bones where automatically fused to accelerate Pluto's transformation selection. Code djacit was simplified to 6 bones, while pcg was simplified at maximum to 21 bones. For the latter, the identity loop transformation was used since the total number of bones remained over the threshold of 15.

To emphasize different features of Apollo, four different inputs were used for spmatmat: (i) a square matrix, (ii) a band matrix, (iii) a randomly distributed matrix, and (iv) a matrix yielding the worst-case scenario for Apollo. Input (i) exhibits a single linear phase, which is conducive to Apollo. Input (ii) yields two different phases: the input band matrix has a constant number of elements per rows, excepting in the very last rows where this number is decreasing. Apollo is successful in optimizing the first large phase where rows have a constant number of elements. But since this number decreases in the last rows, the change of memory accesses and loop bounds yields a rollback, followed by instrumentation and serial loop completion. For input (iii), Apollo is not able to interpolate linear functions to build the prediction model, and continuously switches between instrumented and original executions of the loop. The last input, (iv), represents the worst case scenario for Apollo, consisting of multiple phases of a few iterations. After each instrumented execution for profiling, Apollo successfully builds a prediction model and generates an optimized code; but when executing the speculatively-optimized version, a misprediction is detected, and

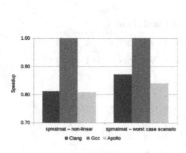

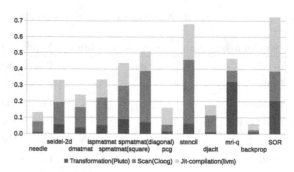

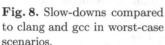

Fig. 8. Slow-downs compared to clang and gcc in worst-case scenarios.

Fig. 9. Total code-generation times.

a rollback occurs. For this last input, the execution yields 6 rollbacks. Figure 8 shows a comparison with inputs (iii) and (iv), where Apollo is not able to execute any optimized slice of iterations successfully. Caption `spmatmat - non-linear` stands for the previously described input (iii), while caption `spmatmat - worst case scenario` stands for input (iv). Even if Apollo imposes an overhead, it does not influence execution time significantly from clang's serial version. Moreover, even if input (iv) yields the worst case scenario – where the framework continuously performs code generation, backup, and fails during the optimized code execution, and rollbacks – the performance impact is weak.

As mentioned, we can distinguish four parts in the code generation process of Apollo: (i) encoding the code bones in an OpenScop object, (ii) determining a polyhedral transformation (Pluto), (iii) generating the scan (Cloog), and finally (iv) generating executable code (LLVM-JIT). In Fig. 9, we only depict the time overheads for the last three stages. Indeed, the time spent encoding the code bones information (i) is negligible when compared to the other parts (0.012 s at maximum, 0.004 s on average). Figure 10 shows the percentage of the total execution time which is spent in the main steps of Apollo. Interpolation, code generation and transformation selection phases are grouped with caption 'code-generation'. The backup time remains lower than 5 % of the execution time for most of the benchmarks. Instrumentation is one of the most time-consuming phases of Apollo, however mostly as fast as executing the original serial code. It is particularly large for `backprop`, since the original serial code is slow due to poor data locality (column-major array access). Since the code generation phase is always executed in parallel with the original code, the percentage corresponding to the original code execution is always higher than the one corresponding to the code generation phase. Figure 11 show a comparison between the code generation strategy using skeletons used by Apollo and the one proposed in this paper with code bones. To compute the speed-ups, we selected the best serial version generated among the gcc-4.8 or clang-3.4 compilers with optimization level 3 (-O3). The code bones approach outperforms the code skeletons approach for

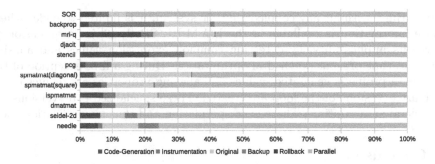

Fig. 10. Overheads of Apollo among the total execution time(24 threads).

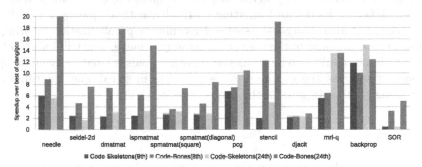

Fig. 11. Speedup against the best of clang/gcc (8 and 24 threads)

codes that benefit from transformations not supported by the code skeleton approach, such as tiling. For benchmarks where the applied transformation is the same with both approaches, similar speed-ups are obtained. For SOR, tiling is required for parallelization.

5 Related Work

The Apollo framework is a major revision of a previous framework called VMAD [9]. To generate parallel code on-the-fly, VMAD builds *code skeletons* at compile-time, whose limitations have been addressed in this paper.

Most TLS systems [11,12,15] are limited to simple parallelization schemes: the outermost loop of the original loop nest is optimistically sliced into speculative parallel threads. Such a scheme does not consider complex reordering of iterations and statements, thus, the implemented code generation mechanisms are reduced to different statically generated and simple code versions, and a runtime system that switches between them. Softspec [4] represents preliminary ideas of our approach. However, no code transformations are performed, only slicing the loop for parallel execution.

Polly [8] may be seen as the static counterpart of our proposal. Polly is a polyhedral compiler built on top of LLVM. However, since Polly operates only at compile-time, without any coupled runtime system, it is limited to codes where precise information is available in the LLVM-IR. SPolly [6] is an extension to enlarge its applicability, by detecting common expression values and aliasing properties that prevent polyhedral optimization. During a first execution of the program, a profile is generated; values and aliasing properties are deduced, and specialized versions of the loop are created. These specialized code versions are not generated at runtime. There is no speculation and thus no verification code.

6 Conclusion

The proposed runtime code generation strategy offers the opportunity of applying any polyhedral loop transformation on-the-fly, without paying a penalizing time overhead. It also enlarges the scope of speculative parallelization by bringing it closer to what a static optimizing compiler may achieve. The CFG abstraction using code bones could also be employed for other goals related to dynamic optimization, as soon as the runtime process consists in scheduling, guarding and instantiating some sub-parts of the target code.

References

1. Bastoul, C.: Code generation in the polyhedral model is easier than you think. In: PACT (2004)
2. Bastoul, C.: Openscop: A specification and a library for data exchange in polyhedral compilation tools. Technical report (2011)
3. Bondhugula, U., Hartono, A., Ramanujam, J., Sadayappan, P.: A practical automatic polyhedral parallelizer and locality optimizer. In: PLDI (2008)
4. Bruening, D., Devabhaktuni, S., Amarasinghe, S.: Softspec: software-based speculative parallelism. In: ACM FDDO (2000)
5. Che, S., Boyer, M., Meng, J., Tarjan, D., Sheaffer, J., Lee, S.H., Skadron, K.: Rodinia: a benchmark suite for heterogeneous computing. In: IISWC (2009)
6. Doerfert, J., Hammacher, C., Streit, K., Hack, S.: Spolly: speculative optimizations in the polyhedral model. In: IMPACT (2013)
7. Feautrier, P.: Some efficient solutions to the affine scheduling problem, part 2: multidimensional time. IJPP 21(6), 389–420 (1992)
8. Grosser, T., Größlinger, A., Lengauer, C.: Polly - performing polyhedral optimizations on a low-level intermediate representation. PPL 22(4), 28 (2012)
9. Jimborean, A., Clauss, P., Dollinger, J.F., Loechner, V., Martinez, J.M.: Dynamic and speculative polyhedral parallelization using compiler-generated skeletons. IJPP 42(4), 1–17 (2014)
10. Lattner, C., Adve, V.: LLVM: a compilation framework for lifelong program analysis & transformation. In: CGO 2004 (2004)
11. Liu, W., Tuck, J., Ceze, L., Ahn, W., Strauss, K., Renau, J., Torrellas, J.: Posh: a tls compiler that exploits program structure. In: PPoPP (2006)
12. Rauchwerger, L., Padua, D.: The lrpd test: speculative run-time parallelization of loops with privatization and reduction parallelization. In: SIGPLAN Not. (1995)
13. van der Spek, H., Bakker, E., Wijshoff, H.: Spark00: A benchmark package for the compiler evaluation of irregular/sparse codes. (2008). arXiv:0805.3897

14. Steffan, J., Mowry, T.: The potential for using thread-level data speculation to facilitate automatic parallelization. In: HPCA 1998 (1998)
15. Steffan, J., Colohan, C., Zhai, A., Mowry, T.: The stampede approach to thread-level speculation. ACM TCS **23**, 253–300 (2005)
16. Stratton, J.A., Rodrigues, C., Sung, I.J., Obeid, N., Chang, L.W., Anssari, N., Liu, G.D., Hwu, W.-M.W.: The Parboil technical report. Techical report, IMPACT (2012)
17. Sukumaran-Rajam, A., Martinez Caamaño, J.M., Wolff, W., Jimborean, A., Clauss, P.: Speculative program parallelization with scalable and decentralized runtime verification. In: Bonakdarpour, B., Smolka, S.A. (eds.) RV 2014. LNCS, vol. 8734, pp. 124–139. Springer, Heidelberg (2014)
18. Upadrasta, R., Cohen, A.: Sub-polyhedral scheduling using (unit-)two-variable-per-inequality polyhedra. In: POPL 2013 (2013)

Piecewise Holistic Autotuning of Compiler and Runtime Parameters

Mihail Popov[1(✉)], Chadi Akel[2], William Jalby[1], and Pablo de Oliveira Castro[1]

[1] Université de Versailles Saint-Quentin-en-Yvelines,
Université Paris-Saclay, Versailles, France
{mihail.popov,william.jalby,pablo.oliveira}@uvsq.fr
[2] Exascale Computing Research, Versailles, France
chadi.akel@exascale-computing.eu

Abstract. Current architecture complexity requires fine tuning of compiler and runtime parameters to achieve full potential performance. Autotuning substantially improves default parameters in many scenarios but it is a costly process requiring a long iterative evaluation.

We propose an automatic piecewise autotuner based on CERE (Codelet Extractor and REplayer). CERE decomposes applications into small pieces called codelets: each codelet maps to a loop or to an OpenMP parallel region and can be replayed as a standalone program.

Codelet autotuning achieves better speedups at a lower tuning cost. By grouping codelet invocations with the same performance behavior, CERE reduces the number of loops or OpenMP regions to be evaluated. Moreover unlike whole-program tuning, CERE customizes the set of best parameters for each specific OpenMP region or loop.

We demonstrate CERE tuning of compiler optimizations, number of threads and thread affinity on a NUMA architecture. On average over the NAS 3.0 benchmarks, we achieve a speedup of 1.08× after tuning. Tuning a single codelet is 13× cheaper than whole-program evaluation and estimates the tuning impact on the original region with a 94.7 % accuracy. On a Reverse Time Migration (RTM) proto-application we achieve a 1.11× speedup with a 200× cheaper exploration.

1 Introduction

The current increase of architecture complexity, multiple cores, out-of-order execution, complex memory hierarchies, and non-uniform memory access (NUMA) complicates the performance characterization. Achieving full efficiency requires fine tuning parameters such as the degree of parallelism, thread placement or compiler optimization. Runtime and compiler standard parameter levels (such as −O3 compiler flag or scatter thread placement) achieve good-enough performance across most of the codes and architectures. But they cannot take advantage of target-specific optimizations since they must correctly work on a large panel of architectures.

Finding the optimal parameters may lead to substantial improvement but is a costly and time consuming process. For example, compilers such as LLVM [1]

© Springer International Publishing Switzerland 2016
P.-F. Dutot and D. Trystram (Eds.): Euro-Par 2016, LNCS 9833, pp. 238–250, 2016.
DOI: 10.1007/978-3-319-43659-3_18

3.4 provide more than sixty passes. Passes have different impact depending on their order of execution and can be executed many times. This leads to a huge exploration space: considering only sequences of 30 passes requires to explore a space over 60^{30} points.

Even worse, some applications may have different optimal parameters for different code regions. For example, compute bound loops and memory bound loops within the same function will not be sensitive to the same compiler optimizations.

There are different approaches to tune parameters. Iterative compilation [2] is a well known automated search method for solving the compiler optimization phase ordering problem. The idea is to apply successive compiler transformations to a program and to evaluate them by executing the resulting code. Similar execution driven studies [3,4] explore the efficiency of different thread placement strategies or frequencies. Smart search algorithms [5,6] through the parameter space reduce the evaluation cost. Genetic algorithms [7,8] or adaptive learning [9, 10] accelerate the search by avoiding unnecessary parameters.

A common point of these execution driven studies is that they require a full program evaluation and execution to quantify the impact of a single parameter value. The problem is that executing application is costly and time consuming, especially if we have thousands of points to evaluate. Also, as regions of code do not benefit from the same parameters, an overall program-evaluation (or *monolithic* evaluation) is not able to achieve the optimal per region optimization. In other words, these studies are expensive to perform and do not necessary lead to the optimal parameters.

In this paper we propose a piecewise exploration framework based on CERE [11] (Codelet Extractor and REplayer) which enhance both the search cost and the search benefits. We partition applications into small pieces called *codelets*. Each independent loop or OpenMP parallel region is extracted as a codelet that can be replayed as a standalone program. Instead of evaluating parameters on the whole application, we separately evaluate them on each codelet (Sect. 3.2). The piecewise evaluation leads to find the best parameters for each region. Combining these regions within a single binary is called *hybridization* and outperforms traditional *monolithic* tuning (Sect. 3.3).

Using codelets as proxies for autotuning requires that codelets faithfully reproduce the application behavior with the exploring parameters. This requires a warmup of the memory state. CERE already implements various warmup strategies. To enable thread placement exploration, we extend these warmups with a new NUMA ownership strategy (Sect. 3.1).

The contributions of this paper are:

- A novel automatic autotuner based on codelets and integrated in CERE.
- A holistic piecewise tuning approach that addresses degree of parallelism, thread placement, NUMA effects, and compiler optimization passes.
- The validation of the codelet tuning over the NAS benchmarks and an industrial proto-application with compiler and runtime parameters.
- A NUMA aware memory page capture and replay.

2   Motivating Example

We will demonstrate how CERE operates on SP, a Scalar Penta-diagonal solver, from the C version of the NPB 3.0 OpenMP benchmarks [12]. CERE autotuning achieved a 1.82× performance speedup over the standard parameters levels. Thanks to the CERE codelet approach, the exploration time was approximately five times cheaper compared to the whole-program iterative compilation.

Table 1. Execution time in megacycles of SP parallel regions across different thread affinities with −O3 optimization. For n threads, we consider three affinities: scatter s_n, compact c_n, and hyperthread h_n. Executing SP with the c8 affinity provides an overall speedup of 1.71× over the standard (s16).

thread	affinity	xsolve	ysolve	zsolve	rhs	total
s2	0;8	32.3	23	28.5	23	106.8
c2	0;1	21.4	17.6	18.1	23.7	80.8
h2	0;16	40	32.6	23	46.1	141.7
s4	0;8;1;9	25.9	20.9	26	12.1	84.9
c4	0;1;2;3	15.5	12.7	13.8	13.2	55.2
h4	0;16;1;17	23.8	17.5	16	24.3	81.5
s8	0;8;1;9;…;11	24.4	21.9	28.6	6.9	81.8
c8	0;1;2;3;…;7	14.4	13.4	14.3	9.1	51.2
h8	0;16;1;17;…;19	17.7	14.2	13.9	13.5	59.3
s16	16 scatter	25.1	21.4	35.5	5.3	87.4
c16	16 compact	17	15	15.5	9.7	57.2
h32	32 scatter	36	31.2	38.9	6.4	112.4

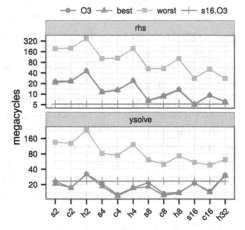

Fig. 1. Tuning exploration for two SP regions. For each affinity, we plot the best, worst, and −O3 optimization sequences. Custom optimization beats −O3 for s2, s4, and s8 on ysolve.

CERE starts by profiling SP and automatically selecting representative OpenMP regions to tune. Xsolve, ysolve, zsolve, and rhs are selected and cover 93 % of SP execution time. CERE extracts these regions as codelets and tunes them with a holistic exploration across three dimensions: thread number, thread placement, and LLVM compiler passes. Once satisfying parameters are found, CERE produces an hybrid application where each region uses the best found parameters.

In this study, we explored the interactions between 12 thread configurations combining different number of threads and affinity mappings and 150 LLVM optimization sequences generated using the random sub-sampling presented in Sect. 4. Combining them produces an exploration space of 1800 points, which gives an insight of how costly it is to simultaneously tune multiple parameters.

Figure 1 shows the performance of two SP parallel regions across this exploration space. We notice that there is a strong interaction between the compiler and the thread parameters as they both significantly impact the performances.

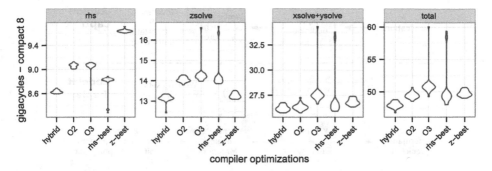

Fig. 2. Violin plot execution time of SP regions using best NUMA affinity. Measures were performed 31 times to ensure reproducibility. When measuring total execution time, Hybrid outperforms all other optimization levels, since each region uses the best optimization sequence available.

Moreover, the best parameters are different for the two regions: scatter placement is best for rhs while compact benefits ysolve.

CERE makes it possible, through codelet replay, to independently explore each region. Moreover, thanks to CERE replay prediction model presented in Sect. 3.2, it is possible to quickly evaluate the impact of each configuration on only a few datasets. CERE evaluates thread affinities and compiler optimizations on SP, respectively 5.84× and 4.52× times faster than a full application evaluation while keeping a low average error of 2.33%.

Custom parameters outperform the standard 16 threads scatter s16 −O3 on SP. Table 1 shows the performance of different thread affinities compiled with −O3. The best custom thread affinity 0;1;2;3;4;5;6;7 (single NUMA socket) achieves a speedup of 1.71× over the standard 16 threads scatter (two NUMA sockets).

We explored with CERE 350 compiler optimization sequences on the best single NUMA configuration found above. Xsolve and ysolve work best at the default −O2 level, but a custom best sequence is found for zsolve and rhs. Figure 2 shows the performance of each region compiled with the default optimization and the best custom sequences. No single sequence is the best for all regions. CERE hybrid compilation produces a binary where each region is compiled using its best sequence, achieving a speedup that cannot be reproduced using traditional monolithic compilation.

3   CERE AutoTuner

CERE [11,12] is an open source framework for code isolation. CERE finds and extracts loops or OpenMP parallel regions from an application as isolated fragments of code, called codelets. Codelets can be modified, compiled, run, and measured independently from the original application.

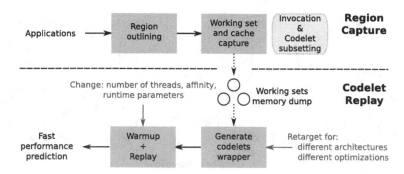

Fig. 3. Codelet capture and replay workflow

Figure 3 presents how a region is captured as a codelet and replayed. Using codelets as a proxy for application characterization requires two steps: *capture* and *replay*. During the capture, the execution state is saved for each region. During the replay, CERE restores the codelet memory and cache state before executing the region. At replay, a cache and NUMA page ownership warmup is necessary to ensure that the replay execution context is close to the original.

CERE extracts regions at the compiler Intermediate Representation (IR) level after clang front-end translation but before LLVM middle-end optimizations. This allows us to re-target the codelet compilation and execution.

3.1   NUMA Aware Warmup

A replay has to faithfully reproduce the original invocation context. CERE already handles two issues: it restores the memory working set of the region and warms up the cache to avoid cold-start bias [13].

It uses a snapshot of the memory at page level granularity. With a memory protection mechanism, the memory pages containing the working set are captured. During replay, pages are remaped to their original addresses. CERE includes different cache warmup approaches [11] that operate by replaying the memory access history at a page granularity before running the codelet.

We outline a new aspect: the placement of the pages across the NUMA nodes. Due to the node local first touch policy, a page is mapped to the core which first attempts to use it. We must ensure that pages are mapped to the same NUMA nodes as they have been in the original run. The problem is that pages are not necessarily bound to the same NUMA nodes across the different thread affinities. Scatter maximizes the number of NUMA nodes while compact minimizes it.

Figure 4 outlines this problem on a 2-NUMA nodes architecture. CERE default warmup uses a single thread to remap the pages to their original addresses: all the pages are bound to a single NUMA node. Replays accurately predict the execution time as long as the affinity binds the threads to the same NUMA node. Otherwise, the replay pays NUMA latencies that do not appear in the original run and which cause prediction discrepancies.

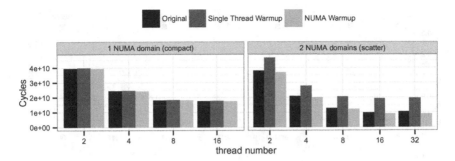

Fig. 4. Prediction accuracy of a single threaded warmup versus a NUMA aware warmup on BT `xsolve`. Only a NUMA aware warmup is able to predict this region execution time on a multi NUMA node configuration.

To solve this issue, we enhance the page capture by saving, for each page, the first thread that touches it. During replay, before replaying the codelet code, each thread touches the pages that it has saved at the capture. Hence, pages are mapped to the NUMA node of the thread which is the first to touch them. To ensure a correct NUMA mapping at replay when we change the number of threads, we must both not exceed the number of threads at capture and spread the pages across the replaying threads.

3.2 Piecewise Optimization with Codelets

Regions within an application may not be sensitive to the same optimizations: SP `rhs` and `zsolve` regions in Sect. 2 have different best compiler optimizations. Unlike monolithic approaches, CERE enables tuning each codelet independently.

The piecewise search not only improves the benefits over a monolithic tuning, but also accelerates the exploration by avoiding the execution of useless compiler sequences (see in experiments Fig. 7) or regions. IS benefits from this as it only times a sorting algorithm included in a region which represents 22 % of the application execution time. Through a codelet, CERE extracts the sorting region and tunes it without executing the rest of the application.

Codelets also accelerate the evaluation of each region. Regions may have performance variations across their different invocations. Using a clustering method, CERE classes these invocations and selects a representative subset of invocations to be replayed. We only execute the subset to predict the region execution time.

We assume that the tuning parameters have a similar impact on the invocations within the same cluster. Figure 5 illustrates this assumption across two parameters on MG `resid`. `Resid` has 42 invocations grouped in 3 performance classes. The invocations remain in the same classes across the parameters. So, by replaying 3 instead of 42 invocations, CERE predicts the region execution for each parameter to explore.

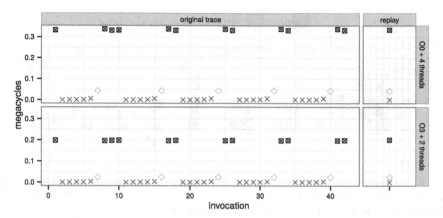

Fig. 5. MG `resid` invocations execution time on Sandy Bridge over −O3 and −O0 with respectively 2 and 4 threads. Each representative invocation predicts its performance class execution time.

3.3 Hybrid Compilation

The piecewise tuning finds the best compiler optimizations for each loop and OpenMP region. Unfortunately, LLVM does not provide a mechanism to select compiler optimizations at the function or loop granularity. To compile each region with a different set of optimizations we must extract each region in its own compilation unit. We leverage the `extract` tool included in LLVM which allows to extract an IR function to a separate IR file.

The first step is outlining each region of interest in its own IR function. Before any middle-end optimization is applied, each region is moved to a separate compilation unit using `LLVM extract`. A special pass changes the visibility of symbols used by the extracted region from internal to global so that they are not removed by the compiler. Then, the best compiler sequence found is applied to each separate IR file and an object file is produced. Finally, all the objects files are linked together producing an hybrid binary.

4 Experiments and Validation

This section validates both usage of codelets as proxies to tune parameters and production of hybrid binaries. Codelets capture most of the application hotspots [11]. Nevertheless, we must demonstrate that codelet tuning helps finding optimal parameters and reducing the search cost. To accurately predict best parameters, codelet replays must capture the original application reaction to the different compiler and thread configurations.

We used two different Intel CPU micro-architectures: a Sandy Bridge E5 with 64 GB of RAM and an Ivy Bridge i7-3770 with 16 GB of RAM. We chose Sandy Bridge to explore thread affinities because it has 2 NUMA sockets and each socket has 8 physical (16 hyper-threaded) cores.

Thread configurations were selected to explore different degrees of parallelism, NUMA and hyper-threading effects. Sandy Bridge has 16 physical cores, so we did not explore configurations beyond 32 threads. We used the Intel kmp affinity [14] notation to characterize the thread placement. Cores ranked between 0 and 7 reference the physical cores of the first NUMA node while cores between 8 and 15 reference the physical cores of the second NUMA node. Similarly, cores from 16 to 23 and from 24 to 31 reference the hyper-threaded cores of respectively the first and the second NUMA node.

The compilation search was performed on LLVM 3.4 using a random pass selection. We use LLVM opt and llc to change respectively middle-end and backend optimizations. Middle-end passes have different impact depending on their order of execution, and can be executed multiple times. −O3 is a manually tuned sequence composed of 65 ordered passes aiming to provide good performances. In this paper, random compilation sequences were generated by down-sampling the −O3 default sequence. Each pass was removed with a 0.7 probability, and the process was repeated four times to explore the impact of pass repetitions. We empirically found that this generation method produces good and diverse candidates. Back-end passes were selected among −O0, −O1, −O2 and −O3.

We performed the experiments on the NAS 3.0 sequential [15] and C OpenMP parallel [16] benchmarks (respectively NAS SER and NPB) with CLASS A datasets and on a Reverse Time Migration [17] (RTM) proto-application.

4.1   Thread Number and Affinity Tuning

This section presents the thread affinity tuning results. CERE page memory capture was performed on a 16 threads scatter run. Table 2 evaluates CERE thread affinities replay accuracy and reduction factor over NAS OpenMP. We focused on regions representing more than 5 % of the application execution time. On average, a region exploration is 6.55× faster with codelets than with whole-program evaluations. Tuning all the SP regions from the motivating example with codelets is five times faster as SP has four regions with an average acceleration of twenty per region. CERE uses an optimistic warmup: it replays four times the codelet over itself. These replays are not amortized on EP and MG: the first executes the main parallel region once in the original execution while the second requires many invocation replays to support the multiple performance classes. As we increase the data sets, the warmup cost overhead becomes smaller compared to the replay execution time. We tested xsolve BT with CLASS B data sets and a single warmup invocation to achieve an acceleration of 9.48×, twice the one achieved in class A, with an accuracy of 98.36%.

The average CERE prediction accuracy is 93.66%. It allows the autotuner to outperform the standard scatter s16 over EP, FT, LU, and SP and to perform an average speedup of 1.40× (see Fig. 6). We note that there is no thread affinity to privilege over the others: h32, s16, and c8 are all optimal on at least two applications.

Table 2. The **accuracy** of the codelet prediction is the relative difference between the original and the replay execution time. The benchmark **reduction factor** or acceleration is the exploration time saved when studying a codelet instead of the whole application. CERE fails to accelerate EP and MG evaluation: EP has a single region with one invocation while MG displays many performance variations.

Benchmarks	Compiler passes			Thread affinity		
	#Regions	Accuracy	Reduction factor	#Regions	Accuracy	Reduction factor
BT	3	98.73	79.63	4	95.24	5.28
CG	2	98.65	3.39	2	79.48	1.23
FT	5	98.3	2.6	5	90.71	2.17
IS	3	96.64	1.26	2	94.85	1.04
SP	6	98.78	68.9	4	97.66	20.07
LU	7	95.04	8.49	2	99.00	12.64
EP	1	83.08	0.36	1	99.31	0.25
MG	4	97.22	0.28	4	93.04	0.45
Average		95.8	20.61		93.66	5.39

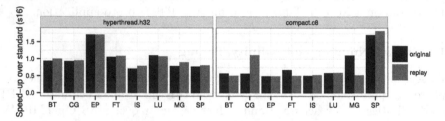

Fig. 6. Original and CERE predicted speedup for two thread configurations. Replay speedup is the ratio between the replayed target and the replayed standard configuration. CERE accurately predicts the best thread affinities in six out of eight benchmarks. For CG and MG, we miss-predict configurations that use all the physical cores.

4.2 Compiler Passes Tuning and Hybridization

Table 2 also presents CERE predictions through compiler optimizations with 3000 compiler sequences for BT, 500 for MG and 1000 for the others NAS SER. The average CERE prediction accuracy and acceleration for a region is 95.8% and 20.61×. Figure 7 presents the number of explored compiler sequences required to achieve a speedup over 1.04× per region. We empirically determined this speedup value. Unlike monolithic approaches which must continue exploration until all regions are optimized, codelets can stop the search over a region once a satisfying speedup is found and focus the exploration on other regions. Here, CERE evaluates BT ysolve 461 times instead of 3000 times. Each evaluation is on average 99 times cheaper than a full application run due to the codelet invocations clustering.

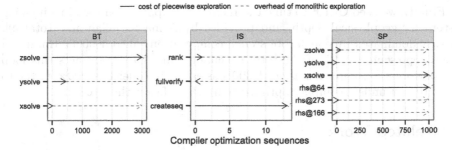

Fig. 7. Compiler sequences required to get a speedup over 1.04× per region. CERE evaluates the sequences in the same order for all the regions. Exploring regions separately is cheaper because we stop tuning a region as soon as the speedup is reached.

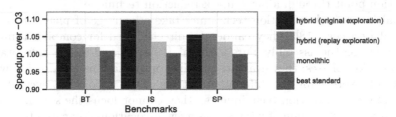

Fig. 8. Speedups over −O3. We only observe speedups from the iterative search over BT, SP, and IS. Best standard is the more efficient default optimization (either −O1, −O2, or −O3). Monolithic is best whole program sequence optimization. Hybrids are build upon optimizations found either with codelets or with original application runs.

The focus of this paper is not on the compiler flag selection, that is why a naive random compiler pass search was used. Nevertheless, CERE results could be improved with more sophisticated techniques for passes selection such as genetic algorithms [6] which would also benefit from the piecewise approach.

CERE outperforms the standard −O3 over BT, SP, and IS with an average speedup of 1.06× (see Fig. 8). IS random generator and sorting algorithm do not benefit from the same optimizations which explains the significant difference between the hybrid and the monolithic approach. Hybrid binaries based on original or replay explorations have the same performances which ensure that we do not miss any optimizations through the codelets.

We make the simplifying assumption that optimizing a region does not affect other regions. This is not always true: due to memory effects, it is possible to have performance interactions between neighbors. We find a compilation sequence which gives a speedup of x1.08× over LU jacu. Unfortunately, optimizing jacu has the side effect of slowing down by 0.92× the neighboring region jacld.

To stress the CERE prediction accuracy model, we performed a simultaneous search of 1000 compiler sequences across the thread affinities on LU ssor. CERE predicted region execution time with a mean accuracy of 99 % across parameters.

Finally, we used CERE to tune the RTM proto-application used in a imaging system for geophysical depth, and provided by Asma Farjallah and Total [9]. RTM is dominated by one Jacobi stencil computation called 3000000 times and which represents 91.1% of the total execution time. CERE extracts this loop and performs a compiler search of 300 passes. This codelet is 200× faster to evaluate and finds a compiler optimization 1.11× faster than -O3.

5   Related Work

While most of the research try to accelerate the iterative compilation by pruning the exploration space [5–8,10], this paper proposes a transverse approach which do not focus on the search space but rather accelerates the evaluation of each exploration point through a benchmark reduction technique.

Usual benchmark reduction techniques take advantage of phases to reduce the simulation cost [18]. They cannot be directly used for compiler tuning as they operate on the assembly. Fursin and al. [19] managed to take advantage of the application phases: they evaluate multiple optimizations for a region with a single run by versioning the different iterations of the region. However, they do not use any code isolation techniques so they cannot focus the search which is problematic when a region of interest has a few invocations compared to the others. Oliveira et al. [20] cluster together codelets that have the same performance behavior, and keep only one representative copy for each group. This benchmark reduction is complementary to the invocations clustering presented in this paper and should accelerate the overall search. We must find clustering metrics that are relevant for compiler optimizations and thread affinities.

Like us, Kulkarni et al. [21] propose a piecewise search at the function level granularity. They propose a per-function compilation using the VPO compiler framework. Yet, they do not use any extraction mechanism during the search: exploring two functions within the same file requires to execute the program many times. Purini et al. [22] find, through LLVM iterative compilation runs, good general sets of compilation sequences that should work well on any given program. They can quickly tune new applications by directly searching passes within the good set instead of exploring the whole optimization space. Codelets could serve proxies to quickly find and test these optimal sequences.

6   Conclusion

In this paper we present an autotuner based on CERE codelets. Codelets serve as proxies for tuning applications holistically, considering the interactions of thread placements, NUMA effects, and compiler passes. CERE proposes a novel piecewise approach that accelerates searching the parameter space and enables an hybrid compilation where each region uses the best set of local parameters. It outperforms traditional monolithic tuning.

CERE codelets predict the impact of thread placement and compiler optimization with a mean accuracy of 94.7% over the NAS 3.0 benchmarks. On the

RTM industrial proto-application, CERE achieved a $1.11\times$ execution speedup through compiler pass selection. The search was $200\times$ faster thanks to codelet tuning. Detailed accuracy and acceleration reports are available at https:// benchmark-subsetting.github.io/autotuning-results/.

Acknowledgments. The research leading to these results has received funding under the Mont-Blanc project from the European Union's Horizon 2020 research and innovation program under grant agreement No. 671697.

References

1. Lattner, C., Adve, V.: LLVM: a compilation framework for lifelong program analysis & transformation. In: International Symposium on Code Generation and Optimization, pp. 75–86. IEEE (2004)
2. Kisuki, T., Knijnenburg, P.M.W., O'Boyle, M.F.P., Bodin, F., Wijshoff, H.A.G.: A feasibility study in iterative compilation. In: Polychronopoulos, C., Fukuda, K.J.A., Tomita, S. (eds.) ISHPC 1999. LNCS, vol. 1615, pp. 121–132. Springer, Heidelberg (1999)
3. Mazouz, A., Touati, S.A.A., Barthou, D.: Performance evaluation and analysis of thread pinning strategies on multi-core platforms: case study of SPEC OMP applications on intel architectures. In: High Performance Computing and Simulation (HPCS), pp. 273–279. IEEE (2011)
4. Rountree, B., Lownenthal, D.K., de Supinski, B.R., Schulz, M., Freeh, V.W., Bletsch, T.: Adagio: making DVS practical for complex HPC applications. In: Proceedings of the Conference on Supercomputing, pp. 460–469. ACM/IEEE (2009)
5. Triantafyllis, S., Vachharajani, M., Vachharajani, N., August, D.I.: Compiler optimization-space exploration. In: International Symposium on Code Generation and Optimization, CGO 2003, pp. 204–215. IEEE (2003)
6. Ladd, S.R.: ACOVEA: Analysis of compiler options via evolutionary algorithm (2007)
7. Cooper, K.D., Schielke, P.J., Subramanian, D.: Optimizing for reduced code space using genetic algorithms. In: SIGPLAN Notices, vol. 34, pp. 1–9. ACM (1999)
8. Hoste, K., Eeckhout, L.: COLE: compiler optimization level exploration. In: Code Generation and Optimization, pp. 165–174. ACM (2008)
9. de Oliveira Castro, P., Petit, E., Farjallah, A., Jalby, W.: Adaptive sampling for performance characterization of application kernels. Concurrency and Computation: Practice and Experience (2013)
10. Fursin, G., et al.: Milepost GCC: machine learning enabled self-tuning compiler. Int. J. Parallel Prog. **39**(3), 296–327 (2011)
11. de Oliveira Castro, P., Akel, C., Petit, E., Popov, M., Jalby, W.: CERE: LLVM based Codelet Extractor and REplayer for piecewise benchmarking and optimization. Trans. Archit. Code Optim. **12**(1), 6 (2015)
12. Popov, M., Akel, C., Conti, F., Jalby, W., de Oliveira Castro, P.: PCERE: fine-grained parallel benchmark decomposition for scalability prediction. In: International Parallel and Distributed Processing Symposium, pp. 1151–1160. IEEE (2015)
13. Kessler, R.E., Hill, M.D., Wood, D.A.: A comparison of trace-sampling techniques for multi-megabyte caches. Trans. Comput. **43**(6), 664–675 (1994)

14. Intel: Reference Guide for the Intel(R) C++ Compiler 15.0. https://software.intel. com/en-us/node/522691
15. Bailey, D., et al.: The NAS parallel benchmarks summary and preliminary results. In: Proceedings of the Conference on Supercomputing, pp. 158–165. ACM/IEEE (1991)
16. Popov, M.: NAS 3.0 C OpenMP. http://benchmark-subsetting.github.io/cNPB
17. Baysal, E.: Reverse time migration. Geophysics **48**(11), 1514 (1983)
18. Sherwood, T., Perelman, E., Calder, B.: Basic block distribution analysis to find periodic behavior and simulation points in applications. In: Parallel Architectures and Compilation Techniques, pp. 3–14. IEEE (2001)
19. Fursin, G.G., Cohen, A., O'Boyle, M., Temam, O.: Quick and practical run-time evaluation of multiple program optimizations. In: Stenström, P. (ed.) Transactions on HiPEAC I. LNCS, vol. 4050, pp. 34–53. Springer, Heidelberg (2007)
20. de Oliveira Castro, P., Kashnikov, Y., Akel, C., Popov, M., Jalby, W.: Fine-grained benchmark subsetting for system selection. In: International Symposium on Code Generation and Optimization, pp. 132–142. ACM (2014)
21. Kulkarni, P.A., Jantz, M.R., Whalley, D.B.: Improving both the performance benefits and speed of optimization phase sequence searches, pp. 95–104. ACM (2010)
22. Purini, S., Jain, L.: Finding good optimization sequences covering program space. Trans. Archit. Code Optim. **9**(4), 56 (2013)

Insights into the Fallback Path of Best-Effort Hardware Transactional Memory Systems

Ricardo Quislant[✉], Eladio Gutierrez, Emilio L. Zapata, and Oscar Plata

Department of Computer Architecture, University of Málaga, 29071 Málaga, Spain
{quislant,eladio,zapata,oplata}@uma.es

Abstract. Current industry proposals for Hardware Transactional Memory (HTM) focus on best-effort solutions (BE-HTM) where hardware limits are imposed on transactions. These designs may show a significant performance degradation due to high contention scenarios and different hardware and operating system limitations that abort transactions, e.g. cache overflows, hardware and software exceptions, etc. To deal with these events and to ensure forward progress, BE-HTM systems usually provide a software fallback path to execute a lock-based version of the code.

In this paper, we propose a hardware implementation of an irrevocability mechanism as an alternative to the software fallback path to gain insight into the hardware improvements that could enhance the execution of such a fallback. Our mechanism anticipates the abort that causes the transaction serialization, and stalls other transactions in the system so that transactional work loss is minimized. In addition, we evaluate the main software fallback path approaches and propose the use of ticket locks that hold precise information of the number of transactions waiting to enter the fallback. Thus, the separation of transactional and fallback execution can be achieved in a precise manner.

The evaluation is carried out using the Simics/GEMS simulator and the complete range of STAMP transactional suite benchmarks. We obtain significant performance benefits of around twice the speedup and an abort reduction of 50 % over the software fallback path for a number of benchmarks.

1 Introduction

Transactional Memory (TM) [8] was first presented in 1993 [9] as a non-blocking synchronization mechanism for shared memory chip multiprocessors (CMPs). TM provides the programmer with the *transaction* construct that executes the code within it atomically and in isolation. Such transactional properties are ensured by the TM system via the cache coherence protocol and dedicated hardware (hardware TM – HTM).

It is not until recently that some processor manufacturers have included HTM support in their commercial off-the-shelf CMPs [4,10,19,21]. Current industry proposals focus on best-effort solutions (BE-HTM) where hardware limits are imposed on transactions. For instance, transactions cannot survive to capacity

© Springer International Publishing Switzerland 2016
P.-F. Dutot and D. Trystram (Eds.): Euro-Par 2016, LNCS 9833, pp. 251–263, 2016.
DOI: 10.1007/978-3-319-43659-3_19

overflows, exceptions, interrupts, page faults, migrations,... To deal with these limitations, BE-HTM systems usually provide a software fallback path to execute a non-transactional version of the code, often comprising a global lock.

In this paper we propose an implementation of a hardware irrevocability mechanism as an alternative to the software fallback path to gain insight into the hardware improvements that could enhance the execution of such a fallback. Irrevocability [3,20] is a transactional execution mode that ensures transaction forward progress since an irrevocable transaction cannot be aborted. Our mechanism anticipates the abort that causes the transaction serialization, and stalls other transactions in the system so that transactional work loss is minimized. In addition, we evaluate the main software fallback path approaches and propose the use of a ticket lock that hold precise information of the number of transactions waiting to enter the fallback. Thus, the separation of transactional and fallback execution can be achieved in a precise manner with the corresponding performance benefits. The result is an enhanced Lemming effect avoidance [6].

The evaluation is carried out using the Simics/GEMS simulator and the complete range of STAMP transactional suite benchmarks. We obtain significant performance benefits of around twice the speedup and an abort reduction of 50 % over the software fallback path for a number of benchmarks.

2   Baseline Architecture

Figure 1 shows the baseline architecture used in this paper. The system relies on the L1 caches to store new transactional values of memory blocks, while old values are kept into the L2 cache. A pair of read and write transactional bits per L1 cache block marks whether the block was read or written within a transaction. Such bits can be flash-cleared on transaction commit and abort. In case of abort, the blocks whose transactional write bit is set are also invalidated. The cache coherence protocol maintains strong isolation [13] and implements an eager conflict detection policy. The conflict resolution policy is requester-wins, where the requesting transaction wins the conflict and the requested one is aborted. The baseline cache coherence protocol is modified to support the execution of transactions:

- *Backup on first transactional store*: If an L1 cache block is in M state and its write transactional bit is not set, the L1 cache has to send the data to the L2 cache before a transactional store is performed. This way the L2 cache holds the last old value for the block.
- *Abort on evictions*: The replacement of a transactional block in an L1 cache implies losing track of transactional loads and stores, which jeopardizes transaction isolation; so transactions must be aborted on these type of evictions. Beside, L2 cache block replacements may abort a transaction because of the inclusion property.
- *L2 cache serves data of aborted transactions*: The L2 cache must send the data of aborted transactions. There are two situations: (i) The requester is already the owner of the block. In such a case, the L2 cache simply responds

with the data; (ii) The requester is not the owner of the block. In this case, the directory forwards the request to the owner, which receives a forward message for a block that is no longer present in its L1 cache. Then, the L1 cache informs the L2 cache and the L2 cache sends the data.

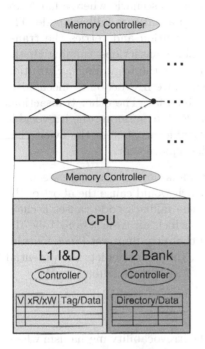

Fig. 1. Baseline architecture of the BE-HTM system.

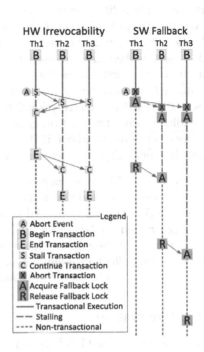

Fig. 2. Execution scenario of hardware irrevocability vs. software fallback.

3   Hardware Irrevocability Fallback Mechanism

A common way to deal with hardware capacity overflows and to ensure forward progress in commercial BE-HTM systems is a software fallback path. The code that Intel suggests as fallback path in its optimization manual [1] comprises a global lock to execute a failed transaction as a non-transactional critical section. Once a transaction aborts a given number of times, the fallback path is taken. In addition, when a transaction is successfully started, the fallback global lock is checked. If the lock was acquired, the transaction aborts. If not, the transaction goes ahead with the lock in its read set so that another transaction acquiring the lock can abort it. The clash of transactions and fallback path sections is thereby avoided.

A hardware irrevocability mechanism provides several benefits over a software fallback code of that kind:

- The programmer is not burdened with the task of writing and tuning a fall-back code, which reduces the programming effort of transactional applications, one of the main goals of transactional memory.
- There is no need for a lock so it is neither cached nor added to the read set of the transaction, thus freeing limited hardware resources.
- Performance benefits: Figure 2 shows an execution scenario where a hardware irrevocability mechanism performs better than a software fallback code. The fallback path version aborts transactional execution and retries the transaction as a locked critical section. The other transactions running in the system abort as well, since they read the lock at the beginning[1]. Execution is rebooted and serialized. However, the hardware irrevocability mechanism does not discard the transactional work done so far. The other transactions are stalled when a transaction gets irrevocable. Furthermore, the irrevocable one does not have to abort if it gets irrevocable just before the event that causes irrevocability, e.g. before an L1 cache replacement.

The scenario in Fig. 2 is optimistic. It considers no contention between the irrevocable transaction and the stalled ones, which would cause the abort of the stalled, conflicting transactions. Additionally, the fallback code causes a chain reaction, also called as *Lemming effect* [6], by which all transactions take the fallback path even if they do not have reached the retry limit yet (Sect. 5 evaluates the Lemming effect problem). Nonetheless, the figure depicts the potential of hardware irrevocability and the weaknesses of a software fallback path.

3.1   Implementation

We propose a token-based implementation of the irrevocability mechanism where only the core that owns the token can run irrevocably. Each core has a flag that indicates whether there is an irrevocable transaction running in the system or not (the I bit). Another flag in the core signals whether the irrevocable transaction belongs to this core or to another core, i.e. whether the core owns the token or not (the T bit). Along with the pair of bits (I,T), each core has a counter (C) that holds the number of transaction retries. The core aks for irrevocability when C is 0.

When a transaction reaches the limit of retries, the L1 cache controller of the core checks its (I,T) bits and acts depending on their value:

- *(I,T) = (0,0)*: There are no irrevocable transactions running in the system and the token is not owned. In this case, the controller broadcasts a token request message that will be responded by the core that owns the token. Should the owner just start irrevocability, the token is not sent and the requester keeps stalling until the owner ends its transaction. If the token is received, the T bit is set to 1 and the controller broadcasts an irrevocability request message

---

[1] The non-transactional write to the lock causes these aborts by means of strong isolation [13]. Correctness is ensured as locks and transactions are not allowed simultaneously.

for the other cores to set the I bit to 1. The requester can safely continue its transaction in irrevocable mode, $(I,T) = (1,1)$, after acknowledgement of the other L1 cache controllers.

- $(I,T) = (0,1)$: The core owns the token, so it can request irrevocability directly.
- $(I,T) = (1,0)$: Someone else is running an irrevocable transaction. Consequently, the transaction stalls. This value for the (I,T) pair can be found on transaction beginning and after receiving an irrevocability message.

Table 1. L1 cache coherence protocol modifications for irrevocability (highlighted in gray).

State	Events					
	L1 Replace ¬(xR∨xW) ∨ (1,1)	L1 Replace (xR∨xW)∧(C>1)	L1 Replace (xR∨xW)∧(C≤1)	L2 Replace ¬(xR∨xW) ∨ (1,1)	L2 Replace (xR∨xW) (1,0)∨(C>1)	L2 Replace (xR∨xW) (0,-)∧(C≤1)
I	–	–	–	ACK	–	–
S	– /I	Abort, C-1 /I	Irre, Z	ACK /I	Abort, C-1 /I	Irre, Zz
E	PUT(no data) /I	Abort, C-1 /I	Irre, Z	ACK /I	Abort, C-1 /I	Irre, Zz
M	PUT+Data /I	Abort, C-1 /I	Irre, Z	ACK+Data /I	Abort, C-1 /I	Irre, Zz

Irre: ask for irrevocability
Z and Zz: recycle mandatory and request queue, respectively
(#,#): pair of bits (I,T)

We have modified the L1 cache controller to implement the anticipation to a block replacement. Table 1 shows the modifications made to the protocol highlighted in gray. L1 cache replacements are left untouched whenever either the block to be replaced is not transactional, $\neg(xR\lor xW)$, or the core is in irrevocable mode and owns the token, $(I,T) = (1,1)$. However, if the block is transactional, $xR\lor xW$, the counter (C) is checked. If C>1 (1 instead of 0 to anticipate the last abort) the transaction aborts and C is decremented. Conversely, if C≤1, the core asks for irrevocability and the mandatory queue is recycled[2] so that the event is triggered later on. Should the core manage to get irrevocable, the L1 replacement is performed safely. If irrevocability is not granted, the core stalls by continuously recycling the message that causes the eviction.

In case of L1 transactional block replacements due to L2 cache evictions (L2 Replace events in Table 1) we have different scenarios. If the core is running an irrevocable transaction, $(I,T) = (1,1)$, the event is treated as a normal L2 cache replacement. However, if the irrevocable transaction is of another core, $(I,T) = (1,0)$, the transaction in this core must be aborted in favour of the irrevocable one. Thus, the only situation in which a transaction asks for irrevocability on

---

[2] The cache controller comprises queues where coherence messages are buffered until they are served by the controller [18]. In this case, there are a mandatory queue that holds the messages from the CPU to the L1 cache, a request queue that holds request messages from/to the L1 cache and a response queue with response messages from/to the L1 cache.

an L2 Replace event is when $C \le 1$ and there is no other irrevocable transaction in the system, $(I,T) = (0,-)$.

The special case in which several transactions ask for the token at the same time is arbitrated by the controller queue of the core that owns the token. The owner of the first token request message found in such a queue is the one that gets the token. The rest of the token request messages are ignored and the requesters stalled. They will ask for irrevocability again after receiving a message of end of irrevocability.

4   Simulation Environment

The simulation environment comprises the full system simulator Simics [12], and the Wisconsin GEMS [14] toolkit that includes Ruby. Ruby is a multiprocessor memory system timing simulator, which we have modified to simulate the best-effort HTM system outlined in Sect. 2, and the proposals described in this paper.

The target system is organized as shown in Fig. 1. It comprises 16 in-order single-issue cores, with a private 32 KB split 4-way L1 cache where the data cache holds two read and write transactional bits per 64B block. The L2 cache is unified, shared and divided into 16 banks of 512 KB each. L2's associativity is 8-way and it does not hold transactional information. The directory keeps a full bit-vector of sharers. Each thread is bound to a core, and so it is the operating system, so that there are not interferences such as migrations and context changes. Consequently, there is a maximum of 15 threads for the use of benchmarks.

The whole Stanford STAMP suite [16] was used for the evaluation. Table 2 shows the parameters and characteristics of the benchmarks. Namely, the number of transactions that successfully commits (# Xact), the percentage of time running transactions (% Time in Xact), and the average RS/WS (read set/write set) cardinality of the transactions, in cache blocks.

Table 2. Workloads: Input parameters and transactional characteristics.

Bench	Input	# Xact	% Time in Xact	$avg|RS|$	$avg|WS|$
Bayes	-v32 -r1024 -n2 -p20 -i2 -e2 -s1	654	94 %	87.64	48.91
Genome	-g512 -s32 -n32768	19496	85 %	23.34	3.58
Intruder	-a10 -l16 -n4096 -s1	54933	92 %	9.87	3.06
Kmeans	-m15 -n15 -t0.05 -i random-n2048-d16-c16	8235	46 %	6.23	1.75
Labyrinth	-i random-x32-y32-z3-n96	222	100 %	139.34	95.12
SSCA2	-s14 -i1.0 -u1.0 -l9 -p9	93721	13 %	3.00	2.00
Vacation	-n4 -q60 -u90 -r16384 -t4096	4095	95 %	63.20	10.16
Yada	-a20 -i633.2	5447	100 %	62.45	38.21

5 Software Fallback Path Evaluation

Figure 3 shows the fallback path code we have evaluated, which includes a variable to specify the number of transaction retries and the Lemming effect[3] avoidance code [6, 11]. The code defines a thread's local retry variable that is initialized to 0 (line 1). The retry limit is defined globally (RETRY_LIMIT). We define two primitives to begin a transaction: (i) TAKE_XACT_CHECKPOINT takes a register checkpoint where we want to resume the transaction on abort, but it does not start transactional bookkeeping; (ii) BEGIN_XACT begins transactional bookkeeping. Then, we can have non-transactional code between the two primitives to check whether we have to take the fallback path or not. The code to begin a transaction (lines 2–13) first takes a checkpoint and then increments the thread's local retry variable. Next, if the number of retries is greater than the retry limit (line 5), the fallback path is taken by acquiring a single spin lock (line 6). If the retry limit is not reached, the code executes transactionally and adds the lock to the read set (line 10). The transaction is explicitly aborted if the lock is taken (line 11). It should be noted that the thread waits for the lock to be released just before beginning the transaction to avoid the Lemming effect (line 8). The code to end a transaction (lines 14–19) checks the number of retries to execute either a transaction commit or a lock release.

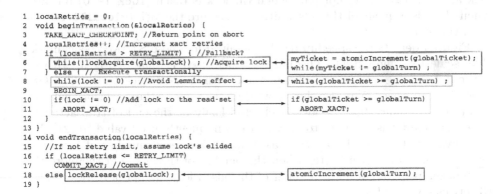

```
1   localRetries = 0;
2   void beginTransaction(&localRetries) {
3       TAKE_XACT_CHECKPOINT; //Return point on abort
4       localRetries++; //Increment xact retries
5       if (localRetries > RETRY_LIMIT) { //Fallback?
6           while(!lockAcquire(globalLock)) ; //Acquire lock          myTicket = atomicIncrement(globalTicket);
7       } else { // Execute transactionally                           while(myTicket != globalTurn) ;
8           while(lock != 0) ; //Avoid Lemming effect                 while(globalTicket >= globalTurn) ;
9           BEGIN_XACT;
10          if(lock != 0) //Add lock to the read-set                  if(globalTicket >= globalTurn)
11              ABORT_XACT;                                           ABORT_XACT;
12      }
13  }
14  void endTransaction(localRetries) {
15      //If not retry limit, assume lock's elided
16      if (localRetries <= RETRY_LIMIT)
17          COMMIT_XACT; //Commit
18      else lockRelease(globalLock);                                 atomicIncrement(globalTurn);
19  }
```

Fig. 3. Fallback code with retry limit and Lemming avoidance. Ticket lock alternative on the right.

On the right hand side of Fig. 3 we show an alternative implementation of the fallback path which replaces the single spin lock by a two-variable *ticket lock* [15]. Each thread takes its own ticket before entering the critical section by atomically incrementing and reading the global ticket variable (line 6). Then, the thread waits for his turn by checking it against the global turn variable. The global

---

[3] If one transaction takes the fallback path, the others abort and wait for the fallback path lock to be released, i.e. a complete serialization of the ongoing transactions is carried out.

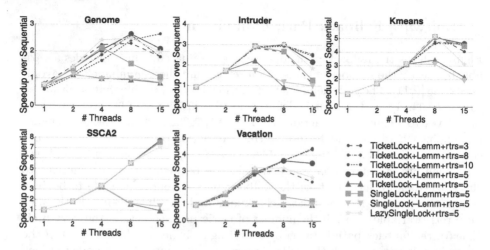

Fig. 4. Speedup over the sequential application for different fallbacks and parameters (Lemm: lemming effect avoidance, rtrs: number of retries).

turn is atomically incremented to release the lock (line 18). The implementation of the Lemming effect avoidance loop (line 8) is more accurate with the ticket lock as the thread waits not only when the lock is taken (`lock != 0`) but also when there is a queue of threads waiting to acquire the lock (`globalTicket >= globalTurn`).

Figure 4 depicts the speedup results obtained for those STAMP benchmarks that scale to some extent. The fallback code used is that of Fig. 3, with or without Lemming avoidance (±Lemm) and with single or ticket lock. The lazy single lock approach [5] is also shown, which is the same as the single lock without Lemming avoidance but the lock is checked lazily at the end of transactions.[4] The retry limit has been set to 5, which is a frequently used value [10,21]. We have evaluated 3, 8 and 10 retries as well. An increased number of retries (8 or 10) seems to perform better when the number of threads, and therefore the contention, is high. For a low number of threads, a low number of retries suffices (3 retries up to 4 threads).

The results show that the fallback path versions with Lemming effect avoidance always beat the ones without it, due to the reduction in unnecessary serializations. As far as the type of lock is concerned, the ticket lock reveals itself as a good option since it reduces lock contention and ensures fairness in lock acquisition. But more importantly, the ticket lock provides the information of how many threads are waiting to enter the critical section and therefore, the Lemming loop waits for them to finish. Conversely, the single lock does not provide such information. Thus, the threads waiting at the Lemming loop may begin a transaction while other threads are contending for acquiring the lock.

---

[4] In this manner, multiple transactions are allowed to execute in parallel with the one in the fallback path, as long as such transactions commit after the lock release and the fallback code does not conflict with them.

Those transaction will be aborted by the eventual lock acquisition. This fact is more probable in those benchmarks that spend a lot of time in transactions such as Genome, Intruder and Vacation (see Table 2), which take advantage of the ticket lock Lemming loop enhancement to avoid unnecessary aborts. SSCA2 and Kmeans are most of the time out of transactions and they are not affected by the type of lock. The lazy single lock yields good results since it encourages parallelism. However, the performance is worse than the ticket lock with Lemming effect avoidance as the number of threads increases, thus increasing the contention (e.g. Intruder, Kmeans and Vacation with 15 threads). The fallback conflicts with the concurrent transactions.

6 Hardware Irrevocability Mechanism Results

Figure 5 shows the speedup of the baseline BE-HTM system with the hardware irrevocability mechanism (Irre) and the software fallback path (Fback) with ticket lock and enhanced Lemming effect avoidance. The hardware irrevocability mechanism counter has been set to 5, as well as the retry counter of the fallback code. From these results we can classify the STAMP benchmarks in the following groups.

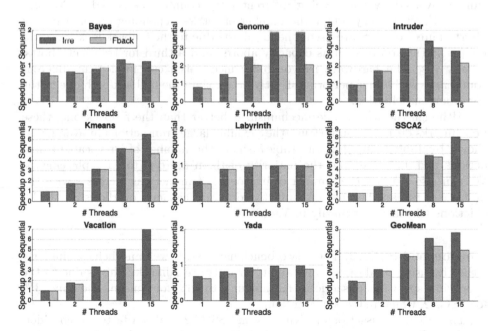

Fig. 5. Speedup of the hardware irrevocability mechanism (Irre) and the software fallback path (Fback) over the sequential application. The geometric mean is also shown (GeoMean).

Table 3. Average number of irrevocable transactions, broken down into those due to L1 or L2 replacements, and those due to conflicts. Average number of aborts of irrevocability and fallback.

Bench	# Irrevocable Xacts (L1/L2/Conflicts)			Aborts(IRRE/FBACK)		
	4 th's	8 th's	15 th's	4 th's	8 th's	15 th's
Bayes	135(86/0/49)	148(68/0/80)	179(37/3/139)	653/728	788/1212	1040/1585
Genome	1203(1120/0/83)	1195(1017/0/178)	1679(1358/21/301)	5022/7942	5582/10821	8217/16321
Intruder	1055(22/0/1033)	3794(36/0/3759)	10562(110/1/10450)	11455/10428	35861/39628	77299/96499
Kmeans	400(0/0/400)	970(0/0/970)	1815(0/0/1815)	2193/2074	5425/6537	10296/19185
Labyrint	97(76/0/21)	122(57/0/64)	160(44/0/116)	435/631	617/783	797/931
SSCA2	127(0/0/127)	283(0/0/283)	515(0/0/515)	657/575	1583/2140	3208/5486
Vacation	249(217/0/33)	347(280/0/68)	433(301/0/132)	1272/2924	1773/6221	2357/9874
Yada	1021(702/0/319)	1245(710/0/535)	1557(628/57/872)	4651/9128	5895/12561	7600/13643

Bayes, Labyrinth and Yada. The speedup obtained for these benchmarks is barely that of the sequential version. And when there is only one thread the results are even worse than the sequential. The problem with performing worse than the sequential when we have only one thread running in the system is the number of retries before getting irrevocable or taking the fallback (set to 5 in this evaluation). With only one thread there is no abort due to conflicts, so all aborts are because of capacity overflows, that are usually persistent. This can be avoided by maintaining different retry counters as stated in Nakaike et al. [17], where they adapt the number of retries depending on the cause of abort. Three counters are used: one for aborts due to the fallback lock, a second for persistent aborts such as capacity aborts, and a third for transient aborts. In any case, the hardware irrevocability mechanism can implement different counters as well and it performs slightly better than the fallback path due to the last abort anticipation.

Although the irrevocable mechanism is better than the fallback one, these benchmarks do not scale because they exhibit large transactions in average, as shown by Table 2. In addition, Table 3 shows the number of irrevocable transactions and its cause, and the majority of them are due to L1 replacements. We can also see how the number of irrevocable transactions increases with the number of threads because of conflict aborts and capacity overflows due to L2 evictions (the latter primarily in Yada).

Kmeans and SSCA2. These two benchmarks scale well and behave similarly either by using hardware irrevocability or software fallback. This is due to the short time spent in transactions that amounts to 46 % for Kmeans and only 13 % for SSCA2, which reduces contention.

The size of transactions in Kmeans and SSCA2 is also a factor to consider. Their small transactions make that the fallback path or hardware irrevocability are barely taken. Actually, Table 3 shows 0 irrevocable transactions due to L1 and L2 replacements. However, contention makes some transactions to abort and take the fallback or the irrevocability mechanism when we have more threads. For this configurations we can see a slight benefit of irrevocability over the fallback version, or not so slight for Kmeans and 15 threads, because the irrevocability mechanism stalls the transactions instead of aborting them. Table 3 shows such an abort reduction that is up to 9000 transactions for Kmeans and 15 threads, which supposes an abort rate of 1.2 with irrevocability in contrast to the 2.32 of the fallback path.

Genome, Intruder and Vacation. For this group of benchmarks we obtain considerable benefits by using the BE-HTM system with hardware irrevocability over the fallback configuration. They are benchmarks with medium and small-sized transactions (Genome and Vacation) or that are more contended (Intruder). These characteristics can be noted in the number of irrevocable transactions that are due to replacements or conflicts in Table 3.

The hardware irrevocability mechanism not only performs better due to the anticipation of the last abort but also reduces the number of aborts by stalling non-irrevocable transactions instead of aborting them. Table 3 shows that the number of transaction aborts for the system with irrevocability is usually lower than that of its software fallback counterpart. The amount of wasted work is larger for the fallback path, specially for Genome and Vacation with 15 threads, where the abort reduction is more than 50 %.

Summarizing, the BE-HTM system with irrevocability speeds up the execution about 2x with respect to the fallback path counterpart for Genome and Vacation, and it is around 20 % better for Intruder and Kmeans, for 15 threads. The rest of the benchmarks yields similar or slightly better speedup by using irrevocability.

7   Related Work

Irrevocability in the context of HTM was first proposed in TCC [7] to deal with overflowed transactions. Blundell et al. [3] introduces OneTM-Serialized as a system where overflowed transactions gets irrevocable and serializes the system to ensure forward progress. Their implementation comprises a log-based HTM where the irrevocable transaction can be aborted as old data can be recovered from the log. They use a shared transaction status word residing in a fixed virtual location that acts as a mutex lock to implement the irrevocability mechanism. We implement irrevocability with a token-based mechanism distributed through the cache controllers, in the context of a best-effort HTM system, comparing its performance with a software fallback path to gain insight into the hardware that could enhance the fallback.

IBM Blue Gene/Q HTM [19] ensures forward progress on capacity overflows and contention scenarios by means of an irrevocable mode. The irrevocability mechanism is implemented in a runtime system, thus freeing the programmer from the task of providing a fallback code. The runtime decides if a transaction gets irrevocable in an adaptive way. However, it has to abort a transaction to run it in irrevocable mode, whereas our hardware irrevocable mechanism anticipates the abort and initiates the irrevocable mode without wasting the work done so far by the transaction.

Afek et al. [2] propose a ticket-lock-based technique to improve the performance of Haswell's hardware lock elision (HLE). It is a different approach to our use of the ticket lock. In this case, the ticket lock guards the HLE lock and is acquired by those transactions that abort due to conflicts. Thus, the conflicting transactions are executed speculatively in turn, in parallel with the non-conflicting ones. After a given number of aborts, the transaction holding the ticket lock acquires the HLE lock and aborts all other transactions in the system. In fact, it is a contention management approach.

8 Conclusions

In this paper we propose a hardware implementation of an irrevocability mechanism to gain insight into the hardware enhancements that may speedup the execution of a fallback path in BE-HTM systems. We find that anticipating the abort that causes the execution of the fallback path and stalling the other transactions running in the system yields a significant improvement over the *abort-all* fallback solution.

On the other hand, we propose an enhanced Lemming effect avoidance loop by means of a ticket lock. A ticket lock provides precise information of how many threads are waiting to acquire the lock, so the separation of transactional and non-transactional execution can be performed more precisely.

We suggest having a hardware accelerated fallback path to retain both hardware benefits and software versatility. However, the possibility of having a hardware alternative to the software fallback path can be interesting for the user due to its simplicity.

Acknowledgement. This work has been supported by the Government of Spain under project TIN2013-42253-P and Junta de Andalucía under project P12-TIC-1470.

References

1. Intel 64 and IA-32 Architectures Optimization Reference Manual. Chapter 12.3:Developing an Intel TSX Enabled Synchronization Library, September 2014
2. Afek, Y., Levy, A., Morrison, A.: Programming with hardware lock elision. ACM SIGPLAN Not. **48**(8), 295–296 (2013)
3. Blundell, C., Devietti, J., Lewis, E.C., Martin, M.M.K.: Making the fast case common and the uncommon case simple in unbounded transactional memory. In: 34th Annual International Symposium on Computer Architecture (ISCA 2007), pp. 24–34 (2007)
4. Cain, H.W., Michael, M.M., Frey, B., May, C., Williams, D., Le, H.: Robust architectural support for transactional memory in the power architecture. In: 40th Annual International Symposium on Computer Architecture (ISCA 2013), pp. 225–236 (2013)

5. Calciu, I., Shpeisman, T., Pokam, G., Herlihy, M.: Improved single global lock fallback for best-effort hardware transactional memory. In: 9th Workshop on Transactional Computing (TRANSACT 2014) (2014)
6. Dice, D., Herlihy, M., Lea, D., Lev, Y., Luchangco, V., Mesard, W., Moir, M., Moore, K., Nussbaum, D.: Applications of the adaptive transactional memory test platform. In: 3rd Workshop on Transactional Computing (TRANSACT 2008) (2008)
7. Hammond, L., Wong, V., Chen, M., Carlstrom, B., Davis, J., Hertzberg, B., Prabhu, M., Wijaya, H., Kozyrakis, C., Olukotun, K.: Transactional memory coherence and consistency. In: 31th Annual International Symposium on Computer Architecture (ISCA 2004). pp. 102–113 (2004)
8. Harris, T., Larus, J., Rajwar, R.: Transactional Memory, 2nd edn. Morgan & Claypool Publishers, San Francisco (2010)
9. Herlihy, M., Moss, J.: Transactional memory: Architectural support for lock-free data structures. In: 20th Annual International Symposium on Computer Architecture (ISCA 1993), pp. 289–300 (1993)
10. Jacobi, C., Slegel, T., Greiner, D.: Transactional memory architecture and implementation for IBM system Z. In: 45th Annual International Symposium on Microarchitecture (MICRO 2012), pp. 25–36, December 2012
11. Liu, Y., Spear, M.: Toxic transactions. In: 6th Workshop on Transactional Computing (TRANSACT 2011). ACM (2011)
12. Magnusson, P., Christensson, M., Eskilson, J., Forsgren, D., Hallberg, G., Hogberg, J., Larsson, F., Moestedt, A., Werner, B., Werner, B.: Simics: A full system simulation platform. IEEE Comput. **35**(2), 50–58 (2002)
13. Martin, M.M.K., Blundell, C., Lewis, E.: Subtleties of transactional memory atomicity semantics. IEEE Comput. Archit. Lett. **5**(2), 17 (2006)
14. Martin, M., Sorin, D., Beckmann, B., Marty, M., Xu, M., Alameldeen, A., Moore, K., Hill, M., Wood, D.: Multifacet's general execution-driven multiprocessor simulator GEMS toolset. ACM SIGARCH Comput. Archit. News **33**(4), 92–99 (2005)
15. Mellor-Crummey, J.M., Scott, M.L.: Algorithms for scalable synchronization on shared-memory multiprocessors. ACM Trans. Comput. Syst. **9**(1), 21–65 (1991)
16. Minh, C., Chung, J., Kozyrakis, C., Olukotun, K.: Stamp: stanford transactional applications for multi-processing. In: IEEE International Symposium on Workload Characterization (IISWC 2008), pp. 35–46 (2008)
17. Nakaike, T., Odaira, R., Gaudet, M., Michael, M.M., Tomari, H.: Quantitative comparison of hardware transactional memory for Blue Gene/Q, zEnterprise EC12, Intel Core, and POWER8. In: 42nd Annual International Symposium on Computer Architecture (ISCA 2015), pp. 144–157 (2015)
18. Sorin, D.J., Hill, M.D., Wood, D.A.: A Primer on Memory Consistency and Cache Coherence, 1st edn. Morgan & Claypool Publishers, San Rafael (2011)
19. Wang, A., Gaudet, M., Wu, P., Amaral, J.N., Ohmacht, M., Barton, C., Silvera, R., Michael, M.: Evaluation of Blue Gene/Q hardware support for transactional memories. In: 21st International Conference on Parallel Architectures and Compilation Techniques (PACT 2012), pp. 127–136 (2012)
20. Welc, A., Bratin, S., Adl-Tabatabai, A.R.: Irrevocable transactions and their applications. In: 20th ACM Symposium on Parallelism in Algorithms and Architectures (SPAA 2008), pp. 285–296, June 2008
21. Yoo, R.M., Hughes, C.J., Lai, K., Rajwar, R.: Performance evaluation of intel transactional synchronization extensions for high-performance computing. In: International Conference on High Performance Computing, Networking, Storage and Analysis (SC 2013), pp. 19:1–19:11 (2013)

Portable SIMD Performance
with OpenMP* 4.x Compiler Directives

Florian Wende[1][(✉)], Matthias Noack[1], Thomas Steinke[1], Michael Klemm[2],
Chris J. Newburn[3], and Georg Zitzlsberger[2]

[1] Zuse Institute Berlin, Takustraße 7, 14195 Berlin, Germany
{wende,noack,steinke}@zib.de
[2] Intel Deutschland GmbH, 85579 Neubiberg, Germany
{michael.klemm,georg.zitzlsberger}@intel.com
[3] Intel Corporation, Santa Clara, USA
chris.newburn@intel.com

Abstract. Effective vectorization is becoming increasingly important
for high performance and energy efficiency on processors with wide SIMD
units. Compilers often require programmers to identify opportunities
for vectorization, using directives to disprove data dependences. The
OpenMP 4.x SIMD directives strive to provide portability. We investigate the ability of current compilers (GNU, Clang, and Intel) to generate
SIMD code for microbenchmarks that cover common patterns in scientific codes and for two kernels from the VASP and the MOM5/ERGOM
application. We explore coding strategies for improving SIMD performance across different compilers and platforms (Intel® Xeon® processor and Intel® Xeon Phi[TM] (co)processor). We compare OpenMP* 4.x
SIMD vectorization with and without vector data types against SIMD
intrinsics and C++ SIMD types. Our experiments show that in many
cases portable performance can be achieved. All microbenchmarks are
available as open source as a reference for programmers and compiler
experts to enhance SIMD code generation.

1 Introduction

On modern CPUs, effective use of SIMD (Single Instruction, Multiple Data) is
essential to approach peak performance. Data parallelism is achieved with a combination of multiple threads and increasingly-wide SIMD units. On the Intel®
Xeon Phi[TM] (co)processor, for instance, the 8-wide double-precision SIMD units
can provide up to one order of magnitude higher performance per core. Exploiting SIMD for codes with complex control flow, leading to masking and execution
overhead, can be difficult. In some cases, unleashing the full performance potential of computational loops can require expertise in language interfaces, compiler
features, and microarchitecture.

Thankfully, the community is converging on a vectorization standard, in
OpenMP 4.x, that eases the programming burden. Just as OpenMP has historically provided a way for users to direct execution to be parallelized across

© Springer International Publishing Switzerland 2016
P.-F. Dutot and D. Trystram (Eds.): Euro-Par 2016, LNCS 9833, pp. 264–277, 2016.
DOI: 10.1007/978-3-319-43659-3_20

threads, it now provides ways to parallelize across SIMD lanes by means of compiler directives. The latter are necessary to disambiguate dependences among loop iterations and communicate vectorization opportunities to the compiler. Our analysis shows that it can yield effective results.

This paper assesses the SIMD capabilities of the current GNU, Clang, and Intel compilers and makes the following contributions: (1) We show limitations of compiler-driven SIMD code generation, and (2) propose coding strategies both to remedy SIMD vectorization issues and to gain SIMD performance in a portable way using OpenMP 4.x. (3) We demonstrate the effectiveness of our SIMD coding strategies for microbenchmarks and two real-world kernels from the VASP and MOM5/ERGOM application.

2 Related Work

The development of techniques to transform (sequential) loop structures into parallelized constructs became a major concern with the arrival of vector and array computer systems in the 1980s. Parafrase [1], KAP [2,3], PFC [4], and VAST [5] are well-known examples of source code transformation tools at that time. To our knowledge, most of the work targets the transformation of complex loop structures into SIMD code, resulting in alternatives to standard compilers. Our approach, however, is to follow a set of SIMD coding schemes that leverage standards and foster portability across compilers. *EXPAND* [6] is a research compiler which implements an expansion approach. It targets a whole C function which is semantically transformed by replacing operators and operands with their SIMD equivalents. For twelve intrinsic functions of the GNU math library, speedups range from 2x to 11x on a PowerPC G5. *Scout* [7] is a configurable C source-to-source translator that generates code with SIMD intrinsics. Vectorization and other optimizations take place at the level of the syntax tree generated by the Clang parser. After basic optimizations, e. g., function inlining, the *unroll-and-jam* technique is applied to vectorize the loop body. For selected CFD kernels, speedups range from 1.5x to 4x on 8 AVX SIMD lanes (Sandy Bridge). Unlike EXPAND and Scout, our vectorization strategy, which includes the expansion step as well, further includes vectorization of nested loops and deep calling trees, as we demonstrate using the VASP and the MOM5/ERGOM codes. Furthermore, our strategy applies high-level code transformations to achieve portability across platforms and applicability across programming languages. The *Whole-Function Vectorization* (WFV) algorithm [8] transforms a scalar function into multiple parallel execution paths using SIMD instructions. WFV can provide a language- and platform-independent code transformation at the level of the LLVM IR. In a rendering system, they achieve up to 3.9x speedup on 4 SIMD lanes.

3   SIMD Vectorization

In this section, we show the impact of using different forms of the OpenMP 4.x constructs. Due to the considered architectures and for simplicity, we interchangeably use the terms SIMD vector and vector in the following.

3.1   Compiler and Library Support for SIMD Vectorization

No matter how aggressive compilers are at SIMD vectorization, they have fundamental theoretical limitations (cf. Rice's Theorem [9]). Compilers perform code analysis to disprove loop dependences [10,11], and attempt vectorization only if code generation for SIMD-capable targets is selected, if aggressive optimization levels are enabled, and if it is safe. We use the following configurations: GNU GCC 5.3 and Clang 3.9 (from SVN repository) compiler: -O3 -ftree-vectorize -ffast-math -mavx2 -mfma; Intel 16.0.3 compiler: -O3 -fp-model fast=2 -xcore-avx2; for the Intel Xeon Phi (coprocessor) -mmic (Knights Corner, KNC) and -xmic-avx512 (Knights Landing, KNL) is used instead of -xcore-avx2, respectively. Optimizing math functions like exp and sqrt is important since they often account for a significant fraction of computational time. Scalar versions of those functions are typically provided by system math libraries like *libm*, and vector versions are commonly available in modern compilers. They extend their scalar equivalents by using SIMD vectors rather than scalars. Their semantics are mostly the same, except for error and exception handling, and accuracy. The availability of such math vector functions depends on the compiler and underlying architecture. The configurations used in this paper are: GNU GCC 5.3 with *libmvec* (GLIBC 2.22) [12], and Intel compiler 16.0.3 with *SVML* (Short Vector Math Library) [13]. *libmvec* and *SVML* contain different sets of vectorized math functions [14,15]. GCC restricts calls into *libmvec* to within OpenMP SIMD constructs only, whereas the Intel Compiler uses *SVML* for any vectorized code.

3.2   OpenMP 4.x SIMD Directives

If compilers fail in vectorizing loop bodies, programmers need to give additional information to the compiler, e. g., by means of directives, which we call *explicit vectorization*, or by going to the extreme of low-level programming using C SIMD intrinsics, to override conservatism induced by language standards. If a loop body contains calls to functions that the compiler cannot directly generate SIMD code for (like *libmvec* or *SVML*), it has to fall back to invoking these functions once per SIMD lane, and it must pack and unpack data into and out of SIMD registers. Providing a SIMD-enabled function reduces calling overhead, increases parallelism, and avoids path length increases, yielding better overall performance. The OpenMP 4.x specification standardizes directives that are consistently implemented across different compilers. The simd construct forces loops to be vectorized, and the declare simd directive creates SIMD-enabled functions, according to a well-defined ABI [16]. A simple example could be:

```
#pragma omp declare simd simdlen(4) uniform(v)
double foo(double x,double v){return v+1.0/sqrt(x);}

void callsite(double* x,double* y,double v){
  #pragma omp simd simdlen(4) aligned(x,y:32)
  for(int i=0;i<N;i++)
    y[i]=foo(x[i],v);}
```

The simd construct instructs the compiler to partition the set of iterations of the loop into chunks of appropriate size so as to match with the SIMD capabilities of the processor. With optional clauses, programmers can specify additional properties and transformations of the loop code, e. g., privatization and reductions, that allow loop iterations to be executed independently in SIMD lanes. In the example, we explicitly define the vector length to be four with clause simdlen(4), and inform the compiler that both arrays x and y are already aligned to 32 bytes.

The declare simd directive augments function definitions and function declarations. Functions tagged with the directive are vectorized by promoting the formal function arguments and the return value (if any) into vector arguments. This was first introduced with Intel® Cilk™ Plus, known as vector_variant or vector attribute [17], but unlike OpenMP 4.x, it is not a standard. Programmers may override this behavior to retain scalar arguments that are broadcast to fill a vector with clause uniform, or express loop-carried dependences that have a linear progression with clause linear. That way, vectorizable loops can invoke suitable vector versions of functions by passing vector registers as arguments.

Table 1 illustrates the effect of adding optional clauses to an OpenMP SIMD (SIMD-enabled) function for our "simple" microbenchmark (see Sect. 5.1), encapsulating the log function. As can be seen, giving the compiler more information about the intended usage and properties of the function arguments helps generate better vectorized code for the function. More information about the full capabilities of the OpenMP 4.x SIMD feature can be found in [18].

Table 1. Gain over a reference execution of a SIMD function in the "simple" microbenchmark.

	Intel, AVX2	Intel, KNC
Reference	1.00x	1.00x
#pragma omp declare simd ..	1.50x	1.48x
.. simdlen(..)	2.03x	1.48x
.. simdlen(..) linear(..)	3.55x	4.15x
.. simdlen(..) linear(..) aligned(..)	3.95x	4.54x

4 Coding Strategies to Gain SIMD Performance

Our observation when writing code for execution on SIMD units is that compiler capabilities vary dramatically when codes are not designed with SIMD in mind from the very beginning. It is obvious that programmers need a (portable) SIMD coding scheme that different compilers can equally understand and generate code for. In this paper, we focus on SIMD vectorization of functions or subroutines as these are the most complex and complicated cases in getting performance. Our premise is that any loop body can be viewed as an inlined call to a function that contains the loop body:

```
for(int i=0;i<N;i++)              for(int i=0;i<N;i++)
   <loop_body>          ⟹           loop_body_func(..)

                                 void loop_body_func(..)
                                    <function_body=loop_body>
```

In the following, our discussion focuses on the OpenMP 4.x declare simd construct introduced in Sect. 3.2, which is simple to use for whole function vectorization. But it lacks a mechanism to explicitly switch back to scalar execution within a SIMD function, e. g., when library calls are present. Furthermore, only the GNU and Intel compiler currently support it. The latest Clang compiler, version 3.9, ignores the directive. The next paragraphs describe how to work around these issues.

(General) SIMD functions: To enhance what is provided by the simd declare directive, and to potentially enable SIMD functions with the Clang compiler in a way that also applies to Fortran code, we propose the following high-level code transformation scheme, given a scalar definition of function foo:

(1) Define vector and mask data types, e. g.,

```
typedef struct{double x[SIMDLEN_LOGICAL_REAL64];}
vec_real64_t __attribute__((aligned(ALIGNMENT)));

typedef struct{bool x[SIMDLEN_LOGICAL_REAL64];}
mask_real64_t __attribute__((aligned(ALIGNMENT)));
```

where SIMDLEN_LOGICAL_REAL64 is a multiple of the hardware's native SIMD vector length or vector width.

(2) Replace all scalar function arguments by vector arguments for the SIMD equivalent vfoo and introduce a masking argument m to allow calling vfoo conditionally, e. g.,

```
void foo(double& x,double& y);
⟹ void vfoo(vec_real64_t& x,vec_real64_t& y,mask_real64_t& m);
```

(3) Within vfoo, vector arguments are processed by extending foo's function body with a SIMD loop with the adjusted trip count. Accesses to scalar arguments in foo are replaced by accessing the components of vfoo's vector arguments instead:

```
void foo(double& x,double& y){y=sqrt(x);}
⟹ void vfoo(vec_real64_t& x,vec_real64_t& y,mask_real64_t& m){
     #pragma omp simd
     for(int ii=0;ii<SIMDLEN_LOGICAL_REAL64;ii++)
       if(m.x[ii]==true) y.x[ii]=sqrt(x.x[ii]);}
```

(4) At the call site, split the SIMD loop candidate into chunks of size
SIMDLEN_LOGI CAL_REAL64 by changing the loop increment accordingly,
and if necessary introduce a prolog and an epilog for vector un-/packing
after/before the call to vfoo, e.g.,

```
for(int i=0;i<N;i++)                for(int i=0;i<N;i+=SIMDLEN_LOGICAL_REAL64){
   foo(x[i],y[i]);       ⟹            for(int ii=0;ii<SIMDLEN_LOGICAL_REAL64;ii++){
                                         m.x[ii]=false;
                                         if((i+ii)<N){m.x[ii]=true; vx.x[ii]=x[i+ii];}}
                                      vfoo(vx,vy,m);
                                      for(int ii=0;ii<SIMDLEN_LOGICAL_REAL64;ii++)
                                         if(m.x[ii]==true) y[i+ii]=vy.x[ii];}
```

Splitting SIMD loops that are inside SIMD functions into parts enables mixing
vectorizable parts and unvectorizable serialized parts. The major differences with
explicit vectorization are (1) we use the "standard" omp simd on a for loop
over the vector length, (2) we use vector data types as arguments to a function
created from the loop body to enable nesting, and (3) we use masking to facilitate
conditional execution for branching and early exits. Hereafter, we refer to this
coding scheme as "enhanced explicit vectorization" to emphasize its distinction
to "explicit vectorization" with OpenMP 4.x SIMD functions.

Branching: SIMD execution with divergent control flow happens because of
branches whose predicates evaluate differently across the SIMD lanes. This can
be handled by using masked vector operations, if supported by the vector instruction set. Masks signal whether each SIMD lane is "active" or "inactive" for a
given execution. Control flow can also be handled with a sequence of consecutive branches, and unmasked operations followed by blend operations. If more
than one branch is taken, the effective speedup due to SIMD execution reduces
with the costs and the number of the branches. Figure 1 illustrates the scalar
and SIMD execution of a loop containing an if-else branch. Throughout iterations 0..3, both the if and the else branch is taken. Assuming they are equally
expensive, the speedup over scalar execution is at most 2x for these iterations.

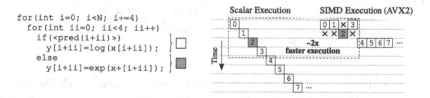

Fig. 1. Scalar and SIMD vector execution of a loop containing an if-else branch
using 4 SIMD lanes. Lanes for which the vector mask is "inactive" are marked by
crosses.

Depending on the compiler's strategy for handling conditional code execution, it can be meaningful to reduce vector masks beforehand to avoid unnecessary operations, e.g., "expensive" math calls. If, for instance, a predicate initiating the execution of a code section evaluates to false on all SIMD lanes, the
entire section can be skipped:

```
void vfoo(..,mask_real64_t& m_0){            // void foo(..){
  bool true_for_any=false; mask_real64_t m_1=false;  //
  #pragma omp simd reduction(|:true_for_any)  //
  for(int ii=0;ii<SIMDLEN_LOGICAL_REAL64;ii++){  //
    ..                                        //   ..
    if(m_0.x[ii] && <pred(ii)>){              //   if(<pred>)
      true_for_any|=true; m_1.x[ii]=true;}}   //   <"expensive code section">
  if(true_for_any)                            //         |
    #pragma omp simd                          //         |
    for(int ii=0;ii<SIMDLEN_LOGICAL_REAL64;ii++)  //     |
      if(m_1.x[ii]) <"expensive code section">  // <——————+
  ..}                                         //   ..}
```

Local copies of frequently-accessed memory references: Loops containing frequent access to the same memory reference(s) should be transformed such that memory loads happen just once into SIMD lanes. These local copies then should be used instead of referencing memory, thereby lowering the number of load operations and potentially increasing the performance.

Arrays on SIMD lanes (SIMD function context): Local arrays with size [d] in the scalar code need to be expanded to two-dimensional arrays of size [d]["SIMD width"] in the SIMD case, where the first dimension is contiguous in main memory. Unfortunately, the (Intel) compiler does not manage to optimize for the desired data layout. Currently, the only high-level approach to solve that issue is manual privatization of the array—Intel specific—or using the enhanced explicit vectorization scheme:

(a) The compiler does not optimize for the desired data layout: scatter is generated.

```
#pragma omp declare simd            #pragma omp declare simd
void foo(double& y){                void bar(double x[],int n){
  double x[5];                        for(int i=0;i<n;i++)
  bar(x,5);                             x[i]=..;}
  for(int i=0;i<5;i++)
    y+=x[i];}
```

(b) Using "enhanced explicit vectorization:" it works out of the box.

```
void vfoo(vec_real64_t& y){         void vbar(vec_real64_t x[],int n){
  vec_real64_t x[5];                  for(int i=0;i<n;i++)
  vbar(x,5);                            for(int ii=0;ii<SIMD..;ii++)
  for(int i=0;i<5;i++)                    x[i].x[ii]=..;}
    for(int ii=0;ii<SIMD..;ii++)
      y.x[ii]+=x[i].x[ii];}
```

Loops on SIMD lanes: Loops within a SIMD context can be challenging to optimize for an effective execution. If the loop trip count is a constant, each SIMD lane performs the same number of loop iterations causing no load imbalances across the lanes. If it is not, and the loop terminates differently from lane to lane, SIMD utilization might be worse. One solution to that issue is implementing dynamic SIMD scheduling, where lanes are dynamically assigned outstanding loop iterations if there are any [19]. However, here, we simply describe how to manage loops on SIMD lanes with enhanced explicit vectorization (the SIMD loop is placed inside the while loop):

```
void foo(..){              void vfoo(..){// called in scalar context
  while(<pred>)      ⟹       while(continue_loop){// to be initialized appropriately
```

```
<loop_body>..}                       continue_loop=false;
                                     #pragma omp simd
                                     for(int ii=0;ii<SIMD..;ii++){
                                         <loop_body with vector replacement>
                                         if(<pred(ii)>) continue_loop=true;}..}}
```

Manual vectorization with SIMD intrinsics and C++ SIMD classes: A very low-level approach to writing SIMD code is using intrinsic functions. This approach gives the programmer maximum control and flexibility without having to rely on the compiler's vectorization capabilities. On the downside, the code is hard to write and not portable, as the intrinsics are specific to a certain SIMD extension (e.g. SSE, AVX, AVX2).

A higher-level approach to manual vectorization is encapsulating intrinsics in C++ classes and overloaded operators. The main abstraction is a vector type, e. g., `double_v`, that offers operators for element-wise operations. Together with a query-able vector width, this allows for portable SIMD code that is valid within the C++ language. All conditional coding paths and potential remainder situations need to be handled through masking, potentially with multiple versions of a function, or blocks of code, that check for an empty, a full or a mixed mask to avoid unnecessary operations. This leads to a trade-off between performance and redundant code. We use Vc [20] for our microbenchmarks, as it provides an API with masking support and vector math-functions.

Masked math calls: We are aware only of the Intel compiler supporting conditional math function calls through SVML. However, an explicit SVML interface is missing. Instead unmasked calls need to be followed by a vector blend operation. For inactive SIMD lanes it is important to input values to these calls that assure fast convergence without causing floating point exceptions. We observed significant performance gains, when assigning, e. g., 1.0 to inactive SIMD lanes before calling `exp` unconditionally.

Source-to-source code translation: The coding scheme and SIMD optimizations presented in this section can be simply integrated into a source-to-source translator. As a proof-of-concept, we built a prototype of such a tool that allows automatic SIMD code generation for our microbenchmarks (Sect. 5.1) using enhanced explicit vectorization.

5 Microbenchmarks and Real-World Codes

To demonstrate the effectiveness of the proposed SIMD coding schemes, we use a combination of microbenchmarks and real-world kernels. The performance has been evaluated for: (1) auto-vectorized reference, (2) explicit vectorization, (3) enhanced explicit vectorization, (4) SIMD intrinsics, and (5) Vc (C++ SIMD class) library.

5.1 Microbenchmarks

Our microbenchmarks are minimal, directed tests that represent common patterns in real-world codes: (a) simple math function call, (b) conditional math function call, (c) conditional return, (d) nested branching with math function calls inside the branches, and (e) loop on SIMD lanes with varying trip count. Besides the reference implementations, we provide versions using OpenMP 4.x SIMD functions, the enhanced explicit vectorization scheme, SIMD intrinsics, and C++ SIMD class vectorization. Additionally, our implementations use OpenMP threading to run multiple instances of the benchmark concurrently. All versions can be found online [21].

Benchmarking setup: All microbenchmarks have been executed on an Intel Xeon E5-2680v3 CPU (Haswell),[1] an Intel Xeon Phi 7120P (KNC) coprocessor, and an Intel Xeon Phi 7210 (KNL) processor.

For all benchmarks we use arrays of 8192 double-precision random numbers chosen uniformly over $[-1.0, +1.0]$—in all cases we use exactly the same random numbers. With x_1, x_2, and y as input and output for the kernels, respectively, this results in 192 KiB data per kernel instance, which on both CPU and KNC/KNL fits into the per-core data cache. That allows us to focus on core performance, rather than diluting vectorization effectiveness results with caching effects. All kernels are called conditionally to reflect a realistic calling context as found in applications, which prevents compiler optimizations on fully occupied masks. The number of OpenMP threads, and hence kernel instances, used for benchmarking has been chosen to use all cores in the package: 12 on Haswell, 120 on KNC (due to 2-cycle decode), and 64 on KNL. Each benchmark setup has been executed 10 times for statistics, where the first 4 runs were skipped (warmup). For all runs, we found the maximum absolute relative error across all array elements to be less than 10^{-14} off compared to the scalar reference, showing the correctness of our implementations. Execution times for all five microbenchmarks calling exp within their function bodies are illustrated in Fig. 2. We only show results for selected runs, where simdlen=x with x=8 on Haswell and x=16 in KNC and KNL, respectively. All benchmark results is available online.

```
a) simple(x1,x2,y){
     y=<mathcall(x1+x2)>;}
```

```
b) conditional_math\
   _call(x1,x2,y){
     if(<pred(x1,x2)>)
       y=<mathcall(x1+x2)>;
     else
       y=1.0;}
```

```
c) conditional\
   _return(x1,x2,y){
     if(<pred(x1,x2)>)
       return;
     y=<mathcall(x1+x2)>;}
```

```
d) nested\
   _branching(x1,x2,y){
     if(<pred_1(x1,x2)>){
       if(<pred_2(x1,x2)>)
         y=<mathcall_1(x1)>;
       else
         y=<mathcall_1(x2)>;
     }else{
       if(<pred_2(x1,x2)>)
         y=<mathcall_2(x1)>;
       else
         y=<mathcall_2(x2)>;
     }}
```

```
e) while_loop(x1,x2,y){
     y=0.0;
     while(y<YMAX)
       y+=<mathcall(x1,x2)>;
   }
```

On the Haswell platform, explicit vectorization gives acceptable performance only with the Intel compiler. GNU achieves only little success and none for Clang. Clang works only with the C++ SIMD classes and has no equivalent of libmvec or SVML, while Vc implements its own SIMD math functions. Using enhanced

---

[1] For CPU benchmarks, the core clock frequency has been set to 1.9 GHz (AVX base frequency) to avoid comparing vector runs against scalar runs with overclocked cores.

explicit vectorization, the performance gets close to SIMD intrinsics with both the Intel and GNU compiler. In some cases it is superior to intrinsics. Even though Clang benefits from enhanced explicit vectorization, it cannot compete with GNU and Intel. On KNC, enhanced explicit vectorization is on a par with explicit vectorization and SIMD intrinsics. On KNL, it ranges between the two.

Except for the while loop kernel, SIMD execution gives speedups between 2x and 4x on Haswell, and up to 2.5x on KNC and KNL. It is somehow surprising that the effect of vectorization is more visible on Haswell than on KNC and KNL. We found that using `simdlen=8` on Haswell results in a major performance gain over using `simdlen=4`, as is the native SIMD width on that platform when 64-bit words are processed. On KNC/KNL, however, going from `simdlen=8` to 16 gives only a slight performance increase. For the while loop kernel the performance gain due to SIMD execution is notable on Haswell only when enhanced explicit vectorization is used together with the Intel compiler, on KNC in all cases except Vc, and on KNL only with intrinsics.

We further found that while Intel successfully vectorized all kernels, it is not able to vectorize through the Vc's C++ SIMD class abstraction layers. With the GNU and Clang compiler, we did not observe this behavior, but found Vc be a bit slower than manual vectorization with intrinsics.

5.2 Real-World Codes

In this subsection, we demonstrate the effectiveness of the enhanced explicit vectorization scheme when used within two real-world applications, VASP and MOM5/ERGOM. Both of these are Fortran codes, where SIMD can only be introduced via high-level programming—SIMD intrinsics and C++ SIMD classes are not directly accessible. Moreover, the community tends to avoid mixing Fortran and C/C++.

VASP 5 [22, 23] is a well-known program for atomic scale materials modeling, e.g., electronic structure calculations and quantum-mechanical molecular dynamics from first principles. One of the computational hotspots is the calculation of electronic properties on a grid by means of a hybrid DFT functional. SIMD vectorization over the grid points includes a calling tree with subroutines which themselves contain a combination of nested branching together with conditional function calls and loops with low trip counts. Vectorization of the grid loop hence means vectorization along the calling tree, which requires the SIMD function feature. The integration of enhanced explicit vectorization mainly consisted of a simple scalar-vector replacement of variables that serve as input and output parameters of the relevant functions. Loops within these functions have been annotated with the `novector` directive, to deactivate compiler vectorization as the execution already happens in the SIMD context. Local arrays within the function bodies have been expanded to two-dimensional arrays as described in Sect. 5.1. As a result of these adaptions, we have the SIMD vector code at hand, which is beneficial for code debugging—this is not the case with explicit vectorization. In fact, we used the enhanced explicit vectorization code to trace

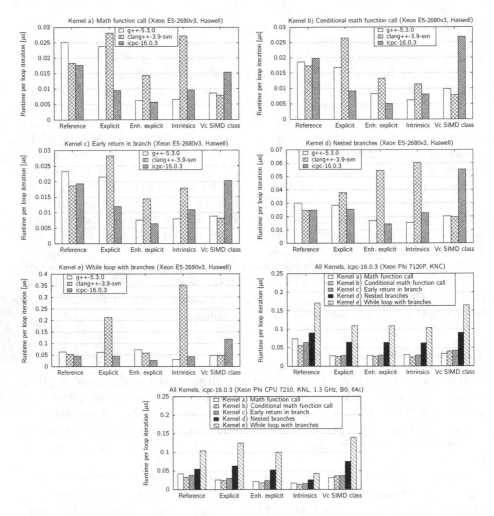

Fig. 2. Runtimes of the five microbenchmarks for different SIMD coding strategies using the GNU, Clang, and Intel compiler on Haswell, and Intel Xeon Phi KNC and KNL. The per-thread work-load is the same for each architecture. Vc on KNL uses AVX2, as AVX512 is not implemented yet.

(and eventually fix) faulty program outputs of the initial code base with explicit vectorization to an incorrect expansion of local array definitions.

   MOM [24] (Modular Ocean Model) is a program to perform numerical ocean simulations that is utilized for research and operations from the coasts to the globe. With *ERGOM* (Ecological Regional Ocean Model), MOM is extended by a bio-geochemical model that incorporates the nitrogen and phosphorus cycle. Within the ERGOM module, this cycle is modeled by numerical integration of a rate equation using an Euler integrator. The latter is at the core of a loop over grid points (hotspot), and happens to converge differently from one grid point to another, potentially causing imbalances among the SIMD lanes. Vectorization of the Euler integration scheme means to vectorize along a "while loop" which is inside the SIMD loop over grid points.

Unlike the microbenchmarks, VASP and MOM5/ERGOM kernels are much more complex and exhibit a combination of the different patterns described in this paper. For VASP, we found that performance with explicit vectorization using the OpenMP 4.x SIMD `declare` construct was inhibited by local array definitions within any of the core routines along the calling tree. We achieved better performance with the enhanced explicit vectorization scheme already in the first instance. For MOM5/ERGOM we achieved success with the enhanced explicit vectorization scheme only. With explicit vectorization we observed faulty simulation outputs. Due to the complexity of the hotspot code section—several hundreds of lines of code within the SIMD context—we were not able to figure out the origin of the faulty outputs.

Performance gains for the hotspot sections (not program performance gain) in VASP and MOM5 are summarized in Table 2. The values show that the enhanced explicit vectorization scheme is an effective high-level approach to gaining SIMD performance.

Table 2. Performance gain in VASP 5 and MOM5/ERGOM within hotspot sections.

	Reference	Explicit	Enhanced explicit	Intrinsics
VASP 5 (Intel 15.0.3, AVX2)	1.0x	1.6x	2.0x	1.9x
VASP 5 (Intel 15.0.3, KNC)	1.0x	1.5x	4.7x	5.7x
MOM5/ERGOM (Intel 15.0.3, AVX2)	1.0x	faulty	1.5x	N/A

6 Summary

We investigated the vectorization capabilities of current compilers that support OpenMP 4.x: GNU, Clang, and Intel. We focused on SIMD functions, because called library functions that are not inlinable are common and they account for a significant fraction of execution time. For a set of microbenchmarks implementing code patterns usually present in scientific codes, we found that only the Intel compiler is able to generate effective vector versions of scalar function definitions, even in case of irregularities within the calling context and the functions themselves, e. g., branching or loops with dynamic trip counts. As an alternative to OpenMP 4.x SIMD functions, we propose a vector coding scheme, "enhanced explicit vectorization," that relies on explicit vector data types together with OpenMP 4.x SIMD loop vectorization to process the vectors. Using this coding scheme, both the GNU and the Intel compiler can close up to or even exceed the performance with manual vector coding using SIMD intrinsics or C++ SIMD classes like Vc. Also the Clang compiler benefits from enhanced explicit vectorization, since it lacks the OpenMP SIMD function feature, but stays behind GNU and Intel, because of missing vector math functions. We showed the effectiveness of our approach for the microbenchmarks, and its portability across compilers. We also showed benchmarking results for two real-world codes, VASP and MOM5/ERGOM, where we successfully integrated the enhanced explicit vectorization scheme. We think that despite various findings and issues which occurred during our analysis, OpenMP 4.x provides a sound base for compiler and language independent SIMD vectorization. Combining it with our enhanced explicit vectorization scheme allows more control over compiler vectorization. We expect future compiler versions to improve support of OpenMP 4.x further.

Acknowledgments. This work is supported by Intel within the IPCC activities at ZIB, and partially supported by the project SECOS—"The Service of Sediments in German Coastal Seas" (Subproject 3.2, grant BMBF 03F0666D). We would like to acknowledge G. Kresse and M. Marsman for collaboration on VASP tuning. (Intel, Xeon and Xeon Phi are trademarks or registered trademarks of Intel Corporation or its subsidiaries in the United States and other countries. * Other brands and names are the property of their respective owners. Performance tests are measured using specific computer systems, components, software, operations, and functions. Any change to any of those factors may cause the results to vary. Intel's compilers may or may not optimize to the same degree for non-Intel microprocessors for optimizations that are not unique to Intel microprocessors. These optimizations include SSE2, SSE3, and SSSE3 instruction sets and other optimizations. Intel does not guarantee the availability, functionality, or effectiveness of any optimization on microprocessors not manufactured by Intel. Microprocessor-dependent optimizations in this product are intended for use with Intel microprocessors. Certain optimizations not specific to Intel microarchitecture are reserved for Intel microprocessors. Please refer to the applicable product User and Reference Guides for more information regarding the specific instruction sets covered by this notice.).

References

1. Kuck, D., Kuhn, R., Leasure, B., Wolfe, M.: The structure of an advanced retargetable vectorizer. In: Tutorial on Supercomputers: Designs and Applications, pp. 163–178. IEEE Press, New York (1984)
2. Davies, I., Huson, C., Macke, T., Leasure, B., Wolfe, M.: The KAP/S-1: an advanced source-to-source vectorizer for the S-1 Mark IIa supercomputer. In: Proceedings of the 1986 International Conference on Parallel Processing, pp. 833–835. IEEE Press, New York (1986)
3. Davies, I., Huson, C., Macke, T., Leasure, B., Wolfe, M.: The KAP/205: an advanced source-to-source vectorizer for the Cyber 205 supercomputer. In: Proceedings of the 1986 International Conference on Parallel Processing, pp. 827–832. IEEE Press, New York (1986)
4. Allen, J., Kennedy, K.: PFC: A program to convert Fortran to parallel form. Report MASC-TR82-6, Rice Univ. Houston, Texas, March 1982
5. Brode, B.: Precompilation of Fortran programs to facilitate array processing. Computer **14**(9), 46–51 (1981)
6. Shin, J.: SIMD Programming by Expansion. Technical report, Mathematics and Computer Science Division, Argonne National Laboratory Argonne, IL 60439 USA (2007). http://www.mcs.anl.gov/papers/P1425.pdf

7. Krzikalla, O., Feldhoff, K., Müller-Pfefferkorn, R., Nagel, W.E.: Scout: a source-to-source transformator for SIMD-Optimizations. In: Alexander, M., et al. (eds.) Euro-Par 2011, Part II. LNCS, vol. 7156, pp. 137–145. Springer, Heidelberg (2012)
8. Karrenberg, R., Hack, S.: Whole function vectorization. In: International Symposium on Code Generation and Optimization. CGO (2011)
9. Rice, H.: Classes of recursively enumerable sets and their decision problems. Trans. Am. Math. Soc. **74**, 358–366 (1953)
10. Bacon, D., Graham, S., Sharp, O.: Compiler transformations for high-performance computing. ACM Comput. Surv. **26**(4), 345–420 (1994)
11. Wolfe, M.: High-Performance Compilers for Parallel Computing. Pearson, Redwood City (1995)
12. Senkevich, A.: Libmvec (2015). https://sourceware.org/glibc/wiki/libmvec
13. Intel: Intrinsics for Short Vector Math Library Operations (2015). https://software.intel.com/en-us/node/583200
14. O'Donell, C.: The GNU C Library version 2.22 is now available (2015). https://www.sourceware.org/ml/libc-alpha/2015-08/msg00609.html
15. Intel: Vectorization and Loops (2015). https://software.intel.com/en-us/node/581412
16. Intel(R) Mobile Computing and Compilers: Vector Function Application Binary Interface, Version 0.9.5 (2013). https://www.cilkplus.org/sites/default/files/open_specifications
17. Intel: Intel Cilk Plus Language Extension Specification (2013). https://www.cilkplus.org/sites/default/files/open_specifications/Intel_Cilk_plus_lang_spec_1.2.htm
18. OpenMP Architecture Review Board: OpenMP Application Program Interface, Version 4.5 (2015). http://www.openmp.org/
19. Krzikalla, O., Wende, F., Höhnerbach, M.: Dynamic SIMD vector lane scheduling. In: Proceedings of the ISC 2016 IXPUG Workshop. LNCS. Springer (2016). www.ixpug.org
20. Kretz, M., Lindenstruth, V.: Vc: a C++ library for explicit vectorization. Softw. Pract. Exper. **42**(11), 1409–1430 (2012)
21. This paper: Code samples hosted on https://github.com/flwende/simd_benchmarks
22. Kresse, G., Furthmüller, J.: Efficient iterative schemes for ab initio total-energy calculations using a plane-wave basis set. Phys. Rev. B **54**, 11169–11186 (1996)
23. Kresse, G., Joubert, D.: From ultrasoft pseudopotentials to the projector augmented-wave method. Phys. Rev. B **59**, 1758–1775 (1999)
24. Griffies, S.M.: Elements of the Modular Ocean Model (MOM). NOAA Geophysical Fluid Dynamics Laboratory, Princeton, USA (2012)

Parallel and Distributed Data
Management and Analytics

Lightweight Multi-language Bindings
for Apache Spark

Luca Salucci[1(✉)], Daniele Bonetta[2], and Walter Binder[1]

[1] Faculty of Informatics, Università della Svizzera italiana (USI),
Lugano, Switzerland
{Luca.Salucci,Walter.Binder}@usi.ch
[2] Oracle Labs, VM Research Group, Lugano, Switzerland
daniele.bonetta@oracle.com

Abstract. Apache Spark has emerged as one of the most prominent frameworks for distributed high-performance data analysis. Among
Spark's most appealing features are its bindings for dynamic languages
such as Python and R. Despite of the great flexibility of such languages,
they often cannot match the performance of statically typed languages
such as Java or Scala. However, this limitation is not only due to the
intrinsic nature of dynamically typed languages. Largely, the performance gap is caused by the way the language runtimes interact with
Spark. In this paper we describe a new approach to integrating Python
and R into data-intensive Spark applications. Our approach significantly
reduces the performance gap between such languages and their statically
typed counterpart, making dynamic languages an attractive alternative
for the implementation of big-data applications.

1 Introduction

In the context of big-data frameworks [22], Apache Spark [3] has emerged as
one of the most popular solutions for writing complex data-analytics applications. The reasons for this success are manifold, spanning from Spark's convenient high-level programming model, to its rich plugins ecosystem and its very
active developer community. Arguably, one of the key aspects of such success
is the extensive support for several different analytics techniques and domains:
started as a research project targeting data-intensive applications in a distributed cluster [29], Spark has rapidly evolved and is now supporting also other,
more complex application domains such as graph analytics [27], stream-based
computations [30], and machine learning [24]. Spark is entirely written in Scala
and runs on the Java Virtual Machine (JVM). Hence, Spark also considers Java
as a first class citizen, and allows developers to write applications in Java, too.
Scala and Java, however, are not popular languages in many of the scientific communities targeted by Spark. For this reason, Spark also offers support for other
languages, providing native built-in support for Python (i.e., PySpark [12]), and
recently also for R (i.e., SparkR [13]). Both languages are very popular among
data scientists, as they provide extensive support and abstractions for specifying

© Springer International Publishing Switzerland 2016
P.-F. Dutot and D. Trystram (Eds.): Euro-Par 2016, LNCS 9833, pp. 281–292, 2016.
DOI: 10.1007/978-3-319-43659-3_21

complex data analyses. Python and R are often considered *complemental* to each other, as they offer extensive support for closely related domains, namely, statistical computing (using R) and numerical computing (using Python with popular libraries such as NumPy [10]). Hence, both languages are often *combined* together to develop complex analyses. The support of high-level, dynamically typed languages (such as Python or R) in Spark, however, comes at the expense of performance. The reasons for this performance difference are twofold. Firstly, dynamically typed languages are inherently slower than statically typed ones. Despite of the progress in dynamic compilation techniques [18,26], languages such as Python still cannot match the performance of Java due to the intrinsic nature of their semantics, which requires the language runtime to perform additional checks for common operations such as function calls and property lookups. Secondly, the runtimes of such languages are not executed within the JVM process of the Spark runtime. If a dynamically typed language is used to express an analysis, several independent language runtimes need to be executed in separate processes, imposing significant additional runtime overheads. In particular, the language runtime driving the analysis (i.e., the JVM) has to exchange data between processes executing another language runtime (e.g., the Python engine). Exchanging data between processes is expensive, as it introduces marshalling, unmarshalling, and communication overheads, and because it prevents the just-in-time (JIT) compiler from optimizing the framework code and the data-analysis code together. Moreover, the overhead imposed by mixing multiple language runtimes grows significantly when developers need to use *more than one* dynamically typed language in their analyses. For example, since Spark does not provide any support for integrating Python with R within the same Spark application, developers need to come up with custom, ad-hoc integration solutions. Typically, such solutions are based on very inefficient integration techniques such as file-based data exchange (using HDFS [9]) between language runtimes. The need to implement such custom integration solutions also imposes a significant loss of developer productivity and may severely limit the benefits that could be gained by combining multiple languages in a data analysis.

In this paper we show how the overheads of integrating multiple dynamic language runtimes with Spark can be drastically reduced by hosting them within the same JVM process. In this way, both the dynamic language runtimes (e.g., Python, R) and the Spark runtime execute in the same JVM, taking advantage from the shared memory of the underlying multicore machine. Our approach is based on a modified version of Spark called TruffleSpark, which supports any language implemented with the Truffle [25] framework; in particular, in this paper we are using the Truffle implementations of Python (ZipPy [14]) and R (FastR [5]). Our approach not only reduces the performance gap between dynamically typed and statically typed languages in Spark, but also allows for an efficient integration of *multiple* dynamically typed language runtimes in the *same* JVM process, enabling an efficient execution of multi-language Spark analyses. Thus, we offer Spark programmers the possibility to directly combine functions written in different languages within the same Spark application without sacrificing performance. This paper makes the following contributions to the state-of-the-art in the field of data analytics:

* We present TruffleSpark and show how Truffle-based languages can be efficiently integrated into Spark. TruffleSpark is based on an earlier research prototype discussed in [15], which we extend to support multiple languages in the same Spark application.
* Thanks to TruffleSpark, we enable multi-language support for Spark, allowing the implementation of data analyses in multiple (statically and dynamically typed) languages. TruffleSpark operates on spark's RDD data types, enabling low-overhead access to RDDs from dynamic languages.
* We evaluate TruffleSpark comparing it against the original Spark framework. On the considered benchmarks TruffleSpark always outperforms the Spark bindings for Python (i.e., PySpark), and has performance close to the ones of equivalent analyses written in a statically typed language (i.e., Scala).

2 Background

Our approach is based on a modified version of the Spark runtime that supports the execution of all the languages developed using the Truffle framework. In this section we provide an overview of Spark and Truffle.

2.1 Apache Spark

Apache Spark [3] is a general-purpose framework for large-scale data processing running on the JVM. One of the main programming abstractions in Spark are Resilient Distributed Datasets (RDDs), a fault-tolerant collection of homogeneous data which can be operated on in parallel. Informally, Spark applications based on RDDs consist of subsequent modifications to such data structures. Modifications can operate either on existing RDDs (e.g., by applying an operator to all elements of an RDD), or on newly created ones. Spark supports the creation of RDDs using multiple data formats, e.g., from a file in an external storage system such as a shared filesystem, HDFS, HBase, or any data source supporting the Hadoop input format. Applications that rely on the RDD abstraction can be developed using both statically typed languages (e.g., Scala) and dynamically typed ones. In the rest of this paper, we will focus on applications using RDDs from dynamic languages such as Python.

At runtime, Spark executes as a set of JVM processes deployed on a cluster in a master-slave fashion: one *driver program* orchestrates the computation from the cluster's master node, while other *worker* processes are responsible for performing the actual distributed computation (i.e., one per cluster node). Upon each Spark's job submission, every worker spawns an *executor* process, which executes the tasks submitted to by the master using a thread pool. Once the executor is started, it communicates directly with the driver program, so that worker processes are not effectively involved in the computation. Spark can be deployed using a distributed and a standalone mode. While in the distributed deployment the driver, workers, and executors are running in distinct JVM processes, in the local standalone deployment they run as threads within a single JVM process. In both configurations, there is always one JVM per cluster node that is responsible for executing all the analyses that have been submitted. Therefore – when running with statically typed languages –, the interactions between the parallel threads running an analysis can happen very efficiently within the same JVM process. This is not the case with dynamically typed languages, as we will discuss in the next section.

2.2 Spark Bindings for Dynamically Typed Languages

Spark features language bindings for Python and R, namely PySpark [12] and SparkR [13]. Both languages are supported via an ad hoc runtime (internal to Spark, but different than the one used when executing Scala or Java applications) which has the role of orchestrating the interaction between the Spark runtime and the external language runtimes. Such an ad hoc runtime mimics the original JVM-based runtime, adopting the same driver-worker architecture, with the notable difference that executors do not use a thread-pool to accomplish their tasks, but instead, additionally spawn external Python processes. As a consequence of this runtime architecture, a Python-based analysis in Spark may require the execution of hundreds of independent processes, each one running a dedicated Python runtime, communicating via inter-process communication channels (e.g., using OS sockets). Such runtime architecture can sometimes be inefficient, as the serialization, transmission and reconstruction of objects that take place in the current system are unnecessary and avoidable, as the entities involved run on the same shared-memory machine. Although such socket-based communications may be optimized by the OS and be highly-efficient, they still incur a cost which is considerably higher than the one of simply sharing an object reference among threads sharing a common memory space. On the contrary, an equivalent analysis using Java or Scala involves in the computation only a single JVM process per cluster node, and can benefit from efficient in-memory communication between JVM threads. More precisely, in the case of the standalone deployment, ZipPy/Spark uses a single JVM process, while PySpark uses one JVM process, plus N Python processes (where N is the number of available cores). In the distributed deployment case, ZipPy/Spark uses one JVM process per cluster node, for a total of M JVM processes (with M being the number of nodes in the cluster). PySpark running in a cluster still requires to spawn one JVM per cluster node, and additionally requires N Python processes per node (resulting in M JVM processes plus $M * N$ Python processes deployed across the cluster). The Spark terminology lacks a term to refer to the threads in the executor's pool and to the Python (respectively, R) processes spawned by PySpark (respectively, SparkR). These two entities fulfill the same role in their runtimes, and we will call them *task runners*. Task runners are threads in the Spark runtime, whereas they are Python (respectively, R) processes in the PySpark (respectively, SparkR) runtime.

2.3 GraalVM and Truffle

Our modified version of Spark relies on Truffle and on the GraalVM runtime. Graal [7] is a state-of-the-art JIT compiler for the JVM, written in Java and focused on performance and language interoperability. Graal implements several optimizations such as method inlining, eliding object allocations, and speculative execution, and can apply them to dynamically typed languages developed with the Truffle [25] language development framework. The GraalVM comes with support for multiple Truffle-enabled languages:

* Graal.js: a JavaScript runtime supporting Node.js applications
* FastR: a fast JIT compiler-enabled R language runtime
* RubyTruffle: a Ruby language runtime
* ZypPy: a Python language runtime

Truffle is a language implementation framework that relies on the notion of self-optimizing Abstract Syntax Tree (AST) interpreters [25]. A Truffle language is defined in terms of AST nodes corresponding to the constructs available in the language. A notable characteristic of such nodes is that they use the information gathered during their execution to specialize (i.e., to rewrite themselves) for the types observed at runtime. In this way, when running with GraalVM, a Truffle AST can be compiled to efficient machine code via partial evaluation [25].

3 TruffleSpark

TruffleSpark is our modified version of Spark that supports Truffle-based languages. In this section we provide an overview of its architecture and of its main components, that is, guest function wrappers and the mechanism for data-type conversions between different language runtimes. TruffleSpark enables the lightweight embedding of dynamically typed guest languages (e.g., Python) in applications developed using the Spark standard API (based on Scala or Java). TruffleSpark extends Spark with the following two main capabilities:

* **New Python and R RDD Bindings**. The two dynamic languages are integrated in the Spark runtime at the level of the JVM, that is, without requiring the integration of external language runtimes. As a consequence, TruffleSpark does not need to execute Python or R code in independent processes, and can execute the entire analysis in the same JVM process. This brings notable advantages (in terms of performance), as the overhead due to inter-process communication is substantially reduced. Moreover, this enables the JVM to perform optimizations (e.g., JIT compilation) of Python code together with the Spark runtime code. Finally, this approach has notable advantages in terms of resource utilization, as Java threads are used instead of more heavy-weight processes.
* **Language Interoperability**. Since the Truffle framework and the GraalVM support several languages, our integration allows one to combine multiple

of them in Spark in the same application, allowing developers to share the infrastructure code and classes necessary to execute analyses that combine multiple language runtimes.

The embedding of guest language functions in the runtime is achieved by means of *function wrappers* (called also *wrappers* in the shorthand form); these wrappers provide the functionality for parsing and compiling the Python or R functions (by interacting with the GraalVM), as well as for invoking them once parsed. Since wrappers are Java objects, they can be optimized together with the Spark runtime and the function they wrap by the JVM's JIT compiler. Being able to execute Python or R code is a necessary but not sufficient condition for integrating foreign languages in Spark. As the dynamically typed language runtime is hosted on the same JVM where Spark runs, it is possible to directly exchange object references between it and the Spark runtime. Exchanged objects, however, have different data type representations and APIs (e.g., when accessed from Python or Java). To support operating on data types belonging to different language runtimes, we provide *object adapters* (called also *adapters*) for each of such objects. In this way, data structures that offer the API enforced by the host language (i.e., Scala or Java) are backed by the original guest object (e.g., in Python). The same approach is used also for objects that are passed from Spark to the dynamic language runtime, with different adapters enclosing a Java object while offering the API expected from, e.g., a Python object. This approach avoids deep copying of the objects exchanged between language runtimes, and greatly reduces overhead.

An example TruffleSpark application combining multiple Truffle languages is depicted in Fig. 1. The application corresponds to a typical word-count benchmark, expressed by combining Java, Python, and R. In the example, each language is used to express a different part of the computation. Java is used to specify how the RDDs are created, and what operations have to be applied to them (e.g., `flatMap`). Python and R are used to express the functional part of the analysis, that is, to split the input and to count words. Like with normal Spark applications, it is the responsibility of the programmer to choose the proper interface for the RDD data types involved in the analysis (e.g., `JavaPairRDD` of strings). Different than plain Spark, expressions in such languages can be directly embedded in the analysis code. To this end, we provide bindings for Truffle-based languages using a function wrapper (e.g., `PyFunction` in Fig. 1). Such wrappers enable the execution of the guest language (e.g., Python) in the same JVM process running Spark, as they implement the methods used internally to construct, serialize, restore, and invoke the guest language function.

3.1   TruffleSpark Implementation

The TruffleSpark framework builds upon the Sparks' Java RDD API, which has been extended to support dynamically typed languages. The two main elements that compose our TruffleSpark framework are guest language wrappers and the object adapters used to perform data-type conversion between the *guest* languages (i.e., Python and R), and the *host* language (i.e., Java).

```
 1 // A file is read from the disk and stored into an RDD
 2 JavaRDD<String> lines = ctx.textFile("file:///wordcount/big-input.txt");
 3 // The first part of the computation is expressed in Python
 4 JavaRDD<String> words = lines.flatMap(new PyFlatMapFunction<String, String>(
 5       "def splitLine(l) : return l.split();");
 6 // The second part of the computation is expressed in R
 7 JavaPairRDD<String, Integer> ones = words.mapToPair(
 8      new RPairFunction<String, String, Integer>(
 9       "tup <- function(x) { return(list(x, as.integer(1))) }"));
10 // Python is used again to accumulate the final result
11 JavaPairRDD<String, Integer> counts = ones.reduceByKey(
12      new PyFunction<Integer, Integer, Integer>(
13       "def reduce(c1,c2) : return c1 + c2;"));
```

Fig. 1. An example word-count computation expressed in TruffleSpark, combining Python, R, and Java code.

Guest Language Function Wrappers. The TruffleSpark runtime is composed of multiple AST interpreters and a single JIT compiler (i.e., Graal [7]), shared among all language runtimes. Each thread in the Spark runtime holds a thread-local instance of a dynamically typed language parser, to avoid data races at the language runtime level. During parsing, an AST for the code provided as input is produced. The AST is implemented as a Java class, and can be used directly from Java by invoking its methods and the API it provides. As described in Sect. 2, when the type for the nodes in the AST stabilizes, the compilation of the AST to machine code is triggered by the Graal compiler, which is shared between threads. The guest language functions are hosted and *embedded* in Java by means of function wrappers that provide the runtime support in order to execute the computation expressed in a dynamically typed language. By embedding we mean that at runtime they resemble standard Java methods, and are therefore optimized by the JVM together with the Spark runtime. Wrappers also implement the parsing and interpretation of guest language functions. This is achieved by interacting with the GraalVM parser and compiler in the background.

Guest-to-host Datatype Conversions. Enabling the execution of guest language code is not sufficient for completing the integration of the different runtimes. The integration of guest languages in TruffleSpark does not require the expensive serialization and reconstruction of objects: objects can be passed from one language to the other freely, since they operate in the same address space. This helps avoiding the transmissions of objects over OS pipes, or other means of IPC, among processes running on the same shared-memory machine. Despite being available and accessible by the other language runtime, such objects cannot be used as they are out of their native context. It is necessary to make possible for the *foreign* language (e.g., Python) to operate on objects generated by another runtime (e.g., R). This has been achieved with a solution that involves object adapters for the foreign language objects, which can consequently be used in the host runtime as if it was a native object. This solution avoids deep copying of the object to its closest equivalent in the target language. In this way, the adapters offer the expected API to the hosting runtime while being backed by the original object. Thus, with respect to the original system which involves the serialization of the data, we only pay for the allocation of an adapter.

Function Wrappers and Guest Object Serializability. In order to be executable by the Spark runtime, wrappers need to be serializable, as it is necessary, together with the input data, to transmit the code to the Spark workers. To meet this requirement we made the AST stored in a function wrapper `transient`, according to the Java terminology (which causes the field and the object contained in it, to be omitted in the serialization process). We transmit only the code of the function over the network. Once restored on the remote node, the remote instance of the GraalVM will translate the code into its AST equivalent and store it back in the newly constructed wrapper, thus restoring the original object. Moreover in Spark, whenever a type is returned as the output of an RDD transformation and has to be stored in the dataset, it is required to be serializable. As in the case of the guest functions, the guest objects passed among the different stages of the computation, could eventually be transmitted over the network if the framework runs in a distributed deployment. To this end, we modified some of the guest language types implementations (e.g., R tuples and lists or vectors, which are frequently used as intermediate objects between stages of the computation), to make them serializable.

4   Evaluation

We performed an initial experimental evaluation of TruffleSpark with the main goal of highlighting the performance of dynamic languages executed using our approach. To this end, we compared our TruffleSpark framework against Spark on common applications that use dynamic languages.

We first focused our evaluation on the Python bindings for Spark, comparing the performance of some well-known Spark benchmarks (included in the Spark distribution) against ZipPy/Spark. The performance of TruffleSpark in this setup is expected to outperform Spark's native Python bindings (i.e., PySpark). We then focused on a single popular benchmark for data-intensive computations (i.e., WordCount), and executed it with different configurations. Specifically, we have re-implemented the benchmark using R, and using a combination of R and Python. In this way, we can highlight the benefits of combining multiple languages in a data-intensive benchmark. We use Spark 1.5.1, PySpark with Python 2.7.3, we build ZipPy using Java 8 and execute on GraalVM 0.9 (the release provided on the OTN website, which includes FastR). All experiments are executed on a server-class machine running Ubuntu 14.04 LTS equipped with two 8-core Intel Xeon CPUs E5-2680 (2.70 GHz) with 2 MB of L2 cache, 20 MB of L3 cache, and 128 GB of RAM. Hyper-threading is enabled on the cores, Intel Turbo Boost Technology is disabled, and the CPU driver is Acpi-cpufreq.

Python Performance. In evaluating the performance of Python, we considered four popular Spark benchmarks: WordCount, a program that counts the distribution of words in a text file; Grep, an analysis that scans a file searching for

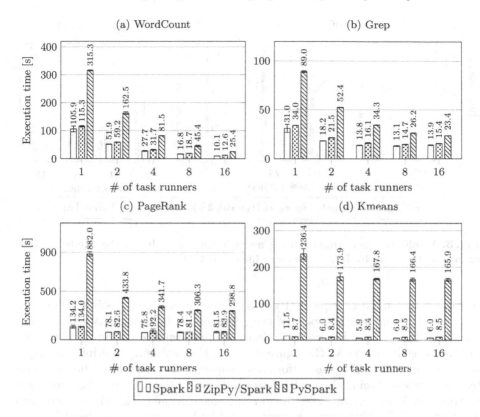

Fig. 2. Python performance. (a) WordCount and (b) Grep use an input file of size 2.7 G; (c) PageRank uses an input size of 42 M (d) KMeans uses a file size of 400 K ($k = 10$)

given patterns; PageRank, a famous graph algorithm [23]; and KMeans, a popular vector quantization method [20]. The performance of TruffleSpark is depicted in Fig. 2, where the average execution time for each benchmark is reported for an increasing number of parallel task runners. As the figure clearly shows, TruffleSpark always outperforms the equivalent implementation in PySpark. Moreover, the performance of Python running with TruffleSpark (i.e., ZipPy/Spark) is close to the same analysis written in Scala. This result suggests that our approach makes Python an efficient alternative to Scala.

Combining R and Python. The R language has performance that often cannot match Python. Still, the language offers very convenient data-analysis functionalities, making it a good choice for certain applications. With the goal of showing the potential of a multi-language solution, we have adapted one of the previous benchmarks (i.e., WordCount) to use R for parts of the computation. The performance for this experiment is depicted in Fig. 3, where the benchmark is executed with different input files. As the picture shows, the performance of

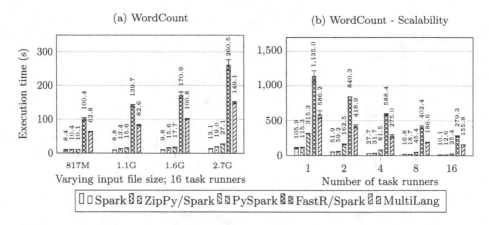

Fig. 3. WordCount performance using multiple languages. In (a) the number of task runnners is fixed at 16. In (b) the input file size is 2.7 G.

R cannot match pure Scala or Python. Nevertheless, it is possible to express the most CPU-demanding part of the computation in another language (Multilang, in the figure). By using Python, the performance of the data analysis improves significantly. As the figure suggests, by combining multiple languages it is possible to mitigate the performance impact of certain (slow) computations by selectively replacing them with faster implementations. Combining languages in this way is not possible in the current Spark framework, where users who want to use R for their analysis must accept its high runtime overhead.

5 Related Work

Other frameworks exist that provide functionalities similar to Spark. Two notable examples are Apache Flink [16], and Google Cloud Dataflow [6]. Both frameworks offer bindings for multiple languages: Flink, for example, enables the development of analyses in Scala and Java, and supports the integration of other languages via inter-process communication. Another example of multi-language integration is Apache Hadoop [1], a well-known open-source implementation of MapReduce [19]. Hadoop's standard way to express computation is to provide map and reduce operations implemented in Java. To enable multi-language integration, Hadoop also enables to execute map and reduce operations as external processes via Hadoop Streaming [8], allowing programmers to express map/reduce jobs using potentially any programming language with basic input-output capabilities. In such configuration, the map and reduce functions are executed by language runtimes that read the input from the standard input and produce results by writing it to the standard output. Another relevant approach is the one of Apache Pig [2], a high-level platform for creating MapReduce programs on top of Hadoop that enables guest language integration through IPC. Multi-language integration is enabled through a language called Pig Latin, which abstracts the

Java map/reduce idiom into a form similar to SQL. Users can extend Pig Latin by writing user defined functions (UDFs in the shorthand form), using Java, Python, Ruby, or other scripting languages, and then call them directly from Pig Latin. Support for such UDFs requires a light-weight serialization/deserialization layer with bindings in the supported languages. Unlike all such frameworks, SparkSQL [17] exemplifies an alternative approach for achieving external language integration, which overcomes IPC and serialization. Built on the previous experience with Shark [28], Spark SQL is the state-of-the-art API for querying structured data in Spark. SparkSQL features a highly extensible optimizer (i.e., Catalyst [4]) and offers much tighter integration between relational and procedural processing. Thanks to Catalyst, Spark SQL uses a code generation approach that involves the translation of SQL queries into equivalent Java byte-code, using a domain-specific language (DSL) similar to R data frames [21] and Python Pandas [11]. SparkSQL corresponds to a relevant improvement towards the integration of foreign languages in Spark. Unlike our approach, it is focused at structured data types (as opposed to RDDs), and supports only a subset of the Python language. Unlike SparkSQL, TruffleSpark supports RDDs and aims at supporting the entire Python language.

6   Conclusion

In this paper we have introduced TruffleSpark, a modified version of the Spark runtime that supports the execution of Truffle-based languages. Thanks to TruffleSpark, it is possible to develop data analyses in Spark that make extensive use of dynamically typed languages, with performance comparable to the ones of statically typed languages. TruffleSpark is based on the tight integration of Truffle-based languages with the Spark runtime, and enables the execution of data analyses that combine more than a single language. Our performance evaluation suggests that dynamically typed languages such as Python or R are a valid alternative to Java or Scala for developing Spark applications.

Acknowledgments. Our research has been supported by Oracle (ERO project 1332) and by the Swiss National Science Foundation (project 200021 153560). We thank the VM Research Group at Oracle for their support. Oracle, Java, and HotSpot are trademarks of Oracle and/or its affiliates. Other names may be trademarks of their respective owners.

References

1. The Apache Hadoop distributed system. http://hadoop.apache.org
2. Apache Pig, high-level platform for MapReduce. https://pig.apache.org/
3. The Apache Spark engine. https://spark.apache.org
4. Catalyst: A Query Optimization Framework for Spark and Shark. https://github.com/apache/spark/tree/master/sql/catalyst
5. FastR, an high performance R runtime. https://bitbucket.org/allr/fastr/overview

6. Google Cloud Dataflow. http://cloud.google.com/dataflow
7. The Graal project. http://openjdk.java.net/projects/graal/
8. Hadoop Streaming. https://hadoop.apache.org/docs/r1.2.1/streaming.html
9. HDFS distributed file system. https://hadoop.apache.org/docs/r1.2.1
10. NumPy, scientific computing with Python. http://www.numpy.org/
11. Pandas, Python Data Analysis Library. http://pandas.pydata.org/
12. PySpark. https://cwiki.apache.org/confluence/display/SPARK
13. Spark on R. https://spark.apache.org/docs/1.6.0/sparkr.html
14. ZipPy, a fast and lightweight Python implementation. https://bitbucket.org/ ssllab/zippy
15. Efficient Embedding of Dynamic Languages in Big-data Analytics. In: Proceedings of the 36th International Conference on Distributed Computing Systems Workshops. DCPerf 2016, IEEE (2016)
16. Alexandrov, A., Kunft, A., Katsifodimos, A., Schüler, F., Thamsen, L., Kao, O., Herb, T., Markl, V.: Implicit parallelism through deep language embedding. In: Proceedings of SIGMOD, pp. 47–61 (2015)
17. Armbrust, M., Xin, R.S., Lian, C., Huai, Y., Liu, D., Bradley, J.K., Meng, X., Kaftan, T., Franklin, M.J., Ghodsi, A., et al.: Spark SQL: relational data processing in spark. In: Proceedings of SIGMOD 2015, pp. 1383–1394. ACM (2015)
18. Bolz, C.F., Cuni, A., Fijalkowski, M., Rigo, A.: Tracing the Meta-level: PyPy's tracing JIT compiler. In: Proceedings of ICOOLPS, pp. 18–25 (2009)
19. Dean, J., Ghemawat, S.: Mapreduce: simplified data processing on large clusters. Commun. ACM **51**(1), 107–113 (2008)
20. Hartigan, J.A.: Clustering Algorithms. Wiley, New York (1975)
21. Ihaka, R., Gentleman, R.: R: a language for data analysis and graphics. J. Comput. Graph. Stat. **5**(3), 299–314 (1996)
22. Nothaft, F.A., Massie, M., Danford, T., Zhang, Z., Laserson, U., Yeksigian, C., Kottalam, J., Ahuja, A., Hammerbacher, J., Linderman, M., Franklin, M.J., Joseph, A.D., Patterson, D.A.: Rethinking data-intensive science using scalable analytics systems. In: Proceedings of SIGMOD 2015, pp. 631–646 (2015)
23. Page, L., Brin, S., Motwani, R., Winograd, T.: The pagerank citation ranking: Bringing order to the web. Technical report 1999–66, November 1999
24. Shanahan, J.G., Dai, L.: Large scale distributed data science using apache spark. In: Proceedings of KDD, pp. 2323–2324 (2015)
25. Würthinger, T., Wimmer, C., Wöß, A., Stadler, L., Duboscq, G., Humer, C., Richards, G., Simon, D., Wolczko, M.: One vm to rule them all. In: Proceedings of Onward! 2013, pp. 187–204. ACM (2013)
26. Würthinger, T., Wöß, A., Stadler, L., Duboscq, G., Simon, D., Wimmer, C.: Self-optimizing AST interpreters. SIGPLAN Not. **48**(2), 73–82 (2012)
27. Xin, R.S., Gonzalez, J.E., Franklin, M.J., Stoica, I.: GraphX: a resilient distributed graph system on spark. In: Proceedings of GRADES, pp. 2:1–2:6 (2013)
28. Xin, R.S., Rosen, J., Zaharia, M., Franklin, M.J., Shenker, S., Stoica, I.: Shark: SQL and rich analytics at scale. In: Proceedings of SIGMOD 2013, pp. 13–24. ACM (2013)
29. Zaharia, M., Chowdhury, M., Das, T., Dave, A., Ma, J., McCauley, M., Franklin, M.J., Shenker, S., Stoica, I.: Resilient distributed datasets: a fault-tolerant abstraction for in-memory cluster computing. In: Proceedings of NSDI 2012, p. 2 (2012)
30. Zaharia, M., Das, T., Li, H., Hunter, T., Shenker, S., Stoica, I.: Discretized streams: fault-tolerant streaming computation at scale. In: Proceedings of SOSP, pp. 423–438 (2013)

Toward a General I/O Arbitration Framework for netCDF Based Big Data Processing

Jianwei Liao$^{(\boxtimes)}$, Balazs Gerofi$^{(\boxtimes)}$, Guo-Yuan Lien, Seiya Nishizawa,
Takemasa Miyoshi, Hirofumi Tomita, and Yutaka Ishikawa

RIKEN Advanced Institute for Computational Science, Kobe 6500047, Japan
{jianwei.liao,bgerofi,guo-yuan.lien,s-nishizawa,takemasa.miyoshi,
htomita,yutaka.ishikawa}@riken.jp

Abstract. On the verge of the convergence between high performance computing (HPC) and Big Data processing, it has become increasingly prevalent to deploy large-scale data analytics workloads on high-end supercomputers. Such applications often come in the form of complex workflows with various different components, assimilating data from scientific simulations as well as from measurements streamed from sensor networks, such as radars and satellites. For example, as part of the next generation flagship (post-K) supercomputer project of Japan, RIKEN is investigating the feasibility of a highly accurate weather forecasting system that would provide a real-time outlook for severe guerrilla rainstorms. One of the main performance bottlenecks of this application is the lack of efficient communication among workflow components, which currently takes place over the parallel file system.

In this paper, we present an initial study of a direct communication framework designed for complex workflows that eliminates unnecessary file I/O among components. Specifically, we propose an I/O arbitrator layer that provides direct parallel data transfer among job components that rely on the netCDF interface for performing I/O operations, with only minimal modifications to application code. We present the design and an early evaluation of the framework on the K Computer using up to 4800 nodes running RIKEN's experimental weather forecasting workflow as a case study.

1 Introduction

With the accelerating convergence between high performance computing (HPC) and a new generation of Big Data technologies, high-end supercomputers are increasingly being leveraged for processing the unprecedented amount of data scientific simulations and sensor networks produce [4]. Consequently, the high-performance computing community has been heavily focusing on how to provide the appropriate execution environment for Big Data workloads on large scale HPC systems.

A motivating example, as well as our case study in this paper, is *SCALE-LETKF* [23], a complex weather forecasting application that is being developed at RIKEN. With the next generation Japanese flagship supercomputer

© Springer International Publishing Switzerland 2016
P.-F. Dutot and D. Trystram (Eds.): Euro-Par 2016, LNCS 9833, pp. 293–305, 2016.
DOI: 10.1007/978-3-319-43659-3_22

(post-K) as its primary target platform, *SCALE-LETKF* is intended to provide high-resolution, real-time weather forecasting of severe guerrilla rainstorms
in Japan. Similar to other operational weather forecasting applications, the
SCALE-LETKF mainly consist of two components developed separately: a
numerical weather prediction (NWP) model and a data assimilation system. The
NWP model used here is the SCALE-LES (Scalable Computing for Advanced
Library and Environment-LES [7]), which simulate the time evolution of the
weather-related atmosphere and land/sea surfaces based on physical equations
(hereafter *"Simulation"*). Meanwhile, the data assimilation method used here
is the Local Ensemble Transform Kalman Filter (*LETKF* [6]), which assimilate
observation data taken from the real world into the simulated state to produce a
better initial condition for the model (hereafter *"Assimilation"*). The two components run in a cyclic way: after the simulation finishes, the data assimilation
starts taking the results from the simulation as its input data, and after the data
assimilation finishes, the simulation of the next cycle follows, depending on the
results from the data assimilation.

Both the simulation (*SCALE*) and data assimilation (*LETKF*) components
in the current workflow rely on *netCDF* for I/O operations using the parallel file
system. *netCDF* is a self-describing, portable, scalable, appendable and share-
able file format, which is widely used to exchange array-oriented scientific data,
such as grids and time-series [1]. Historically, the decision for file based data
exchange was mainly driven by the fact that these models are being developed
and maintained by independent research entities and it's been strongly desired
not to modify either of the component models purely for the purpose of building
a coupled forecasting system.

The prediction of guerrilla heavy rains, however, is a strictly time constrained
procedure, and we identified that file I/O based data transfer between the two
components is one of the hindering factors for acquiring the needed realtimeness.
A large number of coupling tools, targeting effective integration of separately
developed models or applications, have been proposed [13,14,21], nevertheless,
all of them require numerous modifications to the applications.

Our main focus in this paper is to provide an I/O arbitration framework that
can enable high-performance, direct data exchange among workflow components
which process large amounts of data and use *netCDF* for their underlying data
format. Furthermore, we seek to provide a solution that retains the original
netCDF API and requires only minimal changes to existing application code.
Specifically, this paper makes the following contributions:

- *General I/O Arbitration Middleware.* We propose a general I/O arbitration
  middleware, i.e., a software library that enables direct parallel data transfer
  among workflow components that utilize *netCDF* for their data representation. Our library is customizable through configuration files and requires only
  slight modifications to the source code of existing applications.
- *Support for Integration of Existing Models.* Our middleware benefits the integration of existing, separately developed models for solving complicated problems. Individual models or applications are usually developed to tackle specific

scientific issues and easy integration of existing models into complex workflows enables solving more intricate problems.

- *Accelerated Data Exchange in Coupled Systems.* Compared to file based data exchanging the proposed middleware adopts communication pattern-based optimization to efficiently support direct data transfer. It shortens data exchange time among the components so that rigid time constraints of real-time applications can be satisfied.

The remainder of the paper is structured as follows. Related work is described in Sect. 2. The design and implementation of the proposed middleware are explained in Sect. 3. Section 4 shows the evaluation methodology and discusses experimental results. At last, concluding remarks are given in Sect. 5.

2 Related Work

In weather forecasting and geoscientific systems, individual models usually deal with analyzing a single, specific phenomenon. On the other hand, a practical forecasting system takes various aspects into account, and thus it normally employs several models to achieve its final goal. This section introduces related work focusing on coupling existing models or applications, as well as on related work about conducting data transfer among the component models or applications in such systems.

① *Integration mechanisms for individual models.* The intricate global climate problems motivate researchers from different scientific disciplines to integrate existing multi-physics computation models or applications for exhaustive modeling by using a software framework or a coupled system [17]. The Model Coupling Toolkit (*MCT*) [13] is a library providing routines and datatypes for creating a coupled system, and it is mainly used in Community Climate System Model (*CCSM*) [14]. Hereafter, S. Valcke et al. [8] have summarized major coupling technologies used in Earth System Modeling, and their paper shows common features of the existing coupling approaches including the functionality to communicate and re-grid data.

The *OASIS coupler* is another related study (the latest version is *OASIS3*), which is able to process synchronized exchanges of coupling information generated by different components in a climate system, and the coupler mediates communication among the components [16]. But, the *OASIS coupler* has its own interface, and is not a solution for general cases. C. Armstrong et al. [15] have designed an approach to separate the code of models from the coupling infrastructure, but it does not provide coupling functions such as data transfer. However, it enables users to choose the underlying coupling functions from other couplers, such as the *OASIS* coupler. Besides, there are numerous existing middlewares for coupling specific models, such as *ESMF* [9], and *C-Coupler1* [10], which adopt similar integration schemes to the above mentioned solutions, but unfortunately they also require application modifications.

Moreover, G. Waston et al. [19] proposed the scheme to use parallel coupling tool for effectively integrating the existing programming and performance tools, to benefit the development of parallel applications. Dorier et al. [22] have summarized several tools developed by themselves, which can flexibly couple simulations with visualization packages or analysis workflows.

② *Data transfer approaches in coupling or other large-scale systems.* Many integrated approaches employ file-based I/O to exchange data, since the data stored on the global file system can be easily accessed by all participating components [8]. It is worth mentioning the *MCT* framework again, which also enables data transfer among different components via MPI communication [18] rather than file based I/O. For instance, the *CCSM4* system is a single executable implementation, which includes a top-level driver and components integrated via standard *init/run/finalise* interfaces by leveraging *MCT* [11]. From a functionality view point, the *MCT* tool might be the most similar approach to our work, but it requires to compile all individual models or applications together to generate a single executable binary file. The combined binary ensures that all processes can share the same MPI intra-communicator to communicate with each other through MPI function calls. However, this prerequisite is not easy to meet, because it is difficult to combine a large number of separately developed components due to possible collisions on global variables and function names.

However, since all MPI processes share the same MPI_COMM_WORLD communicator in *MCT*, local broadcast operations within a specific (individual) model becomes visible to all other processes belonging to other components. To overcome this limitation, P.A. Browne and S. Wilson [17] have proposed a very similar mechanism for coupling two specified models for the purpose of data assimilation, through a different use of the Message Passing Interface. In their solution, although two models are still compiled together to generate a single MPI job, they split the MPI communicator to enable local MPI communication within individual components. However, this solution implies that the source codes of all involved models have to be modified for enabling usage of split MPI communicators for local communication.

In addition, for supporting flexible communication patterns and better communication efficiency of I/O data transfer, the Adaptable I/O System (*ADIOS*) framework has been proposed [5]. Similarly to the *OASIS coupler*, the users have to modify the models or applications to use *ADIOS*'s specific interfaces. C. Zhang et al. have proposed and implemented a butterfly implementation of data transfer and then developed an adaptive data transfer library for the coupled systems [12]. F. Zhang et al. presented a distributed data sharing and task execution framework to minimize inter-application data exchange [20]. In summary, existing works fail to provide a general framework to integrate separately developed models or applications into a coupled system (so that direct parallel data transfer among all component models could be supported) without modifying the source codes of individual components. To the contrary, our I/O middleware intends to offer a universal communication framework to accelerate data transfer among components in coupled systems in order to meet strict time constraints. Additionally, our framework requires only minimal modifications to existing application codes.

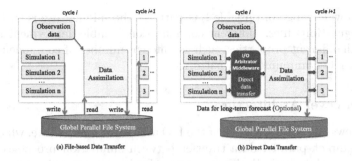

Fig. 1. The communication pattern of one cycle in the *SCALE-LETKF* system utilizing file I/O (a.) or direct data transfer (b.) methods.

3   I/O Arbitrator Middleware

As we mentioned before, our target coupling system of *SCALE-LETKF* repeats a two-step cycle of simulation and data assimilation, performed by two separately developed models, i.e., *SCALE* and *LETKF*. The I/O communication of one cycle in the current *SCALE-LETKF* system is depicted in Fig. 1(a). As seen, the *netCDF* output data of the *Simulation* processes are first written to the global parallel file system, which in turn is read by the *Assimilation* processes. Note that SCALE consists of multiple simulation instances (denoted by *Simulation* 1 to *n* in the figure), which are called "ensemble" simulations and the simulation results are fed to the LETKF processes. Each ensemble member takes slightly different initial condition and outputs different results so the total I/O amount is roughly equal to the I/O amount of one member multiplied by the number of ensemble members. After computation, the ensemble model of *SCALE* generates a large amount of output data, written in *netCDF* format, which are all requested by the subsequent assimilation step of the same cycle. In brief, the output data generated by the simulation processes will be used by the corresponding assimilation processes, which indicates that I/O communication is performed between two separate groups of MPI processes.

To reduce the time needed for data transfer, we have been developing a novel I/O middleware to allow direct parallel data transfer between the two component models. Figure 1(b) illustrates the workflow of the system when the I/O middleware is utilized. As a result, in each cycle, the output data of simulation processes are directly forwarded to the assimilation processes, as well as the analyzed results generated by assimilation processes which can be directly transferred to the simulation processes in the next cycle. Specifically, the I/O middleware connects the two kinds of processes by using MPI communication [18], and consequently, it enables direct communication between the simulation processes and the assimilation processes.

Although only the *SCALE-LETKF* application is detailed in this paper, it is worth emphasizing that our proposed I/O middleware is a general solution for coupled Big Data processing applications on the top of *netCDF*. In order to handle a wide range of possible I/O patterns the middleware is customizable using configurations files. Different configurations enable deployment for applications with different properties, such as different number of component models, or different I/O communication patterns.

3.1 High Level Architecture

Figure 2 shows the software stack of the I/O arbitrator middleware, which is used to support direct parallel data transfer between simulation processes (*SIM* in the figure) and data assimilation processes (*DA* in the figure) in our case study. Except for the application layer itself, the *netCDF*, *POSIX*, and MPI layers are involved in the middleware. Briefly speaking, the mechanism of direct parallel data transfer is transparent to the applications.

As it is shown in the figure, communication is currently performed by using MPI, for which the following subsection will discuss the construction of the communication context between two kinds of processes.

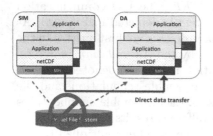

Fig. 2. Architectural overview of the middleware

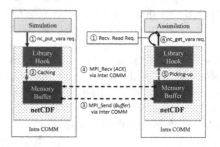

Fig. 3. The internals of direct data transfer among workflow components

3.2 Establishing Communication

Because the simulation and data assimilation models are separately developed applications, and are executed as separate MPI jobs, they do not share the same MPI communicator. To overcome this problem, our prototype implementation currently utilizes the standard MPI intercommunicator family of routines to establish a communication context between the two types of jobs.

We provide an overview of the current MPI based implementation. At initialization time, the *Assimilation* process opens a port using `MPI_Open_port()`, and then publishes it by calling the `MPI_Publish_name()` feature. Subsequently, the *Assimilation* process calls `MPI_Comm_accept()`, to wait for the connection from the simulation processes. The *Simulation* processes, connect to the *Assimilation* processes through `MPI_Comm_connect()` once they successfully obtained the service name by using `MPI_lookup_name()`. As a result, processes of both components can communicate with each other by using standard MPI functions.

Once the data transfer had taken place, the *Simulation* processes proactively disconnect and the *Assimilation* processes can unpublish their connection services with MPI_Unpublish_name().

3.3 Direct Data Transfer Mechanism

The parallel data transfer between the simulation processes and the assimilation processes is carried out when the communication context has been established. Figure 3 depicts the details of direct data transfer in the I/O middleware, where the interaction between two kinds of processes can be described as follows:

1. The *Simulation* process writes the output data to the file system through calling the nc_put_vara() function (i.e. an example write subroutine in the *netCDF* library). We assume that the *Assimilation* process will eventually read the contents of the same file, but the *Assimilation* process is blocked until the requested data is satisfied in *Step 6*.
2. The nc_put_vara() calls are intercepted by *Library Hook* offered by the middleware, and the write contents are cached in the designated *Memory Buffer* instead of flushing them to the global file system.
3. The buffered data is forwarded from the *Simulation* node to the destination node, i.e. the *Assimilation* process, by calling the MPI_Send() routine.
4. The *Assimilation* process responds an *ACK* message, when it has received the data sent by the *Simulation* process, through calling MPI_Recv(). Consequently, the data is cached in the designated memory buffer for satisfying potential future read requests.
5. According to the parameters offered by the nc_get_vara() function (i.e. an example read subroutine in the *netCDF* library), which was blocked because the required data were not yet available, the specified piece of data will be picked up by *Library Hook* from *Memory Buffer*.
6. The *Assimilation* process resumes its execution after it received the data from *Library Hook*.

Both *Simulation* and *Assimilation* processes are able to exchange their data through direct data transfer. Specifically, all nc_put_vara() requests will be fulfilled when the contents have been buffered in the memory, and all cached data are eventually sent to the destination process. On the other side, all nc_get_vara() requests will be satisfied with the data buffered in the memory, which was initially received from the source process.

3.4 Implementation for SCALE-LETKF

To demonstrate the effectiveness of direct parallel data transfer between the simulation and assimilation processes in *SCALE-LETKF*, we have developed a proof-of-concept implementation of the proposed I/O middleware. Besides, since data is exchanged between each *SCALE* process and the corresponding *LETKF* process in *netCDF* format, we have made slight modifications to the *netCDF* library itself (using *ver. 4.2.2.1*), so that it complies with the proposed I/O middleware to enable direct data transfer in an application transparent fashion.

4   Evaluation

This section first describes the experimental setup and experimental methodology for evaluating the proposed I/O middleware. It then presents experimental results and provides the relevant discussion. At last, we summarize the key points of our direct parallel data transfer approach.

4.1   Experimental Setup

Evaluation experiments to assess the advantages of the *SCALE-LETKF* system equipped with our current prototype middleware were conducted on the K computer [2]. The K computer is Japan's flagship supercomputer sporting 88,128 compute nodes (8 CPU cores each), with peak performance more than 10 petaFLOPS. The K computer took the first place of TOP 500 in 2011, and as of June 2015, it is ranked as the fourth fastest machine of the world [3].

As for the input data used in our experiments, we employ real world observations to test the efficiency of *SCALE-LETKF* when equipped with the proposed I/O middleware. In all experiments each MPI process was allocated to one compute node, and we logged the results related to I/O operations during the execution. Three real world test cases for regional weather analysis were used. In each measurement, *SCALE* is composed of up to 100 ensemble instances. *Test Case 1* and *Test Case 2* have 4 processes in each ensemble instance, but there are 48 processes in each ensemble instance of *Test Case 3*. *LETKF* consists of only one instance, but it contains the same number of processes as all *SCALE* instances in total. Note that every MPI process is allocated onto one computing node, and *openMP* is used to explicitly direct multi-threaded parallelism.

Table 1. Total Amount of Transferred Data in the Case Study.

Ensemble Size	Test Case 1	Test Case 2	Test Case 3
10	3, 468 MB	6, 720 MB	53, 328 MB
20	6, 936 MB	13, 440 MB	106, 656 MB
40	13, 872 MB	26, 880 MB	213, 312 MB
60	20, 808 MB	40, 320 MB	319, 968 MB
80	27, 744 MB	53, 760 MB	426, 624 MB
100	34, 680 MB	67, 200 MB	533, 280 MB

Table 1 summarizes the size of transferred data for the cases having different number of ensemble instances. Note that in our current execution model, each application instance corresponds to a separate MPI job.

4.2   Experimental Results

The main limitation of our current proof-of-concept I/O middleware is that we can run only one cycle of the *SCALE-LETKF* system. In other words, each *SCALE* process generates output data after simulation, which will be read by the corresponding *LETKF* process as input for assimilation.

Communication Time for Transferring Data. While running the selected two test cases, we first measured the communication time for transferring data between *SCALE* and *LETKF*, as the function of increasing the number of ensemble instances from 10 to 100. Figure 4 shows the time required for transferring the data from *SCALE* to *LETKF*. The horizontal axis represents the number of ensemble instances and the vertical axis shows the time required for data transmission. As the experimental results imply, the communication time for transferring data between the two components remains essentially unchanged, even with the growing number of involved processes, which is due to the pair-wise communication pattern of the *SCALE-LETKF* system.

Another interesting observation is that while data size between *Test Case 1* and *Test Case 2* differ by a factor of two, transfer time is only increased by approximately 50 %. We believe this is due to the communication protocol that attains higher bandwidth with the increased data size.

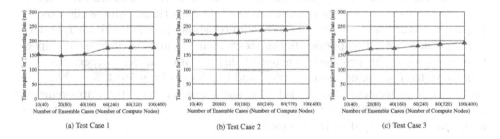

(a) Test Case 1   (b) Test Case 2   (c) Test Case 3

Fig. 4. Communication time needed for transferring data from *SCALE* to *LETKF*.

I/O Acceleration. For comparison, we recorded the time required for I/O operations between the *SCALE* and *LETKF* processes by using both actual file I/O operations and the mechanism of direct data transfer. Figures 5(a), (b) and (c) indicate the time required for carrying out I/O operations between the two types of processes utilizing file I/O and the proposed mechanism, respectively. Note that the I/O time shown by the proposed mechanisms includes the time needed for memory operations issued by both *SCALE* and *LETKF* processes, and the time required for transferring the data from *SCALE* to *LETKF*.

As the Figure depicts, the proposed mechanism can substantially reduce the time needed for I/O operations between *SCALE* and *LETKF* processes compared to the file I/O-based data transfer. For example, when the size of ensemble instances is 100 using the case of *Test Case 3*, the mechanism of direct data transfer can yield over 30× speedup on I/O operations, which in turn implies

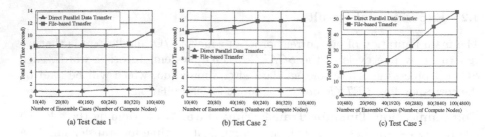

Fig. 5. I/O time contrasting file I/O and direct data transfer.

that more time can be devoted to perform simulation and data assimilation, and that the total execution time can be consequently decreased. Furthermore, the file-based data transfer may require significantly increased I/O time due to contention on the parallel file system. The case of *Test Case 2* required 34.1 % more time for conducting file I/O operations, compared with the case of *Test Case 1*, because the size of transmission data needed by the former case is two times of the size of transmission data of the latter one. In contrast, direct data transfer does not increase the transfer time significantly even for double size data.

Data Throughput. After verifying the proposed mechanism can indeed reduce the time needed for exchanging the data between *SCALE* and *LETKF* in our test cases, this section aims to measure the I/O data throughput while executing various test cases. Figures 6(a), (b) and (c) show the results about I/O data throughput reported by performing the tests with varying ensemble sizes, respectively. As seen, the proposed scheme of direct data transfer outperforms the scheme of file I/O-based data transfer, and it achieves from 758.3 % to 2933.3 % data rate improvements for the selected test cases. Particularly, improvements are getting remarkable while the ensemble size is getting larger that indicates more data are required to be processed.

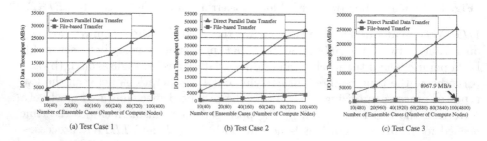

Fig. 6. I/O data throughput utilizing file I/O and direct data transfer.

Another noticeable issue, implied by the figures, is the fact that the larger the amount of data is to be exchanged, the higher the benefits become by utilizing the direct data transfer method.

4.3 Summary

With respect to comparing direct data transfer and file I/O based data transfer, we emphasize the following two key observations. First, with increasing number of processes, direct data transfer yields better relative performance. Second, the more time reduction and higher data throughput can be achieved with the growing size of the involved data. In brief, we conclude that the proposed file I/O middleware is able to significantly reduce the time required by exchanging data between the component models in the *SCALE-LETKF* workflow system.

Furthermore, the implemented I/O middleware offers a general framework for inter-component data exchange in *netCDF*-based workflow systems, where individually developed applications are coupled together. By accelerating the execution of such systems, we believe our middleware, facilitated with the direct data transfer functionality, is particularly important for systems with rigorous time constraints.

5 Concluding Remarks

This paper has proposed a general I/O middleware for Big Data processing, coupled workflows that are comprised of multiple *netCDF*-based and individually developed components. Our framework enables direct parallel data transfer among component models in order to reduce data exchange time, which we applied to the SCALE-LETKF data assimilation based weather forecasting system.

Experimental results on the K computer using up to 4800 nodes have shown that the proposed mechanism can significantly reduce the time spent on I/O operations among SCALE and LETKF. This achievement is useful for real-time weather forecasting in SCALE-LETKF or similar applications, because the I/O time does not noticeably increase while the problem scale is getting larger. Furthermore, we have demonstrated that the benefit of larger data throughput increases with the growing amount of data that is required to be processed.

Enabling asynchronous data transfer so that communication and computation can be efficiently overlapped is an important item on the list of our future work. Furthermore, with asynchronous data transmission we also intend to explore the usage of direct RDMA operations between individual components so that any unnecessary buffering can be eliminated during the data exchange.

Acknowledgment. This work has been partially supported by CREST, JST and the MEXTs program for the Development and Improvement of Next Generation Ultra High-Speed Computer Systems. This research used computational resources of the K computer provided by the RIKEN Advanced Institute for Computational Science through the HPCI System Research project (Project ID: hp150019).

References

1. Network Common Data Form (netCDF) (2013). www.unidata.ucar.edu/netcdf/
2. RIKEN AICS: K computer (2011). http://www.aics.riken.jp/en/k-computer/
3. TOP500 Supercomputer Sites (2015). http://www.top500.org/
4. Reed, D., Dongarra, J.: Exascale computing and big data. Commun. ACM **58**, 56–68 (2015)
5. Podhorszki, N., Klasky, S., et al.: Plasma fusion code coupling using scalable I/O services and scientific workflows. In: Proceedings of the 4th Workshop on Workflows in Support of Large-Scale Science (WORKS 2009), November 2009
6. Hunt, B., Kostelich, E., Szunyogh, I.: Efficient data assimilation for spatiotemporal chaos: a local ensemble transform Kalman filter. Physica D: Nonlinear Phenomena **230**(1), 112–126 (2007)
7. Nishizawa, S., Yashiro, H., Tomita, H., et al.: Influence of grid aspect ratio on planetary boundary layer turbulence in large-eddy simulations. Geosci. Model Dev. **8**(10), 3393–3419 (2015)
8. Valcke, S., Balaji, V., Riley, G., et al.: Coupling technologies for earth system modelling. Geosci. Model Dev. **5**, 1589–1596 (2012)
9. Janjic, Z., Black, T.: An ESMF unified model for a broad range of spatial and temporal scales. Geophys. Res. Abstr. **9**, 05025 (2007)
10. Liu, L., Yang, G., Wang, B., et al.: C-Coupler1: A Chinese community coupler for Earth system modelling. Geosci. Model Dev. Discuss. **7**(3), 3889–3936 (2014)
11. Craig, A., Jacob, R., et al.: A new flexible coupler for Earth system modeling developed for CCSM4 and CESM1. Int. J. High Perform. C **26**(1), 31–42 (2012)
12. Zhang, C., Liu, L., Yang, G., et al.: Improving data transfer for model coupling. Geosci. Model Dev. Discuss. **8**, 8981–9020 (2015)
13. Larson, J., Jacob, R., Ong, E.: The model coupling toolkit: a new Fortran90 toolkit for building multiphysics parallel coupled models. Int. J. High Perform. C **19**(3), 277–292 (2005)
14. Craig, A., Jacob, R., Kauffman, B., He, Y., et al.: CPL6: the new extensible, high performance parallel coupler for the community climate system model. Int. J. High Perform. C **19**(3), 309–327 (2005)
15. Armstrong, C., Ford, R., Riley, G.: Coupling integrated Earth System Model components with BFG2. Concurr. Comp-Pract. E. **21**(6), 767–791 (2009)
16. Valcke, S., Budich, R., Carter, M., et al.: The PRISM software framework and the OASIS coupler. In: Proceedings of the 18 Annual BMRC Modelling Workshop, Melbourne, 28 November - 1 December 2006
17. Browne, P.A., Wilson, S.: A simple method for integrating a complex model into an ensemble data assimilation system using MPI. Environ. Modell. Softw. **68**, 122–128 (2015)
18. MPI: A Message-Passing Interface Standard, Version 2.2, September 2009
19. Watson, G., Frings, W., Knobloch, C., et al.: Scalable control and monitoring of supercomputer applications using an integrated tool framework. In: Proceedings ICPPW 2011, pp. 457–466 (2011)
20. Zhang, F., Docan, C., Parashar, M., et al.: Enabling in-situ execution of coupled scientific workflow on multi-core platform. In: Proceedings of IEEE 26th International Parallel & Distributed Processing Symposium (IPDPS 2012), pp. 1352–1363 (2012)
21. Valcke, S., Craig, A., Dunlap, R., Riley, G.: Sharing experiences and outlook on Coupling Technologies for Earth System Models. Bull. Amer. Meteor. Soc. (2015). doi:10.1175/BAMS-D-15-00239.1

22. Dorier, M., Dreher, M., Peterka, T., et al.: Lessons learned from building in situ coupling frameworks. In: Proceedings of the First Workshop on In Situ Infrastructures for Enabling Extreme-Scale Analysis and Visualization, pp. 19–24. ACM (2015)
23. Miyoshi, T., Kondo, K., Terasaki, K.: Big ensemble data assimilation in numerical weather prediction. IEEE Comput. **48**(11), 15–21 (2015)

High Performance Parallel Summed-Area Table Kernels for Multi-core and Many-core Systems

Angelos Papatriantafyllou[1]([⊠]) and Dimitris Sacharidis[2]

[1] Research Group Parallel Computing, Faculty of Informatics,
Institute of Information Systems, TU Wien, Vienna, Austria
`papatriantafyllou@par.tuwien.ac.at`
[2] E-Commerce Group, Faculty of Informatics,
Institute of Software Technology and Interactive Systems,
TU Wien, Vienna, Austria
`dimitris@ec.tuwien.ac.at`

Abstract. The summed-area table (SAT), also known as integral image, is a data structure extensively used in computer graphics and vision for fast image filtering. The parallelization of its construction has been thoroughly investigated and many algorithms have been proposed for GPUs. Generally speaking, state-of-the-art methods cannot efficiently solve this problem in multi-core and many-core (Xeon Phi) systems due to cache misses, strided and/or remote memory accesses. This work proposes three novel cache-aware parallel SAT algorithms, which generalize parallel block-based prefix-sums algorithms. In addition, we discuss 2D matrix partitioning policies which play an important role in the efficient operation of the cache subsystem. The combination of a SAT algorithm and a partition is manually tuned according to the matrix layout and the number of threads. Experimental evaluation of our algorithms on two NUMA systems and Intel's Xeon Phi, and for three datatypes (int, float, double) by utilizing all system cores, shows, in all experimental settings, better performance compared to the best known CPU and GPU approaches (up to 4.55× on NUMA and 2.8× on Xeon Phi).

1 Introduction

The construction of a summed-area table (SAT) is a well-studied problem in computer graphics and vision [1,13,14], with applications in texture filtering. Since first introduced by Crow [4], several parallel implementations for GPUs [6,7,9] have been proposed. Given a matrix x of size $n \times m$, where n, m are the number of rows and columns, respectively, the problem is to compute the sum of all elements $x(i,j)$, $0 \le i < n$ and $0 \le j < m$, according to the formula:

$$y(i,j) = \sum_{0 \le r < i} \sum_{0 \le c < j} x(r,c) \tag{1}$$

SAT is the 2D generalization of the prefix-sums, or scan, problem, whose parallelization has also been extensively studied; the prefix-sums problem

© Springer International Publishing Switzerland 2016
P.-F. Dutot and D. Trystram (Eds.): Euro-Par 2016, LNCS 9833, pp. 306–318, 2016.
DOI: 10.1007/978-3-319-43659-3_23

(assuming $+$ as the associative operation) for an array x of length n is to compute the sums $y(i) = \sum_{0 \leq r < i} x(r)$. Hence, a straightforward method to parallelize SAT is to scan the matrix row by row, and for each row apply a parallel scan kernel from the literature [5,10–12] to compute its prefix-sums in-place, and then perform in parallel vectorized additions with the prefix-sums of the previous row. Such an approach has throughput bounded by the column size; each process unit (core, warp), henceforth termed thread, operates on a block of size $\frac{m}{p}$, where p is the number of threads. Hence, better performance is achieved with 2D blocks.

In literature, there have been several 2D block-based parallel SAT algorithms for GPUs (Hensley et al. [6], Kasagi et al. [7], Nehab et al. [9] and Yan et al. [15]) and CPUs (Zhang [16]). These can be classified to those that perform within a block prefix-sums along a single dimension [6,15,16], and those along both dimensions, [7,9]. Methods of the former use blocks of size $\frac{n}{p} \times m$ (or $\frac{m}{p} \times n$), resulting in performance degradation when run on NUMA multi-core systems, due to cache misses since blocks can be bigger than the LLC (Last Level Cache), and remote memory accesses imposed by the block assignment, explained later in the related work. On the other hand, algorithms of the latter use square $b \times b$ blocks, and could thus suffer from strided memory accesses when row-major matrix allocation is used on multi-core systems. In addition, the degree of parallelism of [7] is bounded by the number of blocks of each anti-diagonal, which ranges from 1 to $\frac{min(n,m)}{b} - 1$.

Contributions. Our research is motivated by two important facts: (1) the performance limitations of square blocks used in [7,9], and (2) the lack of performance and scalability in current parallel SAT implementations for multi-core systems. Both [7,9] execute row- and column-wise prefix-sums, where the latter are expensive, w.r.t. performance, due to strided memory accesses. To alleviate this overhead, we propose to use non-square blocks of size $br \times bc$, where br, bc are the number of rows and columns, respectively, that are horizontally "stretched" (while being vertically "squeezed"), i.e., $bc > br$. The in-parallel processed blocks must fit into the LLC; thus, $p \times br \times bc$ should not exceed the LLC size.

For the following discussion it helps to conceptually represent the matrix as a sequence of stripes each having size $br \times m$. Depending on the value of bc relative to m, we distinguish two partitionings. In the first called stripe, bc is equal to $\frac{m}{p}$, which means that all blocks processed in parallel lie within a stripe. In the second and more general called tile, $bc \neq \frac{m}{p}$, meaning that a block can span over two consecutive stripes, and thus correspond to a non-rectangular area of the matrix. Figure 1 depicts the two partitionings on the actual matrix, and illustrates that a block in tile can be non-rectangular (shaded).

In addition, we propose three cache-aware parallel SAT algorithms, called psat_cpps, psat_mcstl and psat_sarpps, which process square and non-square blocks, and are generalizations of three parallel block-based prefix-sums algorithms: CPPS [10], MCSTL [12], and the method of Chatterjee et al. [3], which we henceforth call SARPPS (Scan after Reduction Parallel Prefix-Sums). In particular, our algorithms operate in three phases: Phases 1 and 3 are devoted

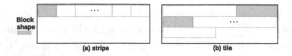

Fig. 1. The 2D matrix partitioning policies: `stripe` (left), and `tile` (right). The grey box indicates the shape of a block according to each partitioning

to in-block computations, and in Phase 2 shared data are propagated across the threads. They differ in the tasks performed in Phases 1 and 3, which are optimized (e.g., using vectorization, loop-unrolling) based on the target system and datatype. Depending on the matrix partitioning employed, certain tasks of Phase 2 are omitted resulting in distinct performance behavior for the same algorithm.

We carefully evaluate the performance and scalability of our kernels compared to the state-of-the-art method for GPU by Kasagi *et al.* [7], and CPU by Zhang [16]. We experiment on two NUMA systems (Westmere- and Opteron-based) and Intel's Xeon Phi, considering different datatypes (int, float, double), while varying number of threads. Our results verify the limitations of existing work when deployed in multi-core and many-core systems. In particular, when utilizing almost the maximum available physical cores, the speedup of [7] drops, and in certain settings down-scale behavior is observed. On the other hand, [16] has the worst performance by a large margin in NUMA systems. In contrast, our proposed kernels have consistent performance behavior across all systems and datatypes, without requiring any modifications to the parallelization code, and in almost all cases outperform the competitors. Specifically, our kernels have up to 4.55x and 3.25x more speedup in NUMA systems, and up to 1.5x and 2.8x in the Xeon Phi system, compared to [16] and [7], respectively. Furthermore, we study the main parameters (e.g., block row and column size, row- and column-wise optimizations, matrix partitioning) and draw conclusions on how they influence the performance of our kernels. In summary, the performance gains we measure are due to the better utilization of the LLC, which is the direct result of larger (non-rectangular) blocks that could not be exploited by previous works.

The rest of this paper is organized as follows. Section 2 describes and reviews related works. Section 3 presents our algorithms in detail. Section 4 shows performance and scalability results. Section 5 concludes the paper.

2   Related Work

Parallel Prefix-Sums Kernels. CPPS [10], MCSTL [12], and SARPPS [3] are block-based algorithms that solve the prefix-sums problem in parallel by splitting the input array in p or $p+1$ blocks. Each block is processed in three phases, where the first and third perform in-block computations, while the second propagates shared data across threads. In Sect. 3, we discuss their generalization to SAT.

Parallel SAT Kernels. There are two algorithmic classes for parallelizing SAT. While, they both split the matrix in blocks assigned to individual threads, they differ in their in-block computations. The first class, termed 1D, is to perform either row- or column-wise prefix-sums, while the second, termed 2D, is to perform prefix-sums in both dimensions, i.e., compute the block's SAT.

Previous 1D approaches on GPUs (Hensley *et al.* [6] and Yan *et al.* [15]) use 2-phase algorithms to compute in parallel first all horizontal and later all vertical prefix-sums. In [6], they use recursive doubling for computing the prefix-sums of both phases, whereas, in [15] they exploit vectorization over a column-major order input matrix by using an auxiliary $n \times m$ matrix. Similarly, Zhang [16] present a 1D method for multi-core systems. Noticeably, the performance of all 1D approaches is expected to be penalized on multi-core systems due to cache-unawareness, because blocks may exceed the LLC. Particularly on NUMA systems, it is expected more performance degradation due to NUMA-unawareness, since the data needed in Phase 2 may reside in remote NUMA nodes.

Nehab *et al.* [9] present a generalization of the SARPPS algorithm for GPUs, which works quite similar to our `psat_sarpps` algorithm. The input matrix is partitioned in blocks of size $b \times b$, where b matches the *warp* size of an SM (i.e. $b = 32$). Each *warp* computes the SAT of a block with the help of the last block row and block column of the other *warp*'s. Their approach could compose and process also rectangular blocks but it is unable to exploit non-rectangular blocks.

Kasagi *et al.* [7] present a different 2D approach, called 1R1W, where each matrix element is read from and written to DRAM only once. To accomplish that, an $n \times n$ matrix is partitioned into $\frac{n}{b} \times \frac{n}{b}$ blocks of equal $b \times b$ size, where $b \leq \frac{n}{p}$, and processed anti-diagonally in $\frac{2n}{b} - 1$ steps. In the kth step, where $0 \leq k < \frac{2n}{b} - 1$, there are $\min(p, k + 1)$ active threads, where each computes its own block by using elements of the adjacent left, top and diagonal blocks. Before entering each step, the threads must synchronize at a global barrier. This approach is expected to exhibit poor performance and scalability in NUMA systems for very large p due to two reasons. First, while increasing p more NUMA nodes are activated and subsequently, more LLCs can be utilized. However, in this approach the block size is bounded by n, implying that the increase of p leads to smaller block sizes and eventually, the LLC footprint is reduced. Exploiting less LLC footprint means that the CPUs process faster the cached blocks and the kernel's performance is exposed to the main memory latency overhead. Second, while increasing p it would need more synchronization steps to reach the maximum degree of parallelism.

In-block Optimizations. Both 1D and 2D approaches benefit from prefix-sums optimizations. Zhang [16] optimizes the row-wise prefix-sums with an algorithm called *enhanced parallelism*, which breaks each row in groups of 6 elements; each group is computed by independent prefix-sums pairs. However, such a technique is not efficient for large vector widths machines like Xeon Phi due to its lack of leveraging vector instructions. Previous works on GPUs [6,9,15] exploit the SIMT (Single Instruction Multiple Thread) model in order to vectorize the row-wise prefix-sums. Unfortunately, SIMT is not compatible with the SIMD (Single Instruction Multiple Data) programming model used in CPUs.

A SIMD-based approach is provided by OpenCV (Open Source Computer Vision) [2], which is a computer vision and machine learning library. This approach uses the SSE instruction set, which has a limited target group and works under specific input constraints (short numbers). Thus, it is not suitable for Xeon Phi systems and it is not applicable to arbitrary datatypes.

From the above discussion, it becomes clear that new optimizations techniques are necessary for multi-core systems. We remark that the sequential (in-block) SAT computation in all 2D methods is based on only one of the three known basic approaches for computing SAT [8]. In Sect. 3, we optimize this approach by further segmenting the assigned block to leverage deep cache hierarchies, investigate prefix-sums optimizations presented in [10], and also explore the optimization space of another basic SAT approach.

3 Parallel Summed-Area Table Kernels (PSAT)

Our algorithms construct the SAT of a matrix x in-place, with an associative operator $+$ over the basic type, by splitting x in 2D blocks. The blocks are formed according to `stripe` and `tile` partitionings, and are further tuned to meet load-balancing requirements according to each PSAT kernel, later explained. Each block is assigned only once to a unique thread and processed in-cache across some or all the algorithmic phases, as illustrated in Figs. 2, 3 and 4. In addition, the threads use two shared buffers, called `blr` and `blc`, to propagate their last row and column, respectively, across all threads. For simplicity and readability, we use square blocks to describe our implementations.

3.1 Implementations

Psat_cpps. This algorithm is the generalization of CPPS [10], which works as follows: (1) Each thread computes the prefix-sums for its assigned block, (2) an inclusive scan is computed over the last element of all the assigned chunks, and (3) each thread, except for thread 0, propagates a previous corresponding cumulative sum to its own elements; thread 0 is assigned and process a new block.

In Phase 1 of `psat_cpps` (Fig. 2), each thread is assigned a $b \times b$ block, on which a sequential SAT is invoked. At the completion of Phase 1, thread 0 is the only thread that has successfully computed its own block. Subsequently, each thread copies the last computed block row and block column to the shared buffers `blr` and `blc`, respectively, and synchronizes with the others at a global barrier. The role of these buffers is crucial. They hold the fragments of the final product computed later at Phase 3. For instance, thread $p - 1$ needs to sum up all the information from the `blc`'s and `blr`'s belonged to the threads accessing the matrix rows and matrix columns which are equal and less than $p - 1$'s.

Phase 2 (Fig. 3) is devoted to process and propagate the shared data from the buffers `blr` and `blc` across threads. The phase is split in three parts, where the first two are responsible to process the `blr`'s, and the third part handles to

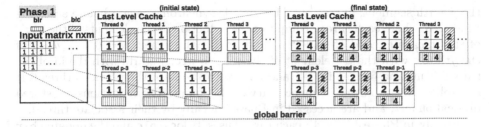

Fig. 2. Phase 1: each of the p threads is assigned a distinct $b \times b$ block of the $n \times m$ matrix and runs a sequential SAT kernel. The computed last block row and block column are copied into the blr and blc shared buffers, respectively. All the blocks fit into the system's LLC. At the end of this phase, thread 0 has completed its block

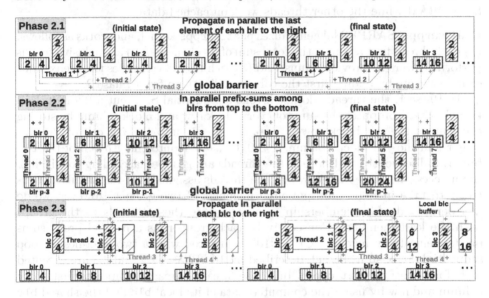

Fig. 3. Phase 2: shared data propagation. Each thread collects the cumulative sum of the last element of each blr at its left and propagates it to its own blr (Phase 2.1). Subsequently, in parallel column-wise prefix-sums are computed by adding the elements of all the blr's, from top to bottom, including the last blr's from the previous group of cached blocks (Phase 2.2). In Phase 2.3, each thread computes and stores locally the cumulative sums of the blc's at its left

process the blc's. Therefore, each thread computes the cumulative sum of the last element of the blr's related to the threads at its left. Subsequently, the sum is used to propagate the elements of its own blr (Phase 2.1). Then, the threads synchronize once again before passing to Phase 2.2, where the aggregation of each shared-row blr's is split in p segments of size $\frac{m}{p}$ and each segment is assigned to each thread. In total, each thread has to process $\frac{pb}{m}$ segments, on which is computed the column-wise inclusive prefix-sums by adding the ith element, $i = 0, 1, ..., \frac{m}{p} - 1$, of the jth segment, $j = 0, 1, ..., \frac{pb}{m} - 1$, with the ith element of the $j + 1$th segment, and so on. The threads synchronize again before accessing Phase 2.3, where each thread computes the reduction of the elements of the blc's of the threads located at its left by adding the ith element, $i = 0, 1, ..., b - 1$ of the jth blc with the ith element of the $j + 1$th blc, j starts with the id of the thread being the first at its left and ends with the id of the previous thread. It is implied that a barrier is unnecessary at this point. The reductions are stored in local buffers, and are accessed in Phase 3 locally by their associate thread.

Finally, in Phase 3 (Fig. 4), each thread uses its local blc and the updated blr (top) to construct the table by propagating their values to its block elements. In addition, thread 0 is assigned a new block, and invokes a sequential SAT kernel by adding at the same time the elements of the left and top buffers. After careful benchmarking, this new block must be a factor of 4 smaller; the block is fetched from DRAM while the other threads work on cached data.

Psat_sarpps. SARPPS [3] generalizes psat_sarpps kernel, and works as follows: (1) each thread first acquires the total sum of its elements, (2) an inclusive scan is performed over the previous sums, and (3) each thread computes the prefix-sums of its elements by adding first the corresponding sum from Phase 2.

In Phase 1, the threads do not invoke a sequential SAT but a 2D reduction kernel, which computes the sums of every block row and block column by storing the sum of the ith block row, $i = 0, 1, ..., b - 1$, into the ith cell of the blc buffer and the jth block column, $j = 0, 1, ..., b - 1$, into the jth cell of the blr buffer. Subsequently, in Phase 2, the threads are grouped according to which of them access the same matrix rows. The threads assigned non-rectangular blocks (refer to tile in Fig. 1) are potential members of two groups. However, we do not permit "double membership", and we place those threads to the groups with the least members. Consecutively, each group runs a parallel prefix-sums on its elements by invoking CPPS [10], configured with a two-level-nested-loop sequential prefix-sums kernel, described in [10]. Phases 2.2 and 2.3 are executed as before (refer to psat_cpps). In Phase 3, each thread updates its first block column and row by using the computed data of its local blc and the shared blr (left), respectively, and subsequently it executes a sequential SAT kernel.

Psat_mcstl. This algorithm is the generalization of the MCSTL [12] algorithm, which resembles SARPPS. However, they are distinguishable from each other.

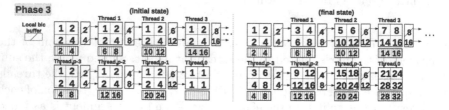

Fig. 4. Phase 3: The adjacent shared top (blr) and local buffers are used by each thread to update its block. Thread 0 is assigned a new block and computes the SAT

In MCSTL's Phase 1, thread 0 computes the prefix-sums of a block, which differs from the block that it computes its prefix-sums in Phase 3.

In Phase 1 of **psat_mcstl**, a block of size $\frac{b}{2} \times \frac{b}{2}$ is assigned to thread 0 and $p-1$ blocks of size $b \times b$ to the other threads. Thread 0 invokes a sequential SAT kernel and the others a 2D reduction kernel for their blocks. The block of thread 0 is a factor of 4 smaller than the other blocks because of the heavyweight computation of the sequential SAT compared to the 2D reduction task. The results of the last computed block row and column are stored again in the `blr` and `blc` buffers, respectively. At that time, thread 0 has produced the final result for its block. Phase 2 can be executed as described in **psat_sarpps**. However, since thread 0 `blc` holds already the final values it should not take part in Phase 2. Instead, we re-segment the input of its group and assign a task to thread 0. Phases 2.2 and 2.3 are executed as before (refer to **psat_cpps**). Consequently, Phase 3 operates as described in **psat_sarrps**, however, thread 0 employs a different block; for better load balancing this block size must be $b \times \frac{b}{2}$.

Optimizations. Due to space constraints, we briefly describe the optimizations that took place for improving the performance of the sequential SAT kernel used by our PSAT kernels. We have optimized two basic SAT approaches, described in [8]; one is used by [7,9]. Since both approaches compute prefix-sums row-wisely, we first tested several optimized sequential prefix-sums kernels, presented in [10]. In addition, both approaches also compute the prefix-sums column-wisely, which can be easily vectorized. However, the size of the vectorized rows can be bigger than the underlying L2/L1 cache size. Thus, we considered performance improvements through further input segmentation. The new formed sub-blocks have 2–4 rows each, and the column size varies according to the architecture. We investigated improving the ILP and the spatial locality of the column-wise prefix-sums. The former through hard-coding the execution of 2–4 independent row-wise prefix-sums of each sub-block, and the latter by storing consecutively in memory the row-wise results (in vector-width chunks).

3.2   Matrix Partitionings

Conceptually, the matrix is composed by a sequence of stripes of size $br \times m$. Accordingly, our PSAT kernels split the stripes in groups of blocks whose cumulative size does not exceed the aggregated system's LLC. Subsequently, each group is processed in parallel by p threads. In `stripe` partitioning (Fig. 1 (left)), the kernels process one stripe at a time by computing groups of blocks located in one stripe, so that, $bc \times p \leq m$. On the contrary, in `tile` (Fig. 1 (right)), the formed groups of blocks may span across several stripes, as long as $bc \leq m$.

`Stripe` affects the functionality of our algorithms. In particular, Phases 2.1 and 2.2 are omitted since all threads operate within the same stripe, where the column-wise propagation is omitted. The reason is that each thread is assigned the same matrix columns across different stripes, and thus can directly access the last block row of the previous stripe without needing the `blr` buffer.

Regarding the scalability of our kernels, we make the following observations. Increasing the number of threads (p) in `stripe` means that the block column size (bc) decreases resulting in the following trade-off: we can increase the br (wider stripe) to improve the LLC utilization at the expense of multiple strided memory accesses, or keep the LLC utilization low, thus avoiding this overhead (strided accesses). In NUMA systems, this effect is amplified since the LLC size increases by activating more NUMA nodes. The benefit of stripe over tile is the decrease of the amount of propagated shared data. Thus, we expect stripe to perform better for small p. On the contrary, tile is not affected by this trade-off and promises better scalability in both multi-core and many-core systems.

4   Evaluation

We implemented our PSAT kernels in C with Pthreads and tested on three systems: two NUMA systems, called Mars and Saturn, and one Intel's Xeon Phi, called MIC. Detailed system information are listed in Table 1. All benchmarks are compiled with Intel's ICC 14.0.1 with -O3 optimization level and executed 30 times. The results show median values. Due to space limitations, we report only results for a specific problem size; similar behavior is observed in other problem sizes. Each kernel is configured and tested with different sequential SAT kernels and block sizes. We only report results for the best configurations. The experiments show the performance and scalability behavior of our PSAT kernels, comparing against 1R1W [7] and Zhang [16]. For a fair comparison, we re-implemented 1R1W for x86 systems according to [7], tested with rectangular blocks, and applied x86-based sequential optimizations. For the testbeds, in-parallel first-touch page placement was applied and consecutive thread pinning.

Table 1. Specifications of the three systems (Mars, Saturn and MIC)

System	CPU (model & freq.)	# cores	# cores/ NUMA	# NUMA nodes	LLC/ NUMA	L2	L1
Mars	Intel Xeon E7-8850 2.0 GHz	80-hyperthreaded	10	8	24576K	256K	32K
Saturn	AMD Opteron 6168 1.9 GHz	48	6	8	5118K	512K	32K
MIC	Intel Xeon Phi 5110P 1.059 GHz	60-hyperthreaded	-	-	only L2 cache	512K	64K

PSATs Performance. Figure 5 (left) depicts the performance (execution time in seconds) of our PSAT kernels (psat_cpps, psat_sarpps, psat_mcstl). Each kernel has been separately tested with tile and stripe partitionings. In addition, Fig. 5 (right) depicts a breakdown phase analysis of our PSATs in stripe mode by reporting the sum of the execution time (in milliseconds) of Phases 1 and 3 when computing p (= #threads) blocks. All the testbeds construct the SAT of a 12K×12K integer matrix on Mars with different number of threads. Due to quantitatively similar results, other systems and datatypes are omitted.

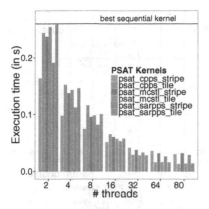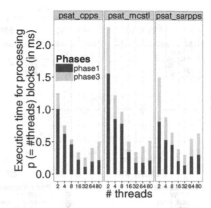

Fig. 5. Performance comparison (left) between our three PSAT kernels (psat_cpps, psat_sarpps, psat_mcstl) configured with the two partitionings (tile, stripe), and breakdown phase analysis (right) of our PSATs in stripe mode, which reports the sum of the execution time of Phases 1 and 3 for computing p (= #threads) blocks. The testbeds run on Mars for different number of threads and a 12K×12K integer matrix

Figure 5 (left) justifies our assumptions about the behavior of both partitionings independently of PSAT kernel. The stripe behaves better for small number of threads due to less computations, and the tile is better for large number of threads due to better cache utilization and less synchronization steps caused by handling bigger block sizes. In most cases, psat_cpps outperforms the other kernels, even though Phase 1 of psat_sarpps is faster, due to the reduction computations being auto-vectorized by gcc and icc compilers, as depicted in the breakdown analysis of Fig. 5 (right). Therefore, our analysis suggests that running a sequential SAT kernel first leads to better performance.

PSATs Speedup. Figure 6 depicts the absolute speedup comparison between 1R1W, Zhang and our best performing PSAT for each of the tile and stripe partitionings. The results are collected by all our systems after composing the SAT of a 12K×12K matrix separately for integers, floats and doubles. In addition, the best performing sequential kernel that we have is selected as a baseline.

Figure 6 shows that in all cases our kernels outperform 1R1W and Zhang for large number of threads (p) in tile. In particular, our kernels run 1.2×–3.25× faster than 1R1W (all datatypes) with all system cores: (1) 3.25× in Mars, (2) 1.8× in Saturn and (3) 2.8× in MIC. Nonetheless, 1R1W behaves slightly better for small p since it spends zero time on processing shared data, and needs fewer synchronization steps until all threads are utilized in parallel. Regarding the performance of Zhang in MIC, it is almost equal to that of our best kernel for integers, but it is 1.5× and 1.2× slower for floats and doubles, respectively. In addition, Zhang proves to be inefficient in NUMA systems, for reasons discussed in Sect. 2. For instance, with all system cores, Zhang is slower by 3.25× (integers and floats) and 3.5× (doubles) on Mars, and 4.55× (integers and floats), and 4.15× (doubles) on Saturn. In conclusion, we observe that the best perfor-

Problem size: 12K×12K elements

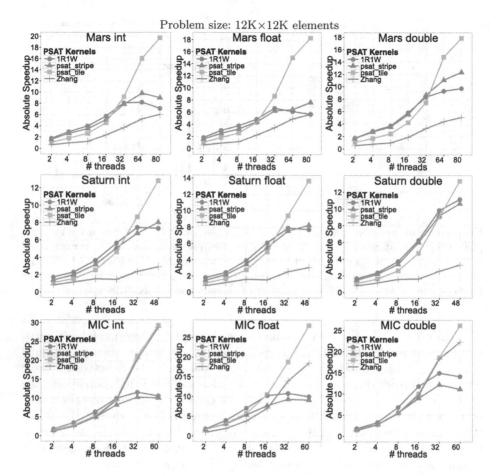

Fig. 6. Absolute speedup comparison between 1R1W kernel [7], Zhang's [16], and our best performing PSAT for each of the two partitionings: `tile` and `stripe` running on all systems. The kernels construct a 12K×12K matrix for three datatypes (int, float, double). The best performing sequential kernel that we have is selected as a baseline

mance of our kernels, in `tile` with all cores, is achieved when the block column size (bc) is smaller than the matrix column size by 1.67× (integers and floats) and 3.3× (doubles) in Mars, 4× (all datatypes) in Saturn and 5× (doubles and floats) and 10× (integers) in MIC. In addition, the block row size is by far smaller than the bc in NUMA (57×–114×), and still quite smaller in MIC (1.2×–4.8×).

5   Conclusions and Future Work

In this paper, we present three new cache-aware parallel SAT (Summed-Area Table) algorithms, called `psat_cpps`, `psat_sarpps` and `psat_mcstl`, for many-core and multi-core systems. Our algorithms are generalizations of three block-based parallel prefix-sums algorithms, and can process rectangular and non-rectangular blocks in order to utilize better the cache subsystem. We provide performance and speedup results after comparing against Kasagi *et al.* [7] and Zhang [16]. Our next step will be designing an auto-tuning mechanism capable of finding the best configurations (parallel and sequential SAT, matrix partitioning, etc.) for different problem sizes in order to increase system efficiency.

Acknowledgements. We would like to thank the members of the TU Wien Research Group Parallel Computing and the anonymous reviewers for their valuable comments.

References

1. Bay, H., Ess, A., Tuytelaars, T., Gool, L.J.V.: Speeded-up robust features (SURF). Comput. Vis. Image Underst. **110**(3), 346–359 (2008)
2. Bradski, G.R., Kaehler, A.: Learning OpenCV - Computer Vision with the OpenCV Library: Software That Sees. O'Reilly, Beijing (2008)
3. Chatterjee, S., Blelloch, G.E., Zagha, M.: Scan primitives for vector computers. In: Proceedings Supercomputing 1990, pp. 666–675 (1990)
4. Crow, F.C.: Summed-area tables for texture mapping. In: Proceedings of the 11th Annual conference on Computer Graphics and Interactive Techniques (SIGGRAPH), pp. 207–212 (1984)
5. Dotsenko, Y., Govindaraju, N.K., Sloan, P., Boyd, C., Manferdelli, J.: Fast scan algorithms on graphics processors. In: Proceedings of the 22nd Annual International Conference on Supercomputing (ICS), pp. 205–213 (2008)
6. Hensley, J., Scheuermann, T., Coombe, G., Singh, M., Lastra, A.: Fast summed-area table generation and its applications. Comput. Graph. Forum **24**(3), 547–555 (2005)
7. Kasagi, A., Nakano, K., Ito, Y.: Parallel algorithms for the summed area table on the asynchronous hierarchical memory machine, with GPU implementations. In: Proceedings of the 43rd International Conference on Parallel Processing (ICPP), pp. 251–260 (2014)
8. Kisacanin, B.: Integral image optimizations for embedded vision applications. In: Proceedings of the 2008 IEEE Southwest Symposium on Image Analysis and Interpretation (SSIAI), pp. 181–184 (2008)
9. Nehab, D., Maximo, A., Lima, R.S., Hoppe, H.: GPU-efficient recursive filtering and summed-area tables. ACM Trans. Graph. **30**(6), 176 (2011)
10. Papatriantafyllou, A.: Energy characterization and optimization of parallel prefix-sums kernels. In: Hunold, S., et al. (eds.) Euro-Par 2015 Workshops. LNCS, vol. 9523, pp. 685–696. Springer, Heidelberg (2015). doi:10.1007/978-3-319-27308-2_55
11. Sengupta, S., Harris, M., Garland, M.: Efficient Parallel Scan Algorithms for GPUs. Technical report, NVIDIA Corporation (2008)
12. Singler, J., Sanders, P., Putze, F.: MCSTL: the multi-core standard template library. In: Kermarrec, A.-M., Bougé, L., Priol, T. (eds.) Euro-Par 2007. LNCS, vol. 4641, pp. 682–694. Springer, Heidelberg (2007)
13. Viola, P.A., Jones, M.J.: Rapid object detection using a boosted cascade of simple features. In: Proceedings of the 2001 IEEE Computer Society Conference on Computer Vision and Pattern Recognition (CVPR), pp. 511–518 (2001)

14. Viola, P.A., Jones, M.J., Snow, D.: Detecting pedestrians using patterns of motion and appearance. Int. J. Comput. Vis. **63**(2), 153–161 (2005)
15. Yan, S., Zhang, Y., Long, G.: Summed-area table algorithm optimization based on the OpenCL. In: Proceedings of the ATIP/A*CRC Workshop on Accelerator Technologies for High-Performance Computing: Does Asia Lead the Way? (2012)
16. Zhang, N.: Working towards efficient parallel computing of integral images on multi-core processors. In: Proceedings of the 2nd International Conference on Computer Engineering and Technology (ICCET), pp. V2-30–V2-34 (2010)

GraphIn: An Online High Performance Incremental Graph Processing Framework

Dipanjan Sengupta[1]([✉]), Narayanan Sundaram[2], Xia Zhu[2],
Theodore L. Willke[2], Jeffrey Young[1], Matthew Wolf[1], and Karsten Schwan[1]

[1] Georgia Institute of Technology, Atlanta, GA, USA
{dsengupta6,jyoung9}@gatech.edu, {mwolf,karsten.schwan}@cc.gatech.edu
[2] Intel Labs, Hillsboro, OR, USA
{narayanan.sundaram,xia.zhu,theodore.l.willke}@intel.com

Abstract. The massive explosion in social networks has led to a significant growth in graph analytics and specifically in dynamic, time-varying graphs. Most prior work processes dynamic graphs by first storing the updates and then repeatedly running static graph analytics on saved snapshots. To handle the extreme scale and fast evolution of real-world graphs, we propose a dynamic graph analytics framework, GraphIn, that incrementally processes graphs on-the-fly using fixed-sized batches of updates. As part of GraphIn, we propose a novel programming model called I-GAS (based on gather-apply-scatter programming paradigm) that allows for implementing a large set of incremental graph processing algorithms seamlessly across multiple CPU cores. We further propose a property-based, dual-path execution model to choose between incremental or static computation. Our experiments show that for a variety of graph inputs and algorithms, GraphIn achieves up to 9.3 million updates/sec and over 400× speedup when compared to static graph recomputation.

Keywords: Graph · Big data · Performance · Incremental processing

1 Introduction

With the increasing interest in many emerging domains such as social networks, the World Wide Web (e-commerce and advertising), and genomics, the importance of dynamic graph processing has grown substantially. This recent trend has given rise to many graph processing frameworks like GraphLab [13], Power-Graph [10], Graphchi [12], and X-Stream [19] that operate on time-varying real-world graphs. However, most current graph analytics on such dynamic graphs follow a store-and-static-compute model that involves storing batches of updates to a graph applied at different points in time and then repeatedly running static graph computations on the "snapshots" of the evolving graph. The key assumption made here is that the dynamic graph changes slower than the static processing rate. Extreme-scale applications like Facebook's representative social graph benchmark [4] reports an update rate of 86,400 objects/second in 2013, while

© Springer International Publishing Switzerland 2016
P.-F. Dutot and D. Trystram (Eds.): Euro-Par 2016, LNCS 9833, pp. 319–333, 2016.
DOI: 10.1007/978-3-319-43659-3_24

Twitter traffic [3] can peak at 143 thousand tweets (and associated updates) per second and emails sent [1] can reach as high as 2.5 millions/sec. This high volume of changes in dynamic graphs is further complicated by the need for soft or hard real-time guarantees for applications like real-time anomaly detection and disease spreading. Both the volume and complexity of queries have outstripped traditional static graph analytics.

To address the challenges of large scale dynamic graph processing, we propose an incremental graph processing framework called GraphIn. The GraphIn framework employs a novel programming model called Incremental-Gather-Apply-Scatter (or I-GAS) to incrementally process a continuous stream of updates (i.e., edge/vertex insertions and/or deletions) as a sequence of batches. The I-GAS programming model is based on the popular gather-apply-scatter (GAS) programming paradigm [10] and it allows an incremental graph problem to be reduced to a sub-problem that operates on a portion, or sub-graph, of the entire evolving graph. This sub-graph abstraction allows I-GAS to substantially outperform traditional static processing techniques.

Furthermore, GraphIn takes into account situations where incremental processing may perform worse than static recomputation, such as with incremental BFS, where updates may affect the entire BFS tree. To handle such scenarios, we introduce the notion of "dual-path execution" or property-based switching between incremental and static graph processing based on built-in and user defined properties (e.g., vertex degree information).

Finally, GraphIn's design allows it to run on top of, and take advantage of performance benefits of any GAS-based static graph analytics framework like X-Stream [19], GraphMat [26] (used in this work) or Graphchi [12].

This paper makes following contributions:

- A high-performance incremental graph processing framework built on top of GraphMat to process time-varying evolving graphs.
- A novel programming model called I-GAS for simplified implementation of incremental versions of many popular graph algorithms using the GraphIn framework that seamlessly generates parallel code for multi-core systems.
- An optimization heuristic to decide between static and dynamic graph execution based on built-in and user defined graph properties. This dual path execution results in speedups of up to 60× over a naïve streaming approach.
- An extensive evaluation of GraphIn for three popular graph algorithms that operate on large scale real-world and synthetic graph datasets. Compared to competitive frameworks such as STINGER [7], GraphIn achieves a speedup of up to 6.6×. Overall, GraphIn achieves up to 9.3 million updates/sec and 400× speedup over the naïve static graph recomputation approach.

The remainder of the paper is organized as follows: Sect. 2 discusses the background on graph analytics. Section 3 introduces our GraphIn framework. Section 4 presents the experimental setup and result analysis. Section 5 discusses the related work and Sect. 6 concludes with future work.

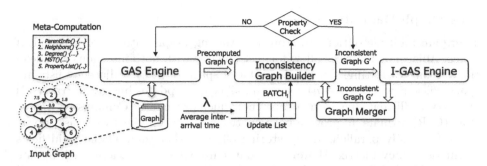

Fig. 1. GraphIn software architecture.

2 Background

Gather-Apply-Scatter (GAS) [10,13,14] is a computational model for graph processing sufficiently general to express a broad set of graph algorithms. With GAS, a problem is described as a directed graph, $G = (V, E)$, where V denotes the vertex set and E denotes the directed edge set and a property/state value is associated with each vertex $v_i \in V$. Graph programs written in GAS typically follow 3-phases: *Gather*, *Apply* and *Scatter*. In the Gather phase, incoming messages are processed and combined (reduced) into one message. In the Apply phase, vertices use the combined message to update their state. Finally, in the Scatter phase, vertices can send a message to their neighbors along their edges.

GraphMat [26] complements the use of the GAS model in GraphIn by taking vertex programs and compiling them into sparse matrix operations (e.g. sparse matrix-vector multiplication). The graph programs are specified by at least 4 user-defined functions (UDFs), SEND_MESSAGE for scatter phase, PROCESS_MESSAGE and REDUCE to specify Gather phase and APPLY for its eponymous phase.

We use 3 algorithms in this paper - Clustering co-efficient (CCof), Connected Components (CC) and Breadth-First Search (BFS). Clustering coefficient C_v for vertex v is defined as $C_v = \frac{T_v}{d_v(d_v-1)}$ where T_v is the total number of triangles in a graph with vertex v as one of the endpoints and d_v is its degree. CC refers to computing the weakly connected components in the graph. BFS assigns a level (equal to the number of edges traversed) to every vertex reachable from a root.

3 GraphIn Framework

The GraphIn framework can efficiently process evolving graphs by dividing the continuous stream of updates (edge or vertex insertions and/or deletions) into fixed size batches processed in the order of their arrival. It simplifies evolving graph analytics programming by supporting a multi-phase, dual path execution model. Figure 1 shows the general software architecture of GraphIn which consists of five major components: *GAS Engine*, *Inconsistency Graph Builder*, *Property Check*, *I-GAS Engine* and *Graph Merger*.

3.1 Graph Data-Structure

There are multiple options to store an evolving graph with n vertices and m edges. Adjacency matrices allow for fast updates with both insertions and deletions taking $O(1)$ time but require a lot of space $(O(n^2))$. Adjacency lists are space efficient with $O(m+n)$ space and allow fast updates but graph traversals are very inefficient due to non-contiguous memory nodes. Compressed Sparse Row (CSR) formats [5] provide both space efficiency combined with fast traversal (often easily parallelized) by storing offsets rather than all the valid fields in the adjacency matrix. However, updates are expensive because each update requires shifting of the graph data throughout the array to match the compressed format.

GraphIn adopts a hybrid data structure involving edge-lists to store incremental updates and compressed matrix format to store a static version of the graph. The edge-list allows for faster updates without adversely affecting the performance of incremental computation. The compressed format allows for faster parallel computation over the entire static version of the graph. The framework merges the update list and the static graph whenever required (see Phase V).

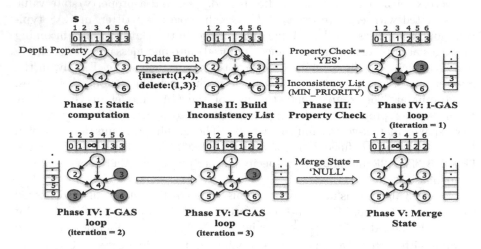

Fig. 2. Incremental BFS phases.

3.2 User Interface

As shown in Table 1, programmers can write a sequential algorithm by simply defining six functions for the different phases in GraphIn. GraphIn will then seamlessly generate parallelized code to incrementally process evolving graph on the target multi-core system. The user-defined functions include *meta_computation()*, *build_inconsistency_list()*, *check_property()*, *activate_frontier()*, *update_- inconsistency_list()* and *merge_state()*, corresponding to the different phases of GraphIn computation described below in detail. Figure 2 shows various incremental BFS phases.

3.3 Phase I: Static Graph Computation (GAS Engine)

This phase is responsible for running (1) Static graph computation and (2) Meta-computation to be used later in the incremental logic. The static graph computation follows the GAS programming model, and therefore, any framework that supports GAS can be used as a Static Engine. We have used GraphMat framework for our static graph computation. Meta computation (*meta_computation()*) involves the computation of graph properties like vertex parents, vertex degree etc., which are needed by incremental algorithms for later phases of GraphIn. As an example, incremental BFS requires parent information during Phase IV.

3.4 Phase II: Inconsistency Graph Builder

Between multiple versions or snapshots of an evolving graph the vertex states for many vertices remain the same over time and therefore their recomputation is essentially redundant. We define an *inconsistent vertex* to be a vertex for which one or more properties are affected when the update batch is applied. For example in BFS, addition of edge (v_i, v_j) can potentially make v_j and all vertices that are downstream from v_j inconsistent. This phase is responsible for marking the portions of the graph that become inconsistent using the user-defined function(UDF) *build_inconsistency_list()*. This UDF takes two parameters: the *update batch* and a *user-defined priority*. The update batch consists of edge or vertex insertions and/or deletions from which a list of inconsistent vertices is built after applying the updates. Vertices in this inconsistency list are assigned the user-defined priorities and by default all the inconsistent vertices have equal priority. This phase also optionally builds a user-defined sub-graph G' to be used in the later phases. By default G' is same as the original graph.

3.5 Phase III: Property Check (Static vs. Dynamic)

Runtime of online graph analytics depends both on the algorithm and the particular choice of updates. There are classes of incremental algorithms that cause large portions of the graph to become inconsistent and hence can result in recomputation over the entire graph. For them, incremental processing will not achieve any performance benefit over static recomputation and may even result in performance degradation due to the overheads associated with incremental execution. To deal with such situations, we allow the user to define a heuristic for determining when one form of computation is used over the other. The user may select from a set of predefined graph properties (e.g. vertex degree) or define their own properties that affect the runtime of the incremental algorithm. The framework will then use the selected properties to decide whether to run incremental or static recomputation by calling *check_property()* for each update batch. This is called "property-based dual path execution". *check_property()* takes four parameters: *inconsistency_list*, *property_list*, *threshold_vector*, and *threshold_fraction*. *Property list* defines the set of graph properties under consideration. *Threshold vector* defines a set of thresholds for properties in the property list, above (or

Table 1. Implementing graph applications in GraphIn

GraphIn APIs and Phases	Graph Algorithms		
	Breadth First Search **(BFS)** [18]	Connected Components **(CC)** [8]	Clustering Coefficients **(CCof)** [6]
Type	All-merge	Delete-only-merge	No-merge
meta_computation() **(Phase I)**	Parent id and vertex degree	Vertex degree	Vertex degree
build_inconsistency_list() **(Phase II)**	1. Inconsistency list contains vertices with incorrect depth values with MIN_PRIORITY.	1. For each edge insertion add an edge in G' if the endpoints belong to different components [8, 15].	1. Inconsistency list contains endpoints of every edge inserted and/or deleted and their respective neighbors.
	2. G' = G	2. G' is also known as component graph	2. G' consists of inconsistent vertices and edge incident on them in G [6]
check_property() **(Phase III)**	Check BFS depth property	Check disjoint component property	Check vertex degree property
activate_frontier() **(Phase IV)**	Activate inconsistent vertices with minimum depth value-Ramalingam and Reps [18]	Activate all the vertices in G'	Activate all the vertices in G'
update_inconsistency_list() **(Phase IV)**	Remove frontier vertices and add inconsistent successors to inconsistency list	Clear inconsistency list	Clear inconsistency list
merge_state() **(Phase V)**	1. Apply all insertions and deletions to G	1. Apply only deletions to G.	1. Applying insertions and deletions to G not required.
		2. Relabel components in G using G'	2. Update triangle counts and degree information in G using G'

below) which the performance of incremental processing will drastically degrade. Finally, *threshold fraction* defines the fraction of inconsistent vertices that are above (or below) the corresponding property threshold. For example, in BFS, the vertex depth could be one of the properties defined in the property list with depth threshold 2 and threshold fraction 0.3. Then, GraphIn would switch to static BFS recomputation if $> 30\%$ of inconsistent vertices have BFS depth < 2, in line with the idea that if updates affect a large number of vertices closer to the root of the BFS tree, its better to run a full static recomputation. Note that these thresholds for static recomputation are algorithm and dataset dependent user-tunable parameters that require training to derive their optimal values.

Algorithm 1. I-GAS computation loop per update batch

1: while(!*inconsistency_list.isempty()*):
2:    *frontier = activate_frontier*(G', *inconsistency_list*)
3:    IGAS(G')
4:    *update_inconsistency_list*(G', *inconsistency_list, frontier*)

3.6   Phase IV: Incremental GAS Computation (I-GAS Engine)

The incremental GAS (or I-GAS) phase ensures that only the inconsistent part of the graph is recomputed incrementally, not the entire graph. The I-GAS Engine identifies the overlap between two consecutive versions of the evolving graph and incrementally processes the graph by opening only the new computational frontier. Here the user implements the I-GAS program as well as the *activate_frontier()* and *update_inconsistency_list()* APIs whose prototypes are: *activate_frontier*(G, inconsistency_list), *update_inconsistency_list*(G, inconsistency_list, frontier).

As shown in Algorithm 1, the I-GAS loop is comprised of three basic steps that are iterated over until the inconsistency list becomes empty. It starts with a set of inconsistent vertices and calls *activate_frontier()* to activate or open the next computational frontier using the vertex priority defined in Phase II and then runs an I-GAS program. An I-GAS program consists of incremental versions of the gather, apply and scatter functions. By default, the I-GAS program is the same as the GAS program for static execution, but the user can choose to override it if necessary. Finally, the new computational frontier information is used to update the vertex inconsistency list.

3.7   Phase V: Merge Graph States (Graph Merger)

This phase is responsible for both merging updated vertex property information (e.g., new vertex depths calculated in incremental BFS) and inserting/deleting edges into the most recent version of the static graph G, thereby generating the next version of the graph. GraphIn can perform the merger in several ways to accommodate different types of graph algorithms and different levels of tolerance for inconsistency in the system.

Some incremental algorithms must accommodate both inserts and deletes from the latest update batch in each iteration of the algorithm. These updates must be applied to G before the next update batch is considered. In general, this is the case for graph algorithms that calculate global properties, like BFS, which needs to consider any added or removed edges before recalculating vertex depths. Other algorithms, often ones that compute properties that are semi-localized within a graph, need to only accommodate deletes e.g. in connected components algorithm, insertions can be handled by creating a component graph [8,15] G' in the Phase II (see Table 1), where each edge insertion (u, v) in G results in an edge in G' if u and v belong to separate components or results in a self-edge, which is ignored in the component graph. Therefore, a batch of insertions has

Table 2. Graph datasets.

Graphs	Type	# Vertices	# Edges
RMAT Scale 20 (G1M16M)	Synthetic	1,048,576	15,700,394
RMAT Scale 21 (G2M32M)	Synthetic	2,097,152	31,771,509
RMAT Scale 22 (G4M64M)	Synthetic	4,194,304	64,155,735
Facebook (FB) [28]	Real World	2,937,612	41,919,708
LiveJournal (LJ) [2]	Real World	4,847,571	68,475,391

no dependency on insertions from prior batches; they can be processed at any point (e.g., deferred) without affecting the accuracy of results. But deletes must be applied before the next update batch is considered. Finally, there are algorithms where each incremental iteration has no dependency on inserts or deletes from the previous batch. This is often the case for algorithms that only utilize local properties. Examples include the computation of clustering coefficients [6], triangle counting, vertex degree, etc. In such cases, both inserts and deletes may be deferred. Based on these observations, we support three merge patterns:

1. **All-Merge:** Both inserts and deletes are merged with static graph G.
2. **Partial-Merge:** Either deletes or inserts are merged with G. The framework defers applying the rest of the updates to the original graphs.
3. **No-Merge:** Neither inserts nor deletes from the update batch are merged with G. The framework defers applying both inserts and deletes.

4 Experimental Evaluation

4.1 Experimental Setup

Evaluation Platform. GraphIn is evaluated[1] on a dual-socket Intel node equipped with two Intel® Xeon®[2]E5-2608L six-core processors running at 2.0 GHz with 64 GB of DDR4 RAM. We used the Intel® C++ Composer XE

---

[1] Software and workloads used in performance tests may have been optimized for performance only on Intel microprocessors. Performance tests, such as SYSmark and MobileMark, are measured using specific computer systems, components, software, operations and functions. Any change to any of those factors may cause the results to vary. You should consult other information and performance tests to assist you in fully evaluating your contemplated purchases, including the performance of that product when combined with other products. For more information go to http:// www.intel.com/performance.

[2] Intel and Xeon are trademarks of Intel Corporation in the U.S. and/or other countries.

2015 Compiler[3] to compile the native and benchmark codes. We used Graph-Mat [26] and STINGER [6–8] for performance comparisons. In order to utilize multiple threads on the CPU, GraphIn, GraphMat and STINGER all use OpenMP. Updates are provided in batches of size ranging from 10,000 up to one million with 1 % of all updates being deletions (except for CC where we use only insertions). The endpoints of the edges used for batch updates are generated randomly.

Graph Datasets. As shown in Table 2, we evaluate the performance of GraphIn using a mix of real-world and synthetic datasets. The synthetic datasets are obtained from the Graph500 RMAT data generator [16] for scales 20, 21 and 22 with an average degree of 16 per vertex.

Evaluated Algorithms. Three widely used graph algorithms are evaluated, including Clustering Coefficient (CCof), Connected Components (CC) and Breadth First Search (BFS). As shown in Table 1, these algorithms are classified as no-merge (CCof), partial-merge (CC) and all-merge (BFS) algorithm defined in the previous section. Algorithms requiring undirected graphs as inputs, e.g., connected components, are stored as pairs of directed edges.

4.2 Evaluation and Analysis

(1) Benefits of Incremental graph computation: From Figs. 3a, 4a, and 5a we can observe that, GraphIn achieves maximum speedups of 107×, 40× and 82× over static computation across all the datasets for CCof, CC and BFS, respectively, with update batch size going as high as 1 million updates. Figures 3b, 4b, and 5b show updates per second versus batch size; GraphIn achieves up to 9.3 million updates/sec. These speedups result from the use of incremental computation in the IGAS execution model to compute the vertex properties/states for only the inconsistent vertices, as opposed to executing the graph algorithm for the entire input graph in the static case. Furthermore, we can draw key inferences as to how the performance of incremental execution varies with the algorithm type and update batch size, which we discuss next in detail.

Effect of graph algorithm. From Figs. 3a–5a we can observe that the maximum speedups occur with no-merge algorithms, such as clustering coefficients, followed by partial-merge and all-merge algorithms like CC and BFS respectively. Note that the maximum speedup for BFS occurs when the update batch

---

[3] Intel's compilers may or may not optimize to the same degree for non-Intel microprocessors for optimizations that are not unique to Intel microprocessors. These optimizations include SSE2, SSE3, and SSE3 instruction sets and other optimizations. Intel does not guarantee the availability, functionality, or effectiveness of any optimization on microprocessors not manufactured by Intel. Microprocessor-dependent optimizations in this product are intended for use with Intel microprocessors. Certain optimizations not specific to Intel micro-architecture are reserved for Intel microprocessors. Please refer to the applicable product User and Reference Guides for more information regarding the specific instruction sets covered by this notice. Notice revision #20110804.

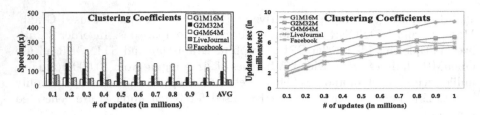

Fig. 3. (a) Incremental speedup over static execution vs. update batch size (b) Update rate vs. batch size for **Clustering Coefficient**.

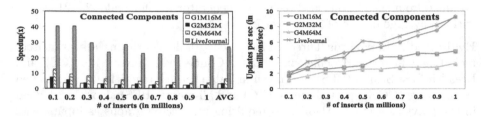

Fig. 4. (a) Incremental speedup over static execution vs. update batch size (b) Update rate vs. batch size for **Connected Components**.

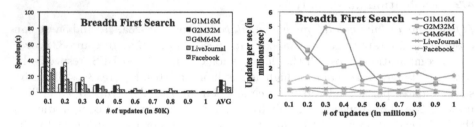

Fig. 5. (a) Incremental speedup over static execution vs. update batch size (b) Update rate vs. batch size for **BFS**.

size is much smaller (∼50k updates) compared to other two incremental algorithms (∼1M updates). This variation in speedups occurs because the fraction of the graph that becomes inconsistent after applying an update batch increases in the order of no-merge, partial-merge, and all-merge. No-merge or partial-merge algorithms affect the graph locally whereas all-merge algorithms like BFS calculate a global property (i.e. depth), so the incremental computation affects a larger portion of the graph and incurs higher incremental processing time.

Effect of Update Batch Size. For a particular incremental algorithm in GraphIn, the speedup achieved falls as the update batch size increases (Figs. 3a–5a) because the problem size and the average incremental runtime increases with the batch size. Also, both runtime (because of decreasing speedup) and update rate increase with the batch size for CCof and CC (see Figs. 3b and 4b), which implies the decrease in speedup changes slower with respect to dynamic update rate increases for larger batches. Whereas in BFS both speedup and update rate

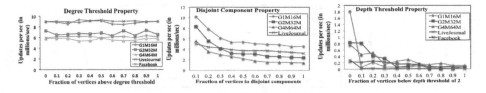

Fig. 6. Effect of (a) vertex degree, (b) disjoint components and (c) depth value from source vertex on the update rate in CCof, CC and BFS respectively

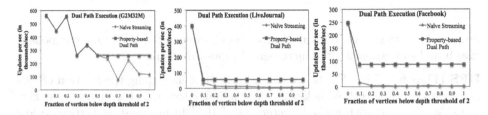

Fig. 7. Dual path execution vs. naïve streaming in incremental BFS using vertex depth property for G2M32M, LiveJournal and Facebook graph

falls with increase in batch size (Fig. 5b) as the incremental computation affects a larger portion of the graph with the increase in batch size causing the update rate to drop.

(2) Performance implications of graph properties: To demonstrate how properties of the inconsistent vertices in an update batch affect the average runtime in GraphIn, we have chosen graph properties: vertex degree (CCof), vertices with disjoint components (CC) and vertex depth from the source (BFS). The rationale behind choosing these properties is that they play a key role in the computation of the corresponding static graph algorithm we are comparing against.

Clustering Coefficient (Vertex Degree): Figure 6a shows the change in the update rate versus the fraction of vertices with degree greater than a certain threshold (e.g. 750 for the Facebook graph) that are affected by the updates. We can observe that the degree property does not have a dramatic effect on the CCof update rate which remains relatively constant. This is because CCof is a no-merge algorithm for both inserts and deletes to the graph and thus the incremental runtime is independent of the vertex degree.

Connected Components (Disjoint Components): As shown in Fig. 6b, the update rate for CC decreases as the fraction of edges inserted whose endpoints belong to different components in the original graph is increased, with a maximum slowdown of 3.97× across all datasets. This happens because such edges have endpoints in multiple components, meaning that there is a corresponding edge in the component graph, as opposed to self edges where endpoints of an edge fall in the same component. As described in Table 1, GraphIn reduces incremental CC to static CC processing on the component graph and increasing the

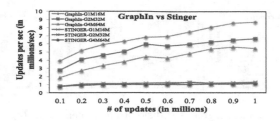

Fig. 8. STINGER comparison for CCof

number of vertices with disjoint components increases the size of the component graph and subsequently, the incremental processing time.

BFS (Depth): From Fig. 6c we can observe that increasing the fraction of vertices with depth threshold below 2 causes a sharp decline in the update rate. E.g. in G1M16M even with a small increment of 10 % in inconsistent vertices below the depth of 2 results in 29x decline in update rate. When more insertions and deletions occur on these lower-depth vertices closer to the root vertex, it results in the I-GAS loop making a much larger portion of the graph inconsistent with each increment and hence the runtime increases sharply. Another observation to make here is that the largest graphs with higher diameter values are affected the most, e.g. LiveJournal, which has the maximum number of edges and the maximum diameter among all input datasets has the largest max to min ratio of update rate . For larger graphs with higher diameter the I-GAS loop iterates through larger number of BFS levels with more work per iteration (large number of edges) until the inconsistency vertex set becomes empty.

Benefits of property-based dual-path execution: Figure 7 shows how GraphIn adapts to situations where the incremental processing performs worse than static recomputation. The performance of incremental BFS for naïve (no property checks) streaming falls relative to static processing beyond a threshold fraction of vertices with depth threshold below 2. For G2M32M, LiveJournal and Facebook this threshold fraction is 0.6, 0.1 and 0.1, respectively. This degradation in incremental performance is because of a larger number of updates to these lower-depth vertices resulting in a larger portion of the graph becoming inconsistent incurring longer processing times. In such scenarios, GraphIn processes the update batch with static recomputation, which ensures that the worst-case performance of GraphIn is no worse than static recomputation. With dual path execution we achieve a maximum speedup of 60× (Facebook input).

(3) Comparison with STINGER: As shown in Fig. 8, clustering coefficients [6] using STINGER shows a max update rate of 1.32 million versus 8.7 million updates per sec with GraphIn for the G1M16M case, which results in 6.6× speedup in throughput. This is because unlike STINGER, which uses a single edge-list based data structures for both static and incremental graph processing, GraphIn's hybrid data structure of edge-list for incremental updates and

compressed matrix format for static versions of the graph enables faster updates as well as fast static computation on the clustering coefficient subproblem.

5 Related Work

Dynamic graph processing can be broken down into offline and online processing; GraphIn is a framework designed to addresses the latter problem.

Offline Processing: Chronos [11], GraphScope [25], and TEG [9] represent recent work in offline graph processing. Chronos supports incremental processing on temporal graphs using a graph representation that places graph vertex data from different versions together leading to good cache locality. Graph-Scope proposes encoding for evolving graphs community discovery and anomaly detection.

Online Processing: Continuous query processing over streaming updates [29] incurs memory constraints and restricts keeping multiple versions of the evolving graph. GIM-V [27] proposes an incremental graph processing model based upon generalized iterative matrix-vector multiplication. STINGER [7] defines an efficient data structure to represent streaming graphs that enables fast, real-time insertions and/or deletions to the graph. Unlike STINGER which uses a single data structure for both static and dynamic graph analysis, GraphIn uses a hybrid data structure that allows for incremental computation on edge lists and a compressed format for static graph computation.

Static Graph Processing: Pregel [14], PowerGraph [10] and GraphLab [13] are some of the distributed frameworks that work by mapping large graphs across the combined memories of multiple machines. GraphChi [12] and X-Stream [19] are recently proposed out-of-memory frameworks that handle graphs that don't fit into a single machine's host memory. GraphReduce [22] framework can efficiently process graphs that cannot fit into the limited GPU memory [20,21] by mapping sub-graphs to the different memory abstractions of slow and fast memory [23]. These projects are complimentary to GraphIn, as they could be used to implement the static component of GraphIn while I-GAS can be leveraged to make these frameworks more dynamic.

6 Conclusion and Future Work

In this paper we present GraphIn, a high performance incremental graph processing framework for time-evolving graphs. Using its novel programming model called I-GAS, GraphIn incrementally computes the required graph properties only for the affected subgraph and thereby eliminates redundant computation. We further propose a user-tunable, property-based dual path execution optimization to choose between an incremental and a static run to achieve the best performance. Extensive experimental evaluations for a wide variety of graph inputs and algorithms demonstrate that GraphIn achieves a throughput of up to 9.3 million updates/sec and over 400× speedup compared to static graph recomputation. Using property-based dual-path execution GraphIn achieves up to 60× speedup compared to a naïve streaming approach. Future work will look at extending GraphIn to take advantage of on-node accelerators like Nvidia GPU [24] and multi-node clusters handling extreme-scale datasets [17].

References

1. Email Statistics: http://tinyurl.com/o7pch5f
2. The University of Florida Sparse Matrix Collection: http://tinyurl.com/me4w55
3. Twitter Statistics: http://tinyurl.com/kcuhdcw
4. Armstrong, T.G., Ponnekanti, V., Borthakur, D., Callaghan, M.: Linkbench: a database benchmark based on the facebook social graph. In: SIGMOD 2013, NY, USA (2013)
5. Bell, N., Garland, M.: Efficient sparse matrix-vector multiplication on CUDA. NVIDIA Technical Report NVR-2008-004, NVIDIA Corporation, December 2008
6. Ediger, D., Jiang, K., Riedy, J., Bader, D.: Massive streaming data analytics: a case study with clustering coefficients. In: IPDPSW 2010, pp. 1–8 (2010)
7. Ediger, D., McColl, R., Riedy, J., Bader, D.: Stinger: high performance data structure for streaming graphs. In: HPEC, September 2012
8. Ediger, D., Riedy, J., Bader, D., Meyerhenke, H.: Tracking structure of streaming social networks. In: IPDPSW 2011, May 2011
9. Fard, A., Abdolrashidi, A., Ramaswamy, L., Miller, J.: Towards efficient query processing on massive time-evolving graphs. In: CollaborateCom, October 2012
10. Gonzalez, J.E., Low, Y., Gu, H., et al.: Powergraph: distributed graph-parallel computation on natural graphs. In: OSDI 2012, Hollywood, CA. USENIX (2012)
11. Han, W., Miao, Y., Li, K., Wu, M., Yang, F., Zhou, L., Prabhakaran, V., Chen, W., Chen, E.: Chronos: a graph engine for temporal graph analysis. In: EuroSys (2014)
12. Kyrola, A., Blelloch, G., Guestrin, C.: Graphchi: large-scale graph computation on just a pc. In: OSDI 2012, Berkeley, CA, USA. USENIX Association (2012)
13. Low, Y., Bickson, D., Gonzalez, J., et al.: Distributed graphlab: a framework for machine learning and data mining in the cloud. Proc. VLDB Endowment 5(8), 716–727 (2012)
14. Malewicz, G., Austern, M.H., Bik, A.J., et al.: Pregel: a system for large-scale graph processing. In: SIGMOD 2010, New York, NY, USA. ACM (2010)
15. McColl, R., Green, O., Bader, D.: A new parallel algorithm for connected components in dynamic graphs. In: HiPC, December 2013
16. Murphy, R.C., Wheeler, K., Barrett, B., Ang, J.A.: Introducing the graph 500. In: Cray Users Group (CUG) (2010)
17. Plimpton, S.J., Devine, K.D.: Mapreduce in mpi for large-scale graph algorithms. Parallel Comput. 37(9), 610–632 (2011)
18. Ramalingam, G., Reps, T.: An incremental algorithm for a generalization of the shortest-path problem. J. Algorithms 21(2), 267–305 (1996)
19. Roy, A., Mihailovic, I., Zwaenepoel, W.: X-stream: edge-centric graph processing using streaming partitions. In: SOSP 2013, New York, NY, USA. ACM (2013)
20. Sengupta, D., Belapure, R., Schwan, K.: Multi-tenancy on gpgpu-based servers. In: VTDC 2013, New York, NY, USA, pp. 3–10. ACM (2013)

21. Sengupta, D., Goswami, A., Schwan, K., Pallavi, K.: Scheduling multi-tenant cloud workloads on accelerator-based systems. In: SC 2014, NJ, USA. IEEE (2014)
22. Sengupta, D., Song, S.L., Agarwal, K., Schwan, K.: Graphreduce: processing large-scale graphs on accelerator-based systems. In: SC 2015, NY, USA. ACM (2015)
23. Sengupta, D., Wang, Q., Volos, H. et al.: A framework for emulating non-volatilememory systemswith different performance characteristics. In: ICPE 2015, NY, USA (2015)
24. Slota, G.M., Rajamanickam, S., Madduri, K.: High-performance graph analytics on manycore processors. In: IPDPS 2015 IEEE International, pp. 17–27, May 2015
25. Sun, J., Faloutsos, C., Papadimitriou, S., Yu, P.S.: Graphscope: parameter-free mining of large time-evolving graphs. In: KDD 2007, New York, NY, USA. ACM (2007)
26. Sundaram, N., Satish, N., Patwary, M.M.A., et al.: Graphmat: high performance graph analytics made productive. Proc. VLDB Endowment **8**(11), 1214–1225 (2015)
27. Suzumura, T., Nishii, S., Ganse, M.: Towards large-scale graph stream processing platform. In: WWW 2014 Companion, Republic and Canton of Geneva, Switzerland (2014)
28. Wilson, C., Boe, B,. Sala, A., Puttaswamy, K.P., Zhao, B.Y.: User interactions in social networks and their implications. In: EuroSys 2009, New York, NY, USA. ACM (2009)
29. Zeitler, E., Risch, T.: Massive scale-out of expensive continuous queries. PVLDB **4**(11), 1181–1188 (2011)

Efficient Large Outer Joins over MapReduce

Long Cheng[1](✉) and Spyros Kotoulas[2]

[1] cfaed, TU Dresden, Dresden, Germany
long.cheng@tu-dresden.de
[2] IBM Research, Dublin, Ireland
spyros.kotoulas@ie.ibm.com

Abstract. Big Data analytics largely rely on being able to execute large joins efficiently. Though inner join approaches have been extensively evaluated in parallel and distributed systems, there is little published work providing analysis of outer joins, especially on the extremely popular MapReduce platform. In this paper, we studied several current algorithms/techniques used in large outer joins. We find that some of them could meet performance bottlenecks in the presence of data skew, while others could be complex and incur significant coordination overheads when applied to the MapReduce framework. In this light, we propose a new algorithm, called POPI (Partial Outer join & Partial Inner join), which targets for efficient processing large outer joins, and most important, is lightweight and adapted to the processing model of MapReduce. We implement our method in Pig and evaluate its performance on a Hadoop cluster of up to 256 cores and datasets of 1 billion tuples. Experimental results show that our method is scalable, robust and outperforms current implementations, at least in the case of high skew.

1 Introduction

In light of the explosion of available data and the increasing connectivity between data systems, the infrastructure for scalable data analytics is as relevant as ever. An essential operation in this domain is the join, which facilitates the combination of records based on a common join key. Since this data-intensive operation can incur significant costs, improving the efficiency of this operation would have a significant impact on the performance of applications.

Outer Join. Although distributed inner join algorithms have been widely studied [1,2], there has been relatively little done on the topic of outer joins. In fact, outer joins are common in complex queries and widely used such as in OLAP applications. For example, in online e-commerce, customer ids are often left outer joined with a large transaction table for analyzing the purchase patterns [3]. In contrast to inner joins, outer joins do **not** discard tuples from one (or both) table(s) that do not match with any tuple in the other table. As a result, the final join results contain not only the matched part but also the non-matched part. This difference makes outer join implementations significantly different from inner joins in a distributed system and challenge current techniques [4].

© Springer International Publishing Switzerland 2016
P.-F. Dutot and D. Trystram (Eds.): Euro-Par 2016, LNCS 9833, pp. 334–346, 2016.
DOI: 10.1007/978-3-319-43659-3_25

MapReduce. As applications grow in scale, joins on multiple CPUs and/or machines is becoming important. Compared to conventional parallel DBMSs, MapReduce (over Hadoop) integrates parallelization, fault tolerance and load balancing in a simple programming framework, and can be easily run in a large computing center or cloud, making it extremely popular for large-scale data processing. In fact, most vendors (such as IBM) provide solutions, either on-premise or on the cloud, to compute on massive amount of structured, semi-structured and unstructured data for their business applications.

In this light, studying analytic techniques on this platform becomes very important. In fact, join operations are sometimes hard in MapReduce [5]. Unlike implementations in DBMS'es, complex designs for joins in MapReduce can easily lead to poor performance: the overhead of starting a communication phase between partitions is very high. Namely, we have to start a new job and re-read (part of) the data. In addition, the MapReduce paradigm is highly sensitive to the presence of data or computation skew: since coordination is infrequent and very costly, there are fewer opportunities to re-balance workloads across nodes.

In comparison to most of current studies focusing on *inner* joins over MapReduce [2,5], in this work, we focus on the design and evaluation of outer joins on this platform. We summarize our contributions as following:

- We introduce several outer join implementations which are applied in MapReduce and discuss their possible performance issues.
- We discuss the possibility to apply some advanced join strategies used in parallel DBMSs to outer joins over MapReduce. We find that, they could either meet performance issues or be complex in implementations and thus bring in high overhead in terms of the number of MapReduce jobs launched.
- We propose a new approach, called POPI (Partial Outer join & Partial Inner join), which targets efficient outer joins adapted to MapReduce.
- We implement the various approaches on Apache Pig. Our experimental results show that our method is robust and can perform better than current implementations in MapReduce, at least in the presence of high skew.

The rest of this paper is organized as follows: In Sect. 2, we shortly introduce the MapReduce framework and describe current outer join implementations over it. In Sect. 3, we discuss some advanced strategies for large data outer joins in MapReduce. We describe our new approach in Sect. 4 and present the evaluation in Sect. 5. We report on related work in Sect. 6 while we conclude the paper in Sect. 7.

2   MapReduce and Outer Joins

Overview. MapReduce [6] is designed to operate over key/value pairs. Specifically, each *Map* function receives a key/value pair and emits a set of key/value pairs. All key/value pairs produced during the map phase are grouped by their key and passed to reduce phase. During the reduce phase, a *Reduce* function is called for each unique key, processing the corresponding set of values.

Though MapReduce has various advantages on large data processing, it entails more overhead compared to traditional DBMSs during execution: This platform sacrifices per-node efficiency, for scalability [5]. Namely, performance loss on a single node can usually be compensated by simply employing more computation resources. Nevertheless, MapReduce has no way of automatically re-balancing load, and any operation that changes the distribution of data should only be performed in the context of a new job (which typically incurs a coordination overhead of tens of seconds, if not minutes). Thus, achieving good load-balancing in data/join processing is critical.

Current Methods for MapReduce. Currently, three outer join methods are commonly applied in MapReduce implementations: hash-based, replication-based and histogram-based outer joins. We focus on *left* outer joins ($\bowtie$) here since they are the most common ones and their implementations would be analogous for right outer joins. In the following, we focus on a single outer join operation between two relations R and S. We assume both R and S are $<k, v>$ pairs with $|R| < |S|$ and k is the join key. For simplicity, we also assume that R is uniformly distributed and S is skewed for all of our examples, unless otherwise specified.

Hash-Based Outer Join. Similarly to inner join implementations, this approach can be done in a single MapReduce job. In the map phase, each map task works on either R or S. To identify which relation an input record is from, each map task tags the record with its originating table, and outputs the extracted join key and the tagged record. For example, for a record $< k_1, v_1 >$ from S, the output will be $< k_1, (s, k_1, v_1) >$ pair, where s is the table tag. Then, the framework brings together records sharing the same key and eventually feed them to a reducer, based on the hash value of their keys. In the reduce phase, the reduce function separates the input records into two sets according to their table tags and then performs a cross-product between each record in these sets and output the final results.

Normally, this scheme can achieve good performance under ideal balancing conditions for distributed systems [1]. However, when the processed records has significant skew, number of records will be flushed to a small part of reducers and cause hotpots. Such issues impact system scalability which will be reduced as employing new nodes[1] cannot yield improvements - the skew records will still be distributed to the same reducers.

Replication-Based Outer Join. Compared to the replication-based *inner* joins containing only a mapreduce job, outer joins within this scheme is significantly different. It is composed by two distinct join stages in an abstract level[2]: (1) A map-side inner join between R and S. Namely, all records of the small table R is retrieved from the DFS and then each map task uses a main-memory hash table to

---

[1] Note that, in terms of terminology, when we talk about a node, we mean a computing unit (e.g., a Reducer in MapReduce) in this work.

[2] The detailed process about how to identity where a record comes from is the same as the hash-based approach described above, thus here we do not present it again.

join S with R, formulating the intermediate results T; and (2) A reduce-side outer join between R and T, which is done in the same way as the hash-based method described above. Namely, all the records of R and T will be grouped based on their keys, and then fed to reducers for the local outer joins.

The *replication* in this method can reduce load imbalance, as each map task has the same workloads in the first phase. Nevertheless, this operation is costly and only suitable for small-large outer joins [3]. Moreover, even if R is small, the cardinality of the intermediate results T could be large when S is highly skewed [4]. This could make tasks in the second stage very costly and consequently decreases the whole performance.

Histogram-Based Outer Join. As data skew is common in most applications, efficient approaches to handle this kind of skew becomes critical for the join performance. Apache Pig has some built-in resistance to skewed joins, a typical method is using histogram [7]. Namely, firstly, a histogram of key popularity is calculated, which can be done with a single MapReduce job. Then, the keys are re-arranged and the jobs are distributed based on that. For instance, if we have the following histogram $k_1 = 19$, $k_2 = 20$, $k_3 = 18$, $k_4 = 60$, and we have two reducers, instead of splitting the keys in a hash-based way (i.e., k_1 and k_3 go to reducer 1, k_2 and k_4 go to reducer 2), the workload will be balanced by sending k_1, k_2 and k_3 to reducer 1 and k_4 to reducer 2. However, this does not work with extreme skew. As shown in [8], if a key is overly popular, the single reducer that it will be sent to will still become a hotspot.

3 Candidate Strategies for MapReduce

In this section, we present some advanced strategies studied in parallel databases and discuss about the possibility to apply them to outer joins in MapReduce.

3.1 The PRPD Method

Xu et al. [9] propose an algorithm named PRPD (Partial Redistribution & Partial Duplication) for inner joins. In their implementation, S is partitioned into two parts: (1) a locally-retained part S_{loc}, which comprises high skew items and which is not involved in the redistribution phase, and (2) the redistributed part S_{redis} which comprises the records with low frequency of occurrence and is redistributed using a common hash-based implementation. The relation R is also divided into two parts: (1) the duplicated part R_{dup}, which contain the keys in S_{loc}, which will be broadcast to all other nodes, and (2) the redistributed part R_{redis} - the remaining part of R that is to be hash redistributed. Then, the final inner join is composed by $R_{redis} \bowtie S_{redis}$ and $R_{dup} \bowtie S_{loc}$.

This method presents an efficient way to process the high skew records (i.e. the ones with keys that are highly repetitive). All these records of S are not transferred at all, instead, a small number of records containing the same keys from R are broadcast. The results for this approach show significant speedup in the presence of data skew. Because PRPD is a hybrid method combining both the hash and duplication-based join scheme, we can simply use outer joins to replace the corresponding inner joins in the case of MapReduce. Namely, we have

$$R \bowtie S = (R_{redis} \bowtie S_{redis}) \bigcup (R_{dup} \bowtie S_{loc}) \qquad (1)$$

However, this implementation could meet the same performance issue as the duplication-based approach described above: the cardinality of the intermediate results of $R_{dup} \bowtie S_{loc}$ could be large, because S_{loc} here is highly skewed, which means that a naive PRPD algorithm cannot be applied to outer joins in MapReduce directly.

3.2   The PRPS Approach

Cheng et al. [10] propose an efficient algorithm for inner joins, named as PRPS (further refined to PRPQ in their work). They use a semijoin-alike way to handle skewed data, inspiring us to apply it to the outer joins of $R_{dup} \bowtie S_{loc}$ in Eq. 1.

In this case, we divide the detailed process into two steps: (1) The unique keys of S_{loc} are extracted and we perform an outer join with R_{dup}; and (2) The matched part of R_{dup} is joined with S_{loc} (inner join), which is union-ed with the non-matching part of R_{dup} to formulate the outputs. Namely,

$$R_{dup} \bowtie S_{loc} = [R_{dup} \bowtie \pi_k(S_{loc})]^\top \bowtie S_{loc}] \bigcup [R_{dup} \bowtie \pi_k(S_{loc})]_\perp \qquad (2)$$

where the symbol $\top$ and $\perp$ means the matched and non-matched results of a outer join respectively.

We can see that this PRPS *outer* join method (referred as PRPS-O in the following) will be efficient on skew handling in MapReduce. The reason is that the large part of skewed records in S is still locally kept and just a small number of unique keys are extracted and transferred, which can be executed with two extra jobs in MapReduce. Nevertheless, as we describe later, we can use a simpler and more efficient method for the outer join implementation.

3.3   Complex Techniques

Other approaches (e.g., [4]) are also very efficient on distributed outer joins. They focus on a fine grained operation of per-node data movement (e.g., peer-to-peer communication based on requirements) to minimizing network communication during implementation. We believe that these algorithms can be coded in MapReduce, however, the number of their execution jobs could be large, more than the PRPS-O method at least. In this case, their implementations could be costly, not only because of their complex data flows, but also the overheads of MapReduce as we described. Actually, in our later evaluation, we have shown that, with two more jobs, PRPS-O takes around 80 s more on runtime, compared to our new method. Thus, we do not consider the detailed implementation and evaluation of these complex techniques in this work.

4 Our Approach

In this section, we present our POPI method and its implementation over Pig [7].

4.1 The POPI Algorithm

The design principles of POPI are: (1) large scale redistribution of skewed records should be limited, so as to avoid load balancing problems; and (2) duplication-based outer join operations should be avoided to the extend possible, in order to simplify the implementation and also reduce possible redundant communication and computation. Based on this, our algorithm adopts the same partitioning approach as PRPD [9]. We process the partitioned records as follows:

$$R \bowtie S = (R_{redis} \bowtie S_{redis}) \bigcup (R_{dup} \bowtie S_{loc}) \tag{3}$$

Namely, the skewed part is executed as an inner join directly. For clarification, we first give a brief proof of the correctness of Eq. 3 here:

Proof sketch: Assume that L is the set of skewed keys of S, then we have that: (1) L is extracted from the skewed part of S, namely, there is $L = \pi_b(S_{loc})$; (2) because the partitioning of R is based on L, namely, a record of R, $< a, x > \subset R_{dup}$ if only if the key meets the condition $a \in L$. Namely, every key of R_{dup} appears in L. In this condition, there will be **no** non-matched results in R_{dup} during its outer join execution with S_{loc}. Therefore, the outer join can be represented as an inner join. Note that, even if a skewed key in S does not appear in R, the inner join between R_{dup} and S_{loc} will still be valid here, since the final left outer join results depend on the match conditions of R only. ■

We can see that our outer join implementation is composed by an outer join and an inner join, which is different from a naive transformation, such as that in Eq. 1, in which there are two outer join operations involved. In the meantime, compared to the PRPS-O as Eq. 2, our approach also greatly simplifies the outer join implementations (with two jobs less over Pig for a left outer join). That is also the motivation behind the naming of our approach, POPI (Partial Outer join & Partial Inner join), since the processing between the skewed part and non-skewed part is different from current approaches and allows us to replace an outer join with an inner join.

Inheriting the advantages of the same data partitioning approach as PRPD, we believe that POPI will be robust and efficient on large outer joins in MapReduce. The reason is that we only need to transfer a small part of keys/records (via DFS), rather than the large number of records in S. Moreover, this method will be more efficient than the PRPS-O algorithm, as the number of MapReduce jobs has been reduced.

Following above, with regard to the case of *skewed-skewed* outer joins (i.e., the relation R is also skewed), we partition R into three parts: the R_{dup} and R_{redis} as we described previous, as well as the locally kept part R_{loc}, which contains all the skewed records in R. Correspondingly, records in S is partitioned into three parts as well, the S_{loc}, S_{redis} and the duplicated part S_{dup}, in which records contains join key belongs to R_{loc}. Then, the final outputs will be composed by three joins: a left outer join for the non-skew part records, namely $R_{redis} \bowtie S_{redis}$, and two inner joins for the skewed records, namely $R_{dup} \bowtie S_{loc}$ and $R_{loc} \bowtie S_{dup}$. In this case, the outer join $R \bowtie S$ can be presented as:

$$(R_{redis} \bowtie S_{redis}) \bigcup (R_{dup} \bowtie S_{loc}) \bigcup (R_{loc} \bowtie S_{dup}) \tag{4}$$

As the uniform-skew join is the core part of a join [9,11], we will focus on such kind of outer joins in our subsequent implementation and evaluation.

4.2 Implementation

We present a general implementation of our method using Pig Latin [12], a language that can be compiled to produce MapReduce programs used with Hadoop. We have three main advantages using this language: (1) It provides a concise notation for algorithms. (2) The outer join methods, such as *hash*, *replicated* and *histogram*, have been integrated in Pig, allowing us for a fair comparison. (3) In a larger pipeline of operations, we can avail of optimisations that are already implemented in Pig, such as performing multiple operation within a job, re-using partitioning of data or executing multiple jobs in parallel.

The detailed implementation of our method in Pig is shown in Algorithm 1. There, R, S, k, t refer to the left table of the outer join, the right side of the outer join, the sampling rate (referred to as *samplingPercentage* later) and the number of chosen top popular keys (refer as *samplingThreshold*) respectively. Initially, we sample the large table S (line 3), group by its join keys (line 4) and count the number of occurrences of sampled key (line 5). Then we order the keys and pick up the most popular keys based on the threshold t (lines 6–7). After that, the tables R and S are partitioned into two parts respectively based on the skewed keys (lines 9–13). With the partitioned data, we then start the outer joins (line 15) and inner joins (line 16). Finally, the outputs of the outer join are composed by the results from both parts (line 18).

5   Experimental Evaluation

5.1   Experiment Setup

Each computation unit of our experimental system has two 8-core Intel Xeon CPU E5-2690 processors running at 2.90 GHz, resulting in a total of 16 cores per physical node. Each node has 32 GB of RAM and a single 128 GB SSD local disk and nodes are connected by Infiniband. The operating system is Linux kernel version 2.6.32-279 and the software stack consists of Hadoop version 1.2.1, Pig version 0.14.0 and Java version 1.7.0_25.

The evaluation is implemented on two relations R and S. We fix the cardinality of R to 64 million records and S to 1 billion records. Because data in

---

Algorithm 1. POPI Outer Joins

1: **DEFINE** Skew_resistant_outer_join(R,S,k,t)
2:   **RETURNS** Result {
3: SS = **SAMPLE** S k;   //sample S
4: SG = **GROUP** SS **BY** S::key;
5: SC = **FOREACH** S2 **GENERATE** group, COUNT(SS) as c;
6: OrderedKey = **ORDER** SC **BY** c **DESC**;
7: SkewedKey = **LIMIT** OrderedKey t;
8:
9: SS = **JOIN** S **BY** key **LEFT**, SkewedKey **BY** group **USING** 'replicated';
10: **SPLIT** SS **INTO** S_loc **IF** SkewedKey::group is not null, S_red **IF** Skewed-Key::group is null;
11:
12: RS = **JOIN** R **BY** key **LEFT**, SkewedKey **BY** group **USING** 'replicated';
13: **SPLIT** RS **INTO** R_dup **IF** SkewedKey::group is not null, R_dis **IF** Skewed-Key::group is null;
14:
15: JA = **JOIN** R_dis **BY** R::key **LEFT**, S_dis by S::key;
16: JB = **JOIN** S_loc **BY** S::key, R_dup **BY** R::key USING 'replicated';
17:
18: $Result = **UNION** JA, JB; }

---

warehouses is commonly stored following a column-oriented model, we set the data format to <*key, value*> pairs, where both the key and value are 8-byte integers. We assume that R and S meet the foreign key relationship and when S is uniform, the tuples are created in such a way that each of them matches the tuples in the relation R with the same probability. Meanwhile we only add skew to S, following the Zipf distribution. The skew factor is set to 0 for uniform, 1 for the low skew (top ten popular keys appear 14 % of the time) and 1.4 for high skew dataset (top ten appear 68 %). Joins with such characteristics and workloads are common in data warehouses and column-oriented architectures [10].

In all experiments, we set the following system parameters: *map.tasks.maximum* to 16 and *reduce.tasks.maximum* to 8 and the rest of the parameters are left to the default values. The implementation parameters of our method are configured as follows: *samplingPercentage* is set to 10, *samplingThreshold* to 4000 as default. We measure runtime as the elapsed time from job submission to the job being reported as finished.

5.2   Experimental Results

Runtime. We focus on examining the runtime of three algorithms: the hash-based algorithm (referred to *Hash*), histogram-based method (referred to as *Skewed*) and the proposed POPI approaches. Since the first two methods have been integrated in Pig, we just simple use them directly. Though Pig also provides the replicated implementation, we do not compare with it here, since it is limited by the fact that the replicated relation needs to fit in memory [4].

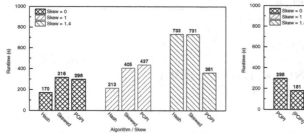 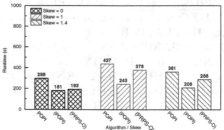

Fig. 1. Runtime of each algorithm.   **Fig. 2.** Compare POPI and PRPS-O.

We implement our tests using over 128 cores (8 nodes) and Fig. 1 shows the runtime of each algorithm. It can be seen that: (1) When S is uniform, Hash is more faster than the other two algorithms. The possible reason is the later two methods have extra-sampling operation and also the overhead of more MapReduce jobs. (2) With low skew, all the runtime increases, which is out of our expectation. As skew handling techniques have been adopted in the later two algorithms, the possible reason could be that not all the highly skewed records were sampled and there remains serious skew in both executions. (3) With high skew, our method becomes the best, which means that, the POPI algorithm can efficiently handle high skew at least.

We also compare the performance of POPI and PRPS-O. The results are shown in Fig. 2. There, the algorithm within "()" means that its sampling operation has been removed. Instead, the top popular keys are stored in a flat file and read as the skewed keys during executions. The reason to do so is for a more precise comparison: the join performance is sensitive to the sampled skew keys and operations like sampling cannot guarantee we always get the skewed keys. In addition, in most data processing pipelines, there is ample opportunity to extract this information as a side-effect of previous jobs. It can be observed that the (POPI) implementation is always faster than the original POPI, which means that the sampling operation could be costly and the sampled skewed keys are also critical for the performance. Moreover, the (POPI) is always faster than (PRPS-O), indicating the more jobs brought by complex implementations in PRPS-O are also costly, about 80 s out of 288 s in a high skewed dataset.

Load Balancing. We also track the detailed time spent on each reducer for each algorithm in the presence of data skew. Under low skew, the results are shown in Fig. 3. It can be seen that there are relatively small discrepancies for all the algorithms. The possible reason is that Hash can not handle data skew and the Skewed and POPI algorithm can not fully catch the skew keys because of their huge number. Furthermore, Fig. 4 shows the results in the condition of high skew. There, Hash is not balanced at all, in comparison, the Skewed method and POPI are much better. We should highlight that POPI achieves excellent load-balancing here. The reason could be that the number of skewed

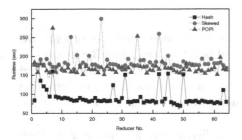

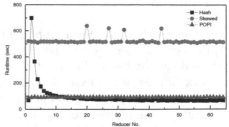

Fig. 3. Runtime of reducer in Skew = 1.   **Fig. 4.** Runtime of reducer in Skew = 1.4.

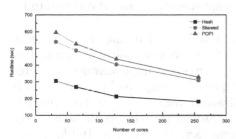

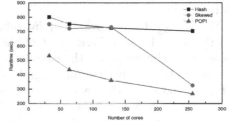

Fig. 5. Scalablity in Skew = 1.   **Fig. 6.** Scalablity in Skew = 1.4.

key is relative small in the condition of high skew and most of the popular keys are extracted, even when we only sample a small part of the input. Moreover, our runtime is much smaller than the Skewed method, which demonstrates the efficiency of this new approach.

Scalability. We finally test the scalability of our algorithm by varying the number of processing cores. We implement our test on the system from 2 nodes (32 cores) to 16 nodes (256 cores) over the skewed datasets. The detailed time-cost is shown in Figs. 5 and 6. We can see that the Skewed method and our algorithms generally scales well with the number of cores under low skew. However, they are slower than Hash. The reason could be the overhead of their implementations on MapReduce since Hash has only a single mapreduce job. This result is greatly different from the conditions when using other programming languages (e.g., X10 in [4]), where the Hash method is slower. In such scenarios, we believe that the hash-based approach could still be a better choice for MapReduce, under low skew. In comparison, with high skew, our method scales well while the other two are not. More importantly, our approach is significantly faster. Combining this with the good load balancing we have illustrated in Fig. 4, it can be seen that our method could be more suitable for the large outer joins in the presence of high skew.

6   Related Work

Several approaches have been proposed to improve the performance of joins over MapReduce [13], regardless, they have modified the basic MapReduce framework and cannot be readily used by existing platforms like Hadoop. Though the work [5] presents an extensive implementation on joins in MapReduce, they focus on execution profiling and performance evaluation, but not for robust join algorithms in the presence of big data.

As data skew has significant impact on distributed join processing, there has been in-depth research on skew handling in parallel and distributed DBMSs [1,3,4,9,10]. However, as we have explained, their methods could either have performance issues or be complex in MapReduce implementations. In comparison, our POPI algorithm is simple on implementation and also shown to be efficient.

Many algorithms have been introduced on skew handling for joins over MapReduce [14], regardless, most of them focus on inner joins, as opposed to consider the challenges on the complexity of outer join implementations. Moreover, several efforts in designing high level query languages on MapReduce, such as Pig [7] and Hive [15], have employed advanced mechanisms on skew handling in outer joins, however, as we have described, sometimes they could be not very efficient. Additionally, though some platforms (e.g., Stratosphere [16]) have provided efficient techniques on big data analytics, they focus on creating optimized plans of executing jobs, in contrast to the detailed implementation of a single operation as we studied in this work.

Recently, Bruno et al. [17] present three *SkewJoin* transformations to mitigate the impact of data skew in a distributed join operation. To prevent an outer join operator from generating *null* values, they partition the skewed tuples (e.g., in S) in a round-robin way so that each node can see at least one such tuple. In comparison, our approach is more light-weighted, since we do not need to repartition the skewed tuples, the number of which is always huge. Even when that some skewed tuples do not appear on some nodes, we will not generate *null*, as we use an inner join operation for the skewed tuples in our approach.

The PRPD algorithm [9] is a very popular method adopted by many companies (e.g., Teradata [9], Microsoft [17] and Oracle [18]). Nevertheless, as we have analyzed, PRPD cannot be applied to outer joins directly. The underlying data partitioning of our method is the same as PRPD, both are based on the skewed keys, therefore, the statistical information of data skew that is collected by the current systems using PRPD can be applied to POPI directly. This means that POPI can be used to extend the join implementations of current systems (or over current platforms like MapReduce [6] and Spark [19]) and consequently simplify the general executions of data queries. For example, skew statistics on the join keys (a, b) for the inner join implementation $R(a, x) \bowtie S(b, y)$ can be applied to the implementation of $R(a, x) \rightthreetimes\!\!\bowtie S(b, y)$ directly, without any modifications for the underlying join patterns.

7   Conclusions

In this paper, we focus on one data-intensive operation - outer joins - over the MapReduce platform. We have described current applied techniques and discussed the potential performance issues in the condition of using current advanced methods from parallel databases. Based on that, we propose our POPI algorithm for efficient large-scale data outer joins over MapReduce. We describe the detailed design and present the evaluation over a Hadoop cluster and Pig. We show that our new method is simple to implement. In the meantime, the experiment results also show that POPI is scalable, robust and can perform better compared with current implementations, at least in the case of high skew.

Acknowledgments. This work is supported by the German Research Foundation (DFG) within the Collaborative Research Center SFB 912 (HAEC) and in Emmy Noether grant KR 4381/1-1 (DIAMOND).

References

1. DeWitt, D., Gray, J.: Parallel database systems: the future of high performance database systems. Commun. ACM **35**(6), 85–98 (1992)
2. Li, F., Ooi, B.C., Özsu, M.T., Wu, S.: Distributed data management using MapReduce. ACM Comput. Surv. **46**(3), 31 (2014)
3. Xu, Y., Kostamaa, P.: A new algorithm for small-large table outer joins in parallel DBMS. In: ICDE, pp. 1018–1024 (2010)
4. Cheng, L., Kotoulas, S., Ward, T.E., Theodoropoulos, G.: Robust and efficient large-large table outer joins on distributed infrastructures. In: Silva, F., Dutra, I., Santos Costa, V. (eds.) Euro-Par 2014. LNCS, vol. 8632, pp. 258–269. Springer, Heidelberg (2014)
5. Blanas, S., Patel, J.M., Ercegovac, V., Rao, J., et al.: A comparison of join algorithms for log processing in Map Reduce. In: SIGMOD, pp. 975–986 (2010)
6. Dean, J., Ghemawat, S.: MapReduce: simplified data processing on large clusters. Commun. ACM **51**(1), 107–113 (2008)
7. Gates, A.F., Natkovich, O., Chopra, S., Kamath, P., Narayanamurthy, S.M., Olston, C., Reed, B., Srinivasan, S., Srivastava, U.: Building a high-level dataflow system on top of Map-Reduce: the Pig experience. PVLDB **2**(2), 1414–1425 (2009)
8. Kotoulas, S., Urbani, J., Boncz, P., Mika, P.: Robust runtime optimization and skew-resistant execution of analytical SPARQL queries on pig. In: Cudré-Mauroux, P., Heflin, J., Sirin, E., Tudorache, T., Euzenat, J., Hauswirth, M., Parreira, J.X., Hendler, J., Schreiber, G., Bernstein, A., Blomqvist, E. (eds.) ISWC 2012, Part I. LNCS, vol. 7649, pp. 247–262. Springer, Heidelberg (2012)
9. Xu, Y., Kostamaa, P., Zhou, X., Chen, L.: Handling data skew in parallel joins in shared-nothing systems. In: SIGMOD, pp. 1043–1052 (2008)
10. Cheng, L., Kotoulas, S., Ward, T.E., Theodoropoulos, G.: Robust and skew-resistant parallel joins in shared-nothing systems. In: CIKM, pp. 1399–1408 (2014)
11. Blanas, S., Li, Y., Patel, J.M.: Design and evaluation of main memory hash join algorithms for multi-core CPUs. In: SIGMOD, pp. 37–48 (2011)
12. Olston, C., Reed, B., Srivastava, U., Kumar, R., Tomkins, A.: Pig latin: a not-so-foreign language for data processing. In: SIGMOD, pp. 1099–1110 (2008)

13. Jiang, D., Tung, A., Chen, G.: Map-Join-Reduce: toward scalable and efficient data analysis on large clusters. TKDE **23**(9), 1299–1311 (2011)
14. Liao, W., Wang, T., Li, H., Yang, D., Qiu, Z., Lei, K.: An adaptive skew insensitive join algorithm for large scale data analytics. In: Chen, L., Jia, Y., Sellis, T., Liu, G. (eds.) APWeb 2014. LNCS, vol. 8709, pp. 494–502. Springer, Heidelberg (2014)
15. Thusoo, A., Sarma, J.S., Jain, N., Shao, Z., Chakka, P., Anthony, S., Liu, H., Wyckoff, P., Murthy, R.: Hive: a warehousing solution over a Map-Reduce framework. PVLDB **2**(2), 1626–1629 (2009)
16. Alexandrov, A., Bergmann, R., Ewen, S., Freytag, J.C., Hueske, F., Heise, A., Kao, O., Leich, M., Leser, U., Markl, V., et al.: The stratosphere platform for big data analytics. VLDB J. **23**(6), 939–964 (2014)
17. Bruno, N., Kwon, Y., Wu, M.C.: Advanced join strategies for large-scale distributed computation. PVLDB **7**(13), 1484–1495 (2014)
18. Bellamkonda, S., Li, H.G., Jagtap, U., Zhu, Y., Liang, V., Cruanes, T.: Adaptive and big data scale parallel execution in Oracle. PVLDB **6**(11), 1102–1113 (2013)
19. Zaharia, M., Chowdhury, M., Das, T., Dave, A., Ma, J., McCauley, M., Franklin, M.J., Shenker, S., Stoica, I.: Resilient distributed datasets: a fault-tolerant abstraction for in-memory cluster computing. In: NSDI, pp. 15–28 (2012)

Cluster and Cloud Computing

Slurm-V: Extending Slurm for Building Efficient HPC Cloud with SR-IOV and IVShmem

Jie Zhang$^{(\boxtimes)}$, Xiaoyi Lu, Sourav Chakraborty,
and Dhabaleswar K. (DK) Panda

Department of Computer Science and Engineering,
The Ohio State University, Columbus, USA
{zhanjie,luxi,chakrabs,panda}@cse.ohio-state.edu

Abstract. To alleviate the cost burden, efficiently sharing HPC cluster resources to end users through virtualization is becoming more and more attractive. In this context, some critical HPC resources among Virtual Machines, such as Single Root I/O Virtualization (SR-IOV) enabled Virtual Functions (VFs) and Inter-VM Shared memory (IVShmem) devices, need to be enabled and isolated to support efficiently running multiple concurrent MPI jobs on HPC clouds. However, original Slurm is not able to supervise VMs and associated critical resources, such as VFs and IVShmem. This paper proposes a novel framework, **Slurm-V**, which extends Slurm with virtualization-oriented capabilities such as job submission to dynamically created VMs with isolated SR-IOV and IVShmem resources. We propose several alternative designs for Slurm-V: Task-based design, SPANK plugin-based design, and SPANK plugin over OpenStack-based design, to manage and isolate IVShmem and SR-IOV resources for running MPI jobs. We evaluate these designs from aspects of startup performance, scalability, and application performance in different scenarios. The evaluation results show that VM startup time can be reduced by up to 2.64X through snapshot scheme in Slurm SPANK plugin. Our proposed Slurm-V framework shows good scalability and the ability of efficiently running concurrent MPI jobs on SR-IOV enabled InfiniBand clusters. To the best of our knowledge, Slurm-V is the first attempt to extend Slurm for the support of running concurrent MPI jobs with isolated SR-IOV and IVShmem resources. The capabilities of Slurm-V can be used to build efficient HPC clouds.

1 Introduction

To meet the increasing demand for computational power, HPC clusters have grown tremendously in size and complexity. Efficient sharing of such HPC resources is becoming more important to achieve faster turnaround time and lower the cost per user. Furthermore, a large number of users experience large variability in workloads depending on business needs, which makes predicting

This research is supported in part by National Science Foundation grants #CNS-1419123, #IIS-1447804, #ACI-1450440, and #CNS-1513120.

P.-F. Dutot and D. Trystram (Eds.): Euro-Par 2016, LNCS 9833, pp. 349–362, 2016.
DOI: 10.1007/978-3-319-43659-3_26

the required resources for future workloads a difficult task. Therefore, virtualized HPC clusters can be an attractive solution that can offer on-demand resource acquisition, high configurability while delivering near bare-metal performance at a low cost.

While virtualization technology has come a long way since its inception, achieving near-native performance for latency-critical HPC application remains a challenge to this date. A significant bottleneck exists in the virtualized I/O subsystem, which is one of the biggest hindrances to large scale adoption of virtualization in the HPC community. The recently introduced Single Root I/O Virtualization (SR-IOV) [3] technology for InfiniBand and High Speed Ethernet is quickly changing the landscape by providing native I/O virtualization capabilities [12]. Through SR-IOV, a single physical device, or a Physical Function (PF), can be presented as multiple virtual devices, or Virtual Functions (VFs). However, our previous studies [10] have shown that SR-IOV lacks support for locality-aware communication, which leads to performance overheads for inter-VM communication within the same physical node. In this context, Inter-VM Shared Memory (IVShmem) [15] has been proposed and can be hot-plugged to a VM as a virtualized PCI device to support shared memory backed intra-node-inter-VM communication. The performance improvements enabled by SR-IOV and IVShmem have contributed to their adoption by the HPC community. For example, the MVAPICH2 MPI library is able to take advantage of SR-IOV and IVShmem to deliver near-native performance for MPI applications [9,10,18].

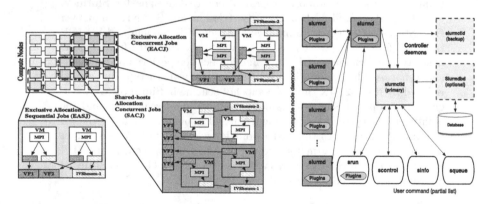

Fig. 1. Different scenarios of running MPI jobs over VMs on HPC cloud

Fig. 2. Slurm architecure

1.1 Motivation

For improved flexibility and resource utilization, it is important to manage and isolate virtualized resources of SR-IOV and IVShmem to support running multiple concurrent MPI jobs. As this requires knowledge of and some level of control over the underlying physical hosts, it is difficult to achieve this with the MPI library alone, which is only aware of the virtual nodes and resources inside. Thus, extracting the best performance from virtualized clusters require support from other middleware like job launchers and resource managers, which have a global view of the VMs and the underlying physical hosts. Figure 1 illustrates three possible scenarios of running MPI jobs over VMs in shared HPC clusters. **Exclusive Allocation for Sequential Jobs (EASJ):** Users exclusively allocate the physical nodes and add dedicated SR-IOV and IVShmem devices for each VM to sequentially run MPI jobs. This scenario requires co-resident VMs select different Virtual Functions, like VF1 and VF2, and add virtualized PCI devices mapping to the same IVShmem region, like IVShmem-1 as shown in Fig. 1. **Exclusive Allocation for Concurrent Jobs (EACJ):** Users get exclusive allocations, but multiple IVShmem devices, like IVShmem-1 and IVShmem-2 in Fig. 1 need to be added to each VM for multiple MPI jobs running concurrently. Because each MPI job at least needs one IVShmem device on one host to support Inter-VM shared memory based communications. **Shared-hosts Allocation for Concurrent Jobs (SACJ):** In shared HPC clusters, different users might allocate VMs on the same physical node. Each VM needs to have a dedicated SR-IOV virtual function, like VF1 to VF4. And IVShmem devices in different users' VMs need to point to different shared memory regions on the physical node, like IVShmem-1 and IVShmem-2 in Fig. 1.

Unfortunately, to the best of our knowledge, none of the currently available studies on resource managers such as Slurm [6,11,16] are SR-IOV and IVShmem aware. Therefore, they are not able to handle the above three scenarios of running MPI jobs. Moreover, one of the major contributors to the increasing popularity of virtual cluster computing is OpenStack [2]. It provides scalable and efficient mechanisms for creation, deployment, and reclamation of VMs on a large number of physical nodes. This offers us with further optimization opportunities - by integrating OpenStack with Slurm, which might be possible to drastically reduce the required interaction and turnaround time for a user attempting to utilize a virtualized cluster. To achieve the above goals, the following challenges need to be addressed:

- Can Slurm be extended to manage and isolate SR-IOV and IVShmem resources for running concurrent MPI jobs efficiently?
- What kind of design alternatives be proposed to achieve better deployment/ job launching times as well as application performance?
- Can Slurm and OpenStack be combined to provide a scalable solution for building efficient HPC clouds?
- Can MPI library running on the extended Slurm with SR-IOV and IVShmem support provides bare-metal performance for end HPC applications on different scenarios?

1.2 Contributions

To address the above challenges, this paper proposes a framework, called **Slurm-V**, which extends Slurm to manage and isolate SR-IOV and IVShmem resources for running MPI applications concurrently on virtual machines. In the proposed Slurm-V, three new components are introduced: VM Configuration Reader, VM Launcher and VM Reclaimer. To support these components, we propose three alternative designs: Task-based design, SPANK plugin-based design, and SPANK plugin over OpenStack-based design. We evaluate these designs from various aspects such as startup time, scalability, and application performance. Our evaluations show that our proposed Slurm-V framework has good deployment performance and scalability. With the proposed designs, VM startup time can be reduced by up to 2.64X through snapshot scheme in Slurm SPANK plugin. The sequential and concurrent MPI jobs can be efficiently executed on shared HPC clusters while maintaining minor overhead.

To the best of our knowledge, our proposed Slurm-V is the first attempt to extend Slurm for the support of running concurrent and sequential MPI jobs with isolated SR-IOV and IVShmem resources. The capabilities of Slurm-V can be used to build efficient HPC Clouds with SR-IOV and IVShmem.

2   Background

2.1   Slurm and SPANK

Simple Linux Utility for Resource Management (Slurm) [17] is an open-source resource manager for large scale Linux based clusters. Slurm can provide users with exclusive and/or shared access to cluster resources. As shown in Fig. 2, Slurm provides a framework including controller daemons (`slurmctld`), database daemon (`slurmdbd`), compute node daemons (`slurmd`), and a set of user commands (e.g. `srun`, `scontrol`, `squeue`) to start, execute and monitor jobs on a set of allocated nodes and manage a queue of pending jobs. Slurm Plug-in Architecture for Node and job (K)control (SPANK) [4] provides a generic interface to be used for dynamically modifying the job launch code. SPANK plugins have the ability to add user options when using `srun`. It may be built without accessing Slurm source code and will be automatically loaded at the next job launch. Thus, SPANK provides a low-cost and low-effort mechanism to change runtime behavior of Slurm.

2.2   SR-IOV and IVShmem

Single Root I/O Virtualization (SR-IOV) [3] is a new PCI Express technology, which specifies the native I/O virtualization capabilities in PCIe adapters. A single Physical Function (PF) can present itself as multiple Virtual Functions (VFs) through SR-IOV. Each VF can be passthroughed to a single VM. However, an efficient management mechanism is required to detect and select an exclusive VF for each VM. Inter-VM Shared Memory (IVShmem) (e.g. Nahanni) [15] provides zero-copy access to data residing on VM shared memory on the KVM platform. The host shared memory region is exposed to VM by serving as a virtualized PCI device in VM. Thus, shared memory based communication can be executed between processes in co-resident VMs. However, the difference from host shared memory is that IVShmem device does not support hierarchical file structure. To support multiple concurrent MPI jobs, multiple IVShmem devices need to be provided accordingly. Therefore, managing and isolating IVShmem devices among different concurrent MPI jobs is critical.

2.3 OpenStack

OpenStack [2] is an open-source middleware for cloud computing that controls large pools of computing, storage, and networking resources. It provides several components, such as Nova, Neutron, Glance, etc. to efficiently manage and quickly deploy cluster resources. OpenStack can work with many available virtualization technologies. It has been widely deployed in many private and public cloud environments.

3 Proposed Design

3.1 Architecture Overview of Slurm-V

This section presents an overview of Slurm-V framework. As we can see in Fig. 3, it is based on the original architecture of Slurm. It has a centralized manager, Slurmctld, to monitor work and resources. Each compute node has a Slurm daemon, which waits for the task, executes that task, returns status, and waits for more tasks [17]. Users can put their physical resource requests and computation tasks in a batch file, submit it by sbatch to the Slurm control daemon, Slurmctld. Slurmctld will respond with the requested physical resources according to its scheduling mechanism. Subsequently, the specified MPI jobs are executed on those physical resources.

In our framework Slurm-V, three new components are integrated into the current architecture. The first component is VM Configuration Reader, which extracts the related parameters for VM configuration. Each time when users request physical resources, they can specify the detailed VM configuration information, such as vcpu-per-vm, memory-per-vm, disk-size, vm-per-node, etc. In order to support high performance MPI communication, the user can also specify SR-IOV devices on those allocated nodes, and the number of IVShmem devices which is the number of concurrent MPI jobs they want to run inside VMs. The VM Configuration Reader will parse this information, and set them in the current Slurm job control environment. In this way, the tasks executed on those physical nodes are able to extract information from job control environment and take proper actions accordingly. The second component is the VM Launcher, which is mainly responsible for launching required VMs on each allocated physical node based on user-specified VM configuration. The zoom-in box in Fig. 3 lists the main functionalities of this component. If the user specifies the SR-IOV enabled device, this component detects those occupied VFs and selects a free one for each VM. It also loads user-specified VM image from the publicly accessible storage system, such as NFS or Lustre, to the local node. Then it generates

XML file and invokes libvirtd or OpenStack infrastructure to launch VM. During VM boot, the selected VF will be passthroughed to VM. If the user enables the IVShmem option, this component assigns a unique ID for each IVShmem device, and sequentially hotplugs them to VM. In this way, IVShmem devices can be isolated with each other, such that each concurrent MPI job will use a dedicated one for inter-VM shared memory based communication. On the aspect of network setting, each VM will be dynamically assigned an IP address from an outside DHCP server. Another important functionality is that the VM Launcher records and propagates the mapping records between local VM and its assigned IP address to all other VMs. Other functionalities include mounting global storage systems, etc. Once the MPI job reaches completion, the VM Reclaimer is executed. Its responsibilities include reclaiming VMs and the critical resources, such as unlocking the passthroughed VFs, returning them to VF pool, detaching IVShmem devices and reclaiming corresponding host shared memory regions.

If OpenStack infrastructure is deployed on the underlying layer, VM Launcher invokes OpenStack controller to accomplish VM configuration, launch and destruction.

3.2   Alternative Designs

We propose three alternative designs to effectively support the three components.

Task-based Design: The three new components are treated as three tasks/steps in a Slurm job. Therefore, the end-user needs to implement corresponding scripts and explicitly insert them in the job batch file. After the job being submitted, **srun** will execute these three tasks on allocated nodes.

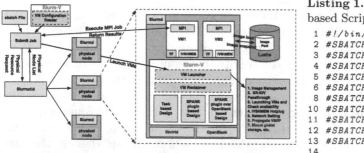

Fig. 3. Architecture overview of Slurm-V

Listing 1.1. SPANK Plugin-based Script

```
 1  #!/bin/bash
 2  #SBATCH -J Slurm-V
 3  #SBATCH -N 2
 4  #SBATCH -p All
 5  #SBATCH --vm-per-node=2
 6  #SBATCH --vcpu-per-vm=2
 8  #SBATCH --disk-size=10G
10  #SBATCH --sriov-ib=1
11  #SBATCH --ivshmem=1
12  #SBATCH --num-ivshmem=1
13  #SBATCH --ivshm-sz=128M
14
15  Slurm-V-run -np 8 a.out
```

The Task-based design is portable and easy to integrate with existing HPC environments without any change to Slurm architecture. However, it is not transparent to end users as they need to explicitly insert the three extra tasks in their jobs. More importantly, it may incur some permission and security issues. VF passthrough requires that VM Launcher connects to the libvirtd instance running with the privileged system account 'root', which in turn exposes security threats to the host system. In addition, the scripts implementation may be varied for different users. This will impact the deployment and application performance. To address these issues, we propose SPANK plugin-based design as discussed below.

SPANK Plugin-based Design: As introduced in Sect. 2.1, the SPANK plugin architecture allows a developer to dynamically extend functions during a Slurm job execution. Listing 1.1 presents an example of a SPANK plugin-based batch job in the Slurm-V framework. As we can see from line5-line13, the user can specify all VM configuration options as inherent ones preceded with #SBATCH. The `Slurm-V-run` on line15 is a launcher wrapper of `srun` for launching MPI jobs on VMs. Also, there is no need to insert extra tasks in this job script. Thus, it is more transparent to the end user compared to the Task-based design. Once the user submits the job using `sbatch` command, the SPANK plugin is loaded and the three components are invoked in different contexts.

Figure 4(a) illustrates the workflow of the SPANK plugin-based design in detail under the Slurm-V framework. Once the user submits the batch job request, SPANK plugin is loaded, and `spank_init` will first register all VM configuration options specified by the user and do a sanity checking for them locally before sending to the remote side. Then, `spank_init_post_opt` will set these options in the current job control environment so that they are visible to all Slurmd daemons on allocated nodes later. Slurmctld identifies requested resources, environment and queues the request in its priority-ordered queue. Once the resources are available, Slurmctld allocates resources to the job and contacts the first node in the allocation for starting user's job. The Slurmd on that node responds to the request, establishes the new environment, and initiates the user task specified by `srun` command in the launcher wrapper. `srun` connects to Slurmctld to request a job step and then passes the job step credential to Slurmds running on allocated nodes.

After exchanging the job step credential, SPANK plugin is loaded on each node. During this process, `spank_task_init_privileged` is invoked to execute VM Launcher component in order to setup VM for the following MPI job. `spank_task_exit` is responsible for executing VM Reclaimer component to tear down VMs and reclaim resources. In this design, we utilize the file-based lock mechanism to detect occupied VFs and exclusively allocate VFs from available VF pool. With this design, each IVShmem device will be assigned a unique ID and dynamically attached to VM. In this way, IVShmem devices can be efficiently isolated to support running multiple concurrent MPI jobs.

In this design, we utilize snapshot and the multi-threading mechanism to speed up the image transfer and VM launching, respectively. This will further reduce VM deployment time.

SPANK Plugin over OpenStack-based Design: This section discusses the design that combines SPANK plugin and OpenStack infrastructure. In this design, the VM Launcher and VM Reclaimer components will accomplish their functionalities by offloading the tasks to OpenStack infrastructure.

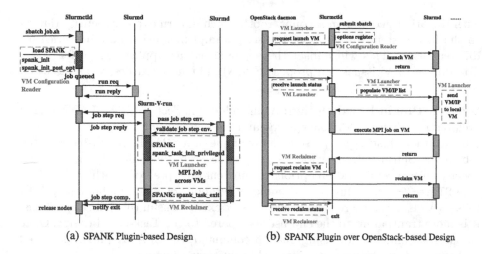

(a) SPANK Plugin-based Design   (b) SPANK Plugin over OpenStack-based Design

Fig. 4. SPANK Plugin-based and SPANK Plugin over OpenStack-based Design

Figure 4(b) presents the workflow of SPANK plugin over OpenStack. When the user submits a Slurm job, SPANK plugin is loaded first. VM configuration options are registered and parsed. The difference is that, on local context, VM Launcher will send a VM launch request to OpenStack daemon on its controller node. The core component of OpenStack, Nova, is responsible for launching VMs on all allocated compute nodes. Upon the launch completes, it returns a mapping list between all VM instance names and their IP addresses to VM Launcher. VM Launcher propagates this VM/IP list to all VMs. The MPI job will be executed after this. Once the result of MPI job is returned, VM Reclaimer in local context sends a VM destruction request to OpenStack daemon. Subsequently, VMs are torn down and associated resources are reclaimed in the way that OpenStack defines. In addition, our earlier work [18] describes in details about VF allocation/release and enabling IVShmem devices for VM under OpenStack framework. In this design, except VM Configuration Reader, the other two components work by sending requests to OpenStack controller and receiving its returning results. There are dedicated services in OpenStack infrastructure to manage and optimize different aspects of VM management, such as identification, image, networking. Therefore, the SPANK plugin over OpenStack-based design is more flexible and reliable.

4 Performance Evaluation

4.1 Experiment Setup

Cluster-A: This cluster has four physical nodes. Each node has dual 8-core 2.6 GHz Intel Xeon E5-2670 (Sandy Bridge) processors with 32 GB RAM and equipped with Mellanox ConnectX-3 FDR (56 Gbps) HCAs. **Chameleon:** [1]

It has eight physical nodes, each with 24 cores delivered in dual socket Intel Xeon E5-2670 v3 (Haswell) processors, 128 GB RAM and equipped with Mellanox ConnectX-3 FDR (56 Gbps) HCAs as well.

CentOS Linux 7 (Core) 3.10.0-229.el7.x86_64 is used as both host and guest OS. In addition, we use KVM as the Virtual Machine Monitor (VMM), and Mellanox OpenFabrics MLNX_OFED_LINUX-3.0-1.0.1 to provide the InfiniBand interface with SR-IOV support. Our Slurm-V framework is based on Slurm-14.11.8. MVAPICH2-Virt library is used to conduct application experiments.

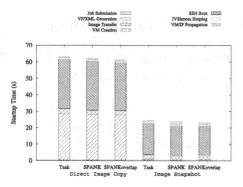

Fig. 5. VM launch breakdown on Cluster-A

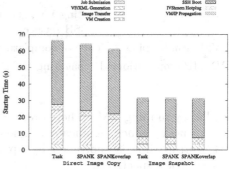

Fig. 6. VM launch breakdown on Chameleon

4.2 Startup Performance

To analyze and optimize the startup performance of the Slurm-V framework, we break down the whole VM startup process into several parts. Table 1 describes the time period of each part.

Overlapping: We found that image transfer is independent of VF/XML generation, so they can start simultaneously after submitting the job. As shown in Figs. 5 and 6, the time spent on direct image copy (2.2 GB) is larger than the time spending on VF selection and XML generation. So it can be completely overlapped. The overlapping effect can be clearly observed between SPANK and SPANKoverlap under direct image copy scheme on Chameleon.

Snapshot: We also observe that direct image copy takes a large proportion of the whole VM startup time for any startup methods on both Cluster-A and Chameleon. In order to shorten the time of image transfer, the external snapshot mechanism is applied. The original image file that user specified will be in a read-only saved state. The new file created using external snapshot will be the delta for the changes and take the original image as its backup file. All the changes from here onwards will be written to this delta file. Instead of transferring a large-size image file, we only create a small-size snapshot file for each VM, which clearly reduces the image transfer time. In addition, the backup file can be read in

Table 1. VM startup breakdown

Part	Time period description
Job submission	From submitting sbatch job to starting VM configuration
VF/XML generation	Reading VM configurations, selecting available VF to generate XML
Image transfer	Transferring VM image from public location to store location of each VM
VM creation	Time between invoking libvirt API to create VM and its return
SSH boot	Booting VM, getting available IP address until starting SSH service
IVShmem hotplug	Time of completing IVShmem hotplug operation
VM/IP propagation	Propagating VMs' hostname/IP records to all VMs

parallel by running VMs. Therefore, the snapshot mechanism enhances the VM startup performance significantly. The evaluation result shows that the whole VM startup time is shortened by up to 2.64X and 2.09X on Cluster-A and Chameleon, respectively.

Total VM Launch Time: We discussed the SPANK plugin over OpenStack-based design in Sect. 3.2. As VM Launcher offloads its task to OpenStack infrastructure as a whole task, we do not breakdown timings within the Open-Stack operations. The evaluation results show that the total VM launch times are 24.6 s, 23.8 s, and 20.2 s for SPANK plugin-based design, SPANK plugin-based design with overlap and SPANK plugin over OpenStack-based design, respectively. Compared to other designs, SPANK plugin over OpenStack has better total VM launch time, which is around 20 s. This is because OpenStack, as a well-developed and relatively mature framework, has integrated optimizations on different steps of VM launch.

4.3   Scalability

In this section, we evaluate the scalability of proposed Slurm-V framework using single-threading (ST) and multi-threading (MT) schemes. In the evaluation, snapshot with overlapping is used for both schemes. In MT case, each thread is responsible for launching one VM. From Figs. 7 and 8, it can be observed that MT scheme significantly improves the VM startup performance, compared to ST scheme on both Cluster-A and Chameleon. For instance, to launch 32 VMs across 4 nodes on Chameleon, ST scheme takes 260.11 s, while MT only spends 34.88 s. Compared with ST scheme, MT scheme reduces the VM startup time by up to 86 % and 87 % on Cluster-A and Chameleon, respectively. As the number of physical nodes increases, we do not see the clear increase for startup time of MT scheme. These results indicate that our proposed Slurm-V framework scales well.

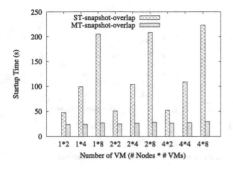

Fig. 7. Scalability study on Cluster-A

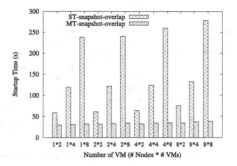

Fig. 8. Scalability study on Chameleon

4.4 Application Performance

The Slurm-V framework extends Slurm to manage and isolate virtualized resources of SR-IOV and IVShmem to support running multiple concurrent MPI jobs under different scenarios. In this section, we evaluate the Graph500 performance under three scenarios (EASJ, EACJ, and SACJ) as indicated in Sect. 1.1 with 64 processes across 8 nodes on Chameleon. Each VM is configured with 6 cores and 10 GB RAM.

For **EASJ**, two VMs are launched on each node. Figure 9(a) shows the Graph500 performance with 64 processes on 16 VMs in this scenario. The evaluation results indicate that the VM launched by Slurm-V with SR-IOV and IVShmem support can deliver near-native performance, with less than 4 % overhead. This is because the Slurm-V framework is able to efficiently isolate SR-IOV VFs and enable IVShmem device across co-resident VMs. Co-resident VMs can execute shared memory based communication through IVShmem device. On the other hand, each VM with the dedicated VF can achieve near-native inter-node communication performance. For **SACJ**, four VMs VM(0–3) are launched on each node. Graph500 is executed across all VM(0–1), while the second MPI job is executed across all VM(2–3) simultaneously. We run NAS as the second MPI jobs. For the native case, we use 8 cores corresponding to VM(0–1) to run Graph500, while another 8 cores corresponding to VM(2–3) to run the second job. As shown in Fig. 9(b), the execution time of Graph500 on VM is similar with the native case with around 6 % overhead. This indicates that the Slurm-V framework is able to efficiently manage and isolate the virtual resource of SR-IOV and IVShmem on both VM and user level, although in the shared allocation. One dedicated VF is passthroughed to each VM and one unique IVShmem device is attached to all co-resident VMs of each user. For **EACJ**, similarly, our Slurm-V framework can also deliver the near-native performance, with around 8 % overhead, as shown in Fig. 9(c). The Slurm-V framework supports the management and isolation of IVShmem on MPI job level, so each MPI job can have a unique IVShmem device to execute shared memory backend communication across the co-resident VMs.

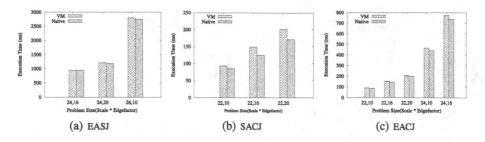

Fig. 9. Graph500 performance with 64 processes on different scenarios

From these application studies, we see that VMs deployed by Slurm-V with appropriately managed and isolated SR-IOV and IVShmem resources are able to deliver high performance for concurrent MPI jobs, which can be seen as promising results for running applications on shared HPC clouds.

5   Related Work

For building cloud computing environments with Slurm, Jacobsen et al. [11] present 'shifter' tightly integrated into Slurm for managing Docker and other user-defined images. Ismael [6] uses VM for dynamic fractional resource management and load balancing in a batch cluster environment. Markwardt et al. [16] propose a solution to run VMs in a Slurm-based batch system. They use a VM scheduler to keep track of the status of Slurm queue on the VMs. For building HPC cloud environments, studies [7,8,13] with Xen demonstrate the ability to achieve near-native performance in VM-based environment for HPC applications. Ruivo et al. [5] explore the potential use of SR-IOV on InfiniBand in an Open Nebula cloud towards the efficient support of MPI-based workloads. Our previous evaluation [10] has revealed that IVShmem can significantly improve intra-node inter-VM communication on SR-IOV enabled InfiniBand clusters. Further, we redesigned MVAPICH2 library [9] to take advantage of this feature and proposed an efficient approach [18] to build HPC clouds by extending OpenStack with redesigned MVAPICH2 library. However, none of these has discussed how to effectively manage and isolate IVShmem and SR-IOV resources in shared HPC cluster under Slurm framework in order to support running MPI jobs in different scenarios as indicated in Sect. 1.1. The initial idea of this work had been presented in Slurm Forum [14], and we further complete the whole Slurm-V design, implementation, and evaluation in this paper.

6   Conclusion and Future Work

In this paper, we proposed a novel Slurm-V framework to efficient support running multiple concurrent MPI jobs with SR-IOV and IVShmem in shared HPC clusters. The proposed framework extends Slurm architecture and introduces three new components: VM Configuration Reader, VM Launcher, and VM Reclaimer. We present three alternative designs to support these components, which are: Task-based design, SPANK plugin-based design and SPANK plugin over OpenStack-based design. We evaluate our Slurm-V framework from different aspects including startup performance, scalability and application performance under different scenarios. The evaluation results indicate that the VM startup time can be reduced by up to 2.64X by using snapshot scheme. Compared with the single-threading scheme, multi-threading scheme reduces the VM startup time by up to 87%. In addition, Slurm-V framework shows good scalability and is able to support running multiple MPI jobs under different scenarios on HPC clouds. In the future, we plan to explore other alternative SPANK-based designs to further extend Slurm framework to have more virtualization support.

References

1. Chameleon. http://chameleoncloud.org/
2. OpenStack. http://openstack.org/
3. PCI-SIG Single-Root I/O Virtualization Specification. http://www.pcisig.com/specifications/iov/
4. SPANK - Slurm Plug-in Architecture for Node and job (K)control. http://slurm.schedmd.com/spank.html
5. De Lacerda Ruivo, T., Altayo, G., Garzoglio, G., Timm, S., Kim, H.W., Noh, S.Y., Raicu, I.: Exploring infiniband hardware virtualization in OpenNebula towards efficient high-performance computing. In: 2014 14th IEEE/ACM International Symposium on Cluster, Cloud and Grid Computing (CCGrid), pp. 943–948 (2014)
6. Estrada, I.F.: Overview of a Virtual Cluster using OpenNebula and SLURM
7. Huang, W., Koop, M.J., Gao, Q., Panda, D.K.: Virtual machine aware communication libraries for high performance computing. In: Proceedings of the 2007 ACM/IEEE Conference on Supercomputing, SC 2007, pp. 9: 1–9: 12. ACM, New York (2007)
8. Huang, W., Liu, J., Abali, B., Panda, D.K.: A case for high performance computing with virtual machines. In: Proceedings of the 20th Annual International Conference on Supercomputing, ICS 2006, New York, NY, USA (2006)
9. Zhang, J., Lu, X., Jose, J., Li, M., Shi, R., Panda, D.K.: High performance MPI library over SR-IOV enabled InfiniBand clusters. In: Proceedings of International Conference on High Performance Computing (HiPC), Goa, India (2014)
10. Zhang, J., Lu, X., Jose, J., Shi, R., Panda, D.K.: Can Inter-VM shmem benefit MPI applications on SR-IOV based virtualized InfiniBand clusters? In: Silva, F., Dutra, I., Santos Costa, V. (eds.) Euro-Par 2014. LNCS, vol. 8632, pp. 342–353. Springer, Heidelberg (2014)
11. Jacobsen, D., Botts, J., Canon, S.: Never Port Your Code Again Docker functionality with Shifter using SLURM. http://slurm.schedmd.com/SLUG15/shifter.pdf
12. Jose, J., Li, M., Lu, X., Kandalla, K., Arnold, M., Panda, D.K.: SR-IOV support for virtualization on infiniband clusters: early experience. In: On 13th IEEE/ACM International Symposium Cluster, Cloud and Grid Computing (CCGrid), pp. 385–392 (2013)

13. Lu, X., Lin, J., Zha, L., Xu, Z.: Vega LingCloud: a resource single leasing point system to support heterogeneous application modes on shared infrastructure. In: Proceedings of the 2011 IEEE Ninth International Symposium on Parallel and Distributed Processing with Applications, ISPA 2011, pp. 99–106. IEEE Computer Society, Washington, DC (2011)

14. Lu, X., Zhang, J., Chakraborty, S., Subramoni, H., Arnold, M., Perkins, J., Panda, D.K.: Supporting SR-IOV and IVSHMEM in MVAPICH2 on Slurm: Challenges and Benefits. http://slurm.schedmd.com/SLUG15/mv2_virt_slug_luxi_osu.pdf

15. Macdonell, A.C.: Shared-Memory Optimizations for Virtual Machines. Ph.D. Thesis. University of Alberta, Edmonton, Alberta, Fall 2011

16. Markwardt, U., Jurenz, M., Rotscher, D., Muller-Pfefferkorn, R., Jakel, R., Wesarg, B.: Running Virtual Machines in a Slurm Batch System. http://slurm.schedmd.com/SLUG15/SlurmVM.pdf

17. Yoo, A.B., Jette, M.A., Grondona, M.: SLURM: simple linux utility for resource management. In: Feitelson, D.G., Rudolph, L., Schwiegelshohn, U. (eds.) JSSPP 2003. LNCS, vol. 2862, pp. 44–60. Springer, Heidelberg (2003)

18. Zhang, J., Lu, X., Arnold, M., Panda, D.K.: MVAPICH2 over OpenStack with SR-IOV: an efficient approach to build HPC clouds. In: 2015 15th IEEE/ACM International Symposium on Cluster, Cloud and Grid Computing (CCGrid), pp. 71–80 (2015)

An Autonomic Parallel Strategy for the Projection of Ecological Niche Models in Heterogeneous Computational Environments

Fernanda G.O. Passos[(✉)] and Vinod E.F. Rebello

Instituto de Computação – Universidade Federal Fluminense (UFF),
Niterói, RJ, Brazil
{fernanda,vinod}@ic.uff.br

Abstract. Ecological Niche Modelling (ENM) is an important process to help ecologists understand and predict the potential geographic distribution of species. In addition to creating correlative models for each species, the projection of the model onto a geographical environment is an essential step in the process to visualize suitable habitats. Given the demand for improved precision and the need to address wider geospatial domains, using larger data sets means that these methods incur increasingly higher processing, memory and I/O demands. This paper proposes a new parallel algorithm for the projection stage of a popular ENM tool. Although the characteristics of ENM already allow this tool to make use of heterogeneous computing environments, a new algorithm has been designed to be autonomic and capable of reconfiguration to respond better to the resource capacities available. The proposal has been compared with the default sequential implementation and a parallel MPI version, both distributed with the ENM tool. An empirical analysis reveals gains from 109 % to 742 % in terms of performance and improved scalability with efficiencies above 81 % in evaluations with up to 128 processors.

1 Introduction

An ecological niche can be defined as the set of environmental conditions for a species to survive and maintain viable populations over time [4]. Ecological Niche Modelling (ENM) is a common procedure most often used in macroecology and biogeography to determine the geographical extension of species distributions. These correlative ecological niche models are generated by relating locations where a given species is known to occur with the environmental conditions at that locale that might influence their distribution [17]. ENM provides a powerful mechanism to predict potential species distribution in distinct geographical and temporal contexts, as well as to study another aspects of evolutionary biology and ecology. ENM has been widely used in various situations such as: searching for rare or endangered species; identifying suitable areas for the (re-)introduction of species; forecasting the impact of climate change on biodiversity; helping in conservation planning and delimitation and evaluation of protected areas; preventing the spread of invasive species; identifying geographical and ecological

© Springer International Publishing Switzerland 2016
P.-F. Dutot and D. Trystram (Eds.): Euro-Par 2016, LNCS 9833, pp. 363–375, 2016.
DOI: 10.1007/978-3-319-43659-3_27

aspects of disease transmission; guiding biodiversity field surveys; among other important applications [12].

ENM applications combine information about the occurrence of species (biotic) with environmental databases (abiotic) in the form of geo-referenced raster layers (such as temperature, rainfall and salinity) to generate potential distribution models. This process includes a combination of 3 dependent steps: model creation, testing and projection. The models are usually generated by statistical techniques, such as maximum entropy, or by machine learning techniques such as artificial neural networks [12]. One of the most widely adopted ENM tools is *openModeller* [7], which offers a choice of 15 modelling algorithms. Although this tool has been adopted by various large scale e-infrastructure projects [1,2,5] to provide ENM services in the cloud, the software tool was designed principally for a single server.

This paper addresses the parallelisation of the costly stage that projects the ecological niche model into the chosen geospatial domain given a set of environmental conditions at the time of interest. *openModeller* is distributed with a default sequential algorithm and an optional parallel version for model projection, but neither implementation takes into sufficient consideration the possibility of having to run on heterogeneous resources or in shared dynamic environments, like clouds. The goal of this paper is to propose an adaptive parallel algorithm for ENM projection that it is able to manage its own execution in any one of these three common types of environments. This algorithm has been integrated with the EasyGrid application management system (EasyGrid AMS) [10] to harness the available environment more efficiently by making the projection autonomic.

The paper is structured as follows. Section 2 presents the openModeller tool and describes the projection of an ecological niche model, as well as the default sequential implementation adopted in openModeller. The existing parallel MPI version is explained in Sect. 3. In Sect. 4, the presentation of the EasyGrid AMS and its programming model is followed by that of the proposed new autonomic algorithm for ENM projection. Section 5 supports the proposal through experimental evaluations, with some conclusions being drawn in Sect. 6.

2   OpenModeller and ENM Projection

OpenModeller (OM), an ENM tool developed by a Brazilian Reference Centre for Environmental Information (*Centro de Referência em Informação Ambiental* – CRIA) together with national and international partners [7], is widely used in the biogeographical and ecological research communities [3,13,18]. OM produces species potential distribution models and includes mechanisms: to read environmental data and species occurrence points; to select environmental layers on which the model may be based; to create a fundamental niche model; project models in an environmental scenario and produce detailed graphical images in several formats.

Several algorithms are available as plug-ins for model creation [7] including: BIOCLIM (Bioclimatic envelopes), GARP (Genetic Algorithm for Rule-set Production), GARP_BS (GARP with Best Subsets), DG_GARP (Desktop GARP), DG_GARP_BS (Desktop GARP_BS), ENVSCORE (Envelope Score), ENVDIST (Environmental Distance), RF (Random Forests), MAXENT (Maximum Entropy), NICHE_MOSAIC (Niche Mosaic), and SVM (Support Vector Machines).

Effort has been invested in distributing ENM workflows consisting of hundreds or more instances of dependent tasks to create, test and project each model [1,2,6]. However, since ENM projection often involves large amounts of data, our work seeks additional gains by exploring parallelism in individual projection tasks. The OM tool presents an optional MPI implementation for the ENM projection [11] using the traditional parallel programming model described in Sect. 3.

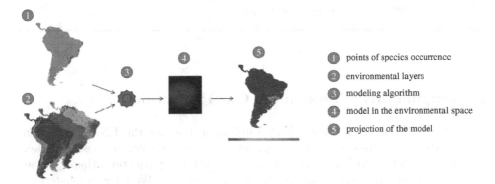

1. points of species occurrence
2. environmental layers
3. modeling algorithm
4. model in the environmental space
5. projection of the model

Fig. 1. Example of modelling and projection of an ecological niche [11]. (Color figure online)

OM project is the OpenModeller mechanism to project the distribution models and generate a high resolution geographical 2D rectangular image of a region bounded by coordinates (a, b) and (c, d) where $a < c$ and $b < d$. A geospatial mask can be selected by the user to project a model in to arbitrary areas within the defined region. The value of each coordinate point represents the probability of the environmental conditions at that locale being hospitable for a given species. Figure 1 shows the modelling and projection of an ecological niche, where item 5 indicates the result of the projection of a model. The colour scale that varies from blue to red represents the suitability of a region (red means high and blue means low). The scientist is responsible for selecting the environmental data layers for which the prediction should be based.

Algorithm 1 briefly presents the sequential implementation of a model projection of a ecological niche. The algorithm inputs are the model, generated previously in a modelling stage by a chosen OM algorithm described in Sect. 2, and the environmental data for the region of interest. The output image contains the map with the predicted probabilities. The following steps are executed by the algorithm for each point (x, y) where $a \leq x \leq c$ and $b \leq y \leq d$: Line 3 obtains the necessary environmental layer data for (x, y); Line 4 applies the model to the coordinate (x, y) considering the environmental data and calculates a probability p; finally, Line 5 converts p to a image value (RGB) and writes it in the file at position (x, y).

Algorithm 1. The sequential algorithm for ENM projection.

Input: *model* - model previously generated by an OM algorithm.
env_data - environmental data.
Output: *map* - file which contain the projection.

```
1  for x ← a to c − 1 do
2      for y ← b to d − 1 do
3          env ← env_data at position (x, y)
4          apply model with env and put the value in p
5          write p in map at position (x, y)
```

---

3 Original MPI Version for OM Projection

The OM tool, currently, has a MPI implementation for the ENM projection. This MPI version uses a parallel programming model similar to master-worker. In addition to the worker processes that compute the projection, there are two master processes instead of one to improve performance. We refer to *masterD* as the process that distributes on-demand ranges of coordinates for each worker to evaluate and *masterR* the process that receives the partial solutions from each worker and writes them to the output image file.

The *masterD* algorithm partitions the total number of points to be projected into fixed-size blocks (the default is 30,000 points). The worker processes request blocks from the *masterD* and compute the species distribution potential for each coordinate within the block. The result is a partial projection block that must be sent to *masterR* which in turn must aggregate the projected blocks and write them to the output file.

4 Autonomic ENM Projection with EasyGrid AMS

The EasyGrid Application Management System (AMS) [10] is an example of an application-centric middleware that is responsible for managing an application's execution on the computational resources available. The EasyGrid philosophy promotes the idea that an AMS should employ autonomic management strategies tuned for each application instead of single resource-centric management approach if one wishes to improve efficiency and performance.

The EasyGrid AMS offers some autonomic features, with different implementations strategies for different classes of MPI applications: a three-level hierarchical dynamic scheduler [8], providing *self-optimization*, and; a fault-tolerant mechanism capable of detecting and recovering application failures [16], providing *self-healing*. *Self-configuration* allows an application to change its configuration dynamically during execution. The implementation of this feature is heavily dependent on characteristics of the application that is to become autonomic. Previous studies with other MPI applications have shown more than satisfactory results using this middleware [9,14–16].

Key to this success is the adoption of an alternative programming model, 1Ptask [9], to the traditional MPI one (referred to here as 1Pproc). The 1Pproc model considers a single process per processor and thus each process is relatively coarse-grained. The 1Ptask model considers each process to be a finer grained task and consequently, the total number of processes tends to be relatively larger than the number of processors. While more processes might mean higher overheads, it also provides greater flexibility to improve load balancing, for example.

4.1   A Self-Configuring Projection Implementation

The original MPI version suffers from two drawbacks: (1) the algorithm is centralized and the messages sent and received by the two masters create a bottleneck, impairing performance; (2) the programming model 1Pproc may not be the best alternative for large-scale distributed environments. Previous work has shown the 1Ptask model to be more appropriate for multi-core servers environments that are typically dynamic, heterogeneous and/or are shared, such as those commonly used in grid and cloud computing. The proposal of an autonomic algorithm with the EasyGrid AMS necessitates: the use of the 1Ptask model; the definition of application tasks that can dynamically self-configure, and; the use of the EasyGrid AMS for the autonomic management of the tasks.

Another issue related to the programming model is the division of the ENM projection domain in fixed-size blocks as presented in the original MPI version. Figure 2 shows the variation in execution times for each block (one map line) of the ENM projection using the BIOCLIM algorithm. One can see a reasonable variation in the execution times, in this case, with an average of 0.28 s and standard deviation of 0.15. The longest block is about to 32 times slower than the fastest. Although the blocks have a fixed size, their processing costs can be different. Furthermore, for example in the case of the first and last lines of the map domain, the processing times can be so low that the communication times may exceed the processing time, causing further losses in performance.

Given that one cannot assume that the processing capacity is the same for all available resources and, from Fig. 2, the task granularities for fixed-sized blocks are not going to be similar amongst them, the proposed EasyGrid AMS version adopts variable block sizes that changes dynamically over time. At first, the workload is divided equally to tasks across all processors cores independently of their processing capacities. Under the 1Ptask programming model, each task is implemented as an MPI process. In this application, the block size is set initially

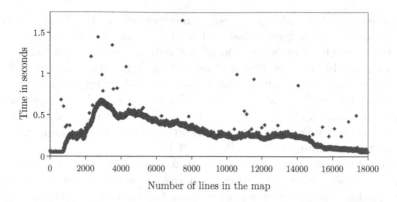

Fig. 2. Execution times of each line of the ENM projection.

to the total number of map lines divided by the number of processor cores. During execution, the tasks calculate the projection of the block in the same way as the sequential algorithm, but with one subtle difference, the task execution time is limited by a timer *timeout*. Should the task's current processing time exceed the *timeout* value, a number of new tasks are created to continue the projection of the remaining portion of the block and the original task then terminates. For each level of task creation (assume the initial tasks to be in level 0, their children belong to level 1 and so on recursively), the task's *timeout* duration is reduced. Effectively, the block size allocated to a task changes dynamically, depending on the resource location and the application's progress. One goal is for later tasks to become finer in granularity so that the EasyGrid AMS can distribute them among the available processors more efficiently [8].

The Algorithm 2 highlights the execution steps of each task of the new ENM projection, called *partition*. As input, this algorithm receives the model, the environmental data, the Y dimension of the original map, the block and the *timeout* value. Each task has its output identified by map_{l_i}, where l_i represents the initial line of the block. The projection of the allocated block is processed line by line (Lines 1) with Lines 2 to 5 evaluating the coordinates in a given block line in the same way as Algorithm 1. Now, in the Line 6, after completing the projection of a map line, the timer *timeout* is verified. If the time limit has been exceeded, the number of new tasks *ntasks* to be created is estimated, in Line 6, based on the processing rate during the last time period. In a total of Dim_x map lines, if x is the current line map that was calculated in *timeout* seconds, to calculate the remaining $Dim_x - x$ we need $\frac{(Dim_x - x)}{x}$ tasks. Line 8 determines a shorter *timeout* value for the new tasks while in Lines 9 to 11, the remaining unprocessed portion of the block is split and sent to the *ntasks* tasks. The new *timeout* value is the quotient of old *timeout* (starting from 10 s) divided by the level plus 1. If the value is lower than 1, the new *timeout* is 1 s. Unfortunately by the end of this process, the parallelisation of ENM projection means that the image is made up of several sub-maps each stored in a separate output file (one

---

Algorithm 2. Algorithm for the *partition* tasks of the ENM projection with EasyGrid AMS.

Input: *model* - model previously generated by an OM algorithm.
env_data - environmental data.
Dim_y - dimension Y of the complete map.
(l_i, l_j) - block to be calculated.
timeout - initial timer.
Output: map_{l_i} - partial maps.

```
1   for x ← l_i to l_j do
2       forall the y ← 0 to Dim_y do
3           env ← env_data at position (x, y)
4           Apply model in env and put the value in p
5           Write p in map_{l_i} at position (x − l_i, y)
6       if timeout is over then
7           Calculate ntasks
8           Calculate the new timeout
9           Split remaining block (x + 1, l_j) into ntasks and put each sub-block in S
10          forall the k ∈ S do
11              Create task with arguments model, env_data, Dim_y, k and timeout
```

---

per task). Thus, a *merge* task is required to combine the files and generate a final image containing the projection map. At present, this is implemented using a sequential algorithm.

5   Experimental Analysis

Three sets of experiments were carried out. The first aims to compare the two parallel implementations (the original MPI version that distributed fixed block sizes on demand with the proposed autonomic version with EasyGrid AMS) using a variety of OM model algorithms at small scales (with up to 24 CPUs). The second aims to evaluate the scalability of the proposal at a larger scale, obviously. The final experiment evaluates the efficiency of the proposal in a dynamic heterogeneous environment where external loads were introduced periodically. The last two experiments used a cluster of 16 8-core processors each, totalling 128 CPUs.

5.1   Experiment 1: Small Scale Performance

In this experiment, p initial worker processes are considered for both projection implementations running on p CPUs. For the original MPI version, there are actually $p + 2$ processes in total (plus 2 masters). The projection modelling algorithms BIOCLIM, ENVSCORE, GARP, DG_GARP, DG_GARP_BS, MAXENT, GARP_BS, SVM, RF and NICHE_MOSAIC are selected for this experiment.

Table 1. Speed-up obtained by the original MPI version and the autonomic version.

Algorithm	Version	Speed-up				
		4	8	12	16	24
BIOCLIM	Original MPI	2.58	4.29	5.54	5.45	5.70
	EasyGrid	3.67	7.15	9.96	13.03	19.21
ENVSCORE	Original MPI	2.65	4.20	5.49	4.86	6.18
	EasyGrid	3.32	6.48	9.35	12.67	18.28
GARP	Original MPI	3.12	4.90	6.85	8.25	8.03
	EasyGrid	3.85	7.57	11.21	14.59	21.58
DG_GARP	Original MPI	2.90	4.74	6.17	7.59	7.49
	EasyGrid	3.67	7.11	10.01	13.10	19.73
DG_GARP_BS	Original MPI	3.11	5.80	7.26	9.68	10.49
	EasyGrid	3.78	7.65	10.87	14.32	21.48
MAXENT	Original MPI	3.46	6.07	7.84	10.38	11.45
	EasyGrid	3.88	7.62	10.99	14.61	21.70
GARP_BS	Original MPI	2.98	5.42	7.50	9.46	10.33
	EasyGrid	3.85	7.60	10.96	14.52	21.77
SVM	Original MPI	3.32	5.63	8.01	10.54	12.59
	EasyGrid	3.90	7.71	11.17	14.85	22.24
RF	Original MPI	3.28	4.99	7.97	10.15	11.30
	EasyGrid	3.59	7.14	10.35	13.78	20.58
NICHE_MOSAIC	Original MPI	3.28	5.68	7.70	7.49	4.63
	EasyGrid	3.94	7.81	10.10	11.48	12.74

Table 2. Execution Time in seconds for *partition* and *merge* operations, when using the ENVDIST and SVM modelling algorithms.

	ENVDIST			SVM		
p	*partition*	*merge*	speed-up	*partition*	*merge*	speed-up
24	70,758.88	75.93	22.85	1,186.28	41.18	24.05
48	35,316.90	85.26	45.72	614.53	59.50	43.80
96	17,758.27	88.44	90.69	313.29	47.33	81.86
128	13,390.62	98.59	119.99	238.36	46.59	103.60

Table 1 presents a comparison of the speed-ups, in relation to sequential algorithm, obtained by original MPI approach and autonomic approach with Easy-Grid AMS for the ENM projection. Using different OM modelling algorithms and varying the number of processors (from 4 to 24), the autonomic version presented significantly better speed-ups and, for the most of modelling algorithms, the value was close to the number of processors used.

Figure 3 provides a better visualisation of the improvements in speed-up. Each bar group with labels 4, 8, 12, 16 and 24 processors represents the gain (in percent) of the autonomic EasyGrid AMS version over the original MPI one. As seen in Table 1, the autonomic version always achieved higher speed-ups. For most algorithms, as the number of processors increases, so does the gain in relation to the original MPI version. For BIOCLIM, with 24 processors, the parallel projection with EasyGrid AMS is more than 3 times better than the existing solution.

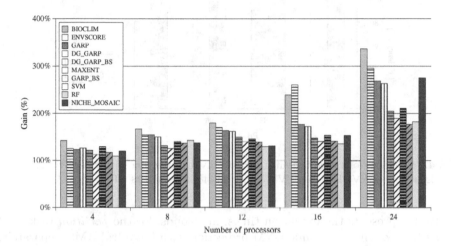

Fig. 3. Performance gain of the autonomic approach over the original MPI versions.

Scalability is a weakness of the original MPI version. For some algorithms, the speed-up values continue to increase as the number of processors grows, while for others, already with 24 CPUs the speed-ups taper off or fall (as in the case of DG_GARP and NICHE_MOSAIC). To obtain good scalability, algorithm designers should aim for near linear speed-up, especially at relatively lower processor counts. In the case of the EasyGrid AMS version, while this property is better it is not perfect, for example, in the case of NICHE_MOSAIC, but in part due to the overhead of the unparallelised merge operation.

5.2 Experiment 2: Larger Scale Homogeneous Performance

From the first experiment, the better performance of the autonomic parallel ENM version with the EasyGrid AMS over the original MPI one was not only clear but also indicated that the former approach was more scalable. This second experiment aims analyse further the scalability of the EasyGrid AMS version. The two modelling algorithms with the longest execution times were selected: ENVDIST and SVM. The sequential implementation of ENM Projection with ENVDIST, for example, has an execution time of almost 20 days. The number of processors varied from 24, 48, 96 to 128 CPUs.

The number of tasks created by the autonomic version during execution can be quite high, relatively speaking. The total amount of tasks for the instance SVM is approximately 5,400 for all CPUs numbers used, with low variance. For the instance ENVDIST, the number is approximately 16,000 for all numbers of CPUs used, again with low variance. For both, the number of tasks keeps more or less static as the processors increase. For this reason, the merge time has a low variability as approximately the same number of sub-maps are merged.

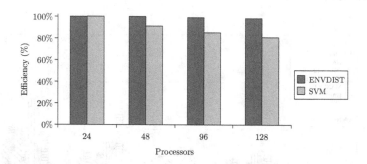

Fig. 4. Efficiency of the autonomic parallel ENM projection with EasyGrid AMS for the algorithms ENVDIST and SVM using 24, 48, 96 and 128 CPUs.

Table 2 presents the execution times, in seconds, for the *partition* tasks and *merge* tasks of the autonomic ENM projection with EasyGrid AMS, separately, for both the ENVDIST and SVM model algorithms. There is also a column that represents the speed-up of the combined *partition* plus *merge* execution. We consider the total execution time to be the sum of the two but are interested in the *partition* execution since it dwarfs the *merge* time. The sequential *merge* algorithm is only used to concatenate the images generated by the *partition* tasks but as in the case of Amdahl's Law will eventually limit the speedup.

The efficiency of the proposal for the 8 ENM projections (i.e. the partition and merge tasks) is indicated in Fig. 4.

The self-configuring and self-optimisation abilities of the autonomic version combined allow significantly better performance and good scalability for the ENM projections with costly OM model algorithms. The self-configuring mechanism ensures an appropriate degree of parallelism by dynamically creating progressively finer grained tasks taking into consideration the application's behaviour. This over-provisioning of tasks is then managed by the self-optimisation through the dynamic scheduler of the EasyGrid AMS. Together, these abilities explore the characteristics of the ENM projection and the execution environment, achieving acceptable efficiency levels for large scale computing.

5.3   Experiment 3: Larger Scale Heterogeneous Performance

The aim of this experiment is to briefly highlight the difference in performance of the original MPI and EasyGrid AMS approaches in a heterogeneous or dynamic

Table 3. Results with a simulated shared environment.

	Original MPI version			EasyGrid AMS version		
Algorithm	Dedicated	With load	Slowdown (%)	Dedicated	With load	Slowdown (%)
SVM	1,156.40	2,446.02	111.5	238.36	329.66	38.30
ENVDIST	24,611.82	46,490.34	88.89	13,390.62	17,840.64	33.23

shared environment. Using the previous dedicated homogeneous 128-core cluster, CPU-intensive loads were introduced to compete with the application. These independent externals loads are not constant – they are effectively active for intercalated periods of processing (20 s) and idleness (20 s) on each CPU-core. Thus, the full capacity of the cluster is available to the OM projection applications for half the time, the rest of the time they will compete, obtaining only half the capacity.

Table 3 presents execution times in seconds for the SVM and ENVDIST modelling algorithms again. For each approach, the execution times in a dedicated environment and in this shared environment were measured. The column *Slowdown* indicates the relative loss in performance due to resource sharing (or dynamic heterogeneity), *i.e.*, how much longer the applications take to execute. These results indicate significantly better performance is obtained by the autonomic version in a shared configuration. Furthermore, while this version is 4.85 and 1.84 times faster in a static homogeneous environment, this shoots up to 7.42 and 2.61 times in this dynamic heterogeneous one, for OM projection with the SVM and ENVDIST modelling algorithms, respectively.

6 Conclusion

This paper proposed a new autonomic approach for the projection of the ecological niches in geospatial domains. This kind of application appears to be easily parallelisable since data processing can be decomposed into independent tasks. However traditional approaches may not be as efficient as expected in current execution environments due to workload granularities of unknown size.

The current MPI approach distributed with the biodiversity tool *openModeller* is based on the master-worker model and uses fixed-block partitioning allowing on-demand load balancing between tasks executing under a 1Pproc model. The new autonomic approach proposes the use of the 1Ptask model with a dynamic partition that generates tasks with variable block sizes. This partitioning is achieved by a simple self-configuring strategy that takes into consideration the execution behaviour. Tasks are created to process blocks on the fly, each being managed by the EasyGrid AMS, which provides self-optimisation.

Results showed that the proposed algorithm presented better performance and scalability in all experiments when compared to the original traditional algorithm. With 24 processors, speed-ups with the EasyGrid AMS version are between 109 % to 337 % higher than those of the original version. The proposed approach scales up to 128 processors, obtaining efficiency values of at least 81 % even though sub-maps are still merged sequentially. Future work includes substituting this with a parallel version. Finally, in heterogeneous or shared environments, the improvements in performance are even greater.

References

1. Amaral, R., Badia, R.M., Blanquer, I., Braga-Neto, R., Candela, L., Castelli, D., Flann, C., De Giovanni, R., Gray, W.A., Jones, A., et al.: Supporting biodiversity studies with the EUBrazilOpenBio hybrid data infrastructure. Concurrency Comput. Pract. Experience **27**(2), 376–394 (2015)
2. EUBrazil Cloud Connect Project: EUBrazil Cloud Connect (2016). http://www.eubrazilcloudconnect.eu. Accessed Feb 2016
3. Geller, G.N., Melton, F.: Looking forward: applying an ecological model web to assess impacts of climate change. Biodiversity **9**(3–4), 79–83 (2008)
4. Grinnell, J.: Field tests of theories concerning distributional control. Am. Nat. **51**(602), 115–128 (1917)
5. Leidenberger, S., De Giovanni, R., Kulawik, R., Williams, A.R., Bourlat, S.J.: Mapping present and future potential distribution patterns for a meso-grazer guild in the baltic sea. J. Biogeogr. **42**(2), 241–254 (2015)
6. Lezzi, D., Rafanell, R., Torres, E., Giovanni, R., Blanquer, I., Badia, R.: Programming ecological niche modeling workflows in the cloud. In: 27th International Conference on Advanced Information Networking and Applications Workshops (WAINA), pp. 1223–1228, March 2013
7. Muñoz, M.E.S., De Giovanni, R., Siqueira, M.F., Sutton, T., Brewer, P., Pereira, R.S., Canhos, D.A.L., Canhos, V.P.: Openmodeller: a generic approach to species' potential distribution modelling. GeoInformatica **15**(1), 111–135 (2011)
8. Nascimento, A., Sena, A., Boeres, C., Rebello, V.E.F.: Distributed and dynamic self-scheduling of parallel MPI grid applications. Concurrency Comput. Pract. Experience **19**(14), 1955–1974 (2007)
9. Nascimento, A., Sena, A., da Silva, J., Vianna, D.Q.C., Boeres, C., Rebello, V.E.F.: On the advantages of an alternative MPI execution model for grids. In: CCGRID 2007, pp. 575–582. IEEE Computer Society, Rio de Janeiro (2007)
10. Nascimento, A., Sena, A., da Silva, J., Vianna, D.Q.C., Boeres, C., Rebello, V.E.F.: Autonomic application management for large scale MPI programs. Int. J. High Perform. Comput. Networking **5**(4), 227–240 (2008)
11. Team, O.: Openmodeller webpage (2016). http://openmodeller.sourceforge.net/. Accessed Feb 2016
12. Peterson, A.T., Soberón, J., Pearson, R.G., Anderson, R.P., Martínez-Meyer, E., Nakamura, M., Araújo, M.B.: Ecological niches and geographic distributions (MPB-49). Princeton University Press (2011)
13. Ramachandra, T., Kumar, U., Aithal, B.H., Diwakar, P., Joshi, N.: Landslide susceptible locations in western ghats: prediction through openmodeller. In: Proceedings of the 26th Annual In-House Symposium on Space Science and Technology, pp. 65–74. Indian Institute of Science, Bangalore, Indian, January 2010
14. Ribeiro, F., Nascimento, A., Boeres, C., Rebello, V., Sena, A.: Autonomic malleability in iterative MPI applications. In: 25th International Symposium on Computer Architecture and High Performance Computing (SBAC-PAD), pp. 192–199, October 2013

15. Sena, A., Nascimento, A., Boeres, C., Rebello, V.: Easygrid enabling of iterative tightly-coupled parallel MPI applications. In: International Symposium on Parallel and Distributed Processing with Applications (ISPA 2008), pp. 199–206, December 2008
16. da Silva, J.A., Rebello, V.E.F.: Low cost self-healing in MPI applications. In: Cappello, F., Herault, T., Dongarra, J. (eds.) PVM/MPI 2007. LNCS, vol. 4757, pp. 144–152. Springer, Heidelberg (2007)
17. Soberón, J., Peterson, A.T.: Interpretation of models of fundamental ecological niches and species distributional areas. Biodivers. Inform. 2, 1–10 (2005)
18. Zarco-González, M.M., Monroy-Vilchis, O., Alaníz, J.: Spatial model of livestock predation by jaguar and puma in mexico: conservation planning. Bio. Conserv. 159, 80–87 (2013)

Towards Network-Aware Service Placement in Community Network Micro-Clouds

Mennan Selimi[1,3]([✉]), Davide Vega[2], Felix Freitag[1], and Luís Veiga[3]

[1] Universitat Politècnica de Catalunya, Barcelonatech, Barcelona, Spain
{mselimi,felix}@ac.upc.edu
[2] University of Bologna, Bologna, Italy
davide.vegadaurelio@unibo.it
[3] INESC-ID Lisboa/Instituto Superior Técnico,
University of Lisbon, Lisbon, Portugal
luis.veiga@inesc-id.pt

Abstract. Cloud services in community networks have been enabled by micro-cloud providers. They form community network micro-clouds (CNMCs), which grow organically, i.e. without being planned and optimized beforehand. Services running in community networks face specific challenges intrinsic to these infrastructures, such as the limited capacity of nodes and links, their dynamics and geographic distribution. CNMCs are used to deploy distributed applications, such as streaming and storage services, which transfer significant amounts of data between the nodes on which they run. Currently there is no support given to users for enabling them to chose better or the best option for specific service deployments. This paper looks at the next step in community network cloud service deployments, by taking network characteristics into account when deciding placement of service instances. We propose a service placement algorithm (PASP) that minimizes the service overlay diameter, while fulfilling service specific criteria. First, we characterize with simulations the potential performance gains of our approach. Secondly, we apply our algorithm to deploy a distributed storage service currently used in Guifi.net, and evaluate it in the real production network, assessing the performance and effects of our algorithm. We find that our PASP algorithm reduces the client reading times by an average of 16 % (with a max. improvement of 31 %) compared to the currently used organic placement scheme. Our results show how the choice of an appropriate set of nodes, taken from a larger resource pool, can influence service performance significantly.

Keywords: Community network micro-clouds · Service placement

1 Introduction

Community networks or Do-It-Yourself networks (DIYs) are bottom-up built decentralized networks, deployed and maintained by their own users. One successful effort of such a network is Guifi.net[1], located in the Catalonia region of

---

[1] http://guifi.net/.

© Springer International Publishing Switzerland 2016
P.-F. Dutot and D. Trystram (Eds.): Euro-Par 2016, LNCS 9833, pp. 376–388, 2016.
DOI: 10.1007/978-3-319-43659-3_28

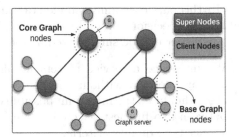

Fig. 1. Guifi.net nodes and links in Barcelona

Fig. 2. Guifi.net topology

Spain. Guifi.net is defined as an open, free and neutral community network built by its members: citizens and organizations pooling their resources and coordinating efforts to build and operate a local network infrastructure. Guifi.net network started in 2004 and today it has more than 30.000 operational nodes, which makes it the largest community network worldwide [1]. Figure 1 shows as example the nodes and links of Guifi.net in the city of Barcelona. Figure 2 shows the topology structure followed in Guifi.net. Client nodes are connected to the super-nodes. These super-nodes interconnect through wireless links different administrative zones.

Until recently, user-oriented local services were not much deployed because of the lack of easy to use mechanisms to exploit the available resources within the community network and due to other technological barriers. Early services include GuifiTV, Graph servers, mail servers, game servers [1]. With the adoption of *community network micro-clouds* (CNMC)², i.e. the platform that enables cloud-based services in community networks, local user-oriented services gained a huge momentum. Community network users started creating their own home-grown services and using alternative open source software for some of today's Internet cloud services, e.g., data storage services, interactive applications such as Voice-over-IP (VoIP), video streaming, P2P-TV, [2,3]. In a CNMC, a server (i.e. a low-power device such as a enhanced home gateway or mini-PC) is connected to a node of the community network.

Since Guifi.net nodes and the connected servers are geographically distributed, it needs to be decided where services should be placed in a network. If the underlying network resources are not taken into account, a service may suffer from poor performance, e.g., by sending large amounts of data across slow wireless links while faster and more reliable links remain underutilized. Therefore, a key challenge in community network micro-clouds is to determine the location of the service deployments, i.e. which servers at a certain geographic points in the network. Due to the dynamic nature of community networks and usage patterns, it is challenging to calculate an optimal placement.

---

² http://cloudy.community/.

In this paper we aim at understanding the impact of network-aware service placement decisions on end-to-end client performance. The main contributions of this paper can be summarized as follows:

- We introduce a service placement algorithm that provides optimal service overlay selection without the need to verify the whole solution space. The algorithm finds the minimum possible distance in terms of the number of hops between two furthest selected resources, and at the same time fulfil different service type quality criteria.
- We extensively study the effectiveness of our approach in simulations using real-world node and usage traces from Guifi.net nodes. From the results obtained in the simulation study, we are able to determine the key features of the network and node selection for different service types.
- Subsequently, we deploy our algorithm, driven by these findings, in a real production community network and quantify the performance and effects of our algorithm with a distributed storage service.

2   System Model

2.1   Network Structure

The Guifi.net community network consists of a set of *nodes* interconnected through mostly wireless equipment that users, companies, administrations must install and maintain in addition to its links, typically on building rooftops. The set of nodes and links are organized under a set of mutually exclusive and abstract structures called administrative *zones*, which represent the geographic areas where nodes are deployed.

We have collected network description data through CNML files (obtained January 2016)[3]. CNML (Community Networks Markup Language) is an XML-based language used to describe community networks. Guifi.net publishes a snapshot of its network structure every 30 min with a description of registered nodes, links and their configurations. In the CNML description, the information is arranged according to different geographical zones in which the network is organised. Furthermore, we used a *Node database*: a dump of the community network database that, in addition to the data described in CNML, includes other details about dates and people involved in the creation and update of the configuration of nodes and links.

The CNML information obtained has been used to build two topology graphs: *base-graph* and *core-graph*. The base-graph of Guifi.net is constructed by considering only operational nodes, marked in *Working* status in the CNML file, and having one or more links pointing to another node in the zone. Additionally, we have discarded some disconnected clusters. All links are bidirectional, thus, we use an undirected graph. We have formed what we call the core-graph by removing the terminal nodes of the base graph (i.e., client nodes). Table 1 summarizes

---

[3] https://guifi.net/en/guifi/cnml/2413.

Table 1. Summary of the used network graphs

	Nodes / edges	Node degree max/ mean/ min	Diameter	Zones
Base-graph	**13636 / 13940**	537 / 2.04 / 1	35	129
Core-graph	**687 / 991**	20 / 2.88 / 1	32	85

the main properties of base and core graphs that we use in our study e.g., number of nodes, node degree, diameter (number of max hops in the sub-graph) and number of zones traversed in core and base-graph.

2.2 Allocation Model and Architecture

In order to generalize the placement model for community services, we made the following assumptions that give to our model the flexibility to adapt to many different types of real services. In our case, a *service* is a set of S generic processes or replicas (with different roles or not) that interact or exchange information through the community network. The service can also be a composite service (e.g., three-tier service) built from simpler parts. These parts or components of a service would create an overlay and interact with each other to offer more complex services. Each of the service replicas or components will be deployed over a node in the network, where each node will host only one process no matter which service it belongs to.

It is important to remark that the services aimed in this work should be at infrastructure level (IaaS), as cloud services in current dedicated datacenters. Therefore the services are deployed directly over the core resources of the network (nodes in the core-graph) and accessed by base-graph clients. Services can be deployed by Guifi.net users or administrators. The architecture that we consider is based on a hybrid peer-to-peer model with three hierarchical levels of responsibility. On each level, members are able to share information among themselves.

The coordination is managed by some peer (i.e., as a super-peer) designated from the immediate upper layer. Three types of peers can be identified:

1. **Community Nodes:** are the computing equipment placed along the wireless community network by users. Besides contributing to the network quality and stability, they share all or part of their physical resources with other community members in an infrastructure as a service (IaaS) fashion. In terms of type and amount of resources, our model assumes the nodes are different. This means that from service point of view there is allocation preference.
2. **Zone Managers:** are single nodes - only one within each zone, selected among all the Community nodes with the extra responsibility to manage local zone services and coordinate inter zones aggregated information. In our model we do not explicitly identify these managers and we assume the existence of at least one of them in each area.

3. **The Controller:** is a unique centralized entity in our system. The role of the controller is to manage all the service allocation requests from the users and update service structures by pulling the configuration information for the zone managers. The allocation algorithms are implemented in the controller.

2.3   Service Quality Parameters

Resource dispersion in a community network scenario can be a drawback or an advantage, as the Nebula [4] authors claim in their research. The overlay created by composite services abstracts from actual underlying network connections. Based on that, services that require intensive inter-component communication (e.g. streaming service), can perform better if the replicas (service components) are placed close to each other in high capacity links [2]. On other side, bandwidth-intensive services (e.g., distributed storage, video on-demand) can perform much better if their replicas are as close as possible to their final users (e.g. overall reduction of bandwidth for service provisioning).

If we have some information about the application SLAs in community networks and node behaviour from the underlying network, decisions can be made accordingly, in order to promote that certain types of applications are executed in certain type of nodes with better QoS.

Our algorithm considers the following network and graph metrics as shown in Table 2, when allocating different type of services.

– **Availability:** The availability of a node is defined as the percentage of ping requests that the node replies when requested by the graph-server system. Graph-servers are distributed in Guifi.net and are responsible for performing network measurements between nodes. This is an important metric for service life-cycle and is considered for two service types. It is measured in percentage (%).
– **Latency:** The latency of a node is defined as the time it takes for a small IP packet to travel from the Guifi.net graph-servers through the network to the nodes and back. It is an important metric for latency-sensitive service in CNMCs. It is measured in milliseconds (ms).

Table 2. Service-specific quality parameters

Type of service	Examples of services	Network metrics	Graph metrics
Bandwidth-intensive	distributed storage, video-on-demand network graph server, mail server	availability **closeness**	**closeness centrality**
Latency-sensitive	VoIP, video-streaming game server, radio station server	availability **latency**	**betweenness centrality**

– **Closeness:** The closeness is defined as the average distance (number of hops) from the solution obtained from the algorithm to the clients. It is an important metric for bandwidth-sensitive services. It is measured in number of hops.

In terms of graph centrality metrics, we consider closeness and betweenness centrality. Closeness centrality is a good measure of how efficient a particular node is in propagating information through the network. Betweenness centrality quantifies the number of times a node acts as a bridge along the shortest path between two other nodes.

3   Service Placement Algorithm

We designed an algorithm that explores different placements searching for the local minimal service overlay diameter while, at the same time, fulfilling different service type quality parameters. Algorithm 1 relies on the method $PASP()$ to evaluate the different service placements in different zones and generate the solutions. The algorithm tries to find a solution in each zone by applying Breadth-First Search (BFS) and utilizing the $IsBetter$ method to choose the best solutions by applying service policies shown at Table 2. In the case of equal diameter allocations, the mean out-degree (the mean boundary of the nodes in the service overlay with the nodes outside of it) is taken. The service allocation with smallest diameter and largest mean out-degree fulfilling different service quality parameters is kept as the optimal. The algorithm iterates using Breadth-First-Search algorithm (BFS) in the network graph, taking as root the given node and selecting the first $S - 1$ closest resources to it. The node with high degree centrality is initially chosen as root. Degree centrality is the fraction of nodes that a particular node is connected to. In the case of several nodes at the same distance, nodes are selected randomly, distributing thus uniformly. Thanks to this feature, our algorithm performs faster than a pure exhaustive search procedure, since size equivalent placements are not evaluated. It is worth noting that the same set of nodes might be obtained from different root nodes, since placements in nearby network areas would involve the exact same nodes. We avoid re-evaluating these placements with a cache mechanism, that improves algorithm efficiency. After the placement solutions for different number of services are returned from BFS, the solutions are compared regarding the service quality parameters.

For each solution set obtained, we check our defined service-specific policies and then accordingly we calculate different scores (e.g., latency or availability score). Once we have the these scores for each solution set, we utilize the $IsBetter$ method to compare the solutions and to choose the new best placement solution according to different service types. Currently, the algorithm has not been optimized regarding the computation time, but it provides near-optimal overlay allocations, as our results show, without need of verifying the whole solution space.

Algorithm 1. Policy-Aware Service Placement (PASP)

Require: $N(V_n, E_n)$ ▷ Network graph
Require: $Z(V_z, E_z)$ ▷ Zones graph
Require: $Zone$ ▷ Search solution zone
Require: S ▷ Number of nodes in the service
Require: $ServicePolicy$ ▷ Service specific policies

```
 1: procedure PASP(N, Z, Zone, S, ServicePolicy)
 2:     Community ← Vₙ ∈ V_{z_i}
 3:     BestSolution ← null
 4:     for all node ∈ Community do
 5:         solution ←BreadthFirstSearch(N, Community, node, S)
 6:         if isBetter(solution, BestSolution, ServicePolicy) then
 7:             BestSolution ← solution
 8:         end if
 9:     end for
10:     return BestSolution
11: end procedure
12: procedure isBetter(currentSolution, bestSolution, ServicePolicy)
13:     for all p ∈ ServicePolicy do
14:         result ←CheckPolicy(currentSolution, bestSolution, p)
15:     end for
16:     return result
17: end procedure
```

---

4 Experimental Results

4.1 Network Behaviour and Algorithmic Performance

Our service placement algorithm proposed in Sect. 3 is used to simulate the placement of different services in Guifi.net. Our goal is to determine the key features of the network and its nodes, in particular to understand the network metrics that could help us to design new heuristic frameworks for smart service placement in CNMCs.

From the data obtained, our first interest is to analyse the availability and latency of Guifi.net nodes. This can be used as an indirect metric of quality of connectivity that new members may expect from the network.

Figure 3 shows that 40 % of the base-graph nodes are reachable from the network 90 % or less of the time. The situation seems to be even worse with the core-graph nodes, which are supposed to be the most stable part of the network (20 % of the core-graphs have availability of 90 % or less). Base-graph nodes have higher availability because they are closer to users, and is of high interest to users to take care of them. It is interesting to note that 20 % of the core-graph nodes have availability between 98–100%, and those are most probably the nodes that comprise the backbone of the network and connect different administrative zones. Since the service placement is done on the core-graph nodes, selecting the nodes with higher availability (e.g., 90–100%) is of high importance.

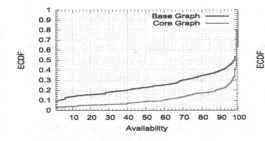

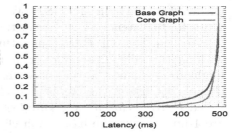

Fig. 3. ECDF of node availability   **Fig. 4.** ECDF of node latency

Figure 4 depicts the Empirical Cumulative Distribution Function (ECDF) plot of the node latency. Similar to the availability case, the latency of base-graph nodes is slightly better. For both cases, 30 % of the nodes have latency of 480 ms or less, which makes the other 70 % of the nodes to have higher latency. The availability and latency graph demonstrate the importance of, and indeed the need for, a more effective, network-aware placement in CNMCs. By not taking the performance of the underlying network into account, applications can end up sending large amounts of data across slow wireless links while faster and more reliable links and nodes remain under-utilized.

In order to see the effects of the network-aware placement in the solutions obtained, we compare two versions of our algorithm. The first version i.e., *Baseline*, allocates services just with the goal of minimizing the service overlay diameter without considering node properties such as availability, latency or closeness. The second version of the algorithm called *PASP*, tries to minimize the service overlay diameter, *while* taking into account these node properties.

The availability and latency of the *Baseline* solutions are calculated by taking the average of nodes in the optimal solutions (after the optimal solution is computed), where the optimal solution is the best solution that minimizes the service overlay diameter, that can only be calculated exhaustively offline.

We allocate services of size 3, 5, 7, 9, 11 and 15. Figures 5 and 6 reveal that nodes obtained in the solutions with *PASP* have higher average availability and lower latency than with *Baseline*, with minimum service overlay diameter.

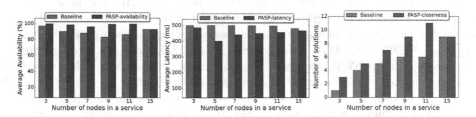

Fig. 5. PASP-availability   **Fig. 6.** PASP-latency   **Fig. 7.** PASP-closeness

On average, the gain of *PASP* over *Baseline* is 8% for the availability, and 45 ms for the latency (5–20% reduction).

We find that our *PASP* algorithm is good in finding placement solutions with higher availability and lower latency, however the service solutions obtained might or might not be very close (in terms of number of hops) to base-graph clients. Because of this we also developed another flavour of PASP algorithm called *PASP − closeness*. Figure 7 shows the number of solutions obtained that are 1-hop close to the base-graph clients. When *PASP − closeness* algorithm allocates three services, on average there are three solutions whose internal nodes (e.g., any of the nodes) are at 1-hop distance to any of the base-graph client nodes, contrary to the *Baseline* where on average there is one solution whose nodes are at 1-hop distance to base-graph clients.

Overall, in the two algorithms, there is a trade-off between latency and closeness. For bandwidth-intensive applications closeness seems to be more important when allocating services (e.g., *PASP−closeness* can be used), while for latency-sensitive applications it is the latency the one that naturally seems to be more important (e.g., *PASP − latency* can be used).

Moreover, from working with the Guifi.net data, we observed some patterns in the node features that conforms optimal allocations. We saw that the solution overlay diameter depends on the nodes degree centrality. Minimum degree centrality can be used to select the first node that composes the service (the solution). We saw that most of the solutions obtained are concentrated on a small set of of average centrality values. Selecting the next nodes in a particular range of closeness centrality (for bandwidth-intensive services) and betweenness centrality (for latency-sensitive services) is specially useful to obtain more optimal overlays.

4.2   Deployment in a Real Production Community Network

In order to understand the gains of our network-aware service placement algorithm in a real production community network, we deploy our algorithm in real hardware connected to the nodes of the QMP[4] network, which is a subset of Guifi.net located in the city of Barcelona. Figure 8 depicts the topoloqy of the QMP network. Furthermore, a live QMP monitoring page updated hourly is available in the Internet[5].

We use 16 servers connected to the wireless nodes of QMP. The nodes and the attached servers are geographically distributed in the city of Barcelona. The hardware of the servers consists of Jetway devices, which are equipped with an Intel Atom N2600 CPU, 4 GB of RAM and 120 GB SSD. They run an operating system based on OpenWRT, which allows running several slivers (VMs) on one node simultaneously implemented as Linux containers (LXC).

The slivers host the Cloudy[6] operating system. Cloudy contains some pre-integrated distributed applications, which the community network user can

---

[4] http://qmp.cat/.

[5] http://dsg.ac.upc.edu/qmpsu/index.php.

[6] http://cloudy.community/.

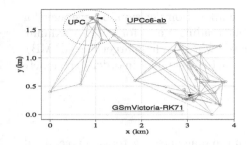

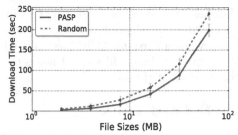

Fig. 8. QMP topology

Fig. 9. Average reading time on clients side

activate to enable services inside the network. Services include a streaming service, a storage service and a folder synchronizing service, among others. For our experiments, we use the storage service, which is based on Tahoe-LAFS[7]. Tahoe-LAFS is an open-source distributed storage system with enforced security and fault-tolerance features, such as data encryption at the client side, coded transmission and data dispersion among a set of storage nodes.

As the controller node we leverage the experimental infrastructure of Community-Lab[8]. Community-Lab provides a central coordination entity that has knowledge about the network topology in real time. Out of the 16 devices used, three of them are storage nodes and 13 of them are clients (chosen randomly) that read files. The clients are located in different geographic locations of the network. The controller is the one that allocates the distributed storage service in these three nodes and clients access this service. On the client side we measure the file reading times. We monitored the network for the entire month of January 2016. The average throughput distribution of all the links for one month period was 9.4 Mbps.

Figure 9 shows the average download time for various file sizes (2–64 MB) perceived at the 13 clients, after allocating services using *Random* algorithm (i.e., currently used at Guifi.net) and using our *PASP* algorithm. The experiment is composed of 20 runs, where each run has 10 repetitions, and averaged over all the successful runs. Standard deviation error bars are also shown.

Regarding the network interferences that may be caused by other users concurrent activities which can impact the results of our experiments, we reference to our earlier work [5] which investigated these issues.

Allocation of services using *Random* algorithm by Controller is done without taking into account the performance of the underlying network. It can be seen for instance that when using our *PASP* algorithm for allocation, it takes around 17 s for the clients on average to read a 8 MB file. In the random case, the time is almost doubled, reaching 28 s for reading a file from the clients side. We observed therefore that when allocating services, taking into account the

---

[7] https://tahoe-lafs.org/trac/tahoe-lafs.
[8] https://community-lab.net/.

closeness and availability parameters in the allocation decision, on average (for all clients) our algorithm reduces the client reading times for 16 %. Maximum improvement (around 31 %) has been achieved when reading larger files (64 MB). When reading larger files client needs to contact many nodes in order to complete the reading of the file.

5   Related Work

Service placement is a key function of cloud management systems. Typically, it is responsible for monitoring all the physical and virtual resources on a system and balance their load through the allocation, migration and replication of tasks.

Data Centers: Choreo [6] is a measurement-based method for placing applications in the cloud infrastructures to minimize an objective function such as application completion time. Choreo makes fast measurements of cloud networks using packet trains as well as other methods, profiles application network demands using a machine-learning algorithm, and places applications using a greedy heuristic, which in practice is much more efficient than finding an optimal solution. In [7] the authors proposed an optimal allocation solution for ambient intelligence environments using tasks replication to avoid network performance degradation. Volley [8] is a system that performs automatic data placement across geographically distributed datacenters of Microsoft. Volley analyzes the logs or requests using an iterative optimization algorithm based on data access patterns and client locations, and outputs migration recommendations back to the cloud service.

Distributed Clouds: There are few works that provides service placement in distributed clouds with network-aware capabilities. The work in [9] proposes efficient algorithms for the placement of services in distributed cloud environment. Their algorithms need input on the status of the network, computational resources and data resources which are matched to application requirements. In [10] authors propose a selection algorithm to allocate resources for service-oriented applications and the work in [11] focuses on resource allocation in distributed small datacenters.

Service Migration: Regarding the service migration in distributed clouds, few works came out recently. The authors in [12,13] study the dynamic service migration problem in mobile edge-clouds that host cloud-based services at the network edge. They formulate a sequential decision making problem for service migration using the framework of Markov Decision Process (MDP) and illustrate the effectiveness of their approach by simulation using real-world mobility traces of taxis in San Francisco. The work in [14] studies when services should be migrated in response to user mobility and demand variation.

Our focus is to perform service placements in community network clouds, which are peer-to-peer clouds formed from low-resource machines and very dynamic and diverse network. Another work in the community network context related to ours is [15] where the authors propose service allocation algorithms that minimize the coordination and overlay cost along the network.

6 Conclusion and Future Work

We addressed the need for network-aware service placement in community network micro-cloud infrastructures. We looked at a specific case of improving the deployment of service instance on micro-servers for enabling an improved distributed storage service in a community network.

As services become more network intensive, the bandwidth, latency etc., between the used nodes becomes the bottleneck for improving performance. In community networks, the limited capacity of nodes and links and an unpredictable network performance becomes a problem for service performance. Network awareness in placing services allows to chose more reliable and faster paths over poorer ones.

In this work we introduced a service placement algorithm that provides improved overlay service selection for distributed services considering service quality parameters, without the need for exploring the whole solution space. For our simulations we employed a topological snapshot from Guifi.net to identify node traits in the optimal service placements. We deployed our service placement algorithm in a real network segment of Guifi.net, a production community network, and quantified the performance and effects of our algorithm. We conducted our study on the case of a distributed storage service. In experiments we showed that by using our service placement algorithm, we were able to improve the total file reading time comparing to the currently used random placement.

In next steps we plan to develop and implement a decentralized version of our investigated service placement algorithm. Service migration should also be addressed to support performance objectives in the case of user mobility and within dynamic changes in the network.

Acknowledgements. This work was supported by the EU Horizon 2020 Framework Program project netCommons (H2020-688768), by the EMJD-DC program and by the Spanish Government under contract TIN2013-47245-C2-1-R. This work was also supported by the national funds through Fundação para a Ciência e a Tecnologia with reference UID/CEC/50021/2013.

References

1. Selimi, M., et al.: Cloud services in the Guifi.net community network. Comput. Netw. **93**, 373–388 (2015). Part 2
2. Selimi, M., et al.: Integration of an assisted p2p live streaming service in community network clouds. In: Proceedings of the IEEE 7th International Conference on Cloud Computing Technology and Science, CloudCom 2015. IEEE, November 2015
3. Selimi, M., et al.: Performance evaluation of a distributed storage service in community network clouds. Concurrency and Computation: Practice and Experience n/a-n/a cpe.3658 (2015)
4. Ryden, M., et al.: Nebula: distributed edge cloud for data intensive computing. In: IEEE International Conference on Cloud Engineering, IC2E 2014, pp. 57–66, March 2014

5. Cerdà-Alabern, L., Neumann, A., Escrich, P.: Experimental evaluation of a wireless community mesh network. In: Proceedings of the 16th ACM International Conference on Modeling, Analysis and Simulation of Wireless and Mobile Systems, MSWiM 2013, pp. 23–30. ACM, New York (2013)

6. LaCurts, K., Deng, S., Goyal, A.: Choreo: network-aware task placement for cloud applications. In: Proceedings of the 2013 Conference on Internet Measurement-Conference, IMC 2013, pp. 191–204. ACM, New York (2013)

7. Herrmann, K.: Self-organized service placement in ambient intelligence environments. ACM Trans. Auton. Adapt. Syst. 5(2), 6:1–6:39 (2010)

8. Agarwal, S., et al.: Volley: automated data placement for geo-distributed cloud services. In: Proceedings of the 7th USENIX Conference on Networked Systems Design and Implementation, NSDI 2010, Berkeley, CA, USA, USENIX Association 2-2 (2010)

9. Steiner, M., Gaglianello, B.G., Gurbani, V., Hilt, V., Roome, W. D., Scharf, M., Voith, T.: Network-aware service placement in a distributed cloud environment. In: Proceedings of the ACM SIGCOMM 2012 Conference, SIGCOMM 2012, pp. 73–74. ACM, NewYork (2012)

10. Klein, A., Ishikawa, F., Honiden, S.: Towards network-aware service composition in the cloud. In: Proceedings of the 21st International Conference on World Wide Web, WWW 2012, pp. 959–968. ACM, New York (2012)

11. Alicherry, M., Lakshman, T.: Network aware resource allocation in distributed clouds. In: 2012 Proceedings of INFOCOM, pp. 963–971. IEEE, March 2012

12. Wang, S., Urgaonkar, R., Zafer, M., He, T., Chan, K., Leung, K.K.: Dynamic service migration in mobile edge-clouds. CoRR abs/1506.05261 (2015)

13. Wang, S., et al.: Dynamic service placement for mobile micro-clouds with predicted future costs. In: IEEE International Conference on Communications, ICC 2015, pp. 5504–5510, June 2015

14. Urgaonkar, R., et al.: Dynamic service migration and workload scheduling in edge-clouds. Perform. Eval. 91, 205–228 (2015). Special Issue: Performance 2015

15. Vega, D., Meseguer, R., Cabrera, G., Marques, J.: Exploring local service allocation in community networks. In: 10th International Conference on Wireless and Mobile Computing, Networking and Communications, WiMob 2014, pp. 273–280. IEEE, October 2014

Heating as a Cloud-Service, A Position Paper (Industrial Presentation)

Yanik Ngoko[✉]

Qarnot Computing, 92120 Montrouge, France
yanik.ngoko@qarnot-computing.com

Abstract. In this paper, we discuss a novel utility computing approach, implemented by the company Qarnot computing in private clouds. The approach promotes a new computing paradigm in which computers are considered as machines that produce both data and heat. It is based on two main technologies: a new model of servers and a new resource manager for servicing both computing and heating as a cloud-service. This paper focuses on the resource manager promoted by this utility computing approach. We summarize the architecture of the middleware and describe the key computational challenges. We also provide a performance characterization on the thermal comfort and processing time. Some preliminary results show that the proposed utility computing approach can lead to distributed systems that are competitive with both traditional cloud solutions and heating systems.

Keywords: Heating as a service · Utility computing · Resource manager

1 Introduction

With the analytical engine [7], Charles Babbage conceptualized the main ideas of a general purpose *computer*. One of the most innovative point of the Babbage machine was the input/output model he defined. This model is still the reference today; indeed, it is commonly admitted that the computers are machines that given an *input data* and a *computer program* will automatically generate *output data* that are the results of the computer program on the input. The Babbage analytical engine was a mechanical machine. Many years after its conceptualization, the electronic design of computers stressed the importance of the heat generated by the run of computer programs. A question was then to know how to consider this heat regarding, computational processes. On this interrogation, one can distinguish two paradigms that we will refer to as the *pure-compute* and *compute-and-heat*.

In the pure-compute paradigm, the heat produced by computing machines is an unexpected outcome. The execution of a computer program must pursue the sole objective to produce the correct output data. On modern computing machines, the pure-compute paradigm led to the usage of fans in personal computers or cooling technologies like chilled water and air conditioning systems.

© Springer International Publishing Switzerland 2016
P.-F. Dutot and D. Trystram (Eds.): Euro-Par 2016, LNCS 9833, pp. 389–401, 2016.
DOI: 10.1007/978-3-319-43659-3_29

The pure-compute paradigm follows the vision that pioneers like Babbage or Alan Turing had about the functioning of computers. It is also supported by the impact of high temperatures on processors aging and reliability [5]. Alternatively to pure-compute, there is the *compute-and-heat* paradigm. Here, the computer programs are not only supposed to produce data. The heat is an expected outcome and computer designers should invest in the mastering of its production. Though heat generation was already observed in successive generation of electronic computers, it is only in recent years that this second vision was clearly set with the development of heat recovery systems in data centers [1] and in the pioneer initiative of Qarnot computing[1].

In his report on the Babbage analytical engine [7], Menabrea suggested to consider two economical gains we could expect from general purpose computers: the economy of time and the economy of intelligence. The conviction that supports this work is that in shifting from the pure-compute to the compute-and-heat paradigm, we could add another gain: the economy of energy. This is because the compute-and-heat paradigm is based on a vision where we can merge a computer and a heater into a single machine. However, this reasoning holds if we are able to establish that the compute-and-heat vision could lead to *satisfactory* solutions for heating and computing. For this purpose, our paper focuses on the implementation of the compute-and-heat paradigm made by Qarnot computing.

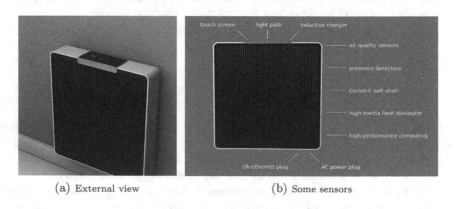

(a) External view   (b) Some sensors

Fig. 1. The Qarnot digital heater.

Qarnot computing promotes a utility computing model in which computing and heating are delivered from a single cloud infrastructure. The model is implemented by means of a geo-distributed cloud plateform based on special *server* nodes named digital heaters or Q.rads (See Fig. 1). Each Q.rad embeds several processors connected to a heat diffusion system. Q.rads are deployed in homes, offices, schools etc. and the network of radiators that they constitute is the physical infrastructure of the distributed data center.

---

[1] http://www.qarnot-computing.com/.

The Qarnot model is based on a new and customized resource manager named Q.ware. In comparison with traditional ones, Q.ware supports a service provisioning model that distinguishes two types of requests: requests for heating and for computing. In addition, the requirements in computations must be balanced with those in heating. Our paper summarizes the design principles of the middleware and presents the new computational problems for which it was built. It also proposes an empirical characterization of the middleware performance on thermal comfort and processing time. These preliminary results show that Q.ware can be used to develop distributed systems that are competitive with traditional clouds and heating systems.

The remainder of this paper is organized as follows. In Sect. 2, we present and discuss the related works. Section 3 describes the Q.ware manager. An empirical characterization of performance is proposed in Sect. 4 before concluding in Sect. 5.

2 Related Works

At first sight, it might not be easy to perceive the interest in developing a new resource manager for geo-distributed computing. Indeed, the study of resource management is a well-investigated topic and several managers already demonstrated their efficiency in the management of huge computer systems. However let us notice that while these managers were mainly developed for the management of grids, clusters, or desktop grids, they are not necessarily adequate for geo-distributed clouds. Indeed, clouds are based on alternative models for service provisioning that introduce specific scheduling problems like the virtual machines consolidation and placement problem.

Several resource managers were proposed for geo-distributed clouds. In [2], the authors showed how to combine a classical grid scheduler with a set of clouds that offers infrastructure as services. In their model, the user requests describe a job to process. The jobs are submitted to the grid scheduler with an additional description of the machines to use for the processing. The requirements in machines is transmitted to an IaaS provider that will then boot the necessary compute nodes with virtual machines. The grid scheduler then deploys the job on the virtual machines that are started. The open stack project[2] developed several solutions for the creation of distributed clouds. This includes distributed scheduling mechanisms implemented in Nova or the open stack cascading solution that can be used for an hierarchical management of several open stack sites. The NebulaEdgeCloud [10] proposes a distributed edge computing. It enables distributed data-intensive computing through a number of optimizations including location-aware data and computation placement, replication, and recovery. The SlapOs cloud [11] introduced a distributed resource manager for cloud that implements the master/slave paradigm. The master nodes are stateful while slaves nodes are stateless. Slaves request to their master node which software they should deploy and which tasks to run. They frequently sent reports to

---

[2] https://www.openstack.org/.

the master about their state. Finally, the distributed cloud management has also been envisionned throughout multi-cloud brokers that interact with several clouds to find the best resources according to user service level agreement [3]. Several other resource managers were proposed for geo-distributed clouds. For an extended state-of-the-art, we invite the interested reader to the report provided in [6]. Despite the existence of such solutions, they do not all implement one of the main feature required in the Qarnot vision, which is the need to define a thermal-aware manager.

We did not find in the literature any geo-distributed cloud resource manager whose specificity is to be thermal-aware. However, let us notice that thermal-aware schedulers were investigated on several computing systems. Thus, one might consider that a thermal manager for geo-distributed cloud would easily be derived in including existing scheduler policies in a geo-distributed manager. However, let us observe that this will not necessarily conduct to the heating as a service model that is targeted. Indeed, existing thermal-aware scheduling policies mainly targeted the minimization of the power consumption and heat generation, the minimization of the flow of hot air or the minimization of the cooling costs of computer systems [8]. This also means that they are formulated in the pure-compute paradigm while with Q.ware, it is the compute-and-heat paradigm that is followed. The consequence is that Q.ware will reduce the temperature only when requested and must manage target of temperatures. In addition, Q.ware can operate in an highly dynamic environment in which it is the requirements in heating that determines the computing power of its nodes. The next section gives a general view of the architecture of the middleware.

3   The Q.ware Resource Manager

3.1   Design Principles

The design of the Q.ware system is based on several guiding principles, including the following:

- **Distributivity:** Q.ware assumes a physical geo-distributed topology for cloud computing. In the case of the Qarnot cloud for instance, the compute nodes consists of digital heaters deployed in buildings of the city of Paris. The management of the geo-distributivity is in particular handled by locality-aware rules in the scheduling.
- **Security:** Q.ware integrates state-of-the-art security modules for encryption and authentication. The REST API is accessible through HTTPS or leased line; All distribution and computing nodes implement TLS/IPSEC with client authentication. In addition, the compute node could be stateless (no storage).
- **Thermal-awarness:** As already mentioned, in addition to traditional cloud computing requests, Q.ware also handles heating requests.
- **Multi-clouds and multi-architectures:** Q.ware assumes a generic view of computing resources. They can consist of a motherboard (part or not of a Q.rad), a connected device, a computer server. Q.ware also supports various types of virtual machines and containers.

- **City-cloud:** The Q.ware is based on a hierarchy of 3 levels of servers: Q.node, Q.box, Q.rad. The organization has been conceived to manage a data center distributed in a city. Thus, the Q.rads support the clouds' abstractions at home level, the Q.boxes aim at managing abstractions at building levels and Q.nodes, abstractions at city level.
- **Autonomy:** In Q.ware, if a top server fails, the servers underneath will autonomously take the appropriate decisions to ensure that the heating is serviced. In addition, if there are no computing requests, the servers will automatically request computations from databases of scientific problems like BOINC (http://sat.isa.ru/pdsat/top_users.php).
- **Fault-tolerance:** A Q.box can be connected to several Q.nodes. This is important for continuing the service provisioning when a Q.node fails. In the same way if a Q.box could not be accessed, the Q.rad will ensure the heating supply.
- **Scalability:** We can easily add new compute nodes to Q.ware by creating new instances of the different servers.

The Q.ware resource manager includes a C# API and several SDK (Python, NodeJs, C#) for the submission of computing requests. In the next section, we will present the architecture of the manager in more details.

3.2   Architecture

The Q.ware is composed of three types of servers: Q.nodes, Q.boxes and Q.rads. The servers are organized in a hierarchical tree where the Q.nodes are root nodes and the Q.rads are leaf nodes. An example of deployment is given in Fig. 2(a). As Q.ware was designed to operate data centers distributed in cities, a typical deployment of the middleware could be seen as a forest of Q.nodes. Here, each home is associated with a set of Q.rads that are controlled at the building level by a Q.box. The Q.boxes are controlled by a Q.node. In practice however, this deployment might not be suitable. For fault tolerance issues, it might be more interesting to link a Q.box to several Q.nodes. Ideally, Q.ware assumes that the computing power of the cloud is in the processors (we will also use the term compute nodes) inside the digital heaters. But, as already mentioned, it can also manage these resources without the digital heater abstraction (See Fig. 2(b)). In particular, Q.ware can exploit concurrently the compute nodes of a heater and container/virtual machines deployed in another data center.

In the server hierarchy of Q.ware, the heating requests are first routed towards the leaf nodes while the computing requests are routed to the top. For computing requests, Q.ware supports a Python and C# client that is used to formulate the computing requests as the submission of a set of tasks. Here, a task refers to the run of a container/virtual machine image and is associated with a set of input files, a virtual disk and an output directory.

Each Q.node runs a scheduler that manages a queue of tasks. The scheduler implements a list scheduling algorithm with priority on tasks. The principle of the algorithm is to iteratively loop on the queue and to select the tasks of

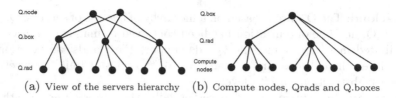

(a) View of the servers hierarchy   (b) Compute nodes, Qrads and Q.boxes

Fig. 2. Possible deployment with Q.ware.

higher priorities to deploy. Once done, several filters are applied to determine the compute nodes that will run the tasks. We will come back on the scheduling model further. For now, it is important to retain that the vision that the Q.ware has of the compute nodes come from data reported by the Q.boxes. Thus, when for instance, in a heater, a request is sent to not produce any heat, the Q.box will state that the corresponding nodes are not exploitable. In the list scheduling algorithm, there are globally three classes of priorities: background, high and low. Tasks of background priorities correspond to those that are mainly deployed to heat. Those with high priorities are associated with a computing request that has a strong SLA. The list scheduler of the Q.node assumes that tasks are preemptable. However, we cannot interrupt a task with a given priority to run another of lower priority.

At the lower end of the Q.ware architecture there are the Q.rads servers, in each digital heater. The Q.rads servers have a direct access to the processors, sensors and the main computing power information. The sensors include humidity, light, CO2, noises, temperature etc. The servers are also connected to a control interface (HMI) that the hosts of the heater can use to control their temperature. The Q.rad is able to negotiate computing loads with Q.nodes, through the secured connexion of its Q.box. In the case, the communication is broken between the Q.box and the Q.nodes, the Q.rads are still able to compute cached tasks. In the case where the communication with the Q.box is broken, are not working, the Q.rads will autonomously ensure that the heating will be serviced in launching a generic benchmarked computer program.

Over the Q.rads and under the Q.node, Q.boxes are acting as local controllers to handle heating and jobs' dispatching, security and caching. The Qboxes manage the storage used in the heaters. The choice to separate the management of the storage was initially driven by noise and integration consideration. Q.boxes are in charge of dynamic container deployment, input, output and session data synchronization with parent Q.nodes. A Q.box can also stop or pause a container of the processors of the heaters it controls. Those decisions are based on collected sensors data. Finally, each Q.box is connected to at least one Q.node. In the same way, a Q.rad can be connected to one or several Q.boxes. These choices were made for being fault-tolerant.

3.3   Scheduling in Q.ware

One main novelty in Q.ware is its scheduling model. To understand why, in this part, we will present here some challenges envisionned in the design of the middleware.

A New Objective Function. Scheduling in Q.ware is naturally a multi-objective problem. This is because there are at least three viewpoints: the one of customers that want to compute (*HPC customers*), the viewpoint of customers that want to be heated (*host*) and the one of the middleware.

In Q.ware, the viewpoint of the HPC customers is what we find in classical distributed scheduling systems: the goal is to get the results of submitted jobs as soon as possible. For this purpose, the current Q.ware implementation focuses on C_{max} minimization. The viewpoint of the hosts completely differs from what we could find in classical scheduling theory. Indeed, let us assume that at date t, a host of the heater i want to be heated at the temperature $target_i(t)$. Let us also assume that $ambiant_i(t)$ is the current ambient temperature observed from the heater. Given n jobs to schedules on m heaters, the hosts expect that the processing of the jobs should be done such as to minimize the difference between the target and ambient temperatures. This is captured with the objective function:

$$\text{minimize} \max_{1 \le i \le m} \int_0^{T_m} |target_i(t) - ambiant_i(t)|.dt$$

Here, T_m is an input estimation of the time required to process the jobs. Finally, the middleware viewpoint is specifically related to one goal of the Qarnot computing business model. It is the one of reducing the energy consumption in the processing of the jobs. This means that if $P_i(t)$ is the power consumption of the heater i at date t, the objective is to minimize:

$$\max_{1 \le i \le m} \int_0^{T_m} P_i(t).dt$$

This objective is related to the Qarnot business model since the company refunds the electricity bill of the hosts. At first sight, it might seem impossible to reduce the energy consumption on a system that produces heat from electricity. However, we can play on the inertia of the heater to not compute all the time.

A general approach for solving multi-objective problems consists of formulating the other objectives as a constraint such as to have a single objective problem [4]. This is done in Q.ware where the hosts and Q.ware viewpoints are defined as constraints in an adaptive scheduling approach.

A New Scheduling Problem. As we have a new objective function, we have a new scheduling problem. But, the novelty of the scheduling problem in Q.ware is not only related to the objective function. The Q.ware scheduling model also introduces two types of constraints related to heat production. To understand why, let us consider the power equation:

$$P = CV^2 f + P_{static} \tag{1}$$

Here, P_{static} is a base power consumption and $CV^2 f$ is the dynamic power where V is the voltage, f the frequency and C the capacitance. In Q.ware, this power equation is used to manage the heat production in Q.ware. The idea is to act on the dynamic part to consume more or less electricity and then generate heat. There are two new types of heat production constraints we consider. The first is on the *velocity* at which we go from a temperature $\Delta(t)$ to $\Delta(t')$ $(t' > t)$. Here, we define the velocity as

$$v(\Delta(t), \Delta(t')) = \frac{\Delta(t') - \Delta(t)}{t' - t}$$

The value of the velocity will depend on the configurations we will set for the load, voltage, frequency of the processors inside the heater. The velocity constraint we consider is that a host could set a minimal speed in heating. In general, all velocities cannot be reached because each heater has a limited power consumption. The second type of constraints is to ensure that the heat diffused by any heater fits within a given interval. The guarantee of a minimal and maximal heat is important since some temperatures are more suitable for the human body. This constraint implies to have an accurate model of the inertia of the heater.

A New Model for Heterogeneity. Since processors frequencies are manipulated to produce heat (see Eq. 1), scheduling in Q.ware must be envisionned in an heterogeneous context. This hypothesis holds even if *the processors layer of the distributed architecture initially have the same characteristics*. Scheduling algorithms for variable speed processors have been investigated in the past. But the main novelty that the Q.ware model introduces is that the variability is dynamic and could be caused by the interaction of the hosts with the heater (modification of the target temperature for instance). In its current implementation, Q.ware supports a simple variability model that calibrates the computational power of the nodes depending on the seasons. But a more elaborated model that uses the sensors embedded in the heaters is in construction. The idea is to to anticipate the available computational power by detecting the presence of host and combining this information with meteorogical parameters like the external temperature, the humidity etc.

An Open Perspective for Non Cooperative Game Scheduling. During the summer, the available computing power in Q.ware will decrease. To increase it, we can however turn the scheduling problem in a game where hosts compete in an ecological perspectives. Given two hosts $host_1, host_2$ let us assume that each could tolerate a small deviation over the temperature they want inside their homes. In a ecological viewpoint, this is interesting because it allows the processing of the tasks in using a free-cooling system. Let us assume that in manipulating the target temperature, each host can either accept or reject a task that was scheduled on his heater. The deviations that $host_1$ could tolerate could differ from the one that $host_2$ could tolerate. In order to make these threshold grow, let us consider a game with four participants: the Q.ware, $host_1$, $host_2$ and a datacenter. Initially in the game, n tasks are submitted to Q.ware. At the round j of the game, the Q.ware pushes a task and chooses a host that will process it. If the host accepts the task, it cumulate the reward associated with it. If not, Q.ware will propose the task to the other host. If none of the hosts accepts, then it is the datacenter that will run the task. Such a game will put in tension two objectives: the thermal confort of each host (private interest) and the ecological benefit for all (public interest). In practice, the winner of such a competition could be the datacenter. To avoid this situation and maximize efficiency, Qarnot gives priority to optimal buildings for deployment (e.g. schools closed in summer) and plan to leverage on others opportunities to exploit heat (e.g. hot water).

Here ends the presentation of the Q.ware middleware. In the next section, we will discuss its performance.

4   Performance Characterization

In this section, we propose an empirical characterization of the performance we could expect from Q.ware. The characterization is based on data collected from the Qarnot cloud. It is important to notice that it is not easy to dissociate the impact of the physical architecture of the cloud on the performance we measured. However, as Q.ware was mainly designed for the Qarnot cloud, we can say that the estimation are significant. We estimated the performance on two criteria: the thermal comfort and processing time. We start with the thermal comfort.

4.1   Thermal Comfort

On thermal comfort, we considered distribution of temperatures inside homes where the Qarnot heater was deployed. The intent was to see whether or not the system was able to produce sufficient heat to people that adopts it.

In Fig. 3(a–d), we present the trends on temperatures we observed. The figures were computed from more than 200 different heaters. The temperatures were taken from November, 01st of 2015 to May 05th of 2016. The measures were collected approximately every 10 min. The interval probabilities were computed for the intervals $[a, a + 1]$ where a is a positive integer. The curves show that the temperatures were concentrated between 17 and 26 ° with an average temperature around 22.5 °. They also show that the heat is guaranteed during the winter with a low probability to observe a temperature under 17 ° (mostly open windows). These results are interesting because as shown by a national french study [9], more than 75 % of french householders have a preference for a reference living room over or equal to 19 °. In addition, this latter study [9]

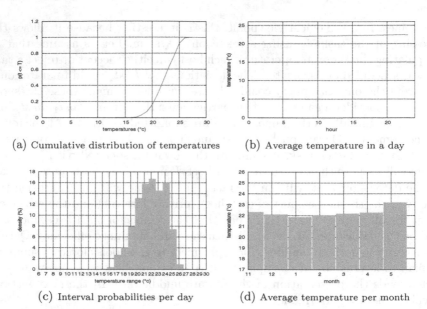

(a) Cumulative distribution of temperatures   (b) Average temperature in a day

(c) Interval probabilities per day   (d) Average temperature per month

Fig. 3. Temperature distributions

reveals a correlation between the increase of this reference temperature and the revenue level of the householders. We can thus conclude that the distributions we observed reflect a vision of the comfort shared by most french householders.

It is important to precise that since these measures were done with sensors embedded in the heater, there might be a biais when we go far from the heater.

4.2 Processing Time

We analyzed the journal of events of the Qarnot cloud scheduler. This journal reports the state of the tasks submitted in the platform. One of these task consists of the execution of the Qarnot parallel rendering service[3]. We focused on it (more than 1500 tasks). In Fig. 4(a), we depict a cumulative distribution of the rendering tasks that were completed (some failed). This figure shows that more than 80 % of the tasks we analyzed were processed in less than 3 h (Makespan). It also shows that some of the processed tasks took more than 80 h. This is very interesting if we consider that we did not get an error even if the processors temperature in these cases often exceeded 80 °. This first curve shows that the idea of provisioning heating from computations can perfectly be applied to long-term and compute intensive tasks.

The tasks we considered were diverse. To show this point, we computed a *load balancing factor* as the ratio between the sum of completion time required for processing a task and the makespan times the number of processors used

---

[3] blender.qarnot.net.

in the run of the task $\left(\frac{\sum c_i}{P.C_{max}}\right)$. Somehow, this factor could approximate the parallel efficiency achieved on the task in the case where the sequential time is the parallel time on one processor. Figure 4(b) gives the interval probability to fall in different range of load balancing. As one can notice, there is not an interval that dominates the other.

In the parallel rendering, any input file is splitted into several sub-files that each is associated with a sub-task to deploy on the heater (Bag of Task parallelism). If these subtasks are deployed on a set S of processors, we considered that the maximal time spent on each processor $p_i \in S$ is the effective computing time. For each task, we computed the ratio between the elapsed time in the processing of the task (the one perceived by HPC customers) and the effective computing time $\left(\frac{elapsed\ time}{C_{max}}\right)$. The goal was to capture the overhead induced by Q.ware (system deployment, remote access to the persistent storage, execution faults, variation of processors speed etc.). Figure 4(b–c) depicts the result we observed. We distinguish two cases: in the former, the number of cores used for the rendering is lower than 64 and in the latter it is greated. We made the distinction to see the impact of the number of processors on the overhead.

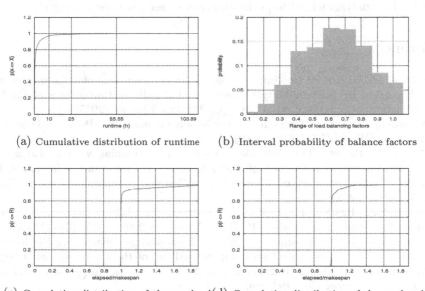

(a) Cumulative distribution of runtime   (b) Interval probability of balance factors

(c) Cumulative distribution of the overhead (d) Cumulative distribution of the overhead
in the release date (1-64 cores)   in the release date (over 64 cores)

Fig. 4. Runtime distributions

As one can notice, the probability for the overhead to be close to 1 was high in all cases. This means that the time taken for the management of the resources is not predominant in Q.ware. However, let us notice that the overhead reached 1.7 in some cases. This phenomenon occurs on short jobs. For instance, the execution of a sub-task in the parallel rendering service always includes a deployment time (operating systems, copy of files etc.). The shorter the execution time, the higher the importance of the deployment time.

5   Conclusion

In this paper, we introduced Q.ware, a new resource manager for a utility computing approach in which heating is considered as a cloud-service. We described the architecture of the middleware and proposed a performance characterization in considering data collected from the Qarnot cloud. The results showed that Q.ware could pave the way of a new vision of distributed computing where cloud-services and heating are provisioned from a unique platform. The results also invite to (1) reconsider the question of cooling in datacenters and servers and (2) to design new scheduling algorithms for the efficient production of heat based on computers.

To continue this work, we mainly envision an extensive benchmarking study of the middleware. The objective is to consider other dimensions of the quality of services like the reliability and the availability. We also intend to make a comparative analysis with the performance of other resource managers.

References

1. Alfonso, C., Giulio, P.: Cooling systems in data centers: state of art and emerging technologies. In: Sustainability in Energy and Buildings: Proceedings of the 7th International Conference, SEB 2015, pp. 484–493. Elsevier (2015)
2. Armstrong, P., et al.: Cloud scheduler: a resource manager for distributed compute clouds. CoRR abs/1007.0050 (2010)
3. Buyya, R., Barreto, D.: Multi-cloud resource provisionning with aneka: a unified and integrated utilization of microsoft azure and amazon ec2 instances. In: International Conference on Computing and Network Communications, pp. 222–235, December 2015
4. Dutot, P.F., Rzadca, K., Saule, E., Trystram, D.: Multi-objective Scheduling, Chap. 9. Chapman and Hall/CRC Press, November 2009. ISBN: 978-1420072730
5. Huang, L., Xu, Q.: Agesim: a simulation framework for evaluating the lifetime reliability of processor-based socs. In: Proceedings of the Conference on Design, Automation and Test in Europe, DATE 2010, pp. 51–56. European Design and Automation Association, Belgium (2010)
6. Jennings, B., Stadler, R.: Resource management in clouds: survey and research challenges. J. Netw. Syst. Manage. **23**(3), 567–619 (2015)
7. Menabrea, L.F., Lovelace, A.: The analytical engine invented by charles babbage. From the Bibliothèque Universelle de Genève, No. 82, October 1842. http://www.fourmilab.ch/babbage/sketch.html
8. Moore, J., Chase, J., Ranganathan, P., Sharma, R.: Making scheduling "cool": temperature-aware workload placement in data centers. In: Proceedings of the Annual Conference on USENIX Annual Technical Conference, ATEC 2005, p. 5. USENIX Association, Berkeley (2005)

9. Penot-Antoniou, L., Zobiri, R., (directeur de la publication), X.B.: Les déterminants de la température de chauffage adoptée par les ménages. Etude et Documents No. 83, Comissariat général du développement durable, April 2013. http://developpement-durable.gouv.fr/IMG/pdf/ED83.pdf
10. Ryden, M., Oh, K., Chandra, A., Weissman, J.: Nebula: Distributed edge cloud for data intensive computing. In: 2014 IEEE International Conference on Cloud Engineering (IC2E), pp. 57–66, March 2014
11. Smets-Solanes, J.P., Cérin, C., Courteaud, R.: Slapos: a multi-purpose distributed cloud operating system based on an erp billing model. In: 2011 IEEE International Conference on Services Computing (SCC), pp. 765–766, July 2011

Distributed Systems and Algorithms

Design and Verification of Distributed Phasers

Karthik Murthy[✉], Sri Raj Paul, Kuldeep S. Meel,
Tiago Cogumbreiro, and John Mellor-Crummey

Rice University, Houston, USA
karthik.murthy@rice.edu

Abstract. A phaser is an expressive barrier-like synchronization construct that supports dynamic task membership. Each task can participate in a phaser as a signaler, a waiter, or both. In this paper, we present a highly concurrent and scalable design of phasers for a distributed memory environment. Our design for a distributed phaser employs a pair of concurrent skip lists augmented with the ability to collect and propagate synchronization signals. To enable a high degree of concurrency, the addition and deletion of participant tasks are performed in two steps: a "fast single-link-modify" step followed by multiple hand-over-hand "lazy multi-link-modify" steps. We verify our design for a distributed phaser using the SPIN model checker. We employ a novel "message-based" model checking scheme to enable a non-approximate complete model checking of our phaser design. We guarantee the correctness of phaser semantics by ensuring that a set of linear temporal logic formulae are valid during model checking. We also present complexity analysis of the cost of synchronization and structural operations.

1 Introduction

Power consumption is now considered to be a very important parameter in the design of future HPC systems. Dynamic voltage and frequency scaling is an essential tool required to operate parallel systems within a tight energy envelope [10]. As a consequence, dynamic task-based programming models are gaining attention as an alternative to static SPMD models. Synchronization between tasks in the dynamic task-based programming models is becoming increasingly important, as noted in the report "Software Challenges in Extreme Scale Systems" [9].

Phasers are a general barrier-like synchronization primitive that supports dynamic registration of tasks. Each task has a choice of participation modes: signal-only, wait-only, and signal-wait. To date, the only phaser design available is for shared memory systems [11,12]. In this paper, we present a highly concurrent and scalable design of phasers for distributed memory parallel systems.

Recent designs for phaser-like synchronization include Alting barriers in Communicating Sequential Processes for Java (JCSP) [13] and Clocks in X10 [8]. While Clocks have been implemented for distributed memory environments, they use a non-scalable design in which a single root task collects information from all

© Springer International Publishing Switzerland 2016
P.-F. Dutot and D. Trystram (Eds.): Euro-Par 2016, LNCS 9833, pp. 405–418, 2016.
DOI: 10.1007/978-3-319-43659-3_30

the participants [7]. Alting barriers similarly maintain global state in a centralized fashion. In contrast, our phaser design uses a scalable distributed protocol.

Synchronization protocols that take time linear in the number of participating tasks are not scalable. Protocols with sub-linear growth in time complexity are necessary. Skip lists [6] have long been used in shared memory environments, providing an expected time complexity of $O(\log n)$ for operations on a skip list containing n items. We make use of a pair of distributed concurrent skip lists as the backbone for a distributed phaser. Insert and delete operations on the skip lists enable a task to dynamically join or abandon a phaser. Additional operations on the skip lists support propagation of synchronization signals.

Proving the correctness of distributed protocols is difficult. The manual enumeration of communication interleavings is infeasible and writing formal proofs is error prone. For these reasons, we employ automated formal verification known as model checking to verify our design. We check whether our design satisfies the required phaser semantics with a quorum of Linear Temporal Logic (LTL) formulae. Model checkers explore all possible paths of execution, verifying the input LTLs at each point along these paths. During this process, the size of the state space needed to completely model check the operations on a distributed phaser is significantly more than a terabyte. However, we employ a novel "message-based" divide-and-conquer strategy to reduce the state space and provide a non-approximate complete model checking of our design. To the best of our knowledge, we are the first to employ a message-based scheme for a non-approximate model checking to prove the correctness of a distributed synchronization protocol.

In this paper, we explore the design of a distributed phaser, complexity of operations and its correctness. Our contributions are as follows:

- We describe a design for distributed phasers that employs a scalable decentralized event-driven approach to synchronize dynamic tasks.
- We prove livelock- and deadlock-freedom, semantic properties about synchronization and structural-modification operations through a novel "message"-based model checking scheme.
- We analyze the time and message complexity of operations on distributed phasers.

Section 2 introduces distributed phasers. Section 3 details the design and operations. Section 3.4 verifies our design using model checking. Section 4 derives the complexity of phaser operations. Section 5 discusses related work. Section 6 presents conclusions.

2   Distributed Memory Phasers

A phaser is a flexible, barrier-like primitive used to synchronize a group of parallel tasks [11]. A phaser enables each task to participate in one of three modes: signal-only, wait-only, signal-wait. This flexibility lets a phaser be used in a spectrum of synchronization patterns ranging from a barrier to a producer-consumer pattern.

A phaser supports five operations: create, register, drop, signal, and wait. create is a collective among a team of tasks that creates a phaser. register adds a task as a participant, while drop lets a task remove its membership. The only way to invoke register is when a task spawns another: the spawner registers the spawnee. Operations create and register indicate whether a task participates as a signaler (signal-only), a waiter (wait-only), or both signaler and waiter (signal-wait). The participation mode affects the two remaining phaser operations signal and wait, explained next.

A phaser synchronization maintains a monotonically increasing global event counter called *phase*. To increment the counter, all signalers that have not dropped from the phaser must invoke signal exactly once. A waiter issues a wait to block until the phaser reaches a certain phase i, effectively observing the i-th collective event. Any task that is both a signaler and a waiter must always signal before waiting. A wait-only task will observe but not affect synchronization. In contrast, a signal-only task contributes to advancing phase, but waits for no other, *e.g.*, a producer in a producer-consumer pattern.

On distributed systems, tasks participating in a phaser may reside on different compute nodes and must interact with each other through messages[1]. Below, we detail the challenges of designing phasers for a distributed memory model and introduce our solutions to address these challenges.

(1) Efficient creation, signal aggregation and diffusion among participants. Communication costs are significantly higher than computation costs in a distributed memory model. Centralized algorithms lack scalability. Decentralized algorithms that grow sub-linearly in the number of communication interactions among participant tasks to perform phaser operations are necessary.

Skip lists [6] have long been used in shared memory environments, providing an expected time complexity of $O(\log n)$ for operations on a skip list containing n items. The items in a skip list participate at one or more levels. Every item participates at level 0. An item at level k participates at level $k + 1$ with probability p. A skip list does not require rebalancing after insertion/removal of items to maintain expected logarithmic time complexity for all operations.

Intuitively, determining the phase of a phaser is equivalent to retrieving the phase information resident on signalers organized as members of a skip list while performing a min-reduce of the phase information along the retrieval path.

(2) Efficient integration of dynamically created participants. The expected cost of including a task into a distributed phaser should be cost effective in terms of the number of communication interactions needed.

In our design, the number of communication interactions to either register or drop a task is sub-linear in the number of phaser participants.

(3) Concurrent synchronization and structural modifications. A distributed phaser design needs to provide a separation of concerns by allowing

---

[1] Our design of distributed phasers is idempotent to whether messages are one-sided (i.e., RDMA) or two-sided.

synchronization signals to propagate through the underlying data structure while structural modifications (adding or deleting a task) are in progress.

We achieve this concurrency by factoring a register/drop into a sequence of sub-operations that can be interleaved with signaling operations. In particular, we factor every register/drop into a "fast single-link-modify" step followed by a "lazy multi-link-modify" step similar to the one presented by Crain et al. [3] to support higher levels of concurrency in a distributed memory environment.

3   Distributed Phaser Design

Our design for a distributed phaser employs a pair of distributed skip lists. Signalers self organize into a signal collection skip list (referred to as $SCSL$), which is used to aggregate signals to a designated signaler at the head of the list. Waiters self organize into a signal notification skip list (referred to as $SNSL$) that is used to diffuse phase information from the head of the list to all the waiters.

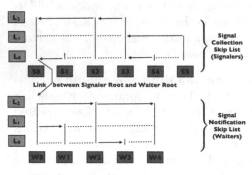

Fig. 1. Phaser synchronization achieved through signal collection and notification skip lists.

In a synchronization round, phase aggregation occurs in a right to left sweep with each signaler communicating the minimum phase of itself and its right neighbors to its highest level left neighbor in the $SCSL$. The designated signaler at the head of the $SCSL$ conveys the aggregated phase to the designated waiter at the head of the $SNSL$, who then initiates a left to right diffusion of the phase to all the waiters.

To support non-blocking `signal` operations, we separate the implementation of a task into actions by a computation and communication thread. The computation thread executes the task, informs the communication thread at `signal`, and proceeds without blocking. The communication thread interacts with other such threads to perform the required $SCSL$ actions. All the task actions described in this paper are those of the communication thread. We explicitly refer to the computation thread where necessary. In this section, we present detailed design descriptions of the creation and operations on $SCSL$. Managing the signal notification skip list - $SNSL$ is similar, but simpler compared to $SCSL$. For lack of space, we omit the design of $SNSL$.

3.1   Distributed Skip Lists Creation

`Create` is a collective operation among a set of tasks that is used to create a phaser. Each task can specify whether it wants to participate in the phaser and if so, its participation mode. Invoking the `create` operation leads to the creation of both *SCSL* and *SNSL*, for which we employ the $O(\log n)$-based recursive doubling algorithm developed by Egecioglu et al. [4] without wrap-around. The algorithm proceeds in $\log n$ rounds of communication. In each round i, a task communicates with its hypercube neighbors at 2^i links away and accumulates left and right "frontiers" that indicate visible neighbors at each level.

3.2   Synchronization Signal Aggregation

Definition 1. `Local_phase` is the number of calls to `signal` by a signaler's computation thread. `Subtree_phase` at a signaler is the minimum `subtree_phase` across all the right neighbors connected to the signaler in the *SCSL* at their highest level and `local_phase` at the signaler itself. These right neighbors are referred to as `from_neighbors`. Messages informing the `subtree_phase` at a signaler to the left neighbor at its highest level is called a synchronization signal. The left neighbor at the highest level is referred to as the `to_neighbor`.

For example, in Fig. 1, s_3 is the `from_neighbor` of s_2 and also the `to_neighbor` for both s_4 and s_5.

Single round of signal aggregation. In a round of signal aggregation, each signaler issues a request (SRQ) to its `from_neighbors` querying whether they can participate in the next phase, i.e., `subtree_phase+1`. On receiving an SRQ, a signaler waits for responses from all its `from_neighbors` and waits for `local_phase` to equal the requested phase. After the wait conditions are satisfied, the signaler increments its `subtree_phase` and sends a response (SRP) to its `to_neighbor` and forming a request-response chain, which begins at the designated `root` signaler task in the *SCSL*. The request-response chain might, however, result in the ripple of requests from the `root` to the farthest signaler participating in the *SCSL*, and the ripple of responses back to the `root` in every synchronization round. To mitigate the latency of such ripples, we require that a signaler issues SRQ for the subsequent synchronization round immediately after sending a response to its `to_neighbor`.

Definition 2 (Synchronization signal invariant). The `to_neighbor` of any signaler aggregates signals for the same or lower synchronization round than the signaler itself, formally:

$$\forall s \in SCSL, s.\text{subtree\_phase} \geq s.\text{to\_neighbor.subtree\_phase}$$

Since every signaler is transitively connected to the `root` in the *SCSL*, this invariant ensures that an increment of the `subtree_phase` at the `root` occurs only after all signalers in the *SCSL* have signaled for that round.

3.3   Registration of a Signaler

Our design supports the dynamic addition of a task into a phaser. Similar to phasers in shared memory, only a task currently participating in a phaser, referred to as parent, registers new tasks, referred to as children, into the phaser. By doing so, we provide the guarantee that a child begins participating in the same phase (local_phase+1) as that of its parent.

Inserting a child into the SCSL is decomposed into multiple steps to enhance concurrency. In each step, the granularity of locking is limited to a maximum of two links at two adjacent levels of the *SCSL*. First, a parent eagerly inserts a child into a link at the lowest level of the *SCSL*, i.e., L_0. Next, the child initiates a lazy hand-over-hand climb from one level to the next until its final level in the list; decision to move to level $k+1$ from k is based on probability p.

There are two modes of signal propagation for a child. After eager insertion, a child still needs the parent to propagate its signals since it might be at a lower phase than the subtree_phase of the left neighbor at L_0 of the *SCSL*. We refer to this state as a *transient* state. Once it reaches the subtree_phase of its left_neighbor at L_0, then it functions as a typical task in the *SCSL* and propagates its signals through its left neighbor. We refer to this state as a *normal* state. In the next two sections, we describe the two steps of eager- and lazy-insertion in detail.

3.3.1   Eager Single-Link Modify

Here, we describe in detail the pattern of communications between tasks in the *SCSL* needed to register a child task with the phaser. When a parent registers a child, the parent's computation thread blocks until the child is linked into the *SCSL* at L_0. The blocking ensures that no signals of the child are lost. To insert a child at L_0 of the *SCSL*, the first step is to find the location where the child should be linked. To do so, we employ the logical rank of the compute node on which a child will execute as a key.

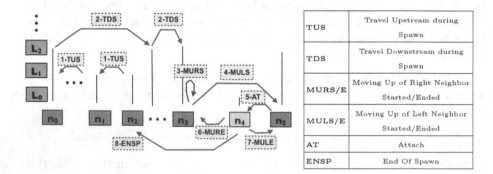

Fig. 2. Message sequence for addition of a task.

Figure 2 illustrates the message sequence for linking a child to L_0 of the *SCSL* based on the child's key. In the figure, n_2 links its child n_4 into the *SCSL*. The first step in this process is to find neighbors n_3 and n_5 such that n_4's key lies between n_3 and n_5. To do so, n_2 initiates an upstream message chain, 1-TUS, that hops from a task to its left_neighbor at its highest level terminating at a task n_0, such that n_4's key lies between n_0 and its right_neighbor or the highest level of the $SCSL$ is reached and n_0's right_neighbor's key is less than the child's key. n_0 then initiates a downstream message chain, 2-TDS, that hops from a task to its right_neighbor until it ends at the L_0 link where the child should be linked. The left_neighbor of this L_0 link, n_3, enqueues 3-MURS on itself because another inclusion/drop might occur concurrently preventing n_3 from handling the 3-MURS immediately. Once, n_3 dequeues 3-MURS, it verifies whether the link with its current right_neighbor, n_5, remains valid for the child's inclusion. If so, n_3 proceeds to lock the link to prevent other structural changes and informs n_5 of the child through 4-MULS. n_5 sets its to_neighbor to n_4, locks its left neighbor at L_0 and sends 5-AT to n_4. n_4 sets its left and right neighbors at L_0 to n_3 and n_5 respectively and sends 6-MURE, 7-MULE and 8-ENSP. n_4 starts in the same phase as n_2. If subtree_phase of n_3 is higher than that of n_2, then n_4 needs to send signals to n_2 till it catches up with the synchronization round of n_3, i.e., *transient* state. On receipt of 6-MURE, n_3 sets its right_neighbor at L_0 to n_4 and unlocks it. On receipt of 7-MULE, n_5 sets its left_neighbor at L_0 to n_4 and unlocks it. The right_neighbor of n_3 and left_neighbor of n_5 are set at the end to ensure that search messages such as 1-TUS and 1-TDS are never blocked and go through a transient task only after its completely linked at L_0. On receipt of 8-ENSP, n_2 determines whether to maintain a signaling link with child task (n_4) and notifies its blocked computation thread to proceed.

3.3.2   Lazy Multi-link Modify

The lazy hand-over-hand movement of a child to its final height in the $SCSL$ does not begin until the child completes transition from *transient* state, i.e., signals through its parent, to *normal* state, i.e., signals through its left_neighbor at L_0 in the $SCSL$. The transition to *normal* state occurs once the child reaches the subtree phase of its left_neighbor at L_0. In *normal* state, at each level k, the child decides to move to level $k+1$ based on probability p until it reaches its final height. To move to level $k+1$, it needs to determine its neighbors at level $k+1$. Using a message chain similar to 1-TUS, the first neighbor on the left of the child with a height of $k+1$ is determined. This neighbor, its right_neighbor at level $k+1$, and the child interact in a hand-shake message sequence exactly like the one for eager insertion to move the child to level $k+1$.

For lack of space, we do not provide details about the drop operation. The message exchanges are similar to the inclusion except the signaler is moving lazily from $k+1$ to k before delinking itself from the $SCSL$ completely.

3.4   Verification of $SCSL$

In this section, we show the correctness of $SCSL$ operations with model checking [2]. In model checking, given a system (specified as a *configuration*) and some properties, a model checker tests these properties in all possible execution paths of the system. The goal of the *SCSL* verification is to show that the signal aggregated at the root is inclusive of signals from all registered signalers who haven't drop'ed; we call this property *root aggregation correctness*. To this end, we define a set of linear temporal logic (LTL) formulae that capture the root aggregation property. We check whether these formulae are satisfied during model checking. We employ a "message"-based strategy that consists of model-checking LTLs against a different configuration for each message type, say 1-TUS. We do so because a naive process-based model checking strategy required more than 1TB of RAM in our experiments.

We realize our verification using the state-of-the-art model checker Spin [5]. The complete set of LTLs and configurations is available online at: http://goo.gl/ypuhaq.

3.4.1   Root Aggregation Correctness

We introduce three categories of properties: synchronization signal, structural consistency, and progress.

Synchronization Signal. Every signaler signals to its to_neighbor only after its from_neighbors and itself have signaled, i.e., *SCSL* maintains the synchronization signal invariant at all times. The synchronization signal invariant, Definition 2, guarantees the integrity of the phase aggregated at the root of the *SCSL*. The LTLs that capture this invariant are as follows:

- $\Box(\forall i, (!\ \texttt{is\_transient}(n_i) \implies$
  $(n_i.\texttt{subtree\_phase} \geq n_i.\texttt{to\_neighbor.subtree\_phase}))$
- $\Box(\forall i, (\texttt{is\_transient}(n_i) \implies$
  $n_i.\texttt{left\_neighbor}[\texttt{cur\_height}].\texttt{subtree\_phase} > n_i.\texttt{subtree\_phase}))$

Structural Consistency. Every signaler is transitively connected to the root, i.e., *SCSL* maintains structural consistency at all times. Every signaler has a single to_neighbor whose identifier is lesser than its own, and every signaler has at most one from_neighbor at each level of *SCSL*. This prevents any independent clusters in the *SCSL* and guarantees eventual connectivity to the root of the *SCSL*, thereby, ensuring that no signal from a signaler is lost.

- $\Box(\forall i, \forall L, (n_i.\texttt{left\_neighbor}[L] < n_i < n_i.\texttt{right\_neighbor}[L]))$ states that for every signaler, its identifier is always between its left_neighbor and right_neighbor at every level in which it participates. This monotonically increasing task-to-to_neighbor chain ensures that there are no independent loops of signalers that are not attached to the *SCSL*.
- $\Box(\forall i, \forall L, (n_i == n_i.\texttt{left\_neighbor}[L].\texttt{right\_neighbor}[L]))$ states that for every signaler, the right_neighbor's left_neighbor is the signaler itself.
- $\Box(\forall i, n_i == n_i.\texttt{to\_neighbor.from\_neighbor}[\texttt{height}(n_i)]))$ states that every signaler always has a to_neighbor and that the from_neighbor of the to_neighbor at the height of the signaler is always the signaler itself.

Progress *SCSL* is deadlock- and livelock-free. This requirement ensures progress.

3.4.2   Message-Based Verification

Every phaser operation in our design is implemented as a series of message exchanges in the *SCSL*, where every message is handled atomically and terminates with the initiation of the next message needed for the operation. For example, if a task processes a 1-TUS then it either sends a 1-TUS or initiates the 2-TDS and does so atomically. Therefore, if each message of an operation can be processed correctly under any possible structural change and every message completes by starting the next message needed for the operation, then the operation is guaranteed to function correctly.

Message-based Modeling and Model Checking. Our scheme uses a quorum of processes, signalers in our case, to undergo structural changes that challenge the successful completion of a single message in an operation on the *SCSL*. The structural changes include the source of the message delinking from the *SCSL* or moving lazily to a higher level, the destination of the message delinking from the *SCSL* or moving lazily to a higher level, and a new signaler linking between the source and destination and later delinking itself. These processes also have to complete a specific set of synchronization rounds. In the presence of such structural changes, if a message successfully completes, the LTL constraints are satisfied, and the specific number of synchronization rounds are complete, then we conclude that the handling of that message is correct.

Verifying 1-TUS message. Consider the 1-TUS message in Fig. 2. n_2 initiates a 1-TUS to n_1 in the *SCSL*. The following structural changes can occur: n_1 can move down from L_1 to L_0, and n_1's new neighbor at L_0, say n_{01}, can drop out of the *SCSL*. To ensure the successful handling of 1-TUS message in these scenarios, we model check a configuration of 6 signalers $n_{0,01,1,2,3,4}$ such that n_2 inserts n_4, n_1 and n_{01} undergo structural changes as mentioned above. This configuration along with others needed to verify eager insertion are present in Table 1. In Table 1, column 1 describes the message while column 2–6 lists configurations of 5 tasks; the root n_0 participates at all levels, does not undergo structural changes, and hence, omitted from the table. Column 7 specifies the memory consumed and Column 8 specifies the number of states explored. A configuration of the task is specified as L:X*, where L indicates the initial level and X* is the sequence of operations comprising of D (drop), M(lazy move up), E[i] (eager insertion with parent task i).

A Model of SCSL in PROMELA. The input specification to Spin is the *SCSL* implemented in PROMELA along with the LTLs. We implement the *SCSL* as a group of processes (`proctypes`), one for each signaler. These signalers interact with each other using `channels`; a `channel` holds messages sent from one process to another. Every signaler is configured to perform a specific number of phase advancements and its probabilistic height is decided a priori based on the configuration needed to verify a specific message. Every signaler executes a message-driven progress engine, which on receipt of a specific message responds with messages as specified in previous sections. We model check our configu-

Table 1. Configurations used to model check the eager insertion of a signaler.

Message	n_{01}	n_1	n_2	n_3	n_4	Mem (GB)	States
TUS	L_0:D	L_1:D	L_1	L_2	:E(2)	135	1.1e10
TDS	L_1:D	L_0:D	:E(0)	L_1	-	23	1.7e9
MURS	:E(0)D	:E(0)	L_0:D	-	-	10	5.6e8
MULS-1	L_0	L_0	:E(01)	:E(0)	-	78	7.4e9
MULS-2	L_0	L_0	:E(01)	L_0:M	-	86	6.7e9
MULS-3	L_0	L_0	:E(01)	L_1:D	-	50	4.3e9
AT	L_1	:E(0)	:E(0)MDD	-	-	6	3.1e8
ENSP	L_1	:E(0)	-	-	-	1	5.4e7

rations on a POWER7 compute node with 256 GB RAM. A few experiments that needed more memory than 256 GB were run on NERSC's Carver system, which had 1TB RAM. In total, we employed 23 configurations to verify all the messages in all operations on the $SCSL$.

Design Influenced by Model Checking: Tagging Messages with Link-sequence Numbers. Monotonically increasing unique integral identifiers are assigned to links between tasks in the $SCSL$ and messages are tagged with them. This design feature avoids problems due to stale messages. Consider the scenario in which n_3 initiates a move into the link between n_2 and n_4 at L_i. Concurrently, n_4 also decides to move into the next level, i.e., L_i to L_{i+1}, and issues an 1-LLNL to n_2. Before the 1-LLNL is processed at n_2, the following events occur: n_3 moves into the link between n_2 and n_4, n_3 processes the move up of n_4, n_3 drops out of the phaser, n_4 drops a level relinking itself to n_2, and n_2 processes the 1-LLNL issued by n_4 prior to these events. Processing the stale 1-LLNL leads to n_2 locking the link n_2-n_4 without n_4 having any intention of moving to the upper level. This led to the introduction of link identifiers.

4   Complexity Analysis

In this section, we present complexity analysis of synchronization and structure modification operations on the signal collection skip list - $SCSL$.

Complexity of Signal Aggregation. The expected critical path length in a skip list from any task to the **root** is logarithmic in the number of tasks in the skip list. Hence, the expected time complexity taken by a signal from any participant in the $SCSL$ to reach the designated **root** is $O(\log n)$, where n is the total number of signalers. The expected time complexity to aggregate signals from all the signaler tasks is also $O(\log n)$ since the aggregation occurs in parallel across all such chains.

Complexity of Participant Addition. Here, we present complexity analysis of the expected number of message hops, i.e., pairwise communications, needed to insert a task to the $SCSL$. Eager insertion requires a skip list search, $O(\log n)$, to find the position to attach and a constant number of operations to finalize attach. Hence, eager insertion has a time and message complexity of $O(\log n)$. The rest of this discussion derives the complexity for moving a task lazily from L_0 to its eventual height.

Let there be a group of tasks that are lazily moving up to the higher levels between two stable tasks; stable tasks are those that have already reached their final height. We use K_i^j to indicate the jth task at L_i and use $|K_i^j|$ to represent the distance between the left stable task and K_i^j. To this end, we abstract our model by making the following assumptions: (1) When considering the movement of tasks from L_i to L_{i+1}, there is a uniform probability distribution over the orders in which they move up. For example, if tasks K_i^1, K_i^2, K_i^3 are moving up, then any of the 6 possible orders are equally likely. (2) The number of hops required for task K_i^j is

(a) $|K_i^j|$, if there is no task $|K_i^l| < |K_i^j|$ that moves to L_{i+1} before K_i^j, and

(b) $|K_i^j| - |K_i^l|$, if K_i^l moves to L_{i+1} before K_i^j and there is no other task K_i^t that reaches before K_i^j and $|K_i^t| > |K_i^l|$.

The key idea in our complexity analysis is to compute the expected number of messages for an arbitrary link, L_i. We then sum up the number of messages across levels and divide by the total number of inserted tasks to obtain per inserted task analysis. Before stating the main result, we prove three helper lemmas. Let m_i denotes the total number of intervals at L_i and m_T denote the total number of intervals at L_0.

Lemma 1. *Let C be the interval contention at L_0 in the $SCSL$ and let the interval contention at L_i be denoted by C^i. Then $C * p^i \leq E[\mathsf{C}^i] \leq C$.*

Proof. Let X be the number of newly inserted tasks that move to L_i and Y is the number of stable tasks excluding root that are present at L_i. X and Y are independent of each other and are binomially distributed with probability p^i. m_T is the total number of intervals at L_0. By definition, $\mathsf{C}^i = X/(Y+1)$ and hence, $E[\mathsf{C}^i] = E[X/(Y+1)]$. Since X and Y are independent and binomially distributed with probability p^i, $E[X] = m_T C p^i$ and $E[1/(Y+1)] = \frac{(1-(1-p^i)^{m_T})}{m_T p^i}$. Since, $E[\mathsf{C}^i] = E[X] * E[1/(Y+1)]$, we have $C * p^i \leq E[\mathsf{C}^i] \leq C$. □

Lemma 2. *Let $K_i = \{K_i^1, \cdots, K_i^{n_i}\}$ be the tasks that move up from L_i to L_{i+1}, then the expected value of total number of hops for K_i, denoted by $E[\mathsf{Cost}(K_i)]$, is $\Sigma_{j=1}^{n_i} \frac{|K_i^j|}{n_i+1-j}$.*

Proof. We first note that $E[\mathsf{Cost}(K_i)] = \Sigma_{j=1}^{n_i} E[\mathsf{Cost}(K_i^j)]$. To compute, $E[\mathsf{Cost}(K_i^j)]$, we further partition the space of different configurations based on the order in which K_i^j moves up and use $M(K_i^j, r)$ to denote the event that K_i^j is r^{th} task to reach the level $i+1$. Note that $\mathsf{Cost}(M(K_i^j, r))$ depends only on the largest $K_i^l < K_i^j$ that reaches L_{i+1} before K_i^j. To this end, we use $MO(K_i^j, r, l)$ to denote the event that K_i^j is r^{th} task to reach L_{i+1} and K_i^l reaches before

K_i^j and there is no other task K_i^t that reaches before K_i^j and $|K_i^t| > |K_i^l|$. We use $MO(K_i^j, 1, 0)$ to denote the event when K_i^j is the first task to reach L_{i+1}. Therefore, $E[\mathsf{Cost}(K_i)] = \Sigma_{j=1}^{n_i} \Sigma_{r=1}^{n_i} \Sigma_{l=0, l \neq j}^{n_i} E[\mathsf{Cost}(MO(K_i^j, r, l))]$. The rest of the proof is completed by first computing $E[\mathsf{Cost}(MO(K_i^j, r, l))]$ and then applying algebraic simplifications to compute $E[\mathsf{Cost}(K_i)]$.

To compute $E[\mathsf{Cost}(MO(K_i^j, r, l))]$, we first note that $E[\mathsf{Cost}\ (MO(K_i^j, r, l))]$ $=$ $Pr(MO(K_i^j, r, l))$ $\times \mathsf{Cost}(MO(K_i^j, r, l))$. Next, $Pr(MO(K_i^j, r, l))$ is (a) $\frac{1}{n_i} \prod_{t=1}^{n_i - l}(\frac{n_i - r - t - 1}{n_i - t})$ for $r \neq 1, j > l - 1$, (b) 0 for $r \neq 1, j <= l - 1$ and (c) $1/n$ for $r = 1$. Also, $\mathsf{Cost}(MO(K_i^j, r, l)) = |K_i^j| - |K_i^l|$ if $r \neq 1$, $|K_i^j|$ otherwise. Therefore, $E[\mathsf{Cost}(K_i^j)] = |K_i^j| - \Sigma_{t=1}^{j-1} \frac{|K_i^{j-t}|}{t(t+1)}$. Summing up over j, we obtain $E[\mathsf{Cost}(K_i)] = \Sigma_{j=1}^{n_i} \frac{|K_i^j|}{n_i + 1 - j}$. To simplify this cost expression, we use the following lemma.   □

Lemma 3. Let $|K_i^*| = \min_{j=1}^{n_i/2} \frac{|K_i^j| + |K_i^{n_i+1-j}|}{2}$, then $E(|K_i^*|) \geq \frac{p}{4} C^i$.

Proof. $|K_i^*| = \min_{j=1}^{n_i/2} \frac{|K_i^j| + |K_i^{n_i+1-j}|}{2} \geq \min_{j=1}^{n_i/2} \frac{|K_i^j|}{2} + \min_{j=1}^{n_i/2} \frac{|K_i^{n_i+1-j}|}{2}$. Therefore, $|K_i^*| \geq \frac{1}{2} + \frac{|K_i^{n_i/2}|}{2} \geq \frac{|K_i^{n_i/2}|}{2}$. Since $E(|K_i^{n_i/2}|) \geq \frac{p}{2} C^i$, $E(|K_i^*|) \geq \frac{p}{4} C^i$.   □

Theorem 1. Let $E[H_C]$ be the expected number of hops consumed by a task inserted at L_0 to reach stable state, then $\Omega(p^3 \log(Cp^3)) \leq E[H_C] \leq \mathcal{O}(\frac{p}{1-p} \log(C \frac{p}{1-p}))$.

Proof. To compute expected number of hops per task, we take the ratio of expected number of hops for all tasks inserted at L_0, denoted by $E[H_C^T]$ and the total number of tasks at L_0. Let $H_C^{T,i}$ denote the total number of hops consumed by tasks moving from L_{i-1} to L_i, then $E[H_C^T] = \Sigma_i E[H_C^{T,i}]$. From Lemma 2, we have $E[H_C^{T,i}] = E[m_i \Sigma_{j=1}^{n_i} \frac{|K_i^j|}{n_i+1-j}]$. Using Lemma 3 and $\forall j, K_i^j < K_i^{n_i}$, we have $E[m_i \Sigma_{j=1}^{n_i} \frac{|K_i^*|}{n_i+1-j}] \leq E[H_C^{T,i}] \leq E[m_i \Sigma_{j=1}^{n_i} \frac{|K_i^{n_i}|}{n_i+1-j}]$. From the proof of Lemma 1, we know that $E[n_i] = E[C^i]p$. Similarly, following the proof of Lemma 1, we have $E[m_i] = m_T p^i$. Since $\Omega(\log n_i) \leq \Sigma_{j=1}^{n_i} \frac{1}{n_i+1-j} \leq \mathcal{O}(\log n_i)$. Next, $E[K_i^{n_i}] \leq C$ and noting the random variables m_i, n_i, K_i are independent, we have $m_T p^i \frac{p}{2} E[C^i] \Omega(\log E[n_i]) \leq E[H_C^{T,i}] \leq m_T p^i C \mathcal{O}(\log E[n_i])$. Hence, $m_T p^i C \frac{p^{i+1}}{4} \Omega(\log(Cp^{i+1})) \leq E[H_C^{T,i}] \leq m_T p^i C \mathcal{O}(\log Cp)$. Therefore, $m_T Cp^3 \Omega(\log(Cp^3)) \leq E[H_C^T] \leq m_T C \frac{p}{1-p} \mathcal{O}(\log(C \frac{p}{1-p}))$. Noting that the total number of tasks inserted at L_0 is $m_T C$ we have, $\Omega(p^3 \log(Cp^3)) \leq E[H_C] \leq \mathcal{O}(\frac{p}{1-p} \log(C \frac{p}{1-p}))$ □

5   Related Work

Agarwal et al. present a distributed version of X10 clocks [1]. In this protocol, each task consults a local snapshot to determine the participant tasks and to make a decision about moving to the next phase. Processes add or drop themselves from these local snapshot. The authors, however, do not depict how this information is exchanged and state that in a basic implementation, one would require $O(n^2)$ messages. Our protocol describes the complete set of actions needed to ensure a total of $O(n)$ messages and $O(\log n)$ time complexity for synchronization using distributed skip lists.

In the non-blocking skip list protocol presented by Crain et al. [3], changes to the skip list structure are divided into two stages: eager abstract modification and lazy structural adaptation. They employ a single adaptive thread with global information to perform the structural changes based on neighborhood information. Our protocol is similar with two stages for insertion and deletion, but does not rely on an adaptive thread to perform the structural changes.

6 Conclusions

In this paper, we present a design for phasers, a general barrier-like synchronization construct that supports dynamic addition and deletion of parallel tasks, for a distributed memory-environment. Our design is based on a pair of distributed concurrent skip lists augmented with the ability to aggregate and diffuse phaser synchronization signals. By employing eager- and lazy-strategies while performing structural operations, our distributed phaser design supports a high-degree of concurrency. We employ a novel "message-based" model checking scheme to prove the correctness of our design. We derive the expected cost of signal aggregation, i.e., $\log n$ and cost for inclusion of a new task in the presence of interval contention C, i.e., $\Omega(p^3 \log(Cp^3)) \leq E[H_C] \leq \mathcal{O}(\frac{p}{1-p} \log(C\frac{p}{1-p}))$.

Acknowledgment. We would like to thank Moshe Y. Vardi for the insightful discussions regarding verification. This research was supported in part by the DOE Office of Science Advanced Scientific Computing Research program through collaborative agreement DE-FC02-12ER26105. This research used resources of the National Energy Research Scientific Computing Center, a DOE Office of Science User Facility supported by the Office of Science of the U.S. Department of Energy under Contract No. DE-AC02-05CH11231.

References

1. Agarwal, S., Joshi, S., Shyamasundar, R.K.: Distributed generalized dynamic barrier synchronization. In: Aguilera, M.K., Yu, H., Vaidya, N.H., Srinivasan, V., Choudhury, R.R. (eds.) ICDCN 2011. LNCS, vol. 6522, pp. 143–154. Springer, Heidelberg (2011)
2. Clarke, E.M., Grumberg, O., Peled, D.: Model Checking. MIT Press, Cambridge (1999)
3. Crain, T., Gramoli, V., Raynal, M.: No hot spot non-blocking skip list. In: Proceedings of ICDCS, pp. 196–205 (2013)
4. Egecioglu, O., Koc, C.K., Laub, A.J.: A recursive doubling algorithm for solution of tridiagonal systems on hypercube multiprocessors. J. Comput. Appl. Math. **27**(1), 95–108 (1989)

5. Holzmann, G.J.: The model checker SPIN. IEEE Trans. Softw. Eng. **23**(5), 279–295 (1997)
6. Pugh, W.: Skip lists: a probabilistic alternative to balanced trees. Commun. ACM **33**(6), 668–676 (1990)
7. Saraswat, V., et al.: Source Distribution of X10 V2.5.1 (2014). http://sourceforge.net/projects/x10/files/x10/2.5.1
8. Saraswat, V., et al.: X10 Language Specification Version 2.5 (2014). http://x10.sourceforge.net/documentation/languagespec/x10-latest.pdf
9. Sarkar, V., Harrod, W., Snavely, A.E.: Software challenges in extreme scale systems. J. Phys. Conf. Ser. **180**(1), 12 (2009)
10. Schöne, R., et al.: Tools and methods for measuring and tuning the energy efficiency of HPC systems. Sci. Program. **22**(4), 273–283 (2014)
11. Shirako, J., Peixotto, D.M., Sarkar, V., Scherer, W.N.: Phasers: a unified deadlock-free construct for collective and point-to-point synchronization. In: Proceedings of ICS, pp. 277–288. ACM (2008)
12. Shirako, J., Sarkar, V.: Hierarchical phasers for scalable synchronization and reductions in dynamic parallelism. In: Proceedings of IPDPS, pp. 1–12 (2010)
13. Welch, P., et al.: Alting barriers: synchronisation with choice in Java using JCSP. Concur. Comput. Pract. Exp. **22**(8), 1049–1062 (2010)

Exploring Partial Replication to Improve Lightweight Silent Data Corruption Detection for HPC Applications

Eduardo Berrocal[1](✉), Leonardo Bautista-Gomez[2],
Sheng Di[2], Zhiling Lan[1], and Franck Cappello[2]

[1] Illinois Institute of Technology, Chicago, IL, USA
{eberroca,lan}@iit.edu
[2] Argonne National Laboratory, Lemont, IL, USA
{loobago,sdi1,cappello}@anl.gov

Abstract. Silent data corruption (SDC) poses a great challenge for high-performance computing (HPC) applications as we move to extreme-scale systems. If not dealt with properly, SDC has the potential to influence important scientific results, leading scientists to wrong conclusions. In previous work, our detector was able to detect SDC in HPC applications to a certain level by using the peculiarities of the data (more specifically, its "smoothness" in time and space) to make predictions. Accurate predictions allow us to detect corruptions when data values are far "enough" from them. However, these data-analytic solutions are still far from fully protecting applications to a level comparable with more expensive solutions such as full replication. In this work, we propose partial replication to overcome this limitation. More specifically, we have observed that not all processes of an MPI application experience the same level of data variability at exactly the same time. Thus, we can smartly choose and replicate only those processes for which our lightweight data-analytic detectors would perform poorly. Our results indicate that our new approach can protect the MPI applications analyzed with 49–53 % less overhead than that of full duplication with similar detection recall.

Keywords: Silent data corruption detection · Partial replication · Data analysis · HPC applications

Government License Section: The submitted manuscript has been created by UChicago Argonne, LLC, Operator of Argonne National Laboratory ("Argonne"). Argonne, a U.S. Department of Energy Office of Science laboratory, is operated under Contract No. DE-AC02-06CH11357. The U.S. Government retains for itself, and others acting on its behalf, a paid-up nonexclusive, irrevocable worldwide license in said article to reproduce, prepare derivative works, distribute copies to the public, and perform publicly and display publicly, by or on behalf of the Government.

© Springer International Publishing Switzerland 2016
P.-F. Dutot and D. Trystram (Eds.): Euro-Par 2016, LNCS 9833, pp. 419–430, 2016.
DOI: 10.1007/978-3-319-43659-3_31

1   Introduction

Silent data corruption (SDC) involves corruption to an application's memory state (including both code and data) caused by undetected soft errors, that is, errors that modify the information stored in electronic devices without destroying the functionality [13]. If undetected, these errors have the potential to be damaging since they can change the scientific output of HPC applications and mislead scientists with spurious results.

External causes of transient faults are usually rooted in cosmic ray particles hitting the electronic devices of the supercomputer [22]. As systems keep scaling up, the increasing number of devices will make these external faults appear more often. Other techniques introduced to deal with excessive power consumption, such as aggressive voltage scaling or near-threshold operation, as well as more complex operating systems and libraries, may also increase the number of errors in the system [7].

Substantial work has been devoted to this problem, both at the hardware level and at higher levels of the system hierarchy. Currently, however, HPC applications rely almost exclusively on hardware protection mechanisms such as error-correcting codes (ECCs), parity checking, or chipkill-correct ECC for RAM devices [10,19]. As we move toward the exascale, however, it is unclear whether this state of affairs can continue. For example, recent work shows that ECCs alone cannot detect and/or correct all possible errors [16]. In addition, not all parts of the system, such as logic units and registers inside the CPUs, are protected with ECCs.

With respect to software solutions, full process replication provides excellent detection accuracy for a broad range of applications. The major shortcoming of full replication is its overhead (e.g., $\geq 100\%$ for duplication, $\geq 200\%$ for triplication). Another promising solution is data-analytic-based (DAB) fault tolerance [2,6,9,26], where detectors take advantage of the underlying properties of the application data (the smoothness in the time and/or space dimensions) in order to compute likely values for the evolution of the data and use those values to flag outliers as potential corruptions. Although DAB solutions provide high detection accuracy for a number of HPC applications with low overhead, their applicability is limited because of an implicit assumption—the application is expected to exhibit smoothness in its variables all the time.

In this work, we propose a new adaptive SDC detection approach that combines the merits of replication and DAB. More specifically, we have observed that not all processes of some MPI applications experience the same level of data variability at exactly the same time; hence, one can smartly choose and replicate only those processes for which lightweight data-analytic detectors would perform poorly. In addition, evaluating detectors solely on overall single-bit precision and recall may not be enough to understand how well applications are actually protected. Instead, we calculate the probability that a corruption will pass unnoticed by a particular detector. In our evaluation, we use two applications dealing with explosions from the FLASH code package [12], which are excellent candidates for testing partial replication. Our results show that our adaptive approach is able to protect the MPI applications analyzed (99.999 % detection recall) replicating only 43–51 % of all the processes with a maximum total overhead of only 52–56 % (compared with 110 % for pure duplication).

The rest of the paper is organized as follows. In Sect. 2 we describe how DAB SDC detectors work. In Sect. 3 we introduce our adaptive method for SDC detection. In Sect. 4 we describe the probabilistic evaluation metric used. In Sect. 5 we present our experimental results. In Sect. 6 we discuss related work in this area. In Sect. 7 we summarize our key findings and present future directions for this work.

2 Data-Analytic-Based SDC Detectors

In this section we describe how DAB detectors work. We also point out their major limitations.

Lightweight DAB SDC detectors are composed of two major parts. The *predictor* component computes a prediction for the next value of a particular data point. The prediction takes advantage of the underlying physical properties of the evolution of the data, since we have observed that this evolution is *smooth* in the time and/or space dimensions for a wide range of variables in HPC scientific applications. After the prediction is done, the *detector* component decides whether the current value of the data point is corrupted.

We have implemented our lightweight DAB SDC detectors inside the Fault Tolerance Interface (FTI) [4], an MPI library for HPC applications to perform fast and efficient checkpoint/restarts (C/R). We can add SDC detection support by taking advantage of the fact that iterative applications already provide (to FTI) the data variables representing their *state*. An HPC application needs to perform only one extra call to FTI: a call at the end of every iteration to allow our detectors to check for SDC in the data.

In our previous work we showed that one can detect a large number of corruptions by using simple and lightweight predictors. For the time dimension, we found that quadratic curve fitting (QCF) outperformed all the other considered options [6] with a memory overhead of less than 90 % for all the applications studied. Another way to do predictions is by using the spatial information instead of the temporal information. In [3], 3D linear interpolation was used succesfully to predict values in a computational fluids dynamics (CFD) miniapplication.

Once a prediction $X(t)$ has been made, our detector decides whether the current value of the data $V(t)$ is a normal value by checking whether it falls inside a particular *confident interval* determined by a parameter δ: $[X(t) - \delta, X(t) + \delta]$. We calculate δ using the maximum prediction error from all data points in a process at $t - 1$ multiplied by some constant: $\delta = \lambda \cdot e_{max}(t - 1)$. This constant λ determines a tradeoff between detection *recall* (how many real corruptions can we actually detect) and *precision* (how many of the detected corruptions are actually real corruptions). In our case, the value for λ is chosen to have zero false positives given a particular execution size (i.e., to maximize *precision*).

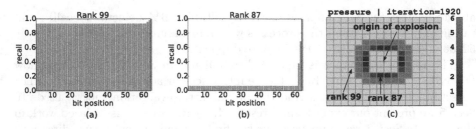

Fig. 1. Detection recall for two different processes in Sedov during 100 iterations.

When data values change too abruptly in a particular process, our δ becomes far too big to detect barely any corruption. Two examples of this kind of application dealing with sharp changes in the data are *Sedov* and *BlastBS*. Sedov is a hydrodynamical test code involving strong shocks and nonplanar symmetry [21]. BlastBS, on the other hand, is a 3D version of the magnetohydrodynamical spherical blast wave problem [27]. Both are part of the FLASH simulation code package.

To illustrate the problem at hand, we show in Fig. 1(a) and (b) detection recall rates for single bit-flips injected on each bit position over two different processes in the variable *pressure* of Sedov during a particular period of time (100 iterations). One can see how the wave of the explosion passing through rank 87 is making detection recall rates decrease substantially for this variable[1]. In contrast, detection recall is high in rank 99. To get a glimpse of how this data looks like, consider Fig. 1(c). Here, we show the state of the maximum of variable *pressure* right after the window of 100 time steps has passed.

3   Adaptive Method

Full replication is generally considered too costly for HPC because of its high overhead both in the time and the space dimensions. Partial replication, however, is worth considering for applications for which sharp changes in the data occur only in a small subset of the processes, such as those involving explosions or collisions (e.g., Sedov). Considering again the example introduced in Sect. 2, we can see that duplicating rank 99 in this situation is a major waste of resources, while rank 87 can surely benefit from replication, making detection recall go from below 10 % in the majority of bits to 100 % in all of them automatically.

One way to detect corruptions efficiently by using replication, proposed by Fiala et al. [11], is by comparing messages in MPI. The idea is that any corruption in the data of a particular process will ultimately produce corrupted messages that will be sent to other process. By comparing messages from replicas of the same process, one can determine whether that process (or any of its replicas)

---

[1] The rank of a process in MPI is its ID inside a group of processes. In this paper we consider only the rank of the general group to which all processes belong. In this sense, we use rank(s) or process(es) interchangeably.

got its data corrupted. In this paper we adopt an adaptive strategy. For some processes (replication set), we use partial replication based on the method of Fiala et al. For the other processes, we use our lightweight DAB detectors.

We implement the following strategy in order to select our replication set and to dynamically adapt it over time. After the first iteration, we choose a subset of processes to replicate, given the maximum prediction error in that first iteration. The number of processes to replicate is determined by the *replication budget B*. During the following w application time steps (where w defines a *window*), we create an array S of size n, which is the number of processes in the application. After every time step, we sort all processes in ascending order given their maximum prediction errors and add their positions in S. For example, if at a particular time step, rank 12 is the one with the highest prediction error and there are 128 processes, then $S[12]$ += 128. When w steps have passed, the score S represents an aggregation of the relative positions of each rank with respect to the others given their prediction errors during the window w. At this point we sort all processes by their score S, pick the top B (which is the allocated budget) as the new replication set, and reset S to start a new window again.

4   Probabilistic Evaluation Metric

In order to understand why this metric is needed, consider the case where we have a mechanism with perfect detection recall for the 22 most significant bits of 32-bit numbers. What is the probability that, in this particular example, a corruption will evade our detector? The answer to this question will actually depend on how many bits can get "flipped" in the memory state of the application. Assuming 1-bit-flip corruptions only, we could say that $10/32 = 0.3125$ (31 % of corruptions will pass undetected). For 2-bit-flips, the probability would be $(10/32) \times (9/31) = 0.0907$ (9.07 % corruptions will pass undetected). For our detector to be unaware of a 2-bit-flip corruption in this case, all flips would always need to hit bits in the 10 less significant positions of the mantissa. We could continue with 3-bit-flips, 4, and so on. An interesting observation from this example is that, generally, the fewer bits that can get "flipped" in a system, the harder it is to detect corruptions using software mechanisms. Furthermore, another interesting question appears: What is the distribution of corruption sizes (in terms of the number of bits) in the system? Is a corruption affecting a large number of bits more or less common than one affecting just a few? The key idea is that protecting the data of simulations at this level is not so much protecting against particular bit-flips as it is protecting against numerical deviations from the original data.

In this work we use an evaluation metric based on the probability that a corruption will pass unnoticed by a particular detector. Since we aim at designing general SDC detectors, we cannot assume that bit-flips in the less significant bits of the mantissa are harmless. For example, we performed a sensitivity study where we injected different corruption sizes on multiple applications (the full study is omitted because of space limitations). In that study we observed that the same exact corruption produces different impacts[2] on different applications, i.e., there are applications that can absorb the corruption effortlesly while there are others that suffer big data deviations.

The evaluation metric, which we also call the *probability of undiscovered corruption*, is defined as

$$P_f = \sum_{i=1}^{N} \left[P(\#\text{bits} = i) \times (1 - (\frac{1}{N} \times \sum_{j=1}^{N} r_j))^i \right], \tag{1}$$

where N is the number of bits per data point (i.e., 64), $\frac{1}{N} \times \sum_{j=1}^{N} r_j$ represents the average recall rate for all bit positions collected during our injection studies, and $P(\#\text{bits} = i)$ represents the probability that the corruption is exactly i bits long.

The distribution $P(\#\text{bits} = x)$ depends on how corruptions in the whole system ultimately affect the numerical data of simulations. Because of the impossibility of calculating this distribution for a system as massive as a supercomputer, we assume four distributions representing the following four cases: (1) the number of bits affected is usually small, with 1 bit being the most common size (for this case, we use a Poisson distribution with $\lambda = 1.0$); (2) all bit sizes are equally probable (i.e., $P(\#\text{bits} = x) = 1/N$); (3) all possible corruptions (2^N) are equally probable (e.g., $P \sim \mathcal{N}(32.5, 13.05)$ for $N = 64$); and (4) the number of bits affected is usually big, with N bits being the most common size (for this case, we use the inverse of distribution (1)).

5   Evaluation

We use two applications from the FLASH code package in our experiments—Sedov and BlastBS—representing two different types of explosions. These applications are excellent candidates for testing the effectiveness of partial replication for data experiencing sharp changes due to explosions and collisions. Implementations of MPI allowing replication at the process level, such as RedMPI [11], do not yet support partial replication; we simulate partial replication by considering precision and recall to be 100 % for those processes that are part of the *replication set*[3]. For the others, we use our lightweight SDC detectors.

Detection recall for each bit position is calculated by averaging the results over hundreds of random injections on the pressure variable in every process over thousands of time steps. For all our experiments, we set our λ parameter, which controls our dynamic *detection range* δ (see Sect. 2) to have exactly zero false positives. In all the experiments, we run the applications using 256 processes,

---

[2] Impact is defined as the rate of deviation over the variable's total data range during the execution. For example, a deviation of 10 on a [0,200] range produces an impact of 0.05.

[3] Of course, this only holds for deterministic applications (which is the case here).

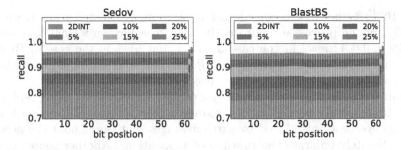

Fig. 2. Single-bit detection recall results from our injection study. We use two applications (Sedov and BlastBS) running 256 processes, and we set $w = 100$. Five partial replication rates (5–25%) are compared with nonreplication (2DINT).

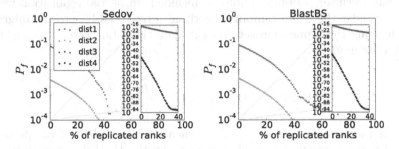

Fig. 3. Probability of undiscovered corruption when replicating a particular percentage of processes using the four distributions $P(\#\text{bits} = x)$ described in Sect. 4. Note that the y-axis is plotted using logarithmic scale. The small subplots represent the data zoomed below 10^{-15}.

while the data domain is configured to be a two-dimensional grid. We report the results only of those experiments using linear interpolation (spatial) as our predictor. Similar results were obtained using our temporal predictor (QCF), so we omit them here.

Figure 2 presents the results of our injection study. Here, we fix the window w (see Sect. 3) to be 100 time steps. One can see the added benefit of using partial replication for improving single-bit detection recall rates. For example, we observe an overall improvement for Sedov from 6 % for 5 % replication to 18 % for 25 % replication. For BlastBS we also see significant gains, with improvements from 5 % for 5 % replication to 24 % for 25 % replication.

Figure 3 presents the results when using the probabilistic evaluation metric P_f with the four distributions presented in Sect. 4. We note that the y-axis is plotted by using a logarithmic scale. We can categorize the distributions in a spectrum from *difficult* (dist1) to *easy* (dist4) (as discussed in Sect. 4, the fewer the number of bits that can get corrupted, the harder it is to detect corruptions at the software level). In this case, only distributions 1 and 2 are of further interest to us, since distributions 3 and 4 represent easy detection cases; that is, the probability of undiscovered corruption using our DAB detectors is already below 10^{-15} without even considering replication. For distribution 1, we need over 43 % of the processes replicated in order to achieve a 99.999 % protection (i.e., $P_f < 0.001$) level in the case of Sedov and 51 % in the case of BlastBS. Recall that distribution 1 is the most *difficult* one, representing an upper bound in the number of replicated processes needed.

Apart from the obvious performance overhead incurred by using replication (i.e., extra hardware needed to run extra processes, or spatial overhead), there is also an overhead introduced by extra messages sent throughout the network, which ultimately enlarges the runtime of applications. Another source of temporal overhead is our own DAB detector, which needs to run on every iteration and check all the data points for all the protected variables. From the system's point of view, both dimensions—temporal and spatial—contribute equally to the overall overhead, so both should be included. In partial replication we also need to consider the extra temporal overhead introduced by process migration when changing the replication set. We calculate the total overhead, then, using the following model:

$$O(r, w) = T(r) \times (r + 1) + \frac{M \times r \times n}{W \times T_w}. \tag{2}$$

where r is the replication rate (e.g., 0.5 when replicating half of the processes) and $T(r)$ is the runtime overhead introduced by the DAB detector and partial replication when running with a replication rate equal to r (e.g., $T(r) = 1.1$ if there is a 10 % increase in running time). The right-hand side of the summation represents the overhead introduced by changing the replication set every w steps[4]. In this part, M is the memory used per process, $r \times n$ is the replication budget (n is the number of processes), W represents the aggregate network bandwidth in the system, and T_w is the time taken to run w steps in the original application. Note that this is an upper bound, since in some cases the number of processes to replicate is less than the budget, namely, when some processes replicated in the previous window are chosen again for the current one. We find that in only a few cases does the replication set change completely.

The temporal overhead $T(r)$ may vary depending on the communication-to-computation ratio of the application. For those cases where computation dominates communication, the extra overhead is usually small. Fiala et al. [11] show that temporal overhead for full duplication (i.e., $r = 1.0$) is not a concern (around 1–2 %) for those applications that can maintain a well-balanced communication-to-computation ratio as they scale (applications exhibiting weak scalability). On the other hand, temporal overheads can reach 30 % for network-bound applications and kernels. Since we are simulating partial replication, we are unable to measure exactly the value of $T(r)$ for the applications used. In this case, we assume the temporal overhead introduced by the extra network messages never to be above 5 %, given that the stencil codes evaluated are not network-bound.

---

[4] Wang et al. [25] show that calculating process migration time as *process_memory/network_bandwidth* is a fairly good estimate.

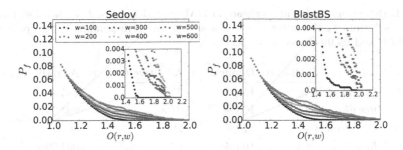

Fig. 4. Total overhead introduced by partial replication using different values of w. Distribution 1 used for $P(\#\text{bits}= x)$; the small subplots represent the data zoomed between $[0.0, 0.004]$.

Moreover, our experiments indicate that the temporal overhead introduced by using our DAB detectors is never above 6 %[5]. Thus, we set $T(r = 1.0) = 1.11$ (i.e., $5+6 = 11$ % temporal overhead introduced by replicating all processes and using our DAB detector on every process). We estimate $T(r)$ for $r < 1.0$ assuming a balanced communication pattern between processes (which is the case in the stencil codes evaluated, where processes communicate mainly with their neighbors): $\hat{T}(r < 1.0) = 1 + r \times 0.05 + 0.06$.

In order to get an idea of how much overhead would be introduced by partial replication, we compute the values of P_f in Fig. 4 based on $O(r)$, for different values of the parameter w. Moreover, we assume distribution 1 for $P(\#\text{bits}= x)$. All the injection experiments are run on the Fusion cluster at Argonne National Laboratory [1], which has an InfiniBand QDR network with a bandwidth of 4 GB/s per link, per direction, arranged on a fat tree topology. Since we are not taking into account network contention issues in our overhead model, we set W to the lowest possible aggregate bandwidth in order to get an upper bound on the effect that the network bandwidth has on the overhead. That is, we set $W=4$ GB/s.

As one can see, a window of a 100 time steps is the best choice among all the considered possibilities. The reason is that the smaller temporal overheads can not compensate for the accuracy loss incurred when using larger window sizes. Our analyses show that we can have a 99.999 % protection (i.e., $P_f < 0.001$) with $w = 100$ with a total overhead of around 1.52 (52 %) for Sedov and 1.56 (56 %) for BlastBS, with a replication rate of 43 % and 51 % respectively. This is an improvement of 53 % and 49 %, respectively, over full duplication (considering 5 % in temporal overhead due to the extra network messages, full duplication has a total overhead of 2.1, or 110 %) with a detection recall close to 100 %. For easy comparison, these results are listed in Table 1.

---

[5] The memory overhead of our DAB detectors is practically 0 % given that we are using spatial-based predictors only in this study. For that reason, we do not include extra memory usage in the overhead calculation.

Table 1. Detection recall and overhead for DAB-only detectors, 2x replication, and our adaptive solution. In the latter, two cases are shown corresponding to two protection levels: 97 % and 99.999 % recall, respectively.

		DAB-only	Duplication	Adaptive (case 1)	Adaptive (case 2)
Sedov	*Overhead*	6 %	110 %	25 %	52 %
	Recall	92 %	100 %	97 %	99.999 %
BlastBS	*Overhead*	6 %	110 %	26 %	56 %
	Recall	91 %	100 %	97 %	99.999 %

6   Related Work

Software solutions for SDC detection can be grouped in four main categories: (1) full replication [11,18], which is the most general but also the most expensive; (2) algorithm-based fault tolerance (ABFT) [14]; (3) approximate computing [5]; and (4) data-analytic-based (DAB) fault tolerance [2,6,9,26]. ABFT and approximate computing are not general enough and have limited applicability, since kernels need to be adapted manually and only a subset of them can be protected. In the case of DAB, detectors take advantage of the underlying properties of the applications' data (their smoothness in the time and/or space dimensions) in order to compute likely values for the evolution of the data and use those to flag outliers as potential corruptions. In this work we combine replication-based and DAB in order to avoid some of their individual shortcomings (i.e., the high cost of replication and the limited applicability of DAB).

Replication mechanisms for fault tolerance have been studied extensively in the past, especially in the context of aerospace and command and control systems [8]. Traditionally, the HPC community has considered replication to be too expensive to be applicable; and, to the best of our knowledge, it has not been implemented in any real production system.

Liu et al. [17] propose partial replication *in time* by taking advantage of the fact that soft errors in the first 60 % of iterations of some iterative applications are relatively tolerable. The idea is to duplicate all processes only during the last 40 % of iterations. Nakka et al. [20], Subasi et al. [24], and Hukerikar et al. [15]—by introducing new programming language syntax—propose to make the programmers responsible for identifying those parts of the code or data that are critical and need to be replicated. In contrast to these solutions, which are application dependent, our work is more general in the sense that we do not require any specific knowledge of tolerability to errors of particular iterations, variables, or code regions.

Partial replication in HPC where processes are chosen at random has also been investigated. Research has shown, however, that such an approach does not pay off [23]. In this work we choose the processes to replicate based on their data behavior.

7   Conclusions and Future Work

In this paper we have shown that combining partial replication along with DAB detectors allows us to get SDC protection levels that are close enough to those achieved by duplication at a lower overhead price. Our results show that we can get an overall SDC protection level, or recall, of 99.999 % replicating only between 43 % and 51 % of all the processes with a maximum total overhead (upper bound) of 52–56 % (compared to 110 % for duplication) for the applications analyzed.

As future steps for this work, we want to consider the situation where the replication budget B is "elastic" during the length of the computation—for example, a situation where we can replicate a small number of processes (say, 10 %) during the majority of the computation but increase the rate to a higher number (say, 60 %), for a short period of time. This strategy can be useful for situations where sharp data changes are concentrated not only in a particular place in space but also in time. One can imagine an scenario in exascale where systems will have spare resources, in our case nodes, which will be allowed to be requested "on the fly" by applications and libraries in order to perform fault tolerance tasks.

Acknowledgments. This material was based upon work supported by the U.S. Department of Energy, Office of Science, Advanced Scientific Computing Research Program, under Contract DE-AC02-06CH11357, and by the ANR RESCUE and the INRIA-Illinois-ANL- BSC-JSC-Riken Joint Laboratory on Extreme Scale Computing. The work at the Illinois Institute of Technology is supported in part by U.S. National Science Foundation grants CNS-1320125 and CCF-1422009.

References

1. Fusion cluster at Argonne National Laboratory. http://www.lcrc.anl.gov/guides/ Fusion
2. Bautista-Gomez, L.A., Cappello, F.: Detecting silent data corruption through data dynamic monitoring for scientific applications. In: PPoPP 2014, pp. 381–382 (2014)
3. Bautista-Gomez, L.A., Cappello, F.: Detecting and correcting data corruption in stencil applications through multivariate interpolation. In: 1st International Workshop on Fault Tolerant Systems (part of Cluster 2015), pp. 595–602 (2015)
4. Bautista-Gomez, L.A., Tsuboi, S., Komatitsch, D., Cappello, F., Maruyama, N., Matsuoka, S.: FTI: high performance fault tolerance interface for hybrid systems. In: SC 2011, pp. 32:1–32:32 (2011)
5. Benson, A.R., Schmit, S., Schreiber, R.: Silent error detection in numerical time-stepping schemes. Int. J. High Perform. Comput. Appl. **29**(4), 1–20 (2014)
6. Berrocal, E., Bautista-Gomez, L., Di, S., Lan, Z., Cappello, F.: Lightweight silent data corruption detection based on runtime data analysis for HPC applications. In: HPDC 2015 (short paper) (2015)
7. Borkar, S.: Major challenges to achieve exascale performance. Intel Corporation, April 2009

8. Briere, D., Traverse, P.: AIRBUS A320/A330/A340 electrical flight controls - a family of fault-tolerant systems. In: Proceedings of the IEEE International Symposium on Fault-Tolerant Computing, pp. 616–623 (1993)

9. Chalermarrewong, T., Achalakul, T., See, S.C.W.: Failure prediction of data centers using time series and fault tree analysis. In: ICPads 2012, pp. 794–799 (2012)

10. Dell, T.J.: A white paper on the benefits of chipkill-correct ECC for PC server main memory. In: IBM Microelectronics Division, pp. 1–23 (1997)

11. Fiala, D., Mueller, F., Engelmann, C., Riesen, R., Ferreira, K., Brightwell, R.: Detection and correction of silent data corruption for large-scale high-performance computing. In: SC 2012, pp. 78:1–78:12 (2012)

12. Fryxell, B., Olson, K., Ricker, P., Timmes, F.X., Zingale, M., Lamb, D.Q., MacNeice, P., Rosner, R., Truran, J.W., Tufo, H.: Flash: an adaptive mesh hydrodynamics code for modeling astrophysical thermonuclear flashes. Astrophys. J. Suppl. Ser. (ApJS) **131**, 273–334 (2000)

13. Hengartner, N.W., Takala, E., Michalak, S.E., Wender, S.A.: Evaluating experiments for estimating the bit failure cross-section of semiconductors using a colored spectrum neutron beam. Technometrics **50**(1), 8–14 (2008)

14. Huang, K.H., Abraham, J.A.: Algorithm-based fault tolerance for matrix operations. IEEE Trans. Comput. **100**(6), 518–528 (1984)

15. Hukerikar, S., Diniz, P.C., Lucas, R.F., Teranishi, K.: Opportunistic application-level fault detection through adaptive redundant multithreading. In: HPCS 2014 (2014)

16. Hwang, A.A., Stefanovici, I.A., Schroeder, B.: Cosmic rays don't strike twice: understanding the nature of dram errors and the implications for system design. In: ASPLOS XVII, pp. 111–122 (2012)

17. Liu, J., Kurt, M.C., Agrawal, G.: A practical approach for handling soft errors in iterative applications. In: Cluster 2015, pp. 158–161 (2015)

18. Mukherjee, S., Kontz, M., Reinhardt, S.: Detailed design and evaluation of redundant multi-threading alternatives. In: ISCA 2002, pp. 99–110 (2002)

19. Mukherjee, S.S., Emer, J., Reinhardt, S.K.: The soft error problem: an architectural perspective. In: HPCA 2005 (2005)

20. Nakka, N., Pattabiraman, K., Iyer, R.: Processor-level selective replication. In: DSN 2007, pp. 544–553 (2007)

21. Sedov, L.I.: Similarity and Dimensional Methods in Mechanics, 10th edn. Academic Press, New York (1959)

22. Snir, M., et al.: Addressing failures in exascale computing. Int. J. High Perform. Comput. **28**(2), 129–173 (2014)

23. Stearly, J., Ferreira, K., Robinson, D., Laros, J., Pedretti, K., Arnold, D., Bridges, P., Riesen, R.: Does partial replication pay off? In: DSN 2012 (2012)

24. Subasi, O., Arias, J., Unsal, O., Labarta, J., Cristal, A.: Programmer-directed partial redundancy for resilient HPC. In: CF 2015 (2015)

25. Wang, C., Mueller, F., Engelmann, C., Scott, S.L.: Proactive process-level live migration in HPC environments. In: SC 2008 (2008)

26. Yim, K.S.: Characterization of impact of transient faults and detection of data corruption errors in large-scale n-body programs using graphics processing units. In: IPDPS 2014, pp. 458–467 (2014)

27. Zachary, A.L., Malagoli, A., Colella, P.: A higher-order godunov method for multidimensional ideal magnetohydrodynamics. SIAM J. Sci. Comput. **15**(2), 263–284 (1994)

Parallel and Distributed Programming, Interfaces, Language

Automatic Verification of Self-consistent MPI Performance Guidelines

Sascha Hunold$^{(\boxtimes)}$, Alexandra Carpen-Amarie, Felix Donatus Lübbe,
and Jesper Larsson Träff

Research Group for Parallel Computing, TU Wien, Vienna, Austria
{hunold,carpenamarie,luebbe,traff}@par.tuwien.ac.at

Abstract. The Message Passing Interface (MPI) is the most commonly used application programming interface for process communication on current large-scale parallel systems. Due to the scale and complexity of modern parallel architectures, it is becoming increasingly difficult to optimize MPI libraries, as many factors can influence the communication performance. To assist MPI developers and users, we propose an automatic way to check whether MPI libraries respect self-consistent performance guidelines for collective communication operations. We introduce the PGMPI framework to detect violations of performance guidelines through benchmarking. Our experimental results show that PGMPI can pinpoint undesired and often unexpected performance degradations of collective MPI operations. We demonstrate how to overcome performance issues of several libraries by adapting the algorithmic implementations of their respective collective MPI calls.

Keywords: MPI · Collectives · Performance guidelines · Benchmarking

1 Introduction

Communication libraries implementing the Message Passing Interface (MPI) are major building blocks for developing parallel, distributed, and large-scale applications for current supercomputers. The performance of parallel codes is therefore highly dependent on the efficiency of MPI implementations. Much research is currently conducted to cope with the problems of exascale computing in MPI.

Assessing the performance of MPI implementations is vital for developers, vendors, and users of the libraries. However, the performance of MPI libraries can be measured in different ways. A common approach is to run a set of MPI micro-benchmarks, such as SKaMPI [12] or ReproMPI [7]. Micro-benchmarks usually report the measured (mean or median) run-time of a given MPI function for different message sizes, e.g., the run-time of MPI_Bcast for broadcasting a 1 Byte message. Developers can gain insights on how the run-time of an MPI function depends on the message size for a fixed number of processes. It is also

---

A. Carpen-Amarie and F.D. Lübbe—This work was supported by the Austrian Science Fund (FWF): P26124 and P25530.

P.-F. Dutot and D. Trystram (Eds.): Euro-Par 2016, LNCS 9833, pp. 433–446, 2016.
DOI: 10.1007/978-3-319-43659-3_32

possible to assess the scalability of MPI functions when the number of processes is increased and the message size stays fixed.

Verifying self-consistent MPI performance guidelines is an alternative, orthogonal method for analyzing the performance of MPI libraries [15]. This approach does not require explicit performance models. Instead, performance guidelines form a set of rules that an MPI library is expected to fulfill. A performance guideline usually defines an upper bound on the run-time behavior of a specialized MPI function. For example, one performance guideline states that a call to MPI_Scatter of n data elements should "not be slower" than a call to MPI_Bcast with n data elements, as the semantics of an MPI_Scatter operation could be emulated using MPI_Bcast [15]. Only minor efforts have been made to systematically test self-consistent performance guidelines for MPI implementations in practice, one example being the mpicroscope benchmark [14]. To close this gap, we introduce the benchmarking framework PGMPI that can automatically verify performance guidelines of MPI libraries.

We make the following contributions: (1) We propose the benchmarking framework PGMPI to detect performance-guideline violations. (2) We present a systematic, experimental verification of performance guidelines for several MPI libraries. (3) We examine different use cases, for which the detection of guideline violations enabled us to tune and improve the libraries' performance.

In Sect. 2, we state the scientific problem and introduce our notation. We continue with summarizing related work and comparing it to our approach in Sect. 3. We introduce the PGMPI framework in Sect. 4 and present an experimental evaluation of different MPI libraries using it in Sect. 5. We summarize our findings and conclude in Sect. 6.

2   Problem Statement and Notation

Träff et al. [15] introduced self-consistent performance guidelines for MPI libraries as follows: The run-time of two MPI functionalities A and B can be ordered using the relation $\preceq$ as MPI_$A(n) \preceq$ MPI_$B(n)$, which means that functionality MPI_$A(n)$ is possibly faster than functionality MPI_$B(n)$ for (almost) all communication amounts n. Performance guidelines are defined for a fixed number of processes p, and thus, they do not mention p explicitly. However, the communication volume per process may vary depending on the semantics of a given MPI function and the number of processes. For example, in the case of MPI_Bcast, the total size n is equal to the message size being transferred to each process. In contrast, the individual message size for an MPI_Scatter is a fraction of the total communication volume n, i.e., each process receives n/p elements.

We examine three types of performance guidelines: (1) *monotony*, (2) *split-robustness*, and (3) *pattern*. The monotony guideline

$$\text{MPI\_}A(n) \preceq \text{MPI\_}A(n + k)$$

ensures that communicating a larger volume should not decrease the communication time.

The split-robustness guideline

$$\mathtt{MPI\_A}(n) \preceq k \; \mathtt{MPI\_A}(n/k)$$

states that communicating a total volume of n data elements should not be slower than sending $\frac{n}{k}$ elements in k steps.

Pattern guidelines define upper bounds on the performance of MPI communication operations. The idea is that a specialized MPI function should not have a larger running time than a combination of other MPI operations, which emulate the functionality of the specialized function. Let us consider the following *pattern* performance guidelines:

$$\mathtt{MPI\_Scatter}(n) \preceq \mathtt{MPI\_Bcast}(n), \text{ and}$$
$$\mathtt{MPI\_Bcast}(n) \preceq \mathtt{MPI\_Scatter}(n) + \mathtt{MPI\_Allgather}(n) \quad .$$

The first states that MPI_Scatter should not be slower than MPI_Bcast. The reason is that the semantics of MPI_Scatter can be implemented using MPI_Bcast, by broadcasting the entire vector before processes take their share depending on their rank. The second guideline states that a call to MPI_Bcast should be at least as fast as a combination of MPI_Scatter and MPI_Allgather, which we call the *mock-up* version of MPI_Bcast, as it emulates its semantics [2].

3 Related Work

Collective communication operations are a central part of the MPI standard, as they are essential for many large-scale applications. Chan et al. [2] provide an overview of typical, blocking collectives and their implementations, as well as lower bounds for the communication cost of each function. For different network topologies, the authors devise algorithms that achieve the lower bounds for either the latency or the bandwidth component. As the model of parallel computation in this paper is rather simplistic, we aim to complement this study by carefully benchmarking MPI collectives on actual hardware.

Träff [14] proposed the MPI benchmark mpicroscope, which can verify two self-consistent performance guidelines: "split-robust" and "monotone". In the present work, we extend this functionality by testing various pattern violations, using the experimental framework that was proposed by Hunold and Carpen-Amarie [7] for better reproducibility of the experimental results. While we focus on performance guidelines for collectives, previous works have also formulated performance guidelines for derived datatypes [4] and MPI-IO operations [5].

As hardware and software factors can influence the performance of MPI collectives, tuning MPI parameters is an essential part for achieving high performance, when installing an MPI library. Yet, optimizing and tuning MPI operations are orthogonal steps compared to the verification of self-consistent performance guidelines, i.e., the latter can help us to verify whether run-times of collectives are consistent in terms of expected performance. For example, the guidelines can be used to ensure that Gather is faster than Allgather for the same problem size. For that reason, if a violation occurs, it usually means that one collective can be tuned. To optimize the latency of collectives at run-time, one can employ the STAR-MPI routines [3]. When a call to a specific MPI function is issued, STAR-MPI selects one of the available algorithms and measures its run-time. When STAR-MPI has enough knowledge about the performance of different algorithms, it is able to pick a good algorithm for a specific case.

Selecting the right algorithm to implement a given MPI function is only one step towards tuning MPI libraries. Another problem is finding the right parameter settings that run-time systems of MPI libraries like OpenMPI or MVAPICH offer. Chaarawi et al. [1] introduced the OTPO tool that can be used to tune OpenMPI run-time parameters. OTPO takes as input the run-time parameters to be tuned as well as their respective ranges, and then starts measuring for all combinations of parameter values. Another approach to tune OpenMPI parameters has been proposed by Pellegrini et al. [10], where the parameter values are predicted using machine learning techniques.

The performance guidelines are formulated as a function of the communication volume. It is also possible to examine the scalability of MPI collectives when increasing the number of processes. Shudler et al. [13] proposed a framework to compare performance characteristics of HPC applications with a theoretical performance model. The framework fits the recorded benchmarking data to analytic speedup functions and compares the experimentally determined scalability behavior to this expected performance model. A model mismatch indicates a scalability problem of the parallel code section.

4   PGMPI: Verifying MPI Performance Guidelines

We now introduce the PGMPI framework to verify self-consistent performance guidelines of MPI libraries. In the first step, PGMPI experimentally determines the number of repetitions needed to obtain stable, reproducible run-time measurements (*Step NREP*). In the second step, the framework performs run-time measurements of all functions for which performance guidelines are formulated (*Step MEASURE*). The data analysis and the statistical verification of performance guidelines is carried out in the last step (*Step ANALYZE*).

4.1   Obtaining Reproducible Results

We start by looking at the main (second) step of PGMPI (*Step MEASURE*), in which the run-times of MPI functions and their emulating counterparts are measured. The guideline-checking program takes as input a set of pattern guidelines, each defined by a pair consisting of an MPI function and its emulating mock-up function. Our PGMPI framework will measure the run-time of one of the specified MPI functions f for all given message sizes m_i and the number of processes p that are given in the input file. Within one call to mpirun, each individual measurement for (f, m_i) is repeated r_i times, where r_i is defined for each m_i. As we expect that mean (or median) run-times vary between different calls to mpirun [7], the PGMPI framework measures the run-time of each MPI function f over R mpiruns.

4.2 Determining the Number of Repetitions

A major problem in MPI benchmarking is the question of how long (how many times) to measure. We need to find the right trade-off between time and measurement stability. One way of dealing with this problem is by executing the experiment sufficiently often, e.g., 1000 times. This would alleviate the problem of low measurement stability, but most often, we cannot afford long-running benchmarking experiments. Therefore, we formulate the following problem:

Definition 1. *The NREP problem is to find a suitable number of repetitions r_i for the tuple (f, m_i, p), such that the obtained run-time metric after r_i repetitions of function f with m_i Bytes on p processes is reproducible between different calls to mpirun. Reproducible in this case means that the distribution of the measured values (for a specific metric) obtained from R mpiruns has a small variance.*

We have experimented with various ways of estimating the number of repetitions needed to obtain reproducible results. One possibility is to monitor the relative standard error of the mean (RSE). SKaMPI, for example, stops the measurements when the RSE falls below a threshold of 0.1 [12]. Although we have tested many different ways to solve the NREP problem, we could not find a generally superior approach. We therefore designed the NREP predictor for *Step NREP* of the PGMPI framework in a flexible manner. The framework currently provides three different methods (metrics) for solving the NREP problem, but new metrics can be added. The NREP prediction may stop

1. when the **relative standard error** (RSE) is smaller than some predefined threshold t_{RSE}; or
2. when the **coefficient of variation of the mean run-time** (COV_{mean}) is smaller than some predefined threshold $t_{COV_{mean}}$. The value of the COV_{mean} is computed over the last $w_{COV_{mean}}$ means (window size); or
3. when the **coefficient of variation of the median run-time** (COV_{median}) is smaller than some predefined threshold $t_{COV_{median}}$ using a window size of $w_{COV_{median}}$.

Users can choose the NREP prediction method on the command line as follows:

```
mpirun -np 4 ./mpibenchmarkPredNreps --calls-list=MPI_Reduce --msizes-list=8
--rep-prediction min=20,max=1000,step=10 --pred-method=rse --var-thres=0.025
```

It is also possible to combine different metrics, i.e., the NREP prediction stops when all selected metrics have been positively evaluated. An example is shown in Fig. 1, in which both the RSE and the COV_{mean} need to be below a specific threshold (marked with horizontal lines). The prediction function for the RSE metric stops after 85 iterations, at which the COV_{mean} value is also below its threshold. As a result, 85 is the number of iterations that will be used when collecting benchmark data in *Step MEASURE*. To cope with the run-time variation between different mpiruns, we perform three NREP predictions for each message size and select the maximum number of repetitions obtained.

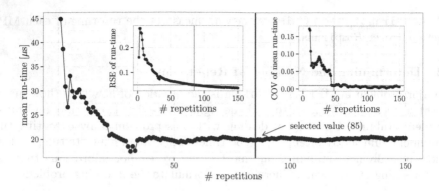

Fig. 1. Example of estimating the required number of repetitions for MPI_Allgather (16 B, 16×1 processes, *Jupiter*, $t_{RSE} = 0.025$, $t_{COV_{mean}} = 0.01$, $w_{COV_{mean}} = 20$)

4.3   Statistically Verifying Performance Guidelines

After gathering the measurement results, PGMPI can proceed to *Step ANA-LYZE*, which consists of the data processing and the verification of performance guidelines. We now explain which statistical methods are applied for guideline verification. For each MPI function, for which guidelines were formulated, the experimental results comprise R (number of mpiruns) data sets for a specific number of processes p. Each data set contains r_i run-time measurements for a specific message size m_i. We first reduce the number of measurements per tuple $(\text{mpirun}_j, m_i, p)$ to a single value, by computing the median run-time over the r_i measurements. In this way, we obtain a distribution of R medians (median run-times) for each message size m_i and processes p. The various performance guidelines will then be verified using these distributions of medians.

Monotony Guideline. PGMPI checks for each pair of adjacent message sizes m_i and m_j, $m_i < m_j$ that the run-time of an MPI function with a message size of m_i is not larger than the run-time with a size of m_j. We use the *Wilcoxon rank-sum test* [6] to test whether the distribution of medians at m_i is smaller or equal than the one at m_j. If the test rejects our hypothesis, we have statistical confidence (at the provided confidence level) that the monotony between message sizes m_i and m_j is violated.

Split-Robustness Guideline. We want to verify that sending a message of size m_j by transferring k packets of size $m_i < m_j$ is not faster than sending only one message of size m_j. We are only given the run-time distribution of one MPI function at m_i. Unfortunately, we have no knowledge about the shape of the run-time distribution when we communicate messages of size m_i in k rounds. As a matter of fact, we cannot simply shift the distribution at m_i by some constant factor, and therefore, we decided to rely on (and to compare) the median values of the distributions.

Table 1. Overview of parallel machines used in the experiments

Name	Hardware	MPI libraries/Compiler
Jupiter	36 × Dual Opteron 6134 @ 2.3 GHz	NEC MPI 1.3.1, MVAPICH2-2.1
	IB QDR MT26428	Open MPI 1.10.1/ gcc 4.4.7
VSC-3	2000 × Dual Xeon E5-2650V2 @ 2.6 GHz	Intel MPI Library 5.0 (Update 3)
	IB QDR-80	gcc 4.4.7

Since we measure the run-time of MPI functions only for a limited number of message sizes, we compute the factor $k = \min_{l \in \mathbb{N}}(lm_i \geq m_j)$, which denotes the smallest multiple of m_i such that the resulting product is at least m_j. Notice that we explicitly allow lm_i to be larger than m_j, which enables us to check whether sending two messages of size 1024 B is faster than sending one message of size 2000 B. The PGMPI framework checks whether the time to communicate messages of size m_i in k rounds is smaller than the run-time for m_j. If we find such a violation for a message size m_j, we only report the largest message size m_i (the smallest factor k) for which the violation occurred; otherwise too many violations would be reported in some cases. It often happens that the predicted run-time for lm_i is very similar to the run-time for m_j. To avoid reporting split-robustness violations for which only marginal relative run-time differences have been measured, we use a 5 % tolerance level to verify this guideline. Currently, PGMPI does not empirically test whether communicating k messages of size m_i is indeed faster than communication a message of size m_j in practice. This additional check would require an additional benchmarking round, and might be added to PGMPI later.

Pattern Guidelines. The verification of pattern guidelines is done similarly to checking the monotony guideline, except that we now compare two run-time distributions of two distinct functions: an MPI function and its mock-up version. We apply the *Wilcoxon rank-sum test* on the two distributions to test whether the run-time distribution of the MPI function is not significantly shifted to the right of the distribution obtained with the mock-up version ("to the right" means larger run-time). If this is the case, PGMPI reports a pattern violation. Alternatively, the *Kolmogorov-Smirnov test* [6] can be employed, as it is less sensitive to ties. Overall, both tests led to similar results in the majority of the considered cases.

5   Experimental Evaluation and Results

We evaluate our proposed PGMPI framework[1] experimentally using the hardware and software setup listed in Table 1. First, we present a summary of detected performance-guideline violations for several MPI libraries. Second, we demonstrate in two case studies that the knowledge about specific guideline violations can help tuning and adapting MPI implementations to parallel systems. Please refer to our technical report [8] for more details about the experimental evaluation.

---

[1] https://github.com/hunsa/pgmpi.

Table 2. Performance-guideline violations of different MPI libraries ($R = 10$); violation types: monotony, split-robustness, pattern; message sizes between 1 B and 100 KiB

(a) *Jupiter*						(b) *VSC-3*		
#processes	type	MVAPICH2-2.1	NEC MPI 1.3.1	Open MPI 1.10.1		#processes	type	Intel MPI 5.0
16x1	m	7/9	6/9	7/9		16x16	m	7/9
16x1	s	1/9	0/9	3/9		16x16	s	6/9
16x1	p	12/15	7/15	9/15		16x16	p	13/15
32x16	m	5/9	4/9	4/9		64x16	m	6/9
32x16	s	3/9	0/9	3/9		64x16	s	7/9
32x16	p	8/15	7/15	7/15		64x16	p	11/15

5.1   Assessing the Guideline Compliance of MPI Libraries

We used the PGMPI framework to verify the performance guidelines listed in Appendix A for different MPI libraries. On *Jupiter*, we evaluated NEC MPI 1.3.1, MVAPICH2-2.1, and Open MPI 1.10.1. The NEC MPI 1.3.1 library was delivered by NEC pre-compiled for our system and we therefore do not know all internals. The other two libraries, MVAPICH2-2.1 and Open MPI 1.10.1, were compiled using the default settings. On *VSC-3*, we recorded guideline violations for the proprietary Intel MPI Library 5.0 (Update 3).

Table 2 presents an overview of the detected guideline violations for several MPI libraries on *Jupiter* and *VSC-3*. For the monotony and the split-robustness guidelines, the table shows the number of MPI functions for which violations occurred, e.g., for MVAPICH2-2.1 using 16×1 processes, PGMPI found seven monotony violations among the nine tested MPI collectives. For the pattern guidelines, we verified the 15 guidelines provided in Appendix A. If a guideline violation is found for any message size of a particular MPI function, we say that this particular guideline is unsatisfied, i.e., a violation is only counted once across all message sizes. We can observe in Table 2 that the monotony and the split-robustness guidelines are violated by approximately 50 % of the collectives. The table also reveals that more than 40 % of the examined pattern guidelines were violated. The guideline violations occurred across different numbers of processes, message sizes, libraries, and machines. We therefore contend that there is a large potential for optimization of the individual MPI libraries on these machines.

Table 3 compares the detected pattern violations for the three libraries on *Jupiter*. Except for two guidelines, we found violations for all other pattern guidelines in at least one MPI library. The experimental results clearly suggest that Reduce-like functions should be improved and tuned in all MPI libraries, which are: MPI_Allreduce, MPI_Reduce, MPI_Reduce_scatter, and MPI_Reduce_scatter_block.

A detailed view on the detected guideline violations for MVAPICH2-2.1 on *Jupiter* is given in Table 4. For this MPI library, we observe a couple of monotony violations. For short message sizes (<32 B), the absolute difference in run-times is very small, and thus, fixing these cases has low priority. Monotony violations occur for larger messages when the message size is not a power of two (e.g., between 10000 B and 16384 B). These cases could be investigated in more

Table 3. Pattern guideline violations of different MPI libraries with 32×16 processes on *Jupiter*, $R = 10$; message sizes between 1 B and 100 KiB

Guideline	MVAPICH2-2.1	NEC MPI 1.3.1	Open MPI 1.10.1
MPI_Allgather $\preceq$ Allreduce	•		
MPI_Allgather $\preceq$ Alltoall			
MPI_Allgather $\preceq$ Gather+Bcast		•	
MPI_Allreduce $\preceq$ Reduce+Bcast	•	•	•
MPI_Allreduce $\preceq$ Reduce_scatter_block+Allgather			•
MPI_Bcast $\preceq$ Scatter+Allgather		•	•
MPI_Gather $\preceq$ Allgather			
MPI_Gather $\preceq$ Reduce	•		
MPI_Reduce_scatter_block $\preceq$ Reduce+Scatter	•	•	•
MPI_Reduce_scatter $\preceq$ Allreduce	•	•	•
MPI_Reduce_scatter $\preceq$ Reduce+Scatterv		•	
MPI_Reduce $\preceq$ Allreduce	•		•
MPI_Reduce $\preceq$ Reduce_scatter_block+Gather	•		
MPI_Scan $\preceq$ Exscan+Reduce_local		•	•
MPI_Scatter $\preceq$ Bcast	•		

detail, as padding up the message to the next power of two could be an option. For split-robustness guidelines we can see potential for improvement only for larger message sizes. When analyzing the pattern guidelines, two cases stand out: MPI_Allreduce is slower than the emulating function using Reduce and Bcast for message sizes up to 2 KiB and MPI_Reduce_scatter exposes a performance degradation compared to Allreduce for almost all message sizes.

As it is impossible for library developers to provide suitable parameters for each individual installation, checking the compliance to performance guidelines can be seen as indicators for programmers and administrators, how to tune MPI libraries. Often, specific MPI libraries already provide efficient algorithms, and violations would not occur if the right algorithm were enabled for a specific case. We therefore show in the next section how violations can guide us to find more suitable algorithms and implementations for collective calls on a specific machine.

5.2   Case Study 1: MPI_Gather $\preceq$ MPI_Allgather, MVAPICH

We consider the violation of this performance guideline that was detected using 32×1 processes and MVAPICH2-2.1 on *Jupiter* and is shown in Fig. 2a. When the *Wilcoxon rank-sum test* reports a violation for a particular message size, we mark this case in the figure with a red background and add asterisks to show the statistical significance. Here, executing MPI_Gather using 32×1 processes (one process per compute node) is slower than performing a Gather using MPI_Allgather. Calling MPI_Gather in the default installation of

Table 4. Performance-guideline violations of MVAPICH2-2.1 using 32×16 processes on *Jupiter* ($R = 10$); violation types: **m**onotony, **s**plit-robustness, **p**attern

type	function	1	2	4	8	16	32	64	100	128	256	512	1024	1500	2048	4096	5000	8192	10000	16384	32768	102400
m	MPI_Allgather				•		•															
m	MPI_Allreduce							•														
m	MPI_Gather																			•	•	
m	MPI_Reduce							•												•		
m	MPI_Scatter	•		•	•																	
s	MPI_Gather													•	•	•	•	•		•		
s	MPI_Reduce																			•		
s	MPI_Reduce_scatter_block											•	•									
p	MPI_Allgather $\preceq$ Allreduce	•	•																			
p	MPI_Allreduce $\preceq$ Reduce+Bcast	•	•	•	•	•	•	•	•	•	•	•	•	•	•							
p	MPI_Gather $\preceq$ Reduce	•	•	•																		
p	MPI_Reduce_scatter_block $\preceq$ Reduce+Scatter	•	•												•	•						
p	MPI_Reduce_scatter $\preceq$ Allreduce	•	•							•	•	•	•	•	•	•	•	•	•	•	•	•
p	MPI_Reduce $\preceq$ Allreduce																			•	•	•
p	MPI_Reduce $\preceq$ Reduce_scatter_block+Gather																			•	•	•
p	MPI_Scatter $\preceq$ Bcast	•	•	•																		

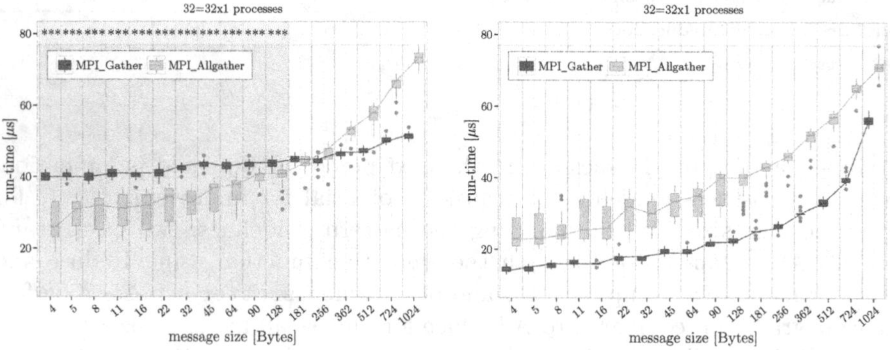

(a) With Violations   (b) No Violations

Fig. 2. Verification of MPI_Gather $\preceq$ MPI_Allgather (a) before and (b) after changing the Gather implementation (MVAPICH2-2.1, *Jupiter*, $R = 30$, $r_i = 1000$)

MVAPICH2-2.1 will use the internal function MPIR_Gather_intra for the first 14 invocations and then switch to MPIR_Gather_MV2_Direct for subsequent calls. The direct implementation of Gather performs $(p - 1)$ MPI_Irecvs on the root process and an MPI_Send on the other processes. We can set the environment variable MV2_USE_DIRECT_GATHER=0 to force MVAPICH to use MPIR_Gather_intra only. The intra-version on our machine uses a binomial tree algorithm to implement Gather, and forcing this algorithm fixes the violation (cf. Fig. 2b). Let us check that the algorithmic change for small message sizes is indeed favorable. We use the Hockney model for MPI_Reduce given by Pjesivac et al. [11], but omit the computational term. As the direct algorithm issues $(p - 1)$ receive operations, we obtain a run-time for small message sizes (we neglect the bandwidth term) of about $52.7\,\mu s$, for a network latency of roughly $1.7\,\mu s$. If we use a binomial tree

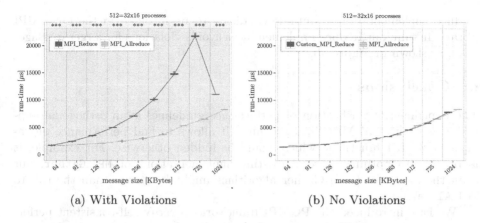

(a) With Violations   (b) No Violations

Fig. 3. Verification of MPI_Reduce $\preceq$ MPI_Allreduce with (a) original and (b) new Reduce implementation (Open MPI 1.10.1, *Jupiter*, $R = 5$, $r_i = 100$)

algorithm instead, the latency cost grows only logarithmically in the number of processes, i.e., $\log 32 \cdot 1.7\,\mu s = 8.5\mu s$. Even though our estimation does not perfectly match the experimental data, it explains why the binomial tree algorithm performs better.

5.3   Case Study 2: MPI_Reduce $\preceq$ MPI_Allreduce, Open MPI

In the second case study, we consider the guideline violations that occurred for MPI_Reduce using Open MPI 1.10.1 on the *Jupiter* system. Here, in contrast to the first case study, violations have only been measured for larger message sizes ($> 2^{16}$ B), but for various numbers of processes: 16×1, 32×1, 16×16, and 32×16. Figure 3 limits the view to message sizes for which violations were detected. Since Open MPI is highly configurable via the MCA parameters, we have tried to find parameter settings for MPI_Reduce, such that executing the latter would be faster than executing MPI_Allreduce. We have tried various segment sizes and fan-outs (where the parameters were applicable). Unfortunately, we failed to tune the parameters in such a way that the violations would disappear on our machine. For that reason, we implemented our own Reduce algorithm, which is based on the MPI_Allreduce algorithm found in Open MPI 1.10.1. Here, MPI_Allreduce is implemented using a Reduce-scatter followed by an Allgatherv on a ring of processes [9]. We modified this algorithm to become an MPI_Reduce by replacing the final Allgatherv by a Gatherv to the root. The Gatherv was realized using a direct Irecv/Send scheme. In the MPI semantics of Reduce, only the root process has a receive buffer. We therefore need to allocate additional buffer space to send and receive data segments in the Reduce-scatter phase. We found that executing malloc in each Reduce call has a severe impact on the performance of Reduce. To overcome this problem, we allocate a temporary buffer outside of Reduce but accessible to the Reduce implementation. This modification helped us to significantly speed up the run-time, and made this

Reduce implementation a suitable candidate to be included in the Open MPI library. In sum, our Reduce implementation avoids violations for larger messages sizes, as shown in Figure 3b.

6   Conclusions

The experimental verification of performance guidelines is an orthogonal approach to traditional MPI library tuning. It allows to find performance degradations of MPI functions, which would be hidden otherwise. For example, it is possible to optimize several existing implementations of MPI_Gather, but even the fastest of these Gather algorithms might be slower than the call to MPI_Allgather.

We have introduced the PGMPI framework to verify self-consistent performance guidelines of MPI functions. Currently, the framework supports blocking MPI collective communication operations, but it can be extended to support MPI point-to-point communication operations and derived datatypes. We have evaluated 17 different guidelines for collective communication operations for several MPI libraries such as MVAPICH and Open MPI. The experimental results reveal that none of the libraries was well adapted to our parallel machines, which might not be surprising. However, by using PGMPI we were able to pinpoint exactly which MPI functions should be tuned and which message sizes should be considered. Thus, PGMPI is a useful tool for MPI developers and system administrators to easily spot tuning potentials.

A   Self-consistent Performance Guidelines in PGMPI

The guidelines are formulated for a variable communication volume n, $n \geq 0$ and fixed number of processes p, $p \geq 1$, which is omitted.

Monotony Guideline

$$\texttt{MPI\_}A(n) \preceq \texttt{MPI\_}A(n + k) \quad , \qquad k \geq 0 \tag{GL1}$$

Split-Robustness Guideline

$$\texttt{MPI\_}A(n) \preceq k\,\texttt{MPI\_}A(n/k) \quad , \qquad k \geq 1 \tag{GL2}$$

Pattern Guidelines

$$\texttt{MPI\_Gather}(n) \preceq \texttt{MPI\_Allgather}(n) \tag{GL3}$$
$$\texttt{MPI\_Gather}(n) \preceq \texttt{MPI\_Reduce}(n) \tag{GL4}$$
$$\texttt{MPI\_Allgather}(n) \preceq \texttt{MPI\_Alltoall}(n) \tag{GL5}$$
$$\texttt{MPI\_Allgather}(n) \preceq \texttt{MPI\_Allreduce}(n) \tag{GL6}$$
$$\texttt{MPI\_Scatter}(n) \preceq \texttt{MPI\_Bcast}(n) \tag{GL7}$$
$$\texttt{MPI\_Reduce}(n) \preceq \texttt{MPI\_Allreduce}(n) \tag{GL8}$$

$$\text{MPI\_Reduce\_scatter}(n) \preceq \text{MPI\_Allreduce}(n) \tag{GL9}$$

$$\text{MPI\_Bcast}(n) \preceq \text{MPI\_Scatter}(n) + \text{MPI\_Allgather}(n) \tag{GL10}$$

$$\text{MPI\_Allgather}(n) \preceq \text{MPI\_Gather}(n) + \text{MPI\_Bcast}(n) \tag{GL11}$$

$$\text{MPI\_Allreduce}(n) \preceq \text{MPI\_Reduce}(n) + \text{MPI\_Bcast}(n) \tag{GL12}$$

$$\text{MPI\_Allreduce}(n) \preceq \text{MPI\_Reduce\_scatter\_block}(n) + \text{MPI\_Allgather}(n) \tag{GL13}$$

$$\text{MPI\_Reduce}(n) \preceq \text{MPI\_Reduce\_scatter\_block}(n) + \text{MPI\_Gather}(n) \tag{GL14}$$

$$\text{MPI\_Reduce\_scatter\_block}(n) \preceq \text{MPI\_Reduce}(n) + \text{MPI\_Scatter}(n) \tag{GL15}$$

$$\text{MPI\_Scan}(n) \preceq \text{MPI\_Exscan}(n) + \text{MPI\_Reduce\_local}(n) \tag{GL16}$$

$$\text{MPI\_Reduce\_scatter}(n) \preceq \text{MPI\_Reduce}(n) + \text{MPI\_Scatterv}(n) \tag{GL17}$$

References

1. Chaarawi, M., Squyres, J.M., Gabriel, E., Feki, S.: A tool for optimizing runtime parameters of Open MPI. In: Lastovetsky, A., Kechadi, T., Dongarra, J. (eds.) EuroPVM/MPI 2008. LNCS, vol. 5205, pp. 210–217. Springer, Heidelberg (2008)
2. Chan, E., Heimlich, M., Purkayastha, A., van de Geijn, R.A.: Collective communication: theory, practice, and experience. Concurrency Comput. Pract. Experience **19**(13), 1749–1783 (2007)
3. Faraj, A., Yuan, X., Lowenthal, D.K.: STAR-MPI: self tuned adaptive routines for MPI collective operations. In: International Conference on Supercomputing (ICS), pp. 199–208. ACM (2006)
4. Gropp, W., Hoefler, T., Thakur, R., Träff, J.L.: Performance expectations and guidelines for MPI derived datatypes. In: Cotronis, Y., Danalis, A., Nikolopoulos, D.S., Dongarra, J. (eds.) EuroPVM/MPI 2011. LNCS, vol. 6960, pp. 150–159. Springer, Heidelberg (2011)
5. Gropp, W.D., Kimpe, D., Ross, R., Thakur, R., Träff, J.L.: Self-consistent MPI-IO performance requirements and expectations. In: Lastovetsky, A., Kechadi, T., Dongarra, J. (eds.) EuroPVM/MPI 2008. LNCS, vol. 5205, pp. 167–176. Springer, Heidelberg (2008)
6. Hollander, M., Wolfe, D.A., Chicken, E.: Nonparametric Statistical Methods, 3rd edn. Wiley, Hoboken (2014)
7. Hunold, S., Carpen-Amarie, A.: Reproducible MPI benchmarking is still not as easy as you think. IEEE TPDS (2016)
8. Hunold, S., Carpen-Amarie, A., Lübbe, F.D., Träff, J.L.: PGMPI: automatically verifying self-consistent MPI performance guidelines. CoRR abs/1606.00215 (2016)
9. Patarasuk, P., Yuan, X.: Bandwidth optimal all-reduce algorithms for clusters of workstations. JPDC **69**(2), 117–124 (2009)
10. Pellegrini, S., Wang, J., Fahringer, T., Moritsch, H.: Optimizing MPI runtime parameter settings by using machine learning. In: Ropo, M., Westerholm, J., Dongarra, J. (eds.) EuroPVM/MPI 2009. LNCS, vol. 5759, pp. 196–206. Springer, Heidelberg (2009)

11. Pjesivac-Grbovic, J., Angskun, T., Bosilca, G., Fagg, G.E., Gabriel, E., Dongarra, J.: Performance analysis of MPI collective operations. Cluster Comput. **10**(2), 127–143 (2007)
12. Reussner, R., Sanders, P., Träff, J.L.: SKaMPI: a comprehensive benchmark for public benchmarking of MPI. Sci. Program. **10**(1), 55–65 (2002)
13. Shudler, S., Calotoiu, A., Hoefler, T., Strube, A., Wolf, F.: Exascaling your library: will your implementation meet your expectations? In: International Conference on Supercomputing (ICS), pp. 165–175 (2015)
14. Träff, J.L.: mpiscope: towards an MPI benchmark tool for performance guideline verification. In: Träff, J.L., Benkner, S., Dongarra, J.J. (eds.) EuroPVM/MPI 2012. LNCS, vol. 7490, pp. 100–109. Springer, Heidelberg (2012)
15. Träff, J.L., Gropp, W.D., Thakur, R.: Self-consistent MPI performance guidelines. IEEE TPDS **21**(5), 698–709 (2010)

ParallelME: A Parallel Mobile Engine to Explore Heterogeneity in Mobile Computing Architectures

Guilherme Andrade[1](✉), Wilson de Carvalho[1], Renato Utsch[1],
Pedro Caldeira[1], Alberto Alburquerque[1], Fabricio Ferracioli[4],
Leonardo Rocha[2], Michael Frank[3], Dorgival Guedes[1], and Renato Ferreira[1]

[1] Department of Computer Science,
Federal University of Minas Gerais, Belo Horizonte, Brazil
gnandrade@dcc.ufmg.br
[2] Department of Computer Science,
Federal University of São João del Rei, São João del Rei, Brazil
[3] LG Electronics, San Jose Lab, Santa Clara, USA
[4] LG Electronics, São Paulo, Brazil

Abstract. Following the evolution of desktops, mobile architectures are currently witnessing growth in processing power and complexity with the addition of different processing units like multi-core CPUs and GPUs. To facilitate programming and coordinating resource usage in these heterogeneous architectures, we present *ParallelME*, a *Parallel Mobile Engine* designed to explore heterogeneity in mobile computing architectures. *ParallelME* provides a high-level library with a friendly programming language abstraction for developers, facilitating the programming of operations that can be translated into low-level parallel tasks. Additionally, these tasks are coordinated by a runtime framework, which is responsible for scheduling and controlling the execution on the low-level platform. ParallelME's purpose is to explore parallelism with the benefit of not changing the programming model, through a simple programming language abstraction that is similar to sequential programming. We performed a comparative analysis of execution time, memory and power consumption between *ParallelME*, OpenCL and RenderScript using an image processing application. ParallelME greatly increases application performance with reasonable memory and energy consumption.

1 Introduction

Mobile phones are no longer devices for basic communication between people, but have become much more sophisticated devices. Current models include a range of sensors that are capable of collecting a wide variety of data about the user and the environment, and many applications are being proposed which involve processing this large data volume either on the device or in the cloud. These application's compute demands have put pressure on the hardware architecture to significantly increase the processing power while also maintaining the power consumption at

© Springer International Publishing Switzerland 2016
P.-F. Dutot and D. Trystram (Eds.): Euro-Par 2016, LNCS 9833, pp. 447–459, 2016.
DOI: 10.1007/978-3-319-43659-3_33

a reasonable level. A recent trend in the desktop scenario is the utilization of different types of processing units (PUs), in the so-called heterogeneous systems - computers which include multi-core CPUs as well as other special purpose processors - GPUs being a favorite among them. Nowadays, many cell phone SoCs are also equipped with multi-core CPUs, such as Qualcomm Snapdragon and NVIDIA Tegra 3, and high-performance GPUs, such as Mali from ARM, and Adreno from Qualcomm.

In this new context, it becomes necessary for applications of different domains to achieve better performance and hence have better user experience by exploring, in a coordinated and efficient way, all the available PUs taking full advantage of their processing capabilities. In this direction it is important to create high-level programming abstractions which allow programmers to define the operations in some simple, intuitive way, while its parallel execution on different PUs can be achieved at run-time with minimal programmer interference. To accomplish that, we propose a specialized run-time framework, providing mechanisms for manipulating the generated low-level tasks. The framework is responsible for creating specific tasks to be executed in specific processing units, for managing the dependencies and for scheduling them across the devices in an efficient way. The proper use of different processing units in a heterogeneous system is a well studied problem in the literature when it comes to conventional computer architectures, for which different run-time environments [8,12] exist. However, it is a new and challenging scenario for mobile architectures.

The challenges in dealing with different processing units on a mobile architecture involve mainly the limited resources available for coding and the absence of facilitating run-time environments. The vast majority of mobile developers use programming languages, frameworks and APIs, supported by the device operating system, with a high level abstraction. In order to access and use different computational resources on a device architecture (i.e. GPU), programmers must be able to work at a lower level of abstraction, which requires advanced knowledge. This major effort to link low-level programming with high-level abstraction leads to a high cost of application programming.

In this work we present *ParallelME*, a *Parallel Mobile Engine* designed for exploring heterogeneity in mobile architectures. *ParallelME* provides a high-level library with a friendly programming language abstraction for developers, enabling programs to easily describe complex operations and translate them into low-level tasks. These tasks are manipulated in a coordinated manner by our proposed run-time framework, which is responsible for scheduling and controlling the execution on the low-level platform. Therefore, in our work, we explore two challenging points in the context of mobile architectures: (1) Expressing high-level parallel operations; and (2) Exploring different processing units in the low-level system. Despite being originally proposed for the mobile scenario, *ParallelME* is a generic parallel Java extension, coupled with a OpenCL run-time framework, which can easily be used for developing to desktop systems as well.

ParallelME's purpose is to explore parallelism with the benefit of not changing the programming model, through a simple programming language abstraction that is similar to sequential programming.

We performed a comparative analysis of execution time, memory and power consumption between *ParallelME*, OpenCL and RenderScript using an image processing application. Our experiments have shown that ParallelME greatly increases application performance up to 32.34 times, with reasonable energy consumption and reducing memory usage significantly, while maintaining a simple programming interface.

The remainder of this paper is organized as follows. In Sect. 2 we present a description of the main parallel programming frameworks currently available for mobile architectures. In Sect. 3 we present all components of ParallelME, on which we highlight the language used and the low level platforms explored. In sequence, in the Sects. 3.1 and 3.3 we explain the features of the proposed high-level library and the details of the low-level execution mechanism, respectively. We evaluate the comprehensiveness and applicability of our abstract language and the run-time performance, presenting the results in Sect. 4. Finally, we conclude and discuss some future work in the last section.

2   Related Works

In this section we present a detailed description of the main parallel programming frameworks available, to the best of our knowledge, for developing efficient application for mobile architectures. The frameworks addressed here are Pyjama [9], Aparapi [11], Rootbeer [14], Paralldroid [7] and RenderScript [6].

Pyjama's [9] focus is on bringing OpenMP's programming model to Java, with optimizations for Graphical User Interface (GUI) applications. Pyjama was built with this requirement in mind: it changes the way threads are generated by its own source to source compiler implementation. In OpenMP, the main thread that triggers parallel computation becomes the master one for the processing, whereas in Pyjama a new master thread is created for that, since the GUI one has to be kept running. It does not exactly present a new parallel programming model, but its changes in benefit of GUI applications are really interesting for developing mobile apps and its source to source compiler model is an effective way of introducing its own changes to the Java language, as it keeps things simple to the user while retaining control over how the parallelization is implemented. The amount of information the user provides and the way it is done is key to the effectiveness of the programming model adopted by the framework. On the other hand, Pyjama does little to address the complexity of OpenMP programming in general: a developer still has to learn his ways through it and is fully responsible for the good execution of his code.

Aparapi [11] is an AMD project, which focuses on making GPU programming easier. Aparapi was presented in 2011 as an API to express data parallel workloads in Java. It avoids changing the basic structure of the language itself, so instead of writing a parallel-for, for example, the developer extends a kernel base class to write his function. The code is compiled to Java bytecode and then parallelized by converting to OpenCL at runtime, reverting back to Java Thread Pool if necessary. It is possible to iterate over objects, not only primitive types. Comparing to programming with OpenCL or CUDA directly, Aparapi offers several advantages, though it does not tackle important problems. It is much easier to create and maintain new code, but debugging is still difficult and searching for execution bottlenecks remains as hard as on other environments. Compared with our own goals, Aparapi is limited in its parallelization scope, regarding both the hardware it optimizes code for and the level of complexity of its functions.

The idea behind Rootbeer [14] is similar to Aparapi's: transform regular Java code into GPU parallel code with minimum effort. Rootbeer is more robust than Aparapi, for it supports a broader set of instructions in Java language, excluding mostly dynamic method invocation, reflection and native methods. Rootbeer was created to make development on GPUs easier, but it never guaranteed performance improvement by treating the Java code written by the programmer. This means that even though it makes it easier to program, the task of code optimization still befalls heavily on the programmer, so much knowledge and time is still required to create efficient code and the programmer has to deal on his own with the performance issues his program's structure creates after the code conversion.

Paralldroid [7] is a framework to ease parallel programming on Android devices, aiming to make source to source translation of OpenMP-like annotated code. The user writes his Java code directing the framework on what should be done in parallel on each occasion, and the framework writes the low level code. Though it has limitations on what it converts, as they only support primitive type variables, conversions are made to RenderScript, OpenCL and Native C code. Akin to Pyjama, Paralldroid also uses OpenMP-like directives as basis for its programming model. These are complex to learn and require knowledge many programmers might not have.

RenderScript [6] is a framework designed by Google to perform data-parallel computation in Android devices, and was formally introduced in Android SDK API Level 11 (aka Android 3.0 Honeycomb). RenderScript is a means of writing performance critical code that can run natively on different processors, selected from those available at runtime. This could be the device CPU, a DSP, or even the GPU. Where it ultimately runs on is a question that depends on many factors that are not readily available to the developer.

3   ParallelME

ParallelME was conceived as a complete infrastructure for parallel programming for mobile architectures, exploring different processing units in a coordinated manner. Our purpose is focused on mobile computing, so we implemented ParallelME through available and widely used platforms for mobile systems. It is composed of three main components: (1) a programming abstraction, (2) a source-to-source compiler and (3) a run-time framework.

The programming abstraction was inspired by ideas found in the Scala collections library and is designed to provide an easy-to-use and generic programming model for parallel applications in Java for Android. The language was extended by means of specialized classes for which we also provided a sequential implementation. This means that the applications can run as they are written using the regular development infrastructure, though they will be very inefficient.

Our source-to-source compiler, however, provides a mechanism for replacing the use of the sequential libraries by extracting the portions of the code that comprise the parallel tasks and generating them either as RenderScript or as OpenCL [5]. The compiler also incorporates the appropriate bindings of the application code with the parallel tasks, thus creating a parallel application and improving the overall performance.

The run-time framework was developed using OpenCL and is responsible for setting up the application to allow several parallel tasks to be specified and queued for execution on different devices. It allows different criteria in deciding the PU in which each task should run at run-time, creating a novel level of control and flexibility which can be explored in order to achieve certain goals as far as resource utilization, which can further improve overall performance.

The first and current version of ParallelME is focused on the mobile operating system (Mobile OS) Android. In addition to being the most used Mobile OS in the world [4], Android is also an open source Linux-based system. In the following subsections we further describe each of the components of our proposed framework.

3.1 Programming Abstraction

General purpose mobile applications are commonly implemented in Java for Android OS through the use of the SDK (Software Development Kit [2]). Even though it is possible to use native Android resources with lower level programming platforms and language through NDK (Native Development Kit [1]), the use of Android SDK is preponderant among programmers. In this sense, ParallelME programming abstraction was designed to be supported by Android SDK in Java. In order to provide a generic and effective programming model for parallel applications, we considered the highest level of abstraction possible regarding the limitations of the current Java version supported by Android.

ParallelME programming abstraction was inspired in ideas found on the Scala collection library [15], which is a library consisted of a set of different data structures with native support for parallel processing. The goal of ParallelME was to create a similar data-structure oriented library that could be used to produce parallel code with the minimum effort by its user. In ParallelME, this set of data structures is called *User-Library*.

The User-Library is composed of a series of classes as part of a collections' package. These collections represent common data structures like arrays, lists, hash sets and hash maps capable of handling an abstract data type. The abstract data type corresponds to the user class which is used to store user data. All user data is stored on the respective collection through one of its data insertion methods.

So far, the User-Library contains three functional classes: Array, BitmapImage and HDRImage. The former class is a generic single-dimensional array, while the last two classes offer support for bitmap and HDR image processing.

Though Scala is a language which differs deeply from Java, ParallelME provides a similar approach for operations that iterate through an entire collection with no restriction of processing order. Given a bitmap object and a *foreach* operation to iterate through all its pixels and perform a simple modification in its color (i.e. add 1 to each RGBA color space parameter), the proposed abstraction for ParallelME is presented in Listing 1.

```
BitmapImage image = new BitmapImage(bitmap);
image.par().foreach(new UserFunction<Pixel>(){
    @Override
    public void function(Pixel pxl) {
        pxl.rgba.red = pxl.rgba.red + 1;
        pxl.rgba.green = pxl.rgba.green + 1;
        pxl.rgba.blue = pxl.rgba.blue + 1;
        pxl.rgba.alpha = pxl.rgba.alpha + 1;
    }
});
bitmap = image.toBitmap();
```

```
BitmapImage image = new BitmapImage(bitmap);
image.par().foreach(pxl -> {
    pxl.rgba.red = pxl.rgba.red + 1;
    pxl.rgba.green = pxl.rgba.green + 1;
    pxl.rgba.blue = pxl.rgba.blue + 1;
    pxl.rgba.alpha = pxl.rgba.alpha + 1;
});
bitmap = image.toBitmap();
```

Listing 1: Programming abstraction for ParallelME

Code in Listing 1 shows two variations of the same operation where left and right side are semantically identical. The first and last lines respectively are the data input and output operations of the User-Library class. They correspond to the class constructor where the user provides the input data (in this case a bitmap object) and the data retrieval method, on which the user gets back the data processed on ParallelME (in this case, the data replaces the original bitmap). The code between data input and output corresponds to the user code with the color modification proposed. It contains a call to method *par* on object *image*, which indicates that a parallel operation will be performed on the next method call. It is followed by a *foreach* iterator containing the user code and where operations have different syntax on each side. Left side code represents a version compatible with Java version supported by Android (Java 7) [2] during the development of the first version of ParallelME, with no support for lambda expressions [10]. Right side code was created using a lambda expression, which is available in Java 8. It demonstrates the simplicity and objectivity of our syntax when applied together with a lambda expression and shows how ParallelME will be used as a more recent version of Java is incorporated in Android.

3.2 Source-to-Source Compiler

The ParallelME source-to-source compiler has been developed with the incorporation of ANTLR, a powerful parser generator [13] widely used to build languages, tools and frameworks. We used an ANTLR Java grammar to create a parser that was used to build and traverse a parse tree, which in turn was the basis for ParallelME compiler. Our compiler takes as input Java code written with the User-Library and translates it into C code compatible with Render-Script or OpenCL frameworks, depending on the chosen runtime. Besides that, the compiler also integrates the translated code with the Java application, performing modifications to the original user code to support the proposed run-time environment.

In this sense, our compiler is composed of 3 macro steps:

- **User-Library detection:** detects where the user code was written with User-Library classes. These locations will point to the code that must be translated to the target runtime.
- **Memory binding:** detects where the user is providing the data that will be processed in order to transmit it to the runtime environment.
- **Conversion to RenderScript/OpenCL:** the user code in the iterators' body of the User-Library classes must be translated to a RenderScript or OpenCL compliant version in order to produce the behaviour specified on the high level programming abstraction.

In favor of creating a first version of ParallelME on a reasonable schedule, we defined some rules and limitations on the input and output code. These rules and limitations are related to the user code provided to the iterator and the user class that is parametrized on User-Library collections. For both user code in iterators and parametrized classes we defined that only Java primitive types, their equivalent wrapper classes and User-Library collections can be used. These restrictions reduced considerably the compiler complexity, yet allowing a high degree of flexibility on user code creation.

Once our User-Library was designed in Java, the user code written with our programming abstraction is 100 % compliant with Java 7 language specification. This allowed us to rely on the Java compiler (javac) for syntactic and semantic evaluation of code, reducing responsibilities of ParallelME compiler and increasing its robustness.

3.3 Run-Time Framework Details

The run-time component of ParallelME is responsible for coordinating, in an efficient way, all processing units available on the mobile architecture. It organizes and manages low level tasks generated by the Compiler component, from definitions expressed by developers in User-Library. The implementation is in C++ and was inserted in the Android system using the NDK toolkit. As previously explained, OpenCL was adopted as a low-level parallel platform to manipulate available processing units. In general, the dynamics of the runtime involves the following phases: (1) Identifying the computing resources available on mobile architecture; (2) Creating tasks and their input and output parameters; (3) Arranging task's data according to their parameters; (4) Submitting them for execution; (5) Instantiating routines of a scheduling policy; (6) Assigning tasks to processing units defined by the scheduling policy. The first four phases correspond to user API phases and must be performed to instantiate the system and the tasks. The run-time internal engine is composed of the two remaining phases. Figure 1 gives an overall picture of the entire framework, which is further described below.

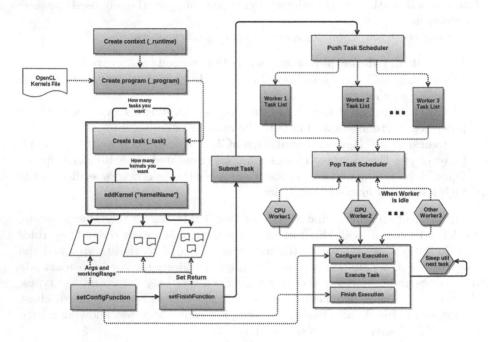

Fig. 1. Execution mechanism dynamics

Run-Time API Phases. The first run-time phase is divided into two steps: (1) detection of the available resources in a specific mobile architecture; and (2) instantiation of all necessary structures related to the framework. In the first step, the framework identifies the available processing units, also called devices. A context is created for each device, which corresponds to specific implementations of OpenCL routines, provided by different suppliers such as NVIDIA, Intel, AMD and Qualcomm. In the second step, the framework instantiates a system thread for each of these contexts. These threads, called *Worker Threads*, are responsible for managing devices using specific OpenCL routines. These configurations are performed by instantiating the run-time constructor.

The second phase corresponds to the creation of tasks that are executed by processing units. In this phase, a source file containing one or more OpenCL kernels is built, and each kernel present in the compiled file composes one or more tasks. When more than one kernel is assigned to a task, they are executed in the order they were instantiated. It is important to emphasize that these tasks are generic and not linked - at this moment - to any device. When submitted to execution, these tasks will be scheduled and executed on a specific device.

The third phase is responsible for preparing the data that will be manipulated by each task. As OpenCL buffers are device-specific, it would be very expensive to set up tasks' data before the scheduler decides where the task will run on. In order to deal with this, we propose a mechanism in which the task configuration is done through callbacks: before sending the task to the scheduler, the user specifies configuration callbacks (that can be lambda functions) that set up task data. Only after the scheduler decides where the task will run, the configuration function is called with the target device specified through a parameter. This avoids the cost of copying task data to multiple devices. Also, in this step, it is possible to configure for each kernel the amount of threads or work units involved (OpenCL work range) in its execution.

Finally, the last phase related to Environment Setup corresponds to submitting tasks to execution. The run-time routine responsible for performing it must be called for each created task.

Execution Engine. After task submission, the run-time engine is responsible for scheduling and executing tasks across the devices. Once a task is submitted, the scheduler routine *Push Task* allocates it to a specific device task list, following a particular scheduling policy. *Worker threads* remain on a sleeping state if there is no task to be executed. However, when a task is submitted, a signal awakens *worker threads* which in turn call the routine *Pop Task*. This routine is responsible for retrieving a task from the task list, following a particular scheduling policy. When a *worker thread* receives the task returned by *Pop task* routine it starts the execution process.

For task execution, first the run-time framework executes the *Configure Execution* callback, responsible for allocating buffers of kernel parameters, assigned in the task, on the specific device. After this allocation, the kernel is also assigned to the device memory in order to start its execution. A *worker thread* waits for the execution to finish and then calls the *Finish Execution* callback, which is responsible for retrieving the output buffers. The *worker thread* then goes back to sleeping state if there is no more tasks in its execution lists.

4   Evaluation

The evaluation of ParallelME was performed on three different metrics: execution time, memory consumption and power consumption. For this purpose, we implemented a HDR tone mapping application using Eric Reinhard's operator [16] with ParallelME. Tone mapping is a technique used in image processing and computer graphics to map one set of colors to another to approximate the appearance of high dynamic range images in a medium that has a more limited dynamic range. Additionally, we implemented and evaluated the same application in different platforms in order to compare and discuss ParallelME performance: (1) Single Threaded Java; (2) OpenCL using CPU; (3) OpenCL using GPU; (4) RenderScript.

In each of the mentioned platforms, along with ParallelME, we ran the application two times. In the first run we measured execution time and power consumption, while in the second we measured memory consumption. We used 5 different HDR images (Clifton Bridge, Clock Building, Crowfoot, Lake Tahoe and Tintern Abbey) [3] and all executions ran each image 1, 2, 5, 10 and 15 times. For ParallelME environment, each tone mapping execution correspond to a task. In this way, in our experiment 1, 2, 5, 10 and 15 tasks were executed concurrently in the ParallelME environment. For these experiments ParallelME used just a simple First Come First Serve scheduler for task distribution.

We also performed each execution from scratch, meaning that after finishing the previous execution, we killed the application process and restarted it anew. We used a LG D855 (LG G3) device set up with Android 5.0, 100 % of screen brightness, flight mode activated and all others applications closed.

4.1 Results and Discussions

To measure execution time we used Java routines to take the time delay between application start and finish and for power consumption evaluation we used a Monsoon Power Monitor configured to 3.9 V. Power consumption was measured in the following way: (1) average consumption of Android OS with 180 s idle (no application running) and the screen on; (2) average consumption during the algorithm execution. Then we subtracted values (1) from (2) to get the application power consumption, that tells us, in milliwatts (mW), the average power consumption required during execution. Additionally we evaluated and discussed the total energy consumption, which was found by multiplying the average power by the application execution time. Taking the single threaded Java implementation as the basis for comparison, we evaluated execution time, memory consumption, power consumption and total energy consumption for the other implementations. Figures 2 and 3 show the mean and median for all executions.

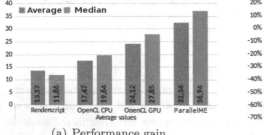

(a) Performance gain.

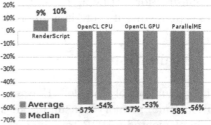

(b) Memory consumption.

Fig. 2. Performance and memory consumption evaluation.

In Fig. 2(a) it is possible to observe that ParallelME presents the best result, increasing application performance, in average, by 32.34 times compared to the single thread Java implementation. The presence of a run-time framework in ParallelME allows tasks to be executed concurrently. Thus all processing units

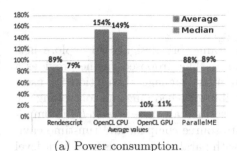

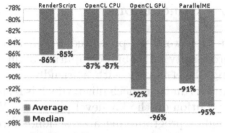

(a) Power consumption.   (b) Total Energy consumption.

Fig. 3. Power and energy consumption evaluation.

in the mobile architecture are being explored in coordination, reflecting a significant performance increase. In addition, analyzing the Fig. 3(a), we note that ParallelME required 88 % more power to achieve the reported performance. This consumption is acceptable given that different resources in mobile architecture are being used in parallel. However, despite ParallelME requiring more power, the total energy consumption, shown in Fig. 3(b), is 94 % lower than the energy consumed by the single threaded Java implementation, since the application execution time using ParallelME is considerably lower than the Java one. This shows that reducing application execution time with ParallelME also causes a significant reduction on the energy consumed by the application. Also, by comparing the results to RenderScript, which required the same amount of power to increase application performance by just 13 times, ParallelME becomes even more satisfactory when it comes to the total energy consumption.

It is necessary to emphasize the results of the OpenCL GPU implementation. In this case, the total energy consumption is only slightly smaller than the ParallelME energy consumption, although power consumption was just 10 % higher than the single thread Java. This also happened because ParallelME runtime is much faster. Also, in this situation, despite the significant increase in performance and low power consumption of OpenCL, GPU resources are fully occupied during application execution. Because of this, the phone screen may freeze, as the screen stops being updated, compromising user experience.

To measure memory consumption we plugged the phone to a computer using a USB cable and adb connection, allowing it to run in debug mode and, in this way, to track memory. Following this strategy, Fig. 2(b) presents in graphical format the average values of the proposed scenario.

RenderScript application consumes more memory to execute; 9 % on average. On the other hand, the improvements achieved by OpenCL version are clear, reducing memory consumption by 57 % on average. ParallelME is based on OpenCL and provides, besides a significant performance improvement and reduction in power consumption, a significant reduction in memory usage compared to RenderScript and single thread Java implementations.

5   Conclusions

In this work we presented *ParallelME*, a framework designed to explore heterogeneity in mobile computing architectures. The proposed engine provides a high-level library with a friendly programming language abstraction for users and a run-time mechanism which is responsible for scheduling and controlling the execution on the low-level platform. We described all three *ParallelME* components - programming abstraction, source-to-source compiler and run-time environment - showing it was designed to deal with parallel operations from high-level implementation to execution on heterogeneous architectures.

We measured execution time, memory and power consumption of *ParallelME* comparing to other implementations of an image processing application. We show that *ParallelME* increases application performance by 32.34 times, which is the best result, with reasonable energy consumption and significantly reducing memory usage.

This work can be extended in different ways. The User-Library can be extended with several different collections and operations to increase ParallelME's usability. The system run-time API makes it possible to insert different scheduling strategies and others schedulers may explore different aspects of the underlying environment, such as reducing energy consumption. It is also possible, at the level of the proposed compiler, to insert features to extract code information that can be used by the scheduler to improve kernel execution assignment in the available processing units.

References

1. Android NDK. http://developer.android.com/tools/sdk/ndk
2. Android SDK. http://developer.android.com/sdk
3. HDR images collection. https://www.cs.utah.edu/~reinhard/cdrom/results.html
4. IDC analyses. http://www.idc.com/prodserv/smartphone-os-market-share.jsp
5. OpenCL by khronos group. https://www.khronos.org/opencl/
6. RenderScript. https://developer.android.com/guide/topics/renderscript
7. Acosta, A., Almeida, F.: Performance analysis of paralldroid generated programs. In: 2014 22nd Euromicro International Conference on Parallel, Distributed and Network-Based Processing (PDP). IEEE (2014)
8. Augonnet, C., Thibault, S., Namyst, R., Wacrenier, P.: StarPU: A Unified Platform for Task Scheduling on Heterogeneous Multicore Architectures (2011)
9. Giacaman, N., Sinnen, O. et al.: Pyjama: OpenMP-like implementation for Java, with GUI extensions. In: Proceedings of the 2013 International Workshop on Programming Models and Applications for Multicores and Manycores. ACM (2013)
10. Gosling, J., Joy, B., Steele Jr., G.L., Bracha, G., Buckley, A.: The Java Language Specification, Java SE 7 Edition. 1st edn. (2013)
11. Gupta, K.G., Agrawal, N., Maity, S.K.: Performance analysis between Aparapi (a parallel api) and Java by implementing sobel edge detection algorithm. In: Parallel Computing Technologies (PARCOMPTECH). IEEE (2013)
12. Kunzman, D.: Charm++ on the cell processor (2006)
13. Parr, T.: The Definitive ANTLR 4 Reference. Pragmatic Bookshelf, 2nd edn. (2013)

14. Pratt-Szeliga, P.C., Fawcett, J.W., Welch, R.D.: Rootbeer: seamlessly using GPUs from java. In: 2012 IEEE 14th International Conference on High Performance Computing and Communication & 2012 IEEE 9th International Conference on Embedded Software and Systems (HPCC-ICESS). IEEE (2012)
15. Prokopec, A., Bagwell, P., Rompf, T., Odersky, M.: A generic parallel collection framework. In: Proceedings of the 17th International Conference on Parallel Processing - Volume Part II, EUROPAR 11 (2011)
16. Reinhard, E., Stark, M., Shirley, P., Ferwerda, J.: Photographic tone reproduction for digital images. ACM Trans. Graph. (TOG) **21**, 267–276 (2002). ACM

CBPQ: High Performance
Lock-Free Priority Queue

Anastasia Braginsky[1(✉)], Nachshon Cohen[2], and Erez Petrank[2]

[1] Yahoo Research, Haifa, Israel
anastas@yahoo-inc.com
[2] Technion - Israel Institute of Technology, Haifa, Israel
{ncohen,erez}@cs.technion.ac.il

Abstract. Priority queues are an important algorithmic component and are ubiquitous in systems and software. With the rapid deployment of parallel platforms, concurrent versions of priority queues are becoming increasingly important. In this paper, we present a novel concurrent lock-free linearizable algorithm for priority queues that scales significantly better than all known (lock-based or lock-free) priority queues. Our design employs several techniques to obtain its advantages including lock-free chunks, the use of the efficient fetch-and-increment atomic instruction, and elimination. Measurements under high contention demonstrate performance improvement by up to a factor of 1.8 over existing approaches.

Keywords: Non-blocking · Priority queue · Lock-free · Performance · Freezing

1 Introduction

Priority queues serve as an important basic tool in algorithmic design. They are widely used in a variety of applications and systems, such as simulation systems, job scheduling (in computer systems), networking (e.g., routing and real-time bandwidth management), file compression, numerical computations, and more. With the proliferation of modern parallel platforms, the need for a high-performance concurrent implementation of the priority queue has become acute.

A priority queue (PQ) supports two operations: insert and deleteMin. The abstract definition of a PQ provides a set of key-value pairs, where the key represents a priority. The insert() method inserts a new key-value pair into the set (the keys don't have to be unique), and the deleteMin() method removes and returns the value of the key-value pair with the lowest key (i.e., highest priority) in the set.

Lock-free (or non-blocking) algorithms [12,13] guarantee eventual progress of at least one operation under any possible concurrent scheduling. Thus, lock-free

This research was supported by THE ISRAEL SCIENCE FOUNDATION (grant No. 274/14).

© Springer International Publishing Switzerland 2016
P.-F. Dutot and D. Trystram (Eds.): Euro-Par 2016, LNCS 9833, pp. 460–474, 2016.
DOI: 10.1007/978-3-319-43659-3_34

implementations avoid deadlocks, live-locks, and priority inversions. Typically, they also demonstrate high scalability, even in the presence of high contention.

In this paper we present the design of a high performance lock-free lineariz-able PQ. The design builds on a combination of three ideas. First, we use a chunked linked-list [3] as the underlying data structure. This replaces the stan-dard use of heaps, skip-lists, linked-lists, or combinations thereof. Second, we use the fetch-and-increment ($F\&I$) instruction for an efficient implementation of deleteMin and insert. This replaces the stronger, but less efficient compare-and-swap (CAS) atomic primitive (used in all other lock-free PQ studies). Third, the resulting design is a great platform for applying an easy variant of elimina-tion [11,18], which resolves the contention of concurrent inverse operations: the insert of a small key and a deleteMin.

Various constructions for the concurrent PQ exist in the literature. Hunt et al. [14] used a fine-grained lock-based implementation of a concurrent heap. Dragicevic and Bauer presented a linearizable heap-based priority queue that used lock-free software transactional memory (STM) [8]. A quiescently consistent skip-list based priority queue was first proposed by Lotan and Shavit [17] using fine-grained locking, and was later made non-blocking [9]. Another skip-list based priority queue was proposed by Sundell and Tsigas [19]. Liu and Spear [16] introduced two concurrent versions of data structure called *mounds* (one is lock-based and the other is lock-free). The mounds data structure aims at very fast $O(\log(\log(N)))$ insert operations. It is built of a rooted tree of sorted lists that relies on randomization for balance. The deleteMin operations have a slower $O(\log(N))$ complexity. Mounds' insert operation is currently the most productive among concurrent implementations of the PQ. Linden and Jonsson [15] presented a skip-list based PQ. Deleted elements are first marked as deleted in the deleteMin operation. Later, they are actually disconnected from the PQ in batches when the number of such nodes exceed a given threshold. Their construction outperforms previous algorithms by $30 - 80\,\%$. Recently, Calciu et al. [5] introduced a new lock-based, skip-list-based adaptive PQ that uses elimination and flat combining techniques to achieve high performance at high thread counts.

Elimination [11,18] provides a method to match concurrent inverse oper-ations so that they exchange values without synchronizing on a centralized data structure. Elimination for PQ was presented in [5], where threads that insert small keys and threads that delete minimum keys post their operations in an elimination array and wait for their request to be processed. Our elimi-nation variant requires no additional similar waiting time and it bears minimal overhead.

We implemented CBPQ in C++ and compared its performance to the cur-rently best performing PQs: the Linden and Jonsson's PQ [15], the lock-free and lock-based implementations of the Mounds PQ [16], and the adaptive PQ (APQ) of Calciu et al. [5]. We evaluated the performance of CBPQ using tar-geted micro-benchmarks: one that runs a mix of insert and deleteMin operations, where each occurs with equal probability, a second one that runs only insert operations, and a third with only deleteMin operations.

The results demonstrate that our CBPQ design performs excellently under high contention, and it scales best among known algorithms, providing the best performance with a large number of threads. Under low contention, our algorithm is not a winner, and it turns out that the LJPQ design performs best. In particular, under high contention and for a mix of deleteMin and insert operations, CBPQ outperforms all other algorithms by up to 80 %. When only deleteMin operations run, and with high contention, CBPQ performs up to 5 times faster than deletions of any other algorithm we compared to. As expected, Mounds perform best with insert only operations, outperforming CBPQ (which comes second) by a factor of up to 2.

2   A Bird's Eye Overview

The CBPQ data structure is composed of a list of chunks. Each chunk has a range of keys associated with it, and all CBPQ entries with keys in this range are located in this chunk. The ranges do not intersect and the chunks are sorted by the ranges' values. To improve the search of a specific range, an additional skip-list is used as a directory that allows navigating into the chunks, so inserting a key to the CBPQ is done by finding the relevant chunk using a skip-list search, and then inserting the new entry into the relevant chunk.

The first chunk is built differently from the rest of the chunks since it holds the smallest keys and supports deleteMin operations. We forbid inserts into the first chunk. Instead, a key in the range of the first chunk is inserted through special handling, as discussed below. The remaining chunks are used for insertions only.

The first chunk consists of an immutable sorted array of elements. To delete the minimum, a thread simply needs to atomically fetch and increment a shared index to this array. All other chunks consist of unsorted arrays with keys in the appropriate range. To insert a key to a chunk other than the first, the insert operation simply finds the adequate chunk using the skip-list directory, and then adds the new element to the first empty slot in the array, again, simply by fetching and incrementing the index of the first available empty slot in the array.

When an insert operation needs to insert a key to the first chunk, it registers this key in a special buffer and requests the first chunk rebuild. Subsequently, a new first chunk with a new sorted array is created from the remaining keys in the first chunk, all keys registered in the buffer, and if needed, more keys from the second chunk. The thread that attempts to insert a (small) key into the buffer yields the processor and allows progress of other threads currently accessing the first chunk, before making everybody cooperate on creating a new first chunk. During this limited delay, elimination can be executed and provide even more progress. By the time a rebuild of the first chunk actually happens, either much progress has occurred, or the buffer has been filled with keys, making the amortized cost of a new first chunk construction smaller. The creation of a new first chunk is also triggered when there are no more elements to delete in it. The creation of a new first chunk is made lock-free by allowing all relevant threads to take part in the construction, never making a thread wait for others to complete a task.

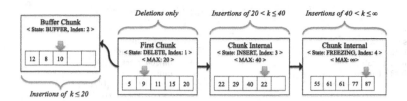

Fig. 1. Overview of the CBPQ data structure

When an internal chunk is filled due to insertions, it is split into two half-full chunks using the lock-free freezing mechanism of [3]. The CBPQ scheme is illustrated in Fig. 1 (with some more algorithmic details shown in text in the angle brackets, to be discussed later). The full description of the algorithm is provided in Sect. 3.

Key Design Ideas: The key design ideas in the proposed priority queue are as follows. First, we aim at using $F\&I$ instructions for the contention bottlenecks. For that, we employ the chunked linked-list as the underlying data structure. This allows most insert and deleteMin operations to be executed with an (atomic) increment of a chunk index. To make the above work, we distinguish the design of the first chunk (that mostly handles deleteMin operations) from the design of the remaining chunks (which handle only insert operations). An extra buffer chunk supports insertions to the first chunk. Finally, we add elimination, which fits this algorithm like a glove with negligible overhead and significant performance benefits.

3   The Full CBPQ Design

In this section we describe the core CBPQ algorithm. In Sect. 4 we describe the skip-list on top of the list and the elimination. The CBPQ linearization points presentation can be found in the full version of this paper [1].

3.1   Underlying Data Structures

Each chunk in the CBPQ has a *status* word, which combines together an array index (shortly denoted *index*), a *frozen index*, and the chunk *state*. The status is always atomically updated. The state of a newly created chunk is INSERT, DELETE, or BUFFER, indicating that it is created for further insertions, for deletions, or for serving as a buffer of keys to be inserted to the first chunk, respectively. In addition, when a chunk has to be replaced by a new chunk, the old chunk enters the FREEZING state, indicating that it is in the process of being frozen. The FROZEN state indicates that the chunk is frozen and thus, obsolete.

Listing 1: Status and Chunk records

```
1   class Status{                    9    class Chunk{
2     uint29_t frozenIdx;           10      Status status;
3     uint3_t state;                11      uint64_t entries[M];
4     uint32_t index;               12      uint32_t max;
5   } // 64 bits, machine word      13      uint64_t frozen[M/63+1];
6                                    14      Chunk *next, *buffer;
7                                    15    }
8
```

The frozen index is used only for freezing the first chunk, as will be explained in Sect. 3.4. In Listing 1, we list the Status and Chunk classes that we use. The relevant state, index, and frozenIdx fields are all held in a single machine word that is called the Status.

In addition to the status, each chunk consists of an array (entries in the Chunk class), which holds the key-value pairs contained in the chunk. The entries are machine words, whose bits are used to represent the key and value. For simplicity, in what follows we will refer to the keys only. Our keys can take any value that can be represented in 31 bits, except 0, which is reserved for initialization. Each chunk has an immutable maximal value of a key that can appear on it, defined when the chunk is created (max in Chunk). A chunk holds keys less than or equal to its max value and greater than the max value of the previous chunk (if it exists). This max field is not relevant for the buffer chunk. Any chunk (except the buffer) uses a pointer to the next chunk (the next field). Finally, only the first chunk uses a pointer to a buffer chunk (the buffer field). The meaning of the frozen array that appears in the Chunk specification of Listing 1 is related to the freeze action and will be explained in Sect. 3.4. Finally, the CBPQ is a global pointer head to the first chunk. In Fig. 1 the chunk's fields are depicted.

3.2   Memory Management

Similar to previous work [15], we use Keir Fraser's epoch based reclamation (EBR) [9] as a (simplified) solution for memory reclamation. The EBR scheme was used in one previous work to which we compare [15]. The other algorithms that we compare against [5,16] did not implement memory reclamation at all. Measurements show that avoiding reclamation buys a performance advantage of up to 4 % (but does not provide a full solution). More advanced lock-free memory reclamation solutions appear in [2,4,6,7].

3.3   Operations Implementation

Insert: The insert pseudo-code is presented in the insert() method in Listing 2. In order to insert a key into the CBPQ, we first need to find the relevant chunk C. Because chunks are ordered by their ranges, a simple search can be used, skipping the chunks with smaller maximums (Line 4). If an insert must be performed to the first chunk, the insertToBuffer() method is invoked (Line 6), as explained in the next paragraph. Otherwise, C is not first. After C is found, its array index is atomically incremented to make room for the new entry (Line 9).

```
1   void insert(int key) {
2     Chunk* cur = NULL, *prev = NULL;
3     while(1) {
4       getChunk(&cur, &prev, key);          // set the current and previous chunk pointers
5       if ( cur==head ) {                   // first chunk
6         if ( insertToBuffer(key, cur, head) ) return;
7         else continue;
8       }
9       Status s = cur->status.aIncIdx();    // atomically increase the index in the status
10      int idx = getIdx(s);
11      if ( idx<M && !s.isInFreeze() ) {    // insert into a non-full and non-frozen chunk
12        cur->entries[idx] = key; memory_fence;
13        if (!cur->status.isInFreeze()) return;
14        if (cur->entryFrozen(idx)) return;  // key got copied
15      }
16      freezeChunk(cur);                    // restructure the CBQP, then retry
17      freezeRecovery(cur, prev);
18    }
19  }
20  bool insertToBuffer(int key, Chunk* cur, Chunk* curhead) {
21    Chunk *curbuf = cur->buffer;  bool result = false;   // PHASE I: key insertion into the buffer
22    if( curbuf==NULL )                                   // the buffer is not yet allocated
23      if ( createBuffer(key,cur,&curbuf) ) goto phaseII;  // key added during buffer creation
24    Status s = curbuf->status.aIncIdx();                 // atomically increase the index in the status
25    int idx = getIdx(s);
26    if ( idx<M && !s.isInFreeze() ) {
27      curbuf->entries[idx] = key; memory_fence;
28      if (!curbuf->status.isInFreeze()) result = true;
29      if (curbuf->entryFrozen(idx)) return true;
30    }
31  phaseII:          // PHASE II: first chunk merges with buffer before insert ends
32    usleep(0);      // yield, give other threads a chance
33    freezeChunk(cur); freezeRecovery(cur, NULL);
34    return result;
35  }
36  int deleteMin() {
37    Chunk* cur, next;
38    while(1){
39      cur = head;
40      Status s = cur->status.aIncIdx();  // atomically increase the index in the status
41      int idx = getIdx(s);
42      if ( idx<M && !s.isInFreeze() )    // delete from not full and non-frozen chunk
43        return curr->entries[idx];
44      freezeChunk(cur); freezeRecovery(cur, NULL);  // Freeze, then restructure the CBPQ and retry
45    }
46  }
```

Listing 2. Common code path: insertion of a key and deletion of the minimum

The aIncIdx() method wraps an $F\&I$ instruction and returns the status with the new value of the chunk index. The index is incremented first and only later we check whether it surpassed the end of the array. However, the number of bits required to represent the chunk size (M) is much smaller than the number of bits in the index, and so even if each thread increments it once after the chunk is full, an overflow would not occur.[1] If C is not frozen and the incremented index does not point beyond the chunk's capacity, we simply write the relevant key to the array entry (Lines 11–15). The write is followed by a memory fence in order to ensure it will be visible to any other thread that may freeze C concurrently.

---

[1] The size of the array plus the number of operating threads limits the index value. In our implementation the array size is less then 2^{10} plus the number of threads (less then 2^6), so 11 bits suffice.

If C is not freezing (Line 13), the insert is completed. Else if C is freezing, but our key was already marked as frozen (Line 14), the insert is also finished. Otherwise (if the index has increased too much or a freeze has been detected), then the freeze is completed and C is split (Lines 16, 17), as will be explained later. After the chunks restructure, the insert is restarted.

Insert to the First Chunk: The lowest range keys are inserted into the buffer pointed to from the first chunk. The pseudo-code of an insertion to the buffer chunk is presented in the insertToBuffer() method in Listing 2. It starts by allocating a new buffer holding the relevant key, if needed (Line 23). The createBuffer() method returns *true* if the new buffer was successfully connected to the first chunk, or *false* if another thread had connected another buffer. In the latter case, a new pointer to the buffer is inserted into curbuf.

Keys are inserted to the buffer in a manner similar to their insertion to other (non-first) chunk: the index is increased and the key is placed. If this cannot be done because the buffer is full or frozen, the insertToBuffer() returns *false* (after the first chunk's freeze and recovery) to signal that the insert operation has to be retried. The insert to buffer operation cannot end until the new key is included in the first chunk and considered for deletion. So after a key is successfully inserted into a buffer, the freeze and merge of the first chunk is invoked. However, if this key is already frozen, the insert to the first chunk can safely return (Line 29), because no deletion can now happen until the new key is taken into the new first chunk. After the first chunk is replaced, the insertion is considered done. The yielding, freeze and merge (Lines 32–33) are explained in Sects. 3.4 and 4.

Delete Minimum: The deletion is very simple and usually very fast. It goes directly to the first chunk, which has an ordered array of minimal keys. The first chunk's index is atomically increased. Unless the need to freeze the first chunk is detected, we can just return the relevant key. The pseudo-code for the deletion operation is presented in the deleteMin() method in Listing 2.

3.4 Split and Merge Algorithms

It remains to specify the case where a freeze is needed for splitting a non-first chunk or merging the first chunk with the buffer and possibly also with the second chunk. This mechanism is developed in [3] and we adopt it with minor modifications. For completeness, we explain this mechanism below.

For splitting or merging chunks, a freeze is first applied on the chunks, indicating that new chunks are replacing the frozen ones. A frozen chunk is logically immutable. Then, a recovery process copies the relevant entries into new chunks that become active in the data structure. Threads that wake up after being out of the CPU for a while may discover that they are accessing a frozen chunk and they then need to take actions to move into working on the new chunks that replace the frozen ones. In [3], the freezing process of a chunk was applied by atomically setting a dedicated freeze bit in each machine word (using a CAS loop), signifying that the word is obsolete. Freezing was achieved after all words were marked in this manner. Applying a CAS instruction on each obsolete word

```
1    void freezeChunk(Chunk* c) {
2      int idx, frozenIdx = 0; Status localS;    // locally copied status
3      while(1){                        // PHASE I: set the chunk status if needed
4        localS = c−>status; idx = localS.getIdx();    // read the current status to get its state and index
5        switch (localS.getState()){
6          case BUFFER:    // in insert or buffer chunks frozenIdx was and remained 0
7          case INSERT: c−>status.aOr(MASK_FREEZING_STATE); break;
8          case DELETE:
9            if (idx>M) frozenIdx=M; else frozenIdx=idx;
10           Status newS; newS.set(FREEZING, idx, frozenIdx); // set: state, index, frozen index
11           if (c−>status.CAS(localS,newS)) break; // can fail due to delete updating the index
12             else continue;
13           case FREEZING: break; // in process of being frozen
14           case FROZEN:    // c was frozen by someone else
15             c−>markPtrs(); return;    // mark the chunk out−pointers as deleted
16         }
17         break;  // continue only if CAS from DELETE state failed
18       }
19       if (c != head) freezeKeys(c);  // PHASE II: freeze the entries
20       c−>status.aOr(MASK_FROZEN_STATE);  // from FREEZING to FROZEN using atomic OR
21       c−>markPtrs();  // set the chunk pointers as deleted
22     }
```

Listing 3. Freezing the keys and the entire chunk

(sometimes repeatedly) may be slow and it turns out that in the context of CBPQ we can freeze a chunk more efficiently.

The freezing mechanism coordinates chunk replacements with concurrent operations. What may come up in the CBPQ is a race between insertion and freezing. An insert operation increments the array index reserving a location for the insert. But such an operation may then be delayed for a long while before actually inserting the item to the array. This operation may later wake up to find that the chunk has been frozen and entries have already been copied to a newer chunk. Since the item's content was not installed into the array, the freezing process could not include it in the new chunk and insertion should be retried. The inserting thread needs to determine whether its item was inserted or not using a freeze bit that is associated with his entry of the frozen chunk. This motivates a freeze bit for each entry, but these bits do not need to reside on the entry.

In CBPQ, all freeze bits are located separately from the entries, with a simple mapping from them to their freeze bits. In the CBPQ Chunk class (Listing 1), the data words (storing the keys and the values) are located in the **entries** array. All freeze bits are compacted in the **frozen** array. Each frozen bit signifies whether the key in the associated entry has been copied into the chunks that replace the current frozen chunk. Assuming a 64-bit architecture, we group each 63 entries together and assign a *freeze word* of 64 bits to signify the freeze state of all 63 entries. We use one bit for each of the 63 entries and reserve the most significant bit (MSB) of the freeze word to make sure that it is written only once (modifying this bit from 0 to 1 when written).

The freezing process reads the actual entries of the 63 entries. Since a value of zero is not a valid key, having a zeroed key word indicates an uncompleted insert. In this case, the entry is not copied into the new chunk. After determining which entries should be copied, the freezing process attempts to set the freeze word accordingly (1 for an existing entry and 0 for an invalid entry) using an atomic CAS of this word in the frozen array. Atomicity is guaranteed, because each freeze word is updated only once, due to the MSB being set only once. The compaction of the freeze bits allows setting them with a single CAS instead of 63 CAS operations, reducing the synchronization overhead for the freeze processing.

Freezing the Chunk: Following the pseudo-code of method freezeChunk() in Listing 3, here is how we execute the freeze for a chunk C. In the first phase of the operation, we change C's status, according to the current status (Lines 4–17). Recall that the status consists of the state, the index and the frozen index. If C is not in the process of freezing or already frozen, then it should be in a BUFFER, an INSERT or a DELETE state, with a zeroed frozen index and an index indicating the current array location of activity. For insert or buffer chunks, we need only change the state to FREEZING; this is done by setting the bits using an atomic OR instruction (Line 7). The frozen index is only used for the first chunk, in order to mark the index of the last entry that was deleted before the freeze. Upon freezing, the status of the first chunk is modified to contain FREEZING as a state, the same index, and a frozen index that equals the index if the first chunk is not exhausted, or the maximum capacity if the first chunk is exhausted, i.e., all entries have been deleted before the freeze (Line 9). Let us explain the meaning of the frozen index.

As the deletion operation uses a $F\&I$ instruction, it is possible that concurrent deletions will go on incrementing the index of the first array in spite of its status showing a frozen state. However, if a thread attempts to delete an entry from the first chunk and the status shows that this chunk has been frozen, then it will not use the obtained index. Instead, it will help the freezing process and then try again to delete the minimum entry after the freezing completes. Therefore, the frozen index indicates the last index that has been properly deleted. All keys residing in locations higher than the frozen index must be copied into the newly created first chunk during the recovery of the freezing process. If all keys in the frozen first chunk have been deleted, then no key needs to be copied and we simply let the frozen index contain the maximum capacity M, indicating that all keys have been deleted from the first chunk.

In Line 11 the status is updated using a CAS to ensure that concurrent updates to the index due to concurrent deletions are not lost. If C is already in the FREEZING state because another thread has initiated the freeze, we can move directly to phase II. If C is in the FROZEN state, then the chunk is in an advanced freezing state and there is little left to do. It remains to mark the chunk pointers buffer and next so that they will not be modified after the chunk has been disconnected from CBPQ. These pointers are marked (in Line 15 or Line 21) as deleted (using the common Harris delete-bit technique [10]). At this point we can be sure that sleeping threads will not wake up and add a link to a new buffer chunk or a next chunk to C and we may return.

The second phase of the freeze assumes the frozen index and state have been properly set and it executes the setting of the words in the frozen array in method freezeKeys() (Line 19). However, in the first chunk no freeze bits are set at all. In the first chunk it is enough to insert the chunk into the FREEZING state. This is so, because no one ever checks the frozen bits on the first chunk. Once we get the index and find that the state is DELETE the relevant minimum is just returned. With the second phase done, it remains to change the state from FREEZING to FROZEN (using the atomic OR instruction in Line 20) and to mark the chunk's pointers deleted as discussed above. The atomic OR instruction is available on the x86 platform and works efficiently. However, this is not an efficiency-critical part of the execution as freezing happens infrequently, so using a simple CAS loop to set the state would be fine.

CBPQ Recovery from a Frozen Chunk: Once the chunk is frozen, we proceed like [3] and replace the frozen chunk with one or more new chunks that hold the relevant entries of the frozen chunk. This is done in the freezeRecovery() method, presented in Listing 4. The input parameters are: cur – the frozen chunk that requires recovery, and prev – the chunk that precedes cur in the chunk list or NULL if cur is the first chunk.[2] The first phase determines whether we need to split or merge the frozen chunk (Line 4). If cur is the first chunk (which serves the deleteMin operation), a merge has to be executed; as the first chunk gets frozen when there is need to create a new first chunk with other keys. If it is not the first chunk, then another chunk (which serves the insert operation) must have been frozen because it got full and we need to split it into two chunks. There is also a corner case in which a merge of the first chunk happens concurrently with a split of the second chunk. This requires coordination that simply merges relevant values of the second chunk into the new first. So if cur is the second chunk and the first chunk is currently freezing, then we should work on a merge.[3]

In order to execute the entire recovery we will need to place the new chunks in the list of chunks following the previous chunk. In the split case, we therefore proceed by checking if prev is in the process of freezing and if it is, we help it finish the freezing process and recover. Namely, we freeze prev, we look for prev's predecessor and then invoke the freeze recovery for prev (Lines 5–12). This may cause recursive recovery calls until the head of the chunk list, Line 8. During this time, there is a possibility that some other thread has helped recovering our own chunk and we therefore search for it in the list, Line 10. If we can't find it, we know that we are done and can return.

In the third phase we locally create new chunks to replace the frozen one (Lines 13, 14). In the case of a split, two new half-full chunks are created from a single full frozen chunk, using the split() method. The first chunk, with the lower-valued part of the keys, points to the second chunk, with the higher-valued part.

---

[2] The freezeRecovery() method is never called with a cur chunk being the buffer chunk.

[3] It is possible that we miss the freezing of the first chunk and start working on a split of the second chunk. In this case a later CAS instruction, in Line 16, will fail and we will repeat the recovery process with the adequate choice of a merge.

```
1   void freezeRecovery(Chunk* cur, Chunk* prev) {
2     bool toSplit = true; Chunk *local=NULL, *p=NULL;
3     while(1) { // PHASE I: decide whether to split or to merge
4       if (cur==head||(prev==head && prev->status.isInFreeze())) toSplit = false;
5       if (toSplit && prev->status.isInFreeze()){ //PHASE II: in split, if prev is frozen, recover it first
6         freezeChunk(prev);   // ensure prev freeze is done
7         if ( getChunk(&prev, &p) )      // search the previous to prev
8           freezeRecovery(prev, p);      // the frozen prev found, p precedes prev; recursive recovery
9         // prev is already not in the list; re-search the current chunk and find its new predecessor
10        if ( !getChunk(&cur, &p) ) return; // the frozen cur is not in the list
11        else {prev = p; continue;}
12      }
13      if (toSplit) local = split(cur);   // PHASE III: apply the decision locally
14      else     local = mergeFirstChunk(cur);
15      if (toSplit) { // PHASE IV: change the PQ accordingly to the previous decision
16        if ( CAS(&prev->next, cur, local) ) return;
17      } else { // when modifying the head, check if cur second or first
18        if (prev==NULL)
19          if( CAS(&head, cur, local) ) return
20          else if( CAS(&head, prev, local) ) return;
21      }
22      if (!getChunk(&cur,&p)) return; // look for new location; finish if the frozen cur is not found
23      else prev = p;
24    }
25  }
```

Listing 4. CBPQ recovery from a frozen chunk

In the case of a merge, a new first chunk is created with M ordered keys taken from the frozen first chunk, the buffer and from the second chunk. This is done using the mergeFirstChunk() method. If there are too many frozen keys, a new first chunk and new second chunk can be created. The new first chunk is created without pointing to a buffer, it will be allocated when needed for insertion.

In phase IV, the thread attempts to attach its local list of new chunks to the chunk list. Upon success the thread returns. Otherwise, the recovery is retried, but before that, cur is searched for in the chunk list. If it is not there, then other threads have completed the recovery and we can safely return. Otherwise, a predecessor has been found for cur in the search and the recovery is re-executed.

4 Optimizations

First, in order to improve the search time for the relevant chunk, we use a simple lock-free skip-list from [13] to index the chunks. Updates to the skip-list are executed during splits and merges of chunks. Second optimization is exploiting the context switch waiting time. When an insert operation needs to insert a key to the buffer and it initiates a freeze of the first chunk. It is worth letting other threads run for a while before executing the freeze. We implemented this by yielding the processor to another thread (using usleep(0)). Third optimization is the elimination. According to the CBPQ algorithm, an insert of a small key k is done by inserting k into the buffer and waiting until the first chunk is rebuilt. During this time, if k becomes smaller than the current minimum key, then the insert operation can be eliminated by a deleteMin operation that consumes its key k. This can be viewed as if the insert and the deleteMin happened instantaneously

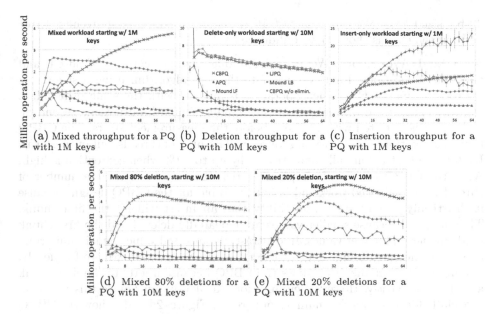

(a) Mixed throughput for a PQ with 1M keys (b) Deletion throughput for a PQ with 10M keys (c) Insertion throughput for a PQ with 1M keys

(d) Mixed 80% deletions for a PQ with 10M keys (e) Mixed 20% deletions for a PQ with 10M keys

Fig. 2. Throughput in different workloads with an increasing threads number

one after the other just at the moment that k was smaller than the minimum key. Due to lack of space, the further details of the optimizations are omitted here and can be found in the full version of this paper [1].

5   Performance Evaluation

We implemented the CBPQ and compared it to the Linden's and Jonsson's PQ [15] (LJPQ), to the lock-free and lock-based implementation of Mounds, and to the adaptive priority queue (APQ) [5]. We chose these implementations, because they are the best performing priority queues in the literature and they were compared to other PQ implementations in [5,15,16]. We thank the authors of [5,15,16] for making their code available to us. All implementations were coded in C++ and compiled with a −O3 optimization level.

We ran our experiments on a machine with 4 AMD Opteron (TM) 6272 16-core processors, overall 64 threads. The machine was operated by Linux OS (Ubuntu 14.04) and the number of threads was varied between 1 and 64. The chunk capacity (M) was chosen to be 928, so one chunk occupies a virtual page of size 4KB. The CBPQ implementation included the skip-list optimization of Sect. 4, we report results with and without elimination, and the results include the EBR memory management (as described in Sect. 3.2). The performance was evaluated using targeted micro-benchmarks: insertion-only or deletion-only workloads and mixed workloads where deletions and insertions appear with equal probability. The keys for insertions were uniformly chosen at random among all

30-bit sized keys. We ran each test 10 times and report average results. Error bars on the graphs show 95 % confidence level.

A Mixed Workload. We considered different percentages of insert and deleteMin operations. First, we evaluate the new algorithm using a stress test micro-benchmark, where each operation is invoked with the same probability. Figure 2a shows the throughput of the 50-50-benchmark during one second on a PQ that is initiated with 1M keys before the measurement starts. The CBPQ (with elimination) is not a winner for a small number of threads, but outperforms LJPQ (the best among all competitors) by up to 80 % when contention is high. Also, the CBPQ is the only implementation that scales for a large number of threads. The balanced workload is not favorable for the CBPQ design, because it drastically increases the probability that an insert will hit the first chunk. This happens because smaller values are repeatedly deleted and the first chunk holds higher and higher values. In contrast, the inserted values remain uniformly chosen in the entire range and hit the range of the first chunk more frequently. Hitting the first chunk often slows the CBPQ because inserts to the first chunk are the most costly. However, elimination excellently ameliorate this problem, especially for an increasing number of threads. Figures 2d and e show the CBPQ with 80 % and 20 % of deleteMin respectively, after the PQ was initiated with 10M keys. In both cases, the CBPQ surpasses the competitors for almost every thread count.

Deletion-Only Workload. Next, we measure the performance of the PQs when only deleteMin is executed. In order to make the deletion measurement relevant with deletion-only workload, we ensured that there are enough keys in the PQ initially so that deletions actually delete a key and never operate on an empty PQ. This requires initiating the PQ with 10M entries for the CBPQ. Elimination is not beneficiary in this case because there exist no pairs to match. Nevertheless, we show the results with and without elimination to highlight the negligible overhead of elimination for the CBPQ. In a deletion-only workloads we see a drastic performance improvement for the CBPQ. Results for the deletion-only workload are reported in Fig. 2b. For a substantial number of threads, the CBPQ deletion throughput is up to 5 times higher than LJPQ throughput, and up to 8 times higher than the rest of the competitors.

Insertion-Only Workload. Similarly to the mixed workload, we start with a PQ that initially contains 1M random keys in it. During the test, we let a varying number of concurrent threads run simultaneously for 1 second, and we measure the throughput. Figure 2c shows the results. Mounds are designed for best performance with inserts of complexity $O(\log(\log(N)))$ and this indeed shows in our measurements. The CBPQ throughput is about 2 times worse than that of lock-based Mound, for a large number of threads. Note that for a smaller amount of threads, the advantage of Mounds is reduced significantly. More over, in spite of the advantage of Mounds with inserts, CBPQ significantly outperforms Mounds on a mixed set of operations. The CBPQ implementation outperforms LJPQ for inserts-only workloads. The performance of the insert operation with

CBPQ is not affected by elimination, therefore the performance of CBPQ on inserts only operations does not change when using or not using elimination.

6 Conclusions

We presented a novel concurrent, linearizable, and lock-free design of the priority queue, called CBPQ. CBPQ cleverly combines the chunked linked-list, elimination technique, and the performance advantage of the *F&I* atomic instruction. We implemented CBPQ and measured its performance against Linden's and Jonsson's PQ (LJPQ) [15], adaptive PQ (APQ) [5] and the Mounds [16] (lock-free and lock-based), which are the best performing priority queues available. Measurements with a mixed set of insert and delete operations show that under high contention CBPQ outperforms all competitors by up to 80 %.

References

1. Braginsky, A., Cohen, N., Petrank, E.: CBPQ: High performance lock-free priority queue (full version). http://www.cs.technion.ac.il/~erez/papers.html
2. Braginsky, A., Kogan, A., Petrank, E.: Drop the anchor: lightweight memory management for non-blocking data structures. SPAA (2013)
3. Braginsky, A., Petrank, E.: Locality-conscious lock-free linked lists. In: Aguilera, M.K., Yu, H., Vaidya, N.H., Srinivasan, V., Choudhury, R.R. (eds.) ICDCN 2011. LNCS, vol. 6522, pp. 107–118. Springer, Heidelberg (2011)
4. Brown, T.A.: Reclaiming memory for lock-free data structures: There has to be a better way. In: PODC (2015)
5. Calciu, I., Mendes, H., Herlihy, M.: The adaptive priority queue with elimination and combining. In: Kuhn, F. (ed.) DISC 2014. LNCS, vol. 8784, pp. 406–420. Springer, Heidelberg (2014)
6. Cohen, N., Petrank, E.: Automatic memory reclamation for lock-free data structures. In: OOPSLA 2015 (2015)
7. Cohen, N., Petrank, E.: Efficient memory management for lock-free data structures with optimistic access, SPAA 2015, pp. 254–263. ACM (2015). http://doi.acm.org/10.1145/2755573.2755579
8. Dragicevic, K., Bauer, D.: Optimization techniques for concurrent stm-based implementations: a concurrent binary heap as a case study. In: IPDPS (2009)
9. Fraser, K.: Practical lock-freedom. In: Ph.D. dissertation, University of Cambridge (2004)
10. Harris, T.L.: A pragmatic implementation of non-blocking linked-lists. In: Welch, J.L. (ed.) DISC 2001. LNCS, vol. 2180, pp. 300–314. Springer, Heidelberg (2001)
11. Hendler, D., Shavit, N., Yerushalmi, L.: A scalable lock-free stack algorithm. J. Parallel Distrib. Comput. **70**, 1–12 (2010)
12. Herlihy, M.: Wait-free synchronization. ACM Trans. Program. Lang. Syst. **13**(1), 124–149 (1991)
13. Herlihy, M., Shavit, N.: The Art of Multiprocessor Programming. Morgan Kaufmann Pub. Inc., San Francisco (2008)
14. Hunt, G., Michael, M., Parthasarathy, S., Scott, M.: An efficient algorithm for concurrent priority queue heaps. In: Information Processing Letters (1996)

15. Linden, J., Jonsson, B.: A skiplist-based concurrent priority queue with minimal . memory contention. In: OPODIS 2013 (2013)
16. Liu, Y., Spear, M.: Mounds: array-based concurrent priority queues. In: Proceedings of the ICpp (2012)
17. Lotan, I., Shavit, N.: Skiplist-based concurrent priority queues. In: Proceedings of the IPDPS (2000)
18. Shavit, N., Touitou, D.: Elimination trees and the construction of pools and stacks. Theory Comput. Syst. **30**, 645–670 (1997)
19. Sundell, H., Tsigas, P.: Fast and lock-free concurrent priority queues for multi-thread systems. J. Parallel Distrib. Comput. **65**, 609–627 (2005)

Multicore and Manycore Parallelism

Redesigning Triangular Dense Matrix Computations on GPUs

Ali Charara[✉], Hatem Ltaief, and David Keyes

Extreme Computing Research Center,
King Abdullah University of Science and Technology,
Thuwal, Jeddah 23955, Saudi Arabia
{Ali.Charara,Hatem.Ltaief,David.Keyes}@kaust.edu.sa

Abstract. A new implementation of the triangular matrix-matrix multiplication (TRMM) and the triangular solve (TRSM) kernels are described on GPU hardware accelerators. Although part of the Level 3 BLAS family, these highly computationally intensive kernels fail to achieve the percentage of the theoretical peak performance on GPUs that one would expect when running kernels with similar surface-to-volume ratio on hardware accelerators, i.e., the standard matrix-matrix multiplication (GEMM). The authors propose adopting a recursive formulation, which enriches the TRMM and TRSM inner structures with GEMM calls and, therefore, reduces memory traffic while increasing the level of concurrency. The new implementation enables efficient use of the GPU memory hierarchy and mitigates the latency overhead, to run at the speed of the higher cache levels. Performance comparisons show up to eightfold and twofold speedups for large dense matrix sizes, against the existing state-of-the-art TRMM and TRSM implementations from NVIDIA cuBLAS, respectively, across various GPU generations. Once integrated into high-level Cholesky-based dense linear algebra algorithms, the performance impact on the overall applications demonstrates up to fourfold and twofold speedups, against the equivalent native implementations, linked with cuBLAS TRMM and TRSM kernels, respectively. The new TRMM/TRSM kernel implementations are part of the open-source KBLAS software library (http://ecrc.kaust.edu.sa/Pages/Res-kblas.aspx) and are lined up for integration into the NVIDIA cuBLAS library in the upcoming v8.0 release.

Keywords: Triangular dense matrix computations · High performance computing · Recursive formulation · KBLAS · GPU optimization

1 Introduction

Most large-scale numerical simulations rely for high performance on the BLAS [13], which are often available through various vendor distributions tuned for their own architecture, e.g., MKL on Intel x86 [3], ACML [5] on AMD x86, ESSL [2] on IBM Power architecture. Indeed, BLAS kernels are considered as

© Springer International Publishing Switzerland 2016
P.-F. Dutot and D. Trystram (Eds.): Euro-Par 2016, LNCS 9833, pp. 477–489, 2016.
DOI: 10.1007/978-3-319-43659-3_35

building blocks, and the library represents one of the last layers of the usual software stack, which is usually where application performance is extracted from the underlying hardware. This drives continuous efforts to optimize BLAS kernels [8]. While performance of Level 1 and 2 BLAS kernels is mainly limited by the bus bandwidth (memory-bound), Level 3 BLAS kernels display a higher flop/byte ratio (compute-bound), thanks to high data reuse occurring at the upper levels of the cache hierarchy. However, BLAS operations on triangular matrix structure, i.e., the triangular matrix-matrix multiplication (TRMM) and the system of linear equations triangular solvers (TRSM), have demonstrated limited performance on GPUs using the highly optimized NVIDIA cuBLAS library [5]. Although in the category of Level 3 BLAS operations, the triangular structure of the input matrix and the in-place nature of the operation may generate many Write After Read (WAR) data hazards. WAR situations usually reduce the level of concurrency due to inherent dependencies, and incur excessive data fetching from memory.

We describe a recursive formulation that enriches the TRMM/TRSM inner structures with GEMM calls and, therefore, enhances data reuse and concurrency on the GPU by minimizing the impact of WAR data hazards and by mitigating the memory transactions load overhead. The idea of casting level 3 BLAS operations into GEMM operations has been promoted by several previous works, including Kågström et al. [16], Goto and van de Geijn [12], Andersen et al. [9], and Elmorth et al. [11]. In this paper, we describe a recursive formulation of the TRMM and TRSM kernels, which suits well the aggressively parallel many-core GPUs architecture. Performance comparisons show up to sixfold and twofold speedups for large dense matrix sizes with our implementation, against the existing state-of-the-art TRMM/TRSM implementations from NVIDIA cuBLAS, respectively, across various GPU generations. After integrating it into high-level dense linear algebra algorithms, such as the Cholesky-based symmetric matrix inversion and the Cholesky-based triangular solvers, the performance impact on the overall application demonstrates up to fourfold and twofold speedups against the equivalent native implementations, linked with NVIDIA cuBLAS TRMM/TRSM kernels.

The remainder of the paper is organized as follows. Section 2 presents related work. Section 3 recalls the TRMM/TRSM kernel operations and identifies the performance bottlenecks seen on NVIDIA GPUs. The implementation details of the high-performance recursive TRMM/TRSM kernels are given in Sect. 4. Section 5 shows TRMM/TRSM performance results on various GPU generations, compares against the state-of-the-art high-performance NVIDIA cuBLAS implementations, and shows the impact of integrating our TRMM/TRSM implementations into the Cholesky-based symmetric matrix inversion and the Cholesky-based triangular solver from MAGMA [7], a high performance dense linear algebra library on GPUs. We conclude in Sect. 6.

2 Related Work

The literature is rich when it comes to BLAS kernel optimizations targeting different x86 and vector architectures [8, 17–19]. We focus only on a few, which are directly related to the topic of the paper.

Kågström et al. [16] proposed a method to express Level 3 BLAS operations in terms of general matrix-matrix multiplication kernel (GEMM). The core idea is to cast the bulk of computations involving off-diagonal blocks as GEMMs, and to operate on the diagonal blocks with Level 1 and 2 BLAS kernels, which increases the byte/flops ratio of the original Level 1 and 2 BLAS operations. Furthermore, recursive formulations for several LA kernels have been promoted by Andersen et al. [9] with packed and general data formats, and by Elmorth et al. [11] with recursive blocking. However, their formulations are designed for the hierarchical memory architecture of x86 multicore processors. Goto and van de Geijn [12] proposed another approach, which casts the computations of Level 3 BLAS operations in terms of a General Panel-Panel multiplication (GEPP). For operations that involve triangular matrices like TRMM and TRSM. The authors customized the GEPP kernel by adjusting the involved loop bounds or by zeroing out the unused elements above the diagonal (for lower triangular TRMM). They show that, with such minimal changes to the main GEPP kernel, this approach enhances the performance of Level 3 BLAS operations against open-source and commercial libraries (MKL, ESSL and ATLAS) on various CPU architectures. Later, Igual et al. [15] extended the same approach to GPUs and accelerated the corresponding Level-3 BLAS operations.

Though these methods are effective for multicore CPU processors, especially when the employed block sizes fits well in the L1 and L2 cache sizes, GPU architecture imposes fundamental changes to the programming model to extract better performance. In general, because the GPU is a throughput-oriented device, Level 3 BLAS kernels perform better when processing larger matrix blocks, which also help mitigate the overhead of extra kernel launches. In particular, GEMM kernel reaches the sustained peak performance with matrix blocks higher than 1024 as shown in Fig. 1(a), an important fact that we will use to explain our method in Sect. 4 to speed up the execution of both TRMM and TRSM operations.

Both Kågström's and Goto's methods require additional temporary buffers to store temporary data, whose size needs to be tuned for better cache alignment. The later method adds the overhead of packing data of each panel before each GEPP operation and unpacking it after GEPP is over. This incurs extra memory allocations and data transfer costs that would prevent an efficient GPU port. Moreover, Kågström's method requires a set of sequentially invoked GEMV kernels to achieve a TRMM for small diagonal blocks, which may be prohibitive on GPUs, due to the overhead of kernel launches. We study the recursive formulations of these kernels in the context of massively parallel GPU devices.

3   The Triangular Matrix Operations: TRMM and TRSM

This Section recalls the general TRMM and TRSM operations and highlights the reasons behind the performance bottleneck observed on GPUs.

3.1   Recalling TRMM and TRSM Operations

As described in the legacy BLAS library [1], TRMM performs one of the triangular matrix-matrix operations as follows: $B = \alpha\ op(A)\ B$, if side is left, or $B = \alpha\ B\ op(A)$, if side is right. On the other hand, TRSM solves one of the matrix equations as follows: $op(A)\ X = \alpha\ B$, if side is left, or $X\ op(A) = \alpha\ B$, if side is right. For both formulas, α is a scalar, B and X are $M \times N$ matrices, A is a unit, or non-unit, upper or lower triangular matrix and $op(A)$ is one of $op(A) = A$ or $op(A) = A^T$. The resulting matrix X is overwritten on B. It is noteworthy to mention that these operations happen *in-place* (IP), i.e., B gets overwritten by the final output of the TRMM or TRSM operations. This in-place overwriting may engender lots of anti-dependencies or WAR data hazards, from successive memory accesses, due to the triangular structure of the input matrix, and may generate lots of memory traffic.

3.2   Current State of Art Performance of TRMM and TRSM

The NVIDIA cuBLAS library currently provides two APIs for TRMM, in-place (IP) and out-of-place (OOP) and a single IP API for TRSM. Figure 1(a) shows the performance of NVIDIA cuBLAS (v7.5) DTRMM in-place (IP) against out-of-place (OOP) using a Kepler K40 GPU, in double precision arithmetic. The theoretical peak performance of the card and the DGEMM sustained peak are given as upper-bound references. The OOP DTRMM runs close to DGEMM performance for asymptotic sizes. However, the IP DTRMM achieves only a small percentage of DGEMM peak and one can notice a factor of six between IP and OOP DTRMM performances. The OOP TRMM removes the WAR data dependencies on B and can be implemented in an embarrassingly parallel fashion, similar to GEMM, at the expense of increasing by a factor of two the memory footprint, besides violating the legacy BLAS TRMM API. On the other hand, MAGMA provides an OOP TRSM implementation, in which the diagonal blocks of matrix A are inverted, followed by a set of GEMM calls. Figure 1(b) shows the performance of NVIDIA cuBLAS (v7.5) IP DTRSM for a low number of right-hand sides (RHS), i.e. 512, as well as a number of RHS equal to the matrix size (square problem), in comparison to the OOP implementation from MAGMA. cuBLAS DTRSM for square problems runs close to DGEMM peak, although it highlights a jaggy performance behavior. However, for low RHS, cuBLAS DTRSM looses computational intensity and seems to further suffer from lack of parallelism. MAGMA DTRSM presents more stable performance for square matrices and much better performance for low RHS cases, at the cost of doubling the memory footprint and thus excessive data transfer.

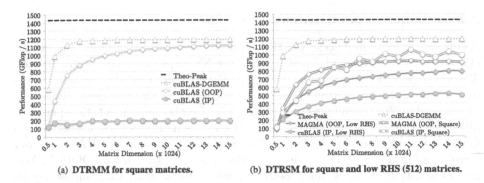

(a) **DTRMM for square matrices.**   (b) **DTRSM for square and low RHS (512) matrices.**

Fig. 1. Performance comparisons of cuBLAS (v7.5) and MAGMA (v2.0.1) in place (IP) against out-of-place (OOP) of TRMM and TRSM using an NVIDIA Kepler K40 GPU, in double precision arithmetic.

3.3 Identifying the Performance Bottlenecks

We can observe limited concurrency for IP TRSM, since updating right columns of matrix B (for a right sided TRSM) needs to read updated left columns of the same matrix, causing a multitude of RAW dependencies. Other variants of TRSM exhibit similar RAW dependencies. The IP TRMM has also limited concurrency due to the fact that updating the left columns of matrix B (for a right sided TRMM) need to read initial values from subsequent right columns of matrix B before the right columns get updated, causing a multitude of WAR dependencies. Other variants of TRMM exhibit similar WAR dependencies. On the other hand, the OOP TRMM/TRSM incurs additional overheads, due to data transfers through the slow PCIe link, as well as extra memory allocation, which would be prohibitive for large matrix sizes, especially since memory is a scarce resource on GPUs.

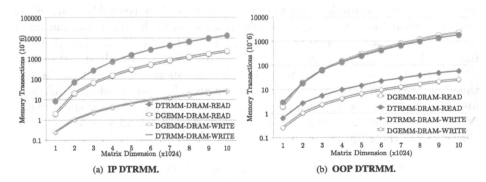

(a) **IP DTRMM.**   (b) **OOP DTRMM.**

Fig. 2. Profiling of memory transactions for cuBLAS IP and OOP DTRMM against that of DGEMM.

3.4 Profiling of NVIDIA cuBLAS TRMM

Assuming square matrices, GEMM performs $2N^3$ floating-point operations (flops) on $3N^2$ data, and TRMM performs N^3 flops on $3/2N^2$ data. In fact, TRMM can be ideally thought of as an IP GEMM; thus, the memory transactions involved are expected to be proportional to the processed data size. However, Fig. 2(a) highlights that cuBLAS IP TRMM implementation performs almost an order of magnitude of DRAM read memory accesses higher than a GEMM for the same input size, and an equivalent number of DRAM memory writes as GEMM for the same input size. On the other hand, OOP TRMM implementation exhibits much better memory traffic load, with almost equivalent DRAM read memory accesses to GEMM for the same input size, but astonishingly more DRAM write accesses as shown in Fig. 2(b). This hints that the cuBLAS implementation of TRMM is not optimal and produces excessive memory accesses, due to its inability to efficiently overcome the WAR dependencies. We note that profiling of cuBLAS and KBLAS memory transactions has been done using the *nvprof* tool available from NVIDIA CUDA Toolkit [4].

In this paper, we propose to further improve the performance of IP TRSM and IP TRMM on NVIDIA GPUs, by running closer to GEMM peak, in addition to staying compliant with the legacy BLAS regarding operating in-place. We refer to IP TRMM and IP TRSM as TRMM and TRSM onward in this paper.

4   Recursive Definition and Implementation Details

In this section, we describe the recursive formulation of the TRMM and TRSM operations, as illustrated in Fig. 3, and their implementation over NVIDIA GPUs. In the following definition, we illustrate the Left Lower Transpose TRMM, and the Left Lower NonTranspose TRSM cases, where $TRMM \rightarrow B = \alpha\ A^T\ B$ and $TRSM \rightarrow A\ X = \alpha\ B$, B is of size $M \times N$ and A of size $N \times N$. All other variants are supported and operate in a similar manner. As illustrated in Fig. 3, we first choose a suitable partition of the number of B rows $M = M_1 + M_2$ as discussed in the next paragraph. We then partition the triangular matrix A into three sub-matrices: two triangular matrices A_1 and A_3 (i.e. two diagonal blocks) of sizes $M_1 \times M_1$ and $M_2 \times M_2$, respectively, and a third rectangular non-diagonal block A_2 of size $M_2 \times M_1$. We correspondingly partition the rectangular matrix B into two rectangular matrices B_1 and B_2 of sizes $M_1 \times N$ and $M_2 \times N$, respectively (such partitioning would be instead along the columns $N = N_1 + N_2$ for right sided TRMM/TRSM, where A is of size $N \times N$). We then re-write these operations as follows:

$$TRMM : \begin{cases} B_1 = \alpha\ A_1^T\ B_1 & \text{recursive TRMM} \\ B_1 = \alpha\ A_2^T\ B_2 + B_1 & \text{GEMM} \\ B_2 = \alpha\ A_3^T\ B_2 & \text{recursive TRMM} \end{cases}$$

$$TRSM : \begin{cases} A_1 \ X_1 = \alpha \ B_1 & \text{recursive TRSM: Solve for } X_1 \text{ over } B_1 \\ B_2 = \alpha \ B_2 - A_2 \ B_1 & \text{GEMM} \\ A_3 \ X_2 = B_2 & \text{recursive TRSM: Solve for } X_2 \text{ over } B_2 \end{cases}$$

Recursion stops when reaching the size where cuBLAS TRMM/TRSM exhibits good performance in comparison to the performance of further splitting, in order to minimize the extra overhead of continuing the recursion on small sub-matrix sizes. At this level we call again the cuBLAS TRMM/TRSM routine, respectively, which launches one kernel with enough thread blocks to process the small sub-matrix in parallel. The stopping size is left as a tuning parameter. In our tests, we used 128 as the stopping size, a value that was effective across various GPU devices.

To maximize the performance of the generated GEMM calls, we need to be careful about the choice of partitioning the rows. Our target is to generate GEMM calls with maximum possible sub-matrix size, which in turn makes the best boost in performance. We choose the partitioning $M_1 = M/2$ if M is a power of 2, otherwise we choose M_1 as the closest power of 2 strictly less than M; in both cases $M_2 = M - M_1$. The same logic applies when partitioning along columns for right sided TRMM or TRSM. Such a strategy for partitioning the matrices is also needed to ensure that the generated GEMM calls operate on suitable matrix sizes and to minimize the number of trailing sub-matrices with odd matrix sizes during the subsequent recursion.

By casting the TRMM and TRSM operations into a set of large GEMMs and a set of small TRMMs and TRSMs, respectively, we benefit not only from the GEMM speed but also from its optimized memory access pattern that fits the GPU memory hierarchy, and thus removing most of the redundant memory accesses observed in cuBLAS TRMM implementation, for instance. The next section details such performance gains.

5   Experimental Results

This section features the performance results of our recursive KBLAS TRMM and TRSM on GPUs and compares it against state-of-the-art implementation from NVIDIA cuBLAS and MAGMA. The experiments have been conducted on various NVIDIA GPUs generations: Fermi, K20, K40, GTX Titan Black, and the latest Maxwell GPU card. Since the Titan and Maxwell are gaming cards, ECC memory correction has been turned off on the other GPUs for consistency. The latest CUDA Toolkit v7.5 has been used. The four precisions for TRMM and TRSM as well as their eight variants are supported in our KBLAS software distribution. It is important to recall the profiling graphs from Sect. 3.4, in which the performance bottleneck of the IP cuBLAS TRMM has been identified, i.e., the excessive amount of memory accesses.

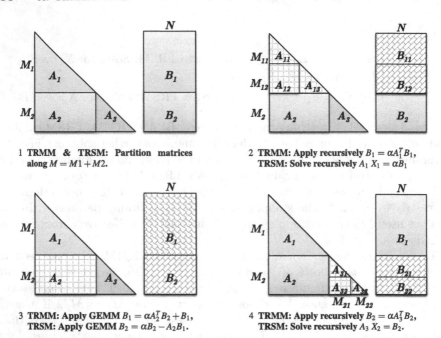

1 **TRMM & TRSM:** Partition matrices along $M = M1 + M2$.

2 **TRMM:** Apply recursively $B_1 = \alpha A_1^T B_1$, **TRSM:** Solve recursively $A_1 X_1 = \alpha B_1$

3 **TRMM:** Apply GEMM $B_1 = \alpha A_2^T B_2 + B_1$, **TRSM:** Apply GEMM $B_2 = \alpha B_2 - A_2 B_1$.

4 **TRMM:** Apply recursively $B_2 = \alpha A_3^T B_2$, **TRSM:** Solve recursively $A_3 X_2 = B_2$.

Fig. 3. Illustrating a Left-Lower-Transpose recursive TRMM, and Left-Lower-NonTranspose recursive TRSM, partitioning along the rows. Steps are to be applied at each recursive call in the order depicted in the picture.

Figure 4 shows how KBLAS TRMM is capable of fetching less data from global memory in favor of reusing already fetched data in the higher cache levels, as opposed to IP cuBLAS TRMM. KBLAS TRMM also performs less data traffic compared to cuBLAS DGEMM, however, this is commensurate to the processed data size. In addition, it performs less data fetching from global memory than the more regular cuBLAS OOP TRMM. Note that the increase in global memory data

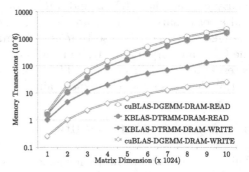

Fig. 4. Profiling of memory transactions for KBLAS DTRMM against that of DGEMM.

writes in KBLAS TRMM is due to the fact that some horizontal panels are updated several times by the recursive nature of the algorithm. The number of updates of a horizontal panel is equal to the number of GEMMs that updates in the left path to the recursion root, plus one for the TRMM leaf. However this increase in the number of global memory writes is much less expensive than the significantly greater number of global memory reads needed by cuBLAS IP TRMM.

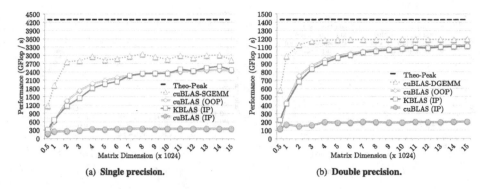

(a) **Single precision.**   (b) **Double precision.**

Fig. 5. Performance comparisons of KBLAS TRMM against that of IP and OOP cuBLAS TRMM running on NVIDIA K40 GPU.

Figures 5(a) and (b) show the performance comparisons of KBLAS TRMM against that of IP and OOP cuBLAS TRMM for single and double precision arithmetics, respectively. KBLAS TRMM achieves similar performance as OOP cuBLAS TRMM, while maintaining the same memory footprint of IP cuBLAS TRMM. For large matrix sizes in double precision, KBLAS DTRMM achieves almost one Tflop/s, six times more than the IP cuBLAS DTRMM from NVIDIA. KBLAS DTRMM attains almost 90 % performance of DGEMM, thanks to the recursive formulation, which inherently enriches TRMM with GEMM calls. Furthermore, Figs. 6(a) and (b) show performance speedup of KBLAS IP TRSM against cuBLAS IP TRSM and MAGMA OOP TRSM for single and double precision arithmetics, respectively, on an NVIDIA K40 GPU. It is noteworthy to mention that although the performance gain with square matrices is not significant, and in fact smooths out the irregular performance from cuBLAS, the performance gain with triangular solves with small number of RHS is important. In practice, it is more common to encounter a triangular solve with only a few RHS. Note also that KBLAS IP TRSM achieves similar (sometimes better) performance as MAGMA OOP TRSM without the overhead of excessive data transfer and extra memory allocations.

Figures 7(a) and (b) show performance speedup of KBLAS TRMM and TRSM against cuBLAS TRMM and TRSM, respectively, on various generations of NVIDIA GPUs. The speedup ranges shown proves the performance portability of the implementation across different NVIDIA GPUs. We note that the double precision capability ratio in reference to single precision capability of the Quadro-M6000 GPU (a Maxwell architecture) has been intentionally fixed by NVIDIA to 1:32, hence its limited double precision computation performance. Although Titan Black GPU is also a graphics card, the user can still control its GPU's double precision performance by switching between 1:3 and 1:24 ratios, which explains its ability to ramp up its double precision performance, as opposed to the Quadro-M6000 GPU.

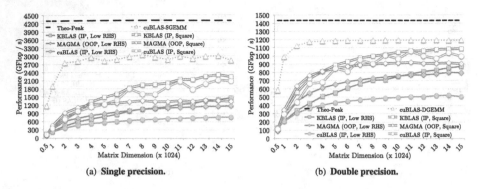

Fig. 6. Performance comparisons of KBLAS IP TRSM against that of cuBLAS IP TRSM and MAGMA OOP TRSM running on NVIDIA K40 GPU, with square and low RHS matrices.

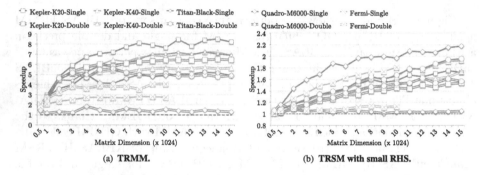

Fig. 7. Performance speedup of KBLAS TRMM and TRSM against cuBLAS TRMM and TRSM, respectively, on various generations of NVIDIA GPUs.

In scientific computations, often data resides on the host memory. Data is shipped to device memory on demand, and as a best practice, one repeatedly operates on the data while it is in the device memory until operations are done or the result is needed back on the CPU memory [6]. It is becoming a common practice within scientific libraries to provide an API that handles the data transfer between the host and the device implicitly and to operate on it in one function call. This practice simplifies the API and puts less burden on programmers who want to replace a CPU call by an equivalent GPU function call, without explicitly handling data transfer. The only way to perform such API with the existing cuBLAS routines is to wrap it within a function call that will initiate the data transfer, wait for it to arrive on device memory, operate on it, then initiate its transfer back to host memory. This results in a severe synchronous communication and computation scheme. In this context, our implementation brings an added value, in that due to the recursive nature of the routine calls, we can overlap communication with computation, i.e., by operating on parts of the data while waiting for the other parts to arrive into device memory and vice

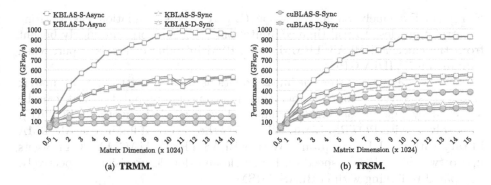

(a) **TRMM.**   (b) **TRSM.**

Fig. 8. Performance impact of asynchronous APIs for TRMM and TRSM.

versa. We provide an asynchronous API that achieves this goal. Figures 8(b) and (a) show the performance gain of such asynchronous TRMM and TRSM API's in comparison to the synchronous cuBLAS and KBLAS API's.

To conclude, we show the impact of KBLAS TRMM and TRSM when integrated into high-level dense linear algebra algorithms. In particular, the Cholesky-based symmetric matrix inversion is one of the main operations for the computation of the variance-covariance matrix in statistics [14]. The inversion algorithm first factorizes the dense symmetric matrix using the Cholesky factorization (POTRF) and then inverts and multiply *in-place* the triangular Cholesky factor by its transpose (POTRI) to get the final inverted triangular matrix. Figure 9 shows the performance impact of linking the MAGMA [7] library with the KBLAS library. With KBLAS TRMM, MAGMA POTRI gets up to twofold and fourfold speedup, for single and double precision, respectively, when compared to using cuBLAS TRMM on a single GPU, as seen in Fig. 9(a). Moreover, the symmetric matrix inversion operation drives a computational astronomy application [10], which designs the future instruments of the

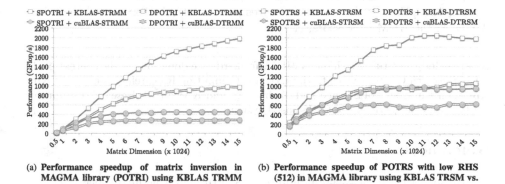

(a) **Performance speedup of matrix inversion in MAGMA library (POTRI) using KBLAS TRMM vs using cuBLAS TRMM.**

(b) **Performance speedup of POTRS with low RHS (512) in MAGMA library using KBLAS TRSM vs. using cuBLAS TRSM.**

Fig. 9. Sample impact of using KBLAS TRMM and TRSM in MAGMA library.

European Extremely Large Telescope (E-ELT). This simulation has a strong constraint on close to real-time computations and, therefore, can highly benefit from the efficient KBLAS TRMM and TRSM introduced in this paper, when running on NVIDIA GPUs.

Finally, solving a system of linear equations with a positive definite matrix (POSV) is a very common operation in scientific computations. It can be achieved by first factorizing the matrix A with Cholesky method (POTRF), then solving with POTRS. Figure 9(b) shows the impact of linking the MAGMA library with the KBLAS library. With KBLAS TRSM, MAGMA POTRS gets up to twofold and 30 % speedup, for single and double precision, respectively, compared to linking with cuBLAS TRSM.

6   Conclusions and Future Work

Recursive formulations of the Level 3 BLAS triangular matrix-matrix multiplication (TRMM) and triangular solve (TRSM) have been implemented on GPUs, which allow reducing memory traffic from global memory and expressing most of the computations in GEMM sequences. They achieve up to eightfold and twofold speedups, respectively, for asymptotic dense matrix sizes against the existing state-of-the-art TRMM/TRSM implementations from NVIDIA cuBLAS v7.5, across various GPU generations. After integrating them into high-level Cholesky-based dense linear algebra algorithms, performance impact on the overall application demonstrates up to fourfold and twofold speedups against the equivalent vendor implementations, respectively. The new TRMM and TRSM implementations on NVIDIA GPUs are available in the open-source KAUST BLAS (KBLAS) library [8] and are scheduled for integration into the NVIDIA cuBLAS library in its future release. Future work includes the performance analysis on tall and skinny matrices, which is critical for some dense linear algebra algorithms (e.g., the generalized symmetric eigenvalue problem), as well as looking at multi-GPU support. The authors would like also to study the general applicability of such recursive formulations on other Level 3 BLAS operations as well as targeting other hardware architectures (e.g., Intel/AMD x86, Intel Xeon Phi, AMD APUs, ARM processors, etc.).

Acknowledgments. We thank NVIDIA for hardware donations in the context of the GPU Research Center Award to the Extreme Computing Research Center at the King Abdullah University of Science and Technology and KAUST IT Research Computing for hardware support on the GPU-based system.

References

1. BLAS: Basic Linear Algebra Subprograms. http://www.netlib.org/blas
2. Engineering and Scientific Subroutine Library (ESSL) and Parallel ESSL. http://www-03.ibm.com/systems/power/software/essl/
3. Intel MKL Library. http://software.intel.com/en-us/articles/intel-mkl

4. NVIDIA CUDA Toolkit. http://developer.nvidia.com/cuda-toolkit
5. The NVIDIA CUDA Basic Linear Algebra Subroutines (CUBLAS). http://developer.nvidia.com/cublas
6. CUDA C best practices guide
7. Agullo, E., Demmel, J., Dongarra, J., Hadri, B., Kurzak, J., Langou, J., Ltaief, H., Luszczek, P., Tomov, S.: Numerical linear algebra on emerging architectures: the PLASMA and MAGMA projects. J. Phys. Conf. Ser. **180**(1), 012037 (2009)
8. Ahmad, A., Ltaief, H., Keyes, D.: KBLAS: an optimized library for dense matrix-vector multiplication on GPU accelerators. ACM Trans. Math. Softw. (2016) (to appear)
9. Andersen, B.S., Gustavson, F.G., Karaivanov, A., Marinova, M., Waniewski, J., Yalamov, P.Y.: LAWRA: linear algebra with recursive algorithms. In: Gebremedhin, A.H., Manne, F., Moe, R., Sørevik, T. (eds.) PARA 2000. LNCS, vol. 1947, pp. 38–51. Springer, Heidelberg (2001)
10. Charara, A., Ltaief, H., Gratadour, D., Keyes, D.E., Sevin, A., Abdelfattah, A., Gendron, E., Morel, C., Vidal, F.: Pipelining computational stages of the tomographic reconstructor for multi-object adaptive optics on a multi-GPU system. In: Damkroger, T., Dongarra, J. (eds.) International Conference for High Performance Computing, Networking, Storage and Analysis, SC 2014, pp. 262–273. IEEE (2014)
11. Elmroth, E., Gustavson, F., Jonsson, I., Kågström, B.: Recursive blocked algorithms and hybrid data structures for dense matrix library software. SIAM Rev. **46**(1), 3–45 (2004)
12. Goto, K., van de Geijn, R.: High-performance implementation of the level 3 BLAS. ACM Trans. Math. Softw. **35**(1), 4:1–4:14 (2008)
13. Hadri, B., Fahey, M.R.: Mining software usage with the automatic library tracking database (ALTD). In: Pfeiffer, H., Ignatov, D., Poelmans, J., Gadiraju, N. (eds.) ICCS. Procedia Computer Science, vol. 18, pp. 1834–1843. Elsevier (2013)
14. Higham, N.J.: Accuracy and Stability of Numerical Algorithms. SIAM, Philadelphia (2002)
15. Igual, F.D., Quintana-Ort, G., van de Geijn, R.A.: Level-3 BLAS on a GPU: picking the low hanging fruit. In: AIP Conference Proceedings, vol. 1504(1), pp. 1109–1112 (2012)
16. Kågström, B., Ling, P., van Loan, C.: GEMM-based level 3 BLAS: high-performance model implementations and performance evaluation benchmark. ACM Trans. Math. Softw. **24**(3), 268–302 (1998)
17. Nath, R., Tomov, S., Dongarra, J.: An improved magma GEMM for fermi graphics processing units. Int. J. High Perform. Comput. Appl. **24**(4), 511–515 (2010)
18. Nath, R., Tomov, S., Dong, T., Dongarra, J.: Optimizing symmetric dense matrix-vector multiplication on GPUs. In: Proceedings of 2011 International Conference for High Performance Computing, Networking, Storage and Analysis, SC 2011, pp. 6:1–6:10. ACM, New York (2011)
19. Volkov, V., Demmel, J.W.: Benchmarking GPUs to tune dense linear algebra. In: Proceedings of 2008 ACM/IEEE Conference on Supercomputing, SC 2008, pp. 31:1–31:11. IEEE, Piscataway (2008)

A Sharing-Aware Memory Management Unit for Online Mapping in Multi-core Architectures

Eduardo H.M. Cruz[1]([✉]), Matthias Diener[1], Laércio L. Pilla[2],
and Philippe O.A. Navaux[1]

[1] Informatics Institute,
Federal University of Rio Grande do Sul (UFRGS), Porto Alegre, Brazil
{ehmcruz,mdiener,navaux}@inf.ufrgs.br
[2] Department of Informatics and Statistics,
Federal University of Santa Catarina (UFSC), Florianópolis, Brazil
laercio.pilla@ufsc.br

Abstract. In modern shared-memory architectures, it is important to map threads and data in a way that increases the locality of their memory accesses, thereby improving performance and energy efficiency. Threads that access shared data should be mapped close to each other in the memory hierarchy, while the data they access should be mapped to their NUMA node, which is called sharing-aware mapping. In this paper, we propose SAMMU, which adds sharing-awareness to the memory management unit in current architectures. SAMMU analyzes the memory access behavior in hardware and provides information to the operating system so it can perform an online mapping of threads and data. In the evaluation with a wide range of parallel applications, performance was improved by up to 35.7 % (13.1 % on average).

1 Introduction

As parallel applications need to access shared data, the memory hierarchy presents challenges for mapping threads to cores, and data to NUMA nodes [24]. Threads that access a large amount of shared data should be mapped to cores that are close to each other in the memory hierarchy, while data should be mapped to the same NUMA node that the threads that access it are executing on [22]. In this way, the *locality* of the memory accesses is improved, which leads to an increase of performance and energy efficiency. This type of thread and data mapping is called *sharing-aware* mapping. For optimal performance improvements, data and thread mapping should be performed together [23]. For the thread mapping, knowledge about how data is shared between the threads is necessary. Data mapping additionally requires information about the memory pages that are accessed by each thread.

Sharing-aware thread and data mapping improve performance and energy efficiency of parallel applications by optimizing memory accesses [11]. Improvements happen for three main reasons. First, cache misses are reduced by decreasing the number of invalidations that happen when write operations are performed

P.-F. Dutot and D. Trystram (Eds.): Euro-Par 2016, LNCS 9833, pp. 490–501, 2016.
DOI: 10.1007/978-3-319-43659-3_36

on shared data [19]. For read operations, the effective cache size is increased by reducing the replication of cache lines on multiple caches [6]. Second, the locality of memory accesses is increased by mapping data to the NUMA node where it is most accessed. Third, the usage of interconnections in the system is improved by reducing the traffic on slow and power-hungry interchip interconnections, using more efficient intrachip interconnections instead.

In this paper, we propose a *Sharing-Aware Memory Management Unit* (SAMMU), which uses the virtual memory implementation to detect the memory access pattern during the execution of a parallel application. SAMMU modifies the memory management unit to analyze the memory access behavior, which is used to perform online thread and data mapping. To the best of our knowledge, SAMMU is the first mechanism that detects the memory access pattern for thread and data mapping completely on the hardware level, considering many more memory accesses than related work to achieve a higher accuracy. It requires no changes to the application or its runtime system, and needs no previous information about application behavior.

2 Related Work

Traditional data mapping strategies, such as *first-touch* and *next-touch* [15], have been used by operating systems to allocate memory on NUMA machines. In the case of first-touch, pages are not migrated during execution. Next-touch can lead to excessive data migrations if the same page is accessed from different nodes. The NUMA Balancing policy [7] was included in more recent versions of Linux. In this policy, the kernel introduces page faults during the execution of the application to perform lazy page migrations, reducing the number of remote memory accesses. However, it does not detect sharing patterns between threads.

Marathe et al. [18] present an automatic page placement scheme for NUMA platforms by tracking memory addresses from the performance monitoring unit (PMU) of Itanium. Their work requires the generation of memory traces to guide data mapping for future executions of the applications, which may lead to a high overhead [3]. A similar technique is used in Marathe and Mueller [17] to perform data mapping dynamically. They enable the profiling mechanism just during the beginning of each application due to the high overhead, losing the opportunity to handle changes in rest of the execution. Data mapping alone is not able to improve locality when more than one thread accesses the same pages, since threads may be mapped to cores of different NUMA nodes.

Azimi et al. [1] map threads based on information from the hardware counters of Power5 processors that sample the memory addresses resolved by remote caches. Accesses resolved by local caches are not considered, generating an incomplete sharing pattern. Cruz et al. [9] detect the pattern by monitoring the invalidation messages of cache coherence protocols. Only thread mapping was performed, which does not improve the locality of memory accesses in NUMA architectures.

The kMAF affinity framework is proposed in [11]. It performs both thread and data mapping and gather information from page faults. Carrefour [10] is a similar mechanism that uses sampling to detect page usage. Due to its over-head, the authors restrict the mechanism to 30,000 pages, which limits its use to applications with a low memory usage. These mechanisms generate mapping information based on a very small number of samples compared to SAMMU, as all memory accesses are handled by the MMU. Some techniques such as Forest-GOMP [4] require annotations in the source code and depend on specific paral-lelization libraries. Similarly, Ogasawara [20] proposes a data mapping method that is limited to object oriented languages.

The usage of the instructions per cycle (IPC) metric to guide thread mapping is evaluated in Autopin [14]. Autopin itself does not detect the sharing pattern, it only verifies the IPC of several mappings fed to it and executes the applica-tion with the thread mapping that presented the highest IPC. The BlackBox scheduler [21], similar to Autopin, selects the best mapping by measuring the performance that each mapping obtained. When the number of threads is low, all possible thread mappings are evaluated. When the number of threads makes it unfeasible to evaluate all possibilities, the authors execute the application with 1000 random mappings to select the best one. These mechanisms that rely on statistics from hardware counters take too much time to converge to an opti-mal mapping, since they need to first check the statistics of the mappings. The convergence is usually not possible because the number of possible mappings is exponential in the number of threads. Also, these statistics do not accurately represent sharing and data access patterns.

3   SAMMU: A Sharing-Aware Memory Management Unit

Computer systems that support virtual memory use a memory management unit (MMU) to translate virtual to physical addresses. To perform the trans-lation, the operating system stores page tables in the main memory, which contain the physical address and metadata of each memory page. A special cache memory, the Translation Lookaside Buffer (TLB), is used to speed up the address translation. A high-level overview of the operation of the MMU, TLB and SAMMU is illustrated in Fig. 1. On every memory access, the MMU checks if the page has a valid entry in the TLB. If it does, the virtual address is trans-lated to a physical address and the memory access is performed. If the entry is not in the TLB, the MMU performs a page table walk and caches the entry in the TLB before proceeding with the address translation and memory access.

The operation of the MMU is extended in two ways, both happening in par-allel to the normal operation of the MMU without stalling application execution:

1. SAMMU counts the number of times that each TLB entry is accessed from the local core. This enables the collection of information about the pages accessed by each thread. We store these saturating *access counters* (AC), one per TLB entry, in a table that we call *TLB access table*, which is stored in the MMU.

2. On every TLB eviction or when an access counter saturates, SAMMU analyzes statistics about the page and stores them in the main memory in two separate

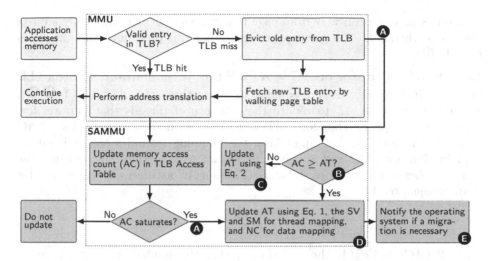

Fig. 1. Overview of the MMU and SAMMU.

structures. The first structure is the *sharing matrix (SM)*, which estimates the amount of sharing between the threads. The second structure is the *page history table*, which contains one entry per physical page in the system with information about the threads and NUMA nodes that accessed them, and is indexed by the physical page address. Each entry of the page history table has three fields: (1) *access threshold* (AT), which defines the minimum number of memory accesses required to modify the statistics; (2) *sharers vector* (SV), which contains the ID of the last threads that accessed the page; (3) *NUMA counters* (NC), which estimate the number of accesses from each NUMA node.

3.1 Gathering Information About Memory Accesses

SAMMU gathers memory access information by counting the number of memory accesses to each page in the TLB of each core. We count the number of accesses to the TLB entry of a page by adding a saturating *access counter* (AC) to the TLB access table. When *AC* saturates or a TLB entry gets evicted, SAMMU collects the information and updates the page history table entry of the page. To filter out threads that perform only few accesses to a page, we use an *access threshold* (AT) in the page history table. *AT* specifies the minimum number of memory accesses required to update the mapping-related information for a page. A number of memory accesses smaller than *AT* means that a thread does not use a page enough to influence its mapping. SAMMU updates the mapping-related statistics of a page only when its *AC* saturates or if the page is evicted from a TLB and the number of memory accesses registered in the *AC* of this TLB entry is greater than or equal to the *AT* of the page (Fig. 1- **Ⓑ,Ⓓ**).

A detailed example of the operation of SAMMU can be found in Fig. 1. The initial value of *AT* is 0. Access thresholds are kept per page because the number of memory accesses can vary from page to page. SAMMU automatically adjusts the access threshold of a given page, separating this procedure into two cases (Fig. 1- **B**):

Case 1 (AC saturates or $AC \geq AT$): When AC saturates or, during a TLB eviction, AC is greater than or equal to its access threshold (AT) (Fig. 1- **B**,**D**), we need to increase AT to reduce the influence of threads that perform few accesses to the page. Therefore, AT is updated with the average value of AC and AT, as illustrated in Eq. 1. Also, the mapping statistics are updated, as explained in Sects. 3.2 and 3.3. It is important to note that, since we use the same number of bits to store AC and AT, when AC saturates, it will be greater than or equal to AT.

$$AT_{new} \leftarrow \frac{AT + AC}{2}, \quad AC \geq AT \tag{1}$$

Case 2 ($AC < AT$): In the second case, when the number of memory accesses registered by AC during a TLB eviction is lower than the access threshold (Fig. 1- **B**,**C**), we update AT in such a way that NUMA nodes with a small number of accesses to the page have a lower influence on the threshold. For that, we use Eq. 2, which guarantees that AT will never be decreased by more than 25 % at each update. In this case, mapping statistics are not updated.

$$AT_{new} \leftarrow AT - \frac{AT - AC}{AT/AC}, \quad AC < AT \tag{2}$$

3.2 Detecting the Sharing Pattern for Thread Mapping

To detect the sharing pattern, SAMMU identifies the last threads that accessed a memory page. To obtain that information, SAMMU adds a small *sharers vector* (SV) to each page history table entry. Each SV stores the IDs of the last threads to access its page. This has the advantage of maintaining temporal locality when detecting which threads share each page. Old entries will be overwritten and not considered as sharers. SAMMU also keeps a *sharing matrix* (SM) in main memory for each parallel application to estimate the number of accesses to pages that are shared between each pair of threads. In the TLB access table, SAMMU stores the ID of the thread that accessed each TLB entry. Control registers containing the memory address and dimensions of the sharing matrix, and the ID of the thread being executed must be added to the architecture and updated by the operating system.

When SAMMU is triggered for a certain page by thread T (Fig. 1- **A**), it accesses the SV of the corresponding page history table entry. If the access counter is greater than or equal to the access threshold (Fig. 1- **B**), SAMMU then increments the sharing matrix in row T, for all the columns that correspond to an entry in the SV (Fig. 1- **D**): $SM[T][SV[i]] \leftarrow SM[T][SV[i]] + 1$.

Each line of SM is accessed by its corresponding thread only, minimizing the impact of coherence protocols. Finally, SAMMU inserts thread T into the SV of the evicted page by shifting its elements, such that the oldest entry is removed.

Table 1. Configuration of the experiments.

System	Parameter	Value
SAMMU	Structure sizes	AC, AT: 32 bits, SV: 2×8 bits, NC: 4×4 bits
	Sharing matrix	256 threads, 4 Byte element size
	Control registers	Support up to 256 threads, $V_{add} = 2$, $NT = 10$
Pin	L1 TLB	64 entries, 4-way, shared between 2 SMT-cores
	L2 TLB	512 entries, 4-way, shared between 2 SMT-cores
Xeon	Processors	4x Xeon X7550 (Nehalem), 8 cores, 2-SMT
	Caches/processor	8x 32 KByte L1, 8x 256 KByte L2, 18 MByte L3
	Main memory	128 GByte DDR3-1333, 4 KByte page size

3.3 Detecting the Page Usage Pattern for Data Mapping

To identify where a memory page should be mapped, SAMMU requires the addition of a vector to each page history table entry. The vector, which we call *NUMA counters* (NC), has N elements for a system with N NUMA nodes. NC employs saturating counters to count a relative number of accesses from different NUMA nodes. The initial value of each NC is 0.

When a TLB entry from a core in NUMA node n is selected for eviction or its AC reaches its maximum value (Fig. 1- **Ⓐ**), SAMMU reads the corresponding page history table entry. If the number of memory accesses stored in AC is greater than or equal to the threshold AT (Fig. 1- **Ⓑ**), SAMMU increments the NUMA counter of node n, and decrements all other NUMA counters (Fig. 1- **Ⓒ**). Since the NUMA counters are saturated, they do not overflow nor underflow.

After updating the values of NC, SAMMU checks if the corresponding page is stored in NUMA node n. If the page is currently mapped to another NUMA node m, SAMMU evaluates if the difference between the NUMA counters of n and m is greater than or equal to a global value *NUMA threshold* (NT) (Fig. 1- **Ⓔ**): $NC[n] - NC[m] \geq NT$. If that is the case, SAMMU notifies the operating system of the page and its destination node n. The NUMA threshold may be configured by a control register. The operating system then chooses how it will handle the migration of the page. The higher the NUMA threshold NT, the lower the number of page migrations.

4 Experiments and Results

In this section, we present the experiments we performed with SAMMU. We describe the methodology and then evaluate the performance and overhead.

4.1 Methodology

The parameters of our experiments are summarized in Table 1. The experiments were performed using a real machine. The machine consists of 4 NUMA nodes with one 8-core, 2-SMT Intel Xeon X7550 processor per node, with a total of 64 virtual cores. It is running version 3.8 of the Linux kernel. Information about the hardware topology is gathered using Hwloc [5]. To generate the thread mappings, we used the EagerMap [8] mapping algorithm, which receives the sharing matrix and a graph representing the memory hierarchy (from Hwloc) as input, and it outputs which core will execute each thread.

As workloads, we used the OpenMP implementation of the NAS parallel benchmarks (NPB) [13], v3.3.1. All experiments were executed 30 times. We show average values as well as a 95 % confidence interval calculated with Student's t-distribution. Results are normalized to the operating system original mapping. We configured the benchmarks to run with one thread per virtual core. Input sizes were chosen to provide similar total execution times and feasible simulation time. Benchmarks BT, LU, SP and UA were executed using input size A. Benchmarks CG, EP, FT, IS and MG were executed using input size B.

Since SAMMU is an extension to the current MMU hardware, we simulate its behavior with the Pin [2] dynamic binary instrumentation tool. The simulated hardware uses the same TLB configuration as the real machines. We used Pin because it is faster than a full system simulator. To make it possible to evaluate SAMMU in real machines, the mapping information generated in Pin is fed into the mapping mechanism in runtime. This is possible because the access pattern of the applications we evaluated and their memory addresses remain the same across different executions, since their memory is statically allocated. Besides performance, we measured L3 cache misses per thousand instructions (MPKI) and QPI interchip interconnection traffic using the Intel PCM tool [12].

4.2 Performance Results

The sharing patterns of a subset of our workloads are illustrated in Fig. 2. The results of execution time can be found in Fig. 3, L3 cache misses per thousand instructions (MPKI) in Fig. 4a, and interchip traffic in Fig. 4b. Lower values are better. In these figures, we also show the average improvements, calculated using the geometric mean function. In this section, we focus on the SAMMU results. The next section presents a comparison to other mapping techniques that are shown in the figures.

In applications whose pages are shared within a small subgroup of threads, mapping presents a high potential for performance improvement. For instance, in SP, most sharing happens between neighboring threads, which is very common in parallel applications that use domain decomposition. In LU and MG, the sharing between more distant threads is more evident than in the other applications. The threads of these applications are able to benefit from the shared cache memories and faster interconnection when mapped nearby in the memory hierarchy, as well as accessing shared pages from their local NUMA node. In general, the effect is a reduction of cache misses and interchip traffic, observed in LU and SP. SP presented the highest improvements, with an execution time reduction of 35.7 %.

To illustrate how thread mapping also affects data mapping, consider MG. MG's sharing pattern indicates that it has a high potential for thread mapping.

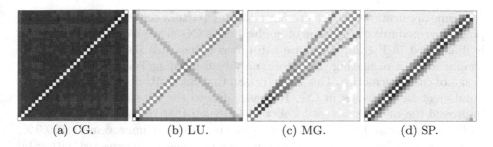

(a) CG.   (b) LU.   (c) MG.   (d) SP.

Fig. 2. Sharing patterns of some applications. Axes represent thread IDs. Cells show the number of accesses to shared pages for each pair of threads. Darker cells indicate more accesses.

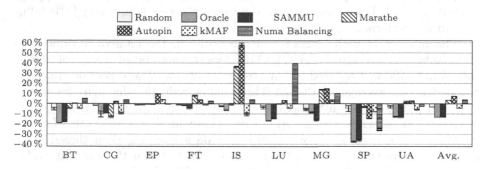

Fig. 3. Execution time normalized to the operating system.

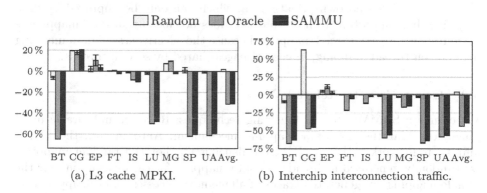

(a) L3 cache MPKI.   (b) Interchip interconnection traffic.

Fig. 4. Performance results, normalized to the operating system.

However, the reduction of interchip traffic is higher than the reduction of cache misses. The reason is that the better fitting thread mapping results in a placement of threads that share data on the same NUMA node, thus reducing interchip traffic. Cache misses were not reduced to the same degree. Therefore, although MG shows a high potential for thread mapping, we are able to observe this by looking at interchip traffic, not at cache misses.

Some applications do not present a sharing pattern suitable for thread mapping. One example of this type of application is CG. The sharing pattern of CG is illustrated in Fig. 2a, where we can observe that each pair of threads has a similar amount of sharing. Therefore, no thread mapping is able to improve the usage of cache memories. This is the reason that SAMMU does not decrease the number of cache misses in CG. However, due to the data mapping, SAMMU improved the memory access locality in CG such that the amount of interchip traffic was decreased by 44.9 %, leading to a performance improvement of 9.0 %.

In some applications, no performance improvements are expected, either by thread or data mapping. For instance, EP is a CPU-bound application [13] with almost no data sharing among its threads. Due to this, there is no thread mapping that is able to optimize the memory accesses. Regarding data mapping, since it is a CPU bound application, the memory accesses have very little influence in the performance of EP.

The number of cache misses and the traffic in the interconnections were reduced by SAMMU significantly. L3 MPKI was reduced by an average of 30.6 %. Interchip traffic was reduced by an average of 39.0 %. The execution time was reduced by an average of 13.1 %. This smaller reduction happens because a better mapping directly influences the number of cache misses and traffic on the interconnections, while the execution time is influenced by several other factors.

Most applications are more sensitive to data mapping than thread mapping, which can be observed in the results by the fact that the interchip traffic presented a higher reduction than cache misses. This happens because, even if an application does not share much data among its threads, each thread will still need to access its own private data, which can only be improved by data mapping. It is important to note that this does not mean that data mapping is more important than thread mapping, because the effectiveness of data mapping depends on thread mapping, in case of pages shared by several threads.

4.3   Comparison to Related Work

We compare SAMMU to the following techniques: Random and Oracle mappings, the Marathe [17] data mapping mechanism, Autopin [14], the kMAF affinity framework [11] and NUMA Balancing [7]. For the random mapping, we randomly generated a thread and data mapping for each execution. For the Oracle mapping, we generated traces of all memory accesses for each application and performed an analysis of the sharing and page usage patterns, similar to [3]. Autopin was executed with 5 mappings: the Oracle mapping and 4 random mappings. We implemented Marathe using a long latency load profile [17] in Pin and fed the information during the execution of the application.

Execution time results of the related work are also shown in Fig. 3. In CG, Marathe presented slightly better results than SAMMU. This happens because, as previously explained, CG is only affected by data mapping, such that SAMMU introduces thread migrations during execution that increase the overhead. Unnecessary thread migrations could be avoided if our mapping algorithm presented features to allow migrations only if the detected sharing pattern has high potential for mapping.

Autopin, in several executions, selected a mapping different from the Oracle, which shows that indirect metrics are not accurate. Also, its performance improvement is lower than ours because it needs to evaluate several other mappings. The results of kMAF are lower than SAMMU for most of the benchmarks. Due to its sampling mechanism, kMAF needs more time to detect the memory access behavior, losing opportunities for improvements. The only application in which NUMA Balancing performed well was SP.

The comparison to the related work shows that mechanisms that perform both thread and data mapping are able to achieve better improvements than mechanisms that perform these mapping separately. It also shows that mechanisms that have access to more accurate information about the memory accesses can provide better performance improvements. SAMMU presented results similar to the Oracle mapping, demonstrating its effectiveness. In most cases, it performed significantly better than the random mapping. This shows that the gains compared to the operating system are not due to the unnecessary migrations introduced by the operating system, but due to a more efficient usage of resources.

4.4   Overhead of SAMMU

SAMMU causes an overhead on the execution of the parallel application on the hardware and software levels. In the hardware level, the additional hardware of SAMMU is not in the critical path, since it operates in parallel to the MMU, such that application execution is not stalled while SAMMU is operating. Therefore, the time overhead introduced by SAMMU consists of the additional memory accesses to update its structures stored in the main memory. To calculate this overhead, we measured the average memory access latency in the Simics full system simulator [16], and multiplied it by the number of additional memory accesses introduced by SAMMU. On the software level, the operating system introduces overhead when calculating the thread mapping, and when migrating threads and pages.

The performance overhead caused by the hardware was 0.41 %, due to the introduction of 1.43 % additional memory transactions, on average. The overhead in the software level was 0.29 %, on average. These results show that SAMMU has only a small performance overhead. Regarding storage overhead, each entry of the page history table would require 8 Bytes, with a total space overhead of 0.2 % relative to the total main memory. The sharing matrix would require 256 KByte, each of its elements with 4 Bytes. We estimate the additional hardware required by SAMMU by counting the amount of transistors required in the implementation. SAMMU would require 143, 000 transistors per core, which results in an increase in transistors of less than 0.05 % in a modern processor.

5   Conclusions and Future Work

In this paper, we presented SAMMU, an extension of the memory management unit to improve locality of memory accesses. SAMMU analyzes the memory accesses of multithreaded applications during execution, such that the operating system can perform a sharing-aware online mapping of threads to cores and data to NUMA nodes. In contrast to previous proposals, it detects the memory access pattern completely in hardware, considering most memory accesses and achieving a higher accuracy. It is independent of the application and its runtime system, and requires no source code modification or previous information about the behavior of the application.

Experiments with the NAS OpenMP benchmarks showed performance improvements of up to 35.7 % (13.1 % on average). L3 cache MPKI and interchip interconnection traffic were reduced by an average of 30.6% and 39.0 %, respectively. Compared to previous work, SAMMU presented the best performance improvements for most applications.

For the future, we will evaluate SAMMU using parallel applications with several processes that do not necessarily share the same virtual address space, as well as running multiple applications simultaneously.

Acknowledgment. This research received funding from the EU H2020 Programme and from MCTI/RNP-Brazil under the HPC4E project, grant agreement n.° 689772. This work was also supported by the STIC-AmSud/CAPES scientific cooperation program under the EnergySFE research project grant 99999.007556/2015-02. Additional funding was provided by CNPq and Capes.

References

1. Azimi, R., Tam, D.K., Soares, L., Stumm, M.: Enhancing operating system support for multicore processors by using hardware performance monitoring. ACM SIGOPS Oper. Syst. Rev. **43**(2), 56–65 (2009)
2. Bach, M., Charney, M., Cohn, R., Demikhovsky, E., Devor, T., Hazelwood, K., Jaleel, A., Luk, C.K., Lyons, G., Patil, H., Tal, A.: Analyzing parallel programs with pin. IEEE Comput. **43**(3), 34–41 (2010)
3. Barrow-Williams, N., Fensch, C., Moore, S.: A communication characterisation of splash-2 and parsec. In: IEEE International Symposium on Workload Characterization, IISWC (2009)
4. Broquedis, F., Aumage, O., Goglin, B., Thibault, S., Wacrenier, P.A., Namyst, R.: Structuring the execution of OpenMP applications for multicore architectures. In: IEEE International Parallel & Distributed Processing Symposium, IPDPS (2010)
5. Broquedis, F., Clet-Ortega, J., Moreaud, S., Furmento, N., Goglin, B., Mercier, G., Thibault, S., Namyst, R.: hwloc: a generic framework for managing hardware affinities in HPC applications. In: Euromicro Conference on Parallel, Distributed and Network-based Processing, pp. 180–186 (2010)
6. Chishti, Z., Powell, M.D., Vijaykumar, T.N.: Optimizing replication, communication, and capacity allocation in CMPs. ACM SIGARCH Comput. Archit. News **33**(2), 357–368 (2005)
7. Corbet, J.: Toward better NUMA scheduling (2012). http://lwn.net/Articles/486858/
8. Cruz, E.H.M., Diener, M., Pilla, L.L., Navaux, P.O.A.: An efficient algorithm for communication-based task mapping. In: International Conference on Parallel, Distributed, and Network-Based Processing (PDP), pp. 207–214 (2015)

9. Cruz, E.H.M., Diener, M., Alves, M.A.Z., Navaux, P.O.A.: Dynamic thread mapping of shared memory applications by exploiting cache coherence protocols. J. Parallel Distrib. Comput. **74**(3), 2215–2228 (2014)
10. Dashti, M., Fedorova, A., Funston, J., Gaud, F., Lachaize, R., Lepers, B., Quema, V., Roth, M.: Traffic management: a holistic approach to memory placement on NUMA systems. In: Architectural Support for Programming Languages and Operating Systems (ASPLOS) (2013)
11. Diener, M., Cruz, E.H.M., Navaux, P.O.A., Busse, A., Heiß, H.U.: kMAF: automatic kernel-level management of thread and data affinity. In: Interntional Conference on Parallel Architectures and Compilation Techniques (PACT) (2014)
12. Intel: Intel Performance Counter Monitor - A better way to measure CPU utilization (2012). http://www.intel.com/software/pcm
13. Jin, H., Frumkin, M., Yan, J.: The OpenMP implementation of NAS Parallel Benchmarks and Its Performance (1999)
14. Klug, T., Ott, M., Weidendorfer, J., Trinitis, C.: Autopin - automated optimization of thread-to-core pinning on multicore systems. High Perform. Embed. Archit. Compil. **3**(4), 219–235 (2008)
15. Löf, H., Holmgren, S.: Affinity-on-next-touch: increasing the performance of an industrial PDE solver on a cc-NUMA system. In: International Conference on Supercomputing (2005)
16. Magnusson, P., Christensson, M., Eskilson, J., Forsgren, D., Hallberg, G., Hogberg, J., Larsson, F., Moestedt, A., Werner, B.: Simics: a full system simulation platform. IEEE Comput. **35**(2), 50–58 (2002)
17. Marathe, J., Mueller, F.: Hardware Profile-guided Automatic Page Placement for ccNUMA Systems. In: ACM SIGPLAN Symposium on Principles and Practice of Parallel Programming (PPoPP) (2006)
18. Marathe, J., Thakkar, V., Mueller, F.: Feedback-directed page placement for ccNUMA via hardware-generated memory traces. J. Parallel Distri. Comput. **70**(12), 1204–1219 (2010)
19. Martin, M.M.K., Hill, M.D., Sorin, D.J.: Why on-chip cache coherence is here to stay. Commun. ACM **55**(7), 78 (2012)
20. Ogasawara, T.: NUMA-aware memory manager with dominant-thread-based copying GC. ACM SIGPLAN Not. **44**(10), 377–389 (2009)
21. Radojković, P., Cakarević, V., Verdú, J., Pajuelo, A., Cazorla, F.J., Nemirovsky, M., Valero, M.: Thread assignment of multithreaded network applications in multicore/multithreaded processors. IEEE Trans. Parallel Distrib. Syst. (TPDS) **24**(12), 2513–2525 (2013)
22. Ribeiro, C.P., Mehaut, J.F., Carissimi, A., Castro, M., Fernandes, L.G.: Memory affinity for hierarchical shared memory multiprocessors. In: International Symposium on Computer Architecture and High Performance Computing (SBAC-PAD) (2009)
23. Terboven, C., an Mey, D., Schmidl, D., Jin, H., Reichstein, T.: Data and thread affinity in OpenMP programs. In: Workshop on Memory Access on Future Processors: A Solved Problem? (MAW) (2008)
24. Wang, W., Dey, T., Mars, J., Tang, L., Davidson, J.W., Soffa, M.L.: Performance analysis of thread mappings with a holistic view of the hardware resources. In: IEEE International Symposium on Performance Analysis of Systems & Software (ISPASS) (2012)

GreenBST: Energy-Efficient Concurrent Search Tree

Ibrahim Umar[✉], Otto Anshus, and Phuong Ha

Department of Computer Science,
UiT The Arctic University of Norway, Tromsø, Norway
{ibrahim.umar,otto.anshus,phuong.hoai.ha}@uit.no

Abstract. Like other fundamental abstractions for energy-efficient computing, search trees need to support both high concurrency and fine-grained data locality. However, existing locality-aware search trees such as ones based on the van Emde Boas layout (vEB-based trees), poorly support *concurrent* (update) operations while existing highly-concurrent search trees such as the non-blocking binary search trees do not consider data locality.

We present GreenBST, a practical energy-efficient concurrent search tree that supports fine-grained data locality as vEB-based trees do, but unlike vEB-based trees, GreenBST supports high concurrency. GreenBST is a k-ary leaf-oriented tree of GNodes where each GNode is a fixed size tree-container with the van Emde Boas layout. As a result, GreenBST minimizes data transfer between memory levels while supporting highly concurrent (update) operations. Our experimental evaluation using the recent implementation of non-blocking binary search trees, highly concurrent B-trees, conventional vEB trees, as well as the portably scalable concurrent trees shows that GreenBST is efficient: its energy efficiency (in operations/Joule) and throughput (in operations/second) are up to 65 % and 69 % higher, respectively, than the other trees on a high performance computing (HPC) platform (Intel Xeon), an embedded platform (ARM), and an accelerator platform (Intel Xeon Phi). The results also provide insights into how to develop energy-efficient data structures in general.

1 Introduction

Recent researches have suggested that the energy consumption of future computing systems will be dominated by the cost of data movement [12,34,35]. It is predicted that for 10 nm technology chips, the energy required between accessing data in nearby on-chip memory and accessing data across the chip, will differ as much as 75× (2 pJ versus 150 pJ), whereas the energy required between accessing on-chip data and accessing off-chip data will only differ 2× (150 pJ versus 300 pJ) [12]. Therefore, in order to construct energy-efficient software systems, data structures and algorithms must not only be concerned with whether the data is on-chip (e.g., in cache) or not (e.g., in DRAM), but must consider also data locality in *finer-granularity*: where the data is located on the chip.

P.-F. Dutot and D. Trystram (Eds.): Euro-Par 2016, LNCS 9833, pp. 502–517, 2016.
DOI: 10.1007/978-3-319-43659-3_37

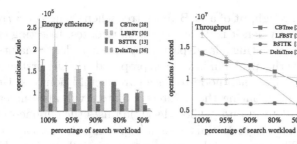

Fig. 1. Result of 5 millions tree operations of decreasing search percentage workloads using 12 cores (1 CPU). DeltaTree's energy efficiency and throughput are lower than the other concurrent search trees after 95 % search workload on a dual Intel Xeon E5-2650Lv3 CPU system with 64 GB RAM.

Concurrent search trees are crucial data structures that are widely used as a backend in many important systems such as databases (e.g., SQLite [24]), filesystems (e.g., Btrfs [32]), and schedulers (e.g., Linux's Completely Fair Scheduler (CFS)), among others. These important systems can access and organize data in a more energy efficient manner by adopting the energy-efficient concurrent search trees as their backend structures.

Devising fine-grained data locality layout for concurrent search trees is challenging, mainly because of the trade-offs needed: (i) a platform-specific locality optimization might not be *portable* (i.e., not work on different platforms while there are big interests of concurrent data structures for unconventional platforms [18,21]), (ii) the usage of transactional memory [20,23] and multi-word synchronization [19,22,27] complicates locality because each core in a CPU needs to consistently track read and write operations that are performed by other cores, and (iii) fine-grained locality-aware layouts (e.g., van Emde Boas layout) poorly support concurrent update operations. Some of the fine-grained locality-aware search trees such as Intel Fast [25] and Palm [33] are optimized for a specific platform. Concurrent B-trees (e.g., B-link tree [28]) only perform well if their B size is optimal. Highly concurrent search trees such as non-blocking concurrent search trees [14,30] and Software Transactional Memory (STM)-based search trees [1,11], however, do not take into account fine-grained data locality.

Fine-grained data locality for *sequential* search trees can be theoretically achieved using the van Emde Boas (vEB) layout [15,31], which is analyzed using cache-oblivious (CO) models [16]. An algorithm is categorized as *cache-oblivious* for a two-level memory hierarchy if it has no variables that need to be tuned with respect to cache size and cache-line length, in order to optimize its data transfer complexity, assuming that the optimal off-line cache replacement strategy is used. If a cache-oblivious algorithm is optimal for an arbitrary two-level memory, the algorithm is also asymptotically optimal for any adjacent pair of available levels of the memory hierarchy [9]. Therefore, cache-oblivious algorithms are expected to be locality-optimized irrespective of variations in memory hierarchies, enabling less data transfer between memory levels and thereby saving energy.

However, the throughput of a vEB-based tree when doing *concurrent* updates is lower compared to when it is doing *sequential* updates. Inserting or deleting a node may result in relocating a large part of the tree in order to maintain the vEB layout. Solutions to this problem have been proposed [7]. The first proposed solution's structure requires each node to have parent-child pointers. Update operations may result in updating the pointers. Pointers will also increase the tree memory footprint. The second proposed solution uses the exponential tree algorithm [3]. Although the exponential tree is an important theoretical breakthrough, it is complex [10]. The exponential tree grows exponentially in size, which not only complicates maintaining its inter-node pointers, but also exponentially increases the tree's memory footprint. Recently, we have proposed a *concurrency-aware vEB layout* [36], which has a higher throughput when doing *concurrent* updates compared to when it is doing *sequential* updates. In the same study, we have proposed DeltaTree, a B+tree that uses the concurrency-aware vEB layout. We have documented that the concurrency-aware vEB layout can improve DeltaTree's *concurrent* search and update throughput over a concurrent B+tree [36].

Nevertheless, we find DeltaTree's throughput and energy efficiency are lower than the state-of-the-art concurrent search trees (e.g., the portably scalable search tree [13]) for the update-intensive workloads (cf. Fig. 1). Our investigation reveals that the cost of DeltaTree's runtime maintenance (i.e., rebalancing the nodes) dominates the execution time. However, reducing the frequency of the runtime maintenance lowers DeltaTree's energy efficiency and throughput for the search-intensive workloads, because DeltaTree nodes will then be sparsely populated and frequently imbalanced. Note that DeltaTree energy efficiency and throughput are already optimized for the search intensive workloads [36,37].

In this paper, we present *GreenBST*, an energy-efficient concurrent search tree that is more energy efficient and has higher throughput for both the concurrent search- and update-intensive workloads than the other concurrent search trees (cf. Table 1). GreenBST applies two significant improvements on DeltaTree in order to lower the cost of the tree runtime maintenance and reduce the tree memory footprint. First, unlike DeltaTree, GreenBST rebalances incrementally (i.e., fine-grained node rebalancing). In DeltaTree, the rebalance procedure has to rebalance *all* the keys within a node and the frequency of rebalancing cannot be lowered as they are necessary to keep DeltaTree in good shape (i.e., keeping DeltaTree's height low and its nodes are densely populated). Incremental rebalance makes the overall cost of each rebalance in GreenBST lower than DeltaTree. Second, we reduce the tree memory footprint by using a different layout for GreenBST's leaf nodes (*heterogeneous* layout). Reduction in the memory footprint also reduces GreenBST's data transfer, which consequently increases the tree's energy efficiency and throughput in both update- and search- intensive workloads. We will show that with these improvements, GreenBST can become up to 195 % more energy efficient than DeltaTree (cf. Sect. 3).

We evaluate GreenBST's energy efficiency (in operations/Joule) and throughput (in operations/second) against six prominent concurrent search trees (cf. Table 1) using parallel micro-benchmarks *Synchrobench* [17] and STAMP

Table 1. List of the evaluated concurrent search tree algorithms.

#	Algorithm	Ref	Description	Synchronization	Code authors	Data structure
1	SVEB	[8]	*Conventional* vEB layout search tree	global mutex	U. Aarhus	binary-tree
2	CBTree	[28]	Concurrent B-tree (B-link tree)	lock-based	U. Tromsø	b+tree
3	Citrus	[4]	RCU-based search tree	lock-based	Technion	binary tree
4	LFBST	[30]	Non-blocking binary search tree	lock free	UT Dallas	binary tree
5	BSTTK	[13]	Portably scalable concurrent search tree	lock-based	EPFL	binary tree
6	DeltaTree	[36]	Locality aware concurrent search tree	lock-based	U. Tromsø	b+tree
7	**GreenBST**	-	Improved locality aware concurrent search tree	lock-based	this paper	b+tree

database benchmark *Vacation* [29] (cf. Sect. 3). We present memory and cache profile data to provide insights into what make GreenBST energy efficient (cf. Sect. 3). We also provide insights into what are the key ingredients for developing energy-efficient data structures in general (cf. Sect. 4).

Our Contributions. Our contributions are threefold:

1. We have devised a new *portable fine-grained locality-aware* concurrent search trees, *GreenBST* (cf. Sect. 2.1). GreenBST are based on our proposed concurrency-aware vEB layout [36] with the two improvements, namely the incremental node rebalance and the heterogeneous node layouts.
2. We have evaluated GreenBST throughput (in operations/second) and energy efficiency (in operations/Joule) with six prominent concurrent search trees (cf. Table 1) on three different platforms (cf. Sect. 3). We show that compared to the state of the art concurrent search trees, GreenBST has the best energy efficiency and throughput across different platforms for most of the concurrent search- and update- intensive workloads.
   GreenBST code and evaluation benchmarks are available at: https://github.com/uit-agc/GreenBST.
3. We have provided insights into how to develop energy-efficient data structures in general (cf. Sect. 4).

2   Design Overview

We devise GreenBST based on the concurrency-aware vEB layout [36], based on the idea that the layout has the same data transfer efficiency between two memory levels as the *conventional* sequential vEB layout [15,31]. Therefore,

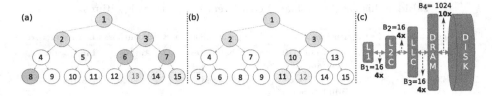

Fig. 2. Illustration of the required data block transfer in searching for (a) key 13 in BFS tree and (b) key 12 in vEB tree, where a node's value is *its address in the physical memory*. Note that in (b), adjacent nodes are grouped together (e.g., (1,2,3) and (10,11,12)) because of the *recursive* tree building. The similarly colored nodes indicates a single block transfer B. An example of multi-level memory is shown in (c), where B_x is the *block transfer* size B between levels of memory.

theoretically, we can use the concurrency-aware layout within a *concurrent* search tree to minimize data movements between memory levels, which can eventually be a basis of an energy-efficient concurrent search tree. This section starts with brief descriptions about the original vEB layout and the concurrency-aware vEB layout for concurrent search tree, followed by detailed description of GreenBST structure and algorithms.

The van Emde Boas (vEB) Layout. The vEB layout has inspired several cache-oblivious (CO) search trees such as the concurrent CO B-trees [5,6] and the CO binary trees [8]. The vEB layout based trees *recursively* arrange related data in contiguous memory locations, minimizing data transfer between any two adjacent levels of the memory hierarchy.

Figure 2 illustrates the vEB layout, where B size is 3. B is the data block transfer between two memory levels (e.g., RAM and disk) in the I/O model [2]. Traversing a complete binary tree with the Breadth First Search layout (or BFS tree for short) with height 4 will need three data block transfers to locate the key at leaf-node 13 (cf. Fig. 2a). The first two levels with three nodes (1, 2, 3) fit within a single block transfer while the next two levels need to be loaded in two separate block transfers that contain nodes (6, 7, 8)[1] and nodes (13, 14, 15), respectively. Generally, the number of data block transfers for a BFS tree of size N is $(\log_2 N - \log_2 B) = \log_2 N/B \sim \log_2 N$ for $N \gg B$.

For a vEB tree with the same height, the required block transfers is only two. As shown in Fig. 2b, locating the key in leaf-node 12 requires only a transfer of nodes (1, 2, 3), followed by a transfer of nodes (10, 11, 12). Generally, the data transfer (or I/O) complexity of searching for a key in a tree of size N is now reduced to $\frac{\log_2 N}{\log_2 B} = \log_B N$, simply by using an efficient tree layout so that nearby nodes are located in adjacent memory locations. If $B = 1024$, searching a BFS tree for a key at a leaf requires $10\times$ (or $\log_2 B$) more I/Os than searching a vEB tree with the same size N, where $N \gg B$.

---

[1] For simplicity, we assume that the memory controller transfers a block of 3 nodes starting at the address of the requested node in memory.

```
 1: Struct Map:                                       10:        return base + map[idx].right
 2:    member fields:                                 11:    else
 3:       left ∈ N, left child pointer address interval   12:        return 0
 4:       right ∈ N, right child pointer address intvl.
                                                      13: function LEFT(p, base)
 5: Map map[UB]                                       14:    nodesize ← SIZEOF(node)
                                                      15:    idx ← (p − base)/nodesize
 6: function RIGHT(p, base)                           16:    if (map[idx].left != 0) then
 7:    nodesize ← SIZEOF(node)                        17:        return base + map[idx].left
 8:    idx ← (p − base)/nodesize                      18:    else
 9:    if (map[idx].right != 0) then                  19:        return 0
```

Fig. 3. Map structure and the *mapping* functions.

On commodity machines with multi-level memory, the vEB layout is even more efficient. So far the vEB layout is shown to have $\log_2 B$ less I/Os for two-level memory. In a typical machine having three levels of inclusive caches (with cache line size of 64B), a RAM (with page size of 4KB) and a disk, a vEB tree search can intuitively give 640× less I/Os than a BFS tree search, assuming the node size is 4 bytes (cf. Fig. 2c). However, the drawback of the vEB layout is in its recursive structure. For example if the tree is full, a new bigger tree needs to be built, recursively in one contiguous block of memory, which also means that the old tree needs to be invalidated and its members copied to the new tree. This drawback prevents an effective way to implement concurrency.

The Concurrency-Aware vEB Layout. Our proposed concurrency-aware vEB layout has been proved to have the same data transfer efficiency between two memory levels as the *conventional* sequential vEB layout [36]. Because of the limited space, we spared the full details of our layout design in this paper, but in brief, a concurrency-aware vEB layout tree (U) is a tree consisting of $|U|$ GNodes $T_i, i = 1, \ldots, |U|$. Nodes of tree T_i are called *internal* nodes in order to distinguish them from GNodes. Each GNode contains a pre-allocated vEB-layout binary search tree (BST) structure that can hold a maximum of *UB internal* nodes. Each GNode's internal leaf nodes may link to another GNode's internal root node, which eventually form a *k-ary* tree of GNodes at the higher level. Note that this *k-ary* tree does not required to have a cache-oblivious layout [36].

2.1 GreenBST

GreenBST and DeltaTree is designed by devising three major strategies, namely it uses a common GNode map instead of pointers or arithmetic-based implicit BST (i.e., a node's successor memory address is calculated *on the fly*) for node traversals, crafting an efficient inter-node connection, and using balanced layouts. In addition to the shared common traits with DeltaTree, GreenBST also employs two new major strategies: (i) GreenBST uses incremental GNode rebalance and (ii) GreenBST uses heterogeneous GNode layouts.

```
 1: function SEARCH(key, GNode, maxDepth)      13:              p ← LEFT(p, base)
 2:     while GNode is not leaf do             14:          else
 3:         rev ← GNode.rev  ▷ Get revision    15:              p ← RIGHT(p, base)
 4:         bits ← 0                           ▷ right child color is 1:
 5:         depth ← 0                          16:      ▷ pad the bits:
 6:         p ← GNode.nodes[0]                 17:          bits ← bits << (maxDepth − depth) − 1
 7:         base ← p                           18:      if (GNode.rev != rev or not even) then
 8:         link ← GNode.link                  19:          Goto 3      ▷ Re-try GNode search
     ▷ continue until leaf node:               ▷ follow nextRight if key ≤ highKey:
 9:         while (p & p.key! = EMPTY ) do     20:      if (GNode.highKey ≤ key) then
     ▷ increment depth:                        21:          GNode ← GNode.nextRight
10:             depth ← depth + 1              22:      else
     ▷ shift one bit to the left in each level 23:          GNode ← link[bits]  ▷ child GNode
11:             bits ← bits << 1               24:      return GNode
12:             if (key < p.key) then
```

Fig. 4. Search within pointer-less GNode. This function will return the *leaf* GNode containing the searched key. From there, an implicit array search using LEFT and RIGHT functions is adequate to pinpoint the key location. The search operations are utilizing both the **nextRight** pointers and **highKey** variables to handle concurrent search even during GNode split.

Data Structures. GreenBST is a collection of GNodes where each GNode consists of an UB internal **nodes** that hold the tree keys and a $1/2\,UB$ **link** array that links the GNode internal leaf nodes to another GNode's root node. Chain of GNodes formed a B+tree (to avoid confusion, from this point onward, we refer the "fat" nodes of GreenBST as GNode and the GNode's internal tree nodes as *internal nodes* or *nodes*). Each GNode also contains a lock (**locked**); a **rev** counter that is used for optimistic concurrency [26]; **nextRight** variable, which is a pointer that points to the GNode's right sibling; and **highKey** variable, which contains the lowest key member of the right sibling GNode. These last four variables are used for GreenBST concurrency control.

Cache-Resident Map Instead of Pointers or Arithmetic Implicit Array. GreenBST does not use pointers to link between its internal nodes, instead it uses a single map-based implicit BST array. This approach is unique to the concurrency-aware vEB layout as it benefits from the usage of the fixed-size GNodes. The usage of pointers and arithmetic-based implicit array in cache-oblivious (CO) trees has been previously studied [8] and both are found to have weaknesses. Pointer based CO tree search operation is slow, mainly because over-heads in every data transfer between memory (although CO tree can minimize data transfers, the inclusion of pointers can lower the amount of meaningful data (e.g., keys) in each block transfer). The implicit array that uses arithmetic calculation for every node traversal may increase the cost of computation, especially if the tree is big.

The cache-resident-maps technique emulates BST's (left and right) child traversals inside a GNode using a combination of a cache-resident GNode *map* structure and LEFT and RIGHT functions (cf. Fig. 3). The LEFT and RIGHT functions, given an arbitrary node v and its GNode's root memory addresses, return the addresses of the left and right child nodes of v, or 0 if v has no children (i.e., v is an internal leaf node of a GNode). The LEFT and RIGHT operations throughout GreenBST share a common cache-resident *map* instance (cf. Fig. 3, line 5). All GNodes use the same fixed-size vEB layout, so only one *map* instance with size *UB* is needed for all traversing operations. This makes GreenBST's memory footprint small and keeps the frequently used *map* instance in cache.

Note that the mapping approach does not induce memory fragmentation. This is because mapping approach applies only for each GNode, and map is only used to point to internal nodes within a GNode. GNode layout uses a contiguous memory block of fixed size *UB* and *update* operations can only change the values of GNode internal nodes (e.g., from EMPTY to a key value in the case of insertion), but cannot change GNode's memory layout.

Inter-GNode Connection. To enable traversing from a GNode to its child GNodes, we develop a new inter-GNode connection mechanism. We logically assign binary values to GNode's internal edges so that each path from GNode root to an internal leaf node is represented by a unique bit-sequence. The bit-sequence is then used as an index in a **link** array containing pointers to child GNodes. As GNode's internal node has only left and right edges, we assign 0 and 1 to the left and right edges, respectively. The maximum size of the bit representation is GNode's height or $\log(UB)$ bits. We allocate a link pointer array whose size is half *UB* length. The algorithm in Fig. 4 explains how the inter-GNode connection works in a pointer-less search function.

Balanced and Concurrent Tree. GreenBST adopts the concurrent algorithms of B-link tree that provides lock-free search operations and adopts the B+tree structure for its high-level structure [28]. However, unlike B-link tree, GreenBST is an in-memory tree and uses optimistic concurrency to handle lock-free concurrent search operations even in the occurrences of the unique "in-place" GNodes maintenance operations.

Similar to B-link tree, GreenBST *insert* operations built the tree from the bottom up, but unlike B-link tree, GreenBST insert operation can trigger *rebalance* operation, a unique GreenBST feature to maintain GNode's small height.

Function REBALANCE(T_i) is responsible for rebalancing a GNode T_i after an insertion. If a new node v is inserted at the *last level* node of a GNode, that GNode is rebalanced to a complete BST. Rebalance sets all GNode leaves node height to $\lfloor \log N \rfloor + 1$, where N is the count of the GNode's internal nodes and $N \leq UB$. Note that this is the default rebalance strategy used by DeltaTree, the incremental rebalance used by GreenBST is explained further in this section.

The *delete* operation in GreenBST simply marks the requested key (v) as deleted. This function fails if v does not exist in the tree or v is already marked. GreenBST does not employ merge operation between GNodes as node reclamation is done by the rebalance and split operations. The offline memory reclamation techniques used in the B-link tree [28] can be deployed to merge nearly empty GNodes in the case where delete operations are the majority. Our new search trees aim at workloads dominated by search operations.

GreenBST concurrency control uses locks and **nextRight** and **highKey** variables to coordinate between search and update operations [28] in addition to **rev** variable that is used for the search's optimistic concurrency. When a GNode needs to be maintained by either rebalance or split operations, the GNode's **rev** counter is incremented by one before the operation starts. The GNode counter is incremented by one again after the maintenance operation finishes. Note that all maintenance procedures happen when the lock is still held by the insert operation and therefore, only one operation may update **rev** counter and maintain a GNode at a time. The usage of **rev** counter is to prevent search from returning wrong key because of the "in-place" GNode maintenance operation.

The *search* operation in GreenBST uses a combination of function SEARCH (cf. Fig. 4) and an implicit tree traversal using map. Function SEARCH traverses the tree from the internal root node of the root GNode down to a leaf GNode, at which the search is handed over to the implicit tree traversal to find the searched key within the leaf GNode. GreenBST *search* operation does not wait or use lock, even in the occurrence of the concurrent updates.

GreenBST *search* uses optimistic concurrency [26] to ensure the operation always returns the correct answer even if it arrives at a GNode that is undergoing the in-place maintenance operation (i.e., *rebalance* and *split*). First, before starting to traverse a GNode, a search operation records the GNode **rev** counter. Before following a link to a child GNode or returning a key, the search operation re-checks again the counter. If the current counter value is an odd number or if it is not equal to the recorded value, the search operation needs to retry search as this indicates that GNodes are being or have been maintained.

Incremental Rebalance. As explained earlier, the rebalance in DeltaTree always involves UB keys, which eventually makes insertions require amortized $\mathcal{O}(UB)$ time. GreenBST borrows the incremental rebalance idea similar to the conventional vEB layout [8] that has the amortized $\mathcal{O}((\log^2 UB)/(1 - \Gamma_1))$ time if used in GreenBST. However, unlike the conventional vEB layout that might have to rebalance the whole tree, we only apply the incremental rebalance to GNodes. To briefly explain the idea, we denote *density(w)* as the ratio of number of keys inside a subtree rooted at w divided by the number of maximum keys that a subtree rooted at w can hold. For example, a subtree with root w that is located three levels away from an internal leaf of a GNode can hold at most $2^3 - 1$ keys. If the subtree only contains 3 keys, then $density(w) = {}^3/_7 = 0.42$. We also denote a *density threshold* $0 < \Gamma_1 < \Gamma_2 < ... < \Gamma_H = 1$, where H is the GNode's height. The main idea is after a new key is inserted at an internal leaf position v, we find the nearest ancestor w of v where $density(w) \leq \Gamma_{depth(w)}$ and $depth(w)$ is the level where w resides, counted from the root of the GNode. If that w is found, we rebalance the subtree rooted at w.

Heterogeneous GNodes. We aim to reduce the overhead of rebalancing and lower the GreenBST height with the usage of different layout for the leaf GNodes (or *heterogeneous*). All DeltaTree's GNodes use the leaf-oriented BST layout, or DeltaTree uses *homogeneous* GNodes. Unlike DeltaTree, leaf GNodes in GreenBST use the internal tree layout instead of the external (or leaf-oriented) tree layout. In the internal tree layout, keys are located in all nodes of a tree, while in the external tree layout, keys are only located in the leaf nodes. The reasoning behind this choice is although leaf-oriented GNodes layout is required for inter-GNode connection (i.e., between parent- and child- GNodes), leaf GNodes do not have any children and therefore, need not to adopt same structure as the other GNodes.

3   Experiments

We run several different benchmarks to evaluate GreenBST throughput and energy efficiency. We combine the benchmark results with the last level cache (LLC) and memory profiles of the trees to draw a conclusion of whether GreenBST improved fine-grained data locality layout (i.e., heterogeneous layout) and concurrency (i.e., lower overall cost of runtime maintenance) over DeltaTree are able to make GreenBST the most energy-efficient tree across different platforms, even when processing the update-intensive workloads. Note that we are not collecting the computation profiles (e.g., Mflops/second) because all the tree operations are data-intensive instead of compute-intensive.

We conduct an experiment on GreenBST and several prominent concurrent search trees (cf. Table 1) using parallel micro-benchmark that is based on Synchrobench [17] (cf. Fig. 5). The trees' LLC and memory profiles during the micro-benchmarks are collected and presented in Fig. 5d and e, respectively. To investigate GreenBST behavior in real-world applications, we implement GreenBST and CBTree as the backend structures in the STAMP database benchmark Vacation [29], alongside the Vacation's original backend structure red-black tree (rbtree) (cf. Fig. 6).

All the experimental benchmarks are conducted on an Intel high performance computing (**HPC**) platform with 24 core 2× Intel Xeon E5-2650Lv3 CPU and 64 GB of RAM, an **ARM** embedded platform with an 8 core Samsung Exynos 5410 CPU and 2 GB of RAM (Odroid XU+E), and an accelerator platform based on the Intel Xeon Phi 31S1P with 57 cores and 6GB of RAM (**MIC** platform). For the parallel micro-benchmark, the trees are pre-initialized with several initial keys before running 5 million operations of 100 % (search-intensive) and 50 % searches (update-intensive), respectively. The initial keys given to both the ARM and MIC platforms are 2^{22} keys and to the HPC platform are 2^{23} keys. All experiments are repeated at least 5 times to guarantee consistent results.

Energy efficiency metrics (in operations/Joule) are the energy consumption divided by the number of operations and *throughput metrics* (in operations/second) are the number of operations divided by the maximum time for the threads to finish the whole operations. Energy metrics are collected from the on-board power measurement on the ARM platform, Intel RAPL interface on the HPC platform, and micras sysfs interface (i.e., /sys/class/micras/power) on the MIC platform.

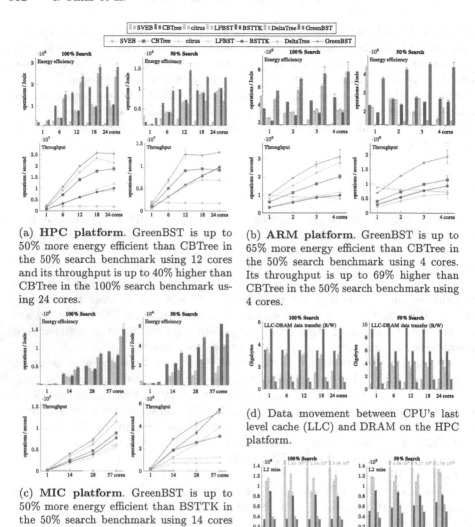

(a) **HPC platform**. GreenBST is up to 50% more energy efficient than CBTree in the 50% search benchmark using 12 cores and its throughput is up to 40% higher than CBTree in the 100% search benchmark using 24 cores.

(b) **ARM platform**. GreenBST is up to 65% more energy efficient than CBTree in the 50% search benchmark using 4 cores. Its throughput is up to 69% higher than CBTree in the 50% search benchmark using 4 cores.

(c) **MIC platform**. GreenBST is up to 50% more energy efficient than BSTTK in the 50% search benchmark using 14 cores and its throughput is up to 20% higher than BSTTK in the 100% search benchmark using 14 cores.

(d) Data movement between CPU's last level cache (LLC) and DRAM on the HPC platform.

(e) L2 cache misses on the MIC platform.

Tree name Memory used (in **GB**)	SVEB	CBTree	citrus	LFBST	BSTTK	DeltaTree	**GreenBST**
	0.1	0.4	0.8	0.7	1.0	0.6	**0.4**

(f) The tree memory footprint after 2^{23} integer keys insertion on the HPC platform.

Fig. 5. (a,b,c) Energy efficiency and throughput comparison of the trees. On the HPC platform, DeltaTree and GreenBST energy efficiency and throughput decreases in the 50 % search benchmark using 18 and 24 cores (i.e., with 2 chips) because of the coherence overheads between two CPUs (cf. Sect. 4). In the 50 % search benchmark using 57 cores (MIC platform), BSTTK energy efficiency and throughput beats GreenBST by 20 % because of the coherence overheads in the MIC platform (cf. Sect. 4). (d) LLC-DRAM data movements on the HPC platform, collected from the CPU counters using Intel PCM. (e) L2 cache miss counter on the MIC platform, collected using PAPI library. (f) The tree memory footprint.

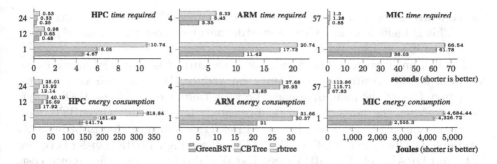

Fig. 6. GreenBST energy efficiency and throughput against CBTree and STAMP's built-in red-black tree (rbtree) for the vacation benchmark. At best, GreenBST consumes 41 % less energy and requires 42 % less time than CBTree (in the 57 clients benchmark on the MIC platform).

Experimental Results. Based on the results in Figs. 5 and 6, GreenBST's energy efficiency and throughput are the highest compared to DeltaTree and the other trees. Because of its *incremental rebalance*, GreenBST outperformed DeltaTree (and the other trees) in the update-intensive workloads. With its *heterogeneous layout*, GreenBST is able to outperform DeltaTree in the search-intensive workloads. GreenBST energy efficiency and throughput are up to 195 % higher than DeltaTree for the update intensive benchmark and up to 20 % higher for the search intensive benchmark (cf. Fig. 5b). Compared to the other trees, GreenBST energy efficiency and throughput are up to 65 % and 69 % higher, respectively. Note that CBTree (B-link tree) is a highly-concurrent B-tree variant that it's still used as a backend in popular database systems such as PostgreSQL.

The reason behind GreenBST good results is GreenBST's data transfer (cf. Fig. 5e) and LLC misses (cf. Fig. 5d) are among the lowest of all the trees. These facts prove that combination of locality-aware layout and the optimizations that GreenBST has over DeltaTree are beneficial to both fine-grained locality and concurrency, of which are the key ingredients of an energy-efficient concurrent search tree.

4   Discussions

Some of the benchmark results showed that besides data movements, efficient concurrency control is also necessary in order to produce energy-efficient data structures. For example, the conventional vEB tree (SVEB) always transferred the smallest amount of data between memory to CPU, but unfortunately, its energy efficiency and throughput failed to scale when using 2 or more cores. SVEB is not designed for concurrent operations and an inefficient concurrency control (a global mutex) had to be implemented in order to include the tree in this study (note that we are unable to use a more fine-grained concurrency because SVEB uses recursive layout in a contiguous memory block). Therefore, even if SVEB has the smallest amount of data transfer during the micro-benchmarks, the concurrent cores have to spend a lot of time waiting and competing for a lock. This is inefficient as a CPU core still consumes power (e.g., static power) even when it is waiting (idle).

Finally, an important lesson that we have learned is that minimizing overheads in locality-aware data structures can reduce the structure's energy consumption. One of the main differences between DeltaTree and GreenBST is that DeltaTree uses the homogeneous (leaf-oriented) layout, while GreenBST does not. Leaf-oriented leaf GNodes increased DeltaTree's memory footprint by 50 % compared to GreenBST (cf. Fig. 5f) and has caused higher data transfer between LLC and DRAM (cf. Fig. 5d). Bigger leaf size also increases maintenance cost for each leaf GNode, because there more data that need to be arranged in every rebalance or split operation, which leads to lower update concurrency. Therefore, DeltaTree energy efficiency and throughput are lower than GreenBST.

Inter-CPU and Many-Core Coherence Issue. Our experimental analysis has revealed that multi-CPU and many-core cache coherence, if triggered, can degrade concurrent update throughput and energy efficiency of the locality-aware trees. Figure 5a shows the "dips" in GreenBST's 50 % update energy efficiency and throughput on the HPC platform (i.e., in the *50 % update/18 cores* and *50 % update/24 cores* cases). Figure 5c also shows that BSTTK beats GreenBST in the *50 % update/57 cores* case on the MIC platform.

Using the CPU performance counters, we have found that the GreenBST concurrent updates frequently triggered the inter-CPU coherency mechanism. In the HPC platform, coherency mechanism causes heavy bandwidth saturation in the CPU interconnect. In the MIC platform, it causes most of the L2 data cache misses to be serviced from other cores and saturates the platform's bidirectional ring interconnect. These facts highlight the challenge faced by the locality-aware concurrent search tree: because of its locality awareness (i.e., related data are kept nearby and often re-used), the tree concurrent update operations might trigger heavy interconnect traffic on the multi-CPU platforms. The coherency mechanisms increase the total number of data transfer and the platform's energy consumption.

5   Conclusions

The results presented in this paper not only show that GreenBST is an energy-efficient concurrent search tree, but also provide an important insight into how to develop energy efficient data structures in general. On single core systems, having locality-aware data structures that can lower data movement has been demonstrated to be good enough to increase energy-efficiency. However, on multi-CPU and many cores systems, data-structures' locality-awareness alone is not enough and good concurrency and multi-CPU cache strategy are needed. Otherwise, the energy overhead of "waiting/idling" CPUs or multi-CPU coherency mechanism can exceed the energy saving obtained by fewer data movements.

Acknowledgments. This work has received funding from the European Union Seventh Framework Programme (EXCESS project, grant no. 611183) and from the Research Council of Norway (PREAPP project, grant no. 231746/F20).

References

1. Afek, Y., Kaplan, H., Korenfeld, B., Morrison, A., Tarjan, R.E.: CBTree: a practical concurrent self-adjusting search tree. In: Aguilera, M.K. (ed.) DISC 2012. LNCS, vol. 7611, pp. 1–15. Springer, Heidelberg (2012)
2. Aggarwal, A., Vitter, J.S.: The input/output complexity of sorting and related problems. Commun. ACM **31**(9), 1116–1127 (1988)
3. Andersson, A.: Faster deterministic sorting and searching in linear space. In: Proceedings of the 37th Annual Symposium on Foundations of Computer Science, FOCS 1996, pp. 135–141, October 1996
4. Arbel, M., Attiya, H.: Concurrent updates with rcu: search tree as an example. In: Proceedings of 2014 ACM Symposium on Principles of Distributed Computing, PODC 2014, pp. 196–205 (2014)
5. Bender, M., Demaine, E.D., Farach-Colton, M.: Cache-oblivious b-trees. SIAM J. Comput. **35**, 341 (2005)
6. Bender, M.A., Farach-Colton, M., Fineman, J.T., Fogel, Y.R., Kuszmaul, B.C., Nelson, J.: Cache-oblivious streaming b-trees. In: Proceedings of the 19th Annual ACM Symposium on Parallel Algorithms and Architectures, SPAA 2007, pp. 81–92 (2007)
7. Bender, M.A., Fineman, J.T., Gilbert, S., Kuszmaul, B.C.: Concurrent cache-oblivious b-trees. In: Proceedings of the 17th Annual ACM Symposium on Parallelism in Algorithms and Architectures, SPAA 2005, pp. 228–237 (2005)
8. Brodal, G.S., Fagerberg, R., Jacob, R.: Cache oblivious search trees via binary trees of small height. In: Proceedings of the 13th ACM-SIAM Symposium on Discrete Algorithms, SODA 2002, pp. 39–48 (2002)
9. Brodal, G.S.: Cache-oblivious algorithms and data structures. In: Hagerup, T., Katajainen, J. (eds.) SWAT 2004. LNCS, vol. 3111, pp. 3–13. Springer, Heidelberg (2004)
10. Cormen, T.H., Leiserson, C.E., Rivest, R.L., Stein, C.: Introduction to Algorithms, 3rd edn. The MIT Press, Cambridge (2009)
11. Crain, T., Gramoli, V., Raynal, M.: A speculation-friendly binary search tree. In: Proceedings of the 17th ACM SIGPLAN Symposium on Principles and Practice of Parallel Programming, PPoPP 2012, pp. 161–170 (2012)
12. Dally, B.: Power and programmability: the challenges of exascale computing. In: DoE Arch-I presentation (2011)
13. David, T., Guerraoui, R., Trigonakis, V.: Asynchronized concurrency: the secret to scaling concurrent search data structures. In: Proceedings of the 12th International Conference on Architectural Support for Programming Languages and Operating Systems, ASPLOS 2015, pp. 631–644 (2015)
14. Ellen, F., Fatourou, P., Ruppert, E., van Breugel, F.: Non-blocking binary search trees. In: Proceedings of the 29th ACM SIGACT-SIGOPS Symposium on Principles of Distributed Computing, PODC 2010, pp. 131–140 (2010)
15. van Emde Boas, P.: Preserving order in a forest in less than logarithmic time. In: Proceedings of the 16th Annual Symposium on Foundations of Computer Science, SFCS 1975, pp. 75–84 (1975)

16. Frigo, M., Leiserson, C.E., Prokop, H., Ramachandran, S.: Cache-oblivious algorithms. In: Proceedings of the 40th Annual Symposium on Foundations of Computer Science, FOCS 1999, p. 285 (1999)
17. Gramoli, V.: More than you ever wanted to know about synchronization: synchrobench, measuring the impact of the synchronization on concurrent algorithms. In: Proceedings of the 20th ACM SIGPLAN Symposium on Principles and Practice of Parallel Programming, PPoPP 2015, pp. 1–10 (2015)
18. Ha, P.H., Tsigas, P., Anshus, O.J.: Wait-free programming for general purpose computations on graphics processors. In: Proceedings of the 2008 IEEE International Symposium on Parallel and Distributed Processing, IPDPS 2008, pp. 1–12 (2008)
19. Ha, P.H., Tsigas, P.: Reactive multi-word synchronization for multiprocessors. In: Proceedings of the 12th International Conference on Parallel Architectures and Compilation Techniques, PACT 2003, pp. 184–193 (2003)
20. Ha, P.H., Tsigas, P., Anshus, O.J.: Nb-feb: a universal scalable easy-to-use synchronization primitive for manycore architectures. In: Proceedings of the 13th International Conference on Principles of Distributed Systems, OPODIS 2009, pp. 189–203 (2009)
21. Ha, P.H., Tsigas, P., Anshus, O.J.: The synchronization power of coalesced memory accesses. IEEE Trans. Parallel Distrib. Syst. **21**(7), 939–953 (2010)
22. Ha, P.H., Tsigas, P., Wattenhofer, M., Wattenhofer, R.: Efficient multi-word locking using randomization. In: Proceedings of the 24th Annual ACM Symposium on Principles of Distributed Computing, PODC 2005, pp. 249–257 (2005)
23. Herlihy, M., Moss, J.E.B.: Transactional memory: architectural support for lock-free data structures. In: Proceedings of the 20th Annual International Symposium on Computer Architecture, ISCA 1993, pp. 289–300 (1993)
24. Hipp, D.R.: Sqlite (2015). http://www.sqlite.org
25. Kim, C., Chhugani, J., Satish, N., Sedlar, E., Nguyen, A.D., Kaldewey, T., Lee, V.W., Brandt, S.A., Dubey, P.: Fast: fast architecture sensitive tree search on modern cpus and gpus. In: Proceedings of 2010 ACM SIGMOD International Conference on Management of Data, SIGMOD 2010, pp. 339–350 (2010)
26. Kung, H.T., Robinson, J.T.: On optimistic methods for concurrency control. ACM Trans. Database Syst. **6**(2), 213–226 (1981)
27. Larsson, A., Gidenstam, A., Ha, P.H., Papatriantafilou, M., Tsigas, P.: Multi-word atomic read/write registers on multiprocessor systems. In: Albers, S., Radzik, T. (eds.) ESA 2004. LNCS, vol. 3221, pp. 736–748. Springer, Heidelberg (2004)
28. Lehman, P.L., Yao, S.B.: Efficient locking for concurrent operations on b-trees. ACM Trans. Database Syst. **6**(4), 650–670 (1981)
29. Minh, C.C., Chung, J., Kozyrakis, C., Olukotun, K.: Stamp: stanford transactional applications for multi-processing. In: IEEE International Symposium on Workload Characterization, IISWC 2008, pp. 35–46, September 2008
30. Natarajan, A., Mittal, N.: Fast concurrent lock-free binary search trees. In: Proceedings of the 19th ACM SIGPLAN Symposium on Principles and Practice of Parallel Programming, PPoPP 2014, pp. 317–328 (2014)
31. Prokop, H.: Cache-oblivious algorithms. Master's thesis. MIT (1999)
32. Rodeh, O.: B-trees, shadowing, and clones. Trans. Storage **3**(4), 2:1–2:27 (2008)
33. Sewall, J., Chhugani, J., Kim, C., Satish, N.R., Dubey, P.: Palm: Parallel architecture-friendly latch-free modifications to b+ trees on many-core processors. Proc. VLDB Endowment **4**(11), 795–806 (2011)

34. Tran, V., Barry, B., Ha, P.H.: RTHpower: accurate fine-grained power models for predicting race-to-halt effect on ultra-low power embedded systems. In: Proceedings of the 17th IEEE International Symposium on Performance Analysis of Systems and Software, ISPASS 2016 (2016) (pages to appear)
35. Tran, V., Barry, B., Ha, P.H.: Supporting energy-efficient co-design on ultra-low power embedded systems. In: Proceedings of the 2016 International Conference on Embedded Computer Systems: Architectures, Modeling, and Simulation, SAMOS XVI (2016) (pages to appear)
36. Umar, I., Anshus, O.J., Ha, P.H.: Deltatree: a locality-aware concurrent search tree. In: Proceedings of the 2015 ACM SIGMETRICS International Conference on Measurement and Modeling of Computer Systems, SIGMETRICS 2015, pp. 457–458 (2015)
37. Umar, I., Anshus, O.J., Ha, P.H.: Effect of portable fine-grained locality on energy efficiency and performance in concurrent search trees. In: Proceedings of the 21st ACM SIGPLAN Symposium on Principles and Practice of Parallel Programming, PPoPP 2016, pp. 36:1–36:2 (2016)

HAP: A Heterogeneity-Conscious Runtime System for Adaptive Pipeline Parallelism

Jinsu Park and Woongki Baek[✉]

School of ECE, UNIST, Ulsan, South Korea
{jinsupark,wbaek}@unist.ac.kr

Abstract. Heterogeneous multiprocessing (HMP) is a promising solution for energy-efficient computing. While pipeline parallelism is an effective technique to accelerate various workloads (e.g., streaming), relatively little work has been done to investigate efficient runtime support for adaptive pipeline parallelism in the context of HMP. To bridge this gap, we propose a heterogeneity-conscious runtime system for adaptive pipeline parallelism (HAP). HAP dynamically controls the full HMP system resources to improve the energy efficiency of the target pipeline application. We demonstrate that HAP achieves significant energy-efficiency gains over the Linux HMP scheduler and a state-of-the-art runtime system and incurs a low performance overhead.

1 Introduction

Heterogeneous multiprocessing (HMP) is rapidly emerging as a promising solution for energy efficient computing [9]. The key idea of HMP is to provide multiple types of cores that are architecturally designed and optimized for different performance and energy efficiency goals. To maximize the energy efficiency of HMP, its system software must be able to dynamically analyze the characteristics of the applications and schedule them on the most efficient cores. Recent work has demonstrated that application-directed dynamic optimization is effective for improving the energy efficiency of parallel applications (e.g., web browser [17] and DBMS [10]).

Pipeline parallelism is an effective software technique to accelerate the execution of tasks, which are difficult to parallelize using conventional techniques (e.g., data parallelism) due to the internal dependence between their subtasks. Pipeline parallelism decomposes the entire task into multiple subtasks and overlaps the execution of the subtasks for different work items to improve the overall throughput on parallel systems. Recent work has shown that various workloads (e.g., data mining [1] and streaming [4]) can be effectively accelerated through pipelining.

To achieve the best possible energy efficiency of pipeline parallelism on HMP, heterogeneity-conscious runtime support is crucial. While there has been prior work that investigates the runtime support for adaptive pipeline parallelism, they have limitations in that they target symmetric multiprocessing (SMP) without

© Springer International Publishing Switzerland 2016
P.-F. Dutot and D. Trystram (Eds.): Euro-Par 2016, LNCS 9833, pp. 518–530, 2016.
DOI: 10.1007/978-3-319-43659-3_38

any support for HMP [14] or they control only a subset of system resources, leaving key HMP system resources unmanaged (e.g., no dynamic voltage and frequency scaling (DVFS) of heterogeneous cores) [8,15] and achieving suboptimal energy efficiency.

To bridge this gap, we propose a heterogeneity-conscious runtime system for adaptive pipeline parallelism (HAP). Unlike the aforementioned prior approaches, HAP dynamically controls the full system resources (i.e., core types, counts, and voltage/frequency levels) of the underlying HMP system to maximize the energy efficiency of the target pipeline application. In addition, HAP provides a simple and easy-to-use application programming interface (API) that programmers can use to exploit the energy-efficient adaptive pipeline parallelism supported by HAP. Through our quantitative evaluation, we demonstrate the effectiveness of HAP. Specifically, this paper makes the following contributions:

- We propose a heterogeneity-conscious runtime system for adaptive pipeline parallelism. HAP manages the full system resources (i.e., core types, counts, and voltage/frequency levels) of the underlying HMP system for energy-efficient adaptive pipeline parallelism.
- We implement and evaluate HAP based on a full HMP system. Prior work is based on architectural simulators that lack the modeling of the entire system software stack such as the operating system [8,15]. We investigate the interaction between the Linux HMP scheduler and pipeline applications, demonstrating its performance and energy inefficiency.
- We quantify the effectiveness of HAP using six pipeline applications and a fully-configurable microbenchmark. Our quantitative evaluation shows that HAP significantly outperforms the Linux HMP scheduler and a state-of-the-art runtime system for adaptive pipeline parallelism [14,15] in terms of energy efficiency. In addition, our experimental results demonstrate that HAP robustly detects and adapts to the phase changes of the target pipeline application and incurs a low performance overhead.

2   Background

Heterogeneous Multiprocessing: A single-ISA heterogeneous multiprocessing system consists of cores that implement the same ISA but exhibit different architectural characteristics such as the instruction issue width [9]. A core cluster is defined as a group of the cores with the same architectural characteristics. In this work, for simplicity, we assume that the HMP system consists of the two types of core clusters, similarly to the ARM's big.LITTLE processor [7]. The core cluster that consists of the cores with higher (or lower) performance and power consumption is referred as the *big* (or *little*) core cluster. We assume that the big and little clusters consist of N_B and N_L cores, respectively.

In addition, we assume that the big and little clusters provide N_{f_B} and N_{f_L} voltage/frequency levels, which can be dynamically controlled in software. While per-core DVFS is promising, we consider cluster-level DVFS in this work because per-core DVFS is not yet widely supported in commodity processors.

Pipeline Parallelism: Pipeline parallelism decomposes a task into subtasks and overlaps their execution to improve the overall throughput. A pipeline application consists of one or more stages, each of which executes its assigned subtask for processing work items. Adjacent stages communicate through the work queues. Each pipeline stage consists of one or more worker threads. Each stage worker thread retrieves a work item from its input queue, processes it, and inserts the processed work item into its output queue, which is used as the input queue of the next stage.

The throughput of a pipeline stage is defined as the number of work items that can be processed by the stage per unit time. If the average processing time of a work item in stage s is t_s and the number of worker threads is N_s, the throughput of the stage s is computed as $\lambda_s = \frac{N_s}{t_s}$. The *limiter* stage of a pipeline application is defined as the stage whose throughput is the minimum among all the stages. The *non-limiter* stages are defined as all the stages except for the limiter stage. The overall throughput of the pipeline application is limited by the throughput of the limiter stage. This indicates that accelerating the non-limiter stages by allocating excessive hardware resources may significantly degrade energy efficiency without achieving any performance gain.

Heterogeneity-conscious runtime support is crucial to achieve the best possible energy efficiency of pipeline parallelism on HMP systems due to the following reasons. First, since the system state space rapidly grows with various characteristics of the target pipeline application (e.g., stage count and worker count) and the underlying HMP system (e.g., core types, counts, and voltage/frequency levels), it is nearly infeasible to develop a profiling-based static system for energy-efficient pipeline parallelism. Second, since the target pipeline application may exhibit widely different behaviors depending on its input data and program phases, it is critical to dynamically adapt its execution in a heterogeneity-conscious and energy-efficient manner, guided by runtime information.

3   Design and Implementation

HAP mainly consists of two components – the application programming interface (API) and the runtime system. For simplicity, we describe the design and implementation of HAP with an assumption that the underlying HMP system consists of two types of core clusters (i.e., big and little). However, we believe that our proposed techniques can be generalized for various HMP systems (e.g., more core types).

3.1   The HAP API

HAP provides the four API functions summarized in Table 1. To exploit the energy-efficient adaptive pipeline parallelism supported by HAP, programmers need to instrument their applications using the API functions. The begin_app function is used to notify the beginning of the target pipeline application. The begin_app function establishes the interprocess communication (IPC) between

Table 1. The HAP application programming interface (API)

Function	Description
begin_app(nStage)	Beginning of the target pipeline application
begin_work()	Beginning of the work-item processing
end_work(stageId)	End of the work-item processing
end_app()	End of the target pipeline application

the target pipeline application and the HAP runtime system and sends the information on the target application (e.g., the number of pipeline stages) to the HAP runtime system through IPC.

The begin_work function is used to mark the beginning of a work-item processing performed by the calling stage worker thread. The begin_work function reads the time (t_B) when the processing of the work item is about to begin. The end_work function reads the time (t_E) when the processing of the work item has ended and sends the data such as the work-item processing time $(t_W = t_E - t_B)$ and the stage ID of the calling thread to the HAP runtime system through IPC. Finally, the end_app function is used to notify the end of the target pipeline application.

The current implementation of the HAP API builds upon the Application Heartbeats framework [6], which provides a well-established interface to communicate messages called *heartbeats* between processes. We extend the Application Heartbeats framework to encode and communicate pipeline-specific information such as the stage ID and work-item processing time (t_W).

3.2 The HAP Runtime System

The HAP runtime system manages the full system resources to significantly enhance the energy efficiency of the target pipeline application. The system resources managed by the HAP runtime system are the core types, counts, and the voltage/frequency levels of each cluster. The HAP runtime system divides the system resources into two groups – the ones allocated to the limiter stage of the target pipeline application and the others allocated to the non-limiter stages. We define the system state space as all the possible combinations of the system resources that can be allocated to the limiter stage. Figure 1 shows the overall architecture of the HAP runtime system, which consists of three components – the performance estimator, power estimator, and runtime manager.

Performance Estimator: For a system state of interest, the performance estimator of HAP estimates the performance of the target pipeline application. The performance estimator assumes that each of the stage worker threads of the target pipeline application is assigned with its dedicated core. The performance estimator employs a linear model that assumes that the performance of each worker thread of the limiter stage is proportional to the computation capacity

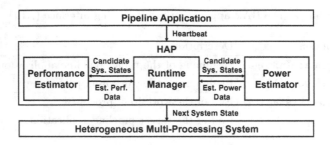

Fig. 1. The overall architecture of the HAP runtime system

of its allocated core. If the performance ratio of the big core to the little core is r_0 at the frequency of f_0, the computation capacities of the big and little cores running at the frequencies of f_B and f_L are $r_0 \cdot \frac{f_B}{f_0}$ and $\frac{f_L}{f_0}$. The performance ratio (r_0) can be either statically determined based on the architectural characteristics (e.g., instruction issue width) of heterogeneous cores or dynamically determined based on the runtime information collected during the execution of the target pipeline application. As discussed later, r_0 is dynamically computed based on the runtime information encoded in the heartbeats.

Power Estimator: For a system state of interest, the power estimator of HAP estimates the power consumption of the underlying HMP system. The power estimator estimates the power consumption of the big and little core clusters based on a linear regression model, which assumes that the power consumption of each cluster is proportional to the sum of the utilization of the cores in the corresponding cluster. The regression coefficient values are determined for every available frequency of each cluster. The regression coefficients are computed based on the data collected through the offline experiments with our microbenchmark that can stress the underlying HMP system with different configurations (e.g., core type, count, frequency, and utilization). Currently, the power estimator assumes that the power consumption of other hardware components (e.g., memory) is constant, which can be extended with more sophisticated models. In summary, the power estimator uses Eq. 1 to estimate the power consumption of the underlying HMP system.

$$P = \alpha_{B,f_B} \cdot \sum_{i=0}^{N_B-1} U_{B,i} + \beta_{B,f_B} + \alpha_{L,f_L} \cdot \sum_{i=0}^{N_L-1} U_{L,i} + \beta_{L,f_L} + \gamma \qquad (1)$$

The power estimator assumes that all the cores allocated to the limiter stage are fully utilized because, by its definition, it is the performance bottleneck among all the stages. To estimate the utilization of the cores allocated to the non-limiter stages, the power estimator sorts the non-limiter stage worker threads in an increasing order of the throughput and the cores allocated to the non-limiter stages in a non-increasing order of the computation capacity, respectively. The power estimator assumes that the runtime manager establishes the one-to-one mapping of each worker thread to core in the sorted order, which is the actual scheduling performed by the runtime manager. Based on the queuing theory [12], the utilization of each core allocated to a non-limiter stage s is approximated to be $U_s = \frac{\lambda_{lim} \cdot t_s}{N_s}$, where λ_{lim}, t_s, and N_s are the throughput of the limiter stage, the work-item processing time, and the worker thread count of the stage s.

We note that the utilization of cores where the non-limiter stages are scheduled can be computed more precisely based on the queuing theory extended for heterogeneous servers [5] because of the different processing capabilities of the stage worker threads scheduled on different clusters. However, we decide to use an approximate solution because the computational complexity of the precise solution is high and the accuracy of the approximate solution is expected to be reasonable [5]. By substituting the estimated utilization of all the cores into Eq. 1, the power estimator estimates the power consumption of the target pipeline application for a system state of interest.

Runtime Manager: The HAP runtime manager explores the system state space to find an efficient system state that significantly reduces the energy consumption of the target pipeline application. The system state space rapidly grows with the worker thread count of the limiter stage and architectural parameters of the underlying HMP system (e.g., core types, counts, and voltage/frequency levels). For instance, if the limiter stage consists of two worker threads and the underlying HMP system consists of four big cores with N_{f_B} voltage/frequency levels and four little cores with N_{f_L} voltage/frequency levels, the number of all the system states is $N_{f_B} + N_{f_B} \cdot N_{f_L} + N_{f_L}$. Due to the large system state space, the runtime manager explores the system state space based on an incremental and greedy algorithm, inspired by the hill-climbing algorithm [13].

The runtime manager executes in two phases – the *adaptation* and *idle* phases. Algorithm 1 shows the pseudocode for the runtime manager. During the adaptation phase, the runtime manager checks if an adaptation period has been reached (Line 7) for every new heartbeat generated when a stage worker thread finishes the processing of a work item. The adaptation phase consists of three sub-phases – the initial, observation, and exploration sub-phases.

The runtime manager runs in the *initial* sub-phase until the first adaptation period is reached. At the end of the initial sub-phase (Line 9), the runtime manager retrieves the information on the target pipeline application such as the thread ID of each stage worker thread. The runtime manager then sets the affinity of all the threads of each stage to each of the available cores, transitioning to the observation sub-phase.

The runtime manager runs in the *observation* sub-phase until the second adaptation period is reached. At the end of the observation sub-phase (Line 11), the runtime manager retrieves the dynamic information on the target pipeline application such as the throughput of each stage using the data encoded in the received heartbeats and the equation discussed in Sect. 2. The runtime manager then identifies the limiter stage, sets the system state to the initial state, and transitions into the exploration sub-phase.

Algorithm 1. The pseudocode for the HAP runtime manager

```
 1: limiter ← invalidStageId; pScore ← 0
 2: firstAdtPeriod ← false; cScore ← 0
 3: pState ← obsState; cState ← obsState
 4: phase ← adaptation; subPhase ← initial
 5: procedure MAIN
 6:   while true do
 7:     if isAdaptPeriod() then
 8:       if phase = adaptation then
 9:         if subPhase = initial then
10:           subPhase ← observation
11:         else if subPhase = observation then
12:           subPhase ← exploration
13:           limiter ← findLimiter()
14:           firstAdtPeriod ← true
15:           cState ← initState
16:         else              ▷ Exploration sub-phase
17:           cScore ← computeScore()
18:           exploreSystemStateSpace()
19:       else                ▷ Idle phase
20:         if triggerReadaptation() = true then
21:           phase ← adaptation
22:           subPhase ← initial
23:           resetVariables()
24:     applyState()
25: procedure EXPLORESYSTEMSTATESPACE

26:   if firstAdtPeriod = false ∧ cScore − pScore
       < δ_c then
27:     cState ← pState
28:     phase ← idle
29:   else
30:     if firstAdtPeriod = true then
31:       firstAdtPeriod ← false
32:     pState ← cState
33:     pScore ← cScore
34:     cState ← getNextState()
35:     if cState = invalidState then
36:       cState ← pState
37:       phase ← idle
38: procedure GETNEXTSTATE
39:   bestState ← invalidState
40:   bestScore ← minScore
41:   candStates ← cState ∪ getNeigh-
       borStates(cState)
42:   for candState in candStates do
43:     estScore ← estimateScore(candState)
44:     if estScore > bestScore then
45:       bestState ← candState
46:       bestScore ← estScore
47:   if bestState = cState then
48:     bestState ← invalidState
49:   return bestState
```

During the *exploration* sub-phase (Line 16), the runtime manager explores the system state space to find an efficient system state that results in the significantly reduced energy consumption of the target pipeline application. The runtime manager explores the system state space based on an incremental and greedy algorithm. At each adaptation period,[1] the runtime manager invokes the `exploreSystemStateSpace` function, which subsequently calls the `getNextState` function to determine the next system state to transition. The runtime manager adapts the system state in an incremental manner in that the candidate system states are generated by incrementally changing the system resources allocated to the limiter stage from the current system state.

Specifically, the runtime manager considers the system states that are within the Manhattan distance d from the current system state in the three dimensional system state space (i.e., the big core count allocated to the limiter stage,[2] the frequencies of the big and little core clusters) as the candidate system states (Line 41). For instance, with the current system state of (n_B, f_{B_i}, f_{L_j}) and $d = 1$, the candidate system states are $(n_B + 1, f_{B_i}, f_{L_j})$, $(n_B - 1, f_{B_i}, f_{L_j})$, $(n_B, f_{B_{i+1}}, f_{L_j})$, $\cdots$, and $(n_B, f_{B_i}, f_{L_{j-1}})$. With larger d, the runtime manager

---

[1] At the end of the first adaptation period of the exploration sub-phase, the runtime manager computes the performance ratio (r_0) of the big core to the little core based on the work-item processing time (t_W) data encoded in the heartbeats generated by the stage worker threads scheduled on the big and little core clusters.

[2] Since the little core count of the limiter stage can be determined from its big core count, we do not consider its little core count when computing d.

explores the system state space more exhaustively at the potential cost of higher performance overheads and instability due to abrupt system state changes.

The runtime manager adapts the system state in a greedy manner in that it chooses the next system state as the one that is estimated to result in the highest energy efficiency among all the candidate system states (Lines 42–46).[3] The energy efficiency (i.e., joules per processed work item) of each candidate system state is estimated to be $\frac{P_{Est}}{\lambda_{Est}}$, where λ_{Est} and P_{Est} are its performance and power consumption estimated through the performance and power estimators.

The runtime manager transitions to the idle phase in the following two cases. First, if the energy efficiency of the current period is lower than the previous period, the runtime manager restores the previous system state and transitions to the idle state (Lines 26 28). The energy efficiency of the current period is computed based on the actual energy consumption data collected using the sensors discussed in Sect. 4 and the throughput data. Second, if none of the candidate states are expected to achieve higher energy efficiency than the current system state (Lines 35 and 47), the runtime manager transitions to the idle phase.

During the idle phase (Line 19), the runtime manager executes the target pipeline application without performing any adaptation but keeps monitoring the application to detect its phase changes. When detecting a program phase change, the runtime manager terminates the idle phase and triggers the entire adaption process again (Lines 20–23). To detect phase changes, the runtime manager computes the work ratio (r_W) of the limiter stage to the non-limiter stages. If the work ratios between the consecutive periods differ by r_{th} and N_R times in a row, the runtime manager determines that the program phase of the target pipeline application has changed and transitions to the adaptation phase to find a new efficient system state. Unless stated otherwise, d, r_{th}, and N_R are set to 5, 25 % and 3.

4 Evaluation

Methodology: To quantify the effectiveness of HAP, we use a full heterogeneous multiprocessing (HMP) system, the ODROID-XU3 embedded development board. The board is equipped with the Exynos 5422 processor based on the ARM's big.LITTLE architecture [7]. The processor consists of the four big cores (i.e., $N_B = 4$) and the four little cores (i.e., $N_L = 4$). The board is installed with Xubuntu 14.04 and the Linux kernel 3.10.69, which implements the HMP scheduler. The configurable frequency ranges of the big and little clusters are 0.2 – 2.0 GHz and 0.2 – 1.4 GHz, respectively. The board is equipped with sensors that periodically sample the power consumption of the big cluster, little cluster, memory, and GPU, which we use to construct the linear regression model of the power estimator and to measure the energy consumption of the HMP system during the execution of the target pipeline application.

---

[3] Note that HAP can be generalized to perform optimizations based on other metrics (e.g., energy-delay product) by customizing the `estimateScore` function (Line 43).

We use the following six pipeline benchmarks, some of which are modified to exploit pipeline parallelism – blackscholes (BL) [3], binomialoptions (BO) [3], bzip2 (BZ) [2], dedup (DD) [1], ferret (FR) [1], montecarlo (MC) [3]. The number of stages of BL, BO, BZ, DD, FR, and MC are 3, 3, 3, 5, 6, and 3, respectively. We also use a microbenchmark that is fully configurable in terms of pipeline parameters such as the number of stages, workers per stage, workload per worker.

Our evaluation aims to investigate the following. First, we quantify how much energy efficiency gain can be achieved through the use of HAP. Second, we evaluate the effectiveness of the re-adaptation functionality of HAP when the target pipeline application has multiple distinct program phases. Third, we investigate the sensitivity of the energy efficiency and performance overhead of HAP to the search distance parameter (d), which controls the exhaustiveness of the system space exploration.

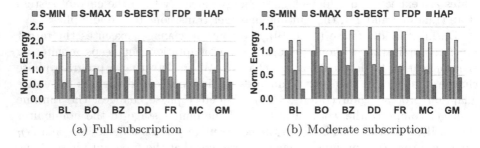

(a) Full subscription   (b) Moderate subscription

Fig. 2. Normalized energy

Energy Efficiency: We evaluate the energy efficiency of HAP. We run the each benchmark with the following five OS or runtime versions – (1) the Linux HMP scheduler with the lowest big and little core frequencies (S-MIN), (2) the Linux HMP scheduler with the highest big and little core frequencies (S-MAX), (3) the Linux HMP scheduler with the big and little core frequencies that result in the best energy efficiency among all the possible combinations of the minimum, medium, and maximum frequencies (S-BEST),[4] (4) feedback-directed pipeline parallelism (FDP), which implements the runtime system proposed in [14,15],[5] and (5) HAP. In addition, to investigate the effectiveness of HAP with different system utilization levels, we configure each benchmark in the following two settings – (1) *full subscription*, in which the worker thread counts of the limiter and each of the non-limiter stages are set to $N_B + N_L - S + 1$ and 1, where S is the number of stages and (2) *moderate subscription*, in which the worker thread counts of the limiter and each of the non-limiter stages are set to 2 and 1.

---

[4] Due to the large system space which requires infeasibly long time for collecting profiled data, we selectively use the most representative frequencies (i.e., min, medium, and max) to determine the configuration for S-BEST.

[5] Due to space limit, we refer to [14,15] for more details on FDP.

Figure 2(a) shows the energy consumption of the five OS and runtime versions normalized to S-MIN with the full subscription setting, demonstrating the following data trends. First, HAP significantly outperforms the Linux HMP scheduler in terms of energy efficiency. Specifically, HAP reduces the energy consumption of the target pipeline applications by 42.4, 64.8, and 20.8 % on average (i.e., geometric mean), compared with the S-MIN, S-MAX, and S-BEST versions. HAP outperforms S-BEST mainly due to the performance inefficiency of the current version of the Linux HMP scheduler. For some benchmarks (e.g., BL), we observe that the Linux HMP scheduler often heavily biases CPU-intensive stage worker threads to the big cores even when the little cores are idle, eventually causing performance and energy efficiency degradation due to the load imbalance.

Second, HAP significantly outperforms FDP, which is a state-of-the-art runtime system for adaptive pipeline parallelism. Specifically, HAP reduces the energy consumption of the target pipeline applications by 63.8 % on average, compared with FDP. This is mainly because FDP lacks the capability of controlling voltage/frequency levels of heterogeneous core clusters, which are critical hardware knobs for achieving high energy efficiency. In contrast, HAP manages the full system resources (i.e., core types, counts, and voltage/frequency levels), significantly improving the energy efficiency of the target pipeline applications.

Figure 2(b) shows the energy consumption of the five OS and runtime versions normalized to S-MIN with the moderate subscription setting. HAP continues to achieve higher energy efficiency gains over the other OS and runtime versions with moderate subscription. Specifically, HAP reduces the energy consumption of the target pipeline applications by 55.3, 67.5, 32.0, and 63.4 % on average, compared with the S-MIN, S-MAX, S-BEST, and FDP versions. Since the system is less utilized with moderate subscription, HAP discovers more opportunities for reducing the energy consumption of the target pipeline application (e.g., setting the frequency of the unused core cluster to the lowest level), achieving higher energy efficiency gains than the case with full subscription. In summary, our experimental results show that HAP is effective in that it significantly outperforms all the other OS and runtime versions in terms of energy efficiency.

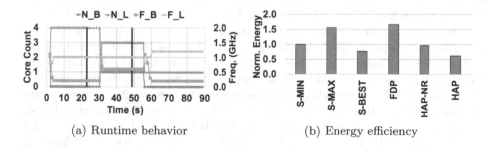

(a) Runtime behavior   (b) Energy efficiency

Fig. 3. Effectiveness of re-adaptation.

Effectiveness of Re-adaptation: To evaluate the effectiveness of the re-adaptation functionality of HAP, we use a microbenchmark, which is config-ured to exhibit three distinct phases. Figure 3(a) shows the runtime behavior of the microbenchmark. At $t = 23.0$, the microbenchmark transitions to the second phase in which the work ratio (r_W) of the limiter to the non-limiter stages signif-icantly changes. HAP robustly detects the phase change and accordingly adapts the system state after observing that the three consecutive samples of r_W are consistent (i.e., $N_R = 3$). At $t = 48.8$, the microbenchmark transitions to the third phase in which one of the non-limiter stages becomes the new limiter stage. HAP also robustly detects the phase change and accordingly performs adapta-tions. Figure 3(b) demonstrates the effectiveness of the re-adaptation functional-ity of HAP in that HAP significantly outperforms all the other OS and runtime versions in terms of energy efficiency, including a variant of HAP (i.e., HAP-NR) with which the re-adaptation functionality is intentionally disabled for illustra-tive purposes.

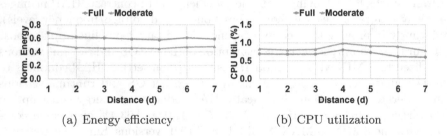

(a) Energy efficiency   (b) CPU utilization

Fig. 4. Sensitivity to the search distance

Sensitivity to the Search Distance: Finally, we investigate the sensitivity of the energy efficiency and performance overhead of HAP to the search distance (d) parameter. Figure 4(a) shows the average (i.e., geometric mean) energy con-sumption of HAP across all the evaluated benchmarks, normalized to OS-MIN when d varies from 1 to 7. With larger d, the energy efficiency of HAP gener-ally improves because it explores the system state more exhaustively. When d is sufficiently large (i.e., $d > 5$), the energy efficiency of HAP slightly decreases as d increases. This is mainly because HAP may converge to a slightly subop-timal system state when the system state changes too abruptly with larger d. Nevertheless, HAP consistently provides significant energy-efficiency gains over the Linux HMP scheduler both in the full and moderate subscription settings.

To quantify the performance overhead of HAP, Fig. 4(b) shows the sensi-tivity of the CPU utilization of HAP to the search distance. With larger d, the CPU utilization of HAP tends to gradually increase because it explores the system state space more exhaustively. However, the CPU utilization of HAP is insignificant (i.e., $< 1.0\%$) across all the configurations. Interestingly, with sufficiently large d (i.e., $d > 4$), the CPU utilization of HAP slightly decreases.

This is mainly because HAP converges faster with sufficiently large d and then consumes significantly less CPU cycles afterward. In summary, our experimental results demonstrate that HAP is an effective runtime system for adaptive pipeline parallelism in that it significantly improves energy efficiency, robustly adapts to program phase changes, and incurs a low performance overhead.

5   Related Work

Prior work has proposed runtime techniques for adaptive pipeline parallelism [8,14,15]. While insightful, the proposed techniques target runtime support for symmetric multiprocessing (SMP) systems [14] or lack the management of full system resources (e.g., no DVFS) of HMP systems [8,15], resulting in suboptimal energy efficiency as quantified by our experimental results. Further, the techniques proposed in [8,15] have been evaluated using architectural simulators without in-depth investigation of the interaction among the target pipeline application, runtime, and OS. Our work differs in that HAP effectively manages the full system resources (core types, counts, and voltage/frequency levels) and is implemented and evaluated based on a real HMP system with the full system software stack.

Prior work has proposed architectural [9] and system software [11,16] techniques to enhance the power and/or energy efficiency of conventional applications on HMP systems. Our work differs in that we propose an energy-efficient runtime system for adaptive pipeline parallelism in the context of HMP. In addition, recent work has investigated application-level techniques to improve the energy efficiency of the web browser [17] and DBMS [10]. While similar in that they utilize the application-level knowledge to enhance energy efficiency, our work differs as HAP targets efficient runtime support for adaptive pipeline parallelism.

6   Conclusions

This work presents HAP, a heterogeneity-conscious runtime system for adaptive pipeline parallelism. HAP dynamically controls the full system resources of the underlying HMP system to maximize the energy efficiency of the target pipeline application. In addition, HAP provides a simple and easy-to-use application programming interface that programmers can use to exploit the energy efficient adaptive parallelism supported by HAP. Our quantitative evaluation demonstrates the effectiveness of HAP in that it significantly outperforms the Linux HMP scheduler and the state-of-the-art runtime system for adaptive pipeline parallelism in terms of energy efficiency, robustly adapts to the phase changes of the target pipeline application, and incurs a small performance overhead. As our future work, we plan to extend HAP by investigating more advanced search algorithms that explore the system state space with higher coverage and efficiency.

Acknowledgements. This research was supported by ICT R&D program of MSIP/IITP (B0101-16-0661).

References

1. Bienia, C., et al.: The PARSEC benchmark suite: characterization and architectural implications. In: PACT 2008 (2008)
2. BZIP2SMP. http://bzip2smp.sourceforge.net/
3. CUDA Samples. http://docs.nvidia.com/cuda/cuda-samples/
4. Gordon, M.I., et al.: Exploiting coarse-grained task, data, and pipeline parallelism in stream programs. In: ASPLOS XII (2006)
5. Gumbel, H.: Waiting lines with heterogeneous servers. Oper. Res. **8**(4), 504–511 (1960)
6. Hoffmann, H., et al.: Application heartbeats: a generic interface for specifying program performance and goals in autonomous computing environments. In: ICAC 2010 (2010)
7. Je, B.: Big.LITTLE system architecture from ARM: saving power through heterogeneous multiprocessing and task context migration. In: DAC 2012 (2012)
8. Joao, J.A., et al.: Bottleneck identication and scheduling in multithreaded applications. In: ASPLOS XVII (2012)
9. Kumar, R., et al.: Single-ISA heterogeneous multi-core architectures: the potential for processor power reduction. In: MICRO 36 (2003)
10. Mühlbauer, T., et al.: Heterogeneity-conscious parallel query execution: getting a better mileage while driving faster! In: DaMoN 2014 (2014)
11. Muthukaruppan, T.S., et al.: Hierarchical power management for asymmetric multi-core in dark silicon era. In: DAC 2013 (2013)
12. Navarro, A., et al.: Analytical modeling of pipeline parallelism. In: PACT 2009 (2009)
13. Skiena, S.S.: The Algorithm Design Manual, 2nd edn. Springer-Verlag, London (2008)
14. Suleman, M.A., et al.: Feedback-directed pipeline parallelism. In: PACT 2010 (2010)
15. Suleman, M.A.: An asymmetric multi-core architecture for efficiently accelerating critical paths in multithreaded programs. Ph.D. thesis. University of Texas at Austin (2010)
16. Yun, J., et al.: HARS: A heterogeneity-aware runtime system for self-adaptive multithreaded applications. In: DAC 2015 (2015)
17. Zhu, Y., et al.: High-performance and energy-efficient mobile web browsing on Big/Little systems. In: HPCA 2013 (2013)

Using Data Dependencies to Improve Task-Based Scheduling Strategies on NUMA Architectures

Philippe Virouleau[1,2]([✉]), François Broquedis[1], Thierry Gautier[2], and Fabrice Rastello[1]

[1] Inria, Univ. Grenoble Alpes, CNRS, Grenoble Institute of Technology, LIG, Grenoble, France
{philippe.virouleau,francois.broquedis,fabrice.rastello}@inria.fr
[2] LIP, ENS de Lyon, Lyon, France
thierry.gautier@inrialpes.fr

Abstract. The recent addition of data dependencies to the OpenMP 4.0 standard provides the application programmer with a more flexible way of synchronizing tasks. Using such an approach allows both the compiler and the runtime system to know exactly which data are read or written by a given task, and how these data will be used through the program lifetime. Data placement and task scheduling strategies have a significant impact on performances when considering NUMA architectures. While numerous papers focus on these topics, none of them has made extensive use of the information available through dependencies. One can use this information to modify the behavior of the application at several levels: during initialization to control data placement and during the application execution to dynamically control both the task placement and the tasks stealing strategy, depending on the topology. This paper introduces several heuristics for these strategies and their implementations in our OpenMP runtime XKAAPI. We also evaluate their performances on linear algebra applications executed on a 192-core NUMA machine, reporting noticeable performance improvement when considering both the architecture topology and the tasks data dependencies. We finally compare them to strategies presented previously by related works.

Keywords: OpenMP · Task dependencies · Benchmark · Runtime systems · NUMA · XKAAPI · Scheduling · Work-stealing

1 Introduction

While non-uniform memory access (NUMA) architectures stand today as one of the most popular design to build large-scale shared memory machines, exploiting them at their full potential remains challenging. On such architectures, the memory is split into several NUMA nodes and both bandwidth and latency depend on which processor accesses specific data: accessing memory allocated locally is

© Springer International Publishing Switzerland 2016
P.-F. Dutot and D. Trystram (Eds.): Euro-Par 2016, LNCS 9833, pp. 531–544, 2016.
DOI: 10.1007/978-3-319-43659-3_39

most of the time faster than accessing data allocated to remotely-located NUMA nodes. Controlling data locality over the application lifetime is one of the key steps to achieving both good performance and scalability on these architectures.

Task-based parallel programming environments like OpenMP have become very popular when it comes to program shared memory machines with hundreds of cores. Indeed, they offer ways of expressing massive fine-grain parallelism with a relatively low overhead. Most of them also come with facilities to dynamically perform load balancing of tasks over the processors. Even if such characteristics fill the need of generating more and more parallelism out of parallel applications, standard parallel programming environments still not explicitly address the problem of data locality on NUMA systems.

The runtime system plays a central role in the execution of a task-based parallel application. For example, it is responsible for assigning ready tasks to the target platforms' processors. It is also in charge of performing load balancing when a processor idles. Both these decisions should take the architecture topology into account in order to avoid NUMA-related performance penalties on the overall application performance.

The recent addition of data dependencies to the OpenMP tasking model provides the runtime system with very precise information about which part of an application accesses which variables. Thanks to these dependencies, the runtime system knows which memory areas are read or written by which task. As shown in this paper, the task scheduler can rely on this information when assigning tasks to processors to implement NUMA-aware strategies.

This paper describes several of these strategies we implemented inside the XKAAPI [9] runtime system. We identified three major steps in the task scheduler workflow that may have an impact on parallel applications on NUMA systems: the data distribution, the assignment of ready tasks to the processors and the way the task scheduler browses the architecture topology to perform load balancing. This paper describes and evaluates them, showing how they impact the application performance on a 192-core NUMA machine. We also compare them to state-of-the-art task scheduling strategies taken from related works and implemented within XKAAPI.

The layout of this paper falls into six sections as follows. In Sect. 2, we first give some background on NUMA architectures and the task programming model with data dependencies. We then describe in Sect. 3 the ideas, strategies and implementation details that we used to improve the runtime performances for these applications. Section 4 is devoted to the presentation performances evaluation. We eventually present some related works in Sect. 5 before concluding.

2   NUMA Architectures Design and Exploitation

2.1   Hardware Background

Most of nowadays parallel shared memory architectures are built according to a NUMA design where the memory is physically split into several banks attached to processors. Many vendors assemble these banks in a hierarchical way, thus building shared memory machines embedding several hundreds of cores. Exploiting such architectures at their full potential requires a fine control of the execution of a parallel application, as accessing local memory is most of the time faster than accessing memory stored in a memory bank attached to a remote processor.

The machine we experimented on is an SGI UV2000 platform made of 24 NUMA nodes. Each NUMA node holds an 8-core Intel Xeon E5-4640 CPU for a total of 192 cores. We refer to this machine as Intel192 in the paper. The memory topology is organized by pairs of NUMA nodes connected together through Intel QuickPath Interconnect. These pairs can communicate together through a proprietary fabric called NUMALink6 with up to two hops.

Table 1 shows the distances advertised by the hwloc library [4] that represents the communication time for different distances normalized to the time of a local communication. Distances named *local* and *peer* form a pair of NUMA nodes (through Intel QPI), other nodes are either one hop away or two hops away (through NUMALink6).

Table 1. NUMA distances from node 0 advertised by the hwloc library on Intel192.

NUMA nodes location	local	peer	one hop away	two hops away
hwloc distances	1.0	5.0	6.5	7.9

2.2 Software Background

To exploit large-scale shared memory architectures, the application programmer needs:

1. to express massive fine grain parallelism to get the most out of the numerous processing units of the platform;
2. to control the execution of the application, especially the way computations and data are distributed over the platform, to prevent the NUMA design to have a negative impact on the overall application performance.

Task-based parallel programming environments provide ways of expressing fine grain parallelism that can be dynamically assigned to processors at runtime. OpenMP [12], the de-facto standard for shared-memory parallel programming, supports task parallelism with dependencies since revision 4.0.

A Glimpse at OpenMP Tasking. An OpenMP *task* can be seen as an independent *unit of work* an OpenMP thread can execute. Tasks can be created by an OpenMP thread and executed by any thread of the same parallel region. As managing tasks at runtime is way cheaper than creating and synchronizing threads, the application programmer can take the parallelization of its application further, as he can now consider portions of code that were too fine grain to be parallelized using only threads. The synchronization of OpenMP 3.0 tasks is performed thanks to the **taskwait** keyword that waits for the completion of all the tasks generated from the current OpenMP parallel region. On one hand, the application programmer is responsible for creating and synchronizing OpenMP tasks explicitly. On the other hand, the runtime system is in charge of correctly assigning tasks to threads during the application execution.

OpenMP 4.0 pushes the concept of task further introducing the **depend** keyword to specify the access mode of each shared variable a task will access during its execution. Access modes can be set to either **in**, **out** or **inout** whether the corresponding variable is respectively read as input, written as output or both read and written by the considered task. This information is then processed by the underlying runtime system to decide whether a task is ready for execution or should first wait for the completion of other ones.

KASTORS Benchmark Suite. Figure 1 shows the implementation of a Cholesky factorization implemented with OpenMP task dependencies. This factorization algorithm comes from the PLASMA library and is very similar to the one implemented in the KASTORS benchmark suite [15]. Task dependencies support comes with several benefits. First, task dependencies involve decentralized, selective synchronization operations that should scale better than the broad-range taskwait-based approaches. In some situations, this way of programming unlocks more valid execution scenarios than explicitly synchronized tasks, which provides the runtime system with many more valid task schedules to choose from. For example, in the Cholesky factorization, many instances of the dtrsm, dsyrk and dgem BLAS computations can legally run concurrently when executing the version with task dependencies. Secondly, information about task dependencies also enables the runtime system to optimize further, such as improving tasks and data placement.

```
1  for (size_t k=0; k < NB; ++k) {        12    for (int m=k; m < NB; ++m) {
2  #pragma omp task shared(A) \           13    #pragma omp task shared(A)\
3     depend(inout: A[k][k])              14       depend(in: A[m][k]) \
4     dpotrf(NB,&A[k][k]);                15       depend(inout: A[m][m])
5                                          16          dsyrk(NB,&A[m][k],  &A[m][m]);
6     for (int m=k; m < NB; ++m)          17
7  #pragma omp task shared(A)\            18       for (int n=k; n < m; ++n)
8     depend(in: A[k][k]) \              19    #pragma omp task shared(A)\
9     depend(inout: A[m][k])             20       depend(in: A[m][k],A[n][k])\
10       dtrsm(NB,&A[k][k],&A[m][k]);     21       depend(inout: A[m][n])
11                                         22          dgemm(NB,
                                           23             &A[m][k],&A[n][k],&A[m][n]);
                                           24    }
                                           25  }
```

Fig. 1. Cholesky factorization with OpenMP-4.0 task dependencies

The Way We Execute Task-Based Applications. Most task-based programming environments rely on a work-stealing execution model, originally introduced in Cilk [8]. Work-stealing is indeed often considered when it comes to dynamically balance the workload among processing units. The work-stealing principle can be summarized as follows. An idle thread, called a thief, initiates a steal request to a randomly selected victim. On reply, the thief receives a copy of one ready task, leaving the original task marked as stolen. Coherency between a thief and its victim is ensured by a variant of Cilk's T.H.E protocol, also described in [8].

The runtime system we develop, called XKAAPI, also implements the work-stealing execution model to execute OpenMP task-based applications. The runtime creates a system thread, called a *kproc*, for each processing unit to be used. On a NUMA multicore machine, a processing unit is a core. A kproc creates tasks and pushes them on its own work queue, which is implemented as a stack. The enqueue operation is very fast: it takes around ten cycles on modern x86/64 processors [3]. As in Cilk, a running XKAAPI task can create children tasks. Depending on the number of tasks per thread, XKAAPI implements two strategies to find a ready task. If the number of tasks is lower than a threshold, XKAAPI follows the Cilk's *work first principle*. In that case, the thief iterates through the victim's stack queue from the least recently pushed task to the most recently one and it computes true data-flow dependencies for each task until a ready task is found. If the number of tasks is greater than the threshold, the data flow graph is built and the thief picks a task from the victim's list of ready tasks [1]. By the nature of our benchmarks, this latter strategy is de facto selected and we developed new NUMA aware strategies.

3 Using OpenMP Tasks Dependencies to Improve Tasks and Data Placement on NUMA Machines

In this section, we describe how the runtime system can have a positive impact on the application execution using the information provided by data dependencies. The following sections describe the way we adapted the behavior of the runtime system to control data placement during the initialization phase, when data will be allocated and accessed for the first time, and how we modified the way tasks that perform the actual computations are dynamically assigned to processors while maximizing data locality.

3.1 Inside the XKAAPI Task-Based Runtime System

This section describes some of the key internal structures and mechanisms of the XKAAPI runtime system.

The Way XKAAPI Models the Architecture. XKAAPI sees the architecture topology as a hierarchy of places. A place is a list of tasks associated with a subset of the machine processing units. XKAAPI's places are very similar to the notion of *shepherd* introduced in [10], or ForestGOMP's *runqueues* [2]. XKAAPI most of the time only considers two levels of places : node-level places, which are bound to the set of processors contained in a NUMA node, and processor-level places, which are bound to a single processor of the platform. This way, at the processor level one place is associated to each of the physical cores, and at the NUMA node level one place is associated to each of the NUMA nodes.

The Way XKAAPI Enables Ready Tasks and Steals Them. The scheduling framework in XKAAPI [1] relies on virtual functions for *selecting a victim*, *selecting a place* to push a ready task and *pushing* a set of initial ready tasks.

When a processor becomes idle the runtime system has to *select a victim* and calls a function, called WSselect for *work-stealing select*, to browse the topology to find a place from which stealing a task from the place task queue.

The completion of a task may unlock the execution of some of its children in the dependency graph. This means marking them as ready for execution and pushing at least one of them to a place. Once again, there are many ways of selecting the place where to push ready tasks, implemented in strategies we refer to as WSpush, for *work-stealing push*.

Before parallel computation begins, the runtime system can distribute (*push*) the set of ready tasks to multiple places, according to the strategy defined by WSpush_init.

These **three functions** are the main entry points to specify a **scheduling algorithm** in XKAAPI. Sections 3.3 to 3.5 describe strategies for these three points, all them were designed to explore the possibilities of the target NUMA architectures, to be able to evaluate which one are worth taking into account.

3.2 Controlling Data Distribution on a NUMA System

Controlling the way data are allocated on a NUMA system requires a good under-standing of the underlying memory architecture. Application programmers can achieve this using dedicated tools or libraries, like libNUMA's numactl [17], which can be used to set a default memory allocation policy for the whole appli-cation. For example, the --interleave=all memory policy spreads out all the memory pages of dynamically allocated variables, over all the NUMA nodes of the machine. This policy is widely used on NUMA systems in conjunction with dynamic parallelism, like task-based programs, as it distributes the memory traf-fic over all the memory controllers, making processors "*all equally bad*" when it comes to memory access. To better control data placement, parallel application programmers are used to relying on the *first-touch* allocation policy, which is the default behavior for memory allocation on most Linux systems. This allows allocating memory pages when they are accessed for the first time.

To better control data distribution on NUMA systems, we propose two dif-ferent approaches:

– either the application programmer explicitly allocates data on specific NUMA nodes of the machine through a dedicated API we provide [7]

(omp_locality_domain_allocate_XXX) where XXX may be a bloc cyclic data distribution for one or two-dimensional arrays over MAMI [2];
- or the application programmer only marks some regions of code that initialize data to give the runtime system the opportunity to map the corresponding tasks to make the first-touch allocation policy indirectly apply the data distribution we target. Indeed, Olivier et al. [11] have shown that specifying affinity for initialization tasks can lead to huge improvement over locality oblivious techniques. To avoid remote memory accesses, the threads must access the data during the computation phase the exact same way it was accessed during the initialization phase, which is very difficult to guarantee with dynamic task-based parallelism. We extend the OpenMP runtime in two ways. First, by adding functions to provide a dedicated API: omp_set_affinity to make the runtime map the next task to a specific NUMA node. Secondly, by extending scheduling heuristics to take into account task's dependencies to better map ready tasks.

During the application's execution, the runtime relies on system's get_mempolicy to determine on which physical node data are allocated. This information is then used to guide the way we perform task creation and load balancing.

3.3   Distribution of Initial Ready Tasks: WSpush_init Strategies

We refer to *initial tasks* when considering the sources of a task dependency graph, usually declared at the beginning of an OpenMP parallel region. These tasks are basically the first ones to be marked as ready and to be distributed over the platforms' places. We have implemented two initial tasks distribution strategies: cyclicnuma which distributes the tasks in a round-robin fashion over the NUMA nodes, and randnuma which randomly distributes the tasks over the NUMA nodes. Note that unlike numactl, the strategies we implemented consider the whole data appearing in the OpenMP task depend clause instead of working at the page level. In other words, while the two memory pages holding an 8K-wide array would be distributed on different nodes by numactl --interleave=all, they are always assigned to the same NUMA node when using one of our data distribution strategies.

3.4   Distribution of Ready Tasks: WSpush Strategies

This section describes four different ways of pushing ready tasks to a NUMA system places. Two of them are data-oblivious while the other two rely on the dependencies expressed using the depend keyword on OpenMP tasks.

The pLoc strategy makes a processor push ready tasks to its own place, while the pLocNum strategy makes a processor push ready tasks to the place of its NUMA node (*local NUMA node*). The pNumaW strategy pushes tasks on the node-level place corresponding to the NUMA node where most of their output data are allocated to (W stands for Write). The last WSpush strategy, called pNumaWLoc, behaves almost the same than pNumaW except that if the data are allocated to the NUMA node of the processor pushing the task, we directly push the task to this processor's place instead of pushing it to the node-level place (Loc stands for Local).

It's important to note that pLoc and pLocNum does not take initial data placement into account, while pNumaW and pNumaWLoc are both aware of where a task's data are physically allocated and which of them are written, thanks to the OpenMP depend keyword.

3.5 Dynamic Load Balancing Using Work-Stealing: WSselect Strategies

Another important step when implementing work-stealing is the selection of the victim processor we want to steal from. This section describes the selection strategies we implemented, that take the architecture memory hierarchy into account. The first two strategies, sRand and sRandNuma are similar to those studied in [10] and distinguish two levels of hierarchy : the processor level and the NUMA node level. sRand selects a random processor's place while sRandNuma selects a random NUMA node's place. We additionally implemented several strategies mixing both levels of hierarchy, described below.

- sProcNuma: First, we browse the processor's place. Upon failure, we browse the topology in the following order: we first browse one of the neighbor processors; when all the neighbors have been visited, we browse the local NUMA place; we continue by browsing all the processors' places from a random remote node and we eventually consider the place of its NUMA node.
- sNumaProc: This strategy is similar, except we always look at the NUMA place before looking at the processors' place.
- sProc: In this strategy the stealer will visit only the processors' places and its own NUMA place.
- sNuma: In this strategy the stealer will visit only NUMA places and its neighbors.

Like proposed in [11], all these strategies come in two versions: a strict version in which we prevent processors from stealing from other NUMA nodes to improve data locality and a loose version where these restrictions do not apply.

4   Evaluation

We ran all our experiments on the Intel192 machine described in Sect. 2.1. We evaluated our strategies using the KASTORS [15][1] benchmark suite. More specifically, we used the dependent tasks version of the blocked QR

---

[1] git available at https://scm.gforge.inria.fr/anonscm/git/kastors/kastors.git, tag "tag-europar16".

factorization (dgeqrf_taskdep), and of the blocked Cholesky factorization (dpotrf_taskdep). These applications rely on kernels from BLAS and
LAPACK libraries provided by OpenBLAS 2.15. We used the OpenMP GCC
compliant runtime libKOMP [3] based on XKAAPI runtime system. We tagged
the version we used on XKAAPI's git repository[2] in the branch *public/eu-
ropar2016*. For all of the above applications, we used the GCC 5.2.0 compiler.
We also made our execution log files public[3], as well as all the scripts we used,
so that anyone may reproduce our data analysis, and look at the other results
we did not put in the figures.

4.1 Impact of the Data Distribution

We first evaluated the impact the initial data distribution has on the application
performance. We did an evaluation for multiple matrix sizes and block sizes, as well
as multiple combinations of WSpush and WSselect strategies. Figure 2 reports the
results we obtained for the Cholesky application, on 32 K-wide matrices divided
into blocks of 512×512 elements. We observed similar behavior running Cholesky
on different matrix sizes (16 K to 64 K) and block sizes (256 to 1024). The lower
double dashed horizontal line is the GCC performance baseline using sequential
initialization. The middle dashed line is the same experiment using numactl. The
upper solid line is the GCC baseline using parallel initialization.

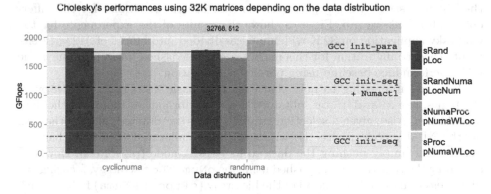

Fig. 2. Evaluating data distributions for multiple strategies (WSselect + WSpush)

Using numactl provides an important performance gain compared to the
sequential initialization. However using a parallel initialization, either controlled
(cyclicnuma, randnuma) or not (GCC init-para), is necessary to significantly
improve the performances, regardless of the strategies used.

The cyclicnuma distribution is the one that works best regardless of the
strategies, and we will use it as the default strategy for the next experiments.

---

[2] https://scm.gforge.inria.fr/anonscm/git/kaapi/xkaapi.git.

[3] https://github.com/viroulep/europar-2016-public.

4.2 Impact of the Stealing Restriction

Given a data distribution, previous works [11] have shown that restricting the task execution to the node where the data are written leads to better data locality which may improve the application performance. However, this is heavily dependent on the algorithm the application implements. For instance, in the case of a Cholesky factorization, many tasks write to the diagonal tiles of the matrix comparatively to other tiles of the matrix. Therefore applying a steal restriction on these tasks will potentially lead to an important number of inactive processors. We evaluated both strict and loose versions of our work-stealing strategies and found out that preventing processors from stealing from other NUMA nodes can lead to a loss of performance by around 25 % to more than 75 % with respect to the same setup without the *strict* restriction. For the sake of brevity we did not include a figure for this, but the results of these experiments are included in the logs publicly published.

4.3 Overview of the Strategies Performances

We took a given data distribution, cyclicnuma, and compared the different strategies, without any steal restriction. The performance obtained running the Cholesky application executed by the libGOMP [4] runtime system (without modification) is considered as a baseline for these experiments. Once again, even if the performances we obtained are obviously not the same, the behavior of the different strategies comparatively to each others are similar for the different applications we ran. Figure 3 shows the results of the experiments for the Cholesky application on 32 K-wide matrices divided into blocks of 512×512 elements (best configuration for this matrix size). The dashed horizontal line is the GCC performance baseline using parallel initialization.

It is first interesting to note that even very basic WSselect and WSpush strategies, like sRand+pLoc, obtained decent performances thanks to the data distribution. Also, given a selection strategy (e.g. sRandNuma), placing the task on the NUMA node where the written data are allocated (pNumaW) behaves better than simply pushing the data to its NUMA node (pLocNum). However, assuming the tasks are being pushed using the pNumaWLoc strategy, focusing the place selection on only one level of the hierarchy (sProc or sNuma) fails to reach the same level of performance we obtained with naive strategies. On the contrary, taking into account both levels of the hierarchy (sProcNuma,sNumaProc) achieve similar performance that outperforms other strategies.

4.4 Strategies Performance Scaling

We eventually selected the three strategy combinations that outperformed the GCC baseline to evaluate their scalability depending on the size of the input matrix. These strategy combinations are:

---

[4] GCC 5.2.0.

Cholesky's performances using 32K matrices depending on the strategy

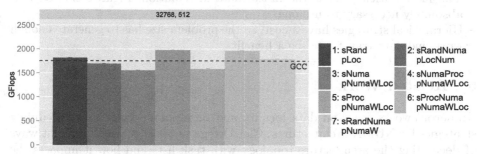

Fig. 3. Evaluating all strategies (WSselect + WSpush), using *cyclicnuma* WSpush_init

Cholesky and QR performances for multiple sizes

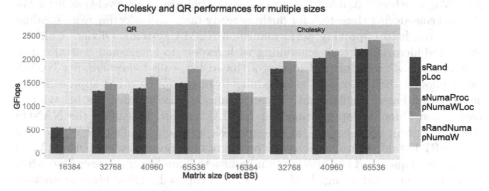

Fig. 4. Evaluating specific strategies on multiple sizes, using *cyclicnuma* WSpush_init

- sRand + pLoc, which is a basic strategy that does not take the architecture topology into account;
- sNumaProc + pNumaWLoc, which was the best strategy in our previous evaluation and is also equivalent to using sProcNuma;
- sRandNuma + pNumaW that performs random selection of node-level places.

Figure 4 reports their performances using a `cyclicnuma` distribution without steal restriction. The figure shows the performances using the best block size for each matrix size (which is, for our setup, 256 for a matrix size of 16384, and 512 for the others). As expected, combinations of strategies taking both the architecture hierarchy and data locality into account (sNumaProc + pNumaWLoc) achieve the best performances. The only exceptions are for small matrix sizes (16384), where there may just be not enough work to be able to take advantage of this strategy. We must note that simply distributing the data over the nodes enables the basic sRand + pLoc combination to achieve satisfying performances.

Finding the appropriate combination is highly application-dependent, therefore it is hard to give a solution for every type of application. However some general guidelines can be followed:

- The most critical part is the initial data distribution. Figure 2 showed it is absolutely necessary to use one.
- Hierarchical strategies have a cost, so the problem size has to generate enough work and data transfer to see a benefit.

5   Related Work

Numerous works focus on data locality and/or topology-aware task scheduling strategies for NUMA architectures. Clet-Ortega et al. [5] studied different ways of decorating the architecture topology with task lists and how it impacts the performance of task-based applications on NUMA systems, promoting private per-threads lists of tasks browsed in a hierarchical way by work-stealing strategies. We somehow extended this work considering also node-level task lists. We showed considering these lists for pushing ready tasks and selecting work-stealing victims can help improving performance on NUMA systems. Olivier et al. [10] evaluated hierarchical task scheduling with respect to traditional centralized or distributed task schedulers. Creating a thread list, called *shepherd*, per NUMA node allowed their hierarchical scheduler to outperforms other approaches on several task-based applications. Tahan et al. [13] also studied the behavior of task-based OpenMP applications on NUMA systems, extending the NANOS runtime system with two NUMA-aware task schedulers called DFWSPT and DFWSRPT, taking into account the notion of task priority when pushing tasks to core-level queues. They also try to minimize the number of *memory hops* when performing load balancing. Drebes et al. [6] proposed similar ideas in another dataflow programming model named OpenStream. This model has a focus on data streaming and has a lot of flexibility on their placement, but does not provide flexibility to the user.

While the same kind of studies have been conducted in other contexts [14, 16,17], none of them takes advantage of the OpenMP depend clause, which precisely indicates which data are read and written by a given task. As advertised by the results obtained by our sNumaProc+pNumaWLoc combined strategy, this information is worth taking into account when choosing a place to push ready tasks to.

6   Conclusion and Future Work

Task-based programming environments like OpenMP have become a standard way to program large-scale NUMA systems. Indeed, they give the programmer ways of expressing massive fine-grain parallelism that can be dynamically mapped to the architecture topology at runtime. OpenMP recently evolved to deal with tasks dependencies describing the data a task reads as input and writes as output.

This paper presented several runtime-level strategies to efficiently assign tasks to processors on any NUMA architecture. We presented strategies assigning ready tasks to lists of tasks, called *places*, attached to processors and NUMA nodes. These strategies define the way a task-based runtime system pushes ready tasks to their initial place and the way idle processors browse the architecture topology to select a place to steal from. We considered several initial data distributions and evaluated different combinations of "push" and "select" strategies on a 192 core NUMA system, on linear algebra applications. We achieved the best performance with strategies taking into account both the architecture topology and the initial data placement obtained through OpenMP tasks dependencies.

A short-term future work will be to extend an OpenMP compiler to be able to identify the initialization tasks in a more OpenMP-friendly manner, like extending the `task` construct with a `init` clause. We also intend to experiment with more OpenMP 4 applications. In a longer term, we intend to move our focus to compile-time techniques able to infer and to attach valuable information on tasks, like an estimation of a task operational intensity, that could guide some of the runtime system's decisions regarding task scheduling and load balancing. We strongly believe a tight cooperation between the compiler and the runtime system is a key step to enhance the performance and scalability of task-based programs on large-scale platforms.

Acknowledgments. This work is integrated and supported by the ELCI project, a French FSN ("Fond pour la Société Numérique") project that associates academic and industrial partners to design and provide software environment for very high performance computing.

References

1. Bleuse, R., Gautier, T., Lima, J.V.F., Mounié, G., Trystram, D.: Scheduling data flow program in XKaapi: a new affinity based algorithm for heterogeneous architectures. In: Silva, F., Dutra, I., Santos Costa, V. (eds.) Euro-Par 2014 Parallel Processing. LNCS, vol. 8632, pp. 560–571. Springer, Heidelberg (2014)
2. Broquedis, F., Furmento, N., Goglin, B., Wacrenier, P.-A., Namyst, R.: Forest-GOMP: an efficient OpenMP environment for NUMA architectures. Int. J. Parallel Program. **38**(5), 418–439 (2010). Special Issue on OpenMP; Guest Editors: Müller, M.S., Ayguade, E
3. Broquedis, F., Gautier, T., Danjean, V.: LIBKOMP, an efficient OpenMP runtime system for both fork-join and data flow paradigms. In: Chapman, B.M., Massaioli, F., Müller, M.S., Rorro, M. (eds.) IWOMP 2012. LNCS, vol. 7312, pp. 102–115. Springer, Heidelberg (2012)
4. Broquedis, F., Clet-Ortega, J., Moreaud, S., Furmento, N., Goglin, B., Mercier, G., Thibault, S., Namyst, R.: hwloc: a generic framework for managing hardware affinities in HPC applications. In: Danelutto, M., Bourgeois, J., Gross, T. (eds.) Proceedings of the 18th Euromicro Conference on Parallel, Distributed and Network-based Processing, PDP 2010, Pisa, Italy, 17–19 February 2010, pp. 180–186. IEEE Computer Society (2010)
5. Clet-Ortega, J., Carribault, P., Pérache, M.: Evaluation of OpenMP task scheduling algorithms for large NUMA architectures. In: Silva, F., Dutra, I., Santos Costa, V. (eds.) Euro-Par 2014 Parallel Processing. LNCS, vol. 8632, pp. 596–607. Springer, Heidelberg (2014)

6. Drebes, A., Heydemann, K., Drach, N., Pop, A., Cohen, A.: Topology-aware and dependence-aware scheduling and memory allocation for task-parallel languages. ACM Trans. Archit. Code Optim. **11**(3), 30:1–30:25 (2014)
7. Durand, M., Broquedis, F., Gautier, T., Raffin, B.: An efficient OpenMP loop scheduler for irregular applications on large-scale NUMA machines. In: Rendell, A.P., Chapman, B.M., Müller, M.S. (eds.) IWOMP 2013. LNCS, vol. 8122, pp. 141–155. Springer, Heidelberg (2013)
8. Frigo, M., Leiserson, C.E., Randall, K.H.: The implementation of the Cilk-5 multithreaded language. SIGPLAN Not. **33**(5), 212–223 (1998)
9. Gautier, T., Besseron, X., Pigeon, L.: Kaapi: a thread scheduling runtime system for data flow computations on cluster of multi-processors. In: PASCO 2007 (2007)
10. Olivier, S., Porterfield, A., Wheeler, K.B., Spiegel, M., Prins, J.F.: Openmp task scheduling strategies for multicore NUMA systems. IJHPCA **26**(2), 110–124 (2012)
11. Olivier, S.L., de Supinski, B.R., Schulz, M., Prins, J.F.: Characterizing and mitigating work time inflation in task parallel programs. In: Proceedings of the International Conference on High Performance Computing, Networking, Storage and Analysis, SC 2012, pp. 65:1–65:12. IEEE Computer Society Press, Los Alamitos (2012)
12. Board, OpenMP Architecture Review: OpenMP application program interface version 4.0, July 2013
13. Tahan, O.: Towards efficient OpenMP strategies for non-uniform architectures. CoRR, abs/1411.7131 (2014)
14. Terboven, C., Schmidl, D., Cramer, T., an Mey, D.: Task-parallel programming on NUMA architectures. In: Kaklamanis, C., Papatheodorou, T., Spirakis, P.G. (eds.) Euro-Par 2012. LNCS, vol. 7484, pp. 638–649. Springer, Heidelberg (2012)
15. Virouleau, P., Brunet, P., Broquedis, F., Furmento, N., Thibault, S., Aumage, O., Gautier, T.: Evaluation of OpenMP dependent tasks with the KASTORS benchmark suite. In: DeRose, L., de Supinski, B.R., Olivier, S.L., Chapman, B.M., Müller, M.S. (eds.) IWOMP 2014. LNCS, vol. 8766, pp. 16–29. Springer, Heidelberg (2014)
16. Weng, T.-H., Chapman, B.M.: Implementing OpenMP using dataflow execution model for data locality and efficient parallel execution. In: Proceedings of the 16th International Parallel and Distributed Processing Symposium, IPDPS 2002, p. 180. IEEE Computer Society (2002)
17. Wittmann, M., Hager, G.: Optimizing ccNUMA locality for task-parallel execution under openmp and TBB on multicore-based systems. CoRR, abs/1101.0093 (2011)

Multicore vs Manycore: The Energy Cost of Concurrency

Martin Groen and Vincent Gramoli[✉]

University of Sydney, Sydney, Australia
mgro7657@uni.sydney.edu.au, vincent.gramoli@sydney.edu.au

Abstract. In this paper, we study the relation between performance and energy in concurrent programs. As energy efficiency became a key challenge of the computing industry, it is crucial to seek solutions that achieve high performance at a reasonable carbon footprint. We show, however, that energy is dramatically impacted by concurrency and it remains difficult to predict the energy consumed even when the application and the thermal design power are given, due to the number of threads running or their level of contention.

To this end, we evaluated concurrent algorithms on a 2.1 GHz multicore and a 1.2 GHz manycore platforms. Our results show that even though the throughput on manycore is lower than the throughput on multicore, we could not find a single concurrent algorithm where the multicore offers consistently a higher performance per watt than the manycore. More importantly, we identified some benchmarks on which the manycore offers up to 4.3× more operations per second per watt than the multicore.

Keywords: Power consumption · Energy · Manycore · Concurrency

1 Introduction

As the complex processing cores require a superlinear growing thermal design power (TDP) to achieve linear performance improvement, increasing concurrency requires to slow down the cores [1]. The limitation stems from the sub-40 nm size of transistors where the Dennard scaling stops applying [2]: the power consumption of a processor is no longer proportional to its area because of current leakage. To lessen the leakage, Intel designed 22 nm Tri-gate transistors, however, this problem will soon limit concurrency by requiring powering off some of the complex cores to let others compute at full speed, a phenomenon known as *dark silicon* [3].

Adopting *manycores*, the concept of placing more simpler cores per chip [4], is thus necessary to keep scaling concurrency within the same power envelope. To understand whether it is worth trying scaling concurrency one has to first answer the question: Does the energy needed to reach some performance on multicore m exceed the energy needed to reach the same performance on manycore M? This

© Springer International Publishing Switzerland 2016
P.-F. Dutot and D. Trystram (Eds.): Euro-Par 2016, LNCS 9833, pp. 545–557, 2016.
DOI: 10.1007/978-3-319-43659-3_40

question is not simple. On the one hand, m is generally known to have higher energy consumption per unit of time than M [5], but on the other hand m is also known to run at higher clock frequencies hence executing more instructions than M over time. Manycores have proved instrumental to run concurrent applications at a high performance per watt—examples include key-value stores [6,7]. These applications have, however, been genuinely re-engineered for the considered manycore platform. Hence, it is hard to compare the original multicore implementations to the resulting manycore implementations.

In this paper, we evaluate the performance per watt under concurrency. In particular, we compare the number of benchmark operations executed per second and per watt on a traditional 32-way Intel Xeon *multicore* platform of 32 nm complex cores running at 2.1 GHz and on a less conventional 36-way Tilera Tile-Gx *manycore* platform of 40 nm simpler cores running at 1.2 GHz.

We ported Synchrobench on Tilera to compare the performance obtained on both platforms. Synchrobench is a benchmark suite executing muti-threaded insert/delete/lookup operations to stress-test concurrent data structures using various synchronization techniques [8]. The C/C++ Synchrobench benchmark-suite [8] was originally designed for x86-64 multicores while the Tilera platform provides a manycore architecture with a reduced instruction set and runs a port of the version 3.10 of the Linux kernel and GCC v4.4.6. Unlike previous manycore applications whose multi-threading was carefully re-factorized to run efficiently on Tilera [7], we only ported the Synchrobench benchmark suite with minimal modifications.

One may think that benchmarking performance is sufficient as the energy can be determined with the Thermal Design Power (TDP) provided by its manufacturers. For example, our Intel multicore consumes more power (95 W TDP for 16 hyperthreaded complex cores) than our Tilera manycore (28 W TDP for 36 simple cores). This comparison is not always easy: in particular, multicore manufacturers offer different definitions of TDP [9] and may even provide a *Configurable TDP* (cTDP) that adapts the performance and the energy consumptions at runtime (AMD offers the Turbo Core technology while Intel offers Turbo Boost).

Moreover, as we will show in this paper, the power consumption of a machine is dramatically affected by the concurrent algorithm. The power consumption depends on the number of cores that run and at which clock frequency but it also depends on whether some simultaneous multithreading technology (like hyperthreading) is enabled on these cores. To accurately report these power measurements, we plugged a hardware power meter on our existing multicore and manycore platforms.

As expected, at similar thread count, all applications run significantly faster on the multicore platform than on the manycore platform. Yet, when looking at the performance per watt attained by both machines, our results are surprising: there is no benchmark where the multicore machine achieves consistently higher performance per watt than the manycore. We also observed that there exist benchmarks where the manycore offers significantly higher performance per watt than the multicore. This is interesting as it shows, for the first time, that the power consumption of state-of-the-art algorithms can compensate the performance advantage of multicores. In other words, even though the highest performance is obtained while running concurrent algorithms on the multicore platform, running them on the manycore platform provides higher performance within the same power envelope.

In Sect. 2, we present the problem of measuring the performance per watt of concurrent applications on multicore and manycore architectures. In Sect. 3, we present our manycore and multicore experimental settings. In Sect. 4, we present the performance and energy consumption of our platforms. In Sect. 5, we relate the energy consumed and the synchronization technique used. In Sect. 6, we discuss the related work and in Sect. 7 we conclude the paper.

2   How to Measure Energy Under Concurrency

To evaluate the performance and energy consumption of the manycore platform, we choose the multicore platform as the baseline.

Figure 1 reports the performance and energy consumed by the 32-way multi-core platform as observed directly on the power socket when running the lock-free linked list Synchrobench benchmark (Algorithm 21 [8]) with 64 K elements and 10 % attempted update, namely the portion of invoked updates (even the ones that return unsuccessfully without writing as described in [8]). The dotted line indicates the throughput T given by Synchrobench when running the benchmark for one minute at different thread count. The bar chart indicates the power consumed in watts E during the experiment as the average over all values read every second on a dedicated power meter (the detailed settings are presented in Sect. 3). The solid line indicates the *performance per watt* $P = \frac{T*1000}{E} ops\,per\,sec/W$ as the number of operations per second divided by the watts. The value reported at thread count 0 corresponds to the machine idle, i.e., not running any experiment.

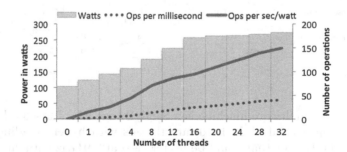

Fig. 1. Power and throughput depending on the level of concurrency

First, we can observe that the power consumption increases with the level of concurrency. The power consumed keeps increasing with the number of threads even when the number of threads exceed the number of cores (16). We can see however that the power increases faster below 16 threads than above 16 threads. This is due to activation of one new core with each new running thread up to 16: we noted a scattered thread pinning strategy, hyperthreading kicking in after 16 threads. Second, the performance increases as the number of hardware threads used increases, confirming the performance scalability of this particular benchmark on multicore as already noted [8]. Finally, we observe that the performance per watt increases also steadily up to the highest hardware thread count, indicating that the multicore machine delivers an energy proportional computation [10] on this particular benchmark. This is not always the case as the performance of several algorithms does not necessarily increase to the highest hardware thread count, as explained in Sect. 4.

As in Fig. 1, we carefully observed that the highest performance per watt for a given workload on both the mutlicore and the manycore platforms was always obtained at the thread count where the performance was the highest. In other words, the energetic overhead is never higher than the performance drop. Hence, in the remainder of the paper and when not explicitly mentioned, we report the performance per watt observed at the thread count that maximizes performance.

3   Energy and Concurrency Settings

In this section, we present the multicore and manycore platforms, the power monitoring tools and the algorithms used. The multicore and the manycore platforms are both 64-bit platforms made available for purchase in 2012. The multicore machine is a 32 nm Xeon platform based on Intel's x86 architecture with a complex instruction set whereas the manycore is a TILExtreme platform with 40 nm manycore Tile-Gx processors based on the Tilera architecture with reduced instruction set. These two platforms use a 3-level cache.

Table 1. The specification summary of our manycore (M) and multicore (m) platforms

Platform	Description	CPU	#CPU	#cores per CPU	#hw threads	Clock frequency	CPU TDP	Power (idle)	Power (loaded)
M	TILExtreme	Tile-Gx	4	36	144	1.2 GHz	28 W	240 W	294 W
m	SandyBridge	Intel Xeon	2	8	32	2.1 GHz	95 W	103 W	277 W

Multicore. The multicore machine is a SandyBridge-EN of 28-core Intel Xeon E5-2450 offering a total of 32 hardware threads with hyperthreading enabled running at 2.1 GHz but that could be overclocked at 2.9 GHz with enabled TurboBoost [11]. Intel offers Turbo Boost 2.0 so that processors may "operate at a power level that is higher than its TDP configuration".[1] Note that this approach is shared by other multicore manufacturers: AMD proposes the Turbo

---

[1] http://www.intel.com/content/www/us/en/architecture-and-technology/turbo-boost/turbo-boost-technology.html.

Core technology to increase similarly the core frequency within the thermal and power limits of the accelerated processing unit.[2] Each processor has a TDP of 95 W.[3] It features transistors of size 32 nm.

Manycore. The manycore machine is a TILExtreme, a four 36-core Tile-Gx processors running at 1.2 GHz. It features 16 fans that run at a speed of 3000 to 16000 rpm that cannot be disabled or tuned individually [12]. The details are summarized in Table 1. There is no coherence across two different TileGx sockets. The 36 cores of each Tile-Gx are organized into a 6 × 6 mesh of tiles where each cache line has a dedicated "home" core. Upon level-2 local cache miss, a core request the cache line from the home core local level-2 cache so that the union of all home core level-2 caches represents an 9 MiB level-3 cache. The cache coherence is maintained through a distributed directory that is more energy efficient than a bus-snooping cache coherency protocol.

3.1 Preliminary Power Measurements

We measured the performance of our platforms using a power metering tool.

Watt Metering. We used the Watts Up? .NET watt meter 100–250 V, 50/60 Hz and 15 amps to perform our power measurements. This device has an accuracy of ±1.5 % when reporting consumption above 60 W like in our case. Note that the same device was previously used to report power consumption in other studies [6]. All power measurements were collected for both the muticore and manycore machines in the same room with a steady temperature of 20.8°C cooled using an independent air conditioning system whose power consumption was not accounted in our measurements.

Power Consumption Under Full Load. The power consumption at full load was measured with the Synchrobench lock-free skip list running with parameters u10-i65536-r132K-d60000[4] with the number of threads set to the maximum number of hardware threads available. Because the fans cannot be disabled and tuned individually on the Tilera [12], we run the full load on the four sockets of the Tilera (144 cores) and divided the energy consumption by four to get an estimate of the energy consumed per socket. It is important to remark that a single socket machine could consume more than a fourth of this overall power due to the consumption of components shared by the four sockets. To confirm that the shared consumption was not impacting our results, we measured the power consumption of the machine with all the sockets shut-down in hardware and observed 87 W. We then confirmed that manycore would still reach higher

---

[2] http://www.amd.com/en-us/innovations/software-technologies/turbo-core.

[3] http://ark.intel.com/products/64611/Intel-Xeon-Processor-E5-2450-20M-Cache-2_10-GHz-8_00-GTs-Intel-QPI.

[4] 10 % update, initial size of 65536, value range of 132 K and a duration of 60 s [8].

performance per watt than multicore even if each fan consumed less than 0.8 W on this heavy workload. We selected the multicore platform as the baseline for our experiments as it is the most common platform of the two. We noticed that the power consumption of this platform when idle was 103 W, which is close to the 95 W TDP announced by the manufacturer.

Table 2. Port of Synchrobench-C/C++ to the Tilera manycore

	Data Structure	Ref.	Synchronization
skip lists	Optimistic skip list	[13]	`spin-lock`, `mutex`
	Rotating skip list	[14]	`lockfree CAS`
	Elastic skip list	[15]	`ESTM`
	Fraser skip list	[16]	`lockfree CAS`
	No hot spot skip list	[17]	`lockfree CAS`
	Sequential skip list	[8]	$\emptyset$
linked lists	Lazy linked list	[18]	`spin-lock`, `mutex`
	Harris's linked list	[19]	`lockfree CAS`
	Reusable linked list	[20]	`ESTM`
	Lock-coupling linked list	[21]	`spin-lock`, `mutex`
	Sequential linked list	[8]	$\emptyset$
hash tables	Lock-free hash table	[22]	`lockfree CAS`
	Elastic hash table	[15]	`ESTM`
	Sequential hash table	[8]	$\emptyset$
binary trees	Speculation-friendly tree	[23]	`ESTM`
	Transactional red-black tree	[24]	`ESTM`
	Sequential binary search tree	[8]	$\emptyset$

3.2 Porting Synchrobench-C/C++ to Manycore

To understand whether the concurrent programs and the synchronization techniques impact energy efficiency, we run the Synchrobench [8] benchmark suite on both the multicore and the manycore machine. Synchrobench is a benchmark suite designed to evaluate the performance of synchronization techniques like compare-and-swap (CAS), spin-lock, mutex and transactional memory (TM), and data structure implementations on multicore machines.

To evaluate the performance on manycore, we ported 17 benchmarks out of the 19 C/C++ benchmarks of Synchrobench-v1.1.0-alpha to the Tilera architecture. We also ported the TM library implementing elastic transactions, ESTM [25]. We restricted our study to C/C++ because the only other available version of Synchrobench is in Java[5] and we know that the experimental

---
[5] https://sites.google.com/site/synchrobench/download.

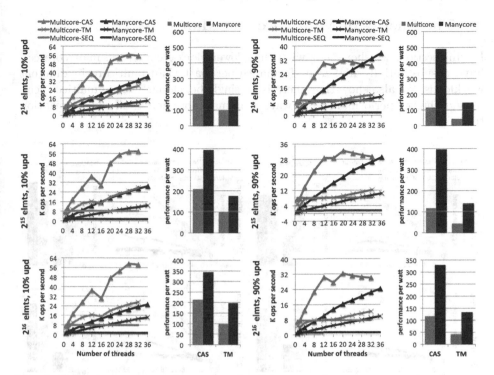

Fig. 2. Operations per second per watt of multicores and manycores running the concurrent hash tables benchmark

measurements are more predictable than in Java especially when running different JVMs [8].[6]

The oldest benchmarks of Synchrobench used the `atomic_ops` library from HP[7], however, this library supports only IA32 and x86-64 and was adapted for SPARC but not for Tilera—we had to manually port some of its operations to the Tilera architecture, as listed in Table 2. Some other benchmarks rely on the recent C/C++11 atomic intrinsics, however, as of today neither `stdatomic` nor the latest versions of GCC are supported on Tilera.[8] We decided not to port the remaining benchmarks because of the changes they would induce: some of the benchmarks of Synchrobench were only designed to run on 64-bit Intel and featured a 128-bit wide compare-and-swap that does not exist on Tilera [26] and adapting them would have affected the veracity of our comparisons on different architectures.

---

[6] The TILExtreme runs a port of Java 1.6.
[7] https://github.com/ivmai/libatomic_ops/.
[8] http://www.tilera.com/scm/.

4   The Energy of Multicore and Manycore

In this section, we show that with their lower clock frequencies, manycore can have a substantially higher performance per watt than multicore in different workloads. We also show that there is no single benchmark where multicore can provide higher performance per watt than manycore across all synchronization techniques.

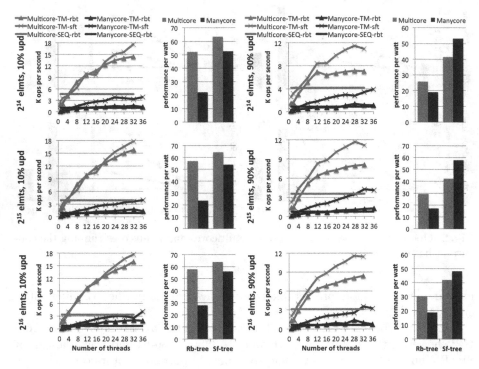

Fig. 3. Operations per second and per watt of multicore and manycore running the concurrent binary search trees benchmark

Raw Performance. Figure 2 (resp. Fig. 3) represents the performance achieved and the energy dissipated when running the hash table (resp. binary search tree) benchmarks of Synchrobench-C/C++ on our multicore and manycore platforms. In these figures, each binary tree or hash table implementation is synchronized with compare-and-swap (denoted by CAS), transactional memory (denoted by TM) or nothing being only able to run sequentially (denoted by SEQ). The different binary search trees algorithms are either of type red-black tree (denoted rbt) or of type speculation-friendly tree [27] (denoted sft). Both figures indicate that concurrent algorithms perform generally better on the multicore than on the manycore. This is expected given the lower clock frequency of the manycore machine (1.2 GHz) compared to the multicore machine (2.1 GHz). However, we can also see that the performance of the manycore can be higher than the one of the multicore in some cases (cf. top-right of Fig. 2). This is due to the contention that induces cross-socket communication on the multicore and that does simply require low-latency network-on-chip communication on the manycore.

Higher Performance per Watt for the Manycore. The hash table benchmark (Fig. 2) clearly shows higher performance per watt on the manycore than on the multicore (across different sizes and update ratios). In particular, on 90 % updates and with 2^{12} elements, the hash table benchmark runs 4.3× more operations per watt than the multicore at maximum thread counts. Note that the speedup is 3.9× when the thread count is 32 on both machines. The reason is probably due to the low contention of hash table and the fact that the manycore platform have high speed core-to-core communication compared to the multicore machine. In addition, the time needed for a core to access the memory or the level-1 cache on the manycore are faster than on the multicore. For example accessing the level-1 cache of the Tilera takes 1.7 µs (2 cycles at 1.2 GHz) while it takes 2.4 µs (5 cycles at 2.1 GHz) on the Xeon. For other data structures, whether the multicore or the manycore is more suitable depends on many parameters, like the synchronization technique used to synchronize the data structure, the level of contention and the size of the data structures. We discuss the impact of the synchronization technique used in Sect. 5.

5   The Energy of Synchronization Techniques

To get a broader view of the performance per watt delivered by the manycore and the multicore, we ran the other Synchrobench benchmarks.

Figure 4 depicts the performance per watt obtained on the multicore and the manycore for the Harris linked list that uses CAS for synchronization. We used this benchmark as an example to illustrate that both manycore and multicore can achieve better performance per Watt results at different thread counts. For clarity and given that we had only 32 hardware contexts on the multicore, we did not represent the performance obtained on the manycore at 36 threads. The other Synchrobench parameters used for this benchmark are -u10-i16384-r32768, indicating an initial size of 2^{14} elements and an attempted update ratio of 10 %.

First, we can observe that the performance per watt delivered by the manycore does not scale up to 32 threads (triangle-dotted line) while the one delivered by the multicore scales with the level of concurrency (square-dotted line). In addition, the peak performance per watt delivered by the manycore is comparatively higher than the peak performance per watt delivered by the multicore. This indicates that the multicore presents some advantage in terms of performance per watt for this particular benchmark. Finally, we observe that the performance per watt delivered by the multicore is, however, not consistently higher than the one delivered by the manycore. In particular, between 1 and 24 threads, the performance per watt obtained from the multicore is higher than the one obtained from the multicore. Although not depicted here, we ran additional experiments and identified some skip list benchmarks with similar differences: the peak performance per watt is higher on manycore whereas, at some thread counts, the manycore delivers higher performance per watt.

We conclude that the multicore does not consistently provide a higher performance per watt than the manycore on a given data structures. This is in contrast with the manycore offering consistently higher performance per watt than the multicore on all the binary search trees evaluated, whether they were synchronized with CAS, TM or simple running sequentially. We also observed, however, that this is not necessarily true when considering a data structure benchmark synchronized with a particular technique. Hence, we observed that the multicore would deliver a higher performance per watt than the manycore on the skip list synchronized with CAS or TM and on the linked list synchronized with TM, but not on the skip list synchronized with locks, the linked list synchronized with CAS and the linked list synchronized with locks.

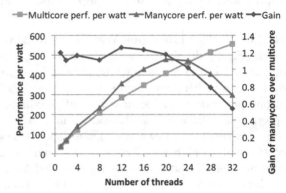

Fig. 4. Performance per watt improvement of the manycore over the multicore

6   Related Work

A study on the impact of concurrency on power consumption [28] shows that running two cores instead of one could, on some workloads, double the power overhead and that simultaneous multi-threading could save energy on recent hardware and in-order processors. The focus of this study is on managed languages showing, for example, how seeminlgy singly-threaded Java applications actually exploit multiple cores through the JVM.

Some research work focuses on algorithms to model theoretically their carbon footprint [29,30]. The first study [30] shows that for matrix multiplication and the n-body problem, the energy consumption remains constant as the number of processors increases and the runtime decreases. The second study raised the question of the relevance of designing algorithms under the constraint of energy efficiency [29]. It does not present experimental measurements but rather exploits energy models applied to graphical processing units. These studies do not model the energy consumed by non-deterministic executions.

A recent work simulated the impact of the MSI cache coherence protocol on the energy consumed by data structure algorithms that experience non-deterministic executions [31]. The authors propose new lease and release instructions to minimize cache invalidation in lock-based and lock-free structures. Simulations of their instructions on the Graphite multi-processor simulator indicate a substantial reduction of the energy consumption, compared to the classic MSI cache coherence protocol without lease/release.

The energy consumption of both simple and complex cores was modelled in the context of distributed heterogeneous platforms [32]. To validate their results, the authors measure the consumption of high performance computing applications on clusters of Intel Xeon and ARM Cortex-A9 nodes. FAWN [6] is an in-memory key-value store well-tuned for running on 21 single-threaded winpy nodes using flash storage to retrieve data that cannot fit in memory. FAWN achieves a peak 350 key-value queries per Joule. With 21 nodes, FAWN achieves 350 key-value queries per Joules. As the goal of our study was to compare concurrent programs running on multicore and manycore platforms released the same year, we minimized the changes of our benchmarks while porting them to manycores.

7   Conclusion

We measured the performance and energy consumption of a multicore and a manycore when running concurrent algorithms. As expected, these algorithms run faster on the multicore but can achieve better performance per watt on the manycore. There are several directions for future work. First, it would be interesting to isolate the power consumption of each individual components, like fans and CPUs, by separating them physically or by using dedicated software toolsets. Second, it would be interesting to broaden the scope of benchmarks to see whether the same results hold for IO-bound applications.

Acknowledgments. This research was supported under Australian Research Council's Discovery Projects funding scheme (project number 160104801) entitled "Data Structures for Multi-Core". Vincent Gramoli is the recipient of the Australian Research Council Discovery International Award.

References

1. Esmaeilzadeh, H., Blem, E., St. Amant, R., Sankaralingam, K., Burger, D.: Power challenges may end the multicore era. Commun. ACM **56**(2), 93–102 (2013)
2. Dennard, R.H., Rideout, V., Bassous, E., Leblanc, A.: Design of ion-implanted mosfet's with very small physical dimensions. IEEE J. Solid-State Circ. **9**(5), 256–268 (1974)
3. Esmaeilzadeh, H., Blem, E. St. Amant, R., Sankaralingam, K., Burger, D.: Dark silicon and the end of multicore scaling. In: ISCA, pp. 365–376, June 2011
4. Borkar, S., Chien, A.A.: The future of microprocessors. Commun. ACM **54**(5), 67–77 (2011)
5. Borkar, S.: Thousand core chips: a technology perspective. In: DAC, pp. 746–749 (2007)

6. Andersen, D.G., Franklin, J., Kaminsky, M., Phanishayee, A., Tan, L., Vasudevan, V.: Fawn: A fast array of wimpy nodes. In: SOSP, pp. 1–14 (2009)
7. Berezecki, M., Frachtenberg, E., Paleczny, M., Steele, K.: Many-core key-value store. In: IGCC, pp. 1–8, July 2011
8. Gramoli, V.: More than you ever wanted to know about synchronization: synchrobench, measuring the impact of the synchronization on concurrent algorithms. In: PPoPP, pp. 1–10 (2015)
9. Intel: Measuring power processor: TDP vs. ACP Intel White paper (2011)
10. Barroso, L.A., Holzle, U.: The case for energy-proportional computing. Computer **40**(12), 33–37 (2007)
11. Rotem, E., Naveh, A., Rajwan, D., Ananthakrishnan, A., Weissmann, E.: Power management architecture of the 2nd generation intel core microarchitecture, formerly codenamed sandy bridge. In: HotChips (2011)
12. Tilera: UG410 - TILExtreme-Gx Platform User's Guide Release 1.1 Doc. N. UG410, May 2013
13. Herlihy, M.P., Lev, Y., Luchangco, V., Shavit, N.N.: A simple optimistic skiplist algorithm. In: Prencipe, G., Zaks, S. (eds.) SIROCCO 2007. LNCS, vol. 4474, pp. 124–138. Springer, Heidelberg (2007)
14. Dick, I., Fekete, A., Gramoli, V.: A skip list for multicore. Concurrency and Computation, Practice and Experience (2016)
15. Felber, P., Gramoli, V., Guerraoui, R.: Elastic transactions. In: Keidar, I. (ed.) DISC 2009. LNCS, vol. 5805, pp. 93–107. Springer, Heidelberg (2009)
16. Fraser, K.: Practical lock-freedom. Ph.D. Thesis, University of Cambridge (2004)
17. Crain, T., Gramoli, V., Raynal, M.: No hot spot non-blocking skip list. In: ICDCS, pp. 196–205 (2013)
18. Hellor, S., Herlihy, M., Luchangco, V., Moir, M., Scherer, W.N., Shavit, N.: A lazy concurrent list-based set algorithm. Parallel Process. Lett. **17**(4), 411–424 (2007)
19. Harris, T.L.: A pragmatic implementation of non-blocking linked-lists. In: Welch, J.L. (ed.) DISC 2001. LNCS, vol. 2180, pp. 300–314. Springer, Heidelberg (2001)
20. Gramoli, V., Guerraoui, R.: Reusable concurrent data types. In: Jones, R. (ed.) ECOOP 2014. LNCS, vol. 8586, pp. 182–206. Springer, Heidelberg (2014)
21. Herlihy, M., Shavit, N.: The Art of Multiprocessor Programming. Morgan Kaufmann Publishers Inc., San Francisco (2008)
22. Michael, M.M.: High performance dynamic lock-free hash tables and list-based sets. In: SPAA, pp. 73–82. ACM, New York (2002)
23. Crain, T., Gramoli, V., Raynal, M.: A contention-friendly binary search tree. In: Wolf, F., Mohr, B., an Mey, D. (eds.) Euro-Par 2013. LNCS, vol. 8097, pp. 229–240. Springer, Heidelberg (2013)
24. Minh, C.C., Chung, J., Kozyrakis, C., Olukotun, K.: Stamp: Stanford transactional applications for multi-processing. In: IISWC, pp. 35–46. IEEE (2008)
25. Felber, P., Gramoli, V., Guerraoui, R.: Elastic transactions. In: Keidar, I. (ed.) DISC 2009. LNCS, vol. 5805, pp. 93–107. Springer, Heidelberg (2009)
26. Intel: Intel 64 and IA-32 architectures software developers manual - vol. 2A: Instruction set reference, A–M (2007)
27. Crain, T., Gramoli, V., Raynal, M.: A speculation-friendly binary search tree. In: PPoPP, pp. 161–170 (2012)
28. Esmaeilzadeh, H., Cao, T., Yang, X., Blackburn, S.M., McKinley, K.S.: Looking back and looking forward: Power, performance, and upheaval. Commun. ACM **55**(7), 105–114 (2012)
29. Choi, J.W., Vuduc, R.W.: How much (execution) time and energy does my algorithm cost? XRDS **19**(3), 49–51 (2013)

30. Demmel, J., Gearhart, A., Lipshitz, B., Schwartz, O.: Perfect strong scaling using no additional energy. In: IPDPS, pp. 649–660 (2013)
31. Haider, S.K., Hasenplaugh, W., Alistarh, D.: Lease/release: architectural support for scaling contended data structures. In: PPoPP (2016)
32. Ramapantulu, L., Loghin, D., Teo, Y.M.: An approach for energy efficient execution of hybrid parallel programs. In: IPDPS, pp. 1000–1009, May 2015

Theory and Algorithms for Parallel Computation and Networking

Work-Efficient Parallel Union-Find with Applications to Incremental Graph Connectivity

Natcha Simsiri[1], Kanat Tangwongsan[2],
Srikanta Tirthapura[3(✉)], and Kun-Lung Wu[4]

[1] College of Information and Computer Sciences,
University of Massachusetts, Amherst, USA
`nsimsiri@umass.edu`
[2] Mahidol University International College,
Phutthamonthon District, Thailand
`kanat.tan@mahidol.edu`
[3] Department of Electrical and Computer Engineering,
Iowa State University, Ames, USA
`snt@iastate.edu`
[4] IBM T.J. Watson Research Center, Yorktown Heights, USA
`klwu@us.ibm.com`

Abstract. On an undirected graph, how can one quickly answer whether two vertices are connected while allowing more edges to be added incrementally? This is the well-studied incremental graph connectivity (IGC) problem, a fundamental problem that can be efficiently solved using solutions to the classical union-find problem. Motivated by the need to handle larger and rapidly-changing graphs, this paper presents the first shared-memory parallel algorithm for IGC and equivalently, Union-Find that is provably work-efficient (i.e., does no more work than the sequential optimal) and has polylogarithmic parallel depth. It performs path compression in parallel without a lock or speculative execution. We also present a simpler algorithm with slightly worse theoretical properties, but which is easier to implement, and has good practical performance.

1 Introduction

The classical *Union-Find* problem is to maintain a collection of disjoint sets, supporting (1) union(u, v): given elements u and v, combine the sets containing u and v into a single set and return (a handle to) the combined set; and (2) find(v): given an element v, return (a handle to) the set containing v.

The problem has many applications, including incremental graph connectivity on undirected graphs. The graph connectivity question asks, *given two vertices, is there a path between them?* The problem of *incremental graph connectivity* (IGC) requires a (quick) answer to the connectivity question as edges are incrementally added. Union-Find is a well-known solution to this problem: To answer whether u and v are connected, check if find(u) and find(v) are equal. To add an edge (w, x), invoke union(w, x) on the union-find data structure.

© Springer International Publishing Switzerland 2016
P.-F. Dutot and D. Trystram (Eds.): Euro-Par 2016, LNCS 9833, pp. 561–573, 2016.
DOI: 10.1007/978-3-319-43659-3_41

With a sharp increase in the amount of linked data, IGC now has to be solved at a much larger scale than before. On the one hand, modern streaming systems (e.g., IBM Streams [1] and Spark Streaming [2]) provide a software platform for using parallelism to achieve high-throughput processing. On the other hand, scalable, parallel, and dynamic algorithms are needed to effectively utilize the platform. This paper makes a step forward in designing parallel algorithms for IGC: the input is a graph stream consisting of a sequence of edges, and the queries are connectivity queries. Our main ingredient is a novel parallel algorithm for Union-Find.

To enable parallelism in stream processing, many systems (e.g., Apache Spark Streaming [2]) use a "discretized streams" input model: a stream is divided into a sequence of *minibatches*, each processed using a parallel algorithm. We adopt this model and seek parallel methods to efficiently process a minibatch of edges. On a vertex set V, a graph stream $\mathcal{A}$ is a sequence of minibatches $A_1, A_2, \ldots$, where each minibatch A_i is a set of edges on V. The graph at the end of observing A_t, denoted by G_t, is $G_t = (V, \cup_{i=1}^{t} A_i)$ containing all the edges up to t. The minibatches A_is may have different sizes. Equivalently, each edge (u, v) can be viewed as a single $\mathtt{union}(u, v)$ operation that merges the sets containing u and v, and a minibatch A_i is a set of $\mathtt{union}$'s that have to be applied in parallel.

We study a *bulk-parallel incremental connectivity problem*, which is to maintain a data structure that provides two operations: $\mathtt{Bulk\text{-}Update}$ and $\mathtt{Bulk\text{-}Query}$. The $\mathtt{Bulk\text{-}Update}$ operation takes as input a minibatch of edges A_i and adds them to the graph—this involves multiple $\mathtt{union}$ operations, one per edge. The $\mathtt{Bulk\text{-}Query}$ operation takes a minibatch of vertex-pair queries and returns for each query, whether the two vertices are connected on the edges observed so far in the stream—this involves multiple $\mathtt{find}$ operations, two per vertex-pair query. On this data structure, the $\mathtt{Bulk\text{-}Query}$ and $\mathtt{Bulk\text{-}Update}$ operations are each invoked with a (potentially large) minibatch of input, each processed using a parallel computation. But a bulk operation, say a $\mathtt{Bulk\text{-}Update}$, must complete before the next operation, say a $\mathtt{Bulk\text{-}Query}$, can begin. By disallowing mixing unions and finds in the same batch, the model provides clean semantics for what constitutes the graph being queried.

1.1   Contributions

We present the first shared-memory parallel algorithm for Union-Find (and hence, for IGC) that is both provably work-efficient and has polylogarithmic parallel depth. We analyze our algorithms in terms of work and depth assuming an underlying concurrent-read, concurrent-write (CRCW) machine[1]. Specifically:

- *Simple Parallel Algorithm.* We first present a simple parallel algorithm for Union-Find (Sect. 2). This is easy to implement and has good theoretical properties. On a graph with n vertices, it makes a single pass through the

---

[1] We use parallel integer sort internally, which assumes CRCW.

stream using $O(n)$ memory, and can process a minibatch of b edges (equivalently, b **union** operations), using $O(b\log n)$ work and $O(\text{polylog}(n))$ parallel depth.

- *Work-Efficient Parallel Algorithm.* We then present an improved parallel algorithm for Union-Find (Sect. 3), with total work $O((m+q)\alpha(m+q,n))$, where m is the total number of **union** operations across all minibatches and q is the total number of **find** operations across all minibatches, and α is an inverse Ackermann's function (see Sect. 1.2). Equivalently, this is a parallel algorithm for IGC with total work $O((m+q)\alpha(m+q,n))$, where m is the total number of edges across all minibatches, q is the total number of connectivity queries across all minibatches. This matches the work of the theoretically time-optimal sequential counterpart (i.e., it is *work-efficient*). Further, processing a minibatch takes $O(\text{polylog}(m,n))$. Hence, the sequential bottleneck in the runtime of the parallel algorithm is very small, and the algorithm is capable of using almost a linear number of processors efficiently. We are not aware of a prior parallel algorithm for Union-Find with such provable properties on work and depth.

We have implemented a variant of the simple algorithm on a 20-core shared memory machine. Our preliminary experimental results suggest that the algorithm achieves good speedups in practice and is able to efficiently use available parallelism. On a 20-core machine, it can process hundreds of millions of edges per second, and realize a speedup of 8–11x over its single threaded performance. Due to space constraints, we are unable to present experimental results in this paper. Experimental results and detailed proofs of our theoretical results can be found in the expanded version of this paper [3].

1.2 Related Work

Let n denote the number of vertices, m the number of operations, and α an inverse Ackermann's function, a very slow-growing function, practically a constant independent of its input. The most efficient sequential data structure for Union-Find (and hence, IGC) is the well-known solution based on path compression, for example, see [4]. The analysis due to Tarjan [5] shows an $O(\alpha(m,n))$ amortized time per **find** and an $O(1)$ time per union, which has been proved to be optimal (see Seidel and Sharir [6] for an alternate analysis). However, current solutions are often unable to take advantage of parallelism and hence are unable to process high-throughput dynamic graphs.

Recent work on streaming graph algorithms has focused on minimizing the memory requirement, with little attention given to the use of parallelism. This line of work has largely focused on the "semi-streaming model" [7], which allows $O(n \cdot \text{polylog}(n))$ space. In the semi-streaming model, the union-find data structure [5] solves IGC in $O(n)$ space and a total time nearly linear in m.

When only $o(n)$ of work space (sublinear) is allowed, interesting tradeoffs are known for multi-pass algorithms. For an allotment of $O(s)$ workspace, an algorithm needs $\Omega(n/s)$ passes [7] to compute the connected components of a graph. Demetrescu et al. [8] consider the W-stream model, which allows the processing of streams in multiple passes in a pipelined manner: the output of the i-th pass is given as input to the $(i+1)$-th pass. They show a tradeoff between the number of passes and the memory required. With s bits of space, their algorithm computes connected components in $O((n \log n)/s)$ passes. Demetrescu et al. [9] present a simulation of a PRAM algorithm on the W-Stream model, allowing existing PRAM algorithms to run *sequentially* in the W-Stream model.

McColl et al. [10] present a parallel algorithm for maintaining connected components in a fully-dynamic graph (addition and deletion), a more general setting than ours. Their work focuses on engineering algorithms that work well on real-world graphs, and gives no theoretical analysis of the parallel complexity. Manne and Patwary [11] present a parallel Union-Find algorithm for distributed memory computers. Unlike these two works, our work aims for algorithms with a provable performance bounds and parallel efficiency.

Berry et al. [12] present methods for maintaining connected components in their parallel graph stream model, called X-Stream, which periodically ages out edges. Their algorithm is essentially an "unrolling" of the algorithm of [8], and edges are passed from one processor to another until connected components are found by the last processor in the sequence. Compared to our work, their input model and notions of correctness differ. Our work views the input stream as a sequence of batches, where each batch is an unordered set of edges or a set of queries. Their work views the input as a totally ordered sequence of interleaved edges and queries and their algorithm must strictly respect this total ordering. Next, they age out edges, while we do not. Finally, they do not give provable parallel complexity bounds while we do.

There exist work-efficient batch parallel algorithms for graph connectivity, such as [13, 14]. These compute the connected components of a graph whose edges are all known in advance, using work linear in the number of edges and parallel depth polylogarithmic in the number of edges. The work of these algorithms is of the same order as that of the best sequential algorithm for connected components of a static graph. However, these algorithms are not efficient for a dynamic graph. If the graph changes due to the addition of a few edges, then the entire algorithm will have to be re-run to update the connected components in the graph. Hence, these are not suitable for parallel ICG.

Shiloach and Vishkin (as presented by Jájá [15]) describe a batch parallel algorithm for graph connectivity based on Union-Find that runs in $O(m \log n)$ work and polylogarithmic depth. Their algorithm, like ours, also relies on a union-find-like forest and linking together trees ("grafting" in their terminology); however, theirs relies on pointer jumping to keep the tree shallow to the point of keeping stars. As far as we know, there is no easy way to port their algorithm to our setting, let alone perform parallel path compression. Our algorithm is simpler as it sidesteps concurrent grafting.

Prior work on wait-free algorithms for Union-Find [16] has focused on the asynchronous model, where the goal is to be correct under all possible interleavings of operations. Unlike us, they do not focus on bulk processing of edges.

In addition, there is a long line of work on sequential algorithms for maintaining graph connectivity on an evolving graph. See the recent work by [17] that addresses this problem in the general dynamic case and the references therein.

1.3 Preliminaries and Notation

Throughout the paper, let $[n]$ denote the set $\{0, 1, \ldots, n\}$. A sequence is written as $X = \langle x_1, x_2, \ldots, x_{|X|} \rangle$, where $|X|$ denotes the length of the sequence. For a sequence X, the i-th element is denoted by X_i or $X[i]$. Following the set-builder notation, we denote by $\langle f(x) : \Phi(x) \rangle$ a sequence generated (logically) by taking all elements that satisfy $\Phi(x)$, preserving their original ordering, and transform them by applying f. For example, if T is a sequence of numbers, the notation $\langle 1 + f(x) : x \in T \text{ and } x \text{ odd} \rangle$ means a sequence created by taking each element x from T that are odd and map x to $1 + f(x)$, retaining their original ordering. Furthermore, we write $S \oplus T$ to mean the concatenation of S and T.

We design algorithms in the work-depth model assuming an underlying CRCW machine: the *work* of an algorithm is the total operation count, and the *depth* (also called parallel time or span) is the length of the longest chain of dependencies within a parallel computation. The gold standard for algorithms in this model is to perform the same amount of work as the best sequential counterpart (work efficient) and to have polylogarithmic depth. At the expense of an extra polylog factor in work and depth, algorithms designed for CRCW can work in other shared-memory models (e.g., CREW).

We use standard parallel operations such as filter, prefix sum, map (applying a constant-cost function), and pack, all of which has $O(n)$ work and at most $O(\log^2(n))$ depth on an input sequence of length n. Given a sequence of m numbers, there is a duplicate removal algorithm removeDup running in $O(m)$ work and $O(\log^2 m)$ depth [15]. We also use the following results to sort integer keys in a small range faster than a typical comparison-based algorithm:

Theorem 1 (Parallel Integer Sort [18]). *There is an algorithm,* intSort, *that takes a sequence of integer keys* $a_1, a_2, \ldots, a_n$, *each* $a_i \in [0, cn]$, *where* $c = O(1)$, *and produces a sorted sequence in* $O(n)$ *work and* polylog(n) *depth.*

Parallel Connectivity: For a graph $G = (V, E)$, a connected component algorithm (CC) computes a sequence of connected components $\langle C_i \rangle_{i=1}^{k}$, where each C_i is a list of vertices in the component. There are algorithms for CC that have $O(|V| + |E|)$ work and $O(\text{polylog}(|V|, |E|))$ depth (e.g., [13,14]), with Gazit's algorithm [14] requiring $O(\log |V|)$ depth.

2 Simple Bulk-Parallel Data Structure

In this section, we describe a simple bulk-parallel data structure for Union-Find and IGC. The data structure is conceptually simple but will be instructive for the theoretical improvements presented in the next section. As before, let n denote the number of vertices in the graph stream, and equivalently, the total number of elements across all disjoint sets. The main result for this section is as follows:

Theorem 2. *There is a bulk-parallel data structure for Union-Find and IGC with the following properties: (1) the total memory consumption is $O(n)$ words; (2) processing a minibatch of b updates takes $O(b \log n)$ work and $O(\log \max(b, n))$ depth; and (3) processing a minibatch of q queries takes $O(q \log n)$ work and $O(\log n)$ depth.*

Our solution is a parallel version of the standard union-find data structure that keeps the height of the union-find forest at most $O(\log n)$. The crux here is to handle concurrent **union** operations efficiently.

Sequential Union-Find Algorithm: We begin by reviewing a basic union-find implementation that uses union by size[2]. The data structure maintains a forest with one tree for each set in the partition. $\mathtt{find}(v)$ returns the element at the root of the tree containing v. If u and v are in different sets $\mathtt{union}(u, v)$ combines the trees containing u and v into a single tree by pointing the root of one to the root of the other. Once $\mathtt{union}(u, v)$ has been applied, $\mathtt{find}(u)$ and $\mathtt{find}(v)$ return the same element.

A tree in a union-find forest is typically represented by remembering each node's parent, in an array **parent** of length n, where $\mathtt{parent}[u]$ is the tree's parent of u or $\mathtt{parent}[u] = u$ if it is the root of its component. The running time of the union and find operations depends on the height of the corresponding trees. To keep the height within $O(\log n)$, a simple strategy, known as *union by size*, is for **union** to always link the tree with fewer vertices into the tree with more vertices. The data structure also keeps an array for the sizes of the trees. The following results are standard (see [6], for example):

Lemma 1 (Sequential Union-Find). *On a graph with n vertices, a sequential union-find data structure implementing the union-by-size strategy consumes $O(n)$ space and has the following characteristics:*

- *Every union-find tree has height $O(\log n)$ and each* **find** *takes $O(\log n)$ sequential time.*
- *Given two distinct roots u and v, the operation* $\mathtt{union}(u, v)$ *implementing union by size takes $O(1)$ sequential time.*

Our data structure maintains an instance of this union-find data structure, called U. The **find** operation here is read-only; unlike some more sophisticated variants, this version of union-find does not perform path compression.

Connectivity Queries in Parallel. Let U denote the union-find data structure stored in shared memory. Connectivity queries can be answered in parallel, using read-only **find** operations on U. To answer whether u and v are connected, we compute $U.\mathtt{find}(v)$ and $U.\mathtt{find}(u)$, and report if the results are equal. Multiple queries can be answered by running multiple copies of this procedure in parallel, as described in $\mathtt{Simple\text{-}Bulk\text{-}Query}$ (Algorithm 1).

Correctness follows from the underlying union-find structure. The parallel cost is simply that of applying q operations of $U.\mathtt{find}$ in parallel. Hence:

---

[2] Other variants, such as union by rank, will also work.

Algorithm 1. Simple-Bulk-Query$(U, \langle (u_i, v_i) \rangle_{i=1}^q)$.

Input: U : union find structure, and (u_i, v_i) is a pair of vertices, for $1 \le i \le q$.
Output: For each i, whether or not u_i is connected to v_i in the graph.
1: **for** $i = 1, 2, \ldots, q$ **do** in parallel
2: $\quad \mid \quad a_i \leftarrow (U.\text{find}(u_i) = U.\text{find}(v_i))$
3: **return** $\langle a_1, a_2, \ldots, a_q \rangle$

Lemma 2. *The parallel depth of* Simple-Bulk-Query *is* $O(\log n)$, *and the work is* $O(q \log n)$, *where* q *is the number of queries input to the algorithm.*

Multiple Updates in Parallel. How to incorporate (in parallel) a minibatch of edges A into an existing union-find structure? Sequentially, this is simple: for each edge, invoke union on the endpoints. But it is dangerous to directly apply different union operations in parallel since union updates the structure.

However, it is safe run multiple unions in parallel as long as they operate on different trees. Because there may be a number of union operations involving the same tree, this is not sufficient in itself—running these sequentially will result in a large parallel depth. For instance, consider adding the edges of a star graph (with a very high degree) to an empty graph. Because all the edges share a common endpoint (the center), this vertex is involved in every union, and hence no two operations can proceed in parallel.

To tackle this problem, our algorithm transforms the minibatch of edges A into a structure that can be connected up easily in parallel. For illustration, we revisit the example when the minibatch is itself a star graph. Suppose there are seven edges within the minibatch: $(v_1, v_2), (v_1, v_3), (v_1, v_4), \ldots, (v_1, v_8)$. By examining the minibatch, we find that all of $v_1, \ldots, v_8$ will belong to the same component. We now apply these connections to the graph.

In terms of connectivity, it does not matter whether we apply the actual edges that arrived, or a different, but equivalent set of edges; it only matters that the relevant vertices are connected up. To connect up these vertices, our algorithm schedules the unions in only three parallel rounds as follows. The notation $X \| Y$ indicates that X and Y are run in parallel:

1. union$(v_1, v_2) \|$ union$(v_3, v_4) \|$ union$(v_5, v_6) \|$ union(v_7, v_8)
2. union$(v_1, v_3) \|$ union(v_5, v_7)
3. union(v_1, v_5)

As we will soon see, such a schedule can be constructed for a component of any size provided that no two vertices in the component are connected previously. The resulting parallel depth is logarithmic in the size of the minibatch.

To add a minibatch of edges Simple-Bulk-Update (Algorithm 2) proceeds in three steps: (1) Relabel edges as links between existing components (2) Discover new connections arising from A (3) Commit new connections to U in parallel, using a divide-and-conquer strategy. We omit further details here. For proof of correctness and analysis of the properties of the algorithm, we refer the reader to the full version of the paper [3].

Algorithm 2. Simple-Bulk-Update(U, A)

Input: U: the union find structure;

A: a set of edges to add to the graph, equivalently viewed as a set of **union** operations to be applied to U

▷ *Relabel each $(u, v) \in A$ with the roots of u and v*

1:   $A' \leftarrow \langle (p_u, p_v) : (u, v) \in A$ where $p_u = U.\mathtt{find}(u)$ and $p_v = U.\mathtt{find}(v) \rangle$

▷ *Remove self-loops*

2:   $A'' \leftarrow \langle (u, v) : (u, v) \in A'$ where $u \neq v \rangle$

3:   $\mathcal{C} \leftarrow \mathtt{CC}(A'')$

4:   **foreach** $C \in \mathcal{C}$ **do** in parallel

5:     | Parallel-Join(U, C)

Algorithm 3. Parallel-Join(U, C)

Input: U: the union-find structure, C: a seq. of tree roots

Output: The root of the tree after all of C are connected

1:   **if** $|C| = 1$ **then**

2:   | **return** $C[1]$

3:   **else**

4:   | $\ell \leftarrow \lfloor |C|/2 \rfloor$

5:   | $u \leftarrow$ Parallel-Join$(U, C[1, 2, \ldots, \ell])$ **in parallel with**

    | $v \leftarrow$ Parallel-Join$(U, C[\ell + 1, \ell + 2, \ldots, |C|])$

6:   | **return** $U.\mathtt{union}(u, v)$

Intuitively, Parallel-Join is correct because the order that the **union** operations are made does not matter and that parallel **union** operations always work on separate sets of tree roots, posing no conflicts.

Lemma 3 (Complexity of Simple-Bulk-Update). *Given a minibatch A with b edges, Simple-Bulk-Update takes $O(b \log n)$ work and $O(\log n)$ depth.*

3   Work-Efficient Parallel Algorithm

Whereas the fastest sequential data structures (e.g., [5]) require $O((m+q)\alpha(m+q, n))$ work to process m edges and q queries, our basic data structure from the previous section needs up to $O((m + q) \log n)$ work for the same input stream. This section describes improvements that make it match the best sequential work bound while preserving the polylogarithmic depth guarantee. The main result for this section is as follows:

Theorem 3. *There is a bulk-parallel data structure for Union-Find and IGC with the following properties:*

(1) The total memory consumption is $O(n)$ words.

(2) The depth of Bulk-Update and Bulk-Query is $O(\log n)$ each.

(3) Over the lifetime of the data structure, the total work for processing m updates (across all Bulk-Update) and q queries is $O((m + q)\alpha(m + q, n))$.

Overview: All sequential data structures with a $O((m + q)\alpha(n))$ bound use a technique called path compression, which shortens the path that `find` takes to reach the root, making subsequent operations cheaper. Our goal in this section is to enable path compression during parallel execution. We present a new parallel `find` procedure called `Bulk-Find`, which answers a set of `find` queries in parallel and performs path compression.

Algorithm 4: `Bulk-Find`(U, S)—find the root in U for each $s \in S$ with path compression.

Input: U: the union find structure. For $i = 1, \ldots, |S|$, $S[i] =$ a graph vertex.
Output: A response array res of length $|S|$ where $res[i]$ is the root of the tree of the vertex $S[i]$ in the input.

▷ **Phase I: Find the roots for all queries**
1: $R_0 \leftarrow \langle (S[k], \textbf{null}) : k = 0, 1, 2, \ldots, |S| - 1 \rangle$
2: $F_0 \leftarrow$ mkFrontier$(R_0, \varnothing)$, $roots \leftarrow \varnothing$, $visited \leftarrow \varnothing$, $i \leftarrow 0$
3: **while** $R_i \neq \varnothing$ **do**

    `def mkFrontier(R, visited):`
        `// nodes to go to next`

4:     $visited \leftarrow visited \cup F_i$
5:     $R_{i+1} \leftarrow \langle (\text{parent}[v], v) \mid v \in F_i \text{ and } \text{parent}[v] \neq v \rangle$
6:     $roots \leftarrow roots \cup \{v \mid v \in F_i \text{ where } \text{parent}[v] = v\}$
7:     $F_{i+1} \leftarrow$ mkFrontier$(R_{i+1}, visited)$, $i \leftarrow i + 1$

    1: $req \leftarrow \langle v : (v, *) \in R \wedge$
        $\textbf{not } visited[v] \rangle$
    2: **return** removeDup(req)

▷ **Set up response distribution**
8: Create an instance of RD with $R_\cup = R_0 \oplus R_1 \oplus \cdots \oplus R_i$
    ▷ **Phase II: Distribute the answers and shorten the paths**
9: $D_0 \leftarrow \{(r, r) : r \in roots\}$, $i \leftarrow 0$
10: **while** $D_i \neq \varnothing$ **do**
11:     For each $(v, r) \in D_i$, in parallel, $\text{parent}[v] \leftarrow r$
12:     $D_{i+1} \leftarrow \bigcup_{(v,r)\in D_i} \{(u, r) : u \in RD.\text{allFrom}(v) \text{ and } u \neq \textbf{null}\}$. That is, create D_{i+1} by expanding every $(v, r) \in D_i$ as the entries of $RD.\text{allFrom}(v)$ excluding **null**, each inheriting r.
13:     $i \leftarrow i + 1$
14: For $i = 0, 1, 2 \ldots, |S| - 1$, in parallel, make $res[i] \leftarrow \text{parent}[S[i]]$
15: **return** res

To understand the benefits of path compression, consider a concrete example in Fig. 1A, which shows a union-find tree T that is typical in a union-find forest. The root of T is $r = 19$. Suppose we need to support `find`'s from $u = 1$ and $v = 7$. When all is done, both `find`(u) and `find`(v) should return r. Notice that in this example, the paths to the root $u \rightsquigarrow r$ and $v \rightsquigarrow r$ meet at a common vertex $w = 4$. That is, the two paths are identical from w onward to r. If `find`'s were done sequentially, say `find`(u) before `find`(v), then `find`(u)—with path compression—would update all nodes on the $u \rightsquigarrow r$ path to point to r. This means that when `find`(v) traverses the tree, the path to the root is significantly shorter: for `find`(v), the next hop after w is already r.

The kind of sharing and shortcutting illustrated, however, is not possible when the `find` operations are run independently in parallel. Each `find`, unaware of the others, will proceed all the way to the root, missing out on possible sharing.

We fix this problem by organizing the parallel computation so that the work on different "flows" of finds is carefully coordinated. Algorithm 4 shows an algorithm Bulk-Find, which works in two phases, separating actions that only read from the tree from actions that only write to it:

▷ *Phase I:* Find the roots for all queries, coalescing flows as soon as they meet up. This phase should be thought of as running breadth-first search (BFS), starting from all the query nodes S at once. As with normal BFS, if multiple flows meet up, only one will move on. Also, if a flow encounters a node that has been traversed before, that flow no longer needs to go on. To proceed to Phase II, we need to record the paths traversed so that we can distribute responses to the requesting nodes.

▷ *Phase II:* Distribute the answers and shorten the paths. Using the transcript from Phase I, Phase II makes sure that all nodes traversed will point to the corresponding root—and answers delivered to all the finds. This phase, too, should be thought of as running breadth-first search (BFS) backwards from all the roots reached in Phase I. This BFS reverses the steps taken in Phase I using the trails recorded. There is a technical challenge in implementing this. Back in Phase I, to minimize the cost of recording these trails, the trails are kept as a list of directed edges (marked by their two endpoints) that are traversed. However, for the reverse traversal in Phase II to be efficient, it needs a means to quickly look up all the neighbors of a vertex (i.e., at every node, we must be able to find every flow that arrived at this node back in Phase I). For this, we use a data structure that takes advantage of hashing and integer sorting (Theorem 1) to keep the parallel complexity low. We discuss this in Lemma 4.

Example: We illustrate how the Bulk-Find algorithm works using the union-find from Fig. 1A. The queries to the Bulk-Find are nodes that are circled. The paths traversed in Phase I are shown in panel B. If a flow is terminated, the last edge traversed on that flow is rendered as ⊢.

Notice that as soon as flows meet up, only one of them will carry on. In general, if multiple flows meet up at a point, only one will go on. Notice also that both the flow $1 \to 2 \to 4$ and the flow $7 \to 8 \to 9 \to 4$ are stopped at 4 because 4 is a source itself, which was started at the same time as 1 and 7. At the finish of Phase I, the graph (in fact a tree) given by $R_\cup$ is shown in panel C. Finally, in Phase II, this graph is traversed and all nodes visited are updated to point to their corresponding root (as shown in panel D).

3.1 Response Distributor

For a sequence $R_\cup = \langle (from_i, to_i) \rangle_{i=1}^\lambda$, we need to answer the query allFrom(f), which returns a sequence containing all to_i where $from_i = f$. To meet the desired overall running bound, we can spend no more than $O(\lambda)$ work and $O(\text{polylog}(\lambda))$ depth on preprocessing $R_\cup$ and cannot generate, say, a sequence of sequences RD where $RD[f]$ stores allFrom(f). This calls for a data structure.

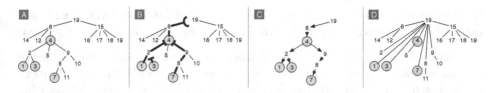

Fig. 1. A: An example union-find tree with sample queries circled; B: Bolded edges are paths, together with their stopping points, that result from the traversal in Phase I; C: The traversal graph $R_\cup$ recorded as a result of Phase I; and D: The union-find tree after Phase II, which updates all traversed nodes to point to their roots.

Lemma 4 (Response Distributor). *There is a response distributor (RD) structure that can be constructed from input* $R_\cup = \langle(from_i, to_i)\rangle_{i=1}^{\lambda}$ *in* $O(\lambda)$ *work and* $O(\mathrm{polylog}(n))$ *depth. Each* `allFrom` *query takes* $O(\log \lambda)$ *depth. If* $\mathbb{F} = \{from_i : i = 1, \ldots, \lambda\}$, *then* $\mathbf{E}\big[\sum_{f \in \mathbb{F}} Work(RD.\mathtt{allFrom}(f))\big] = O(\lambda)$.

Idea Sketch: Each $from_i$ is hashed to a number in, say, $[c\lambda]$, $c = O(1)$. Then, `intSort` places the same keys together. To answer `allFrom`(f), scan the tuples that hash to the same number as f, filtering out those that are not for f. Since two different fs hash to the same place with probability $1/c$, there cannot be too many "collisions," so the overall work across all queries in $\mathbb{F}$ is still $O(\lambda)$.

Hence, `Bulk-Find`(U, S) performs $O(|R_\cup|)$ work and $O(\mathrm{polylog}(n))$ depth.

3.2 Bulk-Find's Cost Equivalence to Serial `find`

To analyze the work bound of the improved data structure, we will show that what `Bulk-Find` does is equivalent to some sequential execution of the standard `find` and requires the same amount of work, up to constants.

To gather intuition, we will manually derive such a sequence for the sample queries $S = \{1, 3, 4, 7\}$ used in Fig. 1. The query of 4 went all the way to the root without merging with another flow. But the queries of 1 and 7 were stopped at 4 and in this sense, depended upon the response from the query of 4. By the same reasoning, because the query of 3 merged with the query of 1 (with 1 proceeding on), the query of 3 depended on the response from the query of 1. Note that in this view, although the query of 3 technically waited for the response at 2, it was the query of 1 that brought the response, so it depended on 1. To derive a sequence of execution, we need to respect the "depended on" relation: if a depended on b, then a will be invoked after b. As an example, one sequential execution order that respects these dependencies is `find(4), find(7), find(1), find(3)`. By applying `find`s in this order, the paths traversed are exactly what the parallel execution does because U.`find` performs full path compression. This idea is formalized as follows:

Lemma 5. *For a sequence of queries S with which `Bulk-Find`(U, S) is invoked, there is a sequence S' that is a permutation of S such that applying U.`Find` to*

S' *serially in that order yields the same union-find forest as* `Bulk-Find`*'s and incurs the same traversal cost of* $O(|R_\cup|)$, *where* $R_\cup$ *is as defined in* `Bulk-Find`.

Finally, to obtain the bounds in Theorem 3, we modify `Simple-Bulk-Query` and `Simple-Bulk-Update` (in the relabeling step) to use `Bulk-Find` on all query pairs. The depth clearly remains $O(\text{polylog}(n))$ per bulk operation. Aggregating the cost of `Bulk-Find` across calls from `Bulk-Update` and `Bulk-Query`, we know from Lemma 5 that there is a sequential order that has the same work. Therefore, the total work is bounded by $O((m + q)\alpha(m + q, n))$.

4   Conclusion

We presented a shared-memory parallel algorithm for incremental graph connectivity in the minibatch arrival model. Our algorithm has polylogarithmic parallel depth and its total work across all processors is of the same order as the work due to the best sequential algorithm for incremental graph connectivity.

We list some natural open questions. (1) The present paper assumes CRCW (Concurrent Read and Concurrent Write) because of parallel integer sorting, which is used internally. Is it possible to design an algorithm with the same work and depth that does not require concurrent write? (2) For large but still $o(n)$ batch size, can the work of our algorithm be improved? Our algorithm, though work-efficient with respect to union-find, still does slightly more than linear work. By contrast, the optimal connectivity in the static setting only requires linear work. Note, however, that for all practical purposes, the work of our algorithm is linear in the number of edges, due to very slow growth of the inverse Ackerman's function. (3) Can these results on parallel algorithms be extended to the fully dynamic case when there are both edge arrivals as well as deletions?

References

1. IBM Corporation: IBM Streams. http://www-03.ibm.com/software/products/en/ibm-streams. Accessed Feb 2016
2. Zaharia, M., Das, T., Li, H., Hunter, T., Shenker, S., Stoica, I.: Discretized streams: fault-tolerant streaming computation at scale. In: Proceedings of ACM Symposium on Operating Systems Principles (SOSP), pp. 423–438 (2013)
3. Simsiri, N., Tangwongsan, K., Tirthapura, S., Wu, K.L.: Work-efficient parallel and incremental graph connectivity. Technical report, arXiv (2016). https://arxiv.org/
4. Cormen, T.H., Leiserson, C.E., Rivest, R.L., Stein, C.: Introduction to Algorithms, 3rd edn. MIT Press, Cambridge (2009)
5. Tarjan, R.E.: Efficiency of a good but not linear set union algorithm. J. ACM **22**(2), 215–225 (1975)
6. Seidel, R., Sharir, M.: Top-down analysis of path compression. SIAM J. Comput. **34**(3), 515–525 (2005)
7. Feigenbaum, J., Kannan, S., McGregor, A., Suri, S., Zhang, J.: On graph problems in a semi-streaming model. Theor. Comput. Sci. **348**(2–3), 207–216 (2005)

8. Demetrescu, C., Finocchi, I., Ribichini, A.: Trading off space for passes in graph streaming problems. ACM Trans. Algorithms **6**(1), 6 (2009)
9. Demetrescu, C., Escoffier, B., Moruz, G., Ribichini, A.: Adapting parallel algorithms to the W-stream model, withapplications to graph problems. Theor. Comput. Sci. **411**(44–46), 3994–4004 (2010)
10. McColl, R., Green, O., Bader, D.A.: A new parallel algorithm for connected components in dynamic graphs. In: 20th Annual International Conference on High PerformanceComputing, HiPC 2013, Bengaluru (Bangalore), Karnataka, India, 18–21 December 2013, pp. 246–255 (2013)
11. Manne, F., Patwary, M.M.A.: A scalable parallel union-find algorithm for distributed memory computers. In: Wyrzykowski, R., Dongarra, J., Karczewski, K., Wasniewski, J. (eds.) PPAM 2009, Part I. LNCS, vol. 6067, pp. 186–195. Springer, Heidelberg (2010)
12. Berry, J., Oster, M., Phillips, C.A., Plimpton, S., Shead, T.M.: Maintaining connected components for infinite graph streams. In: Proceedings of 2nd International Workshop on Big Data, Streams and Heterogeneous Source Mining: Algorithms, Systems, Programming Models and Applications (BigMine), pp. 95–102 (2013)
13. Shun, J., Dhulipala, L., Blelloch, G.E.: A simple and practical linear-work parallel algorithm for connectivity. In: 26th ACM Symposium on Parallelism in Algorithms and Architectures, SPAA 2014, Prague, Czech Republic, pp. 143–153 (2014)
14. Gazit, H.: An optimal randomized parallel algorithm for finding connected components in a graph. SIAM J. Comput. **20**(6), 1046–1067 (1991)
15. JáJá, J.: An Introduction to Parallel Algorithms. Addison-Wesley, Redwood City (1992)
16. Anderson, R.J., Woll, H.: Wait-free parallel algorithms for the union-find problem. In: Proceedings of the 23rd Annual ACM Symposium on Theory of Computing, 5–8 May 1991, New Orleans, Louisiana, USA, pp. 370–380 (1991)
17. Kapron, B.M., King, V., Mountjoy, B.: Dynamic graph connectivity in polylogarithmic worst case time. In: Proceedings of the Twenty-Fourth Annual ACM-SIAM Symposium on Discrete Algorithms, SODA 2013, New Orleans, Louisiana, USA, 6–8 January 2013, pp. 1131–1142 (2013)
18. Rajasekaran, S., Reif, J.H.: Optimal and sublogarithmic time randomized parallel sorting algorithms. SIAM J. Comput. **18**(3), 594–607 (1989)

An Efficient Cache-oblivious
Parallel Viterbi Algorithm

Rezaul Chowdhury, Pramod Ganapathi$^{(\boxtimes)}$, Vivek Pradhan,
Jesmin Jahan Tithi, and Yunpeng Xiao

Department of Computer Science, Stony Brook University, New York, USA
`pramod.ganapathi@stonybrook.edu`

Abstract. The Viterbi algorithm is used to find the most likely path through a hidden Markov model given an observed sequence, and has numerous applications. Due to its importance and high computational complexity, several algorithmic strategies have been developed to parallelize it on different parallel architectures. However, none of the existing Viterbi decoding algorithms designed for modern computers with cache hierarchies is simultaneously cache-efficient and cache-oblivious. Being oblivious of machine resources (e.g., caches and processors) while also being efficient promotes portability. In this paper, we present an *efficient cache- and processor-oblivious Viterbi algorithm* based on *rank convergence*. The algorithm builds upon the parallel Viterbi algorithm of Maleki et al. (PPoPP 2014). We provide empirical analysis of our algorithm by comparing it with Maleki et al.'s algorithm.

Keywords: Viterbi algorithm · Cache-efficient · Cache-oblivious · Recursive · Divide-and-conquer · Parallel · Multi-instance · Rank convergence

1 Introduction

The Viterbi algorithm [36,37] proposed by Andrew J. Viterbi in 1967, is a dynamic programming algorithm that finds the most probable sequence of hidden states, called the "Viterbi path" from a given sequence of observed events in the context of a hidden Markov model (HMM).

Motivation. The Viterbi algorithm has numerous real-world applications. Although it was originally used for speech recognition in CDMA technology, in the last 25 years, it has been heavily used in computational biology and bioinformatics for finding coding and non-coding regions of an unlabeled string of DNA nucleotides (i.e., gene finding) [3], prediction of protein-coding regions in genome sequences modeling families of related DNA or protein sequences and prediction of secondary structure elements in proteins [24], CpG island [17], promoter [29] and conserved elements detection [30]. Apart from computational biology, Viterbi algorithm is used in TDMA system for GSM [15], television sets [28], satellite and space communication [21], magnetic recording systems [23],

© Springer International Publishing Switzerland 2016
P.-F. Dutot and D. Trystram (Eds.): Euro-Par 2016, LNCS 9833, pp. 574–587, 2016.
DOI: 10.1007/978-3-319-43659-3_42

parsing context-free grammars [22], and part-of-speech tagging [16]. Therefore, improving performance of Viterbi algorithm will likely to have impact in these areas as well.

When the input data becomes too large to fit into a cache, between two algorithms that perform the same set of CPU operations, the one that is more *cache-efficient*, i.e., causes fewer block transfers (or IO) between adjacent levels of caches is likely to run faster. Though there have been a lot of efforts and successes in parallelizing the Viterbi algorithm, there is little work in the realm of designing cache-efficient Viterbi algorithms that are also *cache-oblivious* [20], i.e., independent of cache parameters such as cache sizes and block sizes. Similarly, a *processor-oblivious* [12] algorithm does not use the number of processors in the algorithmic description. A cache- and processor-oblivious algorithm is more likely to be portable across machines. To the best of our knowledge, we present the first provably cache-efficient cache-oblivious parallel Viterbi algorithm.

We use *dynamic multithreading model* [14] and *ideal cache model* [20] to measure parallelism and serial cache complexity, respectively.

Related Work. Several efficient cache- and processor-oblivious recursive divide-and-conquer algorithms for solving dynamic programs (DP) have been developed [2,4,7–11,13,31,33,34]. But the approach used in those papers assumes that the set and sequence of DP cell updates to be performed do not depend on the data values in the DP table which is not true in case of Viterbi DP.

One can use auto-parallelizers to parallelize sequential Viterbi programs. Fisher and Ghuloum [19] present a method in which loop body instances are represented in a closed form using function compositions. Reduction is then applied for parallelization. Chin et al. [5,6] use second-order generalization and induction derivation to generate divide-and-conquer parallel programs. None of these methods exploit parallelism across stages. Also the generated parallel programs are not cache-efficient.

The parallel Viterbi algorithm [18] used for homology search in HMMER uses SSE2 instructions and reduces L1 cache misses. Though the phrase "cache-oblivious" appears in the title of the paper, the presented algorithm is not oblivious of the cache parameters as it uses loop-tiling with the tile size determined based on the size of the L1 data cache. Also the algorithm works only for three states, and it is not clear how the method behaves for arbitrarily large number of states as in the case of a general Viterbi algorithm.

The EasyPDP system [32] parallelizes the Viterbi algorithm and also reduces cache misses. However, it requires the user to specify loop tile sizes making it cache-aware. Also the reduction in cache misses is not significant.

The Viterbi algorithm is inherently sequential across stages which constraints parallelism along the time dimension. A parallel Viterbi algorithm presented in [26,27] based on rank convergence is the first to exploit parallelism across stages. However, this algorithm is processor-aware and not cache-efficient.

Our Contributions. Our major contributions are: (1) an *efficient cache-oblivious parallel multi-instance Viterbi algorithm* (Sect. 3), (2) an *efficient cache-oblivious parallel single-instance Viterbi algorithm* (Sect. 5) based on our multi-instance algorithm (Sect. 3) and Maleki et al.'s rank convergence algorithm (Sect. 4), and (3) *experimental results* (Sect. 6) comparing our algorithms with Maleki et al.'s algorithms on modern multicore platforms.

2   Cache-inefficient Viterbi Algorithm

In this section, we formally describe the Viterbi dynamic program (DP), and describe a simple cache-inefficient Viterbi algorithm based on divide-and-conquer.

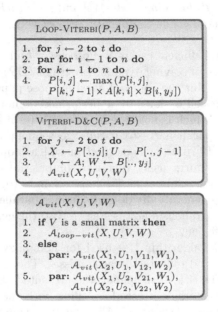

Fig. 1. Iterative and recursive Viterbi algorithms.

Formal Specification. The Viterbi DP is described as follows. We are given an observation space $O = \{o_1, o_2, \ldots, o_m\}$, state space $S = \{s_1, s_2, \ldots, s_n\}$, observations $Y = \{y_1, y_2, \ldots, y_t\}$, transition matrix A of size $n \times n$, where $A[i, j]$ is the transition probability of transiting from s_i to s_j, emission matrix B of size $n \times m$, where $B[i, j]$ is the probability of observing o_j at s_i, and initial probability vector (or initial solution vector) I, where $I[i]$ is the probability that $x_1 = s_i$. Let $X = \{x_1, x_2, \ldots, x_t\}$ be a sequence of hidden states that generates $Y = \{y_1, y_2, \ldots, y_t\}$. Then the matrices P and P' of size $n \times t$, where $P[i, j]$ is the probability of the most likely path of getting to state s_i at observation o_j

and $P'[i, j]$ stores the hidden state of the most likely path (i.e., Viterbi path) are computed as follows. $P[i, j] = I[i] \cdot B[i, y_1]$, and $P'[i, j] = 0$ when $j = 1$. Otherwise (i.e., when $j > 1$):

$$P[i, j] = \max_{k \in [1,n]} (P[k, j - 1] \cdot A[k, i] \cdot B[i, y_j]),$$

$$\text{and } P'[i, j] = \text{argmax}_{k \in [1,n]} (P[k, j - 1] \cdot A[k, i] \cdot B[i, y_j]),$$

Cache-inefficient Algorithm. An iterative parallel and a recursive divide-and-conquer-based parallel Viterbi algorithms are given in Fig. 1. As per the Viterbi recurrence, each cell (i, j) of matrix P depends on all cells of P in column $j - 1$, all cells of A in column i, and the cell (i, y_j) of B. The function $\mathcal{A}_{vit}$ fills jth column of P denoted by X using $(j - 1)$th column denoted by U using a divide-and-conquer approach. To compute each column of P, the entire matrix of A has to be read. Hence the recursive algorithm is cache-inefficient. In both algorithms, the stages are computed sequentially, however, all cells in each stage (or timestep) are computed in parallel.

Complexity Analysis. The serial cache complexity of the iterative algorithm is computed as $\sum_{j=1}^{t} \sum_{i=1}^{n} \mathcal{O}(n/B) = \mathcal{O}(n^2 t/B)$, and that of the divide-and-conquer algorithm is computed as follows. Let $Q_{\mathcal{A}}(n)$ denote the serial cache complexity of $\mathcal{A}_{vit}$ on a matrix of size $n \times n$. Then $Q_{\mathcal{A}}(n) = \mathcal{O}(n^2/B + n)$ if $n^2 \leq \gamma_A M$, and $4Q_{\mathcal{A}}(n/2) + \mathcal{O}(1)$, otherwise; where, γ_A is a suitable constant. Solving, $Q_{\mathcal{A}}(n) = \mathcal{O}(n^2/B + n)$. Thus, the serial cache complexity of the recursive algorithm is $\mathcal{O}(n^2 t/B + nt)$ when n^2 is too large to fit in cache.

Both the iterative and recursive algorithms have spatial locality, but they do not have any temporal locality. Hence, these algorithms are not cache-efficient.

The span (i.e., runtime on a machine with an unbounded number of processors) of the iterative algorithm is $\Theta(nt)$, as there are t time steps and it takes n time steps to fully update a cell of P. The span of the recursive algorithm is computed as follows. Let $T_A(n)$ denote the span of $\mathcal{A}_{vit}$ on a matrix of size $n \times n$. Then $T_A(n) = \Theta(1)$ if $n = 1$, and $2T_A(n/2) + \Theta(1)$, otherwise. Solving, $T_A(n) = \Theta(n)$, which implies that the span of the recursive algorithm is $\Theta(nt)$.

3   Cache-efficient Multi-instance Viterbi

In this section, we present a novel cache-efficient cache-oblivious Viterbi algorithm for multiple instances of the problem.

It is easy to see that a standard recursive divide-and-conquer algorithm has no temporal locality because to compute each column of P ($\Theta(n^2)$ work), we have to scan the entire matrix A ($\Theta(n^2)$ space). We can exploit temporal cache locality by solving multiple instances of the problem simultaneously. The existing method that uses multiple instances [25] is cache-inefficient.

Two problems that have the same transition matrix A and emission matrix B are termed two instances of the same problem. The spoken word recognition problem can be considered as an example of multi-instance Viterbi problem.

```
VITERBI-MULTI-INSTANCE-D&C(P_1, P_2, ..., P_q, A, B, t)

1.  for j ← 2 to t do
2.      X ← [P_1[.., j], P_2[.., j], ..., P_q[.., j]]
3.      U ← [P_1[.., j − 1], P_2[.., j − 1], ..., P_q[.., j − 1]]
4.      V ← A
5.      W ← [B[.., y_{1j}], B[.., y_{2j}], ..., B[.., y_{qj}]]
6.      A_vit(X, U, V, W)
```

```
A_vit(X_{n×q}, U_{n×q}, V_{n×n}, W_{n×q})

1.   if X and V are small matrices then
2.       A_{loop−vit}(X, U, V, W)
3.   else if q > n do
4.       par: A_vit(X_L, U_L, V, W_L), A_vit(X_R, U_R, V, W_R)
5.   else if q < n do
6.       par: A_vit(X_T, U_T, V_{11}, W_T), A_vit(X_B, U_T, V_{12}, W_B)
7.       par: A_vit(X_T, U_B, V_{21}, W_T), A_vit(X_B, U_B, V_{22}, W_B)
8.   else
9.       par: A_vit(X_{11}, U_{11}, V_{11}, W_{11}), A_vit(X_{12}, U_{12}, V_{11}, W_{12}),
             A_vit(X_{21}, U_{11}, V_{12}, W_{21}), A_vit(X_{22}, U_{12}, V_{12}, W_{22})
10.      par: A_vit(X_{11}, U_{21}, V_{21}, W_{11}), A_vit(X_{12}, U_{22}, V_{21}, W_{12}),
             A_vit(X_{21}, U_{21}, V_{22}, W_{21}), A_vit(X_{22}, U_{22}, V_{22}, W_{22})
```

Fig. 2. Cache-efficient multi-instance Viterbi algorithm.

The core idea of the algorithm comes from the fact that by scanning the transition matrix A only once, a particular column of matrix P can be computed for n instances of the problem.

Consider Fig. 2. In the function $\mathcal{A}_{vit}$ (X, U, V, W), the matrix U is an $n \times q$ matrix obtained by concatenating $(j-1)$th columns of q matrices $P_1, P_2, \ldots, P_q$, where P_i is the most likely path probability matrix of problem instance i. The algorithm computes X, which is a concatenation of jth columns of the q problem instances. Each problem instance i has a different observations vector $Y_i = \{y_{i1}, y_{i2}, \ldots, y_{it}\}$. Matrix W W is a concatenation of $B[y_{1,j}], B[y_{2,j}], \ldots, B[y_{q,j}]$. We use X_T, X_B, X_L, and X_R to represent the top half, bottom half, left half, and right half of X, respectively. Executing the divide-and-conquer algorithm once computes the second column of all matrices P_1 to P_q. Executing the algorithm again computes the third column of the q matrices. Executing the algorithm t times, the last column of all problem instances would be filled. Note that for each time step (or observation step), W needs to be reconstructed.

Complexity Analysis. The serial cache complexity of the algorithm in Fig. 2 is computed as follows. Let $Q_{\mathcal{A}}(n, q)$ denote the serial cache complexity of $\mathcal{A}_{vit}$ on a matrix of size $n \times q$, and let n and q be powers of two. Then $Q_{\mathcal{A}}(n, q) = \mathcal{O}\left(n^2/B + n\right)$ when $n^2 + nq \leq \gamma_{\mathcal{A}} M$; $Q_{\mathcal{A}}(n, q) = 8 Q_{\mathcal{A}}(n/2, q/2) + \mathcal{O}(1)$ when $n = q$; $Q_{\mathcal{A}}(n, q) = 2 Q_{\mathcal{A}}(n, q/2) + \mathcal{O}(1)$ when $n < q$; and $Q_{\mathcal{A}}(n, q) = 4 Q_{\mathcal{A}}(n/2, q) + \mathcal{O}(1)$ when $n > q$; where, $\gamma_{\mathcal{A}}$ is a suitable constant. Solving, the cache complexity of the algorithm for t timesteps is $t \times Q_{\mathcal{A}}(n, q) = \mathcal{O}\left(n^2 qt/(B\sqrt{M}) + n^2 qt/M + n(n + q)t/B + t\right)$. As the algorithm exploits temporal locality, it is cache-efficient. The span of the algorithm remains $\Theta(nt)$.

4 Viterbi Algorithm Using Rank Convergence

We briefly describe and improve Maleki et al.'s Viterbi algorithm [26] below.

Preliminaries. We rewrite the Viterbi recurrence using *log-probabilities* (i.e., logarithms of all probabilities) as follows so that we can replace multiplications with additions: $P[i,j] = I[i] + B[i,y_1]$ if $j = 1$, and $P[i,j] = \max_{k \in [1,n]}(P[k,j-1] + A[k,i] + B[i,y_j])$ if $j > 1$.

We rewrite the recurrence above as $s[t-1] = s[0] \odot A_1 \odot A_2 \odot \cdots \odot A_{t-1}$, where $s[j]$ is the jth solution vector (or column vector $P[..,j]$) of matrix P, the $n \times n$ matrix A_i is a suitable combination of A and B, and $\odot$ is a matrix product operation defined between two matrices $R_{n \times n}$ and $S_{n \times n}$ as $(R \odot S)[i,j] = \max_{k \in [1,n]}(R[i,k] + S[k,j])$.

VITERBI-RANK($s[0..t-1], A, B$)

1. $p \leftarrow$ #processors

   ⟨ *Forward phase* ⟩ _____
2. **par for** $i \leftarrow 1$ **to** p **do**
3. $l_i \leftarrow t(i-1)/p$; $r_i \leftarrow ti/p$
4.   **if** $i > 1$ **then** $s[l_i] \leftarrow$ random vector
5.   **for** $j \leftarrow l_i$ **to** $r_i - 1$ **do**
6. $s[j+1] \leftarrow$ VITERBI($s[j], A, B[..,y_{j+1}]$)

   ⟨ *Fix up phase* ⟩ _____
7. *converged* $\leftarrow$ false
8. **while** !*converged* **do**
9.   **par for** $i \leftarrow 2$ **to** p **do**
10. $conv_i \leftarrow$ false; $s \leftarrow s[l_i]$
11.   **for** $j \leftarrow l_i$ **to** $r_i - 1$ **do**
12. $s \leftarrow$ VITERBI($s, A, B[..,y_{j+1}]$)
13.   **if** s is parallel to $s[j+1]$ **then**
14. $conv_i \leftarrow$ true; **break**
15. $s[j+1] \leftarrow s$
16.   *converged* $\leftarrow \wedge_i conv_i$

Fig. 3. Processor-aware parallel Viterbi algorithm using rank convergence as given in Maleki et al. paper [26].

The *rank* of a matrix $A_{m \times n}$ is r if r is the smallest number such that A can be written as a product of two matrices $C_{m \times r}$ and $R_{r \times n}$. Vectors v_1 and v_2 are *parallel* provided they differ by a constant offset. For example, $\langle 1, 2, 3, 4 \rangle$ and $\langle 5, 6, 7, 8 \rangle$ are two parallel vectors.

Original Algorithm. The algorithm, shown in Fig. 3, consists of two phases: (*i*) parallel forward phase, and (*ii*) fix up phase. In the forward phase, the t stages are divided into p segments, where p is the number of processors, each segment having $\lceil t/p \rceil$ stages (except possibly the last stage). The stages in the ith segment are from l_i to r_i. The initial solution vector of the entire problem is the initial vector of the first segment and it is known. The initial solution vectors of all other segments are initialized to non-zero random values. A sequential Viterbi algorithm is run in all the segments in parallel. A stage i is said to converge if the computed solution vector $s[i]$ is parallel to the actual solution vector s_i. A segment i is said to converge if $\text{rank}(A_{l_i} \odot A_{l_i+1} \odot \cdots \odot A_j)$ is 1 for $j \in [l_i, r_i - 1]$.

In the fix up phase a sequential Viterbi algorithm is executed for all segments simultaneously. The solution vectors computed in different segments (except the first) might be wrong. But eventually they will become parallel to the actual solution vectors if *rank convergence* occurs. If rank convergence occurs at every segment then the solution vectors at every stage will be parallel to the actual solution vectors. Otherwise, the fix up phase is run again and again until rank convergence occurs at some point. In the worst case, which rarely happens in practice, the fix up phase will have to be executed a total of $p - 1$ times for rank converngence to happen.

Improved Algorithm. The algorithm described above is processor-aware, and we make it processor-oblivious as follows.

VITERBI-RANK-IMPROVED($s[0..t-1], A, B$)

1. $n \leftarrow 2^k; \; t \leftarrow 2^{k+k'}; \; c \leftarrow 2^8$
    ⟨ *Forward phase* ⟩ ——————————
2. $size \leftarrow c; q \leftarrow t/size$
3. **par for** $i \leftarrow 0$ **to** $q - 1$ **do**
4. $l_i \leftarrow i \times size, \; r_i \leftarrow l_i + size - 1$
5.   **if** $i > 0$ **then** $s[l_i] \leftarrow$ random vector
6.   **for** $j \leftarrow l_i$ **to** r_i **do**
7. $s[j + 1] \leftarrow$ VITERBI($s[j], A, B[.., y_{j+1}]$)
    ⟨ *Fixup phase* ⟩ ——————————
8. $u[0..t - 1] \leftarrow s[0..t - 1]; \; converged \leftarrow$ false
9. **for** $(j \leftarrow \log c$ **to** $(\log t)\text{-}1)$ & !$converged$ **do**
10. $size \leftarrow 2^j; q \leftarrow t/(2 \times size)$
11.   **par for** $i \leftarrow 0$ **to** $q - 1$ **do**
12. $l_i \leftarrow (2i + 1) \times size - 1$
13. $r_i \leftarrow l_i + size; conv_i \leftarrow$ false
14.   **for** $j \leftarrow l_i$ **to** r_i **do**
15. $u[j+1] \leftarrow$ VITERBI($u[j], A, B[.., y_{j+1}]$)
16.   **if** $u[j + 1]$ is parallel to $s[j + 1]$ **then**
17. $conv_i \leftarrow$ true; break
18. $s[j + 1] \leftarrow u[j + 1]$
19.   **for** $i \leftarrow 0$ **to** $q - 1$ **do**
20. $converged \leftarrow converged \wedge conv_i$
21.   **if** $converged =$ true **then** break

Fig. 4. Processor-oblivious parallel Viterbi algorithm using rank convergence.

We chose a suitable segment size c (say 256) that is feasibly large, then use a parallel for loop to solve those t/c segments simultaneously. Unlike Maleki et al.'s algorithm, we need to make sure that the segments are non-overlapping at their boundaries and then adjust the fixup phase accordingly as shown in Fig. 4.

Here is how the algorithm works. Let the initial segment size is c (i.e., c consecutive time steps). For convenience we chose $c = 2^i$ where $i \in [\log c, \log t]$. We divide t time steps into t/c independent segments each of size c. Similar to

Maleki et al.'s algorithm, the first solution vectors of all except the first segment are initialized to non-zero valid random probability values. Then in the forward phase we run serial Viterbi algorithm on all segments simultaneously. At the end of the forward phase solution vectors till the cth column (i.e., all columns in the first segment) will have correct log-likelihood values. Other segments will have values computed from the random values chosen initially which may or may not be parallel to the expected values.

In the fix up phase, we start fixing from the second segment as in the original Maleki et al.'s algorithm. However, in each fix up phase, we work on alternative segments always leaving the first segment of the prior fix up phase. After each fix up phase, the size of each segment being considered doubles and number of segments becomes half with respect to the previous phase. At the end of each fix up phase, we check whether the computed solution vectors are parallel to those in the forward phase, and if the answer is 'yes' for all segments under consideration, the program terminates. Otherwise, the next fix up phase is run. In the worst case, the fix up phase is executed $\lambda \in [1, \log(t/c)]$ times after which all results are guaranteed to be correct since by that time the result from the original input propagates till the end. In the worst case, the program is like a serial Viterbi algorithm with a constant factor overhead.

Complexity Analysis. Let $T_1^F(n,t)$, $Q_1^F(n,t)$, $T_\infty^F(n,t)$, and $S^F(t)$ denote the work, serial cache complexity, span, and the steps for convergence, respectively, of algorithm $F \in \{O, I\}$, where O represents the original rank convergence algorithm and I denotes our modified algorithm. Let $f(t)$ be the number of segments in algorithm O. Note that for Maleki et al.'s original algorithm $f(t) = p$. Let the number of times the fix up phase is executed in O and I be λ_O and λ_I, respectively. Then $\lambda^O \in [1, f(t)]$ and $\lambda^I \in [1, \log(t/c)]$.

Work. $T_1^O(n,t) = \Theta\left(n^2 t \cdot \lambda_O\right)$, and $T_1^I(n,t) = \Theta\left(n^2 t \cdot \lambda_I\right)$. In the worst case, $T_1^O(n,t)$ is $\Theta\left(n^2 t \cdot f(t)\right)$, and $T_\infty^I(n,t)$ is $\Theta\left(n^2 t \cdot \log t\right)$.

Serial Cache Complexity. As there is no temporal locality, $Q_1^O(n,t) = \mathcal{O}\left(T_1^O(n,t)/B\right)$ and $Q_1^I(n,t) = \mathcal{O}\left(T_1^I(n,t)/B\right)$, when n^2 does not fit into the cache.

Span. $T_\infty^O(n,t) = \Theta(n(t/f(t)) \cdot \lambda_O)$, as the number of stages in each segment is $\Theta(t/f(t))$, and the span of executing each stage is $\Theta(n)$. In the worst case, $T_\infty^O(n,t)$ is $\Theta(nt)$. $T_\infty^I(n,t)$ is computed as follows. In the ith fix up phase, the number of stages in each segment is 2^i. Hence, the span of executing all stages for λ_I iterations in the fix up phase is $\Theta\left(\sum_{i=\log c}^{(\log c) + \lambda_I} 2^i\right) = \Theta\left(2^{\lambda_I}\right)$. Then $T_\infty^I(n,t) = \Theta\left(n 2^{\lambda_I}\right)$. In the worst case, $T_\infty^I(n,t)$ is $\Theta(nt)$.

Steps for Convergence. Let the rank of the matrix $A_1 \odot A_2 \odot \cdots \odot A_t$ be k. For the original algorithm, $(S^O(t) - 1) \times (t/f(t)) < k \leq S^O(t) \times (t/f(t))$, which implies $S^O(t) = \lceil kf(t)/t \rceil$. Similarly, for the improved algorithm, $2^{S^I(t)-1+\log c} < k \leq 2^{S^I(t)+\log c}$, which implies $S^I(t) = \lceil k/c \rceil$.

5   Cache-efficient Viterbi Algorithm

In this section, we present an efficient cache- and processor-oblivious parallel Viterbi algorithm based on recursive divide-and-conquer, as shown in Fig. 5. The algorithm is derived by combining ideas from the cache-efficient multi-instance Viterbi algorithm (see Sect. 3) and the improved parallel Viterbi algorithm based on rank convergence (see Sect. 4).

Recall that in the multi-instance Viterbi algorithm works on the ith solution vectors, $s[i]$, of different instances of the problem and generates the $(i + 1)^{th}$ solution vectors, $s[i + 1]$, of the instances cache-efficiently. To develop a cache-efficient Viterbi algorithm, in the forward phase, we divide t time steps into t/c independent segments each of size c as we did in the improved parallel Viterbi algorithm using rank convergence shown in Fig. 4, As before, we chose $c = 2^i$ where $i \in [\log c, \log t]$. Since each segment is independent, we can assume that these segments are different instances of the same Viterbi problem. Therefore, we can use the cache-efficient multi-instance Viterbi algorithm to solve these t/c instances simultaneously. Again, the first solution vectors of all except the first segment are initialized to non-zero valid random probability values.

The fix up phase is similar to that of the VITERBI-RANK-IMPROVED algorithm (see Figs. 4 and 5), except that now we use cache-efficient Multi instance Viterbi algorithms to compute the next solution vector of all segments at once instead of using Viterbi algorithm to compute an entire segment independently. As before, we start fixing from the second segment since the first segment is already fixed after the forward phase.

In each fix up phase, we work on alternative segments always leaving out the first segment of prior fix up phase (already fixed by this time). After each fix up phase, the size of each segment being considered doubles and number of segments halves with respect to the previous phase. For each step, we use multi-instance Viterbi algorithm to compute the $(i + 1)$st solution vector from the ith solution vectors for all segments at once. At the end of each fix up phase, we check whether the computed solution vectors are parallel to those found in the forward phase, and if that is true for all segments under consideration, the program terminates. Otherwise, the next fix up phase is run. In the worse case, the fix up phase is executed for $\lambda \in [1, \log(t/c)]$ times after which all results are guaranteed to be correct.

Complexity Analysis. Let $T_1(n, t), Q_1(n, t)$, and $T_\infty(n, t)$ be the work, serial cache complexity, and span of the cache-efficient Viterbi algorithm, respectively. Let $\lambda \in [1, \log(t/c)]$ be the number of times the fix up phase is executed. $T_1(n, t) = \Theta(n^2 t \cdot \lambda)$. In the worst case, $T_1(n, t) = \Theta(n^2 t \cdot \log t)$. As in Sect. 4, $T_\infty(n, t) = \Theta(n2^\lambda)$. Finally, $Q_1(n, t) = \mathcal{O}\left(\sum_{i=\log c}^{(\log c)+\lambda} \left(Q_A\left(n, t/2^i\right) \cdot 2^i\right)\right) = \mathcal{O}\left(n^2 t \lambda/(B\sqrt{M}) + n^2 t \lambda/M + (n(n2^\lambda + t\lambda))/B + 2^\lambda\right)$. If $n^2, t = \Omega\left(\sqrt{M}\right)$ and convergence happens after $\lambda = \mathcal{O}(1)$ iterations of the fix up phase, $Q_1(n, t)$ reduces to $\mathcal{O}\left(n^2 t \lambda/(B\sqrt{M}) + n^2 t \lambda/M\right)$ which further reduces to $\mathcal{O}\left(n^2 t \lambda/(B\sqrt{M})\right)$ when the cache is tall (i.e., $M = \Omega(B^2)$).

VITERBI-CACHE-EFFICIENT($s[0..t-1], A, B$)

1. $n \leftarrow 2^k; t \leftarrow 2^{k+k'}; c \leftarrow 2^8$

$\langle$ Forward phase $\rangle$ —————————————

2. $size \leftarrow c; q \leftarrow t/size$
3. **par for** $i \leftarrow 0$ **to** $q - 1$ **do**
4. $l_i \leftarrow i \times size, r_i \leftarrow l_i + size - 1$
5.   **if** $i > 0$ **then** $s[l_i] \leftarrow$ random vector
6. VITERBI-MI($s[l_0..r_0], s[l_1..r_1], ..,$
    $s[l_{q-1}..r_{q-1}], A, B, c$)

$\langle$ Fixup phase $\rangle$ —————————————

7. $u[0..t-1] \leftarrow s[0..t-1];$ converged $\leftarrow$ false
8. **for** $(j \leftarrow \log c$ **to** $(\log t) - 1)$ **do**
9. $size \leftarrow 2^j, q \leftarrow t/(2 \times size)$
10.   **par for** $i \leftarrow 0$ **to** $q - 1$ **do**
11. $l_i \leftarrow (2i + 1) \times size - 1$
12. $r_i \leftarrow l_i + size; conv_i \leftarrow$ false
13.   VITERBI-MI($u[l_0..r_0], ..., u[l_{q-1}..r_{q-1}],$
      $A, B, size + 1$)
14.   **par for** $i \leftarrow 0$ **to** $q - 1$ **do**
15. $r_i \leftarrow 2(i + 1) \times size - 1$
16.   **if** $u[r_i]$ is parallel to $s[r_i]$ **then**
17. $conv_i \leftarrow$ true
18.   **else** $s[r_i] \leftarrow u[r_i]$
19.   **for** $i \leftarrow 0$ **to** $q - 1$ **do**
20.   converged $\leftarrow$ converged $\wedge conv_i$

Fig. 5. An efficient cache- and processor-oblivious parallel Viterbi algorithm using rank convergence. VITERBI-MI refers to VITERBI-MULTI-INSTANCE-D&C of Sect. 3.

6   Experimental Results

This section presents our implementation details and performance results.

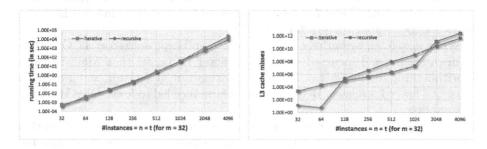

Fig. 6. Running time and L3 miss of our cache-efficient multi-instance Viterbi algorithm along with the multi-instance iterative Viterbi algorithm.

We used a dual socket 16-core (= 2 × 8-cores) 2 GHz Intel Sandy Bridge machine to run all experiments presented in the paper. Each core of this machine was connected to a 32 KB private L1 cache and a 256 KB private L2 cache. All the cores in a socket shared a 20 MB L3 cache, and the machine had 32 GB RAM shared by all cores. We used PAPI 5.2 [1] to count the L3 cache misses (event

PAPI_L3_TCM) and likwid [35] (i.e., likwid-perfctr) to measure energy and power consumption of the program. The matrices A, B, and I were initialized to random probabilities. We used log-probabilities in all implementations and hence used additions instead of multiplications in the Viterbi recurrence. All matrices were stored in column-major order. We performed two sets of experiments to compare our cache-efficient algorithms with the iterative and the fastest known Viterbi (Maleki et al.'s) algorithms. They are as follows.

Cache-efficient Multi-instance Viterbi Algorithm. We compared our cache-efficient multi-instance recursive Viterbi algorithm with the multi-instance iterative Viterbi algorithm. Both algorithms were optimized and parallelized. To construct matrix $W_{n \times q}$ (we chose q to be n in this case), instead of copying all the relevant columns of B, only the pointers to the respective columns were used. Wherever possible, pointer swapping was used to interchange previous solution vector (or matrix) and current solution vector (or matrix).

The running time and the L3 cache misses for the two algorithms are plotted in Fig. 6. The number of stages n, which is also the number of instances was varied from 32 to 4096. Note that in the cache-efficient multi-instance Viterbi algorithm, the number of stages does not need to be the same as the number of instances. The variable m was fixed to 32 and the number of timesteps t was also kept the same as n (hence overall complexity is $O(n^4)$). The recursive algorithm ran slightly faster than the iterative algorithm in most cases when the number of instances increased. When n was 4096, our recursive algorithm ran around 2.26 times faster than the iterative algorithm.

Cache-efficient Viterbi Algorithm. We compared our cache-efficient parallel Viterbi algorithm with Maleki et al.'s parallel Viterbi algorithm. Both implementations were optimized and parallelized and the reported statistics are averages of 4 independent runs. In all our experiments, the number of processors p was set to 16. The plots of Fig. 7 show the graphs of the running time and L3 cache misses for the two algorithms for $n = 4096$.

When $n = 4096$, we varied t from 2^{12} to 2^{18}, and kept m fixed at 32. Our algorithm ran faster and incurred significantly fewer L3 misses than Maleki et al.'s algorithm throughout. For $t = 2^{18}$, our algorithm ran 33 % faster, and incurred a factor of 6 fewer L3 misses. Better cache performance led to lower DRAM energy consumption.

Energy Consumption. We ran experiments to analyze the energy consumption (taking average over three runs) of our cache-efficient recursive algorithm and Maleki et. al.'s algorithm. Our algorithm consumed relatively less DRAM energy compared to the other algorithm.

We used the likwid-perfctr tool to measure CPU, Power Plane 0 (PP0), DRAM energy, and DRAM power consumption during the execution of the programs. The energy measurements were end-to-end, i.e., included all costs during the entire program execution. Note that the DRAM energy consumption is somewhat related to the L3 cache miss of a program as each L3 cache miss results in a DRAM access. Similarly, since CPU energy gives the energy consumed by the

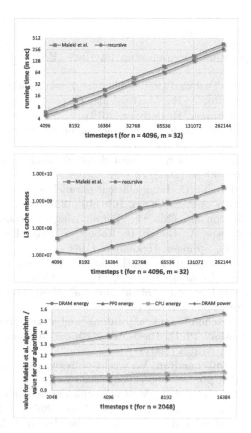

Fig. 7. Running time, L3 miss and energy/power consumption of our cache-efficient Viterbi algorithm along with the existing algorithms.

entire package (all cores, on chip caches, registers and their interconnections), it is related to a program's running time. PP0 is basically a subset of CPU energy since it captures energy consumed by only the cores and their private caches.

For $n = 2048$, t was increased from 2^{11} to 2^{14} while keeping m fixed to 32. Figure 7 shows that the DRAM energy as well as power consumption of our algorithm was significantly less because of the reduced L3 cache misses. When $t = 16384$, Maleki et al.'s algorithm consumed 60 % more DRAM energy and 30 % more DRAM power than ours.

Acknowledgment. Chowdhury and Ganapathi were supported in part by NSF grants CCF-1162196, CCF-1439084 and CNS-1553510.

References

1. Performance Application Programming Interface (PAPI). http://icl.cs.utk.edu/papi/
2. Bille, P., Stöckel, M.: Fast and cache-oblivious dynamic programming with local dependencies. In: Dediu, A.-H., Martín-Vide, C. (eds.) LATA 2012. LNCS, vol. 7183, pp. 131–142. Springer, Heidelberg (2012)
3. Burge, C., Karlin, S.: Prediction of complete gene structures in human genomic DNA. J. Mol. Biol. **268**(1), 78–94 (1997)
4. Cherng, C., Ladner, R.E.: Cache efficient simple dynamic programming. In: Proceedings of AofA, pp. 49–58 (2005)
5. Chin, W., Tan, S., Teo, Y.: Deriving efficient parallel programs for complex recurrences. In: Proceedings of PASCO, pp. 101–110 (1997)
6. Chin, W.N., Darlington, J., Guo, Y.: Parallelizing conditional recurrences. In: Fraigniaud, P., Mignotte, A., Bougé, L., Robert, Y. (eds.) Euro-Par 1996. LNCS, vol. 1123, pp. 579–586. Springer, Heidelberg (1996)
7. Chowdhury, R.A., Ganapathi, P., Tithi, J.J., Bachmeier, C., Kuszmaul, B.C., Leiserson, C.E., Solar-Lezama, A., Tang, Y.: AutoGen: automatic discovery of cache-oblivious parallel recursive algorithms for solving dynamic programs. In: Proceedings of PPoPP, p. 10. ACM (2016)
8. Chowdhury, R.A.: Cache-efficient algorithms and data structures: theory and experimental evaluation. Ph.D. thesis, Department of Computer Sciences, The University of Texas at Austin (2007)
9. Chowdhury, R.A., Ramachandran, V.: Cache-oblivious dynamic programming. In: Proceedings of SODA, pp. 591–600 (2006)
10. Chowdhury, R.A., Ramachandran, V.: Cache-efficient dynamic programming algorithms for multicores. In: Proceedings of SPAA, pp. 207–216 (2008)
11. Chowdhury, R.A., Ramachandran, V.: The cache-oblivious Gaussian elimination paradigm: theoretical framework, parallelization and experimental evaluation. Theory Comput. Syst. **47**(4), 878–919 (2010)
12. Chowdhury, R.A., Ramachandran, V., Silvestri, F., Blakeley, B.: Oblivious algorithms for multicores and networks of processors. J. Parallel Distrib. Comput. **73**(7), 911–925 (2013)
13. Chowdhury, R.A., Le, H.S., Ramachandran, V.: Cache-oblivious dynamic programming for bioinformatics. IEEE/ACM Trans. Comput. Biol. Bioinf. **7**(3), 495–510 (2010)
14. Cormen, T.H., Leiserson, C.E., Rivest, R.L., Stein, C.: Introduction to Algorithms. MIT Press, Cambridge (2001)
15. Costello, D.J., Hagenauer, J., Imai, H., Wicker, S.B.: Applications of error-control coding. IEEE Trans. Inf. Theory **44**(6), 2531–2560 (1998)
16. Cutting, D., Kupiec, J., Pedersen, J., Sibun, P.: A practical part-of-speech tagger. In: Proceedings of ANLC, pp. 133–140. Association for Computational Linguistics (1992)
17. Durbin, R., Eddy, S.R., Krogh, A., Mitchison, G.: Biological Sequence Analysis: Probabilistic Models of Proteins and Nucleic Acids. Cambridge University Press, Cambridge (1998)
18. Ferreira, M., Roma, N., Russo, L.M.: Cache-oblivious parallel SIMD Viterbi decoding for sequence search in HMMER. Bioinformatics **15**(1), 165 (2014)
19. Fisher, A.L., Ghuloum, A.M.: Parallelizing complex scans and reductions. ACM SIGPLAN Notices **29**(6), 135–146 (1994)

20. Frigo, M., Leiserson, C.E., Prokop, H., Ramachandran, S.: Cache-oblivious algorithms. In: Proceedings of FOCS, pp. 285–297 (1999)
21. Heller, J., Jacobs, I.: Viterbi decoding for satellite and space communication. IEEE Trans. Commun. Technol. **19**(5), 835–848 (1971)
22. Klein, D., Manning, C.D.: A* parsing: fast exact Viterbi parse selection. In: Proceedings of NAACL, pp. 40–47 (2003)
23. Kobayashi, H.: Application of probabilistic decoding to digital magnetic recording systems. IBM J. Res. Dev. **15**(1), 64–74 (1971)
24. Krogh, A., Larsson, B., Von Heijne, G., Sonnhammer, E.L.: Predicting transmembrane protein topology with a hidden Markov model: application to complete genomes. J. Mol. Biol. **305**(3), 567–580 (2001)
25. Liu, C.: cuHMM: A CUDA implementation of hidden Markov model training and classification. The Chronicle of Higher Education (2009)
26. Maleki, S., Musuvathi, M., Mytkowicz, T.: Parallelizing dynamic programming through rank convergence. In: Proceedings of PPoPP, pp. 219–232 (2014)
27. Maleki, S., Musuvathi, M., Mytkowicz, T.: Low-rank methods for parallelizing dynamic programming algorithms. ACM Trans. Parallel Comp. **2**(4), 26 (2016)
28. Nam, H., Kwak, H.: Viterbi decoder for a high definition television (1998). http://www.google.com/patents/US5844945 US Patent 5,844,945
29. Ohler, U., Niemann, H., Liao, G.C., Rubin, G.M.: Joint modeling of DNA sequence and physical properties to improve eukaryotic promoter recognition. Bioinformatics **17**(Suppl. 1), S199–S206 (2001)
30. Siepel, A., Bejerano, G., Pedersen, J.S., Hinrichs, A.S., Hou, M., Rosenbloom, K., Clawson, H., Spieth, J., Hillier, L.W., Richards, S., et al.: Evolutionarily conserved elements in vertebrate, insect, worm, and yeast genomes. Genome Res. **15**(8), 1034–1050 (2005)
31. Tan, G., Feng, S., Sun, N.: Locality and parallelism optimization for dynamic programming algorithm in bioinformatics. In: Proceedings of SC, p. 78 (2006)
32. Tang, S., Yu, C., Sun, J., Lee, B.S., Zhang, T., Xu, Z., Wu, H.: EasyPDP: an efficient parallel dynamic programming runtime system for computational biology. IEEE Trans. Parallel Distrib. Syst. **23**(5), 862–872 (2012)
33. Tang, Y., Chowdhury, R.A., Luk, C.K., Leiserson, C.E.: Coding stencil computations using the Pochoir stencil-specification language. In: Proceedings of HotPar (2011)
34. Tithi, J.J., Ganapathi, P., Talati, A., Aggarwal, S., Chowdhury, R.A.: High-performance energy-efficient recursive dynamic programming with matrix-multiplication-like flexible kernels. In: Proceedings of IPDPS (2015)
35. Treibig, J., Hager, G., Wellein, G.: Likwid: a lightweight performance-oriented tool suite for x86 multicore environments. In: Proceedings of ICPPW, pp. 207–216 (2010)
36. Viterbi, A.J.: Error bounds for convolutional codes and an asymptotically optimum decoding algorithm. IEEE Trans. Inf. Theory **13**(2), 260–269 (1967)
37. Viterbi, A.J.: Convolutional codes and their performance in communication systems. IEEE Trans. Commun. Technol. **19**(5), 751–772 (1971)

Gradual Stabilization Under τ-Dynamics

Karine Altisen[1], Stéphane Devismes[1], Anaïs Durand[1(✉)], and Franck Petit[2]

[1] VERIMAG UMR 5104, Université Grenoble Alpes, Saint-Martin-d'Hères, France
anais.durand@imag.fr
[2] LIP6 UMR 7606, INRIA, UPMC Sorbonne Universités, Paris, France

Abstract. In this paper, we introduce the notion of *gradually stabilizing* algorithm as any self-stabilizing algorithm with the following additional feature: if at most τ *dynamic steps* occur starting from a legitimate configuration, it first quickly recovers to a configuration from which a minimum quality of service is satisfied and then gradually converges to stronger and stronger safety guarantees until reaching a legitimate configuration again. We illustrate this new property by proposing a gradually stabilizing unison algorithm.

1 Introduction

Self-stabilization [10] is a general paradigm to enable the design of distributed systems tolerating *any* finite number of transient faults. Consider the first configuration after all transient faults cease. This configuration is arbitrary, but no other transient faults will ever occur from this configuration. By abuse of language, this configuration is referred to as *arbitrary initial configuration* of the system in the literature. Then, a self-stabilizing algorithm (provided that faults have not corrupted its code) guarantees that starting from an arbitrary initial configuration, the system recovers *within finite time*, without any external intervention, to a so-called *legitimate configuration* from which its specification is satisfied. Thus, self-stabilization makes no hypotheses on the nature (*e.g.*, memory corruptions, topological changes) of transient faults, and the system recovers from the effects of those faults in a unified manner. Such versatility comes at a price, *e.g.*, after transient faults cease, there is a finite period of time, called *stabilization phase*, during which safety properties of the system may be violated. Hence, self-stabilizing algorithms are mainly compared according to their *stabilization time*, *i.e.*, the maximum duration of the stabilization phase. Many problem specifications induce a significant stabilization time, *e.g.*, in the context of synchronization tasks [3] and more generally for specifications of non-static problems [13], such as broadcast, the lower bound is $\Omega(\mathcal{D})$ rounds, where $\mathcal{D}$ is the diameter of the network. By definition, the stabilization time is impacted by worst case scenarios, but, in many cases, transient faults are sparse and their effect may be superficial. Recent research thus focuses on proposing self-stabilizing algorithms that also ensure drastically smaller convergence times in favorable cases.

© Springer International Publishing Switzerland 2016
P.-F. Dutot and D. Trystram (Eds.): Euro-Par 2016, LNCS 9833, pp. 588–602, 2016.
DOI: 10.1007/978-3-319-43659-3_43

Defining the number of faults hitting a network using some kind of Hamming distance (minimal number of processes whose state must be changed in order to recover a legitimate configuration), variants of self-stabilization have been defined. A *time-adaptive* self-stabilizing algorithm [21] additionally guarantees a convergence time in $O(k)$ time units when the initial configuration is at distance at most k from a legitimate configuration. *Fault containing* self-stabilizing algorithms [14] ensure that when few faults hit the system, the faults are both spatially and temporally contained. "Spatially" means that those faults cannot be propagated further than a preset radius around the corrupted processes. "Temporally" means quick stabilization when few faults occur. Some other approaches consist in providing convergence times *tailored by the type of transient faults*, e.g., a *superstabilizing algorithm* [11] is self-stabilizing and has two additional properties when transient faults are limited to a *single* topological change: after adding or removing one link or process in the network, it recovers fast (typically $O(1)$ rounds), and a safety predicate, so-called *passage*, should be satisfied meanwhile.

Contributions. We introduce the notion of *gradually stabilizing* algorithm as any *self-stabilizing* algorithm achieving the following additional feature. If at most τ *dynamic steps*[1] occur starting from a legitimate configuration, a gradually stabilizing algorithm first quickly recovers to a configuration from which a specification offering a minimum quality of service is satisfied. It then gradually converges to specifications offering stronger and stronger safety guarantees until reaching a configuration from which its initial (strong) specification is satisfied again, and where it is ready to achieve another gradual convergence in case of up to τ new dynamic steps. Of course, this property makes sense only if convergence to every intermediate weaker specification is fast.

We illustrate this new property by considering three variants of a synchronization problem respectively called *strong, weak,* and *partial* (asynchronous) unison. In these problems, each process maintains a local clock. We restrict our study to periodic clocks, *i.e.*, clocks are integer variables whose domain is $\{0, \ldots, \alpha - 1\}$, where $\alpha \geq 2$ is called the *period*. Each process should regularly increment its clock modulo α (liveness) while fulfilling some safety requirements. The safety of *strong unison* requires that at most two consecutive clock values exist in each configuration of the system. *Weak unison* only requires that the difference between clocks of every two neighbors is at most one increment. Finally, we define *partial unison* as a specification dedicated to dynamic systems which enforces the difference between clocks to remain at most one increment, but only for neighboring processes that do not appear during the dynamic steps.

We propose a self-stabilizing strong unison algorithm which works with any period $\alpha > 4$ in any anonymous connected network. It assumes the knowledge of two values μ and β, where μ is any upper bound on n — the (initial) number of processes, α should divide β, and $\beta > \mu^2$. Our algorithm is designed in the locally shared memory model and assumes the distributed unfair daemon, the most general daemon of the model. Its stabilization time is at most $n + (\mu + 1)\mathcal{D} + 1$ rounds, where $\mathcal{D}$ is the diameter of the network. We then slightly

---

[1] *N.b.*, a dynamic step is a step containing topological changes.

modify this algorithm to make it gradually stabilizing after one dynamic step. In particular, the parameter μ should be now at least $n + \#J$, where $\#J$ is an upper bound on the number of processes that join the system during a dynamic step. This new version is *gradually stabilizing* because after one dynamic step from a configuration which is legitimate for strong unison, it immediately satisfies the specification of partial unison, then converges to the specification of weak unison in at most one round, and finally retrieves, after at most $(\mu+1)\mathcal{D}_1+1$ additional rounds (where $\mathcal{D}_1$ is the diameter of the network after the dynamic step), a configuration from which the specification of strong unison is satisfied and where it is ready to achieve gradual convergence again in case of another dynamic step. This result holds considering dynamic steps which may contain several link and/or process additions and/or removals, however we assume that after a dynamic step, the network stays connected and, if $\alpha > 4$, every new process is linked to at least one process already in the system before the dynamic step. We show that this condition, called UnderLocalControl, is necessary to obtain gradual convergence. However, notice that if the system suffers from arbitrary other kinds of transient fault including, *e.g.*, several dynamic steps that do not satisfy the UnderLocalControl condition, our algorithm still converges to strong unison, yet without intermediate safety guarantees during the stabilization phase.

Related Work. Gradual stabilization is related to two other stronger forms of self-stabilization: *safe-converging self-stabilization* [19] and *superstabilization* [11]. The goal of a safely converging self-stabilizing algorithm is to first quickly ($O(1)$ rounds is the usual rule) converge from an arbitrary configuration to a *feasible* legitimate configuration, where a minimum quality of service is guaranteed. Once such a feasible legitimate configuration is reached, the system continues to converge to an *optimal* legitimate configuration, where more stringent conditions are required. Hence, the aim of safe-converging self-stabilization is also to ensure a gradual convergence, but only for two specifications. However, such a gradual convergence is stronger than ours as it should be ensured after any step of transient faults,[2] while our gradual convergence applies after dynamic steps only. Safe convergence is especially interesting for self-stabilizing algorithms that compute optimized data structures, *e.g.*, minimal dominating sets [19], minimal (f, g)-alliances [8]. However, to the best of our knowledge, no safe-converging algorithm for non-static problems, such as unison, has been proposed until now.

In superstabilization, like in our approach, fast convergence and the passage predicate should be ensured only if the system was in a legitimate configuration before the topological change occurs. In contrast with our approach, superstabilization ensures fast convergence to the original specification. However, this strong property only considers one dynamic step with only *one* topological event. Again, superstabilization has been especially studied in the context of static problems, *e.g.*, spanning tree construction [4,5,11], and coloring [11]. However, there exist few superstabilizing algorithms for non-static problems in particular topologies, *e.g.*, mutual exclusion in rings [16,20].

---

[2] Such transient faults may include topological changes, but not only.

We use the general term *unison* to name several close problems also known in the literature as *phase* or *barrier synchronization* problems. There exist many self-stabilizing algorithms for strong or weak unison problems, *e.g.*, [2,6,7,15,17,18,22,23]. However, to the best of our knowledge, until now there was no self-stabilizing solution for such problems addressing specific convergence properties in case of topological changes, in particular no superstabilizing one. Self-stabilizing strong unison was first considered in synchronous anonymous networks. Particular topologies were considered in [17] (rings) and [22] (trees). Gouda and Herman [15] proposed a self-stabilizing algorithm for strong unison working in anonymous synchronous systems of arbitrary connected topology. However, they considered unbounded clocks. A solution working with the same settings, yet implementing bounded clocks, is proposed in [2]. In [23], an asynchronous self-stabilizing strong unison algorithm is proposed for arbitrary connected rooted networks.

Johnen *et al.* investigated asynchronous self-stabilizing weak unison in oriented trees in [18]. The first self-stabilizing asynchronous weak unison for general graphs was proposed by Couvreur *et al.* [9]. However, no complexity analysis was given. Another solution which stabilizes in $O(n)$ rounds has been proposed by Boulinier *et al.* in [7]. Finally, Boulinier proposed in his PhD thesis a parametric solution which generalizes both the solutions of [9] and [7]. In particular, the complexity analysis of this latter algorithm reveals an upper bound in $O(\mathcal{D}.n)$ rounds on the stabilization time of the Couvreur *et al.*' algorithm.

Roadmap. In the next section, we define the computational model used in this paper. In Sect. 3, we recall the formal definition of self-stabilization, and introduce the notion of gradual stabilization. In Sect. 4, we show that condition Under-LocalControl is necessary to obtain a gradually stabilizing solution. We present our self-stabilizing strong unison algorithm in Sect. 5. The gradually stabilizing variant of this latter algorithm is proposed in Sect. 6. We make concluding remarks in Sect. 7.

Due to the lack of space, proofs are omitted, see the report online [1] for details.

2 Preliminaries

We consider distributed systems made of *anonymous* processes. The system *initially contains $n > 0$ processes and its topology is connected*, however it may suffer from topological changes over time. Each process p can directly communicate with a subset $p.\mathcal{N}$ of other processes, its *neighbors*. In our context, $p.\mathcal{N}$ can vary over time. Communications are assumed to be *bidirectional* and carried out by a finite set of locally shared variables: each process can read its own variables and those of its current neighbors, but can only write into its own variables. The *state* of a process is the vector of values of its variables. We denote by $\mathcal{S}$ the set of all possible states of a process. Each process updates its variables according to a *local algorithm*. The collection of all local algorithms defines a *distributed algorithm*. The local algorithm of p consists of a finite set of *actions* of the following form: $\langle$ label $\rangle::\langle$ guard $\rangle \rightarrow \langle$ statement $\rangle$. *Labels* are used to identify actions in the reasoning. The *guard* of an action is a Boolean predicate involving variables of p and its neighbors. The *statement* is a sequence of assignments on variables of p. If the guard of some action evaluates to true, the action is said to be *enabled* at p. By extension, if at least one action is enabled at p, p is said to be enabled. An action can be executed only if it is enabled. The execution of an action consists in executing its statement, atomically. A *configuration* γ_i is a pair $(G_i, V_i \rightarrow \mathcal{S})$. $G_i = (V_i, E_i)$ is a simple undirected graph, where V_i is the set of processes that exist in γ_i and E_i represents the links between processes in γ_i. $V_i \rightarrow \mathcal{S}$ is a function which associates a state to any process of V_i. We denote by $\mathcal{C}$ the set of all possible configurations.

Executions. The dynamicity and asynchronism of the system are materialized by an adversary, called *daemon*. To perform a *step* from a configuration γ_i, the daemon can (1) activate processes that are enabled in γ_i — each activated process executes one of its enabled actions according to its state and that of its neighbors in γ_i, and/or (2) modify the topology. Activation of enabled processes and/or topology modifications are done atomically, leading to a new configuration γ_{i+1}. The set of all possible steps induces a binary relation $\mapsto$ over configurations (empty steps of the form $\gamma_i \mapsto \gamma_i$ are excluded). Relation $\mapsto$ is partitioned into $\mapsto_s$ and $\mapsto_d$. Relation $\mapsto_s$ defines all possible *static steps* consisting in activation of enabled processes *only*. Relation $\mapsto_d$ defines all possible *dynamic steps* containing topological changes and possibly process activations.

An *execution* is a sequence of configurations $\gamma_0, \gamma_1, \ldots$ such that G_0 is connected and $\forall i \geq 0$, $\gamma_i \mapsto \gamma_{i+1}$. For sake of simplicity, we note $G_0 = G = (V, E)$; we also note $\mathcal{D}$ the diameter of G. Moreover, we note $\mathcal{E}^\tau$ the set of maximal executions which contain at most τ dynamic steps. The set of all possible executions is therefore equal to $\mathcal{E} = \cup_{\tau \geq 0} \mathcal{E}^\tau$. For any subset of configurations $X \subseteq \mathcal{C}$, we denote by $\mathcal{E}_X^\tau$ the set of all executions in $\mathcal{E}^\tau$ that start from a configuration of X.

Dynamic Steps. Any step $\gamma_i \mapsto_d \gamma_{i+1}$ contains a finite number of topological events and maybe some process activations. Each topological event is of the following types. (1) A process p can *join* the system. This event, denoted by $join_p$, triggers the atomic execution of a specific action, called *bootstrap*. This bootstrap is executed without any communication and initializes the variables of p to a particular state, called *bootstate*. We denote by New_k the set of processes which are in bootstate in γ_k. When p joins the system in $\gamma_i \mapsto_d \gamma_{i+1}$, we have $p \in New_{i+1}$, but $p \notin New_i$. Until p executes its bootstrap, say in step $\gamma_x \mapsto \gamma_{x+1}$, it is still in bootstate. Hence, $\forall j \in \{i+1, \ldots, x\}, p \in New_j$, but $p \notin New_{x+1}$. We assume that there are at most #J joins during a dynamic step. (2) A process can also *leave* the system. (3) Finally, some communication links can *appear* or *disappear* between two different processes.

Daemon. We assume the daemon is *distributed* and *unfair*. In a static step, this daemon must select at least one enabled process. In a dynamic step, it can select zero, one, or several enabled processes. It has no fairness constraint, *i.e.*, it might never select a process p during any step unless in the case of a static step from a configuration where p is the only enabled process. Moreover, at each configuration, it freely chooses between making a static or dynamic step, except if no more process is enabled; in this latter case, only a dynamic step containing no process activation can be chosen.

Metrics. We measure the time complexity of our algorithms in *rounds* [12]. The first round of an execution $e = (\gamma_i)_{i \geq 0}$ is the minimal prefix e' of e in which every process that is enabled in γ_0 either disappears, or executes an action, or becomes disabled (due to some changes in its neighborhood). Let γ_j be the last configuration of e', the second round of e is the first round of $e'' = (\gamma_i)_{i \geq j}$, and so on.

Specifications. We define a specification as a predicate over executions. We denote by SP_{su} and SP_{wu} the respective specifications for *strong* and *weak* unison. The specification of *partial* unison, noted SP_{pu}, does not impose any constraint on processes that join the system until they achieve their bootstrap: the safety holds as long as clocks of every two neighboring processes *not in bootstate* differ from at most one increment.

3   Stabilization

Self-stabilization has been defined by only considering executions free of topological changes, yet starting from an arbitrary configuration. Indeed, self-stabilization considers the system immediately after transient faults cease. So, the system is initially observed from an arbitrary configuration reached after occurrences of transient faults (including some topological changes maybe), but from which no faults will ever occur. Below, we recall the definitions of some notions classically used in self-stabilization for a given distributed algorithm $\mathcal{A}$. Let X and Y be two subsets of configurations.

- X is *closed under $\mathcal{A}$ iff* every *static step* of $\mathcal{A}$ starting from a configuration of X leads to a configuration which is also in X.
- Y *converges to X under $\mathcal{A}$ iff* every execution of $\mathcal{E}_Y^0$ contains a configuration of X.
- $\mathcal{A}$ *stabilizes from Y to a specification SP by X iff* X is closed under $\mathcal{A}$, Y converges to X under $\mathcal{A}$, and every execution of $\mathcal{E}_X^0$ satisfies SP. In this case, the *convergence time from Y to X in rounds* is the maximal number of rounds to reach a configuration of X over every execution of $\mathcal{E}_Y^0$.

A distributed algorithm $\mathcal{A}$ is *self-stabilizing* for a specification SP iff $\exists \mathcal{L} \subseteq \mathcal{C}$ such that $\mathcal{A}$ stabilizes from $\mathcal{C}$ to SP by $\mathcal{L}$. $\mathcal{L}$ is said to be a set of *legitimate configurations w.r.t. SP*, and the convergence time from $\mathcal{C}$ to $\mathcal{L}$ is called *stabilization time* of $\mathcal{A}$.

Gradual Stabilization Under τ-Dynamics. This property is a specialization of self-stabilization which additionally requires that after at most τ dynamic steps from a legitimate configuration, the system gradually re-stabilizes to stronger and stronger specifications, until fully recovering its initial (strong) specification. For every execution $e = (\gamma_i)_{i \geq 0} \in \mathcal{E}^\tau$, we note $\gamma_{fst(e)}$ the first configuration of e after the last dynamic step. Formally, $fst(e) = \min\{i : (\gamma_j)_{j \geq i} \in \mathcal{E}^0\}$. For any subset E of $\mathcal{E}^\tau$, let $FC(E) = \{\gamma_{fst(e)} : e \in E\}$ be the set of all configurations that can be reached after the last topological changes in executions of E. Let $SP_1, SP_2, \ldots, SP_k$, be an ordered sequence of specifications. Let $B_1, B_2, \ldots, B_k$ be (asymptotic) complexity bounds such that $B_1 \leq B_2 \leq \cdots \leq B_k$.

A distributed algorithm $\mathcal{A}$ is *gradually stabilizing under τ-dynamics* for $(SP_1 \bullet B_1, SP_2 \bullet B_2, \ldots, SP_k \bullet B_k)$ iff $\exists \mathcal{L}_1, \ldots, \mathcal{L}_k \subseteq \mathcal{C}$ such that

1. $\mathcal{A}$ stabilizes from $\mathcal{C}$ to SP_k by $\mathcal{L}_k$.
2. $\forall i \in \{1, \ldots, k\}$, $\mathcal{A}$ stabilizes from $FC(\mathcal{E}^\tau_{\mathcal{L}_k})$ to SP_i by $\mathcal{L}_i$, and the convergence time in rounds from $FC(\mathcal{E}^\tau_{\mathcal{L}_k})$ to $\mathcal{L}_i$ is bounded by B_i.

The first point ensures that a gradually stabilizing algorithm is still self-stabilizing for its strongest specification. Hence, its performances can be also evaluated at the light of its stabilization time. Indeed, it captures the maximal convergence time of the gradually stabilizing algorithm after the system suffers from an arbitrary finite number of transient faults, *e.g.*, after more than τ dynamic steps.

The second point means that after at most τ dynamic steps from a legitimate configuration *w.r.t.* the strongest specification SP_k, the algorithm *gradually converges* to every specification SP_i with $i \in \{1, \ldots, k\}$ in at most B_i rounds. Note that B_k captures a complexity similar to the *fault gap* in fault-containing algorithms [14]: assume a period of at most τ dynamic steps starting in a legitimate configuration $\mathcal{L}_k$; B_k represents the necessary fault-free interval after this period and before the next period of at most τ dynamic steps so that system becomes ready to achieve gradual convergence again.

4   Necessary Condition

In this section, we establish that Condition UnderLocalControl is necessary to allow the design of a deterministic algorithm $\mathcal{A}$ which is gradually stabilizing under 1-dynamics for $(SP_{PU} \bullet 0, SP_{WU} \bullet 1, SP_{SU} \bullet B)$ (with $B \geq 1$) in any arbitrary anonymous network, assuming the distributed unfair daemon. Below, we assume the existence of $\mathcal{A}$ and denote by $\mathcal{L}^\mathcal{A}_{SU}$ the set of legitimate configurations of $\mathcal{A}$ *w.r.t.* specification SP_{SU}.

UnderLocalControl captures a condition on the network dynamics which is necessary to prevent a notable desynchronization of clocks: the network should stay connected and, if $\alpha > 4$, every process that joins during the dynamic step $\gamma \mapsto_d \gamma'$ should be "under control of" (that is, linked to) at least one process which exists in both γ and γ'. The definition of UnderLocalControl uses the notion of *dominating set* of a graph $G = (V, E)$, *i.e.*, any subset D of V such that every node not in D is adjacent to at least one member of D. Formally, UnderLocalControl holds *iff* $\forall e \in \mathcal{E}^1_{\mathcal{L}^\mathcal{A}_{SU}}$, $G_{fst(e)}$ is connected, and if $\alpha > 4$, then $V_{fst(e)} \setminus New_{fst(e)}$ is a dominating set of $G_{fst(e)}$.

Theorem 1. *An algorithm $\mathcal{A}$ is gradually stabilizing under 1-dynamics for $(SP_{PU} \bullet 0, SP_{WU} \bullet 1, SP_{SU} \bullet B)$ in arbitrary anonymous networks under the distributed unfair daemon only if* UnderLocalControl *holds.*

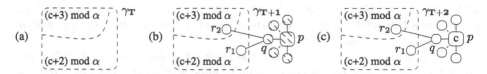

Fig. 1. Proof outline of Theorem 1. The hachured nodes are in bootstate.

Proof Outline. If the graph becomes disconnected after a dynamic step, the distributed unfair daemon can prevent forever all processes of a given connected component from incrementing their clocks, hence violating the liveness of SP_{SU}. Assume, by contradiction, that there is an execution e with $\alpha > 4$ such that $G_{fst(e)}$ is connected but $V_{fst(e)} \setminus New_{fst(e)}$ is not a dominating set. This means that some process p and all its neighbors have been added during the dynamic step. First, to satisfy SP_{WU} after at most one round, p and its neighbors should be enabled to take a clock value immediately after the dynamic step. Let c be the clock value that p would choose in this case. Then, we build another execution e' initiated from a configuration in $\mathcal{L}_{SU}^{\mathcal{A}}$ on another graph containing at least two nodes which are neither p, nor one of its neighbors. As SP_{SU} holds and the execution can be asynchronous, it is possible for the system to eventually reach a configuration γ_T where there are exactly two clock values: $(c+2) \bmod \alpha$ and $(c+3) \bmod \alpha$ (see Fig. 1(a)). Then, assume the daemon chooses to execute, during $\gamma_T \mapsto_d \gamma_{T+1}$, the dynamic step which contains no process activation, but introduces p, its neighborhood, and two links, just as in Fig. 1(b). Then, after this step, SP_{PU} should be satisfied. Finally, assume that the daemon selects no process, except p and its neighbors in the next step. As before, p sets its clock to c, but, as $\alpha > 4$, whatever be the value chosen by q, there is a difference greater than one increment between q and at least one of its neighbors (Fig. 1(c)). Henceforth, the legitimate configurations of SP_{PU} are not closed under $\mathcal{A}$, a contradiction. $\square$

5   Self-Stabilizing Strong Unison

In this section, we propose an algorithm which is self-stabilizing for strong unison in any arbitrary connected anonymous network. This algorithm works for any period $\alpha > 4$ and is based on an algorithm previously proposed by Boulinier in his PhD [6], this latter is self-stabilizing for weak unison and works for any period $\beta > n^2$.

Algorithm $\mathcal{WU}$. We first recall the algorithm of Boulinier [6], noted here Algorithm $\mathcal{WU}$. This algorithm being just self-stabilizing, it only considers executions without any topological change, yet starting from arbitrary configurations.

So, the topology of the network consists in a connected graph $G = (V, E)$ of n nodes which is fixed all along the execution. Each process p is endowed with a clock variable $p.t \in \{0, \ldots, \beta - 1\}$, where β is its period. β should be greater than n^2. The algorithm also uses another constant, noted μ, which should satisfy $n \leq \mu \leq \frac{\beta}{2}$. The algorithm uses the notion of *delay* between two integer values x and y, defined by the function $d_\beta(x, y) = \min\big((x - y) \bmod \beta, (y - x) \bmod \beta\big)$. It also uses the relation $\preceq_{\beta,\mu}$ such that for every two integer values x and y, $x \preceq_{\beta,\mu} y \equiv \big((y - x) \bmod \beta\big) \leq \mu$.

Two actions are used to maintain the clock $p.t$ at each process p. When the delay between $p.t$ and the clocks of some neighbors is greater than one, but the maximum delay is not too big (that is, does not exceed μ), then it is possible to "normally" converge, using Action $\mathcal{WU}$-N below, to a configuration where the delay between those clocks is at most one by incrementing the clocks of the most behind processes among p and its neighbors: $\mathcal{WU}$-N::$\forall q \in p.\,\mathcal{N}, p.t \preceq_{\beta,\mu} q.t \to$ $p.t \leftarrow (p.t + 1) \bmod \beta$

Moreover, once legitimacy is achieved, p can "normally" increment its clock still using Action $\mathcal{WU}$-N when it is on time or one increment late with all its neighbors. In contrast, if the delay is too big (that is, the delay between the clocks of p and one of its neighbors is more than μ) and the clock of p is not yet reset, then p should reset its clock to 0 using Action $\mathcal{WU}$-R: $\mathcal{WU}$-R::$\exists q \in$ $p.\,\mathcal{N}, d_\beta(p.t, q.t) > \mu \wedge p.t \neq 0 \to p.t \leftarrow 0$

Algorithm $\mathcal{SU}$. For this algorithm, we still assume a non-dynamic context (no topological change). Algorithm $\mathcal{SU}$ is a straightforward adaptation of Algorithm $\mathcal{WU}$. More precisely, Algorithm $\mathcal{SU}$ maintains two clocks at each process p. The first one, $p.t \in \{0, \ldots, \beta - 1\}$, is called the *internal clock* and is maintained exactly as in Algorithm $\mathcal{WU}$. Then, $p.t$ is used as an internal pulse machine to increment a second, yet actual, clock of Algorithm $\mathcal{SU}$ $p.c \in \{0, \ldots, \alpha - 1\}$, also called *external clock*.

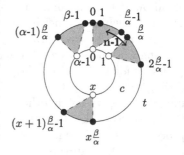

Fig. 2. From t to c.

Algorithm $\mathcal{SU}$ is designed for any period $\alpha > 4$. Its actions $\mathcal{SU}$-N and $\mathcal{SU}$-R are identical to actions $\mathcal{WU}$-N and $\mathcal{WU}$-R of Algorithm $\mathcal{WU}$, except that we add the computation of the external c-clock in their respective statement.

$\mathcal{SU}$-N :: $\forall q \in p.\,\mathcal{N}, p.t \preceq_{\beta,\mu} q.t$ $\to p.t \leftarrow (p.t + 1) \bmod \beta;\ p.c \leftarrow \big\lfloor \frac{\alpha}{\beta} p.t \big\rfloor$
$\mathcal{SU}$-R :: $\exists q \in p.\,\mathcal{N}, d_\beta(p.t, q.t) > \mu \wedge p.t \neq 0 \to p.t \leftarrow 0;\ p.c \leftarrow 0$

Algorithm $\mathcal{WU}$ stabilizes to a configuration from which t-clocks regularly increment while preserving a bounded delay of at most one between two neighboring processes, and so of at most $n - 1$ between any two processes. Algorithm $\mathcal{SU}$ implements the same mechanism to maintain $p.t$ at each process p and computes $p.c$ from $p.t$ as a normalization operation from clock values in $\{0, \ldots, \beta - 1\}$ to $\{0, \ldots, \alpha - 1\}$: each time the value of $p.t$ is modified, $p.c$ is updated to $\big\lfloor \frac{\alpha}{\beta} p.t \big\rfloor$. Hence, we can set β in such way that $K = \frac{\beta}{\alpha}$ is greater than

Algorithm 1. $\mathcal{DSU}$, for every process p

Parameters:

    α: any positive integer such that $\alpha > 4$

    μ: any positive integer such that $\mu \geq n + \#J$

    β: any positive integer such that $\beta > \mu^2$,

    and $\exists K$ such that $K > \mu$ and $\beta = K\alpha$

Variables: $p.c \in \{0, \ldots, \alpha - 1\} \cup \{\bot\}$, $p.t \in \{0, \ldots, \beta - 1\} \cup \{\bot\}$

Predicates:

    $Locked_p$ $\equiv p.t = \bot \vee \exists q \in p.\mathcal{N}, q.t = \bot$

    $NormalStep_p$ $\equiv \neg Locked_p \wedge \forall q \in p.\mathcal{N}, p.t \preceq_{\beta,\mu} q.t$

    $ResetStep_p$ $\equiv \neg Locked_p \wedge (\exists q \in p.\mathcal{N}, d_\beta(p.t, q.t) > \mu \wedge p.t \neq 0)$

    $JoinStep_p$ $\equiv p.t = \bot$

Actions:

    $\mathcal{DSU}\text{-}N$ $:: NormalStep_p \;\rightarrow\; p.t \leftarrow (p.t + 1) \bmod \beta; \; p.c \leftarrow \left\lfloor \frac{\alpha}{\beta} p.t \right\rfloor$

    $\mathcal{DSU}\text{-}R$ $:: ResetStep_p \;\rightarrow\; p.t \leftarrow 0; \; p.c \leftarrow 0$

    $\mathcal{DSU}\text{-}J$ $:: JoinStep_p \;\rightarrow\; p.t \leftarrow MinTime_p; \; p.c \leftarrow \left\lfloor \frac{\alpha}{\beta} p.t \right\rfloor$

    $bootstrap$ $:: join_p \;\rightarrow\; p.t \leftarrow \bot; \; p.c \leftarrow \bot$

---

or equal to n (here, we choose $K > \mu \geq n$ and $\beta > \mu^2$ for sake of simplicity) to ensure that, when the delay between any two t-clocks is at most $n - 1$, the delay between any two c-clocks is at most one, see Fig. 2. The liveness of $\mathcal{WU}$ ensures that every t-clock increments infinitely often, thus so do c-clocks.

Theorem 2. *Algorithm $\mathcal{SU}$ is self-stabilizing for SP_{su} in any arbitrary connected anonymous network assuming a distributed unfair daemon. Its stabilization time is at most $n + (\mu + 1)\mathcal{D} + 1$ rounds.*

We have also proven that, once $\mathcal{SU}$ has stabilized, every process increments its c-clock at least once every $\mathcal{D} + \frac{\beta}{\alpha}$ rounds. This result derives from [6] which states that after stabilization of t-clocks, those ones increment at least once every $\mathcal{D} + 1$ rounds.

6 Gradual Stabilization Under 1-Dynamics for Strong Unison

We now propose Algorithm $\mathcal{DSU}$ (Algorithm 1), a variant of Algorithm $\mathcal{SU}$. $\mathcal{DSU}$ is still self-stabilizing for strong unison, but also achieves a gradual convergence after one dynamic step. This dynamic step may include several topological events (*i.e.* link or process additions or removals). However, according to Theorem 1, it should satisfy Condition UnderLocalControl. Precisely, after any dynamic step which fulfills condition UnderLocalControl, $\mathcal{DSU}$ maintains clocks almost synchronized during the convergence to strong unison since it immediately satisfies partial unison, then converges in at most one round to weak unison, and finally re-stabilizes to strong unison. Remember that, after one dynamic step, the graph contains at most $n + \#J$ processes, by definition, and $\mathcal{D}_1$ denotes the diameter of the new graph.

We first showed a result allowing to simplify proofs and explanations: for every closed set of configurations X, if UnderLocalControl holds, then $\forall \gamma_i \in \mathcal{C}, (\exists \gamma_j \in X \mid \gamma_j \mapsto_d \gamma_i) \Leftrightarrow (\exists \gamma_k \in X \mid \gamma_k \mapsto_{d_{only}} \gamma_i)$, where $\mapsto_{d_{only}}$ is the relation defining all dynamic steps containing no process activa-

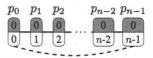

Fig. 3. Link addition.

tion. We apply this result to the set of legitimate configurations w.r.t. strong unison, noted $\mathcal{L}_{SU}^d$ (n.b., $\mathcal{L}_{SU}^d$ is closed, by definition): the set of configurations reachable from $\mathcal{L}_{SU}^d$ after one dynamic step (which may also include process activations) is the same as the one reachable from $\mathcal{L}_{SU}^d$ after one dynamic step made of topological events only. At the light of this result, we only consider this latter kind of dynamic steps in the following.

Consider first link additions only. Adding a link (see the dashed link in Fig. 3) can break the safety of weak unison on internal clocks. Indeed, it may create a delay greater than one between two new neighboring t-clocks. Nevertheless, the delay between any two t-clocks remains bounded by $n - 1$, consequently, no process will reset its t-clock (Fig. 3 shows a worst case). Moreover, c-clocks still satisfy strong unison immediately after the link addition. Besides, since increments are constrained by neighboring clocks, adding links only reinforces those constraints. Thus, the delay between internal clocks of arbitrary far processes remains bounded by $n - 1$, and so strong unison remains satisfied, in all subsequent static steps. Consider again the example in Fig. 3: before the dynamic step, p_{n-1} had only to wait until p_{n-2} increments p_{n-2} in order to be able to increment its own t-clock; yet after the step, it also has to wait for p_0.

Assume now a dynamic step containing only process and link removals. Due to Condition UnderLocalControl, the network remains connected. Hence, constraints between (still existing) neighbors are maintained: the delay between t-clocks of two neighbors remains bounded by one, see the example in Fig. 4: process p_2 and link $\{p_0, p_3\}$ are removed. So, weak unison on t-clocks remains satisfied and so is strong unison on c-clocks.

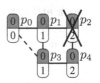

Fig. 4. Removals.

Consider now a more complex scenario, where the dynamic step contains link additions as well as process and/or link removals. Figure 5 shows an example of such a scenario, where safety of strong unison is violated. As above, the addition of link $\{p_1, p_6\}$ in Fig. 5(b) leads to a delay between t-clocks of these two (new) neighbors which is greater than one (here 5). However, the removal of link $\{p_1, p_2\}$, also in Fig. 5(b), relaxes the neighborhood constraint on p_2:p_2 can now increment without waiting for p_1. Consequently, executing Algorithm $\mathcal{SU}$ does not ensure that the delay between t-clocks of any two arbitrary far processes remains bounded by $n - 1$, e.g., after several static steps from Fig. 5(b), the system can reach Fig. 5(c), where the delay between p_1 and p_2 is 9 while $n-1 = 5$. Since c-clock values are computed from t-clock values, we also cannot guarantee that there is at most two consecutive c-clock values in the system, e.g., in Fig. 5(c) we have: $p_1.c = 1$, $p_6.c = 2$, and $p_2.c = 3$.

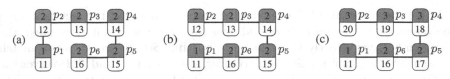

Fig. 5. Execution where links are added and removed ($\mu = 6$, $\alpha = 7$, and $\beta = 42$).

Again, in the worst case scenario, after a dynamic step, the delay between two neighboring t-clocks is bounded by $n - 1$. Moreover, t-clocks being computed like in Algorithm $\mathcal{WU}$, we can use two of its useful properties (see [6]): (1) when the delay between every pair of neighboring t-clocks is at most μ with $\mu \geq n$, the delay between these clocks remains bounded by μ because processes never reset; (2) furthermore, from such configurations, the system converges to a configuration from which the delay between the t-clocks of every two neighbors is at most one. So, keeping $\mu \geq n$, processes will not reset after one dynamic step and the delay between any two neighboring t-clocks will monotonically decrease from at most $n - 1$ to at most one. Consequently, the delay between any two neighboring c-clocks (which are computed from t-clocks) will stay at most one, *i.e.*, weak unison will be satisfied all along the convergence to strong unison.

Consider now a process p that joins the system. The event $join_p$ occurs and triggers the specific action *bootstrap* that sets both the clocks $p.t$ and $p.c$ to a specific *bootstate* value, noted $\perp$. By definition and from the previous discussion, the system immediately satisfies partial unison since it only depends on processes that were in the system before the dynamic step. Now, to ensure that weak unison holds within a round, we add the action $\mathcal{DSU}$-*J* which is enabled as soon as the process is in bootstate. This action initializes the two clocks of p according to the clock values in its neighborhood. Precisely, the value of $p.t$ can be chosen among the non-$\perp$ values in its neighborhood, and such values exist by Condition UnderLocalControl. We choose to set $p.t$ to the minimum non-$\perp$ t-clock value in its neighborhood, using the function $MinTime_p$:

$MinTime_p = 0$ if $\forall q \in p.\mathcal{N}, q.t = \perp$; $= \min\{q.t : q \in p.\mathcal{N} \wedge q.t \neq \perp\}$ **otherwise**.

The value of $p.c$ is then computed according to the value of $p.t$. Notice that $MinTime_p$ returns 0 when p and all its neighbors have their respective t-clock equal to $\perp$. This ensures that Algorithm $\mathcal{DSU}$ remains self-stabilizing (in particular, if the system starts in a configuration where all t-clocks are equal to $\perp$).

To prevent the unfair daemon from blocking the convergence to a configuration containing no $\perp$ values, we should also forbid processes with non-$\perp$ t-clock values to increment while there are t-clocks with $\perp$-values in their neighborhood. So, we define the predicate *Locked* which holds for a given process p when either $p.t = \perp$, or at least one of its neighbors q satisfies $q.t = \perp$. We then enforce the guard of both normal and reset actions, so that no *Locked* process can execute them. See actions $\mathcal{DSU}$-*N* and $\mathcal{DSU}$-*R*. This ensures that t-clocks are initialized first by Action $\mathcal{DSU}$-*J*, before any value in their neighborhood increments.

Finally, notice that all the previous explanation relies on the fact that, once the system recovers from process additions (*i.e.*, once no $\perp$ value remains), the algorithm behaves exactly the same as Algorithm $\mathcal{SU}$. Hence, it has to match the assumptions made for $\mathcal{SU}$, in particular, the ones on α and β. However the constraint on μ has to be adapted, since μ should be greater than or equal to the actual number of processes in the network and this number may increase. Now, the number of processes added in a dynamic step is bounded by $\#J$. So, we require $\mu \geq n + \#J$.

We now consider the example execution of Algorithm $\mathcal{DSU}$ in Fig. 6. This execution starts in a configuration legitimate *w.r.t.* the strong unison, see Fig. 6(a). Then, one dynamic step happens (step (a)$\mapsto$(b)), where a process p_6 joins the system. We now try to delay as long as possible the execution of $\mathcal{DSU}$-J by p_6. In configuration (b), p_3 and p_5, the new neighbors of p_6, are locked. They will remain disabled until p_6 executes $\mathcal{DSU}$-J. p_1 and p_4 execute $\mathcal{DSU}$-N in (b)$\mapsto$(c). Then, p_4 is disabled because of p_5 and p_1 executes $\mathcal{DSU}$-N in (c)$\mapsto$(d). In configuration (d), p_1 is from now on disabled: p_1 must wait until p_2 and p_4 get t-clock value 7. p_6 is the only enabled process, so the unfair daemon has no other choice but selecting p_6 to execute $\mathcal{DSU}$-J in the next step.

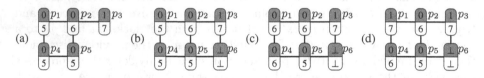

Fig. 6. Execution where the first step of a new process is delayed ($\mu = 6$, $\alpha = 6$, $\beta = 42$).

Theorem 3. *If UnderLocalControl is satisfied then Algorithm $\mathcal{DSU}$ is gradually stabilizing under 1-dynamics for* $(SP_{PU} \bullet 0, SP_{WU} \bullet 1, SP_{SU} \bullet (\mu + 1)\mathcal{D}_1 + 2)$.

After one dynamic step that fulfills Condition UnderLocalControl from any legitimate configuration *w.r.t.* strong unison, the system re-stabilizes to strong unison in at most $(\mu + 1)\mathcal{D}_1 + 2$ rounds. Now, in any other cases (*e.g.*, a dynamic step that does not satisfy UnderLocalControl), the system still recovers to a legitimate configuration within finite time, as the algorithm is self-stabilizing. Nevertheless, in such cases, the stabilization time is slightly bigger: $n + \#J + (\mu + 1)\mathcal{D}_1 + 2$ rounds.

Finally, we have proven [1] that after stabilization to strong unison, every process increments its c-clock at least once every $\mathcal{D}_1 + \frac{\beta}{\alpha}$ rounds, like in Algorithm $\mathcal{SU}$. Moreover, during the convergence from weak to strong unison, the increments are slower, *i.e.*, the c-clocks are guaranteed to increment at least once every $\mu\mathcal{D}_1 + \frac{\beta}{\alpha}$ rounds.

7   Conclusion

The apparent seldomness of superstabilizing solutions for non-static problems, such as unison, may suggest the difficulty of obtaining such a strong property and if so, make our notion of gradual stabilization very attractive compared to merely self-stabilizing solutions. For example, in our unison solution, gradual stabilization ensures that processes remain "almost" synchronized during the convergence phase started after one dynamic step satisfying UnderLocalControl. Hence, it is worth investigating whether this new paradigm can be applied to other, in particular non-static, problems. Concerning our unison algorithm, the graceful recovery after one dynamic step comes at the price of slowing down the clock increments. The question of limiting this drawback remains open. Finally, it would be interesting to address in future work gradual stabilization for non-static problems in context of more complex dynamic patterns.

References

1. Altisen, K., Devismes, S., Durand, A., Petit, F.: Gradual stabilization under τ-dynamics. Technical report (2015). https://hal.archives-ouvertes.fr/hal-01215190
2. Arora, A., Dolev, S., Gouda, M.G.: Maintaining digital clocks in step. Parallel Process. Lett. **1**, 11–18 (1991)
3. Awerbuch, B., Kutten, S., Mansour, Y., Patt-Shamir, B., Varghese, G.: Time optimal self-stabilizing synchronization. In: STOC, pp. 652–661 (1993)
4. Blin, L., Potop-Butucaru, M., Rovedakis, S.: A super-stabilizing log(n)-approximation algorithm for dynamic steiner trees. Theor. Comput. Sci. **500**, 90–112 (2013)
5. Blin, L., Potop-Butucaru, M.G., Rovedakis, S., Tixeuil, S.: Loop-free super-stabilizing spanning tree construction. In: Dolev, S., Cobb, J., Fischer, M., Yung, M. (eds.) SSS 2010. LNCS, vol. 6366, pp. 50–64. Springer, Heidelberg (2010)
6. Boulinier, C.: L'Unisson. Ph.D. thesis, Université de Picardie Jules Vernes, France (2007)
7. Boulinier, C., Petit, F., Villain, V.: When graph theory helps self-stabilization. In: PODC, pp. 150–159 (2004)
8. Carrier, F., Datta, A.K., Devismes, S., Larmore, L.L., Rivierre, Y.: Self-stabilizing (f, g)-alliances with safe convergence. J. Parallel Distrib. Comput. **81–82**, 11–23 (2015)
9. Couvreur, J., Francez, N., Gouda, M.G.: Asynchronous unison (extended abstract). In: ICDCS, pp. 486–493 (1992)
10. Dijkstra, E.W.: Self-stabilizing systems in spite of distributed control. Commun. ACM **17**(11), 643–644 (1974)
11. Dolev, S., Herman, T.: Superstabilizing protocols for dynamic distributed systems. Chicago J. Theor. Comput. Sci. **1997**, 13 (1997)
12. Dolev, S., Israeli, A., Moran, S.: Uniform dynamic self-stabilizing leader election. IEEE Trans. Parallel Distrib. Syst. **8**(4), 424–440 (1997)
13. Genolini, C., Tixeuil, S.: A lower bound on dynamic k-stabilization in asynchronous systems. In: SRDS, p. 212 (2002)
14. Ghosh, S., Gupta, A., Herman, T., Pemmaraju, S.V.: Fault-containing self-stabilizing distributed protocols. Distrib. Comput. **20**(1), 53–73 (2007)

15. Gouda, M.G., Herman, T.: Stabilizing unison. Inf. Process. Lett. **35**(4), 171–175 (1990)
16. Herman, T.: Superstabilizing mutual exclusion. Distrib. Comput. **13**(1), 1–17 (2000)
17. Huang, S., Liu, T.: Four-state stabilizing phase clock for unidirectional rings of odd size. Inf. Process. Lett. **65**(6), 325–329 (1998)
18. Johnen, C., Alima, L.O., Datta, A.K., Tixeuil, S.: Optimal snap-stabilizing neighborhood synchronizer in tree networks. Parallel Process. Lett. **12**(3–4), 327–340 (2002)
19. Kakugawa, H., Masuzawa, T.: A self-stabilizing minimal dominating set algorithm with safe convergence. In: IPDPS, p. 8 (2006)
20. Katayama, Y., Ueda, E., Fujiwara, H., Masuzawa, T.: A latency optimal superstabilizing mutual exclusion protocol in unidirectional rings. J. Parallel Distrib. Comput. **62**(5), 865–884 (2002)
21. Kutten, S., Patt-Shamir, B.: Stabilizing time-adaptive protocols. Theor. Comput. Sci. **220**(1), 93–111 (1999)
22. Nolot, F., Villain, V.: Universal self-stabilizing phase clock protocol with bounded memory. In: IPCCC, pp. 228–235 (2001)
23. Tzeng, C., Jiang, J., Huang, S.: Size-independent self-stabilizing asynchronous phase synchronization in general graphs. J. Inf. Sci. Eng. **26**(4), 1307–1322 (2010)

Parallel Numerical Methods and Applications

High Performance Polar Decomposition on Distributed Memory Systems

Dalal Sukkari, Hatem Ltaief[✉], and David Keyes

Extreme Computing Research Center, Division of Computer, Electrical,
and Mathematical Sciences and Engineering, King Abdullah University
of Science and Technology, Thuwal, Kingdom of Saudi Arabia
{Dalal.Sukkari,Hatem.Ltaief,David.Keyes}@kaust.edu.sa

Abstract. The polar decomposition of a dense matrix is an important operation in linear algebra. It can be directly calculated through
the singular value decomposition (SVD) or iteratively using the QR
dynamically-weighted Halley algorithm (QDWH). The former is difficult to parallelize due to the preponderant number of memory-bound
operations during the bidiagonal reduction. We investigate the latter
scenario, which performs more floating-point operations but exposes at
the same time more parallelism, and therefore, runs closer to the theoretical peak performance of the system, thanks to more compute-bound
matrix operations. Profiling results show the performance scalability of
QDWH for calculating the polar decomposition using around 9200 MPI
processes on well and ill-conditioned matrices of $100\,\mathrm{K} \times 100\,\mathrm{K}$ problem
size. We study then the performance impact of the QDWH-based polar
decomposition as a pre-processing step toward calculating the SVD itself.
The new distributed-memory implementation of the QDWH-SVD solver
achieves up to five-fold speedup against current state-of-the-art vendor
SVD implementations.

1 Introduction

The polar decomposition is a critical numerical algorithm for various applications, including aerospace computations [4], chemistry [8], computation of block
reflectors in numerical linear algebra [15], factor analysis [16], and signal processing [3].

There are rather distinct algorithmic approaches to its calculation: a direct
method based on the singular value decomposition (SVD) [9,18] and an iterative method such as Newton, which requires the explicit matrix inversion at
each iteration, or the inverse-free iterative QR dynamically-weighted Halley
algorithm (QDWH) [12,13]. These approaches present diametrically opposed
computational challenges. The SVD approach is difficult to parallelize due to
the preponderant number of memory-bound operations during the bidiagonal
reduction, where a large portion of the code is characterized by low arithmetic
intensity. The latter can exploit high concurrency effectively and is inherently
compute-bound, and so may run closer to the theoretical peak performance of
the system.

© Springer International Publishing Switzerland 2016
P.-F. Dutot and D. Trystram (Eds.): Euro-Par 2016, LNCS 9833, pp. 605–616, 2016.
DOI: 10.1007/978-3-319-43659-3_44

This comes at the expense of performing extra floating-point operations (flops), which at first glance appear to be a prohibitive handicap. Semiconductor manufacturers are, however, designing future manycore chips to be throughput-oriented and massively parallel, highly populated by floating-point units (e.g., x86 architectures with next generation AVX-512 instruction sets for Knights Landing Xeon Phi processor) at the price of increasingly limited bandwidth per core [7]. Data motion, whether vertically within a node or horizontally across the network interconnect, is becoming as one of the impeding factors for parallel performance and energy efficiency of scientific applications.

To this hostile hardware landscape, the authors offer a new high performance implementation of the polar decomposition on distributed-memory environment systems, based on the inverse-free QDWH. Previous work by the authors [17] has demonstrated the QDWH capability and performance on shared-memory systems equipped with GPUs. This paper presents a comprehensive performance analysis of QDWH on a homogeneous large-scale x86 system, based on the distributed-memory vendor-optimized numerical library ScaLAPACK [5]. Performance scalability and profiling results provide an assessment of the implementation using more then 9200 MPI processes on $100\,K \times 100\,K$ matrix problem size. We look at one of the direct applications for QDWH in the context of singular value decomposition (SVD), a paramount operation in linear algebra. By using the QDWH-based polar decomposition as a pre-processing step toward calculating the SVD itself, a new distributed-memory QDWH-SVD solver has been developed that achieves up to five-fold speedup against current state-of-the-art vendor SVD implementations.

The remainder of the paper is organized as follows. Section 2 presents related work. Section 3 recalls the polar decomposition. Section 4 describes the fundamental design of dense linear algebra algorithms, as implemented in LAPACK [1] and ScaLAPACK [5]. The implementation details of the high performance distributed-memory QDWH are given in Sect. 5. Section 6 shows the numerical robustness of QDWH in distributed memory. Section 7 assesses the QDWH performance scalability and identifies room for improvements. The impact on SVD solvers and a comparison with existing state-of-the-art solvers are given in Sect. 8 and we conclude in Sect. 9.

2   Related Work

Developing a high performance QDWH-based polar decomposition has been a computational challenge due to the high number of flops required within each iteration.

Higham and Papadimitriou [14] introduced the first QDWH implementation on shared-memory systems, for which the early version required the calculation of the matrix inversion. A decade and half later, Nakatsukasa et. al [12] revisited the QDWH algorithm by replacing the expensive matrix inversion kernels with QR factorization and theoretically proved the numerical robustness of the new inverse-free QDWH. Then, Nakatsukasa and Higham [13] presented a new spectral divide and conquer algorithms for the symmetric eigenvalue problem and the singular value decomposition that are backward stable, based on the QDWH polar decomposition. All aforementioned related works focus primarily on the stability of the algorithm and the orthogonality of the polar factor, as a qualitative metric to assess QDWH. However, high performance implementations have not been covered, especially on distributed-memory systems. In previous work [17], we demonstrated a high performance implementation on a shared-memory system equipped with multiple GPUs. The results showed decent scalability and proposed an attractive new QDWH-SVD solver for the scientific community.

We introduce here a high performance QDWH implementation on distributed-memory based on the state-of-the-art vendor-optimized numerical library ScaLAPACK [5]. We then use QDWH as a pre-processing step toward calculating the SVD of a dense matrix, and demonstrate not only the suitability of QDWH for the polar decomposition as opposed to an SVD-based direct method, but also how QDWH can catalyze a new SVD solver, itself.

3 The Polar Decomposition and its SVD Extension

The polar decomposition of the matrix $A \in \mathbb{R}^{m \times n}$ $(m \geq n)$ is written $A = U_p H$, where U_p is an orthogonal matrix and $H = \sqrt{A^\top A}$ is a symmetric positive semidefinite matrix. We use the inverse-free QDWH-based iterative procedure to calculate the polar decomposition as follows [13]:

$$X_0 = A/\alpha,$$

$$\begin{bmatrix} \sqrt{c_k} X_k \\ I \end{bmatrix} = \begin{bmatrix} Q_1 \\ Q_2 \end{bmatrix} R, \ X_{k+1} = \frac{b_k}{c_k} X_k + \frac{1}{\sqrt{c_k}} \left(a_k - \frac{b_k}{c_k} \right) Q_1 Q_2^\top, \ k \geq 0. \quad (1)$$

When, X_k becomes well-conditioned, it is possible to replace Eq. 1 with a Cholesky-based implementation as follows:

$$X_{k+1} = \frac{b_k}{c_k} X_k + \left(a_k - \frac{b_k}{c_k} \right) (X_k W_k^{-1}) W_k^{-\top},$$

$$W_k = \text{chol}(Z_k), \ Z_k = I + c_k X_k^\top X_k. \quad (2)$$

This algorithmic switch at runtime allows to further speed up the overall computation, thanks to a lower algorithmic complexity, while still maintaining numerical stability. In particular, in the subsequent experiments, our implementations switch from Eqs. 1 to 2 if c_k is smaller than 100. In terms of algorithmic complexity, the number of flops depends on the number of iterations required to converge, which is dictated by the condition number of the original matrix problem, but is six at maximum, based on standard double precision. More details on theoretical results can be found in [13]. For instance, assume the matrix is ill-conditioned, QDWH will typically perform an initial condition estimate of the matrix, and then enter the inner iteration loop to perform QR factorization for the first 3 iterations, followed by three iteration of Cholesky, besides executing other expensive Level 3 BLAS operations. The total cost is roughly $43n^3$ from an extreme side of the spectrum, and to help put this flops number in perspective, this corresponds to 43 times the cost of solving a Cholesky-based system of linear equations. A direct method based on the SVD would require $25n^3$ But even with this high flop count, thanks to its high degree of parallelism, the QDWH polar decomposition remains more suitable on distributed-memory system than a direct method based on the SVD itself.

Furthermore, once the QDWH polar decomposition has been calculated, it can be used as a pre-processing stage towards getting the SVD of a general dense matrix as follows: (1) perform the polar decomposition $A = U_p \times H$, (2) calculate the singular values and the right singular vectors $H = V \times \Sigma \times V^T$ using a symmetric dense eigensolver of choice, and (3) compute the left singular vectors $U \times V$. These three successive computational stages represent the crux of the QDWH-SVD on distributed-memory systems.

4   Background on ScaLAPACK

This Section reviews the core methodology used under the hood by ScaLAPACK, the standard dense linear algebra (DLA) library on distributed-memory systems, since it is central to the QDWH implementation presented in this paper.

ScaLAPACK relies on block algorithms, similar to LAPACK [1], and can be expressed by two successive computational stages: the panel factorization and the update of the trailing submatrix. While the former is memory-bound, the latter is rich in compute-bound operations and this is where most of ScaLAPACK (and LAPACK) dense linear algebra algorithms extract parallel performance through calls to Level 3 BLAS. ScaLAPACK employs the bulk synchronous programming model, which may limit performance due to strong synchronization points. Furthermore, ScaLAPACK uses 2D block cyclic data distribution to map the matrix data to the distributed-memory and relies on the Message Passing Interface (MPI) [11] to exchange data within the grid of MPI processes. There are two main categories of DLA algorithms in ScaLAPACK (and LAPACK): the one-sided and two-sided transformations. The one-sided DLA operations mostly regroup solvers of linear equations and their corresponding factorizations (i.e., $QR/LU/LL^T$). The two-sided transformations gather non-symmetric, symmetric eigensolvers, and SVD with their corresponding reductions to condensed forms (i.e., Hessenberg, tridiagonal, and bidiagonal reductions).

It is noteworthy to mention that the panel factorization for the two-sided transformations are much more expensive than the one-sided operations because they operate on the entire unreduced part of the matrix, as opposed to just within the panel. Therefore, two-sided operations require extensive data movement across the network interconnect. This background will later be referenced to help interpret and analyze the performance results.

5   High Performance Implementations

Algorithm 1 describes the pseudo-code of the distributed-memory QDWH-SVD based on ScaLAPACK [5]. We define the MPI process grid configuration $P \times Q$ so that the data mapping can occur following the 2D block-cyclic data distribution (2D-BCDD). Each data structure owns a handle or a descriptor, which is paramount for the computations as it describes which process has which chunk of the original matrix data. The size of the chunk is referred as nb and is used to initialize the various data structures needed in the beginning of the code. The nb parameter is also critical for performance as it trades-off concurrency with arithmetic intensity. We set $nb = 64$ as it seems to be the proper sweetspot for all our experiments. Once the QDWH-based polar decomposition (stage 1) has been calculated, we can plug in a symmetric dense eigensolver of choice (stage 2) (e.g., PDSYEVR from ScaLAPACK based on the MRRR eigensolver [2] or ELPA-EIG [10] which combines a two-stage reduction with a divide-and-conquer eigensolver) to compute the singular values and the right singular vectors, and finally, perform a matrix-matrix multiplication to get the left singular vectors.

6   Numerical Accuracy

This Section highlights the numerical robustness of the distributed-memory QDWH-SVD implementation. We use around 5200 MPI processes on the Cray XC40 system installed at the Swiss National Supercomputing Centre (CSCS) in Lugano, Switzerland. QDWH-SVD relies on the state-of-the-art vendor-optimized ScaLAPACK from the Cray Scientific library (libSCI). Although this has been studied in details on shared-memory systems [17], the deployment on distributed-memory environment may necessitate a change in the algorithm to match the data distribution and/or may require the execution of floating-point operations in a different order, which may be cumbersome for sensitive algorithms due to rounding errors.

6.1   Synthetic Matrix Generation

Since QDWH performance depends strongly on the condition number of the matrix (cond), the numerical robustness of the SVD algorithms is assessed against a well-conditioned synthetic matrix (cond = 1) and an ill-conditioned synthetic matrix (cond = 1e16). Each dense synthetic matrix $A = QDQ^\top$ is generated using the ScaLAPACK routine PDLATMS (matrix of type 4), by initially setting a diagonal matrix $D = diag(\Sigma)$ containing the singular values, which follows a uniform distribution and from an orthogonal matrix Q generated from random entries.

6.2   Accuracy Assessments of SVD Solvers

Figure 1 presents the accuracy of the singular values and the orthogonality of all singular vectors, the backward error of the overall SVD for ill-conditioned and

Algorithm 1. Distributed-memory QDWH Pseudo-Code using ScaLAPACK.

```
1:  // Set the block size
2:  nb = 64
3:  // Initialize data structures using the 2D-BCDD
4:  descinit(nb, nb, A, descA); Fill_in(A, descA)
5:  descinit(nb, nb, B, descB); Fill_in(B, descB)
6:  descinit(nb, nb, C, descC); Fill_in(C, descC)
7:  descinit(nb, nb, X, descX); Fill_in(X, descX)
8:  descinit(nb, nb, H, descH); Fill_in(H, descH)
9:  descinit(nb, nb, U, descU); Fill_in(U, descU)
10: descinit(nb, nb, V, descV); Fill_in(V, descV)
11: // Estimate the condition number
12: pdlacpy(A, descA, B, descB)
13: pdgetrf(B, descB)
14: pdgecon(B, descB, alpha)
15: // Compute the polar decomposition A = U_p H using QDWH
16: pdlacpy(A, descA, X, descX)
17: pdlascl(X, descX, α),  α ≈ ‖A‖_2
18: k = 1, Li = β × α/1.1,  conv = 100
19: while (conv ≥ ∛(5eps) || |Li − 1| ≥ 5eps) do
20:     L2 = Li², dd = ∛((4(1 − L2)/L2²)
21:     sqd = √(1 + dd), a1 = sqd + √(8 − 4 × dd + 8(2 − L2)/(L2 × sqd))/2
22:     a = real(a1), b = (a − 1)²/4, c = a + b − 1
23:     Li = Li(a + b × L2)/(1 + cL2)
24:     pdlacpy(X, descX, B, descB)
25:     if c > 100 then
26:         C = [√c B]
                [ I  ]
27:         pdgeqrf(C, descC, tau)
28:         pdorgqr(C, descC, tau)
29:         // Compute X_k from X_{k−1}
30:         pdgemm(C(1 : m, :), descC, C(m : m + n, :), descC, X, descX)
31:     else
32:         pdlaset(C, descC, 0.0, 1.0)
33:         pdgemm(B, descB, B, descB, C, descC)
34:         pdposv(C, descC, B, descB)
35:         // Compute X_k from X_{k−1}
36:         pdgeam(B, descB, X, descX)
37:     end if
38:     conv ← ‖X_k − X_{k−1}‖_F
39:     k = k + 1
40: end while
41: pdgemm(X_k, descX, A, descA, H, descH)
42: // Compute the singular values and the right singular vectors H = VΣV^T
43: if ScaLAPACK PDSYEVR then
44:     pdsyevr(H, descH, Σ, V, descV)
45: else
46:     ELPA_DSYEVD(H, descH, Σ, V, descV)
47: end if
48: // Compute the left singular vectors X × V
49: pdgemm(X, descX, V, descV, U, descU)
```

well-conditioned matrix of type 4, following the metrics as in [17]. The QDWH-SVD implementations based on the eigensolvers from ELPA [10] or ScaLAPACK (PDSYEVR) provide satisfactory accuracy up to the machine precision across all matrix sizes, regardless on the condition number of the matrix. These extensive numerical tests have in fact helped detect numerical issues for PDGESVD from Cray libSCI library on ill-conditioned matrices for some matrix sizes (Figs. 1(a) and (c)). PDGESVD uses internally a QR iteration to retrieve the singular values and their associated singular vectors. The issues may have to do with the implementation on distributed-memory systems, since this numerical behavior has not been detected on shared-memory systems [17]. The performance numbers corresponding to these matrix sizes have not been reported in the following performance results Section.

7   Performance Results and Analysis

This Section highlights the performance assessment of QDWH on distributed-memory environment system and demonstrates its impact on the SVD solver.

7.1   Environment Settings

The Cray XC40 system codenamed *dora* installed at the Swiss National Supercomputing Centre (CSCS) has been used extensively to generate all experimental data: *dora* has 1256 compute nodes, each with two-sockets Intel Xeon E5-2690 v3 (Haswell) with 12 cores each running at 2.60 GHz. Given our compute-intensive workload, we disabled hyperthreading. Each node has 64GB of DDR3 main memory. The customized network interconnect on the platfrom is Cray Aries, which implements a Dragonfly network topology. The theoretical peak performance of the system is 1.254 Petaflops. We use intel compiler suites v15.0.1.133 and rely on the vendor-optimized ScaLAPACK implementation from Cray scientific library libSCI. We choose a square MPI process grid to support the 2D-BCDD and set nb = 64 for all experimental results as this turns out to be a possible optimal value for QDWH-SVD parallel performance. Given the low number of main memory available per core on each node, we execute 12 MPI processes per node instead of 24. We investigate only pure MPI implementation, since we notice MPI+OpenMP programming model has shown limitations on ScaLA-PACK performance due to its bulk synchronous programming model.

7.2   QDWH Performance in Tflop/s

Although appearing sometime as controversial, the Tflop/s metric is important in the sense that it indicates how well the underlying architecture is used. As depicted in Fig. 2, we show the parallel performance of QDWH in Tflop/s across various matrix sizes and number of MPI processes. QDWH achieves around 90 Tflop/s and 100 Tflop/s for ill and well-conditioned matrix, respectively on 9216 MPI processes. Unfortunately, it is difficult to assess these numbers against the theoretical peak performance of the system, since starting from Haswell processor with the introduction of AVX2 instruction sets, the processor may require more power to run and therefore, may perform at less the marked frequency at runtime to stay within the thermal design power limits.

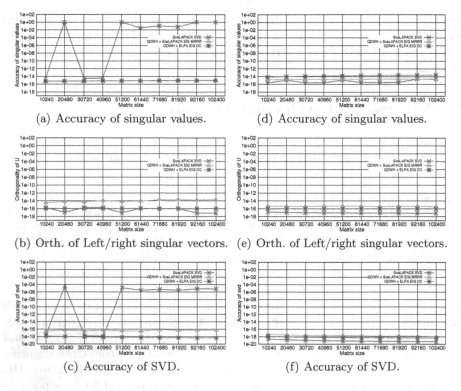

(a) Accuracy of singular values.   (d) Accuracy of singular values.

(b) Orth. of Left/right singular vectors. (e) Orth. of Left/right singular vectors.

(c) Accuracy of SVD.   (f) Accuracy of SVD.

Fig. 1. Accuracy comparison of SVD solvers on matrix of type 4: (a-b-c) for ill conditioned matrix and (d-e-f) for well conditioned matrix.

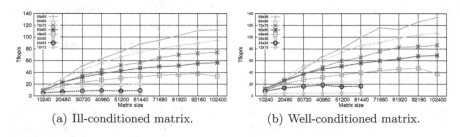

(a) Ill-conditioned matrix.   (b) Well-conditioned matrix.

Fig. 2. QDWH performance results in Tflop/s.

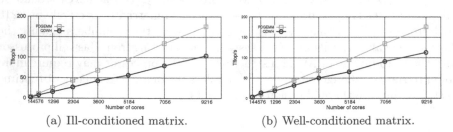

(a) Ill-conditioned matrix.   (b) Well-conditioned matrix.

Fig. 3. QDWH scalability assessment.

7.3   QDWH Performance Scalability

However, we can still assess QDWH performance by looking at weak scaling against the matrix-matrix multiplication PDGEMM from ScaLAPACK as a natural performance upper-bound. Figure 3 highlights the aforementioned experiment for ill and well-conditioned matrix. QDWH achieves around 60 % and 70 % of PDGEMM performance for ill and well-conditioned matrix, respectively. PDGEMM is obviously a loose upper-bound as it is embarrassingly parallel and have less synchronization points and data structures than the sophisticated QDWH implementation. All in all, there is clearly room for further improvements. For instance, for the QDWH iterations for which a QR factorization is

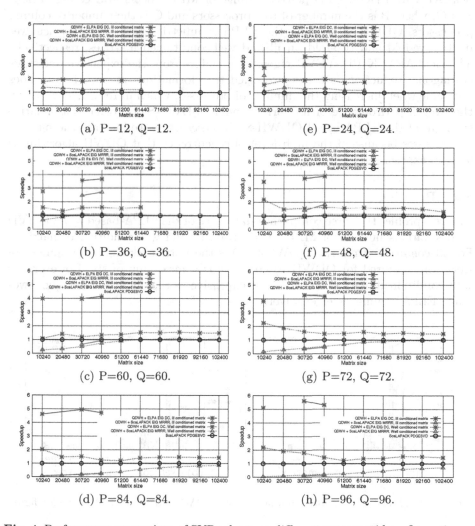

Fig. 4. Performance comparison of SVD solvers on different process grid configurations.

performed, the matrix structure has a size of 2*N rows for N cols (see line 26 in Algorithm 1), for which a square MPI process grid may not be adequate anymore to mitigate the overhead of the sequential panel factorization (see Sect. 4). A rectangular process grid $(P < Q)$ would reduce data movement across the process columns in favor of more concurrency across the process rows.

8   Performance Impact on SVD Solvers

8.1   Performance Speedup

Figure 4 depicts performance comparisons of two QDWH-SVD variants and ScaLAPACK PDGESEVD across a range of MPI process grid configurations $P \times Q$, where P is the number of row processors and Q is the number of column processors. QDWH-SVD using the ELPA symmetric eigensolver [10] achieves up to five-fold and two-fold on ill and well-conditioned matrices, respectively. The ScaLAPACK PDSYEVR eigensolver suffers from the expensive one-stage tridiagonal reduction (see the two-sided transformations in Sect. 4) and does not scale when increasing the number of processes, which in return, slows down the overall perfromance of its corresponding QDWH-SVD solver. The relative speedup of the ELPA-based QDWH-SVD increases as we increase the number of processes because of the high level of concurrency that QDWH algorithm is able to expose.

8.2   Profiling

Figure 5 profiles the three computational stages of the ELPA-based QDWH-SVD across various number of MPI processes on ill and well-conditioned matrices. For ill-conditioned matrix, QDWH-SVD is the bottleneck since the number of iterations required for QDWH to converge is a maximum of six. However, for well-conditioned matrix, QDWH is as expensive as the divide-and-conquer two-stage ELPA symmetric eigensolver.

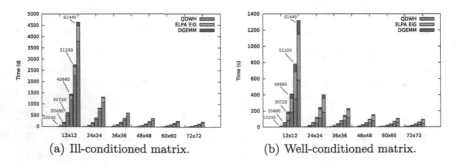

(a) Ill-conditioned matrix.   (b) Well-conditioned matrix.

Fig. 5. Profiling the computational stages of ELPA-based QDWH-SVD (Color figure online).

9 Conclusion and Future Work

A high performance QDWH implementation and its SVD extension has been presented on distributed-memory environment system. Performance analysis shows decent scalability running with around 9200 MPI processes on well and ill-conditioned matrices of $100\,\mathrm{K} \times 100\,\mathrm{K}$ problem size. We have also identified room for improvements to further enhance the scalability. We have also studied the performance impact of using QDWH as a pre-processing step toward calculating the SVD itself. The new distributed-memory implementation of the QDWH-SVD solver achieves up to five-fold speedup against current state-of-the-art vendor-optimized SVD implementations. Moreover, numerical accuracy study highlights the robustness of QDWH-SVD over ScaLAPACK PDGESVD in presence of ill-conditioned matrices. We would like also to investigate task-based programming model to break the ScaLAPACK bulk synchronous programming model in the context of DPLASMA [6] library using dynamic runtime systems for asynchronous task scheduling.

Acknowledgment. For computer time, this research used the resources from the Swiss National Supercomputing Centre (CSCS) in Lugano, Switzerland.

References

1. Anderson, E., Bai, Z., Bischof, C.H., Blackford, L.S., Demmel, J.W., Dongarra, J.J., Croz, J.J.D., Greenbaum, A., Hammarling, S., McKenney, A., Sorensen, D.C.: LAPACK User's Guide, 3rd edn. SIAM, Philadelphia (1999)
2. Antonelli, D., Vömel, C.: PDSYEVR. ScaLAPACK's Parallel MRRR Algorithm for the Symmetric Eigenvalue Problem (168), 18 August 2005. http://www.netlib.org/lapack/lawnspdf/lawn168.pdf, technical Report UCB//CSD-05-1399
3. Arun, K.S.: A unitarily constrained total least squares problem in signal processing. SIAM J. Matrix Anal. Appl. **13**(3), 729–745 (1992)
4. Bar-Itzhack, I.: Iterative optimal orthogonalization of the strapdown matrix. IEEE Trans. Aerosp. Electron. Syst. **AES–11**(1), 30–37 (1975)
5. Blackford, L.S., Choi, J., Cleary, A., D'Azevedo, E.F., Demmel, J.W., Dhillon, I.S., Dongarra, J.J., Hammarling, S., Henry, G., Petitet, A., Stanley, K., Walker, D.W., Whaley, R.C.: ScaLAPACK Users' Guide. Society for Industrial and Applied Mathematics, Philadelphia (1997)
6. Bosilca, G., Bouteiller, A., Danalis, A., Faverge, M., Haidar, A., Hérault, T., Kurzak, J., Langou, J., Lemarinier, P., Ltaief, H., Luszczek, P., YarKhan, A., Dongarra, J.: Flexible development of dense linear algebra algorithms on massively parallel architectures with DPLASMA. In: IPDPS Workshops, pp. 1432–1441. IEEE (2011)
7. Dongarra, J., Beckman, P., Moore, T., Aerts, P., Aloisio, G., Andre, J.C., Barkai, D., Berthou, J.Y., Boku, T., Braunschweig, B., Cappello, F., Chapman, B., Chi, X., Choudhary, A., Dosanjh, S., Dunning, T., Fiore, S., Geist, A., Gropp, B., Harrison, R., Hereld, M., Heroux, M., Hoisie, A., Hotta, K., Jin, Z., Ishikawa, Y., Johnson, F., Kale, S., Kenway, R., Keyes, D., Kramer, B., Labarta, J., Lichnewsky, A., Lippert, T., Lucas, B., Maccabe, B., Matsuoka, S., Messina, P., Michielse, P.,

Mohr, B., Mueller, M.S., Nagel, W.E., Nakashima, H., Papka, M.E., Reed, D., Sato, M., Seidel, E., Shalf, J., Skinner, D., Snir, M., Sterling, T., Stevens, R., Streitz, F., Sugar, B., Sumimoto, S., Tang, W., Taylor, J., Thakur, R., Trefethen, A., Valero, M., Van Der Steen, A., Vetter, J., Williams, P., Wisniewski, R., Yelick, K.: The international exascale software project roadmap. Int. J. High Perform. Comput. Appl. **25**(1), 3–60 (2011). http://dx.org/10.1177/1094342010391989

8. Goldstein, J.A., Levy, M.: Linear algebra and quantum chemistry. Am. Math. Monthly **98**(10), 710–718 (1991). http://dx.org/10.2307/2324422

9. Golub, G.H., Van Loan, C.F.: Matrix Computations. John Hopkins Studies in the Mathematical Sciences, 3rd edn. Johns Hopkins University Press, Baltimore (1996)

10. Marek, A., Blum, V., Johanni, R., Havu, V., Lang, B., Auckenthaler, T., Heinecke, A., Bungartz, H., Lederer, H.: The ELPA library: scalable parallel eigenvalue solutions for electronic structure theory and computational science. J. Phys. Condens Matter 26(21) (2014). http://www.ncbi.nlm.nih.gov/pubmed/24786764

11. Forum, M.P.I.: MPI: a message passing interface. In: Proceedings of Supercomputing 1993, pp. 878–883. IEEE CS Press, Portland, November 1993

12. Nakatsukasa, Y., Bai, Z., Gygi, F.: Optimizing halley's iteration for computing the matrix polar decomposition. SIAM J. Matrix Anal. Appl. **31**, 2700–2720 (2010)

13. Nakatsukasa, Y., Higham, N.J.: Stable and efficient spectral divide and conquer algorithms for the symmetric eigenvalue decomposition and the svd. SIAM J. Sci. Comput. **35**(3), A1325–A1349 (2013)

14. Higham, N.J., Papadimitriou, P.: A new parallel algorithm for computing the singular value decomposition. In: Lewis, J.G. (ed.) Proceedings of the Fifth SIAM Conference on Applied Linear Algebra, pp. 80–84. Society for Industrial and Applied Mathematics, Philadelphia (1994)

15. Schreiber, R., Parlett, B.: Block reflectors: theory and computation. SIAM J. Numer. Anal. **25**(1), 189–205 (1988). http://dx.org/10.1137/0725014

16. Schnemann, P.: A generalized solution of the orthogonal procrustes problem. Psychometrika **31**(1), 1–10 (1966). http://dx.org/10.1007/BF02289451

17. Sukkari, D., Ltaief, H., Keyes, D.: A high performance QDWH-SVD solver using hardware accelerators. Accepted for publication at ACM Trans. Math. Softw. (2016). https://ecrc.kaust.edu.sa/Documents/qdwh-svd.pdf

18. Trefethen, L.N., Bau, D.: Numerical Linear Algebra. SIAM, Philadelphia (1997). http://www.siam.org/books/OT50/Index.htm

A Synchronization-Free Algorithm for Parallel Sparse Triangular Solves

Weifeng Liu[1,2](✉), Ang Li[3], Jonathan Hogg[2], Iain S. Duff[2], and Brian Vinter[1]

[1] Niels Bohr Institute, University of Copenhagen, Copenhagen, Denmark
weifeng.liu@nbi.ku.dk
[2] Scientific Computing Department,
STFC Rutherford Appleton Laboratory, Didcot, UK
[3] Eindhoven University of Technology, Eindhoven, Netherlands

Abstract. The sparse triangular solve kernel, SpTRSV, is an important building block for a number of numerical linear algebra routines. Parallelizing SpTRSV on today's manycore platforms, such as GPUs, is not an easy task since computing a component of the solution may depend on previously computed components, enforcing a degree of sequential processing. As a consequence, most existing work introduces a preprocessing stage to partition the components into a group of level-sets or colour-sets so that components within a set are independent and can be processed simultaneously during the subsequent solution stage. However, this class of methods requires a long preprocessing time as well as significant runtime synchronization overhead between the sets. To address this, we propose in this paper a novel approach for SpTRSV in which the ordering between components is naturally enforced within the solution stage. In this way, the cost for preprocessing can be greatly reduced, and the synchronizations between sets are completely eliminated. A comparison with the state-of-the-art library supplied by the GPU vendor, using 11 sparse matrices on the latest GPU device, show that our approach obtains an average speedup of 2.3 times in single precision and 2.14 times in double precision. The maximum speedups are 5.95 and 3.65, respectively. In addition, our method is an order of magnitude faster for the preprocessing stage than existing methods.

1 Introduction

The sparse triangular solve kernel, SpTRSV, is an important building block in a number of numerical linear algebra routines, such as direct methods [5,7], preconditioned iterative methods [22], and least squares problems [3]. This operation computes a dense solution vector x from a sparse linear system $Lx = b$, where L is a square lower triangular sparse matrix and b is a dense vector.

Compared to a dense triangular solve [9] and other sparse basic linear algebra subprograms (BLAS) [8,14] such as sparse transposition [27], sparse matrix-vector multiplication [16,17] and sparse matrix-matrix multiplication [15], the SpTRSV operation is more difficult to parallelize since it is inherently sequential.

© Springer International Publishing Switzerland 2016
P.-F. Dutot and D. Trystram (Eds.): Euro-Par 2016, LNCS 9833, pp. 617–630, 2016.
DOI: 10.1007/978-3-319-43659-3_45

This means that, for a lower triangular sparse matrix, computing any single component x_k may depend on having first computed a subset of previous components $x_0, \cdots, x_{k-1}$. Therefore, most existing research concentrates on adding a preprocessing stage to divide the entries of x into a number of sets (known as level-sets or colour-sets). Even though the sets have to be executed in sequence, entries in any single set can be computed in parallel. As a result, parallel hardware can be exploited efficiently. This class of methods demonstrates great advantage over the original sequential implementation both on CPUs [10,20,24,28] and on GPUs [12,19,26].

However, the set-based methods have two performance bottlenecks. Firstly, finding a good set partitioning often takes too much time, which may offset or even wipe out the benefits from parallelization. Secondly, the synchronization between consecutive sets reduces parallelization efficiency at runtime. In fact, due to these large overheads, finding an efficient thread synchronization scheme still remains a popular research topic for computer design [11,13,21].

In this paper, we propose a synchronization-free algorithm for parallel SpTRSV on GPUs. Our approach requires only a light-weight preprocessing stage without set partitioning. More importantly, our method completely eliminates the runtime barrier synchronizations among sets. By doing so, our method resolves the bottlenecks and achieves significant performance improvement. Using 11 sparse matrices from the University of Florida Sparse Matrix Collection [6], our method achieves an average speedup of 2.3 times in single precision and 2.14 times in double precision over the vendor provided library on the latest NVIDIA GPU. The maximum speedups are 5.95 and 3.65, respectively. More noticeably, the preprocessing stage of our algorithm is on average 43.7 faster (maximum of 70.5 times) than existing set-based methods in the vendor supplied library.

2   Background

2.1   Serial Algorithm

Without loss of generality, in the paper we assume that the input matrix L is a nonsingular lower triangular matrix and is stored in the *compressed sparse column* (CSC) format composed of three arrays col_ptr, row_idx and val. A typical serial implementation of SpTRSV for solving $Lx = b$ is given in Algorithm 1. This method traverses all columns in ascending order (line 3) and solves a single component of x in each step (line 4). After that, the code updates all the positions corresponding to the nonzero entries of the current column in an intermediate array left_sum (lines 5–7).

As can be seen, the columns in the main *for* loop (lines 3–8) cannot be parallelized as the ith column requires the ith value in left_sum (line 4), which may be affected by previous columns that update left_sum[i] (line 6). To be clear, we give an example. Figure 1 (a) shows a matrix L, for which the underlying dependencies are illustrated in graph form in Fig. 1 (b). Obviously, vertex 5 (i.e., x_5) cannot be solved before vertex 3 is solved, and vertex 3 has to wait for vertex 0.

Algorithm 1. A serial SpTRSV method for $Lx = b$, where L in CSC format.

```
1: MALLOC(*left_sum, n)
2: MEMSET(*left_sum, 0)
3: for i = 0 to n − 1 do
4:     x[i] ← (b[i]-left_sum[i])/val[col_ptr[i]]
5:     for j = col_ptr[i]+1 to col_ptr[i + 1]−1 do
6:         left_sum[row_idx[j]] ← left_sum[row_idx[j]] + val[j] × x[i]
7:     end for
8: end for
9: FREE(*left_sum)
```

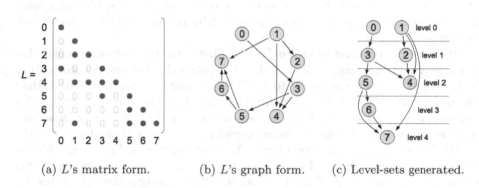

(a) L's matrix form.   (b) L's graph form.   (c) Level-sets generated.

Fig. 1. A lower triangular matrix L and parallel SpTRSV using the level-set method.

2.2 Level-Set Method for Parallel SpTRSV

The motivation of parallel-SpTRSV comes from the observation that some components/vertices are independent and can be processed simultaneously (e.g., vertices 0 and 1 in Fig. 1 (b)). Therefore, the components can be partitioned into a number of sets so that components inside a set can be solved in parallel, while the sets are processed sequentially (i.e., level by level). With this observation, Anderson and Saad [1] and Saltz [23] introduced a preprocessing stage to perform such a partition before the solving stage. Figure 1 (c) shows that five level-sets are generated for the matrix L. Consequently, levels 0, 1 and 2 can use parallel hardware (e.g., a dual-core machine) for accelerating SpTRSV. However, between sets, dependencies still exist so synchronization is required at runtime.

2.3 Motivation for Avoiding Synchronization

Synchronization remains a performance bottleneck for many applications and has long been a classic problem in computer systems research [11,13,21]. To evaluate the synchronization cost in SpTRSV, we run a parallel SpTRSV implemented by Park et al. [20] based on the aforementioned level-set approach. We show the cost of the preprocessing stage and a breakdown of the solving stage

Table 1. Breakdown of naïve level-set method [20] on Intel dual-socket E5-2695 v3.

Matrix name	Preprocessing cost (ms)	SpTRSV cost (ms)	SpTRSV cost breakdown (ms)		#Level-sets
			Synchronization	Compute	
FEM/ship_003	92.46	12.95	10.96	1.99	4367
FEM/Cantilever	47.89	9.60	5.62	3.98	2397
chipcool0	8.74	1.99	1.15	0.84	534
nlpkkt160	484.67	38.30	0.01	38.29	2

execution time (i.e., synchronization cost and floating-point calculations) using four representative matrices[1] from the University of Florida Sparse Matrix Collection [6].

We have two observations from Table 1. Firstly, the preprocessing stage takes much longer than a single call to SpTRSV. Specifically, the preprocessing stage is 4.39 times (matrix *chipcool0*) to 12.65 times (matrix *nlpkkt160*) slower than the main kernel of SpTRSV. This implies that if SpTRSV is only executed a few times, level-set based parallelization is not attractive. Secondly, when the number of level-sets increases, the overhead for synchronization dominates the SpTRSV solving stage execution time. For example, matrix *FEM/ship_003* has 4367 level-sets that implies 4366 explicit barrier synchronizations in the solving stage and accounts for 85 % of the total SpTRSV execution time (10.96 ms out of 12.95 ms). In contrast, the synchronization overhead for matrix *nlpkkt160* is much less as only two level-sets are generated.

Therefore, to improve the performance of parallel SpTRSV, it is crucial to reduce the overhead for preprocessing (i.e., generating level-sets) and to avoid the runtime barrier synchronizations.

3   Synchronization-Free Algorithm

The objective of this work is to eliminate the cost for generating level-sets and the barrier synchronizations between the sets. Due to the inherent dependencies among components, the major task for parallelizing SpTRSV is to clarify such dependencies and to be sure to respect them when solving at runtime.

In this work, we use GPUs as the platform to exploit inherent parallelism when there are many components for a very large matrix. We assign a warp of threads to solve a single component of x (a *warp* is a unit of 32 SIMD threads executed in lock-step for NVIDIA GPUs. For AMD GPUs the warp is 64 threads and is denoted by the term *wavefront*). To respect the partial order of SpTRSV, we need to be sure that the warps associated with dependent entries (if any) must be finished first. Thus thread-blocks of multiple warps are required to be dispatched in ascending order, even though they can be switched and finished in arbitrary order. Since the partial order is essentially **unidirectional** (i.e., any component only depends on previous components but not on later ones, see

---

[1] Similar to [20], the nonsingular matrix L is the lower triangular part of the input matrix, plus a dense main diagonal.

Fig. 1 (b)), we can map entries to warps and strictly respect the partial order of the entries so that no warp execution deadlock will occur.

Therefore, before actually solving for a particular component, we let the processing warp learn how many entries have to be computed in advance (i.e., the number of dependent entries). This number equals the in-degree of a vertex in the graph representation of a matrix (Fig. 1 (b)), which is also identical to the number of nonzero entries of the current matrix row minus one (to exclude the entry on diagonal). Thus, we use an intermediate array **in_degree** of size n to hold the number of nonzero entries for each row of the matrix. This is all we do in the preprocessing stage. Algorithmically, this step is part of transposing a sparse matrix in parallel [27]. Compared with the complex dependency extraction in the set-based methods that have to analyse the sparsity structure, our method requires much less work. Lines 3–7 in Algorithm 2 show the pseudocode of our preprocessing stage.

Algorithm 2. The proposed synchronization-free SpTRSV algorithm.

```
 1: MALLOC(*d_left_sum, *s_left_sum, *d_in_degree, *s_in_degree, n)
 2: MEMSET(*d_left_sum, *s_left_sum, *d_in_degree, *s_in_degree, 0)
 3: function PREPROCESSING-STAGE()
 4:     for i = 0 to nnz − 1 in parallel do
 5:         ATOMIC-ADD(&d_in_degree[row_idx[i]], 1)
 6:     end for
 7: end function
 8: function SOLVING-STAGE()
 9:     th ← SET()                                    ▷ size of diagonal block
10:     for i = 0 to n − 1 in parallel do       ▷ One concurrent warp for one component.
11:         while s_in_degree[i]+1 ≠ d_in_degree[i] do
12:             //busy wait
13:         end while
14:         x[i] ← (b[i]-d_left_sum[i]-s_left_sum[i])/val[col_ptr[i]]
15:         for j = col_ptr[i]+1 to col_ptr[i + 1]−1 in parallel do ▷ One thread for one nonzero.
16:             rid ← row_idx[j]
17:             if rid < i + th − i%th then   ▷ Use on-chip scratchpad for red areas in Figure 3.
18:                 ATOMIC-ADD(&s_left_sum[rid], val[j] × x[i])
19:                 ATOMIC-ADD(&s_in_degree[rid], 1)
20:             else                           ▷ Use off-chip memory for green area in Figure 3.
21:                 ATOMIC-ADD(&d_left_sum[rid], val[j] × x[i])
22:                 ATOMIC-SUB(&d_in_degree[rid], 1)
23:             end if
24:         end for
25:     end for
26: end function
27: FREE(*d_left_sum, *s_left_sum, *d_in_degree, *s_in_degree)
```

Knowing the in-degree information indicating how many warps have to be finished in advance, we can initiate sufficient numbers of warps to fully exploit the irregular parallelism. For an arbitrary warp, after finishing the necessary floating-point computation (line 14 in Algorithm 2) for a component, it notifies all the later entries that depend on the current one, by atomic updating (lines 19 and 22). Note that atomic operations are needed here as multiple updates from different warps may happen simultaneously. Therefore, a warp only has to wait (lines 11–13) until its corresponding in-degrees are all eliminated, implying that all the dependent components are successfully solved and the warp can start processing safely. Due to the warp multi-issuing property of GPUs, a warp can start processing immediately after its dependencies have been satisfied, without any false waiting incurred by the hardware. Also, the first component of x can be solved without any dependencies.

Figure 2 illustrates the procedure of our synchronization-free algorithm[2] using an example. Suppose there are three warps enrolled, tagged as warp0, warp1 and warp2. They follow the same procedure and are context-switched by the hardware scheduler. For an arbitrary warp, the central region contained in the red dotted box (labelled as the critical section protecting the left_sum array) separates the whole procedure into three phases: *lock-wait, critical section* and *lock-update*.

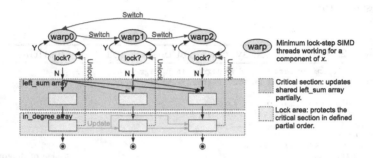

Fig. 2. The basic procedure of our synchronization-free algorithm.

In the **lock-wait** phase, the warp iteratively evaluates the status of the lock protecting the critical section of the current warp. If locked, it waits in the loop (known as *spinning*); otherwise, it stops waiting and enters the next phase. Although the lock here is a spin-lock, it does not have the busy-waiting problem. Based on our observation, if the *clock()* function is invoked inside the waiting loop, the hardware warp scheduler will be signalled to switch to the next warp context. This avoids the execution deadlock. In the **critical section** phase, the warp updates the components in left_sum that have dependencies on the components the warp is currently working on. This is done in an order that depends on the partial dependency defined by the sparsity structure. After that, it aborts the critical section and enters the lock-update phase. In the last **lock-update** phase, the warp updates the dependent in_degree array, in the same order as for the left_sum (so that all the order dependencies are strictly respected). The warp updates the related in-degrees. Depending on the number of components in that column (line 15 in Algorithm 2), it may require one or several updates.

---

[2] Note that hardware-level synchronizations in atomic operations should not be confused with barrier synchronizations in the set-based methods, when we claim that the proposed method is synchronization-free.

When an in-degree is updated to reach the target value (so that all the dependencies of the component are resolved), the lock corresponding to that in-degree is unlocked. Consequently, the warp waiting for that lock can abort the waiting phase and enter its critical section.

Lines 8–26 in Algorithm 2 give the pseudocode for the solving stage of our synchronization-free SpTRSV method. An optimization here is to exploit the GPU on-chip scratchpad memory. The idea is to allocate two sets of intermediate arrays, one on local scratchpad memory (s_left_sum and s_in_degree) and the other on off-chip global memory (d_left_sum and d_in_degree), see line 1 of Algorithm 2. When a warp finds a dependent entry (the later entry that depends on the current one) is in the same GPU thread-block composed of multiple warps, it updates the local arrays (lines 18–19) in the scratchpad memory for faster accessing. Otherwise, it updates the remote off-chip arrays (lines 21–22), to notify warps from other thread-blocks. The sum of the two arrays (line 11) is used to verify if all the dependencies are fulfilled ultimately.

Figure 3 (a) shows an example using 12 warps organized in 3 thread-blocks for solving a system of order 12 × 12. Operations in on-chip scratchpad memory are marked red (lines 18–19 in Algorithm 2), other operations in off-chip memory are marked green (lines 21–22), and the diagonal entries are coloured blue (line 14). Figure 3 (b) plots read/write behaviours for solving the 12 components (presented as 12 columns) of x. We can see that entries 0, 1 and 5 can be solved immediately once the corresponding warps are issued since they have

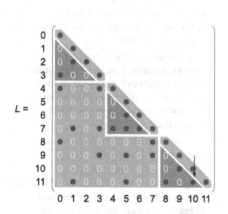

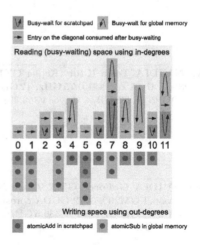

(a) Matrix.   (b) Read/write behaviours.

Fig. 3. An example of the proposed synchronization-free SpTRSV method. The red area performs atomic-adds (lines 18–19 in Algorithm 2) in scratchpad memory, and the green area performs both atomic-adds (line 21) and atomic-subs (line 22) in off-chip memory. (Color figure online)

no in-degree (see the top half of the subfigure), and they update values using their out-degrees (see the bottom half). In contrast, the other entries have to busy-wait until their in-degrees are eliminated.

4   Experimental Results

4.1   Experimental Setup

We have implemented the proposed synchronization-free SpTRSV method both in CUDA and in OpenCL, and have evaluated it on three GPUs: (1) an NVIDIA Tesla K40c GPU of Kepler architecture, (2) an NVIDIA GeForce GTX Titan X GPU of newer Maxwell architecture, and (3) an AMD Radeon R9 Fury X GPU of GCN architecture. As references, we also benchmark the most recent SpTRSV implementations from two libraries cuSPARSE v7.5 and MKL v11.3 Update 1 provided by NVIDIA and Intel, respectively.

Table 2. The testbeds and participating SpTRSV algorithms.

The testbeds	The participating SpTRSV algorithms
A dual-socket Intel Xeon E5-2695 v3 (Haswell, 2×14 cores @ 2.3 GHz, 128 GB ECC DDR4 @ 2×68.3 GB/s).	(1) The `mkl_?csrtrsv` in MKL v11.3 Update 1. Note that this is a highly tuned serial implementation. (2) The parallel executor `mkl_sparse_?_trsv` using the functions `mkl_sparse_set_sv_hint` and `mkl_sparse_optimize` as an inspector in MKL v11.3 Update 1.
An NVIDIA Tesla K40c (Kepler GK110B, 2880 CUDA cores @ 0.75 GHz, 12 GB GDDR5 @ 288 GB/s, driver v352.39).	(1) The latest SpTRSV method `cusparse?csrsv2_solve` using functions `cusparse?csrsv2_bufferSize` and `cusparse?csrsv2_analysis` in its preprocessing stage in the NVIDIA cuSPARSE v7.5. (2) The synchronization-free method proposed in this paper.
An NVIDIA GeForce GTX Titan X (Maxwell GM200, 3072 CUDA cores @ 1 GHz, 12 GB GDDR5 @ 336.5 GB/s, driver v352.39).	(1) The latest SpTRSV method `cusparse?csrsv2_solve` using functions `cusparse?csrsv2_bufferSize` and `cusparse?csrsv2_analysis` in its preprocessing stage in the NVIDIA cuSPARSE v7.5. (2) The synchronization-free method proposed in this paper.
An AMD Radeon R9 Fury X (GCN Fiji, 4096 Radeon cores @ 1.05 GHz, 4 GB HBM @ 512 GB/s, driver v15.12)	(1) The synchronization-free method proposed in this paper.

Because mixed-precision numerical methods have recently attracted much attention, we evaluate all methods in both single and double precision. Information about the platforms and test schemes are listed in Table 2.

Table 3 lists 11 sparse matrices used for our experiments on all platforms. These matrices have also been used in other research on sparse matrix computations [10,14–17,20] and are publicly available from the University of Florida Sparse Matrix Collection [6] (except matrix *Dense*). The selected matrices cover a wide range for the number of level-sets as well as the average parallelism inside a level-set. For example, matrix *nlpkkt160* has only two level-sets so that the computation of most its components can run in parallel, whereas for the matrix *Dense* every component has to wait for earlier components.

4.2 SpTRSV Performance

Figure 4 shows the single and double precision SpTRSV performance on the 11 matrices measured on the four platforms. Overall, the MKL and cuSPARSE libraries show comparable performance, while our synchronization-free method is much faster (in particular on the Maxwell-based Titan X GPU) than the vendor supplied libraries.

Specifically, on the Titan X GPU, our synchronization-free algorithm demonstrates an average speedup over the cuSPARSE library of 2.3 times in single precision and 2.14 times in double precision. The maximum speedups are 5.95 and 3.65, respectively. The best speedups are from a relatively regular matrix *FEM/Cantilever* that has most of its nonzero entries in its diagonal blocks. For this matrix, the optimizing strategy of using both scratchpad and off-chip memory improves the overall performance. Also, our method achieves speedups of 2.69 and 2.52 for single and double precision, respectively, for matrix *hollywood-2009*. This matrix requires 82,735 runtime synchronizations (see Table 3) limiting its

Table 3. The benchmark suite.

Matrix name	#Rows/Columns	#Nonzeros	#Level-sets	Parallelism
nlpkkt160	8,345,600	229,518,112	2	4,172,800
road_central	14,081,816	33,866,826	59	238,675
road_usa	23,947,347	57,708,624	77	311,004
webbase-1M	1,000,005	3,105,536	514	1,946
wiki-Talk	2,394,385	5,021,410	522	4,587
chipcool0	20,082	281,150	534	37
Dense	2,000	4,000,000	2,000	1
FEM/Cantilever	62,451	4,007,383	2,397	26
crankseg_1	52,804	10,614,210	4,056	13
FEM/ship_003	121,728	8,086,034	4,367	28
hollywood-2009	1,139,905	113,891,327	82,735	14

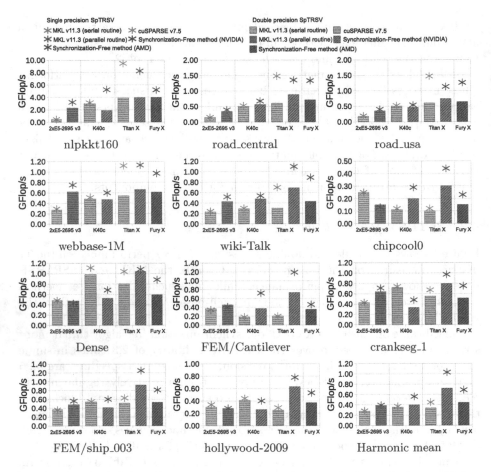

Fig. 4. The SpTRSV performance of the 11 matrices on the four platforms. (Color figure online)

performance from the level-set methods. In contrast, our method avoids synchronizations and thus obtains much superior performance. For the same reason, our method shows comparable performance compared to existing methods on matrix *nlpkkt160*, which requires only two runtime synchronizations.

Compared to the Kepler based K40c GPU, the Titan X GPU offers higher performance. The major reason is that the Maxwell architecture dramatically improves its micro-architectures for faster atomic operations, which are extensively utilized in our approach. Actually, Scogland and Feng [25] also confirmed that atomic operations have been continuously improved in the last generations of modern GPUs. Moreover, although the AMD Fury X GPU has higher bandwidth than the NVIDIA Titan X, it is in general slower for our synchronization-free SpTRSV algorithm. The main reason may be the difference between the warp/wavefront scheduling strategies on the NVIDIA and AMD GPUs.

4.3 Overhead for Preprocessing

Table 4 shows the preprocessing overhead of the parallel SpTRSV implementations from MKL, cuSPARSE and our approach on the four platforms. As can be
seen, our method achieves an average speedup of 43.7 (maximum of 70.5) over
the method in cuSPARSE library on the Titan X card. On the K40c device, the
speedups are on average 58.2 with a maximum of 89.2. The major reason is that
the vendor supplied implementation attempts to find level-sets in the preprocessing phase. Moreover, the AMD Fury X GPU offers lower cost for preprocessing,
due to more cores and higher off-chip memory bandwidth.

Table 4. Preprocessing cost (in millisecond) of the tested methods on the four
platforms.

Matrix name	Intel 2xE5-2695 v3	NVIDIA K40c		NVIDIA Titan X		AMD Fury X
	MKL	cuSPARSE	Sync-Free	cuSPARSE	Sync-Free	Sync-Free
nlpkkt160	64.43	40.58	7.27	19.99	8.91	5.58
road_usa	155.48	160.41	5.06	84.01	3.37	2.31
road_central	92.16	82.01	9.28	42.62	6.98	5.53
wiki-Talk	17.38	16.27	0.33	10.49	0.20	0.16
webbase-1M	7.08	8.53	0.19	5.48	0.13	0.11
chipcool0	1.05	1.48	0.02	1.41	0.02	0.02
FEM/ship_003	9.14	6.41	0.19	4.34	0.26	0.13
FEM/Cantilever	9.52	8.92	0.10	8.28	0.16	0.07
hollywood-2009	223.54	139.98	5.20	204.10	4.82	2.78
crankseg_1	9.30	8.93	0.24	6.14	0.43	0.14
Dense	9.29	3.46	0.08	2.99	0.12	0.05
Harmonic mean	6.80	6.99	0.12	5.71	0.13	0.10

5 Related Work

Existing parallel SpTRSV methods can be classified into two groups: those constructing level-sets and those generating colour-sets.

Anderson and Saad [1] and Saltz [23] proposed that **level-sets** can expose
parallelism in SpTRSV. A few recently developed parallel SpTRSV implementations have improved the level-set method for better data locality and faster
synchronization [10,20,28]. Naumov [19] implemented the level-set method on
NVIDIA GPUs with a tradeoff for decreasing the number of synchronizations.
Li and Saad [12] demonstrated that reordering the input matrix can further
improve parallelism but requires longer preprocessing time. Unlike the above
level-set methods, our synchronization-free SpTRSV algorithm does not analyse
the sparsity structure of the input matrix and thus completely removes costs for
generating sets and executing barrier synchronization. As a result, our method
shows much better performance than level-set methods.

Schreiber and Tang [24] first used graph colouring for constructing **colour-sets** for SpTRSV on multiprocessors. When the input sparse matrix is coloured, it is reorganized as multiple triangular submatrices located on its diagonal. Because all the submatrices can be solved in parallel, this method can be very efficient in practice. Suchoski et al. [26] recently extended the graph colouring method for SpTRSV to GPUs. However, as graph colouring is known to be an NP-complete problem, finding good colour-sets for SpTRSV is in general more time consuming. Thus it may be impractical for real-world applications.

There are also several classes of methods that do not create sets in advance. Mayer [18] pointed out that **2D decomposition** can accelerate SpTRSV but needs to reorganize the data structure of the input matrix. Chow and Patel [4] and Anzt et al. [2] recently developed several **iterative methods** for SpTRSV for use with incomplete factorization. Because iterative methods only give approximate solutions, they should not be used more generally for other scenarios such as using SpTRSV in sparse direct solvers. In contrast, the method we propose in this paper uses the unchanged CSC sparse matrix format and works for general problems.

Some researchers have also utilized atomic operations for **improving fundamental algorithms** such as bitonic sort [29], prefix-sum scan [30], wavefront [11], sparse transposition [27], and sparse matrix-vector multiplication [14, 16,17]. Unlike those problems, the SpTRSV operation is inherently serial and thus more irregular and complex. We also use atomic operations both in on-chip and off-chip memory, and set atomic operations as the central part of the whole algorithm.

6   Conclusions

In this paper, we have proposed a synchronization-free algorithm for parallel SpTRSV. The method completely eliminates the overhead for generating level-sets or colour-sets (in the preprocessing stage) and for explicit runtime barrier synchronization (in the solving stage). Experimental results show that our approach makes preprocessing an order of magnitude faster than level-set methods, and gives average speedups of 2.3 (with a maximum of 5.95) and 2.14 (with a maximum of 3.65) over vendor supplied parallel routines for single and double precision SpTRSV, respectively.

Acknowledgments. The authors would like to thank our anonymous reviewers for their invaluable feedback. We also thank Shuai Che for helpful discussion about OpenCL programming, and thank Huamin Ren for supplying access to the machine with the NVIDIA GeForce Titan X GPU. The research leading to these results has received funding from the European Union's Horizon 2020 research and innovation programme under grant agreement number 671633.

References

1. Anderson, E., Saad, Y.: Solving sparse triangular linear systems on parallel computers. Int. J. High Speed Comput. **1**(1), 73–95 (1989)
2. Anzt, H., Chow, E., Dongarra, J.: Iterative sparse triangular solves for preconditioning. In: Träff, J.L., Hunold, S., Versaci, F. (eds.) Euro-Par 2015. LNCS, vol. 9233, pp. 650–661. Springer, Heidelberg (2015)
3. Björck, Å.: Numerical Methods for Least Squares Problems. Society for Industrial and Applied Mathematics, Philadelphia (1996)
4. Chow, E., Patel, A.: Fine-grained parallel incomplete LU factorization. SIAM J. Sci. Comput. **37**(2), C169–C193 (2015)
5. Davis, T.: Direct Methods for Sparse Linear Systems. Society for Industrial and Applied Mathematics, Philadelphia (2006)
6. Davis, T.A., Hu, Y.: The University of Florida sparse matrix collection. ACM Trans. Math. Softw. **38**(1), 1:1–1:25 (2011)
7. Duff, I.S., Erisman, A.M., Reid, J.K.: Direct Methods for Sparse Matrices. Oxford University Press Inc., New York (1986)
8. Duff, I.S., Heroux, M.A., Pozo, R.: An overview of the sparse basic linear algebra subprograms: the new standard from the BLAS Technical forum. ACM Trans. Math. Softw. **28**(2), 239–267 (2002)
9. Hogg, J.D.: A fast dense triangular solve in CUDA. SIAM J. Sci. Comput. **35**(3), C303–C322 (2013)
10. Kabir, H., Booth, J.D., Aupy, G., Benoit, A., Robert, Y., Raghavan, P.: STS-k: a multilevel sparse triangular solution scheme for NUMA multicores. In: Proceedings of the International Conference for High Performance Computing, Networking, Storage and Analysis, SC 2015, pp. 55:1–55:11 (2015)
11. Li, A., van den Braak, G.J., Corporaal, H., Kumar, A.: Fine-grained synchronizations and dataflow programming on GPUs. In: Proceedings of the 29th ACM on International Conference on Supercomputing, ICS 2015, pp. 109–118 (2015)
12. Li, R., Saad, Y.: GPU-accelerated preconditioned iterative linear solvers. J. Supercomputing **63**(2), 443–466 (2013)
13. Liang, C.K., Prvulovic, M.: MiSAR: minimalistic synchronization accelerator with resource overflow management. In: Proceedings of the 42nd Annual International Symposium on Computer Architecture, ISCA 2015, pp. 414–426 (2015)
14. Liu, W.: Parallel and Scalable Sparse Basic Linear Algebra Subprograms. Ph.D. Thesis, University of Copenhagen (2015)
15. Liu, W., Vinter, B.: A framework for general sparse matrix-matrix multiplication on GPUs and heterogeneous processors. J. Parallel Distrib. Comput. **85**, 47–61 (2015)
16. Liu, W., Vinter, B.: CSR5: an efficient storage format for cross-platform sparse matrix-vector multiplication. In: Proceedings of the 29th ACM International Conference on Supercomputing, ICS 2015, pp. 339–350 (2015)
17. Liu, W., Vinter, B.: Speculative segmented sum for sparse matrix-vector multiplication on heterogeneous processors. Parallel Comput. **49**, 179–193 (2015)
18. Mayer, J.: Parallel algorithms for solving linear systems with sparse triangular matrices. Computing **86**(4), 291–312 (2009)
19. Naumov, M.: Parallel Solution of Sparse Triangular Linear Systems in the Preconditioned Iterative Methods on the GPU. Technical report NVIDIA (2011)

20. Park, J., Smelyanskiy, M., Sundaram, N., Dubey, P.: Sparsifying synchronization for high-performance shared-memory sparse triangular solver. In: Kunkel, J.M., Ludwig, T., Meuer, H.W. (eds.) ISC 2014. LNCS, vol. 8488, pp. 124–140. Springer, Heidelberg (2014)

21. Ros, A., Kaxiras, S.: Callback: efficient synchronization without invalidation with a directory just for spin-waiting. In: Proceedings of the 42nd Annual International Symposium on Computer Architecture, ISCA 2015, pp. 427–438 (2015)

22. Saad, Y.: Iterative Methods for Sparse Linear Systems, 2nd edn. Society for Industrial and Applied Mathematics, Philadelphia (2003)

23. Saltz, J.H.: Aggregation methods for solving sparse triangular systems on multiprocessors. SIAM J. Sci. Stat. Comput. **11**(1), 123–144 (1990)

24. Schreiber, R., Tang, W.P.: Vectorizing the Conjugate Gradient Method. In: Proceedings of the Symposium on CYBER 205 Applications (1982)

25. Scogland, T.R., Feng, W.C.: Design and evaluation of scalable concurrent queues for many-core architectures. In: Proceedings of the 6th ACM/SPEC International Conference on Performance Engineering, ICPE 2015, pp. 63–74 (2015)

26. Suchoski, B., Severn, C., Shantharam, M., Raghavan, P.: Adapting sparse triangular solution to GPUs. In: Proceedings of the 2012 41st International Conference on Parallel Processing Workshops, ICPPW 2012, pp. 140–148 (2012)

27. Wang, H., Liu, W., Hou, K., Feng, W.C.: Parallel Transposition of Sparse Data Structures. In: Proceedings of the 30th ACM International Conference on Supercomputing, ICS 2016 (2016)

28. Wolf, M.M., Heroux, M.A., Boman, E.G.: Factors impacting performance of multithreaded sparse triangular solve. In: Palma, J.M.L.M., Daydé, M., Marques, O., Lopes, J.C. (eds.) VECPAR 2010. LNCS, vol. 6449, pp. 32–44. Springer, Heidelberg (2011)

29. Xiao, S., Feng, W.C.: Inter-block GPU Communication via fast barrier synchronization. In: 2010 IEEE International Symposium on Parallel Distributed Processing, IPDPS 2010, pp. 1–12 (2010)

30. Yan, S., Long, G., Zhang, Y.: StreamScan: fast scan algorithms for GPUs without global barrier synchronization. In: Proceedings of the 18th ACM SIGPLAN Symposium on Principles and Practice of Parallel Programming, PPopp. 2013, pp. 229–238 (2013)

Exploiting Task-Parallelism in Message-Passing Sparse Linear System Solvers Using OmpSs

José I. Aliaga[1], María Barreda[1(✉)],
Matthias Bollhöfer[2], and Enrique S. Quintana-Ortí[1]

[1] Dpto. de Ingeniería y Ciencia de Computadores,
Universidad Jaume I, Castellón, Spain
{aliaga,mvaya,quintana}@icc.uji.es
[2] Institute of Computational Mathematics,
TU Braunschweig, Braunschweig, Germany
m.bollhoefer@tu-bs.de

Abstract. We introduce a parallel implementation of the preconditioned iterative solver for sparse linear systems underlying ILUPACK that explores the interoperability between the message-passing MPI programming interface and the OmpSs task-parallel programming model. Our approach commences from the task dependency tree derived from a multilevel graph partitioning of the problem, and statically maps the tasks in the top levels of this tree to the cluster nodes, fixing the internode communication pattern. This mapping induces a conformal partitioning of the tasks in the remaining levels of the tree among the nodes, which are then processed concurrently via the OmpSs runtime system.

The experimental analysis on a cluster with high-end Intel Xeon processors explores several configurations of MPI ranks and OmpSs threads per process showing that, in general, the best option matches the internal architecture of the nodes. The results also report significant performance gains for the MPI+OmpSs version over the initial MPI code.

Keywords: Programming models · Sparse linear systems · Preconditioned iterative solvers · Task-level parallelism · ILUPACK · MPI · OmpSs

1 Introduction

The solution of large sparse systems of linear equations is a key linear algebra problem arising in many scientific and engineering applications that involve the discretization of partial differential equations (PDEs) [18]. Moreover, the connection between sparse linear algebra and graph algorithms has turned this type of problem into an appealing means to mine the vast amount of information in social networks and other big data analytic processes [13].

ILUPACK[1] (Incomplete LU decomposition PACKage) is a numerical package that contains efficient multilevel ILU factorization solvers, based on Krylov

---

[1] http://ilupack.tu-bs.de.

© Springer International Publishing Switzerland 2016
P.-F. Dutot and D. Trystram (Eds.): Euro-Par 2016, LNCS 9833, pp. 631–643, 2016.
DOI: 10.1007/978-3-319-43659-3_46

subspace methods [18], for large-scale sparse linear systems with up to millions of equations [10, 19, 21]. In previous work, we exploited the task-parallelism exposed by the task dependency graph (TDG) associated with the sparse matrix to develop parallel versions of ILUPACK's preconditioned Conjugate Gradient (PCG) solver. The target platforms for these past efforts included shared-memory multiprocessors via OpenMP [2, 3], multicore architectures with OmpSs[2] [1], and clusters using MPI [1, 4].

Unfortunately, the previous MPI version of ILUPACK [1, 4] could only map one leaf of the TDG to each MPI rank, impeding the exploitation of other types of parallelism internally to the nodes. This is a strong limitation for clusters consisting of "fat" nodes equipped with a significant numbers of cores per node, as the static correspondence between tasks and MPI ranks may result in an unbalanced distribution of the workload and, therefore, inefficiency.

In this work, we present a new implementation of ILUPACK which merges MPI and OmpSs to exploit the benefits of each programming model, and allows the execution of the solver with more than one leaf per MPI process. In addition, we perform an experimental evaluation in order to assess the impact of the MPI+OmpSs configuration, problem dimension, and number of leaves per core on the performance of the iterative solve. Our results on a cluster equipped with 16 Intel Xeon cores per node reveals that the MPI+OmpSs version consistently outperforms the initial MPI code in terms of both strong and weak scaling.

There exist other packages also based on ILU factorizations and Krylov subspace methods. For example, pARMS [15] is an MPI-based library of parallel solvers for solving general sparse linear systems where the preconditioner is based on an algebraic recursive multilevel ILU. In contrast to ILUPACK, it relies on an independent set strategy for partitioning the leading systems into small diagonal blocks and then it re-applies the strategy recursively. In [7] a completely different parallel approach to ILUs was presented treating the error between the ILU and the original matrix as sequences of nonlinear equations to be improved in parallel rather than decomposing the system into a hierarchy of independent blocks. In [9], a parallel incomplete factorization approach uses direct solver techniques based on the level-of-fill and the underlying graph properties Beside ILU-based techniques, efficient parallel direct solvers [11, 12, 14, 17, 20] based on OpenMP/MPI rely on tree–parallelism and a variety of sophisticated techniques. Moreover, there are certainly many further parallel preconditioning methods, e.g., those based on approximate inverses or algebraic multigrid methods just to mention a few of them.

The rest of the paper is organized as follows. Section 2 offers a brief review of ILUPACK and the strategy to extract task-parallelism from this application. Section 3 describes the different parallelization approaches, based on either OmpSs or MPI only, and the new solution that combines both parallel programming interfaces. Section 4 analyzes the performance and scalability of the different parallel versions. Finally, Sect. 5 summarizes our work and offers several concluding remarks.

---

[2] https://pm.bsc.es/ompss.

2   Exposing Task-Parallelism in ILUPACK

Introduction to ILUPACK. The C and Fortran routines included in ILU-PACK can be leveraged to solve sparse linear systems of the form $Ax = b$ via Krylov subspace methods [18]. This library provides multilevel preconditioners that improve the numerical properties of the linear system, reducing the number of steps of the iterative solver. Concretely, the procedure obtains an efficient preconditioner from the ILU factorization of the system matrix, dropping the small entries of the factors, while relying on pivoting to bound the norm of the inverse triangular factors, yielding a numerical multilevel hierarchy of partial inverse-based approximations [5,6].

S1: Compute the preconditioner $A \to M \approx LU$	
S2: Initialize $x_0, r_0, z_0, d_0, \beta_0, \tau_0$	
S3: $k := 0$	
S4: **while** $(\tau_k > \tau_{\max})$	**Iterative PCG solve**
S5: $w_k := A d_k$	(SPMV)
S6: $\rho_k := \beta_k / d_k^T w_k$	(DOT product)
S7: $x_{k+1} := x_k + \rho_k d_k$	(AXPY)
S8: $r_{k+1} := r_k - \rho_k w_k$	(AXPY)
S9: $z_{k+1} := M^{-1} r_{k+1} \approx U^{-1} L^{-1} r_{k+1}$	Apply preconditioner
S10: $\beta_{k+1} := r_{k+1}^T z_{k+1}$	(DOT product)
S11: $\alpha_k := \beta_{k+1} / \beta_k$	
S12: $d_{k+1} := z_{k+1} + \alpha_k d_k$	(AXPY-like)
S13: $\tau_{k+1} := \| r_{k+1} \|_2$	(2-norm)
S14: $k := k + 1$	
S15: **endwhile**	

Fig. 1. Algorithmic formulation of the PCG method. Here, $\tau_{\max}$ is an upper bound on the relative residual for the computed approximation to the solution.

For the particular case of a symmetric positive definite (s.p.d.) linear system, Fig. 1 illustrates a simplified version of the PCG solver underlying ILUPACK. The most challenging operations in this algorithm are the computation of the preconditioner (S1), before the iteration commences, and its application at each iteration (S9). We will describe in detail the task-parallelism implicit in these two operations.

Nested Dissection. Exploiting the relationship between sparse matrices and adjacency graphs, nested dissection can be recursively applied to permute a sparse matrix, yielding a collection of diagonal blocks that are linked to certain subgraphs and separators [3]. Moreover, the hierarchy of subgraphs and separators fixes the order in which the diagonal blocks have to be factorized. This process renders a TDG with the structure of a tree, where the subgraphs occupy the leaves and the separators correspond to the internal nodes. For example,

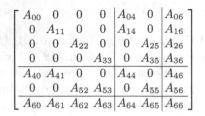 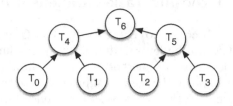

Fig. 2. Partitioning (left) and task dependency tree of the diagonal blocks (right). Task T_j is in charge of processing the diagonal block A_{jj}.

Fig. 2 (left) reflects the structure of a sparse matrix on which two nested dissection steps have been applied, yielding 4 subgraphs and 3 separators. Figure 2 (right) shows the TDG of the permuted matrix, where the edges of this directed acyclic graph define the dependencies between the diagonal blocks (tasks).

Computation of the Preconditioner. In order to improve the concurrency of this computation, the permuted matrix can be disassembled into one submatrix per leaf of the TDG. For instance, the submatrices for the graph in Fig. 2 are decomposed as

$$\begin{bmatrix} A_{00} & A_{04} & A_{06} \\ A_{40} & A_{44}^0 & A_{46}^0 \\ A_{60} & A_{64}^0 & A_{66}^0 \end{bmatrix}, \begin{bmatrix} A_{11} & A_{14} & A_{16} \\ A_{41} & A_{44}^1 & A_{46}^1 \\ A_{61} & A_{64}^1 & A_{66}^1 \end{bmatrix}, \begin{bmatrix} A_{22} & A_{25} & A_{26} \\ A_{52} & A_{55}^2 & A_{56}^2 \\ A_{62} & A_{65}^2 & A_{66}^2 \end{bmatrix}, \begin{bmatrix} A_{33} & A_{35} & A_{36} \\ A_{53} & A_{55}^3 & A_{56}^3 \\ A_{63} & A_{65}^3 & A_{66}^3 \end{bmatrix}, \quad (1)$$

where

$$A_{44} = A_{44}^0 + A_{44}^1, \ A_{55} = A_{55}^2 + A_{55}^3, \ A_{66} = A_{66}^0 + A_{66}^1 + A_{66}^2 + A_{66}^3. \quad (2)$$

Thus, the factorizations of the leading blocks of these four submatrices can proceed in parallel, while the modified blocks $A_{ij}^{0/1/2/3}$ are needed to solve the dependencies of the ancestor tasks. This process continues traversing the dependency tree, until the root task factorizes its local submatrix.

Application of the Preconditioner. The application of the preconditioner requires the solution of two triangular systems, corresponding to the lower and the upper incomplete triangular factors. The TDG for the former triangular system presents the same structure and dependencies as that associated with the computation of the preconditioner. In the latter triangular solve, the structure is preserved but dependencies are reversed, pointing top-down from the root to the leaves. Therefore, concurrency increases/decreases as we move towards/away from the leaves.

Other Kernels in the PCG Iteration. The remaining operations of PCG conform the computation/application of the preconditioner. The matrix is disassembled following (1) and the vectors are partitioned in a conformal manner, but the operation on vectors does not always fulfill (2). With this formulation all these computations only involve the leaves of the TDG and, therefore, can be computed fully in parallel, except for the dot products, which require an atomic addition (reduction) of the values locally computed in each leaf.

Degree of Concurrency. The number of leaves of the TDG grows exponentially with the number of nested dissection steps, so that the degree of concurrency can be easily increased by expanding additional levels. However, each dissection step introduces additional numerical levels in the computation yielding both a different TDG and a distinct preconditioner. While the numerical properties of all these preconditioners are similar, in practice the number of iterations of the PCG solver increases significantly after a few levels (8 and more) are expanded.

3   Exploiting Task-Parallelism with OmpSs and MPI

In this section, we first briefly review how to exploit the task-parallelism explicitly exposed by the TDG, using either OmpSs or MPI, to then introduce our approach that combines both parallel programming models to yield a task-parallel MPI+OmpSs solution.

3.1   Parallelization Using OmpSs

OmpSs is a task-based parallel programming model developed at Barcelona Supercomputing Center (BSC) [8,16]. At execution time, the runtime system underlying OmpSs detects data dependencies between tasks, with the help of OpenMP-like compiler directives (pragmas) annotated with clauses that indicate the task operands' directionality (input, output or input/output). OmpSs then generates a task graph during the execution, which is leveraged to schedule the tasks to the cores, exploiting the inherent task-level parallelism while fulfilling the dependencies embedded in the graph.

   The opportunities to exploit task-level parallelism in ILUPACK's PCG method lie within the computations that involve the preconditioner (computation and application) as well as the vector operations. The introduction of OmpSs in the operations with the preconditioner is quite intricate, mostly due to the complexity of ILUPACK itself. Nonetheless, it is possible to create a "skeleton" structure that explicitly exposes/governs the dependencies associated with the TDG while requiring only minor modifications in the routines included in the OpenMP version of ILUPACK [1,2]. In contrast, as the sparse matrix and the vectors in the PCG iteration are disassembled conformally, according to (1), the operations on the latter can be decomposed into a number of independent vector suboperations, which are easily parallelized using OmpSs. The only exception are the dot products which, after the reduction of the subvectors local to each thread, involve an atomic addition and, therefore, a synchronization/barrier [1,2]. Although nested parallelism could be applied to optimize the operations related to each node, our experience with this technique is negative.

3.2   Parallelization with MPI

The original MPI-based parallel version of ILUPACK, introduced in [4], spawns one MPI rank per leaf (task) of the TDG, with a one-to-one static mapping between leaves and ranks. This task-rank correspondence is fixed before the preconditioner computation, by the root process, which sends the information for each leaf to the appropriate MPI rank. The same mapping is then maintained during the complete execution, for all computations and iterations, including the preconditioner computation/application and vector operations.

The operations with the preconditioner potentially transform the dependencies of the TDG into communications among MPI ranks. To reduce the number of transfers, an inner task is always mapped to one of the two MPI ranks where the two "children" tasks were mapped to. For example, consider a TDG consisting of 4 leaves mapped to 4 MPI ranks: R0–R3. Then, in order to collapse the first level when the graph is traversed bottom-up during the lower triangular system solve, ranks R0, R2 send their data to R1, R3, respectively. Next, the receivers accumulate this information with the results from their own computations, and process the tasks in the next higher level, while the senders block till the top-down traversal of the TDG during the upper triangular system solve. Following this strategy, traversing the TDG only requires a communication between "sibling" tasks/"neighbour" MPI ranks.

Disassembling the matrix and the vectors, according to (1), allows all the other computations in PCG to operate with the leaves, avoiding any communication, except for the dot operations, which require an MPI reduction (`MPI_Reduce`) to accumulate the values computed in each node.

3.3   Combining MPI+OmpSs

In general, a strong motivation for mixing OmpSs with MPI is to unleash a higher level of asynchronism, for example in order to overlap communication with computation reducign the number of global synchronizations. In this particular work, the major advantage of combining both programming models is to exploit dynamic scheduling within the cluster nodes via OmpSs.

The first step to obtain an MPI+OmpSs solution is to develop a *new* MPI version of ILUPACK where an MPI rank can handle a subtree of the TDG comprising several leaves and the related inner tasks. With this version, OmpSs can then be used to process the tasks mapped to each MPI rank, dynamically distributing the work between several OmpSs threads. For example, consider a 2-level TDG composed of one root task and two leaves to be executed on a processor with two cores. If the computational cost associated with the leaves is unbalanced, this can be tackled by expanding an additional level of the TDG, yielding a 3-level tree consisting of four leaves. Now, if the parallelization is based on MPI only, an optimal mapping of the tasks to MPI ranks requires a prior knowledge of the computational costs of the tasks. Compared with this, an OmpSs parallel version with 2 threads features a dynamic mapping of tasks to threads that is more flexible and can use the resources more efficiently by, e.g., prioritizing the execution of the more expensive tasks.

The MPI+OmpSs version still requires an initialization where the root process distributes the data corresponding to (the leaves of) the subtrees among the MPI ranks. The MPI+OmpSs version of ILUPACK is then divided into a sequence of interleaved OmpSs and MPI stages, with the former ones computing the tasks internal to the subtrees local to the MPI ranks, and the latter requiring communication between MPI ranks. In particular, the computation of the preconditioner comprises only one stage of each type, but its application in the loop body of PCG has two OmpSs stages per iteration because the TDG is traversed twice. Figure 3 illustrates the initial distribution for a TDG with 8 leaves, together with a scheme of the execution of the two stages in the preconditioner computation. In that example, the OmpSs threads process the tasks within the bottom two levels, with no MPI communication involved. For the top two levels, the OmpSs threads remain inactive and it is the MPI ranks that are in charge of processing the tasks. The dot operations also exhibit the same two stages: On the leaves, the OmpSs threads accumulate their local subvectors, and an atomic reduction is then applied to compute the reduction inside each MPI rank. These local values are then reduced using an MPI collective primitive. The remaining vector computations of the PCG iteration operate in the bottom level only and, therefore, are computed by OmpSs threads with no MPI communication involved.

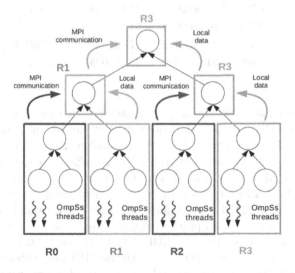

Fig. 3. Mapping of a TDG to 4 MPI ranks (R0–R3) with 2 OmpSs threads per rank.

4 Experimental Results

4.1 Setup and Preliminaries

The experiments in this section were performed using IEEE754 double-precision arithmetic on *MareNostrum*, a large-scale computing infrastructure at BSC. This platform connects 3,056 compute nodes via an Infiniband Mellanox FDR10 network. Each node contains two Intel Xeon E5-2670 processors for a total of 16 cores per server (2.6 GHz). The nodes employed in our experiments were also equipped with 64 Gbytes of DDR3 RAM.

For the experimental analysis, we employed an s.p.d. linear system arising from the finite difference discretization of a 3D Laplace problem, with instances of different size; see Table 1. In the experiments, all entries of the right-hand side vector b were initialized to 1, and the PCG iterate was started with the initial guess $x_0 \equiv 0$. For the tests, the parameters that control the fill-in and convergence of the iterative process in ILUPACK were set as droptool $= 1.0E-2$, condest $= 5$, elbow $= 10$, and restol $= 1.0E-6$.

Table 1. Matrices employed in the experimental evaluation, where n_z only includes the non-zeros in the upper triangular part.

	Matrix	Dimension n	#non-zeros n_z	Density (%)
Laplace	A159	4,019,679	16,002,873	9.90E-7
	A200	8,000,000	31,880,000	4.98E-7
	A252	16,003,008	63,821,520	2.49E-7
	A318	32,157,432	128,326,356	1.24E-7
	A400	64,000,000	255,520,000	6.23E-8

In the following we analyze the performance of two parallel versions of the PCG solver in ILUPACK: one based on MPI that can handle several leaves per MPI rank, with no intervention of OmpSs (hereafter, referred to as MPI-only); and an alternative variant that combines MPI+OmpSs also capable of processing several leaves per MPI rank, but which does so via OmpSs threads internally to each node. The MPI+OmpSs code was compiled using Mercurium C/C++ (1.99.8), with the OpenMPI (1.8.1) flags -showme:compile and -showme:link. The MPI-only variant was compiled with the same version of OpenMPI. Other software included OmpSs (15.06), ILUPACK (2.4), and ParMetis[3] (4.0.2) for the graph reorderings. In the executions with the MPI-only version, we spawned one MPI rank per core (i.e., 16 per node). For MPI+OmpSs, we tested distinct combinations of MPI ranks and OmpSs threads, with the numbers of ranks multiplied by the number of threads always being equal to 16 per node.

In the following, we consider the behaviour of the iterative PCG solver only, without the preconditioner computation, because the computational cost of the

---

[3] http://glaros.dtc.umn.edu/gkhome/metis/parmetis/download.

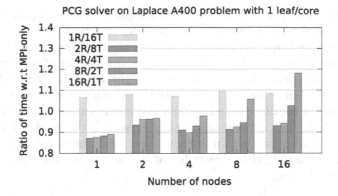

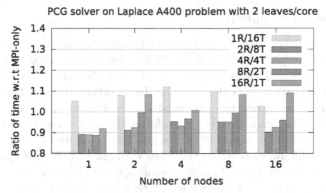

Fig. 4. Execution time per PCG iteration for the Laplace A400 problem for different configurations, using 1 leaf per core (top) and 2 leaves per core (bottom). (Color figure online)

latter is in general smaller and we observed no significant performance differences between the MPI-only and MPI+OmpSs parallel versions of this procedure. In addition, several previous experiments (for brevity, not shown here) revealed that the best performance was obtained when splitting the sparse matrix via nested dissection to generate a TDG with a number of leaves that equals or doubles the number of cores. Therefore, for simplicity, in the following we analyze only these two cases.

4.2   Analysis of Configurations

In order to assess the performance of the parallel MPI+OmpSs version of ILU-PACK, we first evaluate different combinations of MPI ranks and OmpSs threads per node (configurations). Given the node target architecture, with 2 sockets/8 cores per socket, we employ 1, 2, 4, 8 or 16 MPI ranks per node and the corresponding number OmpSs threads that fill all cores per node: 16, 8, 4, 2 or 1, respectively. We will denote these configurations as 1R/16T, 2R/8T, 4T/4T, 8R/2T, and 16R/1T (#Ranks/#Threads). Figure 4 reports the ratio of execution time of these configurations normalized with respect to the MPI-only implementation for the A400 problem, splitting the problem to obtain one leaf per core and two leaves per core. Both graphs reveal that, for almost all cases, the best option is 2R/8T, which mimics the internal socket/core architecture of the servers. Furthermore, we also notice that the extreme configurations, 1R/16T and 16R/1T, deliver the lowest performance. In the case that employs 1 rank and 16 OmpSs threads, this is due to the intersocket implicit communications. In the alternative with 16 MPI ranks and 1 thread per rank the reason is the overhead introduced by the OmpSs runtime system. In order to avoid this, when exploiting the hardware concurrency using MPI ranks only, we will not employ the OmpSs runtime system in the following.

4.3   Analysis of Scalability

We first evaluate the strong scalability of the parallel solvers. Figure 5 shows the execution time per iteration of the PCG solve for the A400 problem as the resources are increased from 16 cores/1 node to 256 cores/16 nodes. In general, as expected, there is a decrease in the iteration time as the number of cores grows. If we compare the two versions, the results demonstrate that the MPI+OmpSs variant consistently outperforms the MPI version (with no underlying OmpSs runtime system), by a margin that is around 5–10 %. Moreover, there is a slight difference between the cases with one or two leaves per core that is enlarged with the number of cores, revealing the TDG with one leaf per core as the best choice for 32 or more cores. The reason is that, as the amount of computational resources grows, the additional concurrency explicitly exposed by further splitting the computational load (sparse matrix/adjacency graph) does not compensate the overhead that is introduced for this particular (moderate) problem dimension.

The next experiment aims to provide an evaluation of weak scaling for the parallel solvers. Unfortunately, for ILUPACK's PCG solve it is not possible to

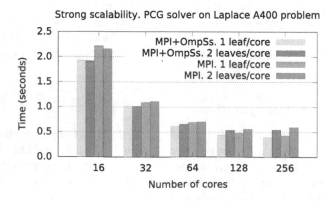

Fig. 5. Execution time per PCG iteration for the Laplace A400 problem. (Color figure online)

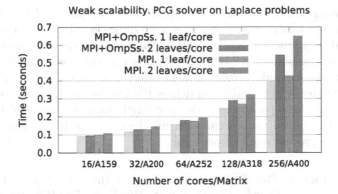

Fig. 6. Execution time per PCG iteration for different Laplace problems. (Color figure online)

generate an instance of the Laplace problem with a computational complexity that grows exactly in proportion to the number of resources. To approximate this scenario, we set the number of non-zeros of the sparse matrix (n_z) to be roughly proportional to the number of cores. However, we emphasize that n_z only offers an estimation of the computational cost, as other factors such as the fill-in/quality of the preconditioner may play a relevant role. Figure 6 reports the performance of the parallel implementations of the PCG solve (per iteration) for the different matrices in Table 1. These results show that the execution times grow with the number of cores/problem dimension. The reason is that the number of actual floating-point arithmetic operations per iteration increases faster than n_z. Comparing both implementations, the MPI+OmpSs version outperforms the MPI variant; and the difference between the cases with one or two leaves per core also grows with the number of cores.

5   Concluding Remarks

We have presented a new parallel version of the complete method underlying ILUPACK for solving symmetric positive definite linear systems on clusters of multicore processors. The approach extracts task-parallelism by splitting the sparse matrix into multiple levels, yielding a directed acyclic graph, with the form of a binary tree, where the nodes represent tasks, the arrows indicate data dependencies, and most computational work is performed in the leaf tasks. This graph is then traversed from bottom-up for the computation of the preconditioner and one of the triangular solves during its application, and top-down for the second triangular solve. In principle, the tree can be expanded into further levels to expose any number of tasks and, therefore, degree of concurrency. However, doing so yields different preconditioners and, from a certain depth, incurs into a significant overhead. In general, the best compromise is to generate up to two leaves per core, to allow the OmpSs scheduler optimize the computation. The experimental results confirm this assert for configurations with a reduced number of nodes, where the overhead is compensated by the OmpSs optimization. For unstructured matrices, the OmpSs runtime system accelerates the computation in most scenarios, due to the irregularity of the node sizes.

The solver combines the MPI and OmpSs programming models, with the best solution corresponding to a configuration that maps one MPI rank and eight OmpSs threads per socket, mimicking the internal architecture of the cluster nodes. With these parameters, the new MPI+OmpSs version of ILUPACK outperforms the initial implementation for clusters, which was based on MPI and could only process one leaf per rank.

Acknowledgements. This work was supported by the CICYT project TIN2014-53495-R of the MINECO and FEDER, and the H2020 EU FETHPC Project 671602 "INTERTWinE". María Barreda was supported by the FPU program of the *Ministerio de Educación, Cultura y Deporte*. The authors thankfully acknowledge the computer resources provided by BSC-CNS (*Centro Nacional de Supercomputación*).

References

1. Aliaga, J.I., Badia, R.M., Barreda, M., Bollhöfer, M., Dufrechou, E., Ezzatti, P., Quintana-Ortí, E.S.: Exploiting task and data parallelism in ILUPACK's preconditioned CG solver on NUMA architectures and many-core accelerators. Parallel Comput. **54**, 97–107 (2016)
2. Aliaga, J.I., Badia, R.M., Barreda, M., Bollhöfer, M., Quintana-Ortí, E.S.: Leveraging task-parallelism with OmpSs in ILUPACK's preconditioned cg method. In: 26th International Symposium on Computer Architecture and High Performance Computing (SBAC-PAD 2014), pp. 262–269 (2014)
3. Aliaga, J.I., Bollhöfer, M., Martín, A.F., Quintana-Ortí, E.S.: Exploiting thread-level parallelism in the iterative solution of sparse linear systems. Parallel Comput. **37**(3), 183–202 (2011)
4. Aliaga, J.I., Bollhöfer, M., Martín, A.F., Quintana-Ortí, E.S.: Parallelization of multilevel ILU preconditioners on distributed-memory multiprocessors. In: Jónasson, K. (ed.) PARA 2010, Part I. LNCS, vol. 7133, pp. 162–172. Springer, Heidelberg (2012)
5. Bollhöfer, M., Grote, M.J., Schenk, O.: Algebraic multilevel preconditioner for the Helmholtz equation in heterogeneous media. SIAM J. Sci. Comput. **31**(5), 3781–3805 (2009)
6. Bollhöfer, M., Saad, Y.: Multilevel preconditioners constructed from inverse-based ILUs. SIAM J. Sci. Comput. **27**(5), 1627–1650 (2006). Special issue on the 8-th Copper Mountain Conference on Iterative Methods
7. Chow, E., Patel, A.: Fine-grained parallel incomplete *lu* factorization. SIAM J. Sci. Comput. **37**(2), C169–C193 (2015)
8. Duran, A., Ferrer, R., Ayguadé, E., Badia, R.M., Labarta, J.: A proposal to extend the OpenMP tasking model with dependent tasks. Int. J. Parallel Program. **37**(3), 292–305 (2009)
9. Gaidamour, J., Hénon, P.: A parallel direct/iterative solver based on a schur complement approach. In: 11th IEEE International Conference on Computational Science and Engineering, CSE 2008, pp. 98–105. IEEE (2008)

10. George, T., Gupta, A., Sarin, V.: An empirical analysis of the performance of preconditioners for SPD systems. ACM Trans. Math. Softw. **38**(4), 24:1–24:30 (2012)

11. Hénon, P., Ramet, P., Roman, J.: PaStiX: a high-performance parallel direct solver for sparse symmetric definite systems. Parallel Comput. **28**(2), 301–321 (2002)

12. Irony, D., Shklarski, G., Toledo, S.: Parallel, fully recursive multifrontal supernodal sparse Cholesky. Future Gener. Comput. Syst. **20**(3), 425–440 (2004). Special issue: Selected numerical algorithms archive

13. Kepner, J., Gilbert, J. (eds.) Graph Algorithms in the Language of Linear Algebra. SIAM (2011)

14. Li, X.S.: An overview of SuperLU: algorithms, implementation, and user interface. ACM Trans. Math. Software **31**(3), 302–325 (2005)

15. Li, Z., Saad, Y., Sosonkina, M.: pARMS: a parallel version of the algebraic recursive multilevel solver. Numerical Lin. Alg. W. Appl. **10**, 485–509 (2003)

16. The OmpSs programming model. http://pm.bsc.es/ompss

17. Amestoy, J.K.P.R., Duff, I.S., L'Excellent, J.-Y.: A fully asynchronous multifrontal solver using distributed dynamic scheduling. SIAM J. Matrix Anal. Appl. **23**(1), 15–41 (2001)

18. Saad, Y.: Iterative Methods for Sparse Linear Systems. SIAM (2003)

19. Schenk, O., Bollhöfer, M., Römer, R.A.: On large scale diagonalization techniques for the Anderson model of localization. SIAM Rev. **50**, 91–112 (2008)

20. Schenk, O., Gärtner, K.: On fast factorization pivoting methods for symmetric indefinite systems. Electr. Trans. Num. Anal. **23**(1), 158–179 (2006)

21. Schenk, O., Wächter, A., Weiser, M.: Inertia-revealing preconditioning for large-scale nonconvex constrained optimization. SIAM J. Sci. Comput. **31**(2), 939–960 (2009)

Lightweight and Accurate Silent Data Corruption Detection in Ordinary Differential Equation Solvers

Pierre-Louis Guhur[1,2]([⊠]), Hong Zhang[1], Tom Peterka[1], Emil Constantinescu[1], and Franck Cappello[1]

[1] Argonne National Laboratory, Lemont, USA
pierre-louis.guhur@ens-cachan.fr,
{hongzh,tpeterka,emconsta,cappello}@mcs.anl.gov
[2] ENS de Cachan, Cachan, France

Abstract. Silent data corruptions (SDCs) are errors that corrupt the system or falsify results while remaining unnoticed by firmware or operating systems. In numerical integration solvers, SDCs that impact the accuracy of the solver are considered significant. Detecting SDCs in high-performance computing is necessary because results need to be trustworthy and the increase of the number and complexity of components in emerging large-scale architectures makes SDCs more likely to occur. Until recently, SDC detection methods consisted in replicating the processes of the execution or in using checksums (for example algorithm-based fault tolerance). Recently, new detection methods have been proposed relying on mathematical properties of numerical kernels or performing data analysis of the results modified by the application. None of those methods, however, provide a lightweight solution guaranteeing that all significant SDCs are detected. We propose a new method called *Hot Rod* as a solution to this problem. It checks and potentially corrects the data produced by numerical integration solvers. Our theoretical model shows that all significant SDCs can be detected. We present two detectors and conduct experiments on streamline integration from the WRF meteorology application. Compared with the algorithmic detection methods, the accuracy of our first detector is increased by 52 % with a similar false detection rate. The second detector has a false detection rate one order of magnitude lower than these detection methods while improving the detection accuracy by 23 %. The computational overhead is lower than 5 % in both cases. The model has been developed for an explicit Runge-Kutta method, although it can be generalized to other solvers.

Keywords: Resilience · Fault tolerance · Runge-kutta · Numerical integration solvers · HPC · SDC

1 Introduction

Ensuring trustworthy results has always been a critical challenge for scientists. In numerical simulations, results can be impaired by silent data corruptions

© Springer International Publishing Switzerland 2016
P.-F. Dutot and D. Trystram (Eds.): Euro-Par 2016, LNCS 9833, pp. 644–656, 2016.
DOI: 10.1007/978-3-319-43659-3_47

(SDCs). Because of an ever increasing number of processes, exascale reports [18] project an increase in the SDC rate in future systems. The origins of SDC are diverse. Examples of SDC sources are electromagnetic interferences [15], ionizing radiation [1], and aging of hardware components.

Replication [10] can detect SDCs by duplicating the same program (or other versions in n-version programming [7]) and comparing their results. In a deterministic program, all duplications must provide exactly the same result; otherwise the results are considered corrupted. The protection of linear algebra results in algorithm-based fault tolerance (ABFT) [12] and the error-correction code memory (ECC) [9] are both based on checksums: ABFT computes and performs detection inside the software, and ECC memory is inside the hardware. All these methods are generic (although ABFT is limited to certain numerical kernels), and only a few percents of SDCs are undetected. However, these methods may be too computationally expensive (replication), or they do not protect each component. ABFT covers only the data used in the kernel and not the other data used by the application. ECC protects only memory, caches, and registers; it usually does not protect the CPU control logic or its functional units.

In the context of iterative, time-stepping methods, new detection techniques compare the results of the numerical method with results produced by a surrogate function. Because previous steps of the numerical method have already been validated, the surrogate function can use these values as trusted references to compute its own results for the current step. In the adaptive impact-driven (AID) detector [8], the surrogate function computes value predictions for the current step by extrapolation from several past steps of the numerical method. If the difference between the numerical method results and the surrogate function predictions is outside a certain confidence interval, an SDC is reported. AID uses different extrapolation methods from order 0 to 2 and selects dynamically the one that minimizes the prediction error. The confidence interval is built from the acceptable bound (given by the user) upon which SDCs are considered as impacting the results, the number of false positives, and the maximum error of extrapolation. Following a different direction, Benson et al. [3] propose a more complex surrogate function by computing an error estimate. As with the AID detector, the estimate is compared to a predicted value. If the estimate for the current step is not similar to previous estimates, an SDC is reported. Their detector is called *BSS14*. While the two approaches are different, they both use extrapolation and thus rely on the smoothness property of the data set (AID) or of the estimate (BSS14) to perform accurate detection. As shown in Sect. 4, they do not guarantee that all SDCs impacting the accuracy of the iterative methods are detected, in particular when the data set (AID) or the estimate (BSS14) presents stiff variations.

Our objective is to design and develop a new SDC detection technique that presents a high detection accuracy (also called recall or true positive rate, TPR) and a low false detection rate (also called false positive rate, FPR), and does not rely on extrapolation. *We mathematically show that all significant SDCs are detected.* In the context of numerical integration solvers, a solver is chosen because its approximation error is acceptable with respect to the required accuracy of the results.

We consider that an SDC is significant when the introduced error is bigger than the approximation error of the solver. We built two new detectors relying on mathematical properties of the ordinary differential equation (ODE) integration method. Our detection technique compares two estimates of this approximation error. We chose estimates that are similar if and only if no significant SDC occurs. A confidence interval on the similarity, established from a simple machine learning algorithm, controls the SDC detection. If an SDC is detected, the correction is done by recomputing the step. The two detectors present different tradeoffs. One has a high accuracy (we call it *Hot Rod HR*, for resilient ODE high recall) and small false detection rate. The other has a false detection rate lower than 1 % but also a lower accuracy (we call it *Hot Rod LFP*, for low false positive). We designed the two detectors for Cash-Karp's method [5], a fourth-order Runge-Kutta method with a fixed-step size. However, our technique can be applied to other ODE integration methods as discussed in Sect. 5.

Because all significant SDCs are detected, our detectors improve the trustworthiness of the results while avoiding wasting of resources to recover from insignificant SDCs. We performed experiments on a streamline integrator used for visualizing of WRF meteorology application results [17].

Section 2 explains background. In Sect. 3, our method for detecting SDC is detailed, and proof is given that all significant SDC are detected. In Sect. 4, our SDC detectors are tested in a meteorology application and compared with replication and the AID and BSS14 detectors.

2   Background

An ODE is a differential equation of one independent variable and its derivatives. Because numerical integration solvers are widely used, trust in their results is critical. An initial value problem can be formulated as

$$x'(t) = f(t, x(t)), \quad x(t_0) = x_0,$$

with $t_0 \in \mathbb{R}, x_0 \in \mathbb{R}^m, x : \mathbb{R} \to \mathbb{R}^m$, and $f : \mathbb{R} \times \mathbb{R}^n \to \mathbb{R}^m$; f is L-Lipschitz continuous.

2.1   Runge-Kutta Methods

For each $n = 1, ..., N$ with N the total number of steps, Runge-Kutta methods (RKMs) provide an approximation x_n of $x(t_n)$, where $t_n = t_0 + nh$, $h \in \mathbb{R}_+^*$ the step size, and $x(t_n)$ the exact solution of the ODE at time t_n. An s-stage explicit RKM is defined by

$$\forall i \leq s, k_i = f\left(t_n + c_i h, x_n + h \sum_{j=1}^{i-1} a_{ij} k_j\right); \quad x_{n+1} = x_n + h \sum_{i=1}^{s} b_i k_i.$$

Here $(k_i)_i$ are called the stage slopes and represent the most computationally expensive part of the method. The local truncation error (LTE) is the approximation error introduced at a step $n + 1$, and it can be defined by $LTE_{n+1} = x_{n+1} - \tilde{x}(t_{n+1}, x_n)$, with $\tilde{x}(t, x_n)$ the exact solution of the problem $\tilde{x}'(t, x_n) = f(t, \tilde{x}(t, x_n)), \tilde{x}(t_n, x_n) = x_n$. The global truncation error (GTE) is the absolute difference between the correct value $\tilde{x}(t_n, x_0)$ and the approximated value x_n. An ODE integration method is said to have an order p if $LTE_n = O(h^{p+1})$ and $GTE_N = O(h^p)$, where N is the last step.

In the following, we focus on a RKM called Cash-Karp's method [5], while generalization is discussed in Sect. 5. Cash-Karp's method has an order 4 and computes six stages. Although two more stages than the classical fourth-order Runge-Kutta method are required, Cash-Karp's method allows us to compute the embedded method that is used in BSS14, in our detectors, and in the adaptive integration scheme.

2.2 Embedded Methods

LTE can be estimated with *embedded methods*. These methods compute two results x_n^p and x_n^q from two RKMs with orders p and q (in general $|q - p| = 1$). The solution is propagated by one of these results, while its stages (and possibly extra stages) are reused to compute the other result in order to achieve a low overhead. In the case of Cash-Karp's method, $p = 4$ and $q = 5$. If LTE_n^q has a higher order than does LTE^p, the difference between x_n^p and x_n^q estimates LTE^p:

$$
\begin{aligned}
\mathcal{E}_n &= x_n^p - x_n^q \\
&= x_n^p - \tilde{x}(t_n, x_{n-1}) - (\tilde{x}(t_n, x_{n-1}) - x_n^q), \\
&= LTE_n^p - LTE_n^q, \\
&= LTE_n^p + O(h^{q+1}).
\end{aligned}
$$

2.3 Radau's Quadrature

Another way of estimating LTE is suggested by Stoller and Morrison [19] and extended by Ceschino and Kuntzmann [6]. Relying on Radau's quadrature and Taylor's expansion, Ceschino and Kuntzmann give an expression of the LTE of a method given its order $p \leq 5$. The estimate $\mathcal{R}_n$, called here *Radau's estimate*, does not require the computation of any extra stage, but it checkpoints previous stages and solutions. Therefore, it has a memory overhead, rather than a computational overhead as does the embedded method. Since $\mathcal{E}_n$ is a sixth-order estimate, we use the following estimate $\mathcal{R}_n$ presented by Butcher and Johnston [4]:

$$
\begin{aligned}
\mathcal{R}_n &= \frac{h}{10} \left[3f(t_{n-2}, x_{n-2}) + 6f(t_{n-1}, x_{n-1}) + f(t_n, x_n) \right] \\
&\quad - \frac{1}{30} \left[x_{n-3} + 18x_{n-2} - 9x_{n-1} - 10x_n \right] \\
&= LTE_n^p + O(h^{p+2}).
\end{aligned}
$$

3   Proposed Hot Rod Method

Our method relies on a surrogate function Δ_n that is the difference between two estimates: $\Delta_n = \mathcal{A}_n - \mathcal{B}_n$. For Cash-Karp's method, we use the embedded estimate $\mathcal{A}_n = \mathcal{E}_n$ and Radau's estimate $\mathcal{B}_n = \mathcal{R}_n$. In the absence of SDC, the surrogate function becomes $O(h^{p+2})$:

$$\Delta_n = \left(LTE_n + O(h^{p+2})\right) - \left(LTE_n + O(h^{p+2})\right) = O(h^{p+2}).$$

In Hot Rod HR, the surrogate function is compared with a certain confidence interval centered over zero. When the surrogate function is outside the confidence interval, an SDC is reported. We show that all significant SDCs are detected. However, Hot Rod HR may have a false positive rate of a few percents. In Hot Rod LFP, we chose a larger confidence interval, and its false positive rate remains below 1 percent.

3.1   First Detector: Hot Rod HR

In regular cases, our surrogate function is one order higher than that of the LTE. In presence of an SDC, Δ_n is outside the confidence interval, as shown in the following paragraph. Hence, SDCs whose introduced errors are even smaller than the LTE are expected to be detected. We show that all significant SDCs are detected by Hot Rod HR.

Detection of Significant SDCs. An SDC is detected when $|\Delta_n^c| \geq C_n$ with C_n the half-length of the confidence interval at step n. It is all the more difficult to detect when $\Delta_n^o = 0$. We show that the minimum injected error ϵ_{min} that can be detected is of the same order as that of the approximation error.

We study the case of a corrupted stage k_i; the case of a corrupted result x_n itself is similar. Here, $k_i^c = \epsilon - k_i^o$, where ^{c} (resp. ^{o}) denotes corrupted (resp. uncorrupted) data:

$$\Delta_n^c - \Delta_n^o = \mathcal{E}_n^c - \mathcal{E}_n^o - (\mathcal{R}_n^c - \mathcal{R}_n^o) = h\epsilon \left[\hat{b}_i + b_i \left(\frac{1}{30} - \frac{3\delta_{i,1}}{10}\right)\right],$$

where δ_{ij} is defined by $\delta_{ij} = 1$ if $i = j$; otherwise $\delta_{ij} = 0$, (b_i) (resp. $(\hat{b}_i)$) are the coefficients of the order 4 (resp. 5) in Cash-Karp's method.

The minimum error ϵ_{min} that we can detect corresponds to the case $|\Delta_n^c - \Delta_n^o| = C_n - 0$. We note that $B = \hat{b}_i + b_i \left(\frac{1}{30} - \frac{3\delta_{i,1}}{10}\right)$. This leads to

$$\epsilon_{min} = \frac{C_n}{hB} = O\left(\frac{C_n}{h}\right).$$

When x_n is corrupted instead of a stage, one can derive that $\epsilon_{min} = O(C_n)$. If C_n has the same order as Δ_n, then (1) $\epsilon_{min} = O(h^{p+1})$ when an error is injected inside a stage and (2) $\epsilon_{min} = O(h^{p+2})$ when an error is injected inside a result. In other words, the threshold of detection has the same order as (or better than) the LTE of Cash-Karp's method. This guarantees that all significant SDCs are detected.

Confidence Interval. Because $\Delta_n = O(h^{p+2})$, one can assume that Δ_n acts as a random variable, with a zero-mean in the absence of SDC. Its standard deviation can be estimated from a training set T composed of N_s samples with the unbiased sample standard deviation

$$\sigma = \sqrt{\frac{1}{N_s - 1} \sum_{n=1}^{N_s} \Delta_n^2}. \tag{1}$$

Assuming that $(\Delta_n)_n$ follows a normal distribution, the "three sigma rule" [14] suggests choosing $C_n = 3\sigma$. Thus, we expect that 99.7 % of uncorrupted $(\Delta_n)_n$ fall within the confidence interval, or in other words a false positive rate of 0.3 %. The normal distribution is a natural choice for modeling the repartition of training samples.

Because items from T are not labeled as trusted or untrusted samples, the evaluation of σ might be corrupted. It thus would jeopardize the confidence interval and thus the SDC detector. To improve reliability, we weighted each Δ_n with its own value. Equation (1) becomes

$$\Sigma = \sum_{n=1}^{N_s} \exp\left(-\Delta_n^2\right); \qquad \sigma = \sqrt{\frac{1}{(N_s - 1)\Sigma} \sum_{n=1}^{N_s} \exp\left(-\Delta_n^2\right)\Delta_n^2}.$$

Adaptive Control. The hypothesis of a normal distribution may be invalidated. We therefore developed a correction of the confidence interval based on false positives.

When an SDC is reported, the current step is recomputed. If the result has the same value, we can assume that it was a false positive and not an SDC. Because of the "three sigma rule," the FPR is expected to be 0.3 %. If the FPR is an order of magnitude higher, at 3 %, for k times, the confidence interval is increased with a certain coefficient $1 + \alpha$. C_n becomes $C_n = (1 + \alpha)^k \times 3\sigma$, where α fixes the rate of the adaptive control. Because $(1 + \alpha)^k = 1 + \alpha k + O(\alpha^2)$, α is taken as $1/(\max(FPR) \times N)$, where N is the number of steps in the application and $\max(FPR)$ is the maximum acceptable false positive rate. Because a false positive requires the recomputation of a noncorrupted step, we suggest setting $\max(FPR)$ at 5 % to limit the computational overhead. In our experiments, we have $N = 1000$; thus $\alpha = 0.02$.

Thanks to the adaptive control, the training set requires only a few steps. In our experiments, we have found that $N_s = 5$ samples are sufficient to initialize the confidence interval.

3.2   Second Detector: Hot Rod LFP

If the cost of a false positive is too high, Hot Rod HR is not suitable. Hence, we designed a second detector with a larger confidence interval. Nonetheless, all significant SDCs must still be detected.

This new confidence interval is defined by $\mathfrak{C}_n = 10C_{99}(|\Delta| \in T)$, with C_{99} the 99th percentile of the training set. The interval can be interpreted as a threshold that is an order of magnitude bigger than the surrogate functions in the training set. Because this threshold is higher than the previous one, this detector's recall is lower. Because the estimates are at order $p = 4$ for Cash-Karp's method, the LTE at step n can be expressed as $LTE_n = Ch^{p+1} + O(h^{p+2})$. We show that the GTE at the last step N is still an order p, since it used to be without corruption. We assume the probability that an SDC occurs and is accepted as small enough to guarantee that at most only one SDC will be accepted. The worst case is when this SDC is accepted at the first step, $n = 1$, and when $\mathfrak{C}_n = \Delta_n$. Hence, the introduced error is $LTE_1 = 10Ch^{p+1} + O(h^{p+2})$. Because $GTE_1 = LTE_1$, $GTE_1 = 10Ch^{p+1} + O(h^{p+2})$.

With $\tilde{x}(t, x_n)$ the notation in Sect. 2.1, $x(t) = \tilde{x}(t, x_0)$, and one can write that the GTE at a step $0 < n < N$ is

$$|\text{GTE}_{n+1}| = |x(t_{n+1}) - \tilde{x}(t_{n+1}, x_n) + \tilde{x}(t_{n+1}, x_n) - x_{n+1}|,$$
$$\leq |x(t_{n+1}) - \tilde{x}(t_{n+1}, x_n)| + |x_{n+1} - \tilde{x}(t_{n+1}, x_n)|.$$

Because f is L-Lipschitz continuous, the Gronwall's inequality simplifies the first term to

$$|x(t_{n+1}) - \tilde{x}(t_{n+1}, x_n)| \leq |\tilde{x}(t_n, x_0) - \tilde{x}(t_n, x_{n-1})| e^{Lh} = |GTE_n| e^{Lh}.$$

The second term, $|x_{n+1} - \tilde{x}(t, x_{n+1})|$, is the LTE at step $n+1$ and so is evaluated at $Ch^{p+1} + O(h^{p+2})$. Denoting $\gamma = e^{Lh}$, we obtain

$$\frac{|\text{GTE}_{n+1}|}{\gamma^n} \leq \frac{|\text{GTE}_n|}{\gamma^{n-1}} + \frac{Ch^{p+1}}{\gamma^n} \leq \dots \leq |GTE_1| + Ch^{p+1} \sum_{i=1}^{n} \frac{1}{\gamma^i}.$$

Because $\sum_{i=1}^{N} 1/\gamma^i = (\gamma^N - 1)/\gamma^N(\gamma - 1)$ and $\gamma - 1 \geq Lh$, noting $\tau = Nh$, we obtain

$$|\text{GTE}_{n+1}| \leq 10Ch^{p+1} + \frac{Ch^p}{L}\left(e^{L\tau} - 1\right) + O(h^{p+2}).$$

At the last step, we have verified that $GTE_N = O(h^p)$. The order of GTE is unchanged: the SDC is insignificant.

3.3   Algorithm

We presented two detectors and showed their efficiency. They differ in their tradeoffs: Hot Rod HR has a higher TPR, and Hot Rod LFP has a lower FPR. We saw that undetected SDCs have no impact on the accuracy of the ODE method. They require fixing the parameter α, but simple indications are given. We can thus derive two scenarios. If an SDC is likely to happen (it could be the case when the processor is not protected from SDC by ECC memory or other protection system), then Hot Rod HR is employed. Otherwise, employing Hot Rod LFP allows us to detect all significant SDCs with fewer false positives. The schema is illustrated in Algorithm 1 for a given detector.

```
while learning do
    step ← simulation(prev. step) ;
    Δ ← |A(step, prev.steps) − B(step, prev.steps)| ;
    TraininigSet.push(Δ) ;
end
while new step do
    step ← simulation(prev. step) ;
    Δ ← |A(step, prev.steps) − B(step, prev.steps)| ;
    if (Detector == Hot Rod HR and Δ ≤ Cₙ) or (Detector == Hot Rod
    LFP and Δ ≤ ℭₙ) then
        report("no error") ;
        accept step ;
    end
    else
        step ← simulation(prev. step) ;
        Δ' ← |A(step, prev.steps) − B(step, prev.steps)| ;
        if Δ' = Δ then
            report("false positive") ;
            if FPR > 3% then
                k++ ;
            end
        end
        accept step ;
    end
end
```

Algorithm 1. Pseudocode for the execution of our detectors

4 Experiments and Results

We have shown theoretically that all significant SDCs are detected with Hot Rod. In this section, we evaluate the SDC detectors with a meteorology application.

4.1 Environment

Experiments were computed on a machine with four Intel Xeon E5620 CPUs (each with 4 cores and 8 threads), 12 GB RAM, and one NVIDIA Kepler K40 GPU with 12 GB memory. It was programmed in C++11 using CUDA. The application is particle tracing for streamline flow visualization [11,16,17]. The solver integrates a velocity field to compute the streamline. It stops when the streamline goes outside the velocity field. Uncorrected streamlines can thus be shorter than they were supposed to be (Fig. 1).

4.2 SDC Injection Methodology

An SDC can arise from many sources in hardware and software [2,13], and these sources may change with new versions and generations of hardware and software.

Fig. 1. Streamlines computed by the application. The color gradient starts in red at seeds; 1,408 streamlines are computed (Color figure online).

We do not attempt to evaluate exhaustively the coverage of our approach because of space limitations. SDCs are simulated by flipping bits in data items. SDCs affect one or several bits in the same data item, called respectively singlebit and multibit corruption. We experimented with both cases. In multibit corruption, we chose the number of bit-flips $Nflips$ from a uniform distribution. Other distributions such as normal and beta distributions were tested with several different parameters, but the results were not significantly different from those reported below. Corruption can affect data items in any stage (or even directly in the result). The position of a bit-flip is drawn from a uniform distribution. In multibit corruption, we have forced the $Nflips$ bit-flips to be applied on $Nflips$ different positions. Some SDC have no impact on the results. In a third scenario, we inject only significant singlebit corruptions. We considered that an SDC is significant when the difference between the corrupted result and the safe result is higher than the mean LTE.

4.3 Benchmark

We compared our approach with similar methods presented in Sect. 1: replication, AID and BSS14 detectors. Those methods need to be parametrized. We compared results with a set of parameters and selected the parameters that provide the best results in our application. Using the same notation as in [8], we configure AID with $\theta r = 1$. Results were improved if the confidence interval is taken as $(1 + \alpha)^k(\epsilon + \theta r)$ with $\alpha = 0.2$ and k defined in Sect. 3.1. Concerning BSS14, five parameters should be set, but no indication is detailed in [3] about two of them. With the notation of [3], the considered values are $\tau_j = 1e^{-5}$, $\tau_v = 0.02$, $\Gamma = 1.4$, $\gamma = 0.95$, and $p = 10$.

4.4 Results

Table 1 presents results from our benchmark. We did not compare each detector with a solver with no detector. We compared each detector with a perfect detector that returns the ground truth. For computational overhead, we divided the execution time of each detector with that of the perfect detector. Our detectors have a computational overhead lower than 5 %, as do the BSS14 and AID detectors. It is 20 times less computationally expensive than replication. But

Table 1. Benchmark of our detectors Hot Rod (H.R.) LFP and HR, replication, AID and BSS14. Values in the column "IRE 95 %" are the injected relative errors (IRE) that were detected 95 % of the time.

Detector	TPR (%)			FPR (%)	IRE 95 %	Overheads (%)	
	Singlebit	Multibit	Significant			Comp	Memory
Replication	100.0	100.0	100.0	100.0	0.0	+100	+100
AID	14.3	43.2	86.7	1.6	$7e^{-6}$	+4.6	+50
BSS14	18.8	49.5	91.2	0.6	$4e^{-6}$	+3.7	+13
H.R. LFP	23.1	64.6	99.9	0.01	$7e^{-8}$	+3.8	+50
H.R. HR	28.6	69.6	99.9	1.2	$5e^{-9}$	+4.4	+50

unlike the AID detector, our detectors have to employ an embedded integration method that computes more stages than does another Runge-Kutta method of the same order.

Our detectors have a higher memory cost than does the BSS14 detector, but a smaller memory cost than does replication. For estimating memory overheads, we counted the number of stored vectors, such as solutions $(x_n)_n$, stage slopes $(k_i)_i$ and estimates. Cash-Karp's method requires computing and storing two additional stage slopes than does Runge Kutta 4, but the same number as the other embedded fourth-order methods. Cash-Karp's method requires storing 6 $(k_i)_i$ (among them $f(x_{n-1})$), and x_n; x_{n-1} is stored to allow a rollback in case of SDC detection; when $f(x_{n-1})$ is employed in the Radau estimation, $f(x_n)$ can be computed at the position (the result is employed at the next step if the step is accepted). Thus in total, 8 data elements are stored by the perfect detector, whereas $\mathcal{E}$ ($\mathcal{R}$ can use the same storage as $\mathcal{E}$), $f(x_{n-2}), x_{n-2}$ and x_{n-3} are stored for our detectors; AID stores x_{n-2}, x_{n-3}, x_{n-4}, and the extrapolated solution; and BSS14 stores $\mathcal{E}$.

The true positive rate (TPR) shows that our detectors detect perfectly (at 99.9 %) significant SDCs. Replication does as well, but the BSS14 and AID detectors have a TPR of 91.2 % and 86.7 % of significant SDCs, respectively. For BSS14 and AID, some SDCs can thus be undetected while affecting the accuracy of the solvers. Moreover, the "IRE 95 %" value of our detectors is smaller than the mean local error estimate ($1.5e^{-6}$) by a factor of 100. Because all significant SDCs are detected, SDCs undetected by Hot Rod are sure to have no impact. The undetected 76.9 % of SDCs by Hot Rod LFP are thus insignificant and do not need to be corrected: correcting these insignificant SDCs would not improve results and would demand extra computation. Figure 2 shows the LTE of the solver in the confidence interval in the absence of SDC. It represents the approximation error. As defined in Sect. 1, significant SDCs inject errors that are higher than this error. Because the streamlines of the AID and BSS14 detectors are pushed outside the confidence interval at SDC injections, they do not detect those SDCs. On the other hand, Hot Rod HR and LFP's streamlines are not affected by SDCs: these detectors protected the solver. This result is consistent

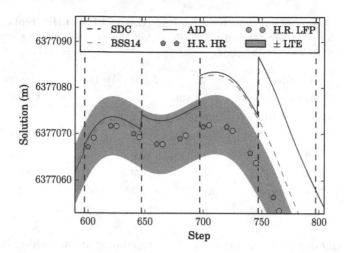

Fig. 2. One streamline computed by the different detectors. Singlebit injection is made every 50 steps. In the window, the position of the bit-flip varies from 31 to 35 in IEEE754 doubleprecision. The interval "$\pm LTE$" represents the approximation error. Significant SDCs shift the solution outside this interval. In the application, the origin is the center of the Earth. (Color figure online)

with the fact that the IRE 95 % of Hot Rod is two orders of magnitudes less than the approximation error.

5   Conclusion

This study presented our SDC detection method Hot Rod for ODE integration solvers. Both experimental and theoretical results show that all significant SDCs are detected. Except for replication, no other tested SDC detectors achieve these results. More specifically, compared with the algorithmic detection SDC detectors, the true positive rate is improved by 52 % for singlebit corruptions; whereas compared with replication, the computational overhead is reduced by 20 times. Moreover, users need only to fix the maximum false positive rate, as explained in Sect. 3.

Our detectors were employed for one of the ODE integration methods. Other embedded Runge-Kutta methods can be directly employed. Radau's estimates have a general expression in the case of adaptive step size; see the work of Butcher and Johnston [4]. For implicit methods or linear multisteps, Richardson's estimates can also be used. In future work, we plan to investigate detection in partial differential equation solvers.

Acknowledgments. We express our gratitude to Julie Bessac for assistance with the algorithm and Gail Pieper for comments that greatly improved the manuscript. We also gratefully acknowledge the use of the services and facilities of the Decaf project at

Argonne National Laboratory, supported by U.S. Department of Energy, Office of Science, Advanced Scientific Computing Research, under Contract DE-AC02-06CH11357, program manager Lucy Nowell. We also thank the anonymous reviewers for their helpful comments.

References

1. Bagatin, M., Gerardin, S.: Ionizing Radiation Effects in Electronics: From Memories to Imagers. Devices, Circuits, and Systems. CRC Press, Cleveland, Boca Raton (2015)
2. Bairavasundaram, L.N., Goodson, G.R., Pasupathy, S., Schindler, J.: An analysis of latent sector errors in disk drives. ACM SIGMETRICS Perform. Eval. Rev. **35**, 289–300 (2007)
3. Benson, A.R., Schmit, S., Schreiber, R.: Silent error detection in numerical time-stepping schemes. Int. J. High Perform. Comput. Appl. **29**, 403–421 (2014)
4. Butcher, J., Johnston, P.: Estimating local truncation errors for Runge-Kutta methods. J. Comput. Appl. Math. **45**(1), 203–212 (1993)
5. Cash, J.R., Karp, A.H.: A variable order Runge-Kutta method for initial value problems with rapidly varying right-hand sides. ACM TOMS **16**(3), 201–222 (1990)
6. Ceschino, F., Kuntzmann, J.: Numerical solution of initial value problems (1966)
7. Chen, L., Avizienis, A.: N-version programming: A fault-tolerance approach to reliability of software operation. In: Digest of Papers FTCS-8, pp. 3–9 (1978)
8. Di, S., Cappello, F.: Adaptive impact-driven detection of silent data corruption for HPC applications. In: IEEE Transactions on Parallel and Distributed Systems (2016)
9. Ghosh, S., Basu, S., Touba, N.A.: Selecting error correcting codes to minimize power in memory checker circuits. J. Low Power Electron. **1**, 63–72 (2005)
10. Guerraoui, R., Schiper, A.: Software-based replication for fault tolerance. Computer **4**, 68–74 (1997)
11. Guo, H., He, W., Peterka, T., Shen, H.W., Collis, S.M., Helmus, J.J.: Finite-time lyapunov exponents and lagrangian coherent structures in uncertain unsteady flows. In: IEEE TVCG (Proceedings of the PacificVis 16) 22, to appear (2016)
12. Huang, K.H., Abraham, J., et al.: Algorithm-based fault tolerance for matrix operations. IEEE Trans. Comput. **100**(6), 518–528 (1984)
13. Hwang, A.A., Stefanovici, I.A., Schroeder, B.: Cosmic rays don't strike twice: understanding the nature of DRAM errors and the implications for system design. ACM SIGPLAN Not. **47**, 111–122 (2012)
14. Krishnamoorthy, K., Mathew, T.: Statistical tolerance regions: theory, applications, and computation, vol. 744. Wiley, Hoboken (2009)
15. Lapinsky, S.E., Easty, A.C.: Electromagnetic interference in critical care. J. Crit. Care **21**(3), 267–270 (2006)
16. McLoughlin, T., Laramee, R.S., Peikert, R., Post, F.H., Chen, M.: Over two decades of integration-based, geometric flow visualization. In: Eurographics 2009 State of the Art Report, pp. 73–92. Munich, Germany (2009)
17. Peterka, T., Ross, R., Nouanesengsy, B., Lee, T.Y., Shen, H.W., Kendall, W., Huang, J.: A study of parallel particle tracing for steady-state and time-varying flow fields. In: IPDPS, pp. 580–591. IEEE (2011)

18. Snir, M., Wisniewski, R.W., Abraham, J.A., Adve, S.V., Bagchi, S., Balaji, P., Belak, J., Bose, P., Cappello, F., Carlson, B., et al.: Addressing failures in exascale computing. Int. J. High Perform. Comput. Appl. **28**, 129–173 (2014)
19. Stoller, L., Morrison, D.: A method for the numerical integration of ordinary differential equations. Math. Tables Other Aids Comput. **12**, 269–272 (1958)

Accelerator Computing

High-Performance Matrix-Matrix Multiplications of Very Small Matrices

Ian Masliah[2(✉)], Ahmad Abdelfattah[1], A. Haidar[1], S. Tomov[1],
Marc Baboulin[2], J. Falcou[2], and J. Dongarra[1,3]

[1] Innovative Computing Laboratory, University of Tennessee, Knoxville, TN, USA
[2] University of Paris-Sud, Orsay, France
ian.masliah@lri.fr
[3] University of Manchester, Manchester, UK

Abstract. The use of the general dense matrix-matrix multiplication (GEMM) is fundamental for obtaining high performance in many scientific computing applications. GEMMs for *small matrices* (of sizes less than 32) however, are not sufficiently optimized in existing libraries. In this paper we consider the case of many small GEMMs on either CPU or GPU architectures. This is a case that often occurs in applications like big data analytics, machine learning, high-order FEM, and others. The GEMMs are grouped together in a single *batched* routine. We present specialized for these cases algorithms and optimization techniques to obtain performance that is within 90% of the optimal. We show that these results outperform currently available state-of-the-art implementations and vendor-tuned math libraries.

Keywords: GEMM · Batched GEMM · Small matrices · HPC · Autotuning

1 Introduction

Parallelism in todays computer architectures is pervasive not only in systems from large supercomputers to laptops, but also in small portable devices like smartphones and watches. Along with parallelism, the level of heterogeneity in modern computing systems is also gradually increasing. Multicore CPUs are combined with discrete high-performance GPUs, or even become integrated parts with them as a system-on-chip (SoC) like in the NVIDIA Tegra mobile family of devices. To extract full performance from systems like these, the heterogeneity makes the parallel programming for technical computing problems extremely challenging, especially in modern applications that require fast linear algebra on many independent problems that are of size $\mathcal{O}(100)$ and smaller. According to a recent survey among the Sca/LAPACK and MAGMA [17] users, 40% of the responders needed this functionality for applications in machine learning, big data analytics, signal processing, batched operations for sparse preconditioners, algebraic multigrid, sparse direct multifrontal solvers, QR types of factorizations

© Springer International Publishing Switzerland 2016
P.-F. Dutot and D. Trystram (Eds.): Euro-Par 2016, LNCS 9833, pp. 659–671, 2016.
DOI: 10.1007/978-3-319-43659-3_48

on small problems, astrophysics, and high-order FEM. At some point in their execution, applications like these must perform a computation that is cumulatively very large, but whose individual parts are very small; when such operations are implemented naively using the typical approaches, they perform poorly. To address the challenges, we designed a standard for Hybrid Batched BLAS [6], and developed innovative algorithms [10], data and task abstractions [1], as well as high-performance implementations based on the standard that are now released through MAGMA 2.0 [5,9]. Figure 1 illustrates how the need for batched operations and new data types arises in areas like linear algebra (Left) and machine learning (Right). The computational characteristics in these cases are common to many applications, as already noted: the overall computation is very large but is made of operations of interest that are in general small, must be batched for efficiency, and various transformations must be explored to cast the batched small computations to regular and therefore efficient to implement operations, e.g., GEMMs. We note that applications in big data analytics and machine learning target higher dimension and accuracy computational approaches (e.g., ab initio-type) that model mutilinear relations, thus, new data abstractions, e.g., tensors, may be better suited vs. the traditional approach of flattening the computations to linear algebra on two-dimensional data (matrices). Indeed, we developed these tensor data abstractions and accelerated the applications using them significantly [1] compared to other approaches.

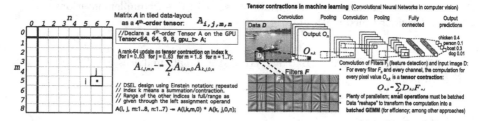

Fig. 1. Left: Example of a 4th-order tensor contractions design using Einstein summation notation and a Domain Specific Embedded Language (or *DSEL*). **Right**: Illustration of batched computations needed in machine learning.

There is a lack of sufficient optimizations on the batched GEMMs needed and targeted in this paper. We show this in comparison to vendor libraries like CUBLAS for NVIDIA GPUs and MKL for Intel multicore CPUs. Related work on GEMM and its use for tensor contractions [1] target only GPUs and for very small sizes (16 and below). Batched GEMM for fixed and variable sizes in the range of $\mathcal{O}(100)$ and smaller were developed in [2]. The main target here is multicore CPUs and GPUs for sizes up to 32.

2   Contributions to the Field

The evolution of semiconductor technology is dramatically transforming the balance of future computer systems, producing unprecedented changes at every level of the platform pyramid. From the point of view of numerical libraries, and the myriad of applications that depend on them, three challenges stand out: (1) the need to exploit unprecedented amounts of parallelism; (2) the need to maximize the use of data locality and vectorized operations; and (3) the need to cope with component heterogeneity. Below, we highlight our main contributions related to the algorithm's design and optimization strategies aimed at addressing these challenges on multicore CPU and GPU architectures:

Exploit Parallelism and Vector Instructions: Clock frequencies are expected to stay constant, or even decrease to conserve power; consequently, as we already see, the primary method of increasing computational capability of a chip will be to dramatically increase the number of processing units (cores), which in turn will require an increase of orders of magnitude in the amount of concurrency that routines must be able to utilize as well as increasing the computational capabilities of the floating point units by extending it to the classical Streaming SIMD Extensions set (SSE-1, to SSE-4) in the earlier 2000, and recently to Advanced Vector Extensions (AVX, AVX-2, AVX-3). We developed specific optimization techniques that demonstrate how to use the many cores (currently multisocket 10–20 cores for the Haswell CPU and 15×192 CUDA cores for the K40 GPU) to get optimal performance. The techniques and kernels developed are fundamental and can be used elsewhere.

Hierarchical Communication Techniques that Maximizes the Use of Data Locality: Recent reports (e.g., [7]) have made it clear that time per flop, memory bandwidth, and communication latency are all improving, but at exponentially different rates. So computation on very small matrices, that can be considered as computation-bound on old processors, is, –today and in the future– communication-bound and depends from the communication between levels of the memory hierarchy. We demonstrate that, performance is indeed harder to get on new manycore architectures unless hierarchical communications and optimized memory management are considered in the design. We show that, only after we developed multilevel memory design, our implementations reach optimal performance.

Performance Analysis and Autotuning: We demonstrate the theoretical maximal performance bounds that could be reached for computation on very small matrices. We studied various instructions and performance counters, as well as proposed a template design with different tunable parameters in order to evaluate the effectiveness of our implementation and optimize it to reach the theoretical limit.

3   Experimental Hardware

All experiments are done on an Intel multicore system with two 10-cores Intel Xeon E5-2650 v3 (Haswell) CPUs, and a Kepler Generation Tesla K40c GPU. Details about the hardware are illustrated in Fig. 2. We used gcc compiler 5.3.0 for our CPU code (with options -std=c++14 -O3 -avx -fma), as well as the icc compiler from the Intel suite 2016.0.109, and the BLAS implementation from MKL (Math Kernel Library) 16.0.0 [12]. We used CUDA Toolkit 7.5 for the GPU. For the CPU comparison with the MKL library we used two implementations: (1) An OpenMP loop statically or dynamically unrolled among the cores (we choose the best results), where each core computes one matrix-matrix product at a time using the optimized sequential MKL dgemm routine, and (2) The batched dgemm routine that has been recently added to the MKL library.

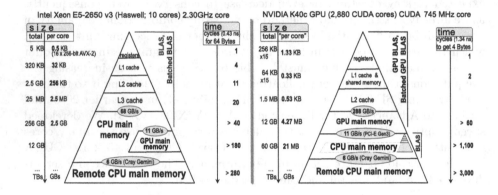

Fig. 2. Memory hierarchies of the experimental CPU and GPU hardware

4   Methodology, Design, and Optimization

To evaluate the efficiency of our algorithms we derive theoretical bounds for the maximum achievable performance $P_{max} = F/T_{min}$, where F is the number of operations needed by the computation and T_{min} is the fastest time to solution. For simplicity, consider $C = \alpha AB + \beta C$ on square matrices of size n. We have $F \approx 2n^3$ and $T_{min} = min_T(T_{Read(A,B,C)} + T_{Compute(C)} + T_{Write(C)})$. Note that we have to read/write $4n^2$ elements, or $32n^2$ Bytes for double precision (DP) calculations. Thus, if the maximum achievable bandwidth is B (in Bytes/second), and we assume $T_{Compute(C)} \to 0$ for very small computation, then $T_{min} = T_{Read(A,B,C)} + T_{Write(C)} = 4n^2/B$ in DP. Note that this time is theoretically achievable if the computation totally overlaps the data transfer and does not disrupt the maximum rate B of read/write to the GPU memory. Thus,

$$P_{max} = \frac{2n^3 B}{32n^2} = \frac{nB}{16} \text{ in DP.}$$

The achievable bandwidth can be obtained by benchmarks. For our measures, we used the STREAM benchmark [16] and the Intel memory latency checker 3.0 tool for CPU, and the NVIDIA's `bandwidthTest` for GPU. Our tests show that the practical CPU bandwidth we are able to achieve using different benchmarks is about 44 GB/s per socket. On the K40 GPU with ECC on the peak is 180 GB/s, so in that case P_{max} is 2.75 n GFlop/s per socket for the CPU and 11.25 n GFlop/s for the K40 GPU. The curve representing this theoretical maximal limit is denoted by the "upper bound" line on Figs. 5 and 8. Thus, when $n = 16$ for example, we expect a theoretical maximum performance of 180 GFlop/s in DP on the K40 GPU.

4.1   Programming Model, Performance Analysis, and Optimization for CPUs

The design of our code is done using new features of C++ for better re-usability and adaptability of the code. By using advanced template techniques we can create high-level interfaces [15] without adding any cost even for small matrix-matrix products. To do so, we have designed a batch structure which will contain a C++ vector for the data and static dimensions. By using the C++ constexpr keyword and integral constants we can make a generic batched code that will dispatch at compile time the correct version depending on the size of matrices. We use this environment for each code sequence we generate.

The implementation of a matrix-matrix products kernel for very small matrices for CPUs requires specific design and optimisations. As we can store three double precision matrices of size up to 32×32 in the L1 cache of an Intel Xeon E5-2650 v3 processor, one can expect that any implementation will not suffer from data cache misses. This can be seen on Fig. 5b where the performance of an ijk implementation, which is not cache-aware and cannot be vectorized, is pretty close to the ikj one. For smaller sizes, the ijk implementation is even more efficient than the ikj one, as it optimizes the number of stores (Fig. 3a). To obtain a near optimal performance, we conduct an extensive study over the performance counters using the PAPI [18] tools. Our analysis concludes that in order to achieve an efficient execution for such computation, we need to maximize the occupancy and minimize the data traffic while respecting the underlying hierarchical memory design. Unfortunately, today's compilers cannot introduce highly sophisticated cache/register based loop transformations and, consequently, this kind of optimization effort should be studied and implemented by the developer [13]. This includes techniques like reordering the data so that it can be easily vectorized, reducing the number of instructions so that the processor spends less time in decoding them, prefetching the data that will be reused in registers, and using an optimal blocking strategy.

Data Access Optimizations and Loop Transformation Techniques. In our design, we propose to order the iterations of the nested loops in such a way that we increase locality and expose more parallelism for vectorization.

The matrix-matrix product is an example of perfectly nested loops which means that all the assignment statements are in the innermost loop. Hence, loop unrolling, loop peeling, and loop interchange can be useful techniques for such algorithm [3,4]. These transformations improve the locality and help to reduce the stride of an array based computation. In our approach, we propose to unroll the two inner-most loops so that the accesses to matrix B are independent from the loop order, which also allows us to reorder the computations for continuous access and improved vectorization. This technique enables us to prefetch and hold some of the data of B into the SIMD registers. Here, we manage to take advantage from the knowledge of the algorithm, and based on the principle of locality of references [11], to optimize both the temporal and spatial data locality.

Register Data Reuse and Locality. Similarly to the blocking strategies for better cache reuse in numerically intensive operations (e.g., large matrix-matrix products), we focus on register blocking to increase the performance. Our study concludes that the register reuse ends up being the key factor for performance. The idea is that when data is loaded into SIMD register, it will be reused as much as possible before its replacement by new data. The amount of data that can be kept into registers becomes an important tuning parameter. For example, an 8 × 8 matrix requires 16 256-bit AVX-2 registers to be completely loaded. As the targeted hardware consists of only 16 256-bit AVX-2 registers, one can expect that loading the whole B will not be optimal as we will have to reload the vectors for A and C. However, if we load only 8 registers for B, which is equal to 4 rows, we can compute a row of C at each iteration and reuse these 8 registers for each iteration. We propose an auto-tuning process to check all the possible scenarios and provide the best option. This reduces the number of load, store, and total instructions from $\mathcal{O}(n^2)$ to $\mathcal{O}(n)$, compared to a classical ijk or ikj implementation as depicted in Figs. 3a, b, and 5a, respectively.

Algorithmic Advancements. Algorithm 1 is an example of our methodology for a matrix-matrix product of 16 × 16 matrices. In this pseudo-code, we start by loading four 256-bit AVX-2 registers with values of B which correspond to the first row. These registers are reused throughout the algorithm. In the main loop (Lines 4–14), we start by computing the first values of every multiplication (stored into a register named M = A×B) based on the prefetched register in line 1. Then, we iterate on the remaining rows (Lines 7–11) loading B, multiplying each B by a value of A, and adding the result into M. Once the iteration over a row is accomplished, the value of M is the final result of A×B and thus, we can load the initial values of C, multiply by α and β, and store it back before moving toward the next iteration such a way to minimize the load/store as shown in Fig. 3. Each C ends up being loaded/stored once. We apply this strategy to matrix sizes ranging from 8 to 32 as for smaller sizes the whole matrix can fit in registers. Different blocking strategies (square versus rectangular) have been studied through our auto-tuning process in order to achieve the best performance. We generate each matrix-matrix product function at compile time with

C++ templates. The matrix size is passed as a function parameter using C++ integral constants.

```
 1: Load B0, B1, B2, B3
 2: Load α, β
 3: S = 16
 4: for i = 0, 1, ... , S-1 do
 5:     Load A[i*S]
 6:     Mi0 = A[i*S] * B0; ... Mi3 = A[i*S] *B3
 7:     for u = 1, 2, ... , S-1 do
 8:         Load A[i*S + u]
 9:         Load Bu0, Bu1, Bu2, Bu3
10:         Mi0 += A[i*S+u] * Bu0; ... Mi3 += A[i*S+u] *Bui3
11:     end for
12:     Mi0 = α Mi0 + β (Load Ci0); ... Mi3 = α Mi3 + β (Load Ci3)
13:     Store Mi0, Mi1, Mi2, Mi3
14: end for
```

Algorithm 1: Generic matrix-matrix product applied to matrices of size 16×16

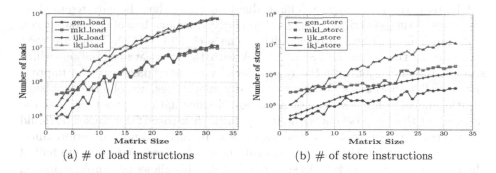

(a) # of load instructions   (b) # of store instructions

Fig. 3. CPU Performance counters measurement of the memory accesses

Effect of the Multi-threading. As described above, operating on matrices of very small sizes is memory-bound computation and thus, increasing the number of CPU cores may not always increase the performance since the performance will be limited by the bandwidth which can be saturated by a few cores. We performed a set of experiments towards clarifying this behaviour and illustrate our findings in Fig. 4b. As shown, the notion of perfect speed-up does not exist for a memory-bound algorithm, and adding more cores increases the performance slightly. We performed a bandwidth evaluation when varying the number of cores to find that a single core can achieve about 18 GB/s while 6 and 8 cores (over the available 10 cores) can reach about 88 % and 93 % of the practical peak bandwidth, which is about 44 GB/s.

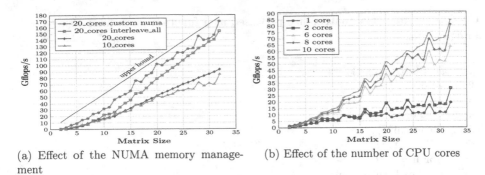

(a) Effect of the NUMA memory management

(b) Effect of the number of CPU cores

Fig. 4. CPU Performance analysis

Effect of the NUMA-Socket and Memory Location. We also studied NUMA-socket (non-uniform memory access) [8] when using two Xeon sockets as seen in Fig. 4a. A standard memory allocation puts all of the data in the memory slot associated to the first socket until it gets filled, then starts filling the second socket. Since the problem size we are targeting is very small, most of the data is allocated on one socket, and thus using extra 10 cores of the second socket will not increase the performance. This is due to the fact that the data required by the cores of the second socket goes through the memory bus of the first socket, and thus is limited by the bandwidth of one socket (44 GB/s). There are ways to overcome this issue. By using NUMA with the interleave=all option, which spreads the allocation over the two sockets by memory pages, we can improve the overall performance. However, for very small sizes, we observe that such solution remains far from the optimal bound since data is spread out over the memory of the two sockets without any rules that cores from socket 0 should only access data on socket 0, and vice versa. To further improve performance, we use a specific NUMA memory allocation, which allows us to allocate half of the matrices on each socket. As shown in Fig. 4a, this allows our implementation to scale over the two sockets and to reach close to the peak bound.

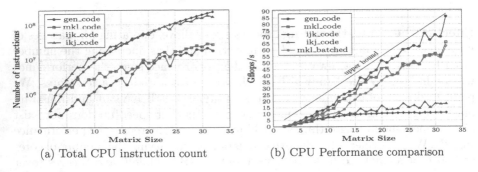

(a) Total CPU instruction count

(b) CPU Performance comparison

Fig. 5. Experimental results of the matrix-matrix multiplication on CPU's

4.2 Programming Model, Performance Analysis, and Optimization for GPUs

Our goal is to minimize coding effort and to design one kernel that can be easily adapted for very small matrix size computations, providing very efficient execution. To design a GEMM kernel in CUDA to take advantage of the available threads, thread blocks, and streaming multiprocessors (SMs) of a GPU, the computation must be partitioned into blocks of threads (also called thread blocks, or simply TBs) that execute independently from each other on the multiprocessors of the GPU. We use a hierarchical blocking model of both communications and computations, similarly to the MAGMA batched GEMM kernel [2] for medium and large sizes. We designed CUDA C++ templates to enable unified code base for all the small sizes. Templates enable an easy instantiation of a kernel with a specific precision and tuning parameters.

A Cache-Based Approach. Unlike multi-core CPUs, the L1 cache (per SM) is not intended for global memory accesses, which are cached only in the L2. The L2 cache is shared among all SMs, which makes it difficult to use for cache-based optimizations, since all TBs will be sharing it (L2 cache is up to 1.5 MB). However, a modern Kepler GPU has a 48 KB per SM of a read-only cache (rocache), which can be used for global memory reads. A possible implementation that takes advantage of this is to read the input matrices A and B through the read-only cache. Each matrix computation is associated to one TB that is configured with M × N threads, where each thread is responsible for computing one output element of the resulting matrix C. Thus, each thread reads an entire row of A and entire column of B. This cache-based design ideally assumes that most of the global memory accesses hit in the rocache. This kernel does not use the shared memory, and so it does not need any synchronization points.

A Shared Memory Based Approach. Another approach is to use shared memory (shmem) for data reuse rather than rocache. We refer to this implementation as the MAGMA kernel, since it is distributed within the MAGMA library. We performed an extensive set of auto-tuning and performance counter analysis to optimize and improve this implementation. The matrices A and B are loaded by block into the shared memory, and the corresponding block of the matrix C is held into registers. Prefetching can also be used to load the next blocks of A and B. The prefetching can be done through either the shared memory or the register, and is controlled by a tunable parameter. This implementation is very well parametrized, and can work for any dimension with tunable block sizes for A, B, and C.

Instruction Mix. We performed a detailed performance study based on the collection and analysis of hardware counters. Counter readings were taken using performance tools (Nvidia's CUPTI and PAPI CUDA component [14]). Our analysis shows that it is important to pay attention to the instruction mix of

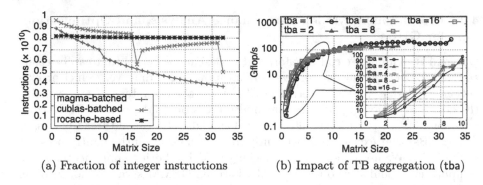

(a) Fraction of integer instructions   (b) Impact of TB aggregation (tba)

Fig. 6. Performance counters measurement on the K40 GPU

the GPU kernel, in particular when operating on matrices of such very small sizes. Integer instructions, which are used for loop counters and memory address calculations, can be quite an overhead in such computations. Moreover, our study showed that a loop with predefined boundary can be easily unrolled and optimized by the Nvidia compiler. We adopt an aggressive approach to produce a fully unrolled code for every size of interest. We add the sizes M, N, and K to the template parameters such a way to use a unified code base to produce a fully unrolled and optimized implementation for any of these very small sizes. Figure 6a shows the ratio of integer instructions to the total number integer and floating point instructions, the MAGMA kernel has the smallest ratio for most sizes. An interesting observation of the CUBLAS implementation, for this range of matrices, is that it uses a fixed blocking size of 16×16. This explains the drops at sizes 16 and 32, where the problem size matches the internal blocking size.

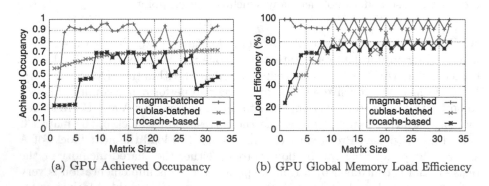

(a) GPU Achieved Occupancy   (b) GPU Global Memory Load Efficiency

Fig. 7. Performance counters measurement on the K40 GPU

Thread Block-Level Aggregation. We further improved the proposed design by another optimization that helps significantly increase the performance for the tiny sizes (e.g. less than 12). Multiple TBs, each is assigned for one problem, are

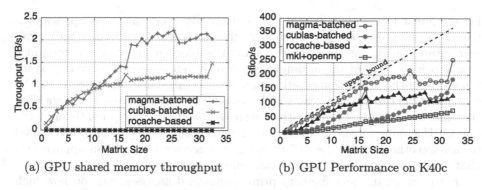

(a) GPU shared memory throughput   (b) GPU Performance on K40c

Fig. 8. Performance counters measurement and efficiency of our design for the matrix-matrix multiplication on the K40 GPU

aggregated together into one larger TB. The motivation behind this technique is to increase the number of threads, especially when the TB configuration has few warps or even less than a warp. Aggregation is controlled through an additional parameter tba, which controls the number of TBs to be fused together. Figure 6b shows the impact of tba on performance. For example, we achieve a speed-up of 6.8× for size 2 and 3.8× for size 3. The performance improvement reaches 24 % at size 8. Beyond size 10, setting tba larger than 1 does not achieve any gains because the resources required by one fused TB become expensive, which affects the number of residing TBs per SM. Some curves look incomplete, since a large value of tba sometimes requires more threads than the hardware-defined maximum number of threads allowed per TB.

Performance Counter Analysis. Figure 7 shows two of the key factors to high performance on a GPU: the achieved occupancy and the efficiency of global memory reads. The first one is the ratio between the number of active warps per active cycles and the maximum number of warps that can run on an SM. The second is defined as the ratio between the load throughput requested by the kernel, and the actual required throughput needed to fulfil the kernel load requests. Our proposed MAGMA implementation achieves more than 75 % occupancy in most cases, which is nearly the upper limit for the other design. It can also achieve very high occupancy (≈90 %) even for very small matrices, thanks to the TB-level aggregation. On the other hand, the MAGMA approach is at least 90 % efficient in reading from global memory, which means that the kernel encounters very little overhead in terms of load instructions replays.

5   Conclusions and Future Directions

We presented work motivated by a large number of applications, ranging from machine learning to big data analytics, that require fast linear algebra on many independent problems that are of size 32 and smaller. The use of batched GEMM for small matrices is fundamental for obtaining high performance in applications like these. We presented specialized algorithms for these cases – where the overall computation is memory bound but still must be blocked – to obtain performance that is within 90 % of the optimal, significantly outperforming currently available state-of-the-art implementations and vendor-tuned math libraries. Here, the optimal is the time to just read the data once and write the result, disregarding the time to compute. The algorithms were designed for modern multi-core CPU and GPU architectures. The optimization techniques and algorithms can be used to develop other batched Level 3 BLAS and to accelerate numerous applications that need linear algebra on many independent problems.

Future work includes further optimizations and analyses, e.g., on how high performance can go using CUDA. It is known that compilers have their limitations in producing top performance codes for computations like these, thus, requiring the use of lower level programming languages. Current results used intrinsics for multi-core CPUs and CUDA for GPUs, combined with auto-tuning in either case, to quickly explore the large algorithmic variations developed in finding the fastest one. Future work includes also use in applications, development of application-specific optimizations, data abstractions, e.g., tensors, and algorithms that use them efficiently.

Acknowledgments. This material is based in part upon work supported by the US NSF under Grants No. CSR 1514286 and ACI-1339822, NVIDIA, the Department of Energy, and in part by the Russian Scientific Foundation, Agreement N14-11-00190.

References

1. Abdelfattah, A., Baboulin, M., Dobrev, V., Dongarra, J., Earl, C., Falcou, J., Haidar, A., Karlin, I., Kolev, T., Masliah, I., Tomov, S.: High-performance tensor contractions for GPUs. In: International Conference on Computational Science (ICCS 2016). Elsevier, Procedia Computer Science, San Diego, CA, USA, June 2016

2. Abdelfattah, A., Haidar, A., Tomov, S., Dongarra, J.: Performance, design, and autotuning of batched GEMM for GPUs. In: Kunkel, J.M., Balaji, P., Dongarra, J. (eds.) ISC High Performance 2016. LNCS, vol. 9697, pp. 21–38. Springer, Heidelberg (2016). doi:10.1007/978-3-319-41321-1_2

3. Ahmed, N., Mateev, N., Pingali, K.: Tiling imperfectly-nested loop nests. In: ACM/IEEE 2000 Conference Supercomputing, p. 31, November 2000

4. Bacon, D.F., Graham, S.L., Sharp, O.J.: Compiler transformations for high-performance computing. ACM Comput. Surv. **26**(4), 345–420 (1994)

5. Dong, T., Haidar, A., Luszczek, P., Harris, A., Tomov, S., Dongarra, J.: LU Factorization of small matrices: accelerating batched DGETRF on the GPU. In: Proceedings of 16th IEEE International Conference on High Performance and Communications, August 2014

6. Dongarra, J., Duff, I., Gates, M., Haidar, A., Hammarling, S., Higham, N.J., Hogg, J., Valero-Lara, P., Relton, S.D., Tomov, S., Zounon, M.: A proposed API for batched basic linear algebra subprograms. MIMS EPrint 2016.25, Manchester Institute for Mathematical Sciences, The University of Manchester, UK, April 2016. http://eprints.ma.man.ac.uk/2464/

7. Fuller, S.H., Millett, L.I., Committee on Sustaining Growth in Computing Performance; National Research Council: The Future of Computing Performance: Game Over or Next Level? The National Academies Press, Washington (2011). http://www.nap.edu/openbook.php?record_id=12980
8. Hager, G., Wellein, G.: Introduction to High Performance Computing for Scientists and Engineers. CRC Press, Boca Raton (2011)
9. Haidar, A., Dong, T., Luszczek, P., Tomov, S., Dongarra, J.: Batched matrix computations on hardware accelerators based on gpus. Int. J. High Perform. Comput. Appl. **29**(2), 193–208 (2015). http://hpc.sagepub.com/content/early/2015/02/06/1094342014567546.abstract
10. Haidar, A., Dong, T.T., Tomov, S., Luszczek, P., Dongarra, J.: A framework for batched and GPU-resident factorization algorithms applied to block householder transformations. In: Kunkel, J.M., Ludwig, T. (eds.) ISC High Performance 2015. LNCS, vol. 9137, pp. 31–47. Springer, Heidelberg (2015). http://dx.doi.org/10.1007/978-3-319-20119-1_3
11. Hennessy, J.L., Patterson, D.A.: Computer Architecture: A Quantitative Approach, 5th edn. Morgan Kaufmann Publ. Inc., San Francisco (2011)
12. Intel Math Kernel Library (2016). http://software.intel.com
13. Loshin, D.: Efficient Memory Programming, 1st edn. McGraw-Hill Professional, New York (1998)
14. Malony, A.D., Biersdorff, S., Shende, S., Jagode, H., Tomov, S., Juckeland, G., Dietrich, R., Poole, D., Lamb, C.: Parallel performance measurement of heterogeneous parallel systems with gpus. In: Proceedings of ICPP 2011, pp. 176–185. IEEE Computer Society, Washington, DC (2011)
15. Masliah, I., Baboulin, M., Falcou, J.: Metaprogramming dense linear algebra solvers applications to multi and many-core architectures. In: 2015 iIEEE TrustCom/BigDataSE/ISPA, Helsinki, Finland, vol. 3, pp. 69–76, August 2015
16. McCalpin, J.D.: Memory bandwidth and machine balance in current high performance computers. IEEE Computer Society Technical Committee on Computer Architecture (TCCA) Newsletter, pp. 19–25, December 1995
17. Tomov, S., Dongarra, J., Baboulin, M.: Towards dense linear algebra for hybrid GPU accelerated manycore systems. Parallel Comput. **36**(5–6), 232–240 (2010)
18. Weaver, V., Johnson, M., Kasichayanula, K., Ralph, J., Luszczek, P., Terpstra, D., S.: Measuring energy and power with PAPI. In: 41st International Conference on Parallel Processing Workshops, September 2012

Effective Minimally-Invasive GPU Acceleration of Distributed Sparse Matrix Factorization

Anshul Gupta[1](✉), Natalia Gimelshein[2], Seid Koric[3], and Steven Rennich[2]

[1] IBM Research, Yorktown Heights, NY, USA
anshul@us.ibm.com
[2] NVIDIA Corporation, Santa Clara, CA, USA
srennich@nvidia.com
[3] NCSA, University of Illinois, Urbana, IL, USA
koric@illinois.edu

Abstract. Sparse matrix factorization, a critical algorithm in many science and engineering applications, has had difficulty leveraging the additional computational power afforded by the infusion of heterogeneous accelerators in HPC clusters. We present a minimally invasive approach to the GPU acceleration of a hybrid multifrontal solver, the Watson Sparse Matrix Package, which is already highly optimized for the CPU and exhibits leading performance on distributed architectures. The novel aspect of this work is to demonstrate techniques for achieving substantial GPU acceleration, up to 3.5x, of the sparse factorization with strategic, but contained changes to the original, CPU-only, code. Strong scaling results show that performance benefits scale to as many as 512 nodes (4096 cores) of the Blue Waters supercomputer at NCSA. The techniques presented here suggest that detailed code reorganization may not be necessary to achieve substantial acceleration from GPUs, even for complex algorithms with highly irregular compute and data access patterns, like those used for distributed sparse factorization.

1 Introduction

The solution of large sparse linear systems is central to many problems in science, engineering, and optimization. Direct methods, for which the major computational task is factoring the coefficient matrix, are often the solution method of choice due to their generality and robustness. Performance optimization, and particularly parallelization, of sparse factorization has been the subject of intensive research.

A growing portion of the computational capability of supercomputer clusters is now being provided by accelerators, particularly GPUs [22]. The characteristics of these distributed heterogeneous systems, most particularly the separate CPU and GPU memories and the limited bandwidth of the PCIe bus over which they communicate, makes leveraging their computational power a challenge for irregular algorithms such as sparse factorization. As a result, there is a need to adapt the implementation of sparse factorization algorithms to such heterogeneous architectures and there is active work in this area [6,14,19,21,23].

© Springer International Publishing Switzerland 2016
P.-F. Dutot and D. Trystram (Eds.): Euro-Par 2016, LNCS 9833, pp. 672–683, 2016.
DOI: 10.1007/978-3-319-43659-3_49

Previous work in GPU acceleration of sparse factorization has taken the approach of offloading dense math to the GPU and making accommodation for:

- only sending appropriately sized work to the GPU,
- asynchronous operations to overlap CPU computation, GPU computation and PCIe communication,
- minimizing communication by re-using data on the GPU as much as possible.

Most previous work has been limited to single-node, shared-memory systems with the notable exception of [20], which showed performance on 8 nodes.

This work seeks GPU acceleration of sparse factorization on large heterogeneous, distributed systems. The current scope is limited to Cholesky factorization ($A = LL^T$) [8] as that is the simpler case, but we expect many of the techniques will be applicable to other factorization algorithms as well. While we leverage the techniques itemized immediately above, this implementation is novel in that it:

- accelerates the factorization in the Watson Spare Matrix Package (WSMP) [9] which shows leading performance on distributed systems [12,13,18],
- uses a minimally-invasive approach promoting maintainability and portability of the underlying CPU code,
- demonstrates improved performance that scales to an order of magnitude more nodes and cores than previous work, and
- identifies techniques for achieving further performance improvements.

We describe the implementation in detail, present experimental results, analyze the results to gain insights about performance bottlenecks, and suggest concrete and feasible avenues for further performance improvement. We show that the use of GPUs can more than double the performance of WSMP's sparse Cholesky factorization on up to 128 nodes of NCSA's Blue Waters supercomputer [3] and can strong scale beneficially up to 512 nodes.

2   WSMP Cholesky Factorization

WSMP uses a highly scalable distributed-memory parallel sparse Cholesky factorization algorithm [11] based on the multifrontal method [5,16]. A multifrontal algorithm expresses the entire sparse matrix factorization in terms of partial factorizations of smaller dense matrices called frontal matrices of the type illustrated in Fig. 1. The rows and columns of the sparse coefficient matrix A are divided into contiguous blocks called supernodes [2] to aid the construction of the dense frontal matrices. The control and data flow in a typical multifrontal algorithm follows a dependency graph known as the elimination tree [15], which can be computed inexpensively from the structure of a symmetric sparse matrix. Each vertex of the elimination tree corresponds to a supernode.

In WSMP's distributed-memory parallel factorization utilizing p MPI processes, the supernodal tree is binary in the top $\log_2 p$ levels. The portions of this binary supernodal tree are assigned to the processes using a subtree-to-subcube strategy [17] illustrated in the top two levels of the tree in Fig. 2.

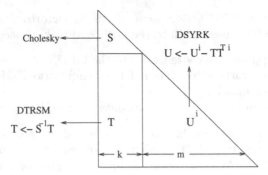

Fig. 1. Computation in the frontal matrix F^i of a typical supernode in sparse multifrontal Cholesky factorization.

The frontal matrix of each supernode is distributed among a logical mesh of a subset of processes using a bit-mask based block-cyclic scheme [11]. The parallel partial factorization operation at each supernode is a pipelined implementation of dense block Cholesky factorization and update operations shown in Fig. 1. Most of the flops corresponding to a supernode are performed by Basic Linear Algebra Subprograms (BLAS) [4]. The three key operations are: (1) dense Cholesky on the top left matrix S, (2) the DTRSM operation $T = S^{-1}T$, and (3) the DSYRK operation $U^i = U^i - TT^T$. Hereafter, we will refer to a factorization task like this performed on multiple nodes as **cooperative factorization**.

Usually, each MPI process is multithreaded, and the portion of multifrontal factorization assigned to each process is further parallelized [10]. Just like the message-passing portion, tasks at each subroot of the elimination tree are assigned to independent groups of threads until each thread ends up with its own subtree. However, mapping between tasks and threads is more flexible than mapping between tasks and MPI processes. Furthermore, a strict block-cyclic mapping based on the binary representation of the indices is not used because all processors can access all rows and columns with the relatively small overhead. Hereafter, we will refer to factorization tasks performed on a single MPI rank as **individual factorization**.

Figure 2 shows one hypothetical mapping of an elimination tree among the 16 CPUs of a 4-node cluster with 4-way shared-memory parallel nodes. The symbolic computation preceding the numerical phase ensures that the tree is binary at the top $\log_2 p$ levels. At each of these levels, multithreaded BLAS calls are used to utilize all the CPU's on each node during cooperative factorization. In the individual factorization region of the tree, which lies below the top $\log_2 p$ levels, the subtrees are assigned to groups of threads until they are mapped onto single threads.

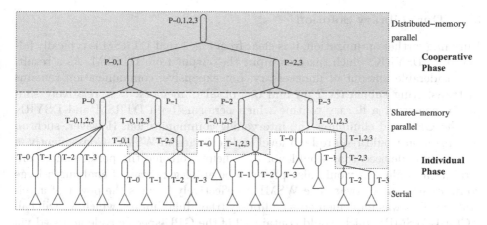

Fig. 2. A mapping of a hypothetical elimination tree on a 4-node cluster with 4-way shared-memory nodes.

3   Minimally Invasive Approach

When factorization involves sufficient large dense blocks, where the computation can be tiled and the communication costs hidden behind computation, then satisfactory acceleration can be achieved by offloading large level-3 BLAS computations to the GPU [21]. However, when dense blocks are fewer or smaller (as might occur in shell models), or at scale (when the dense blocks are distributed across many nodes), communication can't be hidden and opportunity for easy GPU acceleration is limited. Further, for best PCIe performance, and to support transfers which are asynchronous or concurrent with GPU computation, the data on the host must reside in 'pinned' (page-locked) memory. Copying data in or out of pinned memory buffers adds to the communication overhead. This BLAS-offloading approach was investigated for use with WSMP, and as expected, did not yield significant benefit.

However, as WSMP is inherently multithreaded and performs multiple BLAS operations simultaneously in different threads, the same idea behind tiling a single large BLAS operation can be used to hide communication between many simultaneous small BLAS operations. That is, computation for a BLAS operation in one thread call can proceed simultaneously with PCIe communication for a BLAS operation in another thread and with host memory copy (in or out of pinned buffers) of a third. This parallel offload approach did not involve any changes to the base WSMP code and permitted GPU acceleration of the numerical factorization by a factor of 1.7 on average [7] on a single multicore node. Unfortunately, at scale, to promote load-balancing, WSMP decomposes BLAS calls into smaller blocks than can be accelerated in this fashion.

3.1   Dual Library Solution

Pursuing further optimization, it is clear from Fig. 1 that DTRSM is typically followed by DSYRK, which takes as input the output from DTRSM. As a result, a considerable amount of unnecessary, but expensive, communication remains in transferring submatrix T in and out of GPU memory. It is relatively simple to construct a library routine which performs both DTRSM and DSYRK on the GPU and eliminates the intervening communication. However, such an optimization would require high-level code changes to WSMP. Consequently, a scheme was devised that would not only help overcome the performance constraints at scale, but would also prove to be an elegant way of enabling a large complex existing software like WSMP to effectively harness the power of accelerators. This was implemented as a library (the 'dual' library, which we will call ACCEL_WSMP) which would contain all of the GPU-specific code accessed via high-level (BLAS or above) routines. Only small changes would need to be made to WSMP to call these routines, if available, at a high level. In this way, WSMP could be linked with a driver code and, if ACCEL_WSMP was not provided, it's CPU-only behavior would be unchanged. However, if ACCEL_WSMP were provided during the linking, then GPU acceleration would be obtained. In this way, we consider the GPU acceleration to be 'minimally-invasive'.

   In addition to the acceleration techniques described so far, additional optimizations were to:

1. limit the minimum tile size by which the problem can be decomposed to a size that still permits acceleration on the GPU,
2. create high-level API which would provide parallel, tiled GPU BLAS operations and combined GPU BLAS operations to minimize communication,
3. define a protocol for flagging matrices on the GPU for re-use (and later indicating when they can be deleted) to further minimize communication.

   ACCEL_WSMP consists of a total of 10 functions for performing computations on dense matrices and/or moving them between host and device memories, and two administrative functions, including one-time initialization that performs the relatively expensive task of allocating pinned buffers. Note that all computational routines in ACCEL_WSMP have a CPU equivalent (pre-existing in WSMP). If the GPU is occupied (by other WSMP threads) or the operation is too small, then the computation remains on the CPU.

3.2   Individual Factorization

In the initial individual factorization phase of WSMP's sparse Cholesky factorization, the computationally-intensive portion consists of DTRSM followed by symmetric rank-k update DSYRK, with DTRSM output used as input for DSYRK.

   In Fig. 3, we show the benefit of caching the output matrix T from DTRSM for subsequent use in DSYRK. The figure shows performance of DTRSMSYRK

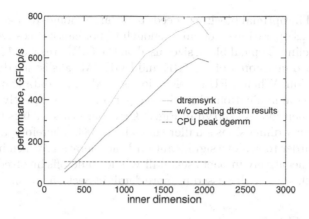

Fig. 3. Performance of DTRSMSYRK with and without caching of DTRSM output on the GPU.

routine on a single node of our target cluster (see Sect. 4) as a function of dimension k of the triangular matrix S. The number of rows of matrix T is set to an empirically derived value of $3k$. Separately, performance achieved when DTRSM output is not cached in GPU memory (and is therefore reloaded for DSYRK) is shown. Caching appears to improve DTRSMSYRK performance by about 30 % over individual uncached DTRSM and DSYRK. GPU performance increases with k, with a slight decrease at $k = 2048$ due to switch to tiled version of the algorithm. GPU performance overtakes the highest possible CPU performance at about $k = 384$, which is the cutoff that we use to determine if DTRSMSYRK operation is performed on the CPU or on the GPU.

3.3   Cooperative Factorization

In the cooperative phase of WSMP's Cholesky factorization, the dense Cholesky panel factorization illustrated in Fig. 1 is performed with S, T, and U distributed among multiple nodes. The dense level-3 BLAS calls in this phase are DTRSM, DSYRK and matrix-matrix multiply (DGEMM). Generally, the caching strategy described in Sect. 3.2 is used to hide PCIe communication.

There are two additional challenges in this phase. First, unlike DTRSM-SYRK, where DTRSM output could be immediately reused as DSYRK input, this reuse is spread over multiple DGEMM calls during cooperative factorization. To expose this data reuse, WSMP signals when a particular matrix is likely to be used across many DGEMM and DSYRK calls. Such flagging of reusable data is a key part of the interface between WSMP and ACCEL_WSMP libraries. ACCEL_WSMP attempts to retain flagged matrices until WSMP signals that a previously flagged matrix is no longer needed.

The second challenge is related to load balancing. In the cooperative factorization phase, WSMP distributes frontal matrices in a block-cyclic fashion among participating nodes in order to facilitate pipelined load-balanced dense panel factorization. The optimal block size needs to be large enough to take advantage of the level-3 BLAS operations, yet small enough to not cause excessive load imbalance in the pipeline. Typical block sizes used on the CPU are 64 or 128. This block size is the inner dimension k of DSYRK and DGEMM calls issued during cooperative factorization. When GPU acceleration is used, since data transfer must be hidden behind communication, the achieved performance is directly proportional to the inner dimension. As shown in Fig. 3, GPU performance lags the CPU performance for inner dimensions smaller than about 384. Therefore, a block size of 512 is used in order to achieve significant GPU acceleration, which increases load imbalance and negatively impacts scalability. The trade-offs involved in the selection of the block size are discussed in more detail in Sects. 5 and 6.

4   Testing Configuration

The matrices used in this study are extracted via a DMAP procedure [1] from a commercial FEA software NX Nastran (2012) from Siemens PLM. They represent industrial CAD geometries and are automatically meshed with 10-noded tetrahedral elements. Figure 4 shows the geometries used for this paper: a symmetric machine part cutter with asymmetrical loads, and a header part of a Charge Air Cooler (CAC) with complex geometry and whose elements have higher aspect ratios and therefore a higher condition number. For the cutter model, the element size control has produced different levels of automatic mesh refinement. Table 1 summarizes the test matrices.

Table 1. Test matrices

Matrix	Source model	Dimension	Nonzeros	Condition number
M2	Cutter	2,246,022	175,360,626	5.9E + 06
M6	Cutter	6,418,305	512,711,831	7.5E + 06
M11	CAC	11,562,627	937,454,416	6.8E + 09
M20	Cutter	20,056,050	1,634,926,088	2.7E + 07

Fig. 4. Finite element discretization of the cutter model (left) used for matrices M2, M6, and M20, and the Charge Air Cooler (CAC) model used for M11 (right).

The sustained peta-scale Blue Waters [3] system, hosted at the University of Illinois National Center for Supercomputing Applications (NCSA) was used in our experiments. Accelerated XK7 nodes of Blue Waters, each containing a single AMD Interlagos processor with 8 floating point cores (32 GB node memory) and a single NVIDIA Kepler K20X GPU (8 GB of GPU memory) were used. The Intel compiler version 2015 was used for compilation, and sequential Intel MKL for BLAS routines on the host. The CUDA runtime and cuBLAS libraries from the NVIDIA CUDA Toolkit version 5.5 were used for operations on the GPU. All runs used a single MPI rank per node and utilized all 8 cores on each node by using 8 threads per MPI rank.

5 Scaling and Acceleration Results

The numerical factorization performance for each of the four matrices has been benchmarked against the number of XK7 nodes used, spanning from the minimum number of nodes necessary to accommodate the problem to where the GPU-accelerated performance drops below the CPU-only performance. Parallel speed-up is defined as the ratio of the wall clock time on one cluster node to the wall clock time on p nodes. However, since most matrices do not fit on one node, the speed-up is computed with respect to the performance on the minimum number of nodes that a problem fits on.

Figure 5 shows the parallel speed-up obtained both on the CPU and the GPU for the numerical factorization of the four test matrices. Solid blue and green lines show the CPU and GPU results, respectively. Plain dashed lines of the corresponding colors show ideal speed-up from the starting point of each curve. The actual Cholesky factorization times on which these speed-up curves are based are shown in Table 2.

Table 2. Numerical factorization times (seconds) of the default CPU code, CPU code with block size of 512, and GPU-enabled code with block size of 512.

XK7 Nodes	Matrices											
	M2			M6			M20			M11		
	4.5E12 flops			*3.7E13 flops*			*3.6e14 flops*			*3.5E13 flops*		
	CPU default	CPU 512	GPU 512	CPU default	CPU 512	GPU 512	CPU default	CPU 512	GPU 512	CPU default	CPU 512	GPU 512
1	82.5	83.5	28.7									
2	47.0	48.0	19.1									
4	24.0	26.5	11.5	180.0	190.0	51.2						
8	14.9	15.8	7.6	98.0	103.0	34.6				95.0	102.5	36.0
16	9.0	10.5	5.0	55.8	63.4	24.0				52.0	57.2	25.0
32	6.0	7.5	4.0	34.0	39.5	17.0	249.1	252.7	78.2	32.0	35.3	17.0
64	4.0	5.7	3.3	22.5	26.5	11.6	137.5	152.5	55.5	18.7	21.6	10.8
128	2.7	4.6	2.8	12.6	17.0	9.0	77.2	91.1	38.4	12.0	15.6	7.9
256				8.8	13.0	7.4	44.9	57.1	25.2	8.5	11.8	6.4
512							26.3	40.8	20.7	5.1	10.5	5.3
1024							16.7	30.2	18.0			

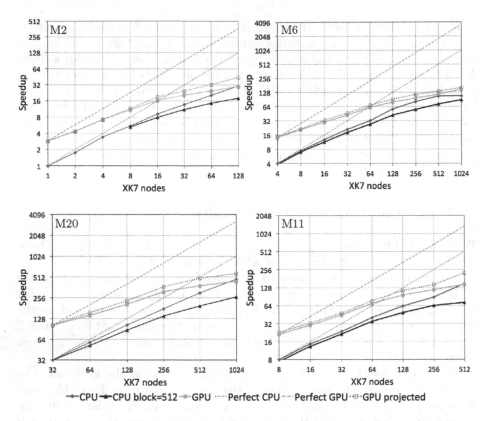

Fig. 5. Numerical factorization speed-ups vs. number of XK7 nodes for the four matrices studied. (Color figure online)

In Sect. 3.3, it was noted that a block size of 512 was used in the cooperative factorization phase on the GPU runs. In CPU-only runs, WSMP automatically chooses the block size, which is typically 64 or 128. To estimate the effect of increased block size on load imbalance, and hence on performance, the CPU runs were also performed using a block size of 512 and the results are shown as black lines in Fig. 5. In all the cases, the performance penalty due to larger block size grows larger as the number of nodes increases. This happens because, in wider runs, more levels of the elimination tree are cooperatively factored, and the number of XK7 nodes among which the frontal matrices of a given size are distributed increases. This creates a longer pipeline of dense operations on smaller submatrices for a given size of a frontal matrix. A large block size results in fewer stages in the pipelines, and therefore greater load imbalance. Figure 5 also shows the projected performance (dotted red line) that could be achieved through further optimizations discussed in Sect. 6.

In all four cases, on the smaller number of nodes, the GPU-accelerated version shows significant performance gains (2.5–3.5x) vs. the CPU-only case. These gains become smaller for wider runs.

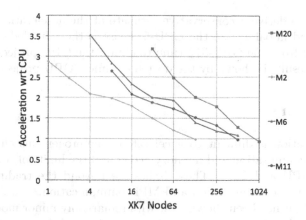

Fig. 6. Acceleration achieved by the GPU-enabled library over the CPU-only factorization vs. number of XK7 nodes.

Figure 6 shows the acceleration achieved by the GPU-enabled library over the CPU-only version for the cutter and CAC matrices on the different numbers of nodes. This plot directly demonstrates the benefit of GPU-acceleration.

6   Performance Analysis

The results in Sect. 5 show significant GPU acceleration with respect to the CPU-only performance for complex problems on hundreds of nodes. In this section, we study the issues limiting further GPU acceleration and show that one of the key causes can be remedied to a large extent by a simple enhancement to our implementation.

We have already touched upon the need for using a large block size of 512 in the GPU implementation and the resulting load imbalance in cooperative factorization. Splitting submatrix T into column blocks results in same blocks of U being updated multiple times. Therefore, if ACCEL_WSMP can retain the result matrix of DGEMM in GPU memory, then the result matrix, U, could be transferred back to the CPU only once, after the last DGEMM. In this way, the total amount of PCIe traffic would become independent of the block size - in fact it would be the same as in the unblocked case, permitting effective use of smaller block sizes and improving both performance and load-balancing. The current implementation of ACCEL_WSMP includes the capability to cache the input matrices of DGEMM, but not the output.

To estimate the effect of caching U, we first calculated the time lost on the CPU due to the added load imbalance when running with a block size of 512. This is the difference between CPU-512 and CPU-default times (Table 2). We assume that this overhead is reflected in the GPU-512 time, but scaled by a factor related to GPU acceleration of the imbalanced computation. We used the ratio of CPU-512 to GPU-512 time on the smallest number of nodes that a matrix ran on to derive a conservative estimate of the acceleration factor. The time lost to load imbalance in the CPU-512 case is then divided by this factor and subtracted from the GPU-512 time to obtain the GPU-projected time and speedup. The results is shown by the dotted red lines (GPU-projected) in Fig. 5.

7   Conclusions

Sparse factorization is difficult to accelerate on heterogeneous clusters due to irregularity in computation and memory access and the limits of communication between the GPU and CPU. These issues are beyond the traditional cluster hurdles of problem decomposition and MPI communication.

In this work, it has been shown that with remarkably minor modifications to the original CPU code, the numerical factorization performance of WSMP can be accelerated on a large number of GPU-enabled nodes of a supercomputer. Our approach relied upon identifying computations that could be beneficially accelerated, defining an accelerator API at an appropriate level of abstraction to capture these computations, and designing a separate library to implement the API. Benchmarking on industrially-derived matrices shows that on 128 nodes, the GPU-accelerated code is about 2 times faster than the CPU-only code using the same number of XK7 nodes, but without using the GPUs. A high degree of acceleration, up to 3.5x, is observed for lower node counts. Due to issues with load-balancing and difficulty maintaining sufficient computational intensity to hide CPU↔GPU transfers, the observed speed-up is reduced at larger node counts. A way to overcome a significant portion of the performance loss on large numbers of nodes has been identified as caching U on the GPU.

The observed scaling represents a unique demonstration that GPUs can be effectively applied to sparse factorization on large clusters and has the potential to permit application of these GPU-accelerated clusters to new types of analysis.

A relatively unique feature of our approach is that performance portability across different accelerator platforms can be achieved by simply tuning or reimplementing the well-contained accelerator library.

Acknowledgments. The authors would like to thank the Private Sector Program and the Blue Waters sustained-petascale computing project at NCSA. Blue Waters is supported by NSF awards OCI-0725070 and ACI-1238993, and by the state of Illinois.

References

1. NX Nastran User's Manual, Version 8.0 (2012). http://support.industrysoftware. automation.siemens.com/general/nxn.shtml
2. Ashcraft, C., Grimes, R.G.: The influence of relaxed supernode partitions on the multifrontal method. ACM Trans. Math. Softw. **15**(4), 291–309 (1989)
3. Bode, B., Butler, M., Dunning, T., Hoefler, T., Kramer, W., Gropp, W., mei Hwu, W.: The Blue Water Super-System for Super-Science, pp. 339–366. Chapman and Hall/CRC (2013) (02 Oct 2016)

4. Dongarra, J.J., Croz, J.D., Hammarling, S., Duff, I.S.: A set of level 3 Basic Linear Algebra Subprograms. ACM Trans. Math. Softw. **16**(1), 1–17 (1990)
5. Duff, I.S., Reid, J.K.: The multifrontal solution of indefinite sparse symmetric linear equations. ACM Trans. Math. Softw. **9**(3), 302–325 (1983)
6. George, T., Saxena, V., Gupta, A., Singh, A., Choudhury, A.: Multifrontal factorization of sparse spd matrices on gpus. In: 2011 IEEE International Parallel Distributed Processing Symposium (IPDPS), pp. 372–383 (2011)
7. Gimelshein, N.E., Gupta, A., Rennich, S.C., Koric, S.: GPU acceleration of WSMP. In GPU Technology Conference 2015. Nvidia Corp. (2015)
8. Golub, G.H., Van Loan, C.F.: Matrix Computations, 3rd edn. Johns Hopkins University Press, USA (1996)
9. Gupta, A.: WSMP: Watson sparse matrix package (Part-I: direct solution of symmetric sparse systems). IBM TJ Watson Research Center, Yorktown Heights, NY, Technical report, RC, 21886 (2000)
10. Gupta, A.: A shared- and distributed-memory parallel sparse direct solver. In: Dongarra, J., Madsen, K., Waśniewski, J. (eds.) PARA 2004. LNCS, vol. 3732, pp. 778–787. Springer, Heidelberg (2006)
11. Gupta, A., Karypis, G., Kumar, V.: Highly scalable parallel algorithms for sparse matrix factorization. IEEE Trans. Parallel Distrib. Syst. **8**(5), 502–520 (1997)
12. Gupta, A., Koric, S., George, T.: Sparse linear solvers on massively parallel machines. In: Proceedings of the ACM/IEEE Conference on High Performance Computing, SC 2009, Portland, Oregon, USA, 14–20 November. IEEE (2009)
13. Koric, S., Lu, Q., Guleryuz, E.: Evaluation of massively parallel linear sparse solvers on unstructured finite element meshes. Comput. Struct. **141**, 19–25 (2014)
14. Lacoste, X., Faverge, M., Bosilca, G., Ramet, P., Thibault, S.: Taking advantage of hybrid systems for sparse direct solvers via task-based runtimes. In: 2014 IEEE International Parallel & Distributed Processing Symposium Workshops (IPDPSW), pp. 29–38. IEEE (2014)
15. Liu, J.W.-H.: The role of elimination trees in sparse factorization. SIAM J. Matrix Anal. Appl. **11**, 134–172 (1990)
16. Liu, J.W.-H.: The multifrontal method for sparse matrix solution: theory and practice. SIAM Rev. **34**(1), 82–109 (1992)
17. Mu, M., Rice, J.R.: A grid-based subtree-subcube assignment strategy for solving partial differential equations on hypercubes. SIAM J. Sci. Stat. Comput. **13**(3), 826–839 (1992)
18. Puzyrev, V., Koric, S., Wilkin, S.: Evaluation of parallel direct sparse linear solvers in electromagnetic geophysical problems. Comput. Geosci. **89**, 79–87 (2016)
19. Rennich, S.C., Stosic, D., Davis, T.A.: Accelerating sparse Cholesky factorization on GPUs. In: Proceedings of the Fourth Workshop on Irregular Applications: Architectures and Algorithms, IA3 2014, pp. 9–16 (2014)
20. Sao, P., Vuduc, R., Li, X.S.: A distributed CPU-GPU sparse direct solver. In: Silva, F., Dutra, I., Santos Costa, V. (eds.) Euro-Par 2014 Parallel Processing. LNCS, vol. 8632, pp. 487–498. Springer, Heidelberg (2014)
21. Schenk, O., Christen, M., Burkhart, H.: Algorithmic performance studies on graphics processing units. J. Parallel Distrib. Comput. **68**(10), 1360–1369 (2008)
22. Strohmaier, E., Dongarra, J., Simon, H., Meuer, M.: The Top 500 list. http://www.top500.org
23. Yeralan, N., Davis, T.A., Ranka, S.: Sparse QR factorization on the GPU. Submission to ACM Trans. Math. Softw

Automatic OpenCL Task Adaptation
for Heterogeneous Architectures

Pierre Huchant$^{(\boxtimes)}$, Marie-Christine Counilh, and Denis Barthou

Inria/LaBRI, University of Bordeaux, Bordeaux INP, Bordeaux, France
{pierre.huchant,denis.barthou}@inria.fr, counilh@labri.fr

Abstract. OpenCL defines a common parallel programming language
for all devices, although writing tasks adapted to the devices, managing
communication and load-balancing issues are left to the programmer.

In this work, we propose a novel automatic compiler and runtime technique to execute single OpenCL kernels on heterogeneous multi-device
architectures. The technique proposed is completely transparent to the
user, does not require off-line training or a performance model. It handles communications and load-balancing issues, resulting from hardware
heterogeneity, load imbalance within the kernel itself and load variations between repeated executions of the kernel, in an iterative computation. We present our results on benchmarks and on an N-body
application over two platforms, a 12-core CPU with two different GPUs
and a 16-core CPU with three homogeneous GPUs.

1 Introduction

Heterogeneous parallel architectures are ubiquitous, from supercomputers to cell
phones. Developing an application for a heterogeneous, multi-devices system,
taking advantage of all available devices is extremely challenging. OpenCL is a
standard language for the development of code on heterogeneous architectures. It
leverages part of this difficulty by defining one language for all platforms, and by
structuring parallelism into a task graph, where tasks are parallel computations
to be mapped onto one device. However, this implies that the developer has
to design as many tasks as there are devices, with tasks adapted in terms of
parallelism and memory granularity: There should be enough parallelism for all
devices, and communications between devices have to be explicit.

OpenCL kernels describe tasks as parallel work-groups. To transform one
kernel into as many kernels as devices, these work-groups have to be partitioned among devices. This raises load balancing issues, stemming from device
heterogeneity and from workload variation between work-groups. As many kernels are executed in iterative computation, the workload may change from one
iteration to the other, requiring a constant adaptation of the work-group partitioning. Moreover, data has to be split among partitioned work-groups in order
to reduce communication time and written data has to be merged upon kernel completion. Achieving adaptive work-group partitioning, with no training,
handling load balancing and data movements has never been conducted before.

© Springer International Publishing Switzerland 2016
P.-F. Dutot and D. Trystram (Eds.): Euro-Par 2016, LNCS 9833, pp. 684–696, 2016.
DOI: 10.1007/978-3-319-43659-3_50

We propose in this paper a static/dynamic approach for the execution of any given OpenCL kernel on a multi-device heterogeneous architecture. The method tackles, without training, load balancing issues coming from device heterogeneity and from varying computational intensity inside the kernel, when the kernel is called multiple times. The load-time analysis computes how to partition data for the execution of work-groups as a function of the work-group partitioning. The dynamic method evaluates from previous runs how to partition work-groups and splits/merges data accordingly. We show with two different runtime methods that only a few iterations are required to reach the optimal granularity. Finally, when the kernel computational intensity changes with each execution, the method dynamically adapts the load and stays close to the optimal load.

Section 2 shows causes for load balancing issues when splitting a kernel into subkernels. Section 4 presents our method, generating partition-ready kernels and instantiating with the appropriate granularity, for each device. Computation of granularities is described in Sect. 5. Related works and experimental results are given in Sects. 6 and 7.

2   Motivating Example

Given an OpenCL kernel, we define a subkernel as a code executing only part of the kernel computation. As OpenCL kernel executions are characterized by the number of parallel work-groups, the ratio of work-groups of a subkernel over the total number of work-groups is called *granularity*, a granularity of 1 meaning the whole kernel is executed. We study the performance variation of a kernel on one device, decreasing manually its granularity. The granularity is indicated as a percentage of the total number of work-groups, and performance is indicated as the mean time per work-group (lower is better). Figure 1 shows performance of AESEncrypt and EP from SNU NPB Suite [13] for different granularities on a 16-core Intel Xeon E5-2650 2.00 GHz with 64GB (CPU) and on an Nvidia Tesla M2075 (GPU). For AESEncrypt, the average time per work-group is nearly constant for all granularities, and very different on CPU and on GPU. For EP, we observe large performance drops (higher average time/work-group) at regular intervals of granularities on both CPU and GPU. This may come from compiler optimizations (such as unrolling), cache effects, and inefficient occupancy of the parallel resources due to a low number of work-groups within subkernels.

Work-groups are indexed in OpenCL by a vector of indices among a rectangular space (from 1D to 3D) called the NDRange. Selecting a granularity boils down to defining a subvolume of indices. In this paper, the subvolumes we consider are obtained by selecting one smaller interval in one dimension of the NDRange. The *offset* is the first index of this interval of indices, the granularity defining the size of this interval. Figure 1c shows for a Sparse Matrix Vector Multiply (SpMV) the influence of the offset on performance when, for a kernel of 1/4 granularity, the offset is changed. In the chosen sparse matrix, rows with a high index have more non-0 elements than those with a low index. This accounts for the execution time increase for large offsets, more than 7× the time of a 0-offset. When splitting a kernel into subkernels, this is a possible source of load-imbalance.

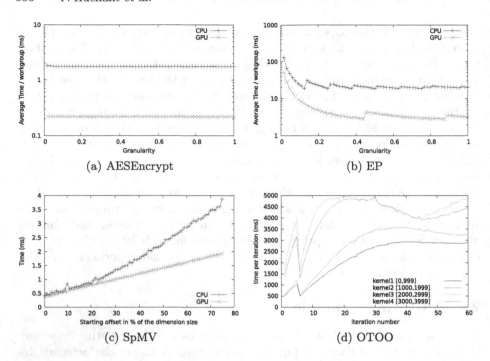

Fig. 1. (a) and (b): Impact on performance of architectural heterogeneity and granularity on AESEncrypt and EP benchmarks. Performance is given as an average time per work-group, granularity as a percentage of the total number of work-groups. (c) Impact on performance of the offset (starting index) for SpMV kernel, with a fixed granularity of 1/4. (d) Impact of iteration count on performance for OTOO application. Granularities are set to 1/4 for all devices, and offset is fixed on all devices. (Color figure online)

Many OpenCL kernels are executed in iterative computations. For instance, OTOO [11] is an astrophysics particle N-Body simulation and the same kernel is called repeatedly to compute forces and move the different particles. Figure 1d shows how the execution time changes for different iterations, for different offsets, for each iteration of the computation. The kernel is split into 4 subkernels, each one is given a granularity of 1/4 and executed on one GPU. The input set corresponds to a non-uniform distribution of the masses in space. As this space is partitioned among the work-groups, this results in a non-homogeneous load distribution among the work-groups, changing with iteration number.

These results advocate for a method able to cope with the heterogeneity of the hardware, but also with the performance variations associated to different granularities, depending on the offset and varying with each execution of the kernel. Adapting a single OpenCL kernel to an heterogeneous architecture with any number of devices, taking into account these four sources of imbalance has never been tackled before.

3   Principle of Adaptive Granularity

The method proposed is threefold. First the kernel is analyzed and a new version, partition-ready, is generated at compile time. Then each time the original kernel has to be executed, a granularity is chosen for each device, based on previous executions if any, and the partition-ready kernel is instantiated on each device with the chosen granularity. More precisely: (i) When the kernel code is first loaded, it is analyzed. This step is more thoroughly described in Sect. 4. The objective is the generation of a parametric and partition-ready kernel, executing only a slice of the NDRange space. The analysis is performed once on the OpenCL code (no host code analysis) but the code generated can be instantiated at run-time for many different granularities and offsets. The slicing of the rectangular volume corresponding to the NDRange is done in any of the dimensions of the volume and does not require to flatten it. The memory region accessed by each work group is computed, parametrically w.r.t. the work-group id and the scalar parameters of the kernel; (ii) Each time the original kernel is launched, a granularity for each device is determined. This granularity determines the number of work-groups to execute on a particular device. The array regions are instantiated with the granularities and the actual parameters of the kernel. Depending on the result, all arrays are communicated to the devices or only the region they require. The same occurs for bringing back data from the devices. This kernel instantiation is described in Sect. 4.2; (iii) The execution time of each kernel execution is collected for refining the granularity in the possible following runs. This iterative granularity optimization is described in Sect. 5.

4   Automatic Adaptation of Data and Parallelism

We describe in this section how a kernel is analyzed and transformed into a parametric partition-ready kernel, function of the granularity.

4.1   Static Analysis and Transformation

The analysis determines for each array passed to the kernel how this array can be split among the different devices. For arrays that are read-only, a safe over-approximation is to broadcast the whole array to all devices. A more precise analysis can determine a finer partition, allowing shorter communication times and for some extreme case, may be the only possible way to execute the kernel if the initial array is too large for any of the devices. Finding how to partition arrays written by the kernel is essential: When the written region is precisely known and there is no overlap with other device regions, bringing back this data to the host can be done in parallel. On the contrary, if the analysis is not able to precisely determine which region has been written, a merge operation is necessary to build the output array [8, 12]. The analysis only handles arrays (buffer objects) but could be extended to OpenCL images. We describe in this section how to determine precisely the array regions accessed by each work-item. The case where the analysis fails is discussed in the next section.

We first identify in the kernel all statements accessing arrays passed as a parameter. In OpenCL, arrays can be cast into other types (from 1D to 3D for instance), with possible offsets. All accesses through the cast arrays are also accesses to the initial array. Likewise, array accesses may occur inside functions called by the kernel. We therefore resort to an inter-procedural alias analysis, following assignments and use-def chains on arrays. In the following example,

```
void KERNEL(float *A) {
   ...
S1: double(* B)[3][5][5] =(double (*) [3][5][5])&A[ offset ];
   ...
S2: B[1][0][0]=..
}
```

the analysis detects that statement S1 defines the 3D array B, aliasing A. S2 accesses B[1][0][0], corresponding to A[25+offset]. Only constant offsets are handled so far. It generates a list of statements accessing input arrays, with their mapping function turning the index into an index of the input array.

For each array that is a parameter of the kernel and for each statement, we compute the array region accessed by this statement. The idea is to consider the index expression and to replace the variables in it by their values, repeatedly. Assuming the code is in SSA-form, this repeated substitution may lead to several cases: The resulting expression only uses scalar parameters of the kernel and work-item ids (local, group or global). In such case, the substitution process stops and the exact region will be evaluated dynamically, when these ids and parameter values are known. When a variable is defined by a Φ-function, if this is an induction variable and its interval of values can be determined, then the variable is replaced by its interval, the index expression becoming an interval expression. For other cases of Φ-functions or variables that are defined by loads, the region accessed is assumed to be *unknown*. Therefore, the array regions computed are interval expressions of the scalar inputs of the kernel and of the work-item ids, or *unknown*. We compute additionally the conditions on the ids for which this region is accessed. The abstraction we use for these conditions is an interval on ids. A similar analysis is therefore conducted on the conditionals (or loop bounds) governing the execution of the statement considered. Out of simplicity, only uniform expressions on ids are kept, *i.e.* inequalities of the form: $\pm id \leq expr$ where $expr$ is an expression independent of the work-item ids. Conditionals that are not uniform expressions are assumed to be true. The conjunction of such conditionals define intervals of ids. To wrap-up, the array region accessed by a statement is either *unknown*, or represented by a guarded region of the form:

$$\textbf{if } id \in [lb(s), ub(s)] : [expr_1(id, s), expr_2(id, s)]$$

with s the scalar parameters of the kernel and id a work-item id. Note that the expressions $expr_1$, $expr_2$, lb and ub have no restriction and can use any operator allowed by the language. For a given array, the region accessed by the kernel is defined by the union of array regions accessed by all statements.

Kernel Modification. When executing a kernel with a fraction of the original NDRange, some syntactic modifications are needed in order to keep the correct

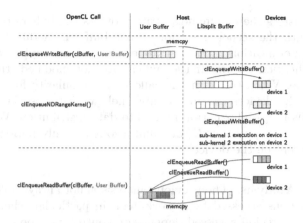

Fig. 2. Buffer management

semantics. Indeed, the global size is different from the original kernel, the number of work-groups has changed and their id has changed too. Two additional parameters are added to the partition-ready kernel: `splitdim` (1, 2 or 3) accounts for the dimension of the NDRange that is split, and `numgroups` is the number of work-groups in this dimension. Then the following function calls are changed:

Expression	Rewritten into
`get_global_size(expr)`	`(expr == splitdim?numgroups*get_local_size(expr):get_global_size(expr))`
`get_num_groups(expr)`	`(expr == splitdim?numgroups:get_num_groups(expr))`
`get_group_id(expr)`	`(get_global_id(expr)/get_local_size(expr))`

All analyses and transformations are performed once at compile-time within the LLVM compiler [6]. The granularities are determined later at runtime.

4.2   Kernel Instantiation and Communication Generation

OpenCL function calls from the host are intercepted by our runtime library `Libsplit`. When the kernel is called in the OpenCL code, the values of the scalar parameters and the size of the NDRange are known. The array regions can then be evaluated and the range of work-item ids is determined according to the chosen granularity and the size of work-groups. The runtime kernel instantiation evaluates the dimension of the NDRange that allows to distribute written data among devices, with no need for a merge operation, if possible.

Array regions are defined as union of guarded intervals. In order to reduce the number of communications, we evaluate for a given interval of ids and for each array an interval including the array region.

Figure 2 shows the different steps, assuming a copy is performed with a `WriteBuffer` command. Calls to `WriteBuffer` are deferred communications:

Our library registers the commands, write protects the buffers to prevent any modification until the kernel execution. If one buffer is modified before kernel execution, the modification access is trapped, so that the copies occur first and then the modification occurs. If the buffers are not modified, this additional copy is not done. When the kernel is called, the granularity for each device is computed and the devices execute a subkernel, with its associated data. The runtime analysis keeps information related to data distribution. When multiple kernel executions are performed, communications are only performed for data not already present on the device.

Limits of the Analysis. The previous analysis is not always able to precisely compute the regions accessed by a work-item, in particular when indirections occur or when the region accessed depends on control flow too complex for the analysis. When this happens, the array is not split between devices. If the array is written, a merge operation is required after the kernel execution in order to fuse the different contributions computed by each device. The operation we propose is based on a diff, similarly to [12]. Such operation degrades the overall performance for communication/memory bound kernels.

5   Adapting Granularity

This section proposes a method to dynamically adapt granularity to the devices and to the kernel, assuming the same kernel is executed multiple times.

5.1   Formalization

Given a kernel and n devices, the problem consists in determining how to split the computation among the devices so as to minimize the execution time. Each device executes the same kernel, but possibly with a different number of work-groups and different data. We formally define the granularity as a value x_i in $[0, 1]$ corresponding to the ratio between the number of work-groups allocated to the device i and the total number of work-groups (numgroups). numgroups is known when the kernel is called. We define $f_i(x_i, \text{offset}_i, t)$ as the mean time to execute one work-group on device i, when a subkernel of granularity x_i is executed at time step t, with an offset offset$_i$. The total execution time of this subkernel is therefore $f_i(x_i, \text{offset}_i, t) * x_i * \text{numgroups}$.

The solution to the problem consists in finding the time T and the granularities x_i and the offsets offset$_i$ such that the system in Fig. 3a is fulfilled.

The functions f_i are not known precisely but they can be measured for a given x_i, offset$_i$ and t. We arbitrarily order the offsets by increasing id of device. Thus, with offsets defined by values in $[0, 1]$, offset$_1 = 0, \ldots$, offset$_n = \sum_{k<n} x_k$ and f_i no longer depends on offsets in Fig. 3b but on x_i.

We generalize this formulation by introducing a new set of variables, $y_i \in [0, 1]$, as shown in Fig. 3b. Now it is possible to define a function F such that

$$\min T$$
$$f_1(x_1, \text{offset}_1, t) * x_1 * \text{numgroups} \leq T$$
...
$$f_n(x_n, \text{offset}_n, t) * x_n \text{numgroups} \leq T$$
$$\sum_i x_i = 1$$

(a) Initial formulation

$$\min T$$
$$f_1(x_1, t) * y_1 * \text{numgroups} \qquad \leq T$$
...
$$f_n(x_1 \ldots, x_n, t) * y_n * \text{numgroups} \leq T$$
$$\sum_i x_i = 1, \qquad\qquad\qquad \sum_i y_i = 1$$

(b) Generalized formulation

Fig. 3. Formulations of the granularity problem

$F_t(\mathbf{x}) = (\mathbf{y})$, with $\mathbf{x} = (x_i)_i$ the vector of all x_i and $\mathbf{y} = (y_i)_i$ the vector of all y_i, satisfying conditions from Fig. 3b:

$$F_t(\mathbf{x}) = \left(\frac{T}{f_i(x_1, \ldots, x_i, t) * \text{numgroups}} \right)_i,$$

with $T = \frac{\text{numgroups}}{\sum_i 1/f_i(x_1, \ldots, x_i, t)}$. The evaluation of $F_t(\mathbf{x})$ requires $O(n)$ basic arithmetic operations with n the number of devices. A solution to the problem of Fig. 3a can be found by computing a fixed point of the function F: $F_t(\mathbf{x}) = \mathbf{x}$ or similarly, by finding the 0 of the function G_t: $G_t(\mathbf{x}) = \mathbf{x} - F_t(\mathbf{x})$.

5.2   Resolution Method

First assume the function F_t does not depend on t, the iteration count. Several methods have been proposed in the literature for solving such problem, when the function is not known analytically: The fixed point method consists in computing the suite of granularity vectors $\mathbf{x}_k = F(\mathbf{x}_{k-1}), k \geq 1$ from some initial value $\mathbf{x}_0$. The evaluation of $F(\mathbf{x}_{k-1})$ requires to execute the kernel with the granularities $\mathbf{x}_{k-1}$. When the suite converges, it converges linearly towards a vector of optimal granularities satisfying the initial problem and achieving perfect load balance. The convergence depends on F and on the initial value $\mathbf{x}_0$ but is in general linear. The secant method, or its generalization for n-D space the Broyden's method, uses an approximate gradient to converge to the 0 of a function with a near quadratic convergence rate. We implemented both methods to refine the granularity assigned to each device. F is evaluated at each kernel instantiation and provides the input necessary to instantiate the partition-ready kernel.

Finally, when the functions f_i also depend on the iteration count t, the fixed point equation becomes $F_{t-1}(\mathbf{x}_{k-1,t-1}) = \mathbf{x}_{k,t}$ with F changing for each term of the suite. For real applications, a good approximation of the solution at step t remains a good approximation at step $t+1$. As the fixed point method converges quickly when the approximation is close to the solution, we believe this approach can be used for many real cases. We demonstrate for a N-Body application that the fixed point method is able to stay close to the optimal, even when the optimal granularity is varying with the iteration count (see Sect. 7).

6   Related Works

Several works focus on kernel splitting. Kim *et al.* [3] are making one OpenCL device unifying multiple uniform GPUs. The OpenCL NDRange and array regions are split according to the values recorded by sampling. They do not handle conditionals as we do and assume that array indices are all linear in the kernel parameters. Moreover, they assume subkernels have the same load. Luk *et al.* [10] propose an heterogeneous programming system that provides an adaptive mapping technique based on execution-time projections stored in a database during training runs. The technique we propose does not require training runs. Li *et al.* [9] present STEPOCL, a tool which takes as input kernels along with a configuration file and generates automatically an OpenCL multi-devices application. The configuration file describes how to split data, the control flow of the program, and allow to have specialized kernels for different architectures. The work partitioning between devices is based on offline profiling. Grewe and O'Boyle [2] propose a pure static task partitioning method, based on predictive modeling and program features. They do not collect data dynamically however and cannot adapt to differences in terms of computing efficiency, depending on granularity, as shown in the motivating example. Kim *et al.* [4] propose to solve the load imbalance problem on CPUs by dynamically assigning sets of work-groups with decreasing sizes to an idle compute unit thread. It manages data across different devices in a cluster, however mapping tasks and data to these devices is left to the programmer. Seo *et al.* [14] propose an automatic work-group size selection technique for OpenCL kernels on multicore CPUs. Their method uses a profiling-based algorithm. Heterogeneity is not handled however and kernels are assumed to be work-group size independent. Kofler *et al.* [5] present a method for OpenCL task partitioning relying on offline generated model. This model is based on artificial neural networks, relying on the features of the kernels, including their input sizes. In [1], the authors propose a dynamic method to partition OpenCL tasks and perform load balancing. Their approach generates chunks of work-groups with increasing size to execute on different devices and selects the best partition of these chunks on the devices. The chunks are manually generated and there is no automatic scheme to partition data for the OpenCL kernels.

In [12], the authors propose an OpenCL runtime that takes a single device kernel and executes it on CPU and GPU. Load balancing is managed at run-time. While their approach dynamically balance work between one CPU and one GPU, it cannot be easily generalized for any number of devices. Besides, data is not split between subkernels, all arrays are transferred to all devices. Finally, the kernel transformation is achieved by hand, and not with an automatic compiler optimization. In [15], Shen *et al.* present a method for heterogeneous platforms and imbalanced applications. They propose a model integrating both the workload of the application, determined by sampling, and the architecture. When the workload has an irregular shape, it is reshaped by sorting to obtain a regular shape with a peak and a bottom part. Based on this model and after profiling, a predictor determines the optimal partitioning between CPUs and GPUs. The work described in [8] relies on complex training, resorting to linear regression techniques in order to predict the correct load balance. They do not handle dynamic load changes, such as the SpMV or OTOO case shown in the motivating example. In [7], the authors extend the previous work to complete task graphs.

It is still based on offline training and assumes performance per work-group is constant, while we have shown there can be large variations.

7   Performance Evaluation

Experiments are conducted on two platforms: conan — 16-core Intel Xeon E5-2650 2.00 GHz with 64 GB, 3 Nvidia Tesla M2075; happyCL — 12-core Intel Xeon E5-2680 2.80 GHz with 64 GB, Nvidia Tesla K20c, Nvidia Quadro K5000.

Detailed Load Balancing: Figure 4 shows the speed-up obtained on conan with AESEncrypt and EP compared to the best single device performance, when these kernels are repeated 10 times. The speed-up shown here are per iteration. For the first iteration, the granularity is the same for all devices (uniform hypothesis), explaining poor performance compared to the GPU performance. For EP, Fig. 4a, the fixed point method requires 6 iterations to reach a maximum speed up of 2.8, whereas the Broyden's method converges in only 4 iterations leading to a better global speedup. For AESEncrypt, Fig. 4b shows there is no such difference and both methods are similar, reaching a peak speed-up of 2.15 in 3 iterations only. The variations are due to the fact that the optimal granularity does not correspond to a round number of work-groups, hence there are granularity adjustments and communications at each step.

Figure 5 shows how our method behaves when the load changes over 60 iterations. Figure 5a illustrates the time taken by each subkernel for OTOO when the granularity is the same for all devices (Uniform strategy). From one device to the other, the execution time differs by more than a factor 3 (iteration 15 for instance). Figure 5b shows how the same load is shared among the four devices when it is continuously adapted by our technique (Adaptive strategy). As the 4 plots are close to each other, this shows the execution time is nearly optimal. We observe that convergence to the optimal only requires 2 iterations.

Overall Speedups: Figure 6 presents speed-ups compared to the best single device performance on the two target architectures, for a large number of benchmarks when they are repeated 100 times. We observe the results of our method

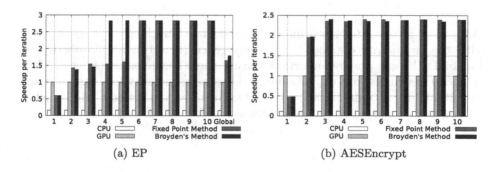

(a) EP   (b) AESEncrypt

Fig. 4. Speedup per iteration of EP and AESEncrypt

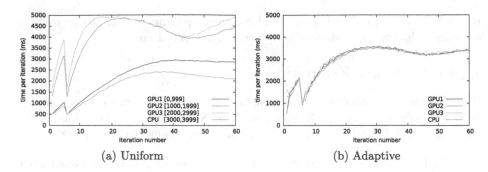

(a) Uniform   (b) Adaptive

Fig. 5. Performance of OTOO executed on conan (3GPUs+CPU) for 60 iterations. (a) shows the execution time of each subkernel with the Uniform splitting, same granularity for all devices. (b) shows the execution time of each subkernel with the Adaptive splitting. (Color figure online)

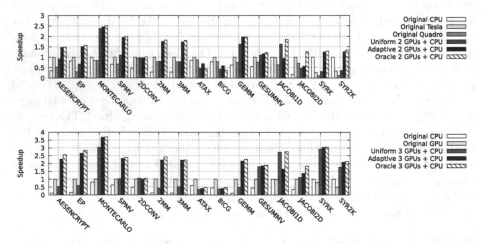

Fig. 6. Performance of AESEncrypt, EP, MonteCarlo, OTOO, SPMV and some Polybench on happyCL (top) and conan (bottom). Original codes run only on one device. Uniform and Adaptive are using subkernels automatically obtained by our method.

(`Adaptive`) are close to the optimal, obtained when launching the kernel directly with the granularity obtained after convergence (`Oracle`). For Jacobi1D and Jacobi2D, the gap is more important because these benchmarks consist in 2 kernels, one stencil and one copy. Defining the same granularity for both copy and stencil minimizes communication time. Our method handles only one kernel at a time, and does not find the same granularity for both kernels.

8   Conclusion

We proposed in this paper the design and implementation of a method that simplifies the development of OpenCL applications for heterogeneous, multi-device
systems. Our technique splits computation and data automatically across the
computing devices, handling all load-balancing issues, including load variations
when the kernel is executed iteratively. We have shown the optimal granularity
is obtained in a few iterations and the technique does not require profiling or
training. The approach is completely transparent to the user, and the same code
can be executed without modification on different machines.

References

1. Boyer, M., Skadron, K., Che, S., Jayasena, N.: Load balancing in a changing world:
   dealing with heterogeneity and performance variability. In: Computing Frontiers
   Conference (2013)
2. Grewe, D., O'Boyle, M.F.P.: A static task partitioning approach for heterogeneous
   systems using OpenCL. In: Knoop, J. (ed.) CC 2011. LNCS, vol. 6601, pp. 286–305.
   Springer, Heidelberg (2011)
3. Kim, J., Kim, H., Lee, J.H., Lee, J.: Achieving a single compute device image in
   OpenCL for multiple GPUs. In: Principles and Practice of Parallel Programming,
   PPopp 2011, pp. 277–288. ACM, New York (2011)
4. Kim, J., Seo, S., Lee, J., Nah, J., Jo, G., Lee, J.: SnuCL: an OpenCL framework for
   heterogeneous CPU/GPU clusters. In: ACM International Conference on Supercomputing, ICS 2012, pp. 341–352. ACM, New York (2012)
5. Kofler, K., Grasso, I., Cosenza, B., Fahringer, T.: An automatic input-sensitive
   approach for heterogeneous task partitioning. In: International Conference on
   Supercomputing, pp. 149–160. ACM, New York (2013)
6. Lattner, C., Adve, V.: LLVM: a compilation framework for lifelong program analysis and transformation, San Jose, CA, USA, pp. 75–88, March 2004
7. Lee, J., Samadi, M., Mahlke, S.: Orchestrating multiple data-parallel kernels on
   multiple devices. In: Parallel Architecture and Compilation Techniques. IEEE
   (2015)
8. Lee, J., Samadi, M., Park, Y., Mahlke, S.: SKMD: single kernel on multiple devices
   for transparent CPU-GPU collaboration. ACM Trans. Comput. Syst. **33**(3),
   9:1–9:27 (2015)
9. Li, P., Brunet, E., Trahay, F., Parrot, C., Thomas, G., Namyst, R.: Automatic
   OpenCL code generation for multi-device heterogeneous architectures. In: International Conference on Parallel Processing, pp. 959–968 (2015)
10. Luk, C.K., Hong, S., Kim, H.: Qilin: exploiting parallelism on heterogeneous multi-
    processors with adaptive mapping. In: Symposium on Microarchitecture, MICRO
    42, pp. 45–55. ACM, New York (2009)
11. Nakasato, N., Ogiya, G., Miki, Y., Mori, M., Nomoto, K.: Astrophysical Particle
    Simulations on Heterogeneous CPU-GPU Systems. CoRR abs/1206.1199 (2012)
12. Pandit, P., Govindarajan, R.: Fluidic kernels: cooperative execution of OpenCL
    programs on multiple heterogeneous devices. In: Code Generation and Optimiza-
    tion, pp. 273–283. ACM (2014)
13. Seo, S., Jo, G., Lee, J.: Performance characterization of the NAS parallel bench-
    marks in OpenCL. In: Workload Characterization, pp. 137–148 (2011)

14. Seo, S., Lee, J., Jo, G., Lee, J.: Automatic OpenCL work-group size selection for multicore CPUs. In: Parallel Architectures and Compilation Techniques, pp. 387–397 (2013)
15. Shen, J., Varbanescu, A.L., Sips, H., Arntzen, M., Simons, D.G.: Glinda: a framework for accelerating imbalanced applications on heterogeneous platforms. In: Computing Frontiers Conference, p. 14. ACM (2013)

Author Index

Printed in the United States
By Bookmasters